36th UIT (Italian Union of Thermo-Fluid-Dynamics) Heat Transfer Conference (UIT 2018)

Journal of Physics: Conference Series Volume 1224

Catania, Italy
25 - 27 June 2018

Editors:

Alberto Fichera
Arturo Pagano
Rosaria Volpe

ISBN: 978-1-5108-8789-3
ISSN: 1742-6588

Printed from e-media with permission by:

Curran Associates, Inc.
57 Morehouse Lane
Red Hook, NY 12571

Some format issues inherent in the e-media version may also appear in this print version.

Institute of Physics
Dirac House, Temple Back
Bristol BS1 6BE UK

Phone: 44 1 17 929 7481
Fax: 44 1 17 920 0979

techtracking@iop.org

Additional copies of this publication are available from:

Curran Associates, Inc.
57 Morehouse Lane
Red Hook, NY 12571 USA
Phone: 845-758-0400
Fax: 845-758-2633
Email: curran@proceedings.com
Web: www.proceedings.com

TABLE OF CONTENTS

COMPUTATIONAL FLUID DYNAMICS AND HEAT TRANSFER

CONDUCTION

HEAT TRANSFER AND EFFICIENCY IN ENERGY SYSTEM, ENVIRONMENTAL TECHNOLOGIES, AND BUILDINGS

FLOW VISUALIZATION

HEAT TRANSFER IN NUCLEAR PLANTS

IR THERMOGRAPHY

MULTI-PHASE FLUID DYNAMICS, HEAT TRANSFER AND INTERFACE PHENOMENA

IOP Conf. Series: Journal of Physics: Conf. Series **1224** (2019) 011001 doi:10.1088/1742-6596/1224/1/011001

Preface

Organized by

UIT Unione Italiana Termofluidodinamica

UNIVERSITÀ degli STUDI di CATANIA

IOP Conf. Series: Journal of Physics: Conf. Series **1224** (2019) 011001 doi:10.1088/1742-6596/1224/1/011001

36th UIT Heat Transfer Conference

Catania, 25-27 June, 2018

The annual UIT Heat Transfer Conference of the "Unione Italiana di Termofluidodinamica" (UIT) aims at endorsing the collaboration among scientists and engineers working in the field of heat transfer and thermal sciences.

The 36th UIT Conference was held in Catania, Italy, 25-27 June, 2018 at the Four Points by Sheraton Hotel. The coverage of topics reflects the current interests of the international and Italian research communities, such as natural, forced and mixed convection, conduction, radiation, computational fluid dynamics, multi-phase fluid dynamics and interface phenomena, heat transfer in nuclear plants, flow visualization, heat transfer and efficiency in energy system, environmental technologies and buildings. With 72 abstracts, 36 oral and 22 poster contributions, the Conference provided not only the opportunity for authors to present and discuss their research development, but also to stimulate the debate on their respective research areas. Keynote Lectures were given by Professor Giovanni S. Barozzi (University of Modena and Reggio Emilia), Professor Michele Ciofalo (University of Palermo) and Professor Marco Marengo (University of Brighton). These Proceedings contain 50 full papers, selected from among the scientific contributions discussed during the Conference.

The Editors thank the authors for their participation and contributions, the Scientific Committee and the reviewers for their efforts. Special thanks are extended to the Organizing Committee. The Editors gratefully acknowledge the Italian Union of Thermal-Fluid Dynamics (UIT) and its President, Professor Alfonso Niro, for all efforts made for promoting the Conference. Finally, they acknowledge the financial support of the University of Catania.

Alberto Fichera, Arturo Pagano and Rosaria Volpe – Editors

IOP Conf. Series: Journal of Physics: Conf. Series **1224** (2019) 011001 doi:10.1088/1742-6596/1224/1/011001

Scientific Committee of the 36th UIT Heat Transfer Conference

Alberto Fichera, *Chair*	Università degli Studi di Catania, Italy
Paolo Baggio	Università degli Studi di Trento, Italy
Giovanni S. Barozzi	Università degli Studi di Modena e Reggio Emilia, Italy
Gian Paolo Beretta	Università degli Studi di Brescia, Italy
Neima Brauner	Tel Aviv University, Israel
Gianni Cesini	Politecnica delle Marche, Italy
Gian Piero Celata	Centro Ricerche ENEA Casaccia, Rome, Italy
Wilson K.S. Chiu	University of Connecticut, USA
Michele Ciofalo	Università degli Studi di Palermo, Italy
Luigi P.M. Colombo	Politecnico di Milano, Italy
Davide Del Col	Università degli Studi di Padova, Italy
Paolo Di Marco	University of Pisa, Italy
Walter Grassi	University of Pisa, Italy
Adriano Lezzi	Università degli Studi di Brescia, Italy
Giovanni Longo	Università degli Studi di Padova, Italy
Oronzio Manca	Seconda Università degli Studi di Napoli, Italy
Luigi Marletta	Università degli Studi di Catania, Italy
Rita Mastrullo	Università degli Studi di Napoli Federico II, Italy
Gianluca Morini	Università degli Studi di Bologna, Italy
Vincenzo Naso	Università degli Studi di Napoli Federico II, Italy
Hisashi Ninokata	Professor Emeritus of the Tokyo Institute of Technology, Japan
Alfonso Niro	Politecnico di Milano, Italy
Enrico Nobile	Università degli Studi di Trieste, Italy
Carlo Nonino	Università degli Studi di Udine, Italy
Bruno Panella	Politecnico di Torino, Italy
Sara Rainieri	Università degli Studi di Parma, Italy
Gianfranco Rizzo	Università degli Studi di Palermo, Italy
Luisa Rossetto	Università degli Studi di Padova, Italy
Peter C. Stephan	Technische Universität Darmstadt, Germany
Giovanni Tanda	Università degli Studi di Genva, Italy
John R. Thome	École Polytechnique Fédérale de Lausanne (EPFL), Switzerland
Akio Tomiyana	Kobe University, Japan

Organizing Committee

Alberto Fichera, *chair*
Gianpiero Evola
Antonio Gagliano
Luigi Marletta
Francesco Nocera
Arturo Pagano
Rosaria Volpe

IOP Conf. Series: Journal of Physics: Conf. Series **1224** (2019) 011002 doi:10.1088/1742-6596/1224/1/011002

Peer review statement

All papers published in this volume of *Journal of Physics: Conference Series* have been peer reviewed through processes administered by the proceedings Editors. Reviews were conducted by expert referees to the professional and scientific standards expected of a proceedings journal published by IOP Publishing.

IOP Conf. Series: Journal of Physics: Conf. Series **1224** (2019) 012001 doi:10.1088/1742-6596/1224/1/012001

Bioheat transfer in a spherical biological tissue: a comparison among various models

Assunta ANDREOZZI[1,*], Luca BRUNESE[2], Marcello IASIELLO[1], Claudio TUCCI[2], Giuseppe Peter VANOLI[2]

[1]DII, Università di Napoli FEDERICO II, Piazzale Tecchio 80, 80125, Napoli, ITALY
[2]DIMES "Vincenzo Tiberio", Università del Molise, Via Francesco De Sanctis 1, 86100, Campobasso, ITALY

[*]Corresponding author:assunta.andreozzi@unina.it

Abstract. The investigation of bioheat transfer is a difficult issue because it entails a mixture of many mechanisms to take into account, such as thermal conduction in tissues, convection and blood perfusion, metabolic heat generation, vascular structure, changing of tissue properties depending on physiological condition and so on. This topic has a key role to predict accurately the temperature distribution in tissues, especially during biomedical applications. In this paper, different bioheat transfer models are resumed and compared. The biological tissue is modelled as a porous sphere and liver tissue properties are used. Governing equations are averaged over a Representative Elementary Volume (REV) of the living tissue. Transient bioheat equations based on models like, for example, Pennes model, Local Thermal Non-Equilibrium equations (LTNE model), are employed. In the employed equations, radiative heat transfer is also considered. Governing equations with the appropriate boundary conditions are solved with the finite-element code COMSOL Multiphysics®. The effects of hyperthermia on the living tissue are included with a source term in the tissue energy equation. Results are presented in terms of temperature profiles in the biological tissue; the aim is to appreciate differences due to the various bioheat models.

Nomenclature

a	volumetric transfer area, m^{-1}	*Greek symbols*	
c_p	specific heat, J kg^{-1} K^{-1}	β	extinction coefficient, m^{-1}
d	diameter of blood vessel, m	ε	porosity
h	heat transfer coefficient, W m^{-2} K	ρ	density, kg m^{-3}
I_0	intensity of radiation, W m^{-2}	σ	Stefan-Boltzmann constant, W m^{-2} K^{-4}
k	thermal conductivity, W m^{-1} K^{-1}	ω	blood perfusion rate, s^{-1}
k_R	"radiative" thermal conductivity, W m^{-1} K^{-1}	*Subscripts*	
n	refractive index	a	arterial
q_R	radiative heat flux, W m^{-2}	t	tissue
Q	power density, W m^{-3}	b	blood
t	time, s	*dis*	dispersion
T	temperature, K	*ext*	external
u	velocity, m s^{-1}	f	fluid
R	external radius of the sphere, m	v	venous
r	heating zone radius, m	*Pennes*	pertinent to the Pennes' model
r,z	spatial coordinates, m		

IOP Conf. Series: Journal of Physics: Conf. Series **1224** (2019) 012001 doi:10.1088/1742-6596/1224/1/012001

1. Introduction

Bioheat transfer is a difficult issue because it entails a mixture of many mechanisms to take into account, such as thermal conduction in tissues, convection and blood perfusion, metabolic heat generation, vascular structure, changing of tissue properties depending on physiological condition and so on. This topic has a key role to accurately predict the temperature distribution in tissues, especially during biomedical applications.

Hyperthermia is a human body situation caused by a lack of thermoregulation, in which the temperature goes beyond its normal values. Hyperthermia can be caused by natural reasons, as in fever, or it can be induced. Induced hyperthermia is a solution for many diseases [1]. For example, heat can be applied to destroy cancer cells, as in radiofrequency or microwave ablation [2 - 5]. With this technique, an antenna is placed into a tissue, and necrosis is induced by applying a certain thermal dose. Other examples are radiofrequency cardiac ablation, in which heat is used to destroy abnormal conduction pathways through the myocardium caused by arrhythmia [6, 7], or laser angioplasty [8-11], in which heat is employed for cardiovascular diseases.

When designing engineering tools for biomedical applications, modelling has a primary role, because of the difficulty of carrying out experiments, due to ethical reasons and small phenomena scale. Many models have been proposed through the years to describe heat transfer in biological tissues. This paper is an overview of some of the most relevant models used through the years. Four models (Pennes [12], Local Thermal Equilibrium [13], Local Thermal Non-Equilibrium [14] and Nakayama and Kuwahara [15]) are applied on spherical human tissue, in order to determine temperature profiles. Heating is supplied to the tissue by a spherical internal heating source. Various geometrical and fluid-dynamics conditions are analyzed, together with an analysis on radiative effects in the semi-transparent medium. The scope is to compare such models to appreciate when differences are significant, and results are presented in terms of temperature profiles in the biological tissue.

2. Geometry and properties

The transient temperature of a biological tissue during hyperthermia treatment with different models of bioheat equations is analyzed. The vascular structures of tissue have been supposed to be uniformly distributed in order to consider the physical model as a uniform porous medium.

The entire computational domain is taken to be a sphere with a radius of 3.10 cm, and the heating zone within the biological tissue is a centrally located sphere with a radius of 0.62 cm, as shown in figure 1. All the blood vessels are assumed to be straight in the blood flow direction and to merge in the porous medium, and both the entrance blood and the boundary temperatures are equal to 37 °C. In addition, the thermal properties of tissue and blood are isotropic, and the heat transfer coefficient and blood velocity are supposed to be constant throughout the domain.

Furthermore, metabolic heat generation is neglected because it is much smaller than the power density released during hyperthermia treatment.

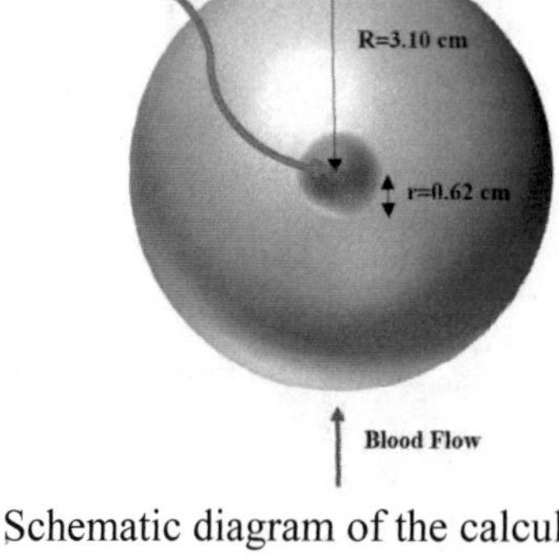

Figure 1. Schematic diagram of the calculation domain.

In this study a vessel diameter of 15 μm and an intervessel distance of 140 μm are considered to yield an estimated porosity of 0.01 [16] and two values of porosities (0.005 and 0.05) are selected.
The relationship between volumetric heat transfer areas and diameters of blood vessels is given by Yuan [16], and resumed in table 1, in which the different blood velocities related to the sizes of the vessels are included, according to the literature [17-19].

Table 1. Volumetric transfer areas for different porosities (ε) and vessels diameters (d).

	Volumetric transfer area a (m^2 m^{-3})	
	ε=0.005	ε=0.05
d=8 μm (u=0.07 cm s^{-1})	2500	25000
d=20 μm (u=0.3 cm s^{-1})	1000	10000
d=30 μm (u=0.4 cm s^{-1})	667	6667
d=50 μm (u=2 cm s^{-1})	400	4000
d=100 μm (u=3 cm s^{-1})	200	2000
d=140 μm (u=3.4 cm s^{-1})	143	1429

Thermal properties of tissue and blood are chosen according to the work by Kou et al. [20]. Thermal conductivities of tissue and blood are 0.5 W m^{-1} K^{-1}, densities of tissue and blood are 1050 kg m^{-3}, and specific heat capacities of tissue and blood are 3770 J kg^{-1} K^{-1}. To notice that the common input parameters are considered the same for the different models, because the paper is focused on the comparison among the different bioheat equations and not on the sensitivity of the models to the input parameters. However, it could be interesting to consider this analysis in further works. Moreover, two heating conditions are assumed: 2 s heating with a power density of 50 x 10^6 W m^{-3} and 50 s heating with a power density of 2 x 10^6 W m^{-3}. The absorbed power density of blood is estimated to be one-tenth of that of tissue [21].

3. Mathematical models

Here the different four bioheat models applied to the porous biological sphere are described in detail. They are: the Pennes bioheat equation, the Local Thermal Non-Equilibrium equations (LTNE model), the Local Thermal Equilibrium equation (LTE model), and the three-energy equation model.

3.1 The Pennes' bioheat equation

The bioheat equation developed in 1948 by Harry Pennes [12] is based on an experimental analysis of human forearm. Pennes' model is a modified form of the transient heat conduction equation, including the effect of blood perfusion and metabolic heat generation rate on heat transfer within the living tissue. This model has been widely used by many researchers for numerous biological and medical applications, but it shows some shortcomings because of its simplicity; in fact, it does not take into account the blood flow direction, neglecting also important anatomical features of the circulatory network system such as the artery-vein countercurrent arrangement. Besides Pennes' model considers only the venous blood stream as the one equilibrated with the tissue. In this case, as mentioned before, metabolic heat generation is neglected because it is much smaller than the power density released during hyperthermia treatment. The Pennes' bioheat equation, considering the external heat source and neglecting the metabolic heat source, can be written as:

$$\left(\rho c_p\right)_t \frac{\partial T_t}{\partial t} = \nabla \cdot \left(k_t \nabla T_t\right) + \left(\rho c_p\right)_b \omega_{Pennes} \left(T_a - T_t\right) + Q_{ext} \tag{1}$$

where t is the time and T is the temperature, the subscripts t, b and a stand for tissue, blood, and arterial blood, respectively, ρ is the density, c_p is the specific heat, k is the thermal conductivity, ω_{Pennes} is mean

IOP Conf. Series: Journal of Physics: Conf. Series **1224** (2019) 012001 doi:10.1088/1742-6596/1224/1/012001

blood perfusion rate, that is selected to be 0.0036 s^{-1} for all cases [22], and Q_{ext} is the heat generation due to hyperthermia treatment. The blood temperature is assumed to be uniform throughout the tissue and it is taken as body temperature equal to 37 °C. The tissue temperature T_t is equal to 37 °C at the boundary of the sphere R=3.10 cm and at the starting time t=0 s.

3.2 The Local Thermal Non-Equilibrium equations (LTNE)

In 1998, Roetzel and Xuan [14] introduced a two-equation bioheat model that considers the heat transfer in porous media. They modelled the biological tissue by dividing it into two different regions, namely, the tissue region and the blood region (i.e. solid phase consists of muscle, vascular tissues and other solid compounds, while the fluid phase is made up by the blood flow that streams), without considering local thermal equilibrium between the two media and introducing an equivalent effective thermal conductivity in the energy equations of blood and tissue. Furthermore they proposed an interfacial convective heat transfer term instead of perfusion one. Based on the study of Yuan [16], the value of the heat transfer coefficient is selected to be 170 W m^{-2} K^{-1} for all cases. Considering the conservation of energy to the tissue and blood, the simultaneous differential equations can be formulated as follows.

For the tissue phase:

$$(1-\varepsilon)(\rho c_p)_t \frac{\partial T_t}{\partial t} = (1-\varepsilon)k_t \nabla^2 T_t + ha(T_b - T_t) + (1-\varepsilon)Q_t \tag{2}$$

For the blood phase:

$$\varepsilon(\rho c_p)_b \left(\frac{\partial T_b}{\partial t} + \mathbf{u}_b \cdot \nabla T_b\right) = \varepsilon k_b \nabla^2 T_b + ha(T_t - T_b) + \varepsilon Q_b \tag{3}$$

where T_t and T_b are temperatures averaged over the tissue and blood volumes, ε is the porosity, h is the heat transfer coefficient, $\mathbf{u}_b$ is the blood velocity vector, a is the volumetric transfer area between tissue and blood, and Q is the absorbed power density.

The initial and boundary conditions are:

$T_t(t,r,z)$ = 37 °C at r=R=3.10 cm;
$T_t(0,r,z)$ = $T_b(0,r,z)$ =37 °C

3.3 The Local Thermal Equilibrium equation (LTE)

When the Local Thermal Equilibrium hypothesis is maintained, the temperature of the tissue is the same of the blood temperature (T_t=T_b=T), thus the equations (2) and (3) can be combined into a single equation as follows:

$$\left[(1-\varepsilon)(\rho c_p)_t + \varepsilon(\rho c_p)_b\right]\frac{\partial T}{\partial t} + \varepsilon(\rho c_p)_b \mathbf{u}_b \cdot \nabla T = \left[(1-\varepsilon)k_t + \varepsilon k_b\right]\nabla^2 T + (1-\varepsilon)Q_t + \varepsilon Q_b \tag{4}$$

The initial and boundary conditions are the same of the LTNE model.

3.4 The three-energy equation model

In 2008, Nakayama and Kuwahara [15] developed a generalized two-equation bioheat models for vascular and extravascular space in local thermal non-equilibrium condition and they incorporated blood perfusion term within the two sub-volume equations. Then they extended the two-equation model to three-equation model to consider the effect of heat transfer in closely spaced countercurrent artery-vein pair, illustrated in figure 2. The three equations are derived for arterial blood phase, venous blood phase and tissue phase distinctively with three different temperatures as follows.

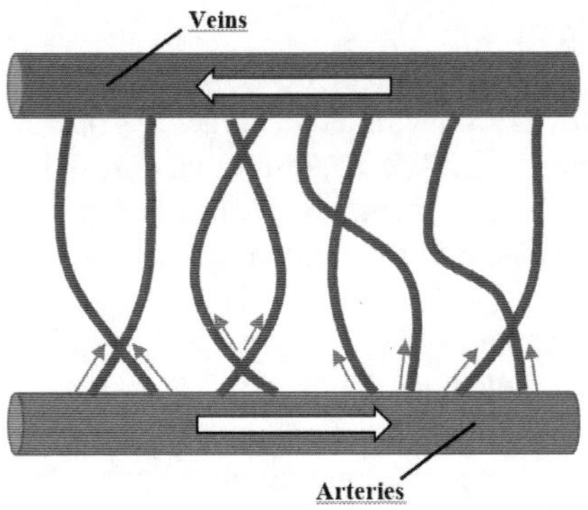

Figure 2. Schematic view of countercurrent heat exchange.

For the arterial blood phase:

$$\varepsilon_a\left(\rho c_p\right)_f\frac{\partial T_a}{\partial t}+\varepsilon_a\left(\rho c_p\right)_f\mathbf{u}_a\cdot\nabla T_a=\left[\varepsilon_a k_f+\varepsilon_a k_{dis,a}\right]\nabla^2 T_a-a_a h_a\left(T_a-T_t\right)-\omega_a\left(\rho c_p\right)_f T_a+\varepsilon_a Q_b \quad (5)$$

For the venous blood phase:

$$\varepsilon_v\left(\rho c_p\right)_f\frac{\partial T_v}{\partial t}+\varepsilon_v\left(\rho c_p\right)_f\mathbf{u}_v\cdot\nabla T_v=\left[\varepsilon_v k_f+\varepsilon_v k_{dis,v}\right]\nabla^2 T_v-a_v h_v\left(T_v-T_t\right)-\omega_v\left(\rho c_p\right)_f T_v+\varepsilon_v Q_b \quad (6)$$

For the tissue phase:

$$\begin{aligned}\left(1-\varepsilon\right)\left(\rho c_p\right)_t\frac{\partial T_t}{\partial t}&=\left(1-\varepsilon\right)k_t\nabla^2 T_t+a_a h_a\left(T_a-T_t\right)+a_v h_v\left(T_v-T_t\right)+\left(\rho c_p\right)_f\left(\omega_v T_v+\omega_a T_a\right)+\\&+\left(1-\varepsilon_a-\varepsilon_v\right)Q_t\end{aligned} \quad (7)$$

The subscripts t, a, v, represent tissue, arterial and venous blood, respectively, while k_{dis} is the thermal dispersion conductivity, which can be estimated according to the relationship in [23]:

$$k_{dis}=\frac{3\varepsilon_a\left[\left(\rho c_p\right)_f\right]^2\left|\mathbf{u}_b\right|^2}{14 a_a h_a} \quad (8)$$

Following the studies of Nakayama et al [15, 24], some assumptions have to be considered:

- $\varepsilon_a=\varepsilon_v$ and $\varepsilon_a+\varepsilon_v=\varepsilon$;
- $a_a h_a=a_v h_v$;
- $u_a=u_v$;
- $\omega_a=-\omega_v$.

In this model the convective heat transfer coefficients ah considered in the LTNE equations, are replaced by the convection-perfusion terms, namely $(a_a h_a \pm (\rho c_p)_f\omega)$ [25]; furthermore, the perfusion rate ω should vary locally, unlike that of Pennes, but it is assumed that its local value is provided everywhere, and its value varies in the range from 2 x 10^{-4} s^{-1} to 5 x 10^{-4} s^{-1} [15], so in this paper the value of blood perfusion is selected to be 0.0005 s^{-1} for all cases, considering the value that reduces mostly the heat transfer coefficient; specifically, h=156.1 W m^{-2} K^{-1} in the worst case, for porosity ε=0.005 and blood velocity modulus u=3.4 cm/s. Also, for this bioheat model, the initial and boundary conditions are the same of those used for the models above.

IOP Conf. Series: Journal of Physics: Conf. Series **1224** (2019) 012001 doi:10.1088/1742-6596/1224/1/012001

4. Numerical approach and validation

The bioheat transient models have been implemented by using the finite-element commercial code COMSOL Multiphysics software. A 2D axisymmetric model has been used to minimize computational efforts and consequently computing time. The mesh chosen for all the models has 11746 triangular elements; thicker meshes have been tested on temperature profiles, but without particular improvements. As regards the transient solver, the absolute tolerance used is 0.001, the time stepping method is the intermediate BDF with initial and maximum steps of 0.001 s and 1 s respectively for the 50 s heating case. For the 2 s heating condition, the maximum step considered is 0.1 s. In order to validate the presented different mathematical models, simulation results of tissue temperature distributions at the centre of the sphere are then compared against the results obtained by Yuan [16], in which LTNE and LTE models are implemented for different conditions of porosity, blood velocities and heating conditions for a cubical tissue. In this work the sphere has the same volume of the Yuan's cube, in order to have the same heating power density; moreover the properties of blood and tissue, the blood velocities and the heating conditions chosen in this work are identical in order to validate the results obtained in COMSOL. The LTNE and LTE models have been simulated in COMSOL for all porosities, blood velocities and heating conditions and they fit the results obtained by Yuan very well in all cases.
In figure 3 only the extreme cases of LTNE equations have been presented, considering three blood velocities, i.e. figure 3(a) shows the results for ε=0.005, Q_t= 2 x 10^6 W m^{-3} and the blood velocities u=0.07 cm s^{-1}, u=0.4 cm s^{-1}, u=3.4 cm s^{-1}; figure 3 (b) regards the outcomes for ε=0.05, Q_t= 50 x 10^6 W m^{-3} and the same three blood velocities. An excellent agreement with literature data has been found.

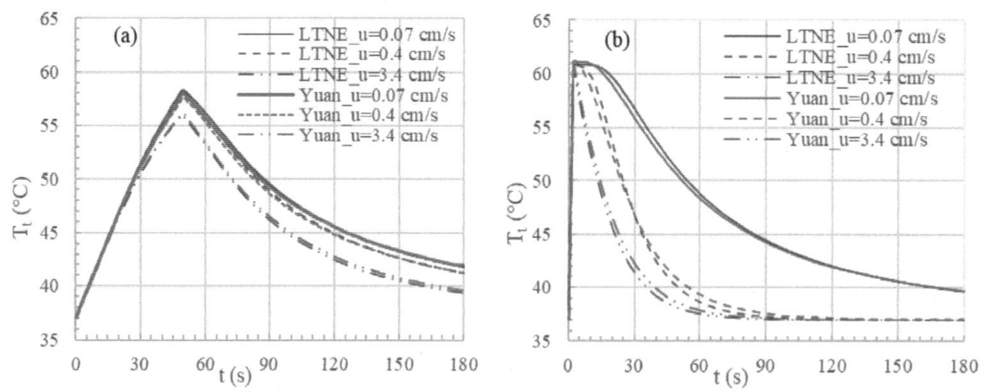

Figure 3. Temperature distributions of LTNE model compared with Yuan's LTNE equations: (a) ε=0.005, Q_t= 2 x 10^6 W m^{-3}; (b) ε=0.05, Q_t= 50 x 10^6 W m^{-3}.

5. Results and discussion

In this section all the tissue temperatures at the central point of the sphere obtained from the simulations for the different analyzed bioheat models are presented and compared at the same blood velocities, considering four separate cases as described in figures 4-7.

Comparing figure 4 to figure 6 and figure 5 to figure 7, the first result to underline is that for all the models the maximum temperature increases in the case of shorter heating duration, but in this condition of quick heating the temperature drop rate occurs more rapidly than in the longer heating because of the enhanced heat transfer. Another common conclusion for all the models is that the temperature decreases with an increase in porosity because higher porosity means having more blood to carry the heat away from the tissue, see figure 4 vs. figure 5 and figure 6 vs. figure 7. Moreover the Peclet number, that is equal to $\varepsilon u \rho c_p 2R/k$, increases as the porosity increases too, so the advection term becomes more important than the conductive one and consequently the temperature decreases more quickly. Obviously this is not valid for the Pennes equation that is not referred to a porous medium.

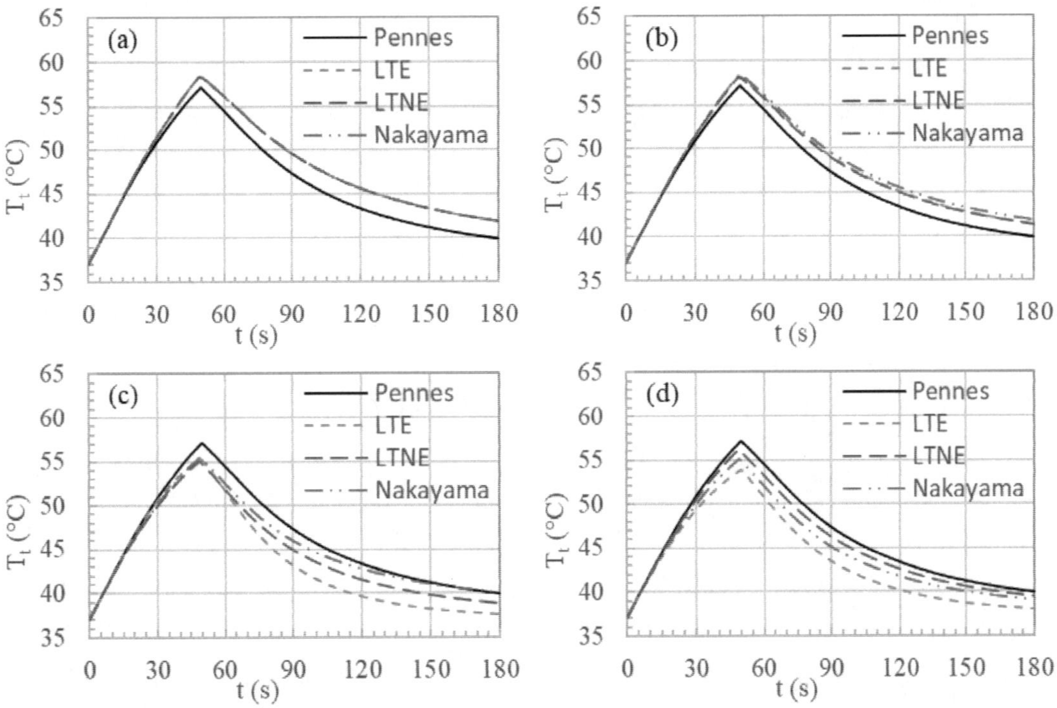

Figure 4. Tissue temperature distributions for ε=0.005, Q_t= 2 x 10^6 W m^{-3} when heating time is 50 s: (a) u =0.07 cm/s, (b) u=0.4 cm/s, (c) u =2 cm/s, (d) u =3.4 cm/s.

Comparing LTNE and LTE models for the two porosities considered, they yield the same results for lower velocities, specifically for u<0.4 cm s^{-1}, which means that the LTE equation is suitable for predicting the temperature during hyperthermia treatment when the blood vessels distributed in the tissue have a diameter less than 30 µm (capillaries, arterioles and terminal arteries); for higher velocities (and consequently vessels diameters), the LTE model overrates the heat carried out by the blood flow; in particular, the increase of velocity corresponds to a decrease of the Stanton number, that is equal to $h/(u\rho c_p)$, so the LTNE equations have to be used. In addition, for the LTNE equations the tissue temperature drops as the blood velocity rises, until u=2 cm/s; for higher velocities, the temperature restarts to increase because the higher velocity is compensated by the growth of the volumetric transfer area a. These considerations on different velocities are not valid for the Pennes' model, which does not take into account the blood velocity, but only the blood perfusion.

As regards the Nakayama three-energy equation model, the results are in good agreement with the LTNE model for the lowest porosity and both the heating conditions, while for ε=0.05 it yields higher temperatures than the LTNE ones, except for the highest velocity u=3.4 cm/s, when LTNE equations result in higher temperatures. However, this more complex model requires more detailed anatomical data compared to others, so its application could be useful only in particular case, such as when artery and vein temperatures have necessarily to be considered different.

To notice that typical values of tissue temperatures are higher than 55 °C because the goal of thermo ablation therapy is the necrosis of tumoral tissue, and actually this can be obtained with different combinations of input power and time of application depending also on the tumor dimension, for example from 32 W to 180 W for a total duration of the corresponding treatment of 15 min and 6 min as described in [26]. On the other hand, the surrounding healthy tissue has to be preserved and the medical probe usually cannot support temperatures higher than 120 °C, so it is preferable to not overcome 100 °C. For these reasons, the differences observed are important especially taking into account higher input powers and application times.

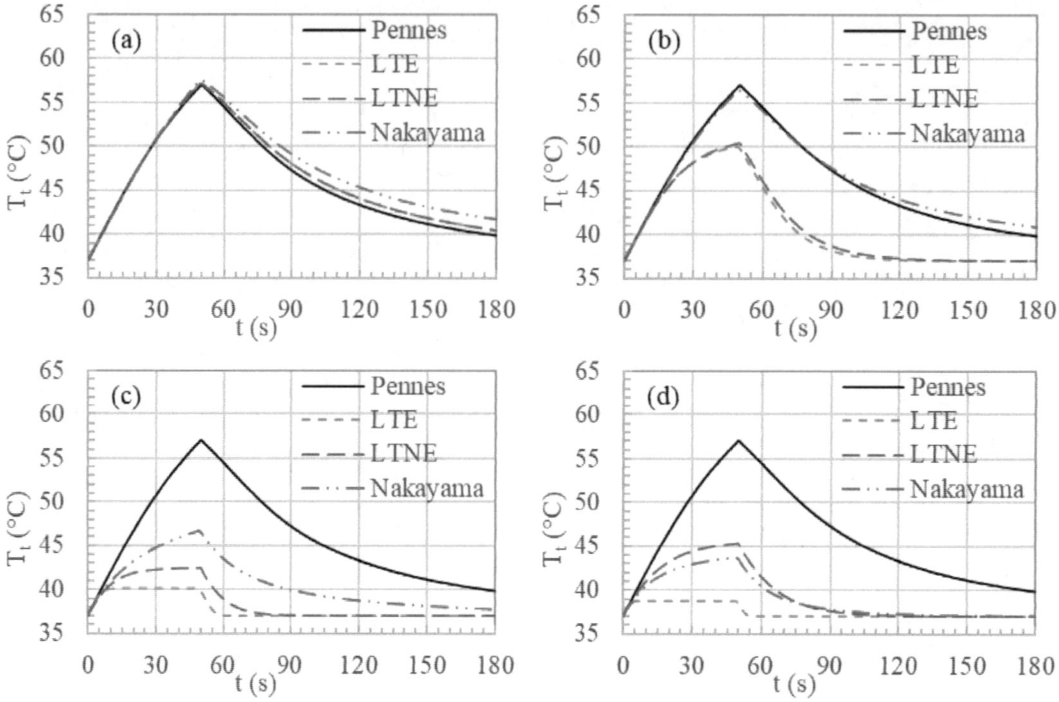

Figure 5. Tissue temperature distributions for ε=0.05, Q_t= 2 x 10^6 W m^{-3} when heating time is 50 s: (a) u =0.07 cm/s, (b) u =0.4 cm/s, (c) u =2 cm/s, (d) u =3.4 cm/s.

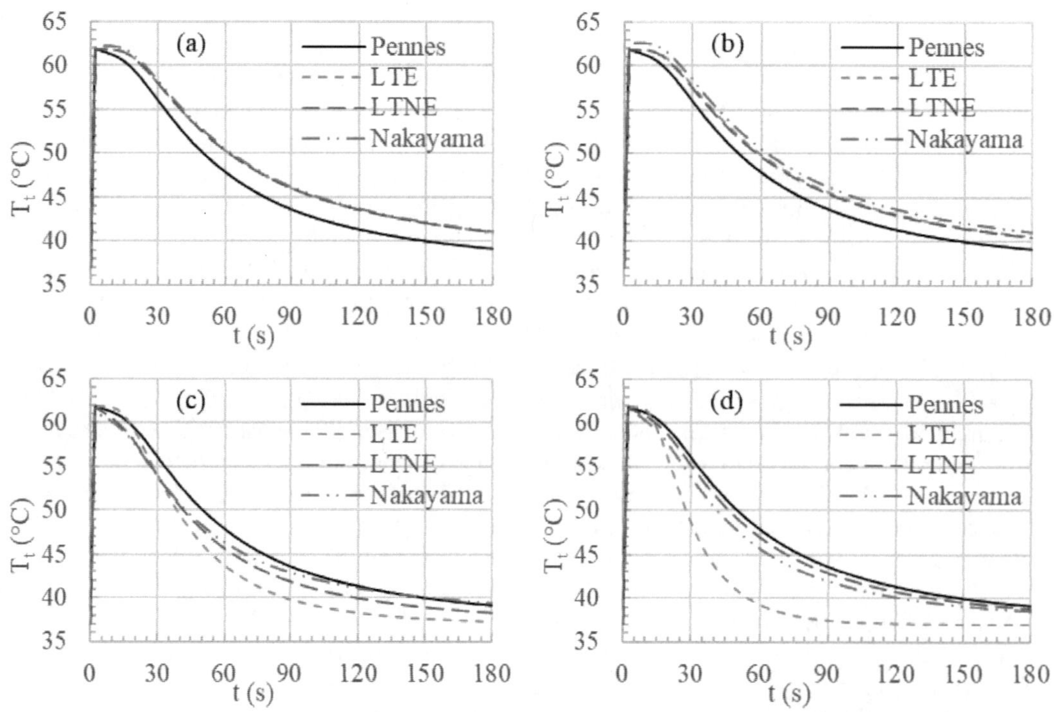

Figure 6. Tissue temperature distributions for ε=0.005, Q_t= 50 x 10^6 W m^{-3} when heating time is 2 s: (a) u=0.07 cm/s, (b) u=0.4 cm/s, (c) u=2 cm/s, (d) u=3.4 cm/s.

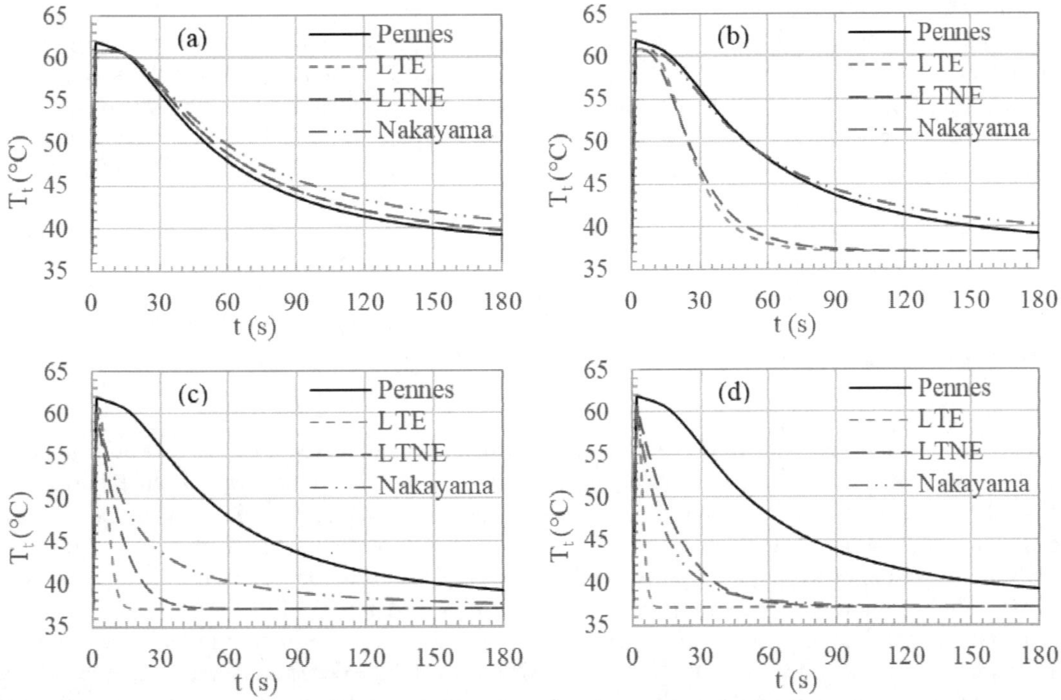

Figure 7. Tissue temperature distributions for ε=0.05, Q_r= 50 x 10^6 W m^{-3} when heating time is 2 s: (a) u=0.07 cm/s, (b) u=0.4 cm/s, (c) u=2 cm/s, (d) u=3.4 cm/s.

In addition, it is interesting to show the computational times of the different models simulations, more in particular, it has been chosen the most onerous case, which is referred to Q_r= 50 x 10^6 W m^{-3} when heating time is 2 s, for ε=0.05, and u=3.4 cm s^{-1}. In this situation, using a 2.50 GHz Intel Core i7-4710MQ CPU and a 16 GB 799 MHz DDR3 RAM, the computational times for Pennes' equation, LTE model, LTNE equations and three-energy equation model are 30'19'', 33'37'', 45'23'', and 59'50'' respectively. These results confirm the simplicity of Pennes' bioheat model, which is also the less onerous computationally, but in this case, it produces not accurate outcomes. In the other cases, the computational times are comparable to the Pennes' ones, so it has not been reported.

6. Radiative heat transfer

Radiation effects on the biological medium considered in this paper can be analysed with two simplified approaches, the Rosseland diffusive approximation and the Beer-Lambert-Bouguer law; the first assumes the medium to be optically thick and allows to define a "radiative conductivity" as follows:

$$k_R = \frac{16n^2\sigma T^3}{3\beta} \tag{9}$$

where n is the refractive index, assumed to be 1.4 [27], σ is the Stefan-Boltzmann constant and β is the extinction coefficient, considered equal to 5550 m^{-1}, according to the study of optical properties of ex vivo human tissues by Simpson et al. [28]. The Beer-Lambert-Bouguer law, instead, is a simplified form for the Radiative Transfer Equation in which the radiation is assumed to be collimated, and consequently the radiative heat flux divergence (heat source term for the governing equations) could be written as:

$$\nabla \cdot q_R = -\beta I_0 e^{-\beta\sqrt{r^2+z^2}} \tag{10}$$

where I_0 is the intensity of radiation measured in W m^{-2}. Since one can assume that the energy irradiates from the internal sphere to the external (the internal sphere acts as a catheter with a spherical tip), the heat exchange area for the radiation contribution is the internal/external spheres contact surface area. This because if one assumes that the radiative source is pointwise, for $r = 0$ one can have an infinite value of the radiative source term in Eq. 10. For this reason, the tissue temperature is evaluated at r=0.62 cm in the model that accounts for the Beer-Lambert-Bouguer law and consequently it is compared with tissue temperature at r=0.62 cm for LTNE. Radiative contribution effects are presented in figure 8 for both Rosseland approximation and Beer-Lambert-Bouguer law. It is shown that in both cases the radiative contribution does not affect the tissue temperature in all the models considered and can be neglected, even in the case of the maximum temperature reached (i.e. for ε=0.005, Q_t= 50 x 10^6 W m^{-3}, u=0.07 cm/s and when heating time is 2 s).

7. Conclusions

The study of bioheat transfer is a challenging task because of the complexity of the involved phenomena. Modelling heat transfer in tissues has a key role because of the difficulty of carrying out experiments. In this paper some of the most relevant bioheat transfer models proposed through the years are applied on a spherical human tissue modelled as a porous medium, in order to compare them. This comparison is performed in terms of temperature profiles at equal conditions. More in detail, from the simplest to the most complex, the four models considered are the Pennes equation, the Local Thermal Equilibrium equation, the Local Thermal Non-Equilibrium equations and the three-equation model. By using the finite-element commercial code COMSOL Multiphysics software, the tissue temperature profiles versus time are presented for each model under different porosities, blood velocity, vessels diameters and heating conditions. Numerical results are validated by means of results from literature. According to the results, the Pennes' model shows its inaccuracy due to its simplicity, while the LTE model is suitable to predict temperatures during hyperthermia therapy when the diameter of the blood vessels is less than 30 μm, which corresponds to the low blood velocity of 0.4 cm s^{-1}. For higher velocities, and consequently vessels diameters, the LTE model overrates the heat carried out by the blood flow, so the LTNE equations have to be used to take into account interfacial convective heat transfer. As regards the Nakayama three-energy equation model, the results are in good agreement with the LTNE model for the lowest porosity and both the heating conditions, even if this model requires more detailed anatomical data compared to others, so its application could be useful only in particular case, such as when artery and vein temperatures have necessarily to be considered different. This means that the LTNE model is preferable.

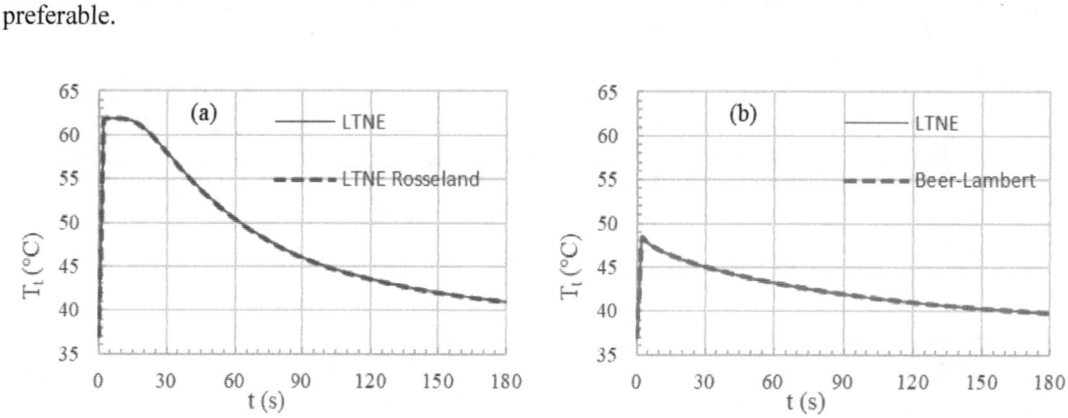

Figure 8. Tissue temperature profiles calculated for LTNE model compared with the tissue temperatures reached taking into account the radiative contributions from the Rosseland approximation (a) and Beer-Lambert-Bouguer law (b).

Furthermore, for ε=0.05 the three-energy equation model yields higher temperatures than the LTNE ones, except for the highest velocity u=3.4 cm/s, when LTNE equations result in higher temperatures. Finally, radiative heat transfer effects have been analyzed by employing either Rosseland approximation or Beer-Lambert-Bouguer law, showing that radiation through biological tissues is negligible for the conditions herein presented. In conclusion, we can state that LTNE model is preferable because it is a good compromise between accuracy and complexity. For further developments, it is clear that these models will require more comparisons with experiments in order to understand which one is better depending on the application.

References

[1] Moros E 2012 *Physics of thermal therapy: fundamentals and clinical applications* (Boca Raton: CRC Press).
[2] Chu K F and Dupuy D E 2014 Thermal ablation of tumours: biological mechanisms and advances in therapy *Nature Reviews Cancer* **14** 199-208.
[3] Diederich C J 2005 Thermal ablation and high-temperature thermal therapy: overview of technology and clinical implementation *International journal of hyperthermia* **21** 745-53.
[4] Wang K, Tavakkoli F, Wang S and Vafai K 2015 Analysis and analytical characterization of bioheat transfer during radiofrequency ablation *Journal of biomechanics* **48** 930-40.
[5] Rattanadecho P and Keangin P 2013 Numerical study of heat transfer and blood flow in two-layered porous liver tissue during microwave ablation process using single and double slot antenna *International Journal of Heat and Mass Transfer* **58** 457-70.
[6] Yee R, Connolly S and Noorani H 2003 Clinical review of radiofrequency catheter ablation for cardiac arrhythmias *The Canadian journal of cardiology* **19** 1273-84.
[7] González-Suarez A and Berjano E 2016 Comparative analysis of different methods of modeling the thermal effect of circulating blood flow during RF cardiac ablation *IEEE Transactions on Biomedical Engineering* **63** 250-9.
[8] Karsch K R and Haase K K 1991 *Coronary laser angioplasty: an update* (Berlin: Springer-Verlag Berlin Heidelberg).
[9] Vincent G M, Fox J, Johnson M D, Strickland R, Garry S Land Hammond E 1990 Thermal laser probe angioplasty: Influence of constant tip temperature, plaque composition, and probe/vessel diameter ratio *Lasers in surgery and medicine* **10** 420-6.
[10] Iasiello M, Vafai K, Andreozzi A, Bianco N and Tavakkoli F 2015 Effects of external and internal hyperthermia on LDL transport and accumulation within an arterial wall in the presence of a stenosis *Annals of biomedical engineering* **43** 1585-99.
[11] Iasiello M, Vafai K, Andreozzi A and Bianco N 2016 Low-density lipoprotein transport through an arterial wall under hyperthermia and hypertension conditions–An analytical solution *Journal of biomechanics* **49** 193-204.
[12] Pennes H H 1948 Analysis of tissue and arterial blood temperatures in the resting human forearm *Journal of applied physiology* **1** 93-122.
[13] Vafai K 2015 *Handbook of porous media* (Boca Raton: CRC Press).
[14] Roetzel W and Xuan Y 1997 Bioheat equation of the human thermal system *Chemical engineering & technology* **20** 268-76.
[15] Nakayama A and Kuwahara F 2008 A general bioheat transfer model based on the theory of porous media *International Journal of Heat and Mass Transfer* **51** 3190-9.
[16] Yuan P 2008 Numerical analysis of temperature and thermal dose response of biological tissues to thermal non-equilibrium during hyperthermia therapy *Med Eng & Phy* **30** 135-43.
[17] Crezee J and Lagendijk J J W 1992 Temperature uniformity during hyperthermia: the impact of large vessels *Phys Med Biol* **37** 1321-37.
[18] Chato J C 1980 Heat transfer to blood vessels *J Biomech Eng-T ASME* **102** 110-8.
[19] Weinbaum S and Jiji L M 1985 A new simplified bioheat equation for the effect of blood flow on local average tissue temperature *J Biomech Eng-T ASME* **107** 131-9.

[20] Kou H S, Shih T C and Lin W L 2003 Effect of the directional blood flow on thermal dose distribution during thermal therapy: an application of a Green's function based on the porous model *Phys Med Biol* **48** 1577-89.

[21] Duck F A, Baker A C and Starrit H C 1998 *Ultrasound in medicine* London: Institute of Physics Publishing 57-88.

[22] Vera A and Leija L 2008 Microcoaxial Double slot antenna for interstitial hyperthermia: design, modeling and validation, in: International Conference on Advances in Electronics and Micro-electronics ENICS2008 138-143.

[23] Yang C and Nakayama A 2010 A synthesis of tortuosity and dispersion in effective thermal conductivity of porous media *International Journal of Heat and Mass Transfer* **53** 3222-30.

[24] Nakayama A, Sano Y and Yoshikawa K 2010 A rigorous derivation of the bioheat equation for local tissue heat transfer based on a volume averaging theory *Heat Mass Transfer* **46** 739-46.

[25] Nakayama A, Kuwahara F and Liu W 2009 A Macroscopic Model for Countercurrent Bioheat Transfer in a Circulatory System *Journal of Porous Media* **12** 289-300.

[26] Hoffmann R, Rempp H, Erhard L, Blumenstock G, Pereira P L, Claussen C D, Clasen S 2013 Comparison of four microwave ablation devices: an experimental study in ex vivo bovine liver *Radiology* **268** 89-97.

[27] Tsenova V and Stoykova E 2003 Refractive index measurement in human tissue samples *Proc. SPIE* **5226** 413-17.

[28] Simpson RC, Kohl M, Essenpreis M and Cope M 1998 Near-infrared optical properties of ex vivo human skin and subcutaneous tissues measured using the Monte Carlo inversion technique *Phys Med Biol* **43** 2465–78.

IOP Conf. Series: Journal of Physics: Conf. Series **1224** (2019) 012002 doi:10.1088/1742-6596/1224/1/012002

Projection algorithm for simulation of fluid flow around moving objects with immersed boundary method

Alessia Abbati*, Andrea Chierici*, Leonardo Chirco*, Roberto Da Vià*† and Sandro Manservisi*

*University of Bologna, DIN, Lab. of Montecuccolino, Via dei Colli 16, Bologna 40136, Italy

E-mail: †roberto.davia2@unibo.it

Abstract. The simulations of fluid flow around moving objects are usually done using body conforming meshes, but recently there is growing interest in algorithms using non conforming grids, known as immersed boundary methods (IB). In particular the resolution of the flow around objects with moving boundary may involve a high meshing work that has to be repeated at each time step. In the present work the immersed boundary method is applied to an in-house finite element code, in order to simulate transient applications involving the movement of immersed solid interfaces. The distinguishing feature of this method is that the entire simulation is carried out on a grid which cover the entire domain and do not conform to the geometry of the rigid objects. A second mesh independent from the first one and built on the immersed body is considered. The presented method consists in the projection of fields, including boundary conditions, from the second mesh mentioned to the first one, that cover the whole domain. The projection is carried out considering the movement of the immersed objects, through the handling of the second mesh. The proposed algorithm is built in order to satisfy the mass conservation in the entire fluid region and to avoid any numerical instability. For the projection of such fields we develop a multiphysics open-source coupling platform, based on the open-source SALOME platform. MED (Modèle d'Échange de Donné) data structure has been used for all field operations, allowing the algorithm to be used with all the numerical codes where a MED duplicate of the field solution can be created. Applications of this approach will be presented and we provide results that show the consistency of the proposed algorithm.

1. Introduction

Over the last decades, numerical simulation has become an increasingly widespread tool to investigate complex engineering applications. Numerous computational fluid dynamics (CFD) problems involve fluid flow around moving objects within the framework of finite element approaches. Such systems are characterized by internal boundaries between fluid and solid regions with dynamically changing geometry. In this regard, one of the challenges to devising numerical techniques is to track the moving boundary and simulate its evolving interaction with the system. Even though considerable efforts have been made, simulating moving boundaries is still not a trivial problem. Methods for handling such problems can broadly be classified into Lagrangian, Eulerian and combined formulations according to their computational grid representation. Lagrangian methods feature a moving computational grid that is dynamically adapted to the interface. Boundary conditions can be applied at the exact location of the

IOP Conf. Series: Journal of Physics: Conf. Series **1224** (2019) 012002 doi:10.1088/1742-6596/1224/1/012002

interface since its evolution is explicitly tracked. These methods are widely used Fluid Structure Interaction problems solved with monolithic approach, allowing a better treatment of the stress continuity along solid-fluid interface [1, 2], for example in [3], where Fluid Structure Interaction problem is used to perform a preliminary investigation of a ventricle contraction, or [4] where a dual grid approach is used to couple an FSI problem with multiphase advection. The major drawback to this approach is the continuous grid rearrangement over the changing computational domain. Moreover large displacements are difficult to tackle since mesh element deformations can lead to stability issues due to excessive mesh element stretching. Conversely, Eulerian methods do not track the interface explicitly but reconstruct it by means of appropriate field variables. This allows a fixed grid formulation where the interface is not treated as a sharp discontinuity, which implies smearing of boundary information. In Eulerian methods topological changes are easier to handle since they do not require mesh rearrangement work and thus they are particularly suited for free surfaces applications. These methods are widely used in the case of multiphase flows [5, 6] and multiphase flows with solid particles, for example in [7], where an innovative dual grid approach is used to couple the multiphase CFD problem to the Discrete Element Method one. It is clear that for these applications an adaptive mesh algorithm cannot be used. The present work specifically address rigid boundaries problems by introducing a combined approach that aims to overcome the drawbacks resulting from both Eulerian and Lagrangian formulations.

Recent years have seen a significant interest in solution algorithms for fluid flow around objects based on non-conforming grids, namely meshes that are not configured to conform to the fluid-solid interface. Several numerical models referred to as immersed boundary methods (IB) have received attention due to their remarkable advantages in simulating fluid flow around moving objects with complex boundaries. The IB method was originally introduced by Peskin as a means of investigating cardiovascular flow patterns [8]. Although originally intended for biological fluid dynamics, the IB method has been successfully applied to a wide range of CFD problems and some noteworthy works have been proposed in the literature [9, 10]. Such methods share as a common core feature the ability to simulate the presence of boundaries without fitting the computational grid to the immersed body. The grid generation process is not altered and spatial discretization is performed over the whole computational domain, resulting in mesh elements being cut by the fluid-solid interface. IB methods constitute a promising alternative to conventional body-conformal grid approaches. The process of generating body-fitted grids requires extensive meshing work that may become prohibitively expensive for complex geometries. In order to handle moving boundaries problems, the grid needs to be regenerated at each time step, adding up to the computational costs. By contrast, IB methods allows simplified spatial discretization that results in more uniform computational grids obviating the need for mesh rearrangement. Because tackling complex moving boundaries is easier, non-conforming grid approaches experience less deterioration in grid quality which affects accuracy and convergence properties of the numerical scheme. Another common point of concern underlying IB methods is the imposition of boundary conditions on the fluid-solid interface. Conventional body-conformal meshes allows boundary conditions to be specified directly on the interface grid points with relative ease. Implementation of boundary condition is not as straightforward in IB methods, which in turn may negatively impact the robustness and conservation properties of the solution algorithm. IB methods account for the fluid-solid interface and its effects through several different strategies, which are extensively reviewed in [11]. The presence of the immersed body is mimicked by means of additional source terms and modified boundary conditions which may be imposed both by indirect and direct techniques.

In the present work, we present a finite element based IB method intended for simulation of fluid flow around moving objects. Numerical solution of fluid flow is computed on an Eulerian grid which spans over the entire domain whereas moving boundaries are tracked by a Lagrangian reference frame. This combined formulation employs a fixed unstructured grid in conjunction with a moving unstructured grid built exclusively on top of the immersed body. The solid mesh accounts for the movement of the object without altering the fixed computational grid and consequently simplifying the discretization process. Coupling between fixed and moving mesh is tackled by appropriate projections of fields and boundary conditions. Thus, effects of moving objects on fluid flow may be imposed through indirect means by introducing an adequate projection algorithm. The proposed IB method succeeds in decoupling the equations for the fluid domain from those imposed on the moving solid and consequently no numerical solution is required for governing equations inside the immersed body. The pressure discontinuity associated with the fluid-solid interface requires modifying the computational stencil near the immersed body to impose an additional boundary condition. Effects of additional parameters introduced in this formulation are not straightforward and may negatively affect computational stability and numerical solution conservation properties.

We develop a multiphysics coupling tool based on the open-source SALOME platform to implement advanced interpolation techniques [12, 13]. MED (Modèle d'Échange de Donné) data structure has been used as a common format to define information exchange between codes. Manipulation of meshes and fields in numerical schemes involved in a simulation process is greatly simplified by adopting a common exchange format. Reduced code coupling complexity allows sharing advanced functions and preserving existing routines. Several interpolation methods have emerged in recent years and a detailed discussion on this issue is presented in [4]. Interpolation techniques projects numerical fields from a source computational grid to a target domain. In the proposed method we use consistent or node-wise interpolation and we evaluate its performance potential. A key factor in consistent interpolation is imposing equality between the node field values on the target domain and the corresponding source mesh. In this work we define a piece-wise indicator function on the support of the solid moving grid, which is set to be the source mesh. The algorithm then projects this field to the target fixed grid spanning over the entire computational domain in order to define a discretized indicator function. The resulting function allows to represent the immersed body and its location in the fluid domain. Additionally, we specify a cell-wise volume fraction field to evaluate the solid body portion of each computational cell. A different interpolation scheme, denoted Galerkin projection, is then implemented to determine a point-wise target field. This information allows to explicitly track the fluid-solid interface through linear reconstruction. The proposed method represent an efficient approach to interface reconstruction since it enables the computation of interface variables without being prohibitively complex and time consuming. Numerical solution for fluid flow around a moving object is then accurately computed on the fixed grid. At each time iteration the solid computational grid position is changed using an arbitrary law, involving a fixed movement or an interaction with fluid through stress calculation on fluid-solid interface. Consequently indicator function is interpolated and the volume fraction field updated on the target mesh (fluid computational grid). Numerical solution accuracy strictly relates to the robustness of the projection algorithm, thus it is worthwhile to investigate the interpolation process. We assess the performance of the proposed IB approach with two significant test cases with the intent of validating the numerical scheme.

2. Numerical Modeling

Let us consider a moving rigid body $\Omega_s(t)$ completely immersed in a fluid domain $\Omega_f(t)$ and separated by a time dependent interface $\Gamma_{f,s}(t) = \partial\Omega_f(t) \cap \partial\Omega_s(t)$, as sketched in Figure 1 a). Any topological changes in the solid region affect in turn the fluid domain $\Omega_f(t)$ in such a way that the entire computational domain $\Omega = \Omega_f(t) \cup \Omega_s(t)$ does not vary over time. The governing equation considered here is the unsteady Navier-Stokes equation for incompressible laminar flow

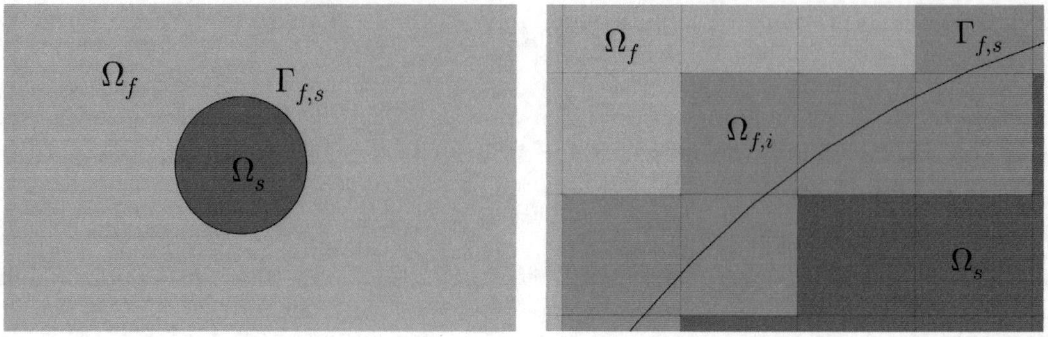

Figure 1. Sketch representation a) of a computational domain made of a fluid (Ω_f) and solid (Ω_s) sub domains which interact through a fluid-solid interface ($\Gamma_{f,s}$) and close-up view b) of computational grid cut by the fluid-solid interface.

given by, namely

$$\nabla \cdot \mathbf{u} = 0\,, \tag{1}$$

$$\frac{\partial \mathbf{u}}{\partial t} + (\mathbf{u} \cdot \nabla)\mathbf{u} = -\frac{1}{\rho}\nabla P + \nu(\nabla^2 \mathbf{u}) + \mathbf{g}\,. \tag{2}$$

where $\mathbf{u}$ is the velocity vector, ρ the density, P the pressure, ν the kinematic viscosity and $\mathbf{g}$ the gravity acceleration vector.

2.1. Field projection

Let us introduce a source field Ψ_s defined on the computational domain Ω_s and a target field Ψ_t in Ω_t. Let N_s^e be the number of elements and n_e the number of nodes that compose Ω_s. The projection operator that interpolates the source field Ψ_s to Ψ_t is defined by

$$\Psi_t = \prod_{\Omega_t}^{\Omega_s}(\Psi_s) = \sum_{e=1}^{N_s^e}\sum_{j=1}^{n_e} \Psi_{s,j}^e \varphi_j^e(\pi^{-1}(\mathbf{x}_j^{t,e}))\,. \tag{3}$$

where φ_j are the approximation functions, π^{-1} the linear transformation from global coordinates to canonical ones and $\mathbf{x}^t$ the target field node coordinates. The projection operator may be denoted by $|P|$ and written in the following compact form

$$\Psi_t = |P|\Psi_s \tag{4}$$

More specifically, the proposed projection scheme implements in turn a search point algorithm, reverse mapping and an interpolation algorithm. A more detailed description of the point-wise interpolation method can be found in [4].

2.2. Immersed boundary formulation

In the present formulation of Immersed Boundary algorithm we will label with $\Omega_s(t)$ and $\Omega_f(t)$ the solid and fluid physical regions, respectively, as sketched in Fig. 1 a). The union of these two regions will be denoted as Ω. We remark that Ω is not a function of time and that this region will be used as computational domain. When dealing with immersed boundary algorithms it is important to define a method for identifying fluid and solid regions. In literature many methods

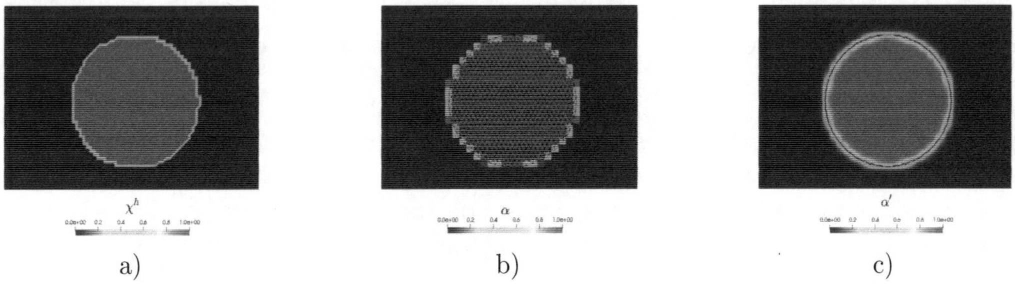

Figure 2. Example of discretized indicator function χ^h a) of volume fraction α, together with a mesh representation of solid body b), and interpolated volume fraction α' together with a fluid-solid interface reconstruction c).

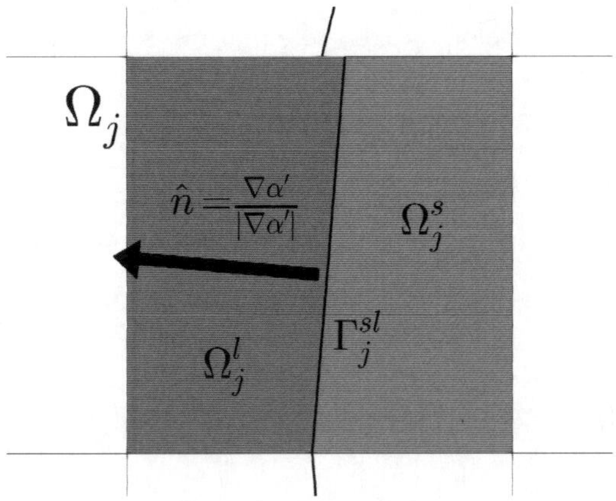

Figure 3. Close-up view of interface reconstruction on a generic mesh element.

are proposed, involving, for example, level set methods based on wall distance between a mesh node and the fluid-solid interface [9, 10]. In the present work we adopt an indicator function $\chi(\mathbf{x}, t)$ defined as a multidimensional Heaviside function, meaning

$$\chi(\mathbf{x}, t) = \int_{\Omega_s(t)} \delta(\mathbf{x}' - \mathbf{x})\, d\mathbf{x}' \quad \forall \mathbf{x} \in \Omega\,, \tag{5}$$

where δ is the Dirac delta function. The possible values of $\chi(\mathbf{x}, t)$ over Ω are then

$$\chi(\mathbf{x}, t) = \begin{cases} 1\,, & \forall \mathbf{x} \in \Omega_s(t)\,. \\ 0\,, & \forall \mathbf{x} \in \Omega_f(t)\,. \end{cases} \tag{6}$$

A discretized indicator function $\chi^h(\mathbf{x}, t)$ is computed using the projection method described in Section 2.1. In particular we define a source field ψ_s on the solid body mesh, with a uniform value $\psi_s(\mathbf{x}) = 1 \forall \mathbf{x} \in \Omega_s$. The field $\chi^h(\mathbf{x}, t)$ is then calculated as a projection of ψ_s over the computational domain, meaning $\chi^h = |P|\psi_s$. Alongside the point-wise indicator function χ^h we define a cell-wise volume fraction field α that, for each computational cell, indicates the fraction of volume occupied by the solid body. Examples of both χ^h and α are reported in Fig. 2 a) and b), respectively. We finally introduce a point-wise field, denoted as α', which is calculated with a Galerkin projection of volume fraction α, namely

$$\int_\Omega \left(\alpha'\phi + \gamma \nabla\alpha' \cdot \nabla\phi\right) d\Omega = \int_\Omega \alpha\phi d\Omega \quad \forall \phi \in H^1(\Omega)\,, \tag{7}$$

where γ is an arbitrary diffusion coefficient. Field α' can be used to perform a linear reconstruction of real solid-fluid interface $\Gamma_{f,s}$ as labeled in Fig. 1 a) and b). The interpolated values of the α field shown in Fig. 2 b) are reported in Fig. 2 c), together with the interface linear reconstruction. The linear reconstruction is performed as follows. As presented in Fig. 3, a generic mesh cell, labeled as Ω_j, having a volume fraction value $0 < \alpha_j < 1$, can be divided into sub-elements Ω_j^l and Ω_j^s by means of a straight edge. The interface normal unit vector $\hat{\mathbf{n}}$ is calculated as the ratio of local α' field gradient and the gradient modulus. The interface edge Γ_j^{sl} inclination is computed in accordance to $\hat{\mathbf{n}}$, while its position is determined from an area conservation equation, namely

$$\int_{\Omega_j^s} d\Omega = \alpha_j \int_{\Omega_j} d\Omega\,. \tag{8}$$

The reconstructed interface can be used for evaluating quantities of interest, like the stress exerted from the fluid on solid body surface.

The weak formulation of the solved system of equation is then

$$\int_\Omega \frac{\partial \mathbf{u}}{\partial t}\cdot\boldsymbol{\varphi}\,d\Omega + \int_\Omega (\mathbf{u}\cdot\nabla)\mathbf{u}\cdot\boldsymbol{\varphi}\,d\Omega =$$

$$= -\frac{1}{\rho}\int_\Omega \nabla P\cdot\boldsymbol{\varphi}\,d\Omega + \int_\Omega \nu(\nabla^2\mathbf{u})\cdot\boldsymbol{\varphi}\,d\Omega + \int_\Omega \mathbf{g}\cdot\boldsymbol{\varphi}\,d\Omega\,, \qquad \forall\boldsymbol{\varphi}\in H_0^1(\Omega) \tag{9}$$

$$\begin{cases} \int_{\Omega_j}(\nabla\cdot\mathbf{u})\,\psi\,d\Omega = 0\,, & \forall\psi\in L^2(\Omega) \qquad \forall\Omega_j : \alpha_j < \tilde{\alpha} \\ \int_{\Omega_j}\left(\lambda\nabla^2 P - \nabla\cdot\mathbf{u}\right)\,\psi\,d\Omega = 0\,, & \forall\psi\in L^2(\Omega) \qquad \forall\Omega_j : \alpha_j > \tilde{\alpha} \end{cases} \tag{10}$$

$$\mathbf{u}_i = \tilde{\mathbf{u}}_s \qquad \forall\mathbf{x}_i : \chi(\mathbf{x}_i, t) = 1\,, \tag{11}$$

where λ is a constant value and $\tilde{\alpha}$ is a threshold value. In cells where $\alpha > \tilde{\alpha}$ we solve a pressure equation that is similar to the ones used in velocity projection methods [14]. It will be later shown that this equation allows to obtain a smoother pressure field. As reported in (11), the fluid velocity field in the solid-occupied computational region is modeled with the solid velocity field. In particular, the values $\tilde{\mathbf{u}}_s$ imposed in (11) are obtained by a point-wise projection of the velocity field $\mathbf{u}_s$ that is defined on the solid body mesh, namely $\tilde{\mathbf{u}}_s = |P|\mathbf{u}_s$. As can be seen from the solved system of equations, in the present immersed boundary formulation we don't use forcing terms for the imposition of boundary conditions along the solid-fluid interface. The algorithm used for the realization of the numerical simulation with the present immersed boundary method can be summarized, for a generic n-th time step, as:

(i) Solid position on computational grid: creation of volume fraction α, projection of solid velocity $\tilde{\mathbf{u}}_s$ and indicator function χ^h
(ii) System of equations solution
(iii) Calculation of interpolated volume fraction α'
(iv) Reconstruction of solid-fluid interface $\Gamma_{f,s}$
(v) Stress calculation
(vi) Update solid position and velocity field

As explained above, the present Immersed Boundary method is formulated similarly to a Volume Of Fluid method for multiphase problems since in both cases an indicator function and a volume fraction field are used to distinguish one phase from another and to reconstruct the interface between the two phases. Differently from the multiphase problem, where the evolution of the indicator function is studied with the solution of an appropriate advection equation, the simulation of non deformable solid bodies allows to compute the indicator function field with a projection algorithm. With the projection operator arbitrarily moving solid objects can be handled without the need of computing a wall distance field between mesh nodes and the fluid-solid interface, which could be a cumbersome task when dealing with complex solid shapes [9, 15]. Moreover, if the solid body position is fixed in time, in the above described algorithm steps (i), (iii), (iv) and (vi) can be performed just once.

The present Immersed Boundary method has been implemented into Finite Element code FEMuS [16].

3. Results

In the present section we describe the results obtained for two different test cases, involving both a fixed ad a moving obstacle.

3.1. Flow around a cylinder at Re=20

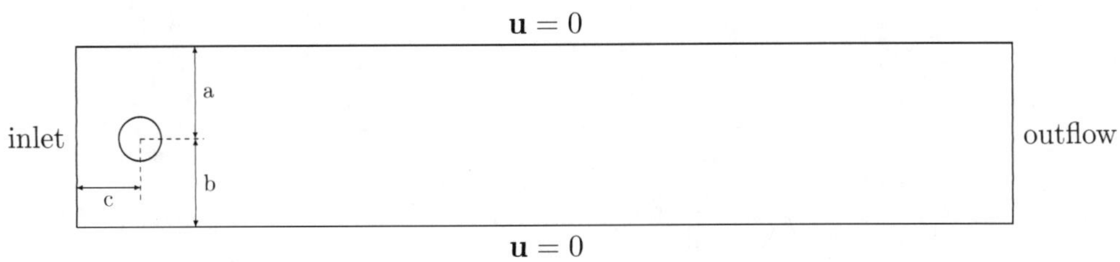

Figure 4. Sketch of the physical domain.

a $[m]$	b $[m]$	c $[m]$	d $[m]$	L $[m]$
0.21	0.2	0.2	0.1	2.2

Table 1. Geometrical parameters of the simulated domain.

	REF0	REF1	REF2
$d/\delta x = d/\delta y$	7.2	14.5	29
ε	-1.5×10^{-5}	-6×10^{-6}	-1.4×10^{-6}

Table 2. Grid resolution and mass loss coefficient for the three different used computational grids.

The present test consists of simulating a stationary fluid flow around a fixed cylinder. This benchmark case is well studied in [17]. Let U_0, d and ν be the mean inlet velocity, the cylinder diameter and the fluid kinematic viscosity, respectively. The resulting Reynolds number is $Re = U_0 d/\nu = 20$. A sketch of the simulated domain, together with the imposed boundary conditions, is reported in Fig. 4, while the geometrical parameters are described in Tab. 1, where L is the length of the channel. A parabolic velocity profile is imposed on the inlet section, namely

$$U(x, y) = 4U_{max} \frac{(y_{max} - y)(y - y_{min})}{(y_{max} - y_{min})^2}, \tag{12}$$

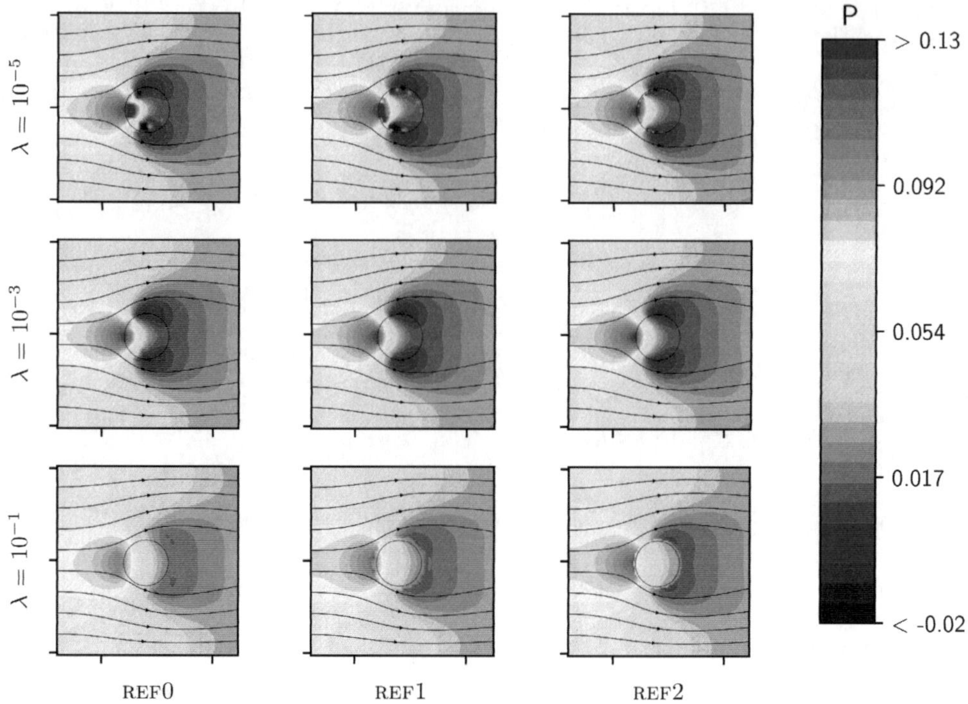

Figure 5. Influence of diffusion coefficient λ on pressure field and flow streamlines in the near solid region.

where y_{max} and y_{min} are the channel maximum and minimum y coordinate values, and U_{max} is the maximum velocity value, $U_{max} = 1.5U_0$. One major concern when dealing with immersed boundary methods is mass conservation. For the present case we use three different computational grids with increasing levels of mesh refinements, in order to demonstrate that the proposed method allows to obtain a good mass conservation. The three different meshes are parameterized as a function of grid resolution with respect to the obstacle diameter and will be hereafter labeled as REF0, REF1 and REF2, where the number stands for the number of refinements. The grid resolution for the three different meshes is reported in Tab. 2 as a function of the ratio $d/\delta x$, where d is the obstacle diameter and δx is the mesh size along x direction. In order to evaluate the behavior of the proposed immersed boundary method regarding mass conservation, we report the values of mass loss fraction ε defined as $(U_0 - U_{m,out})/U_0$, where $U_{m,out}$ is the mean outlet velocity. As can be seen from the values reported in Tab. 2, the mean outlet velocity is slightly higher than the mean inlet velocity value and mass loss fraction decreases by an order of magnitude using a double mesh refinement. In the present test we studied the influence of the diffusion coefficient λ that is used in (10) for the pressure equation in the solid-occupied cells. In Fig. 5 we report a set of close-up views of the pressure fields obtained with the three different mesh refinements and with three different values of λ, namely 10^{-5}, 10^{-3} and 10^{-1}. We see that pressure spots arise in the vicinity of fluid-solid interface when using a small value of λ. Although λ value has not a sensible impact on mass conservation (the values of Tab. 2 refer to $\lambda = 10^{-3}$ but are quite the same for each of the tested values of λ), we see that λ has a great impact on fluid behavior in the near solid body region, as can be seen from the flow streamlines reported in Fig. 5. Finally we compare the obtained results for pressure difference ΔP between front and rear side of the cylinder and drag coefficient c_d with

	ΔP			c_d		
Case	10^{-5}	10^{-3}	10^{-1}	10^{-5}	10^{-3}	10^{-1}
REF0	0.1072	0.1122	0.0741	5.051	5.185	5.351
REF1	0.1105	0.1116	0.0671	4.897	5.014	5.224
REF2	0.1162	0.1120	0.0532	4.912	5.019	5.060
Reference		0.1175			5.579	

Table 3. Computed values of pressure difference ΔP between front and rear side of the cylinder and of drag coefficient c_d, together with reference data [17].

the reference ones, discussed in [17]. The computed values for all simulated cases, in terms of computational grid and λ values, are reported in Tab. 3 along with reference values. The drag coefficient is here calculated as

$$c_d = \frac{2F_x}{\rho d U_0^2} \quad \rightarrow \quad \mathbf{F} = \int_S \left[\mu\left(\nabla\mathbf{u} + \nabla\mathbf{u}^T\right) - p\mathbf{I}\right] \cdot \hat{n} dS\,, \tag{13}$$

where $\hat{n}$ is the normal to the solid-fluid interface, pointing from solid to fluid region, S is the area of solid-fluid interface and F_x the x component of the stress $\mathbf{F}$. In the present case the integral is calculated over the reconstructed interface. With small values of λ we observe an underestimation of pressure difference, ranging from 1 to 8 %, while with $\lambda = 0.1$ we obtain much higher errors (37 to 54 %). If we exclude the case $\lambda = 0.1$, the percentage difference between the computed drag coefficient and the reference one lays in the range 7 to 12 %.

3.2. Flow around an impulsively started cylinder

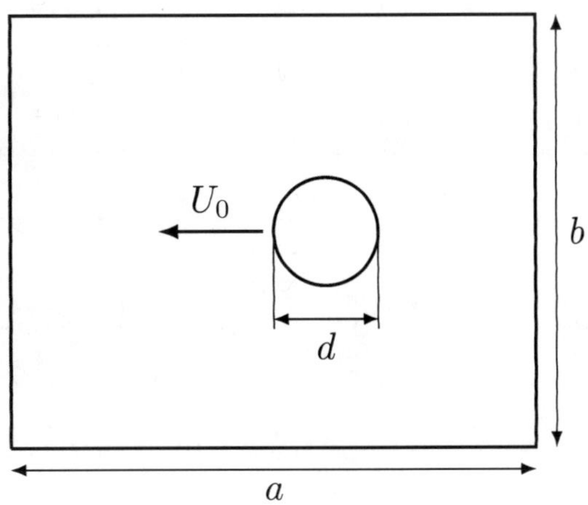

Figure 6. Sketch of the simulated domain for the case of flow around an impulsively started cylinder.

In the present test we simulate the case of a cylinder that is impulsively put in motion with a constant and uniform velocity $\mathbf{U}_0$. For this particular case an experimental study was performed in [20], as well as numerical simulations with vortex method [18, 21, 19] and immersed boundary techniques [22, 23] for various Reynolds numbers. A sketch of the simulated domain is reported

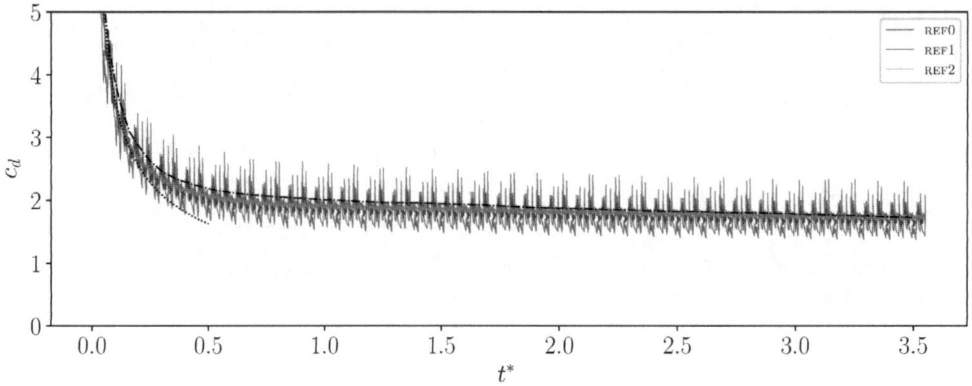

Figure 7. Comparison of drag coefficient evolution, for cases REF0, REF1 and REF2, with analytical law (dotted line) [18] and numerical reference values (dash-dotted line) [19].

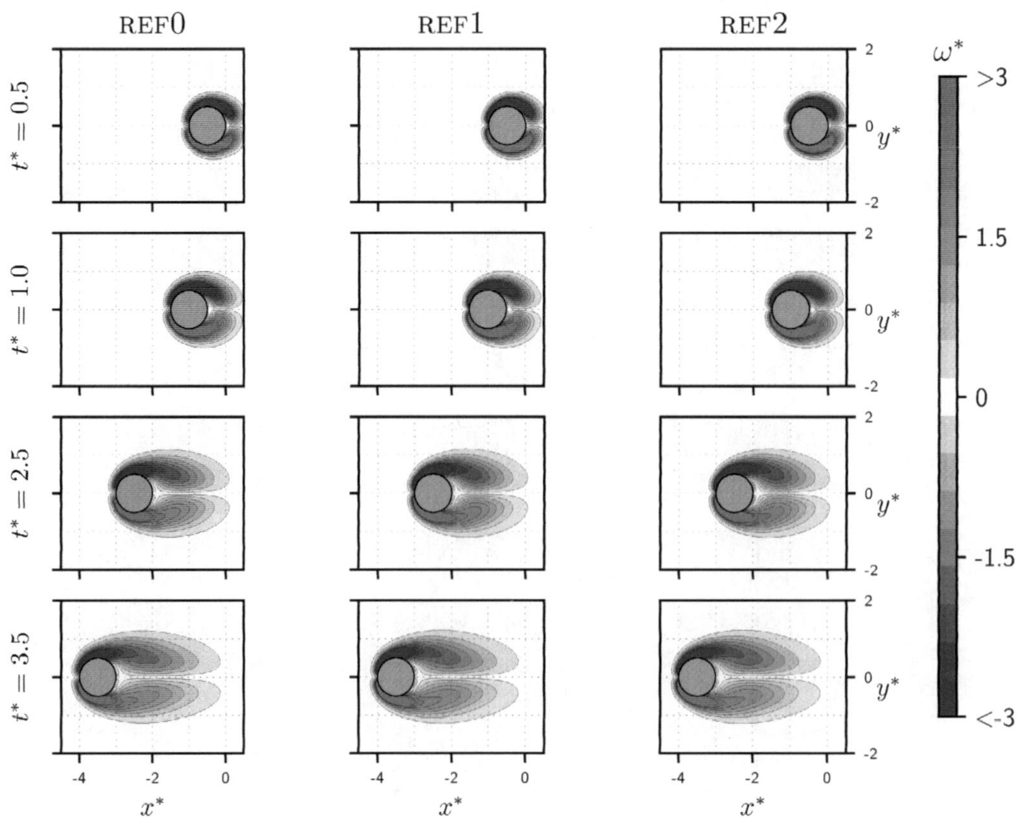

Figure 8. Equi-vorticity contours for the three simulated cases at three different non-dimensional time steps. Vorticity values range from -3 to 3 with a step of 0.4.

in Fig. 6, together with parameters a, b and d, denoting the domain width, height and cylinder diameter respectively. The computational box is characterized by the values $a = 12d$ and $b = 10d$. A no-slip boundary condition is imposed on all boundaries . The Reynolds number, based on cylinder velocity modulus U_0, cylinder diameter and fluid kinematic viscosity, is equal to 40. At time $t = 0$ the cylinder is put into motion with a constant velocity, moving from right to left. The simulated time interval is $t \in [0, 3.5\tilde{t}]$, where $\tilde{t}$ is the reference time constant, $\tilde{t} = d/U_0$. Numerical results will be presented as a function of non-dimensional time $t^* = t/\tilde{t}$, in order to compare them with literature values [18, 19, 22, 23]. We use two different computational grids that are characterized by the following mesh sizes δx and δy, along x and y direction: $\delta x = \delta y = d/9$ for case REF0, $\delta x = \delta y = d/18$ for case REF1 and $\delta x = \delta y = d/36$ for case REF2. In Fig. 7 we report the time evolution of the drag coefficient c_d for the three simulated cases and compare them with the reference numerical results [19] and with the analytical relation proposed in [18], namely

$$c_d = \frac{4\sqrt{\pi}}{\sqrt{t^*\ Re}} + \frac{\pi}{Re}\left(9 - \frac{15}{\sqrt{\pi}}\right), \tag{14}$$

which is valid for $t^* < 0.5$. As in the previous test, the drag coefficient is computed using (13) on the reconstructed solid-fluid interface. We observe that the computed values of c_d are highly oscillating for case REF0, while for the other two cases oscillations amplitude decrease with mesh refinement. Drag coefficient values are slightly underestimated, with respect to the reference values [19], in the interval $t^* \in [0, 2]$, while for higher values of the non-dimensional time we observe a good agreement. A comparison between the simulated cases of the obtained vorticity ω values is reported in Fig. 8 for three different time steps, namely $t^* = 0.5$, $t^* = 1.0$, $t^* = 2.5$ and $t^* = 3.5$. In particular, non-dimensional vorticity ω^* contours are shown in a range $\omega^* \in [-3, 3]$ with increments of 0.4, in order to compare them with the reference results [22, 23]. Comparing case REF0 and REF1 we see that grid spacing has a major impact on the evaluation of vorticity in the fluid area behind the cylinder and also in the estimation of a higher vorticity region. An increase of mesh resolution from REF1 to REF2 leads to less considerable changes in the predicted values of ω^*. We observe a very good agreement with the reference profiles even though the maximum mesh resolution used here is much smaller, namely $\delta x/d = 36$ instead of $\delta x/d = 50$ and $\delta x/d = 100$ [22, 23]. As a final test we study the evolution of the

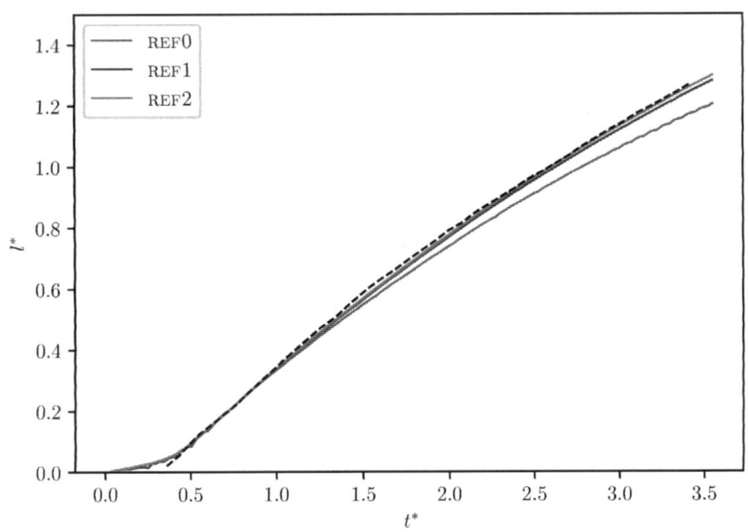

Figure 9. Time evolution of recirculation length l^* and comparison with reference data [20].

recirculation zone length l. Non-dimensional values of this variable, $l^* = l/d$, are plotted against non-dimensional time t^* in Fig. 9 for all the simulated cases and compared with reference values [20]. The length l^* is calculated by examining the velocity field $\mathbf{U} - \mathbf{U}_0$, i.e. from a reference frame defined on the moving cylinder. For the first part of the simulated time interval, namely $t^* < 1$, we observe a good agreement between our results and the reference ones. For higher values of non-dimensional time the results obtained from case REF0 increasingly underestimated l^*, while the ones of case REF1 and REF2 are still close to reference data.

4. Conclusions

In the present paper we presented a simple immersed boundary method implementation. The proposed method is based on a computational grid for the fluid flow and one for the solid body. By using numerical fields projection, we can define a velocity and an indicator function field on the solid body mesh and then project them on the fluid computational grid in order to define solid-occupied regions and to calculate velocity values on those particular grid nodes. A key aspect of this procedure is that the method can be applied on arbitrarily shaped objects that can either be static or move. In mesh cells occupied by solid, velocity values are imposed with Dirichlet boundary condition while an equation similar to those used for Navier-Stokes projection methods [14] is solved for pressure, in order to obtain both mass conservation and a smooth pressure field. We report two test cases, involving respectively a static solid body and a moving one. In the first test we study mass conservation, pressure difference between front and rear sides of solid obstacle and drag coefficient values on a reconstructed solid-fluid interface. We use three different domain discretizations with increasing mesh refinements. The mass conservation error rapidly decreases with the increase of mesh resolution and also pressure difference is in good agreement with reference data [17]. Higher discrepancy between the present calculations and reference values is found for the drag coefficient. The second test involves a cylinder that is suddenly set in motion with uniform and constant velocity in a fluid that is initially quiescent. Even for this case we use three different domain discretizations and we study drag coefficient values, recirculation zone length and vorticity field. The results are compared with reference data [18, 21, 19, 22, 23] and a very good agreement is found. The results presented for both test cases are very promising and positively contribute to the validation of the numerical approach. Future developments will involve simulations with higher Reynolds number values and three-dimensional objects.

References

[1] Michler C, Hulshoff S, van Brummelen E and de Borst R 2004 *Comput. Fluids* **33** 839 – 848
[2] Causin P, Gerbeau J and Nobile F 2005 *Comput. Methods. Appl. Mech. Eng.* **194** 4506 – 4527
[3] Cerroni D, Giommi D, Manservisi S and Mengini F 2018 Preliminary monolithic fluid structure interaction model for ventricle contraction *Biomedical Technology: Modeling, Experiments and Simulation* ed Wriggers P and Lenarz T (Springer International Publishing) pp 217–231
[4] Cerroni D, Da Vià R and Manservisi S 2018 *J. Comput. Phys.* **354** 646–671
[5] Rider W J and Kothe D B 1998 *J. Comput. Phys.* **141** 112 – 152
[6] Tryggvason G, Scardovelli R and Zaleski S 2011 *Direct numerical simulations of gas–liquid multiphase flows* (Cambridge University Press)
[7] Pozzetti G and Peters B 2018 *Int. J. Multiph. Flow* **99** 186 – 204
[8] Peskin C S 1972 *J. Comput. Phys.* **10** 252–271
[9] Ilinca F and Hétu J F 2011 *Int. J. Numer. Methods Fluids* **65** 856–875
[10] Jendoubi A, Yakoubi D, Fortin A and Tibirna C 2014 *Int. J. Numer. Methods Fluids* **75** 63–80
[11] Mittal R and Iaccarino G 2005 *Annu. Rev. Fluid Mech.* **37** 239–261
[12] Cerroni D 2015 *Multiscale multiphysics coupling on a finite element platform* Ph.D. thesis University of Bologna
[13] Chierici A, Chirco L, Da Vià R, Manservisi S and Scardovelli R 2017 *J. Phys. Conf. Ser.* **923** 012025
[14] Guermond J, Minev P and Shen J 2006 *Comput. Methods Appl. Mech. Eng.* **195** 6011 – 6045
[15] Hétu J F and Ilinca F 2013 *Comput. Fluids* **87** 2 – 11
[16] Code FEMuS URL `https://github.com/FemusPlatform/femus/`

[17] Schäfer M *et al.* 1996 *Benchmark Computations of Laminar Flow Around a Cylinder* (Wiesbaden: Vieweg+Teubner Verlag) pp 547–566
[18] Bar-Lev M and Yang H T 1975 *J. Fluid Mech.* **72** 625647
[19] Koumoutsakos P and Leonard A 1995 *J. Fluid Mech.* **296** 1–38
[20] Coutanceau M and Bouard R 1977 *J. Fluid Mech.* **79** 257272
[21] Chloe M, Federico G, GeorgesHenri C and Iraj M 2015 *Int. J. Numer. Methods Fluids* **79** 55–83
[22] Taira K and Colonius T 2007 *J. Comput. Phys.* **225** 2118 – 2137
[23] ShangGui C, Abdellatif O, Julien F and Yannick H 2017 *Int. J. Numer. Methods Fluids* **85** 288–323

IOP Conf. Series: Journal of Physics: Conf. Series **1224** (2019) 012003 doi:10.1088/1742-6596/1224/1/012003

Distributed optimal control applied to Fluid Structure Interaction problems

A Chierici*, **L Chirco***, **R Da Vià***, **M Magnani*** **and S Manservisi***

*University of Bologna, DIN, Lab. of Montecuccolino, Via dei Colli 16, Bologna 40136, Italy

E-mail: leonardo.chirco2@unibo.it

Abstract. In this paper we focus on an optimal control problem based on Lagrangian multipliers and adjoint variables applied to the steady state Fluid-structure interaction problem. FSI systems consist of one or more solid structures that deform by interacting with a surrounding fluid flow. We consider the Fluid-structure interaction problem written in variational monolithic form in order to balance automatically solid and liquid forces at the interface. The objective of the control problem is the minimization of a velocity matching functional and it is obtained with a distributed force control that deforms the shape of the solid domain. This distributed control for Fluid-structure interaction systems can be easily used to control the shape of the solid domain and the flow effects of the fluid-solid interactions. The optimality system is derived from the first order optimality conditions by taking the Fréchet derivatives of the Lagrangian with respect to all the variables involved. The optimal solution is then obtained through a steepest descent algorithm applied to the optimality system. To support this approach we report a few numerical tests where the distributed force deforms the solid structure in order to match the desired velocity profile. The results obtained show the feasibility and robustness of this approach and its possible use in many industrial relevant applications.

1. Introduction

Fluid structure interaction (FSI) systems are commonly studied in many engineering application such as the design of floating structures, wind turbines, parachutes, injection systems, valves and in the study of biological systems [1, 2, 3, 4, 5]. Usually those kind of problems are solved in a forward approach in order to evaluate the tensional state of a mechanical component due to a surrounding fluid flow taking into account the effects of the solid deformations on the motion of the interior flows. However, in recent years optimization has become a key aspect in the field of engineering, in order to improve the performance of devices or reduce the cost of production processes. Nowadays several approaches to optimization are available, such as single and multi-objective, adjoint or sensitivities based method, evolutionary algorithms and many others, see [6, 7, 8, 9, 10, 11]. In this paper we would like to study the inverse FSI problem by using an optimal control approach based on adjoint variables which have been proven to be a good approach for the optimal control of complex systems [12, 8]. We adopt a monolithic approach to solve the FSI system to obtain a stable and well defined solution in a finite element setting.

Inside this framework the objective of our control problem is the minimization of a velocity field tracking functional in a specific region of the fluid domain with a distributed force that acts in the solid domain and changes the location of both the solid and fluid domains. Through Lagrange multipliers formalism we reformulate the physically constrained minimization problem into an unconstrained one. The optimality system is then obtained by imposing the first order necessary conditions which consists in setting to zero the Fréchet derivatives of the Lagrangian taken with respect to all the variables involved. In order to solve the optimality system we use a simple steepest descent algorithm. In the next section we first recall some basic notation about the functional spaces of interest. Then we present the FSI system together with the boundary and interface conditions. Finally the optimality system is obtained and the iterative algorithm used is presented. In order to support the proposed approach we perform some numerical tests with different values of the regularization parameter.

2. Mathematical formulation

In this section we discuss the mathematical formulation of the FSI problem and we derive the optimality system. We briefly recall some notations about functional spaces used in this paper, for a detailed description one can consult [13, 14]. We use standard notation $H^s(\Omega)$ for the Sobolev spaces with norm $\|\cdot\|_s$ ($H^0(\Omega) = L^2(\Omega)$ and $\|\cdot\|_0 = \|\cdot\|$). Let $H_0^s(\Omega)$ be the space of all functions in $H^s(\Omega)$ that vanish on the boundary of the bounded open set Ω and $H^{-s}(\Omega)$ be the dual space of $H_0^s(\Omega)$. The trace space for the functions in $H^1(\Omega)$ is denoted by $H^{1/2}(\Gamma)$.

Let us consider a domain $\Omega \subset R^N$ which consists of a structure domain Ω_s and a fluid domain Ω_f so that $\Omega = \Omega_s \cup \Omega_f$. The deformed solid domain $\Omega_s(\mathbf{l})$ is expressed through the solid displacement field $\mathbf{l}$ as

$$\Omega_s(\mathbf{l}) = \{\mathbf{x} \in R^3 \,|\quad \mathbf{x} = \mathbf{x}_0 + \mathbf{l}\}\,, \tag{1}$$

where the vector $\mathbf{x}_0$ defines the initial solid domain $\Omega_s(\mathbf{0})$, taken as the reference domain. For our steady state FSI problem we consider the interaction of a Newtonian fluid, whose behavior is described by the Navier-Stokes equations, with a St.Venant-Kirchhoff material. The mathematical model in strong form of the FSI problem is then the following

$$\nabla \cdot \mathbf{v} = 0 \qquad on \quad \Omega_f\,, \tag{2}$$

$$\rho_f(\mathbf{v}\cdot\nabla)\mathbf{v} - \nabla\cdot\mathbf{T} = 0 \qquad on \quad \Omega_f\,, \tag{3}$$

$$\nabla\cdot\mathbf{S}(\mathbf{l}) = \mathbf{f} \qquad on \quad \Omega_s\,, \tag{4}$$

with $\mathbf{v}$ being the fluid velocity field and ρ_f its density. We denoted with $\mathbf{f}$ the distributed force that acts in the solid region only. The constitutive relations for the fluid stress tensor $\mathbf{T}$ in the Newtonian incompressible case and for the solid Cauchy stress tensor $\mathbf{S}$ read

$$\mathbf{T}(p_f, \mathbf{v}_f) = -p_f\mathbf{I} + \mu_f(\nabla\mathbf{v} + \nabla\mathbf{v}^T)\,, \tag{5}$$

$$\mathbf{S}(\mathbf{l}) = \lambda_s(\nabla\cdot\mathbf{l})\mathbf{I} + \mu_s\nabla\mathbf{l}\,, \tag{6}$$

where p_f is the fluid pressure, μ_f the dynamic viscosity of the fluid while λ_s and μ_s are the solid Lamé parameters. To complete the system (2-4) it is necessary to define the appropriate boundary and interface conditions, which are

$$\begin{aligned}
\mathbf{v} &= \mathbf{v}_0 && on \quad \Gamma_{fd}\,,\\
\mathbf{l} &= \mathbf{l}_0 && on \quad \Gamma_{sd}\,,\\
\mathbf{T}\cdot\mathbf{n} &= 0 && on \quad \Gamma_{fn}\,,\\
\mathbf{S}\cdot\mathbf{n} &= 0 && on \quad \Gamma_{sn}\,,\\
\mathbf{T}\cdot\mathbf{n} &= \mathbf{S}\cdot\mathbf{n} && on \quad \Gamma_i\,,\\
\mathbf{v} &= 0 && on \quad \Gamma_i\,,
\end{aligned} \tag{7}$$

IOP Conf. Series: Journal of Physics: Conf. Series **1224** (2019) 012003 doi:10.1088/1742-6596/1224/1/012003

where Γ_{fd} and Γ_{sd} are the surfaces where Dirichlet boundary conditions are imposed for the fluid velocity and solid displacement, Γ_{fn} and Γ_{sn} are the surfaces where standard homogeneous outflow boundary conditions are imposed for both the fields. By Γ_i we denote the interface between the solid and fluid domain, $\Gamma_i = \Omega_s \cap \Omega_f$.

In this work we study a fluid velocity matching profile problem in a region of the liquid domain by using a distributed force that moves and deforms the solid. This goal is expressed in terms of the following objective functional

$$\mathcal{J}(\mathbf{v},\mathbf{f}) = \frac{1}{2}\int_{\Omega_d} w||\mathbf{v}-\mathbf{v}_d||^2 d\Omega + \frac{1}{2}\beta\int_{\Omega_c} ||\mathbf{f}||^2 d\Omega \tag{8}$$

where $\mathbf{v}_d$ is the desired velocity on the controlled domain region $\Omega_d \subset \Omega_f$ and w a weight function of the coordinates $\mathbf{x}$ used to give more importance to some parts of Ω_d. The functional is completed by a regularization term which is needed to obtain a control function $\mathbf{f}$ in the space of square integrable functions $L^2(\Omega_c)$, with Ω_c being the solid region where the control can act. The regularization parameter β plays a fundamental role for the numerical solution of the minimization problem. If a too high value of β is chosen the regularization contribution dominates over the objective one and the objective cannot be achieved well, while a lack of regularization leads to singular solutions or convergence issues in the numerical solution of the problem.

To obtain the optimality system we write the full Lagrangian of the problem which is composed of the functional (8) and state equations (2-4) multiplied by the appropriate Lagrangian multipliers, the so-called *adjoint variables*. By doing so we transform a constrained minimization problem into an unconstrained one.

$$\begin{aligned}\mathcal{L}(p,\mathbf{v},\mathbf{l},\widehat{\mathbf{l}},p_a,\mathbf{v}_a,\widehat{\mathbf{l}}_a,\widehat{\mathbf{s}}_a,\beta_a) = \mathcal{J}(\mathbf{v},\mathbf{f}) &+ \int_{\Omega_f}(\nabla\cdot\mathbf{v})p_a\,d\Omega + \int_{\Omega_s}(\nabla\cdot\mathbf{v})p_a\,d\Omega + \\ &+ \int_{\Omega_f}\left[\rho^f(\mathbf{v}\cdot\nabla)\mathbf{v} + \nabla p - \nabla\cdot(\mu_f\nabla\mathbf{v})\right]\cdot\mathbf{v}_a\,d\Omega + \\ &+ \int_{\Omega_s}\left[-\nabla p + \nabla\cdot(\mu_s\nabla\mathbf{l}) + \mathbf{f}\right]\cdot\mathbf{v}_a\,d\Omega + \int_{\Omega_s}\nabla^2\widehat{\mathbf{l}}\cdot\widehat{\mathbf{l}}_a\,d\Omega + \\ &+ \int_{\Gamma_i}\widehat{\mathbf{s}}_a\cdot\left[(\widehat{\mathbf{l}}-\mathbf{l}) - \frac{\mathbf{v}}{h}\right]d\Gamma + \int_{\Omega_s}\beta_a\cdot\left[\mathbf{v} - h(\mathbf{l}-\widehat{\mathbf{l}})\right]d\Omega\,. \end{aligned} \tag{9}$$

In (9) we have introduced the auxiliary mesh displacement $\widehat{\mathbf{l}}$ defined as

$$\nabla^2\widehat{\mathbf{l}} = 0 \qquad \textit{on } \Omega_s \tag{10}$$

$$\widehat{\mathbf{l}} = \mathbf{l} - \frac{\mathbf{v}}{h} \qquad \textit{on } \Gamma_i \tag{11}$$

$$\widehat{\mathbf{l}} = 0 \qquad \textit{on } (\Gamma_s - \Gamma_i)\,, \tag{12}$$

through which we can extend the velocity field in the whole domain Ω, with h positive and constant,

$$\mathbf{v} = \begin{cases} h(\mathbf{l}-\widehat{\mathbf{l}}) & \text{on } \Omega_s, \\ \mathbf{v} & \text{solution of (2)-(3) on } \Omega_f. \end{cases} \tag{13}$$

It is clear that at steady state $\Omega^f(\mathbf{l}) = \Omega^f(\widehat{\mathbf{l}})$ and $\Omega^s(\mathbf{l}) = \Omega^s(\widehat{\mathbf{l}})$ with $\mathbf{v}=0$ on the interface Γ_i.

In order to derive the optimality system we impose the minimization necessary condition

$$\delta\mathcal{L} = 0\,. \tag{14}$$

Since the total variation of the Lagrangian has to be zero and each variation is independent from the others then each variation has to be zero. By taking the Fréchet derivatives of the Lagrangian (9) with respect to the adjoint variables the weak form of the state system (2-4) is obtained, together with the correct boundary and interface conditions. When the derivatives are taken with respect to the state variables, after some term rearrangement, the following adjoint system in weak form is recovered

$$\int_{\Omega_f} (\nabla \cdot \mathbf{v}_a)\delta p \, d\Omega = 0 \qquad \forall\, \delta p \in L^2(\Omega_f) \tag{15}$$

$$\begin{aligned}
&\int_{\Omega_f} \left[\left(\rho^f (\delta \mathbf{v} \cdot \nabla)\mathbf{v}\right) \cdot \mathbf{v}_a + \left(\rho^f (\mathbf{v} \cdot \nabla)\delta \mathbf{v}\right) \cdot \mathbf{v}_a + \mu^f \nabla \delta \mathbf{v} : \nabla \mathbf{v}_a \right] d\Omega + \\
&+ \frac{1}{h} \int_{\Omega^s} \left[\mu_s \nabla \boldsymbol{v}_a : \nabla \delta \mathbf{v} + \lambda_s (\nabla \cdot \boldsymbol{v}_a)(\nabla \cdot \delta \mathbf{v}) \right] d\Omega + \\
&+ \int_{\Omega_f} (\nabla \cdot \delta \mathbf{v}) p_a \, d\Omega + \int_{\Omega_d} w(\mathbf{v} - \mathbf{v}_d)\delta \mathbf{v} \, d\Omega = 0 \qquad \forall\, \delta \mathbf{v} \in H^1_{\Gamma_{fd} \cup \Gamma_{sd}}(\Omega)
\end{aligned} \tag{16}$$

$$\int_{\Omega_s} \nabla \widehat{\mathbf{l}}_a : \nabla\, \delta \widehat{\mathbf{l}} \, d\Omega + \int_{\Omega_s} \mu^s \nabla \mathbf{v}_a : \nabla \delta \widehat{\mathbf{l}} \, d\Omega = 0 \qquad \forall\, \delta \widehat{\mathbf{l}} \in H^1_{\Gamma_{sd}}(\Omega_s) \tag{17}$$

The shape derivatives with respect to the fluid and solid domain have been taken into account and simplified. Furthermore the equation for the adjoint displacement $\mathbf{l}_a$ can be neglected since we do not need it to determine the force $\mathbf{f}$. Finally, when considering the Fréchet derivatives of the Lagrangian (9) with respect to the control parameter $\mathbf{f}$ we obtain the control equation

$$\mathbf{f} = \frac{\mathbf{v}_a}{\beta}. \tag{18}$$

By performing integrations by parts on the system (15-17) it is possible to determine the strong form of the adjoint system together with the boundary conditions for the adjoint variables. After performing the integration by parts, we recover the adjoint state $(\mathbf{v}_a^f, \mathbf{v}_a^s, p_a) \in \mathbf{H}^1_{\partial\Omega^f - \Gamma_i}(\Omega^f) \cap \mathbf{H}^2(\Omega^f) \times \mathbf{H}^1_{\partial\Omega^s - \Gamma_i}(\Omega^s) \cap \mathbf{H}^2(\Omega^s) \times L^2_0(\Omega^f) \cap \mathbf{H}^1(\Omega^f)$, by solving

$$\nabla \cdot \boldsymbol{v}_a^f = 0\,, \tag{19}$$

$$-\rho_l (\nabla \mathbf{v})^T \boldsymbol{v}_a^f + \rho_l [(\mathbf{v} \cdot \nabla)\boldsymbol{v}_a^f] + \nabla p_a - \nabla \cdot (\mu_l \nabla \boldsymbol{v}_a^f) = w(\mathbf{v} - \mathbf{v}_d)\,, \tag{20}$$

$$\nabla \cdot \mathbf{S}(\boldsymbol{v}_a^s) = 0\,. \tag{21}$$

with boundary conditions defined as

$$\begin{aligned}
&\boldsymbol{v}_a^s = \boldsymbol{v}_a^f && && on \quad \Gamma_i\,, \\
&\mathbf{S}(\boldsymbol{v}_a^s) \cdot \mathbf{n} = \mathbf{T}(\boldsymbol{v}_a^f) \cdot \mathbf{n} && && on \quad \Gamma_i\,, \\
&\mu_l (\nabla \boldsymbol{v}_a) \cdot \mathbf{n} = -(\mathbf{v} \cdot \mathbf{n})\, \boldsymbol{v}_a\,, && p_a = 0 && on \quad \Gamma_n^f\,, \\
&\boldsymbol{v}_a = 0 && && on \quad \Gamma_{fd} \cup \Gamma_{sd}\,.
\end{aligned} \tag{22}$$

The optimality system is then composed by the state (2-4) and adjoint (15-17) systems coupled with the control equation (18). Due to the strong non-linearity of the problem a one-shot solution of the optimal system can not be performed. In this work we use a segregate approach for the solution of the optimality system, solving the state system, the adjoint system and the control equation separately and iteratively, see for example [15]. By doing so we can also use the same solver with only minimal modifications to solve the adjoint and state systems. The

Algorithm 1 Description of the Steepest Descent algorithm.

1. Set a state $(\mathbf{v}^0, p^0, \mathbf{l}^0)$ satisfying (2-4) ▷ *Setup of the state - Reference case*
2. Compute the functional $\mathcal{J}^0$ in (8)
3. Set $r^0 = 1$

for $i = 1 \rightarrow i_{max}$ **do**
  4. Solve the system (15)-(17) to obtain the adjoint state $(\boldsymbol{v}_a^i, p_a^i)$
  5. Set the control update $\delta\mathbf{f}^i = (\mathbf{f}^{i-1} - \mathbf{v}_a^i/\beta)$
  6. Set $r^i = r^0$
  while $\mathcal{J}^i(\mathbf{f}^{i-1} + r^i\delta\mathbf{f}^i) > \mathcal{J}^{i-1}(\mathbf{f}^{i-1})$ **do** ▷ *Line search*
    7. Set $r^i = \rho r^i$
    8. Solve (2-4) for the state $(\mathbf{v}^i, \mathbf{f}^i, \mathbf{l}^i)$ with $\mathbf{f}^i = \mathbf{f}^{i-1} + r^i\delta\mathbf{f}^i$
    if $r^i < toll$ **then**
      Line search not successful ▷ *End of the algorithm*
    end if
  end while
  if $||\mathcal{J}^i(\mathbf{f}^{i-1} + r^i\delta\mathbf{f}^i) - \mathcal{J}^{i-1}(\mathbf{f}^{i-1})||/\mathcal{J}^{i-1}(\mathbf{f}^{i-1}) < \tau$ **then**
    9. Convergence reached ▷ *End of the algorithm*
  end if
end for

iterative algorithm adopted to minimize the objective functional is the simple *Steepest Descent method* described in Algorithm 1.

It starts solving the state system with no control obtaining the reference case $(\mathbf{v}^0, p^0, \mathbf{l}^0)$. Then it is determined the gradient direction from the solution of the control equations (18), in which $\mathbf{v}_a$ is known once that the adjoint system (15-17) has been solved. The core of the algorithm consists of the backtracking line search process, where the state system is solved iteratively reducing the value of the control parameters until a functional reduction is obtained. The step length r determines how far from the current state solution we are moving along the gradient direction given by $\delta\mathbf{f}$. This algorithm comes to an end either when two consecutively computed functionals are almost equal and no further improvement can be achieved or when r becomes lower than a tolerance value $toll = 10^{-7}$. In this case the optimal solution is the one obtained at the previous iteration step.

This algorithm requires several solutions of the state and adjoint systems in order to find the optimal control, however it does not need a great amount of memory which is limited to a standard CFD simulation. We implemented this algorithm in an in-house finite element code which is parallelized by using openMPI libraries and uses a multigrid solver with mesh-moving capability. We have used standard quadratic-linear elements for all the variables except the pressure which is assumed linear to satisfy the *BBL inf-sup* condition. The displacements are approximated with standard quadratic elements.

3. Numerical results

In this section we present the results obtained by applying Algorithm 1 to a two-dimensional test case. In Figure 1 we have reported the domain considered with the geometrical properties. The origin of the reference system is the bottom left corner (point A). The solid region Ω_s is colored with darker gray while the fluid region Ω_f with lighter gray. In red is the fluid controlled region $\Omega_d = [0.465, 0.5]m \times [0.16, 0.2]m$ which is located near the channel exit. We impose pressure boundary conditions with vanishing tangential velocity on the left and right surfaces, $p_{AD} = 10000Pa$ and $p_{BC} = 9750Pa$, respectively. On the lower boundary AB we prescribe a no-slip condition, while the upper surface CD is a symmetry axis and we set a homogeneous Neumann boundary condition for all the state and adjoint variables apart from

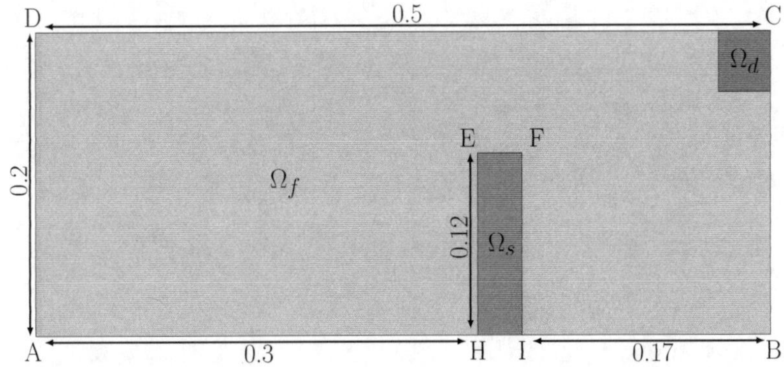

Figure 1. Case study. Domain overview with the solid region (Ω_s), the fluid region (Ω_f) and the controlled region (Ω_d).

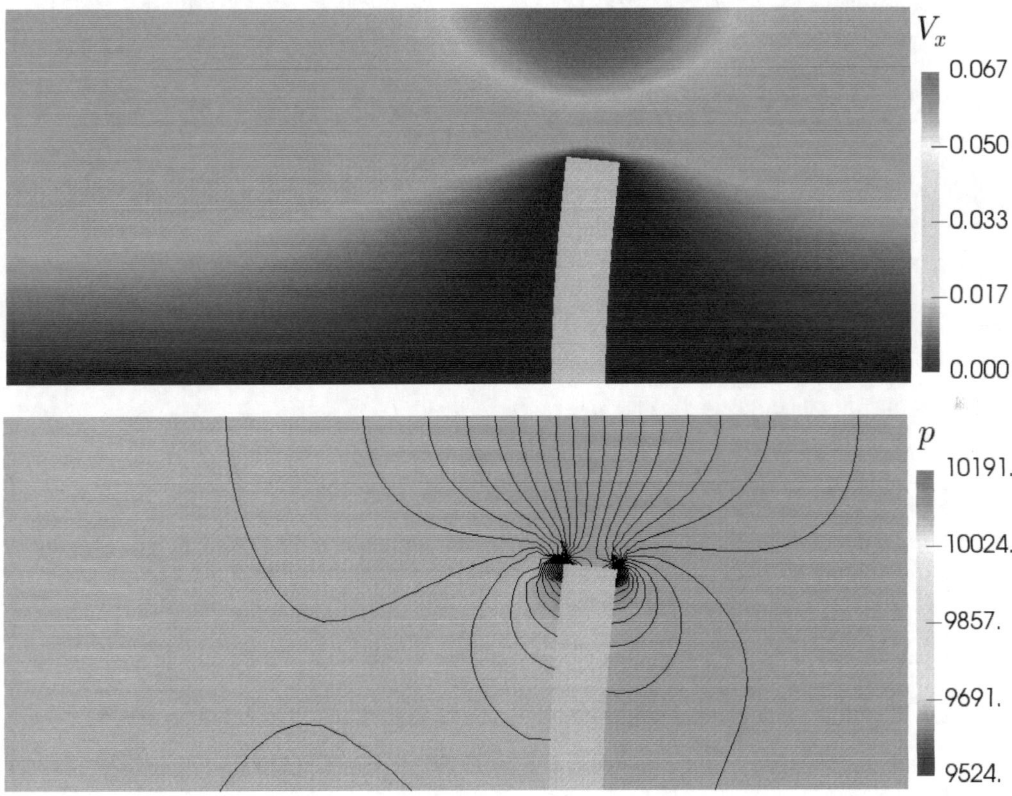

Figure 2. Reference geometry with velocity (m/s) profile in the fluid region (on the top). Pressure (Pa) profile and iso-lines in the fluid region (on the bottom).

u_y and u_{ay} that vanish. All the boundaries are fixed with the exception of the interface Γ_i, that may move with fixed endpoints H and I. The physical properties are $\rho_s = \rho_l = 1000Kg/m^3$ with $\mu_l/\rho_l = 0.07Pa\,m^3s/Kg$ and $\mu_s = 76250Pa$ so that the flow is not turbulent and the solid can bend easily. Young's modulus E_s and Poisson's coefficient ξ are equal to $183000Pa$ and $\xi = 0.2$, respectively.

According to Algorithm 1, we start solving the state system (2-4) assuming $\mathbf{f} = 0$ in order to

Figure 3. Velocity (m/s) profile in the fluid region and force (N) vectors (on the top), $\beta = 10^{-9}$. Pressure (Pa) field and iso-lines in the fluid region (on the bottom), $\beta = 10^{-9}$.

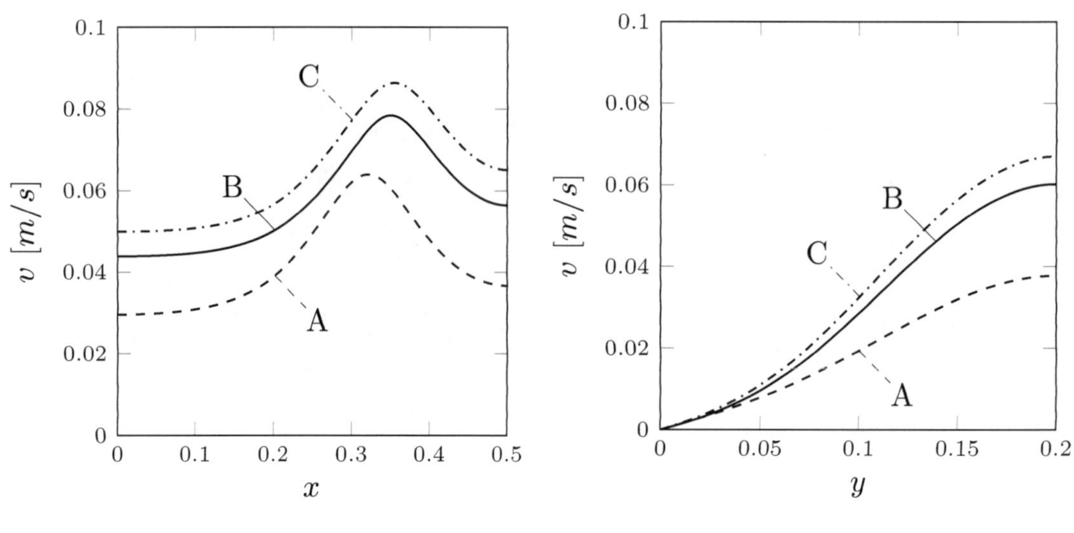

Figure 4. On the left axial velocity u_x, profile on a line at $x = 0.49$. On the right u_x on a line at $y = 0.18$. Result (A) obtained with no control, (B) with $\beta = 10^{-7}$ and (C) with $\beta = 10^{-9}$.

obtain the reference case shown in Figure 2. The velocity profile is reported on the top of this Figure, while on the bottom the pressure field with iso-lines is shown. Since the solid obstacle

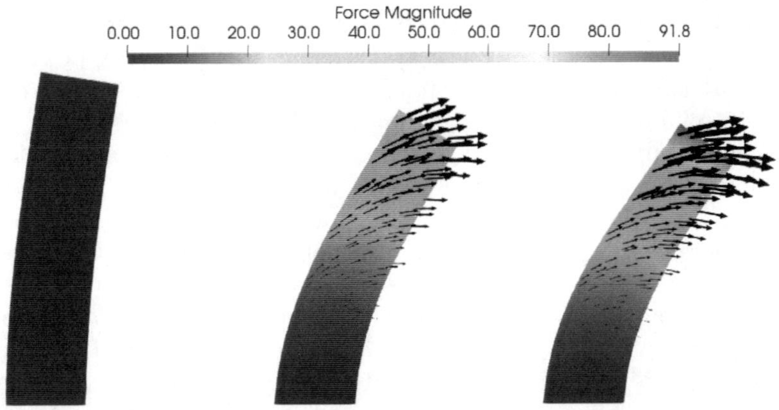

Figure 5. Force vectors in black and magnitude in colors on the solid domain for $\beta = 10^{-5}$, 10^{-7}, 10^{-9} from left to right.

Figure 6. Adjoint velocity profile and vectors in the fluid domain, $\beta = 10^{-9}$.

reduces the fluid domain cross section the fluid has to accelerate to satisfy the mass conservation equation. Furthermore, the solid object is responsible for the majority of the pressure losses as it can be seen on the bottom of Figure 2.

Our optimal control problem consists in finding the optimal solid deformation so that the axial component of the velocity (v_x) matches a desired profile over the controlled domain Ω_d. The solid is deformed by a distributed force $\mathbf{f}$ which is the control parameter acting in the whole solid domain, see equation (4). In the reference case with no control we obtained a mean value of v_x over Ω_d equal to $\bar{v}_x = 0.037\ m/s$. We then choose a constant target value $v_{xd} = 0.065\ m/s$ in the whole controlled domain, thus requiring a higher fluid velocity near the channel outlet. The objective functional becomes

$$\mathcal{J}(v_x, \mathbf{f}) = \frac{1}{2} \int_{\Omega_d} w(v_x - 0.065)^2 d\Omega + \frac{1}{2}\beta \int_{\Omega_s} ||\mathbf{f}||^2 d\Omega \tag{23}$$

in which we set

$$w = \begin{cases} 1 & \mathbf{x} \in \Omega_d, \\ 0 & \text{otherwise.} \end{cases}$$

Once that the reference case has been set up it is possible to solve the optimality system with the technique described in Algorithm 1.

IOP Conf. Series: Journal of Physics: Conf. Series **1224** (2019) 012003 doi:10.1088/1742-6596/1224/1/012003

We now report the results obtained for different values of the regularization parameter $\beta = 10^{-5}$, 10^{-7}, 10^{-9}. On the top of Figure 3 we reported the velocity profile and the force acting in the solid domain, and on the bottom of the same Figure is shown the pressure profile. The force bends the solid to the right so that the channel cross section is larger than that of the reference case. Since the total pressure drop between the channel inlet and outlet is fixed by the boundary conditions ($250Pa$), if the pressure losses induced by the obstacle decrease then the fluid velocity has to increase. In particular it is worth noticing that, due to the regularization term in (23), it is more effective to apply an intense force near the tip of the solid instead of a weaker but more distributed one. In Figure 4 it is shown the axial component of the velocity field (v_x) on a vertical line at $x = 0.49$ on the left and on a horizontal one at $y = 0.18$ on the right, for different values of the regularization β. Both lines cross the controlled region Ω_d. By reducing the value of β the control can act more strongly and the velocity is higher than that of the reference case. In Figure 5 is reported the control force for different amount of regularization. The force acting on the solid is very small in the case $\beta = 10^{-5}$, while becomes more relevant when reducing β, leading to a higher solid deformation. The adjoint velocity profile is reported in Figure 6 with arrows and colors related to the $\mathbf{v}_a$ magnitude. The source term of the adjoint velocity, which is the difference between the actual and the desired velocity in the area Ω_d is the driving force that deforms the solid.

β	∞	10^{-5}	10^{-7}	10^{-9}
$\mathcal{J}(\mathbf{u}) \cdot 10^8$	113.69	111.68	32.75	1.10
v_x	0.0367	0.0375	0.0584	0.0651

Table 1. Objective functionals $\mathcal{J}$ and average axial velocity over the controlled region Ω_d computed with no control ($\beta = \infty$) and different β values.

Finally, in Table 1 we reported the functionals and the average axial velocity over the controlled region as computed in the reference and in the controlled case with different β. The functional has been reduced in every case and in particular with $\beta = 10^{-9}$, in which the axial velocity is very close to the desired value 0.065.

4. Conclusions

In this work we have presented an optimal control method for the fluid structure interaction problem based on adjoint variables. The objective of the optimal control problem is the velocity matching in a specific region of the fluid domain through a distributed control acting in the solid to change the shape of the solid domain. We have adopted a monolithic variational formulation to satisfy automatically the coupling conditions at the fluid-solid interface. Furthermore, we have extended the velocity field to the solid domain to couple adjoint variables and forces on the interface. The optimality system has been derived by imposing the first order necessary conditions to the full Lagrangian and solved iteratively with a steepest descent algorithm in order to use the same solver for state and adjoint system. The results obtained show the feasibility and robustness of this approach. In future works we plan to assess this kind of control to more realistic geometries such as wing shape optimization to show the feasibility of this approach in real complex cases.

References

[1] Bazilevs Y, Takizawa K and Tezduyar T 2013 *Computational fluid-structure interaction* (John Wiley & Sons)
[2] Formaggia L, Quarteroni A and Veneziani A 2009 *Cardiovascular Mathematics, vol. 1* (Springer, Heidelberg)
[3] Cerroni D, Da Via R and Manservisi S 2018 *J. Comput. Phys.* **354** 646–671
[4] Cerroni D, Giommi D, Manservisi S and Mengini F 2018 *Lecture Notes* **84** 217–231
[5] Aulisa E, Cervone A, Manservisi S and Seshaiyer P 2009 *Commun. Comput. Phys.* **6** 319–341
[6] Gunzburger M D 2003 *Perspectives in flow control and optimization* vol 5 (Siam)
[7] Nocedal J and Wright S 2006 *Numerical optimization* (Springer Science & Business Media)
[8] Gunzburger M and Manservisi S 2000 *Comput Methods Appl Mech Eng* **189**(3) 803–823
[9] Rowley C W and Williams D R 2006 *Annu. Rev. Fluid Mech.* **38** 251–276
[10] Kim J and Bewley T R 2007 *Annu. Rev. Fluid Mech.* **39** 383–417
[11] Manservisi S and Menghini F 2016 *Comput. and Fluids* **125** 130–143
[12] Yan Y and Keyes D E 2015 *Journal of Computational Physics* **281** 759–786
[13] Adams R A 1975 *Sobolev spaces* (Academic Press, New York)
[14] Brenner S and Scott R 2007 *The mathematical theory of finite element methods* vol 15 (Springer Science)
[15] Cerroni D and Manservisi S 2016 *J. Computat. Phys.* **313** 13–30

A new surface tension VOF evaluation by using variational representation and Galerkin interpolation projection

L. Chirco*, A. Chierici*, R. Da Vià* and S. Manservisi*

*University of Bologna, DIN, Lab. of Montecuccolino, Via dei Colli 16, Bologna 40136, Italy

Abstract. In this work we propose a new algorithm for studying two-phase interface advection problems dominated by surface tension. We use a Volume Of Fluid (VOF) algorithm for studying the evolution of the two-phase interface on a Cartesian grid and a finite element numerical scheme for the velocity-pressure state. The velocity field that drives the evolution of this interface is obtained by solving the weak form of the Navier-Stokes equation where the surface tension force is not defined in a singular way. With standard numerical approaches that solve the strong form of the Navier-Stokes equations the surface force is determined by taking the divergence of the surface tension tensor. The computation of the divergence term results in a force which is non-convergent when the grid is refined since the tensor is computed in a discontinuous cell-by-cell way. In the past this approach was proposed with artificial different smoothing schemes in order to compute such a singular force. In this work we use the variational formulation of the Navier-Stokes equation and avoid differentiation. The tensor which is a function of the unit normal is evaluated by a Galerkin projection over regular Sobolev spaces. This allows the piece-wise continuous representation of the surface tensor and the unit normal based on the VOF reconstruction. Tests on convergence for two and three-dimension in the static and dynamical cases are reported to show the correct representation in the desired spaces. This method is also natural for coupling non uniform grid computation of the fluid with Cartesian grid of the VOF algorithm.

Keywords: Navier-Stokes equations; finite element methods; VOF; two-phase flows; capillary force.

1. Introduction

Two-phase flows are of great interest in many engineering applications ranging from environmental sciences to the oil and nuclear industries. In recent years this interest has inspired a great number of algorithms for the solution of problems involving moving interfaces and surface tension forces, see [1, 2, 3, 4, 5] and references therein. All these methods provide good solutions to the problem of interface advection but an accurate representation of surface forces remain a a challenging problem with fixed grid representation. As a matter of fact, a surprising feature of the simulation of a single droplet in equilibrium in another phase without any volume force, is the appearance of the so-called spurious currents resulting from an unbalanced representation of capillary forces due to surface tension and the related pressure jump across the interface. This leads to the problem of an accurate velocity field description at the interface which otherwise may lead to numerical instabilities and artificial interface breakups.

IOP Conf. Series: Journal of Physics: Conf. Series **1224** (2019) 012004 doi:10.1088/1742-6596/1224/1/012004

In this paper the Navier-Stokes equations are discretized by using a variational formulation in order to cope with the discontinuities of the fluid properties. The variational formulation of the Navier-Stokes equations allows the treatment of the capillary force in its original form. The derivative of the surface tension tensor, which is inherently singular in its standard classic representation, is avoided. The surface tension tensor is computed by extending the unit normal and projecting the indicator phase into a more regular functional spaces by a Galerkin projection with the purpose of reproducing this force more accurately on fixed grids. Spurious currents are greatly reduced and drop dynamics can be numerically simulated with great accuracy. Furthermore the proposed approach does not need an explicit computation of the curvature which is a very challenging task. The basics of statics and dynamics of the surface tension force is reproduced with well known tests in two and three-dimensional cases.

In this paper we consider two-phase flows over a two-dimensional open bounded domain Ω. Let the reference phase 1 be contained in Ω_f and $\chi(\mathbf{x}, t)$ be the indicator function for the reference phase defined in such a way that is one for all $\mathbf{x} \in \Omega_f$ and zero on $\Omega - \Omega_f$. Over the interval of time $[0, \mathsf{T}]$ the indicator function χ behaves like a passive scalar and satisfies the following advection equation

$$\frac{\partial \chi}{\partial t} + \mathbf{u} \cdot \nabla \chi = 0 \qquad \text{in } \Omega \times [0, \mathsf{T}] \,, \tag{1}$$

where the velocity field $\mathbf{u}$ obeys the incompressibility constraint

$$\nabla \cdot \mathbf{u} = 0 \qquad \text{in } \Omega \times [0, \mathsf{T}] \tag{2}$$

and the Navier-Stokes equations over the whole domain Ω are given by the momentum conservation equation [6]

$$\rho \left(\frac{\partial \mathbf{u}}{\partial t} + \mathbf{u} \cdot \nabla \mathbf{u} \right) = -\nabla p + \nabla \cdot \left[\mu (\nabla \mathbf{u} + \nabla \mathbf{u}^T) \right] + \mathbf{F} + \mathbf{F}_s \qquad \text{in } \Omega \times [0, \mathsf{T}] \,. \tag{3}$$

$\mathbf{F}$ and $\mathbf{F}_s$ are the external body (e.g., gravity) and surface tension forces, respectively. The force $\mathbf{F}_s$ per unit of volume is defined as $\mathbf{F}_s(\mathbf{x}) = \mathbf{f}_s \, \delta_s$ where δ_s is one on the surface and zero otherwise. The surface force $\mathbf{f}_s$ can be written clearly on the surface points $\mathbf{x}_s$ as

$$\mathbf{f}_s(\mathbf{x}_s) = -\sigma \kappa \mathbf{n} \quad \text{on } \Gamma_f \tag{4}$$

where $\mathbf{n}$ is the unit external normal, κ the interface mean curvature and σ the surface tension coefficient. The coefficient σ is assumed to be constant. The normal $\mathbf{n}$, the mean curvature κ and the interface line Γ_f are all geometrical quantities.

The geometrical quantities can be computed in a standard geometrical way but it should be more convenient a direct computation from the indicator function χ. It is clear that the function χ is a discontinuous function and its derivative is a distribution. In fact we can relate the normal direction vector $\mathbf{d}$ (non unitary vector) with the gradient of χ through a distribution function. By using the generalized definition of derivatives we have

$$\int_\Omega \nabla \chi \cdot \mathbf{w} d\mathbf{x} = -\int_\Omega \chi \nabla \cdot \mathbf{w} d\mathbf{x} = -\int_{\Omega_f} \nabla \cdot \mathbf{w} d\mathbf{x} = -\int_{\Gamma_f} \mathbf{n} \cdot \mathbf{w} d\mathbf{x}_s =$$
$$= -\int_\Omega \delta_s \mathbf{n} \cdot \mathbf{w} \, d\mathbf{x} \quad \forall \mathbf{w} \in \mathbf{C}_0^\infty(\Omega) \,. \tag{5}$$

Formally this implied the extension of the normal direction vector $\mathbf{d}$ from the surface Γ_f ($\mathbf{d}$) to the whole domain Ω ($\widehat{\mathbf{d}}$) in the following form

$$\widehat{\mathbf{d}} = \mathbf{n} \, \delta_s = -\nabla \chi \quad on \, \Omega \,. \tag{6}$$

We remark that the extended normal direction $\widehat{\mathbf{d}}$ is not a unitary vector and takes values equal to zero on all points except those on the surface Γ_f. This normal direction extedended function is a distribution and it can be formally written only by introducing the surface Dirac distribution δ_s defined in (5). In the rest of the paper we use the symbol $\widehat{\cdot}$ to denote the extension of functions from interface/phase to the rest of the domain.

The evaluation of the curvature κ through the geometric normal is standard. For each point on the surface Γ_f we have

$$k = \nabla_s \cdot \mathbf{n} \quad on\, \Gamma_f \,, \tag{7}$$

where $\nabla_s = (\mathbf{I} - \mathbf{nn})\nabla$. Therefore we can write

$$\mathbf{F}_s(\mathbf{x}_s) = -\sigma\kappa\mathbf{n} = -\sigma\mathbf{n}\nabla_s \cdot \mathbf{n} \quad \mathbf{x}_s \in \Gamma_f \,. \tag{8}$$

It is easy to note that the surface divergence on the tensor $\mathbf{T}_s = \sigma(\mathbf{I} - \mathbf{nn})$ gives the same result

$$\nabla_s \cdot \mathbf{T}_s = -\sigma(\nabla_s \cdot \mathbf{n})\mathbf{n} - \sigma(\mathbf{n} \cdot \nabla_s)\mathbf{n} = -\sigma(\nabla_s \cdot \mathbf{n})\mathbf{n} = -\sigma\mathbf{n}\kappa = \mathbf{F}_s(\mathbf{x}_s) \quad on\, \Gamma_f \,. \tag{9}$$

The evaluation of the curvature κ through the χ indicator is rather difficult and could be understood only in a distributional sense. In fact, since $\nabla_n \cdot \mathbf{T}_s = 0$, we can write

$$\int_{\Gamma_f} (\nabla_s \cdot \mathbf{T}_s) \cdot \mathbf{w} d\mathbf{x} = \int_{\Gamma_f} \mathbf{T}_s : \nabla_s \mathbf{w}\, d\mathbf{x} = \int_{\Gamma_f} \mathbf{T}_s : \nabla \mathbf{w}\, d\mathbf{x} = \int_{\Omega} \mathbf{T}_s \delta_s : \nabla \mathbf{w}\, d\mathbf{x} \tag{10}$$

for all $\mathbf{w} \in \mathbf{C}_0^\infty(\Omega)$ with $\nabla_n \cdot \mathbf{w} = 0$ on the surface Γ_f. Therefore we can write formally the extended force over the whole domain $\widehat{\mathbf{F}}$ as

$$\widehat{\mathbf{F}} = \mathbf{T}_s \delta_s = \sigma(\mathbf{I} - \mathbf{nn})\delta_s \quad on\, \Omega \,, \tag{11}$$

where we have again introduced the surface distribution δ_s. The extended force $\widehat{\mathbf{F}}$ can be computed only by (10) where the derivative on discontinuous functions are considered in their generalized form.

The two fluids, the reference phase 1 and the secondary phase 2, have both constant densities, ρ_1 and ρ_2, and dynamical viscosities, μ_1 and μ_2, respectively. The density ρ and viscosity μ in (3) are functions of the indicator function χ

$$\rho = \chi\rho_1 + (1 - \chi)\rho_2 \,, \tag{12}$$

$$\mu = \chi\mu_1 + (1 - \chi)\mu_2 \quad on\, \Omega \,. \tag{13}$$

Since the derivatives of the discontinuous function χ are distributions, a treatment of the problem in its weak formulation, which will be introduced in the next section, is more appropriate.

2. Variational formulation of the two-phase Navier-Stokes equations

Let $\Omega \in R^2$ be an open set with boundary Γ. We denote with $H^d(\Omega)$ the Sobolev space of all functions having square integrable derivatives up to order d on Ω. We also denote with $\|\cdot\|_d$ and $(\cdot,\cdot)_d$ the standard Sobolev norm and inner product, respectively. When $d = 0$, we write $L^2(\Omega)$ instead of $H^0(\Omega)$ and drop the index from the inner product designation. For vector-valued functions and spaces, we use boldface notation. For example, $\mathbf{H}^d(\Omega)$ denotes the space of $\Re^2$-valued functions such that each of the two components belongs to $H^d(\Omega)$. We recall that $H^1(\Omega) \subset L^2(\Omega) = H^0(\Omega)$ [7].

In order to define a weak form of the Navier-Stokes equations, we multiply by the test functions $\mathbf{w}$ and integrate over the domain in order to write [3, 4, 7]

$$\int_\Omega \rho\left(\frac{\partial \mathbf{u}}{\partial t} + \mathbf{u}\cdot\nabla\mathbf{u}\right)\cdot \mathbf{w}\, d\mathbf{x} = -\int_\Omega \left[\mu(\nabla\mathbf{u} + \nabla\mathbf{u}^T)\right] : \nabla\mathbf{w}\, d\mathbf{x} + \tag{14}$$

$$+\int_\Omega p\nabla\cdot\mathbf{w}\, d\mathbf{x} + \int_\Omega \mathbf{F}\cdot\mathbf{w}\, d\mathbf{x} + \int_\Omega \widehat{\mathbf{T}} : \nabla\mathbf{w}\, d\mathbf{x}$$

$$\int_\Omega \nabla\cdot\mathbf{u}\psi = 0\, d\mathbf{x} \qquad \text{in } \Omega\times[0,\mathsf{T}] \tag{15}$$

for all test functions $(\mathbf{w},\psi) \in \mathbf{H}^1(\Omega)\times L^2(\Omega)$. The symbol ":" is the double product operator, namely $\mathbf{A} : \mathbf{B} = \sum_i\sum_j A_{ij}B_{ji}$. The system (14-15) includes vanishing velocity or vanishing pressure boundary conditions along the domain boundary Γ. It is important to remark that in (14-15) no curvature values are involved and only the extension of $\widehat{\mathbf{T}}$ with no derivatives is required.

The variational form of the two-phase equation can be written as [4, 8]

$$\frac{\partial}{\partial t}\int_\Omega \chi\varphi d\mathbf{x} = -\int_\Omega \varphi\nabla\cdot\left(\mathbf{u}\chi\right)d\mathbf{x} \qquad \text{in } \Omega\times[0,\mathsf{T}]\,, \tag{16}$$

for all test functions $\varphi \in L^2(\Omega)$. The problem is completed once the extended tensor $\widehat{\mathbf{T}}$ is known. From the extended normal direction $\widehat{\mathbf{d}}$, defined by

$$\int_\Omega \widehat{\mathbf{d}}\cdot\mathbf{w}\, d\mathbf{x} = -\int_\Omega \nabla\chi\cdot\mathbf{w}d\mathbf{x} = \int_\Omega \chi\nabla\cdot\mathbf{w}d\mathbf{x} \quad \forall\mathbf{w}\in\mathbf{C}_0^\infty(\Omega)\,, \tag{17}$$

we should compute the normal unit vector and the extended tensor $\widehat{\mathbf{T}}$. Formally we can write $\widehat{\mathbf{n}} = \widehat{\mathbf{d}}/|\widehat{\mathbf{d}}|$ and $\widehat{\mathbf{T}} = \sigma(\widehat{\mathbf{I}} - \widehat{\mathbf{n}}\widehat{\mathbf{n}})$ but all these terms are meaningful only in distributional spaces.

It clear that it is difficult to work with distributions and it seems reasonable to regularize the problem in more regular spaces by projecting the function χ from $L^2(\Omega)$ into $H^1(\Omega)$ where all the derivatives are well defined. Therefore it is convenient to compute the normal and surface tension by projecting χ from $L^2(\Omega)$ into $\chi^* = P\chi \in H^1(\Omega)$. For example we can use a simple Galerkin projection as

$$\delta\nabla^2\chi^* + \chi^* = \chi \tag{18}$$

where the derivatives of χ^* are well defined for $\delta > 0$. From χ^* we can compute the extended normal direction $\widehat{d} = -\nabla\chi^*$, an extension of the normal $\widehat{n} = \widehat{d}/|\widehat{d}|$ and the extended tensor $\widehat{T} = \widehat{I} - \widehat{n}\widehat{n}$.

3. Finite element and VOF spatial discretization

Let Ω_h be an open bounded domain with boundary Γ_h. The reference phase is contained in the region $\Omega_{fh} \subset \Omega_h$ whose boundary Γ_{fh} consists of a chain of connected segments. Let $X^h \subset H^1(\Omega_h)$ and $S^h \subset L^2(\Omega_h)$ be two families of finite dimensional subspaces parameterized by h that tends to zero. Also we donote by S_0^h the family of finite dimensional subspaces that contains piecewise constant functions. We make the usual approximation assumptions on X^h and S^h including the *LBB* condition (see, e.g.,[7]), and use standard cubic Taylor-Hood elements. Under these conditions, the velocity field $\mathbf{u}_h \in \mathbf{X}^h$ and the pressure field $p_h \in S^h$

satisfy the following discrete Navier-Stokes equations

$$\frac{1}{\Delta t}\int_{\Omega_h} \rho\, \mathbf{u}_h \mathbf{v}_h d\mathbf{x} + \int_{\Omega_h} (\rho\, \mathbf{u}_h \cdot \nabla \mathbf{u}_h) \cdot \mathbf{v}_h d\mathbf{x} + \int_{\Omega_h} \mu\, (\nabla \mathbf{u}_h + \nabla \mathbf{u}_h) : \nabla \mathbf{v}_h d\mathbf{x} + \tag{19}$$
$$- \int_{\Omega_h} p_h \nabla \cdot \mathbf{v}_h d\mathbf{x} = \frac{1}{\Delta t}\int_{\Omega_h} \rho\, \mathbf{u}_h^{(n-1)} \mathbf{v}_h d\mathbf{x} + \sigma \int_{\Omega_h} \left(\widehat{\mathbf{I}}_h - \widehat{\mathbf{n}}_h \widehat{\mathbf{n}}_h\right) : \nabla \mathbf{v}_h d\mathbf{x}$$

$$b(q_h, \mathbf{u}_h) = 0\,, \tag{20}$$

for all test functions $(\mathbf{v_h}, q_h) \in \mathbf{H}^h(\Omega_h) \times S^h(\Omega_h)$. The term with the time derivative has been discretized with a simple backward Euler scheme, then the velocity $\mathbf{u}_h^{(n-1)}$ is defined at the previous time step $n-1$, while all the other quantities are at time step n. Furthermore the Navier-Stokes equations are discretized by using SUPG and GLS discretizazion schemes. The GLS is not necessary here since Taylor-Hood finite elements are considered but it may be necessary to reduce pressure errors from the surface tension term which is not discretized in agreement with the LBB condition.

For the phase equation we use a VOF piecewise linear approximation. Let $C_h \in S_0^h$ be the discrete color function defined by

$$C_h = \frac{1}{\Omega_h}\int_{\Omega_h} \chi_h(\mathbf{x}, t)\, d\mathbf{x}\,, \tag{21}$$

where Ω_h is the volume of the finite element. Clearly, the color function C_h is one for all elements located in the interior of Ω_{1h}, zero for all external elements and between zero and one in the elements cut by the interface. The density $\rho_h \in S^h$ and the viscosity $\mu_h \in S^h$ become

$$\rho_h = \rho_1\, C_h + \rho_2\, (1 - C_h)\,, \qquad \mu_h = \mu_1\, C_h + \mu_2\, (1 - C_h) \tag{22}$$

The color function C_h solves [4]

$$\frac{\partial}{\partial t}\int_{\Omega} C_h\, \varphi_h d\mathbf{x} = -\int_{\Omega} \varphi_h\, \nabla \cdot \left(\mathbf{u} C_h\right) d\mathbf{x} \qquad \text{in } \Omega_h \times [0, \mathrm{T}]\,, \tag{23}$$

for all test functions $\varphi_h \in S_0^h$. In order to reguralize the computation of the normal and surface tension we project C_h from S_0^h into C_h^* in X_h^1 by using the simple Galerkin projection. We set

$$\alpha \nabla^2 C_h^* + C_h^* = C_h \tag{24}$$

for a small positive δ. We can compute the extended normal direction $\widehat{d}_h^* = -\nabla C_h^* \in S^h$ and the unit normal as $\widehat{n}^* = \widehat{d}^* / |\widehat{d}^*| \in S^h$. Once the extended normal is known in S^h also the extended tensor $\widehat{\mathbf{T}} = \left(\widehat{\mathbf{I}}_h - \widehat{\mathbf{n}}_h \widehat{\mathbf{n}}_h\right)$ is known in S^h.

4. Numerical tests

4.1. Steady solutions and stability tests

The numerical discretization of the surface tension term generates parasitic or spurious currents. These spurious currents are generated in the vicinity of fluid interfaces whose magnitude can be estimated by the simulation of the static case, i.e. a static spherical droplet/bubble immersed in another fluid in zero gravity conditions. The analytical solution of the static case is characterized by a zero velocity field and a pressure jump between the two phases $\Delta p = \sigma\, \kappa$, where κ is the mean curvature. However, due to the singularity of the force, the forcing term results in small pressure-velocity fluctuations near the interface. Depending on the numerical representation, on

IOP Conf. Series: Journal of Physics: Conf. Series **1224** (2019) 012004 doi:10.1088/1742-6596/1224/1/012004

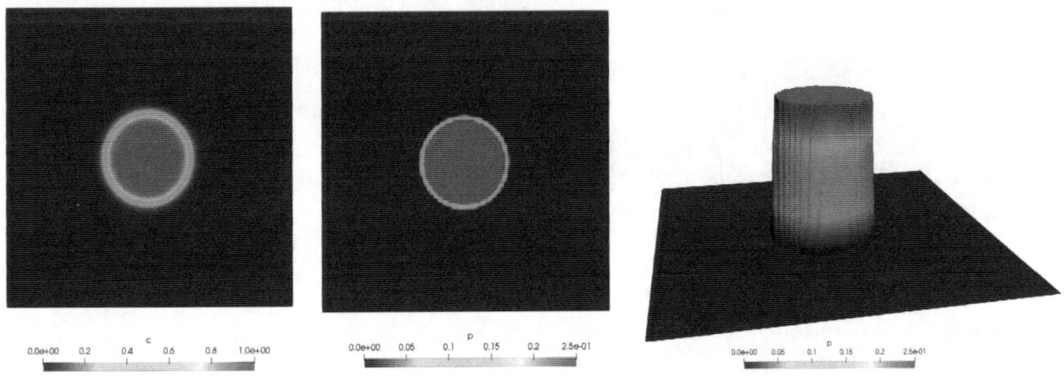

Figure 1. Two-dimensional steady case (A). The projected color function C_h^* (left) and the pressure distribution p_h (center and right) for VOF grid resolution 160×160.

viscous damping and on the value of the surface tension coefficient σ, these fluctuations give rise to flows that, in many cases, do not decrease with mesh refinement. In this work we introduce a numerical scheme that does not need the explicit curvature computation and represents the

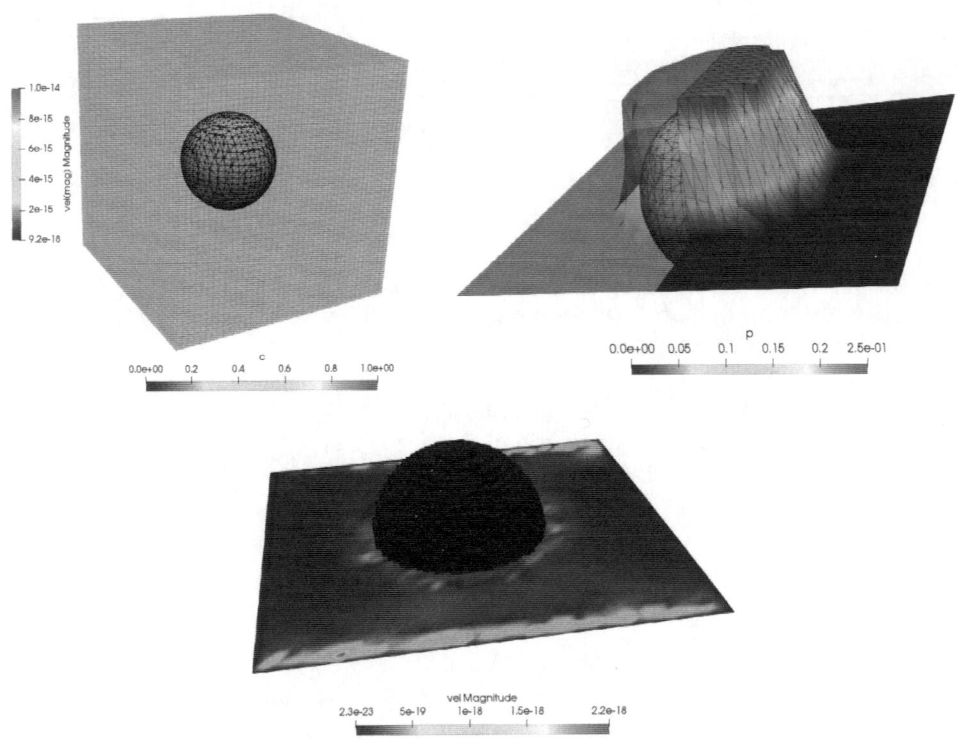

Figure 2. Three-dimensional steady case (A). Plane reconstruction of the interface (top left), the pressure p_h (top right) and the velocity field $\mathbf{u}_h$ on a plane (bottom) for a $40 \times 40 \times 40$ VOF grid resolution.

IOP Conf. Series: Journal of Physics: Conf. Series **1224** (2019) 012004 doi:10.1088/1742-6596/1224/1/012004

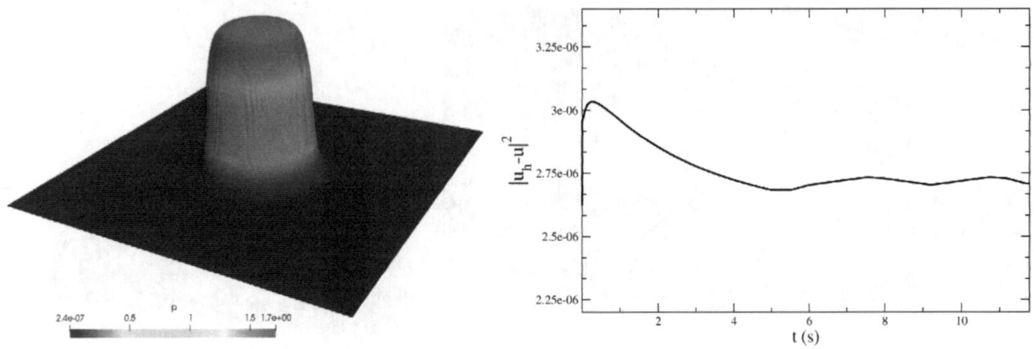

Figure 3. Two-dimensional steady case (B). Pressure for VOF resolution 320×320 (on the left) and square norm $\|\mathbf{u}_h - \mathbf{u}\|_{L^2}^2$ of the velocity error as a function of time for VOF grid resolution 160×160 .

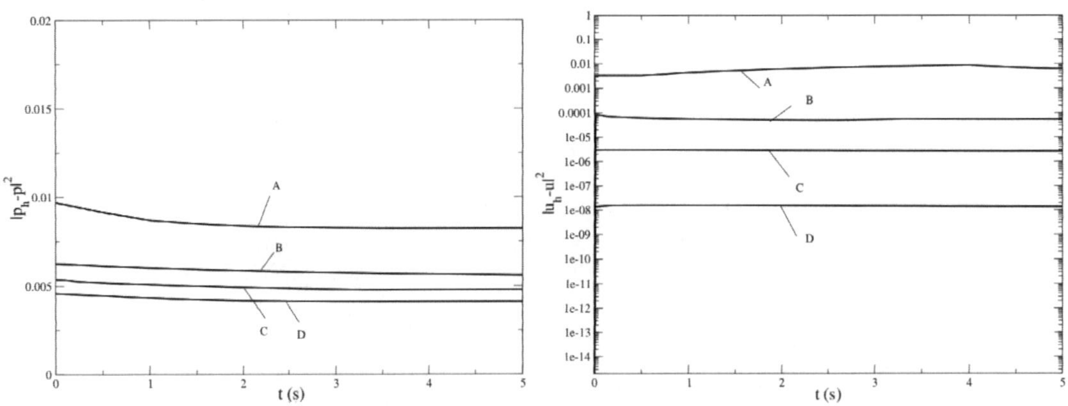

Figure 4. Two-dimensional steady case (B). Velocity $\|\mathbf{u}_h - \mathbf{u}\|_{L^2}^2$ and pressure $\|p_h - p\|_{L^2}^2$ errors as a function of time for different VOF grid resolutions 40×40 (A), 80×80 (B) 160×160 (C) and 320×320 (D) as a function of time.

surface tension force in an appropriate space.

We present two different tests: case A, where we consider known the exact value of the curvature and case B where the surface tension term is computed directly from the numerical representation of the interface. In order to show the robustness of the approach we show computation in two and three-dimensional spaces.

First we consider Case A with assigned curvature. If the curvature $\kappa = \kappa_0$ is constant we can extend this value outside the interface $\widehat{\kappa} = \kappa_0$ and, for the surface tension term in (14), we have

$$\sigma \int_{\Omega_h} \left(\widehat{\mathbf{I}}_h - \widehat{\mathbf{n}}_h \widehat{\mathbf{n}}_h\right) : \nabla \mathbf{v}_h d\mathbf{x} = \sigma \int_{\Omega_h} \left(C_h \widehat{\kappa}\right) \widehat{\mathbf{I}} : \nabla \mathbf{v}_h d\mathbf{x} . \tag{25}$$

For the two-dimensional steady case A we consider a unitary box with a two-dimensional droplet radius of 0.15. In Figure 1 the two-dimensional steady case A with assigned constant curvature is shown. On the left one can see the projected color function C_h^*, while the pressure distribution p_h is shown in the center and on the right for VOF grid resolution 160×160. In this case we set $\mu_1/\mu_2 = \rho_1/\rho_2 = 100$ with $\mu_2 = 1.e-4$ and $\rho_2 = 1$. If the curvature is constant and assigned as

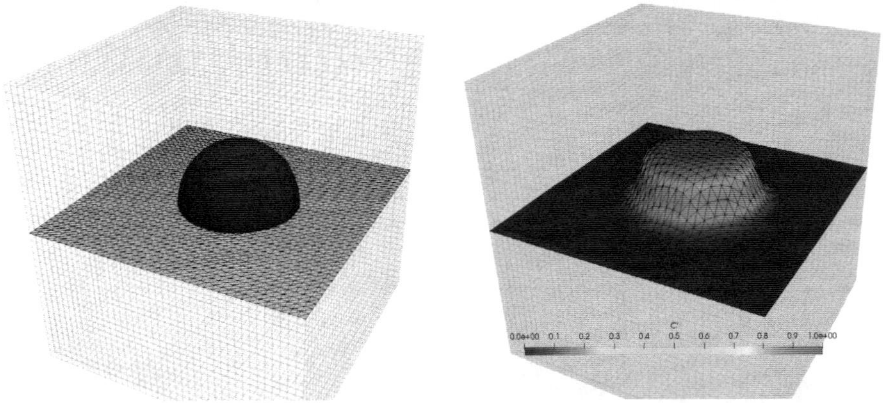

Figure 5. Three-dimensional steady case (B). Surface plane with PLIC-VOF reconstruction (left) and the projected color function C_h^* for $40 \times 40 \times 40$ VOF grid resolution.

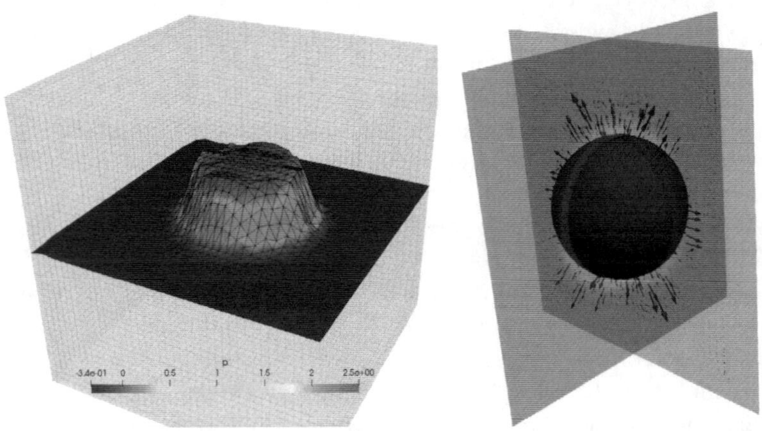

Figure 6. Three-dimensional steady case (B). Pressure p_h (left) and velocity field $\mathbf{u}_h$ (right) for $40 \times 40 \times 40$ VOF grid resolution.

in (25) the velocity field is zero ($\approx 10^{-18}$) and no spurious currents are present.

For the three-dimensional steady case A we consider again a unitary box with a droplet radius of 0.15. In Figure 2 the three-dimensional steady case A with assigned constant curvature is shown. On the left top one can see the reconstruction of the interface of the color function C_h^* and on the right top the pressure distribution p_h for VOF grid resolution $40 \times 40 \times 40$. In this case we set $\mu_1/\mu_2 = \rho_1/\rho_2 = 100$ with $\mu_2 = 1.e-3$ and $\rho_2 = 1$. On the bottom of Figure 2 it is possible to see the velocity field over the central plane. If the curvature is constant, and assigned as in (25), the velocity field is in the range of 10^{-23}-10^{-10} and no spurious currents are present. In fact, since the pressure term has the form in (25), $p_h = \sigma C_h \widehat{\kappa}$ when the interpolation space for p_h and C_h is the same and respects the LBB condition.

This test aims to check the correct distribution of the interface forces. However, if the curvature κ_h is not available analytically the numerical noise alters the interface force balance and spurious currents are generated. The computation of the surface tension force can be seen in a great variety of implementations, see for example [9]. Almost all VOF methods for different numerical techniques, which are based on the same discrete color function C_h, do not usually

IOP Conf. Series: Journal of Physics: Conf. Series **1224** (2019) 012004 doi:10.1088/1742-6596/1224/1/012004

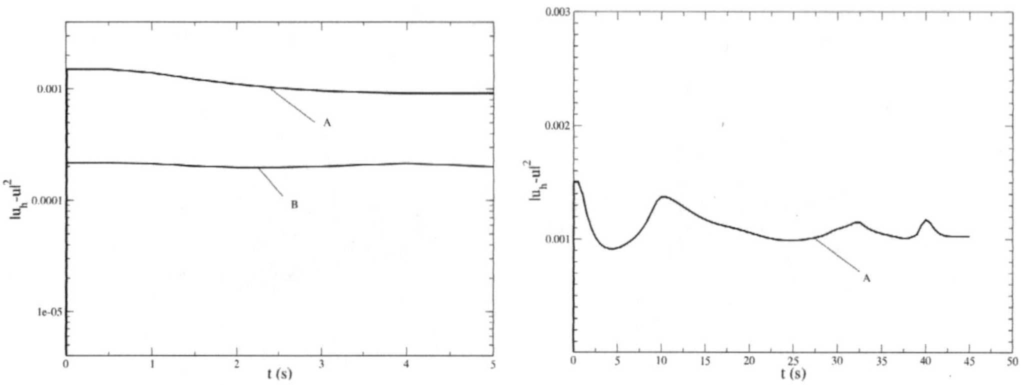

Figure 7. Three-dimensional steady case (B). Velocity $\|\mathbf{u}_h - \mathbf{u}\|^2_{L^2}$ error as a function of time for different VOF grid resolutions $20 \times 20 \times 20$ (A) and $40 \times 40 \times 40$ (B) as a function of time.

approximate the interface local curvature in an accurate way.

Now we consider Case B where the curvature is not known and the surface tension term is computed numerically as in (14). We propose an approach where the color function field $C_h \in S_h$ is projected into $C^*_h \in X_h \subset H^1(\Omega)$ where its gradient is well defined. In order to project the function we use a Galerkin type interpolation, for example as in (24) with $\alpha = 5 \times 10^{-4}$. We remark that the curvature is not computed explicitly but only the unit normal is needed both in two and the three-dimensional cases.

Figures 3-4 show the results for the pressure and velocity for the two-dimensional steady case with numerical surface tension term. The geometry is the same as in Case A. On the left and on the right of Figure 3 the pressure for VOF resolution 320×320 and square norm $\|\mathbf{u}_h - \mathbf{u}\|^2_{L^2}$ of the velocity error as a function of time for VOF grid resolution 160×160 are shown, respectively. The error is evaluated in L^2 norm in agreement with the functional spaces considered. It is important to remark that the spurious currents do not vanish or increase but show an oscillating behavior as a function of time. The numerical solution is a non linear solution that takes into account the numerical reconstruction of the interface and the pressure field. From Figure 3 on the right one can see that the corrections on the interface position move the pressure field and this adjusts back the interface position with oscillating behavior. If one improves the resolution the behavior remains the same as can be seen in Figure 4. On Figure 4 velocity $\|\mathbf{u}_h - \mathbf{u}\|^2$ and pressure $\|p_h - p\|^2$ errors are shown for different VOF grid resolutions 40×40 (A), 80×80 (B) 160×160 (C) and 320×320 (D) as a function of time. The error decreases with higher resolutions but does not vanish in time. The steady solution error oscillates depending from many factors such as temporal step and location of the interface.

For the three-dimensional case B the results are shown in Figures 5-7. The data and the geometry are the same as in the previous case A. In Figure 5 one can see surface plane with PLIC-VOF reconstruction and the projected color function C^*_h for $40 \times 40 \times 40$ VOF grid resolution. In Figure 6 the pressure p_h and the velocity field $\mathbf{u}_h$ for $40 \times 40 \times 40$ VOF grid resolution are shown on the left and right, respectively. Finally the velocity $\|\mathbf{u}_h - \mathbf{u}\|^2_{L^2}$ error for different VOF grid resolutions $20 \times 20 \times 20$ (A) and $40 \times 40 \times 40$ (B) as a function of time is reported in Figure 7. As in the previous case the L^2-norm error decreases with the resolution but oscillates in time as the reconstruction algorithm moves back and forward the plane inside the cell.

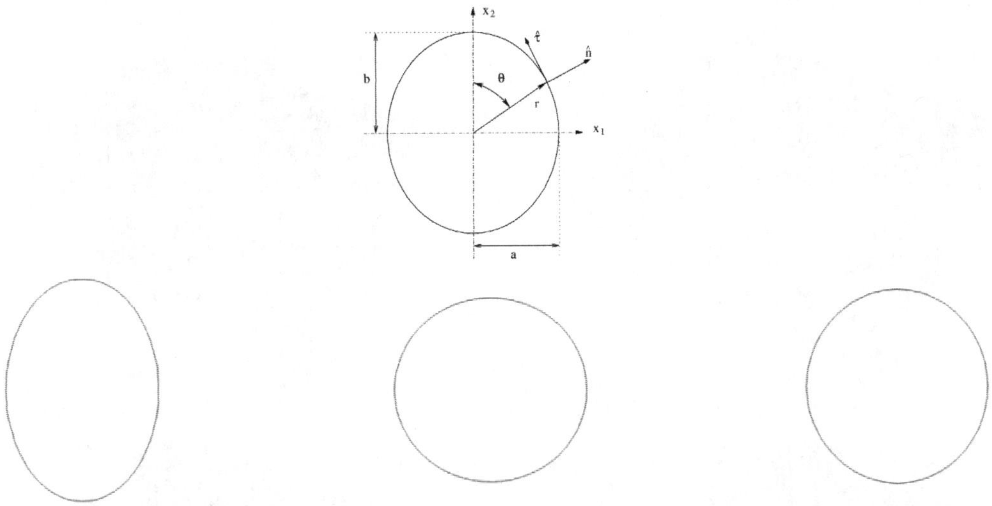

Figure 8. Interface evolution for $t = 0, 3.5, 6$ for VOF resolution $80 \times 80 \times 80$ and $\rho_1/\rho_2 = 100$, $\mu_1/\mu_2 = 100$, $\mu_1 = 0.002$.

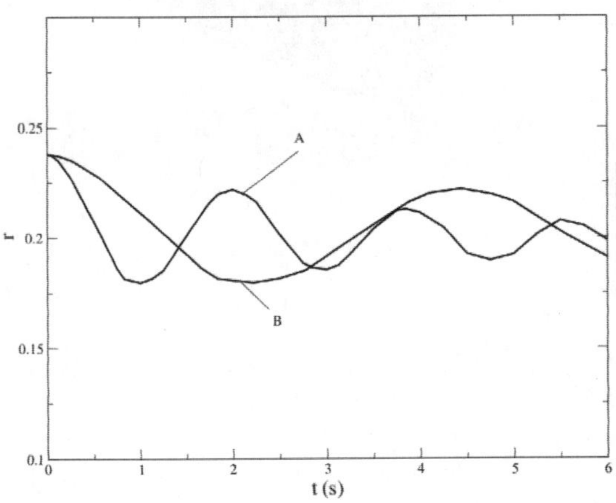

mesh	analytic T	C T	D T
64×64	4.437	4.201	4.333
128×128	4.437	4.383	4.405
256×256	4.437	4.455	4.440

Figure 9. Radius oscillation amplitude as a function of time for different density ratio $\rho_1/\rho_2 = 10$ (A), 100(B) for VOF resolution 64×64 and table of the time period T for different resolutions and time step $dt = 0.025$ (C) and $dt = 0.01$ (D) in comparison with the analytic value for $\mu_2 = \mu_1 = 0.002$.

4.2. Two-dimensional droplet oscillations

In the previous section we have considered a simulation based on the Laplace's law and test the behavior of our surface tension algorithm with respect to the generation of parasitic currents. The results for two and three-dimensional pressure-velocity field can be considered satisfactory. However, even if the spurious currents remain bounded for small time steps in the equilibrium solution for different mesh resolution and density ratio, we need to study the system during its evolution in time. In this section we consider the two-dimensional case. We consider a two-dimensional geometry, as shown in figure 8, with cylindrical coordinates (r, θ). The interface

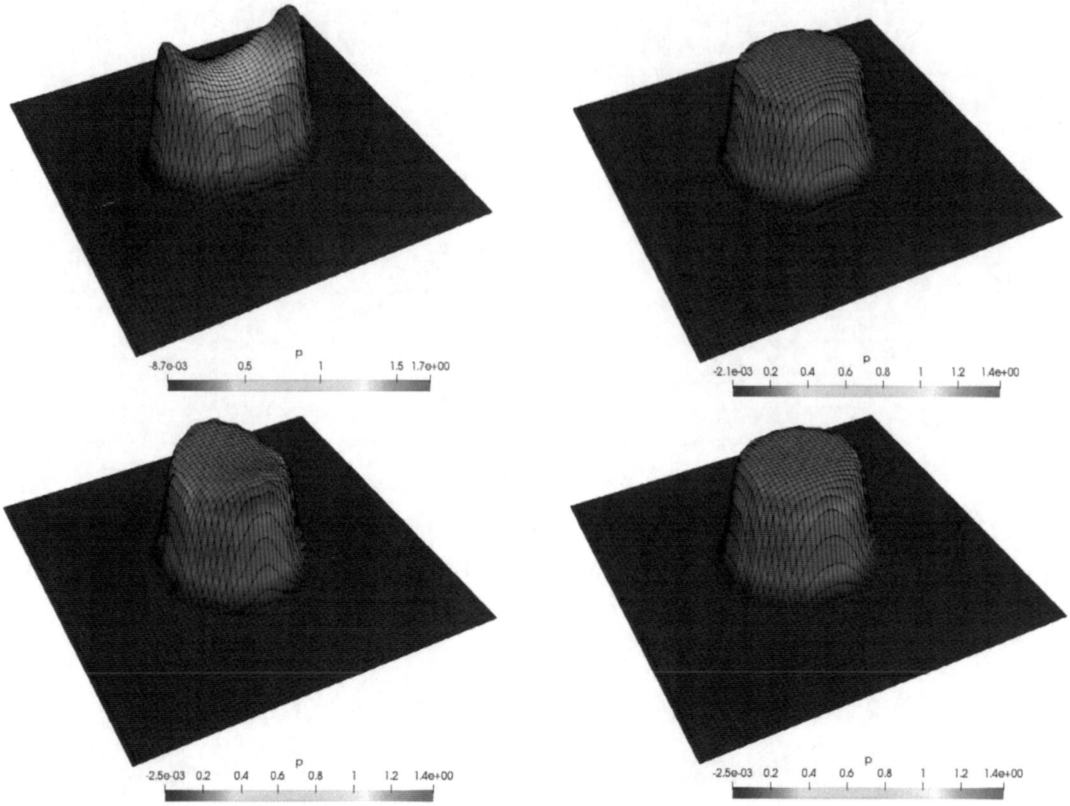

Figure 10. Pressure time evolution for $t = 0$, 2., 3.5, 6 for VOF resolution $80 \times 80 \times 80$ and $\rho_1/\rho_2 = 100$, $\mu_1/\mu_2 = 100$ $\mu_1 = 0.002$.

line $r = r(\theta)$, in its initial configuration, is given by

$$r = R(1 - \epsilon/4 + \epsilon P_2(cos\theta))\,,$$

where R is the droplet radius at equilibrium, ϵ the relative deformation factor and $P_2(x)$ the Legendre Polynomial of order 2 [10]. This test is focused on small oscillations released by the second spherical harmonic. In particular, the oscillations are characterized by an angular frequency ω_2 given by the expression [11]

$$\omega_2^2 = \frac{6\,\sigma}{(\rho_1 + \rho_2)\,R_0^3}\,. \tag{26}$$

Our simulations are performed over a two-dimensional unit square domain. Homogeneous Neumann boundary conditions are enforced over the rest of the boundary. The droplet with density ρ_1 and viscosity μ_1 is surrounded by a medium with density ρ_2 and viscosity μ_2. Different density ratios are obtained by fixing the reference phase density $\rho_1 = 1$ and changing the density ρ_2. The viscosity for reference phase is set to 2×10^{-3}. In this test we assume $R = 0.195$, $\epsilon = 0.02$, $\sigma = 0.25$, $\rho_1/\rho_2 = 10$-100, $\mu_1/\mu_2 = 1$ and $\alpha = 0.0002$. We also consider different time steps with $dt = 0.025$-0.01. In Figure 4.2, on the left, we plot the oscillating radial position of the interface point at $\theta = 0$ as a function of time for VOF resolution 64×64 and different density ratios. One can see the results obtained for $\rho_1/\rho_2 = 10$ (case A) and for $\rho_1/\rho_2 = 100$ (case B).

On the right of Figure 4.2 we have the table of the time period $T = 2\pi/\omega_2$ with $\rho_1/\rho_2 = 100$ for different resolutions and time step $dt = 0.025$ (C) and $dt = 0.01$ (D) in comparison with the analytic value.

By changing the viscosity the oscillation are damping quickly. We consider the same geometry and we set $\rho_1/\rho_2 = 100$, $\mu_1/\mu_2 = 100$, $\mu_1 = 0.002$. The reconstructed interface evolution for $t = 0, 3.5, 6$ for VOF resolution $80 \times 80 \times 80$ is shown in Figure 8. In Figure 10 the time evolution of the pressure field for $t = 0$, 2., 3.5, 6 for VOF resolution $80 \times 80 \times 80$ is reported

4.3. Three-dimensional droplet oscillations

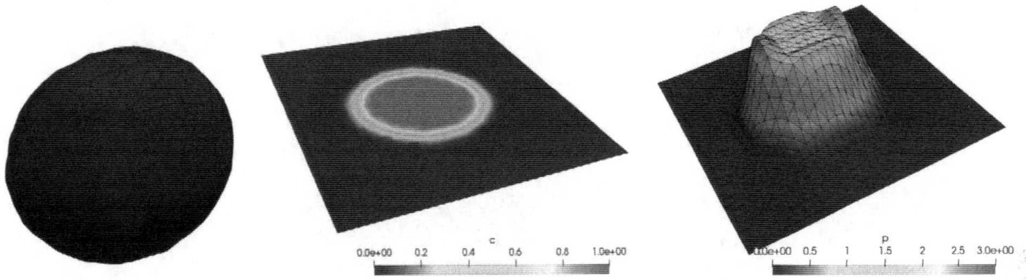

Figure 11. Interface reconstruction at $t = 0$.

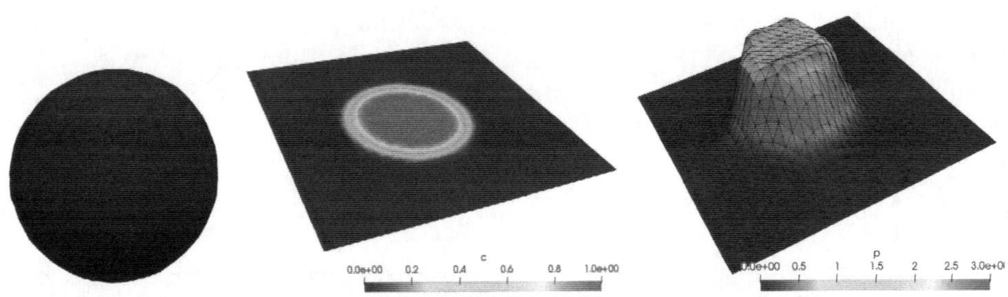

Figure 12. Interface reconstruction at $t = 0.4$.

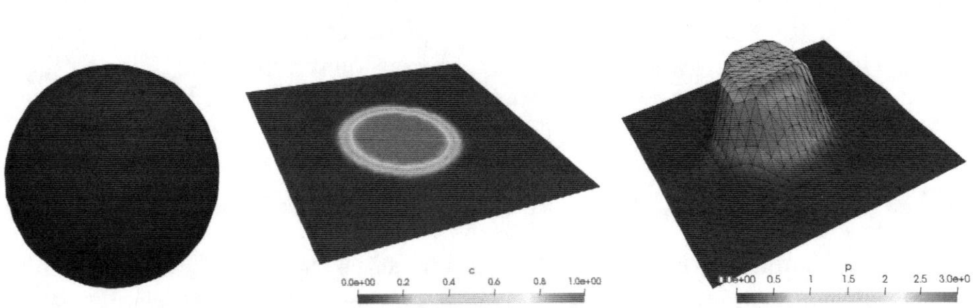

Figure 13. Color function on the central plane at $t = 1$.

In this section we consider the two-dimensional case. We consider, in axisymmetric coordinates (r, θ), the initial ellipsoidal shape of figure 8, but now $r = R(1 + \epsilon P_2(cos\theta))$, where R is the radius of the spherical droplet at equilibrium, ϵ the relative deformation and $P_2(x)$ the Legendre Polynomial of order 2 [10]. In [11], an analytical solution for the small amplitude oscillations of an incompressible inviscid droplet in the vacuum is obtained. The author wrote the fundamental mode of oscillation associated to the Legendre polynomial of order 2 a with the corresponding angular frequency

$$\omega_2^2 = 8\frac{\sigma}{\rho R^3} \,. \tag{27}$$

Our simulations are performed over a three-dimensional cubic unit domain. Homogeneous Neumann boundary conditions are enforced over the rest of the boundary. As in the two-dimensional case the droplet with density ρ_1 and viscosity μ_1 is surrounded by a media with density ρ_2 and viscosity μ_2. The density and viscosity for reference phase are set to 1 and 2×10^{-3}, respectively. In this test we assume $R = 0.195$, $\epsilon = 0.01$, $\sigma = 0.25$, $\rho_1/\rho_2 = 10$-100, $\mu_1/\mu_2 = 1$ and $\alpha = 0.0002$.

mesh	analytic T	$\Delta t = 0.025$ T
$32 \times 32 \times 32$	3.182	3.361
$64 \times 64 \times 64$	3.182	3.255

Table 1. Oscillation period T for different mesh resolutions.

In Table 1 we show the oscillation period T for different mesh resolutions in comparison with the analytical one. By changing the viscosity the oscillation are damping quickly. We consider the same geometry and we set $\rho_1/\rho_2 = 100$, $\mu_1/\mu_2 = 100$ and $\mu_1 = 0.002$.

The results, obtained with a $32 \times 32 \times 32$ mesh resolution and a time step $dt = 0.025$ are shown in Figures 11-13. For each Figure we report form the left to right the reconstructed interface, the color function and the pressure field on the central plane.

5. Conclusions

A new numerical approach for computing the surface tension term for two-phase incompressible flows in two and three-dimensional geometries has been proposed. This algorithm computes the surface tension tensor directly from the normal vector and a Galerkin projection of the indicator function. In the framework of a variational formulation of the Navier-Stokes equations it is possible to evaluate the surface tension force without the explicit evaluation of the local mean curvature. Classical tests, reproducing the solution of the static case and the dynamic of the droplet oscillations, are simulated in two and three-dimensional geometries. This method can be classified as continuous surface stress (CSS) method. For these methods the direct computation of the curvature is not necessary. Our proposed method has several advantages over the other CSS methods: the numerical implementation of the two-dimensional and three-dimensional cases is basically the same and no derivative of the tensor stress is required. On the contrary continuous surface force (CSF) methods, such as methods of smoothing the discontinuity of surface tension forces and convolution with a suitable kernel, need the curvature computation. This is a very hard task when the second derivative of the color function should be evaluated since the accuracy and convergence of the numerical evaluation of a distribution function remain an open question. The main weakness of our approach is the cost of computation: the extended field should be computed over all the domain and the projection in H^1 need the solution of a linear system. At the same time this projection can be used easily in non-Cartesian grids with moving meshes [5, 12].

References

[1] Osher S, Fedkiw R, 2001 J. Comput. Phys. **169** 463-502
[2] Aulisa E, Manservisi S, Scardovelli R, 2004 J. Comput. Phys. **197** 555-584
[3] Aulisa E, Manservisi S, Scardovelli R, 2003 J. Comput. Phys. **188** 611-637
[4] Aulisa E, Manservisi S, Scardovelli R, 2006 Comput Methods Appl Mech Eng **195**(44-47) 6239-6257
[5] Cerroni D, Da Vià R, Manservisi S, 2018 J. Comput. Phys. **354** 646-671
[6] G. Tryggvason G, Brunnen B, Esmaeli A, Juric D, Al-Rawahi N, Tauber W, Han J, Nas S, Jan YJ, 2001 J. Comput. Phys. **169** 708-759
[7] V. Girault V, Raviart PA, 1986 Springer Series in Comput. Math. 5
[8] Cervone A, Manservisi S, Scardovelli R, 2011 Comput. Fluids **47**(1) 101-114
[9] Kothe DB, Williams MW, Puckett EG, 1998 Fluid Dynamics at Interfaces, Cambridge University Press
[10] Bnà S, Manservisi S, Scardovelli R, Yecko P, Zaleski S, 2016 Comput. Phys. Commun. **200** 291-299
[11] A. Prosperetti A, 1980 J. Fluid Mech. **100** 333-347.
[12] Aubert F, Aulisa E, Manservisi S, Scardovelli R, 2006 Comput. Fluids **35**(10) 1332-1343

IOP Conf. Series: Journal of Physics: Conf. Series **1224** (2019) 012005 doi:10.1088/1742-6596/1224/1/012005

Tunnel fire active protection: improving ventilation system

Laura Cirrincione[1], Gianluca Scaccianoce[1], Michael Scrudato[1] Matilde Pietrafesa[2], Gianfranco Rizzo[1]

[1]Dipartimento di Energia, ingegneria dell'Informazione e modelli Matematici, Università degli Studi di Palermo, Italy
[2] Dipartimento di Ingegneria Civile, dell'Energia, dell'Ambiente e dei Materiali, Università Mediterranea di Reggio Calabria, Italy

Laura Cirrincione, email address: laura.cirrincione@unipa.it

Abstract. The main aim of this work is to evaluate how the use of an active protection method, consisting in an improved forced longitudinal ventilation system, can determine a positive impact on heat extraction and people evacuation time in a tunnel fire scenario, by considering a case study which simulations are based on a real gallery.

1. Introduction

Tunnel fire safety has the ultimate goal of reducing the risk of injury and/or mortality for users, also by reducing damage to the tunnel structure and facilitating rescue operations.

With increasing number of fire accidents [1] it is necessary to ancillary improve the current fire protection methods and technologies. Tunnel fires, however, are very complex phenomena because of the interactions among the general causes of fire (combustion, radiation, etc.), the tunnel geometry (tunnel shape, vehicle geometry, arrangement of vehicles, etc.) and the factors that might cause an increase in fire power (presence and kind of ventilation, back layering effect, presence of bituminous conglomerate causing pyrolysis, etc.) [2].

For this reason numerous *Computational Fluid Dynamics* (CFD) numerical investigations have been carried out in different tunnel scenarios to predict the smoke movement and temperature distribution [3, 4], by considering various aspects such as: different kind of ventilation systems (longitudinal or naturally ventilated tunnels with roof openings) [5, 6, 7, 8, 9], multiple fire sources, the presence of obstacles [10, 11], extinguishing systems [12], passive fire protection layers on the concrete, etc..

The advantage in using CFD is that such complex physical interactions can be modelled at the same time, thus studying their relative influence on fire behavior [13]. Small scale [14, 15] and full-scale [16, 17] experiments were also performed to assess smoke removal efficiency, temperature distribution and compression strength of concrete tunnel lining during a fire.

The main aim of this work is the use of CFD to evaluate how the use of an active protection method, forced longitudinal ventilation, can have an impact on the heat extraction and people evacuation time during a tunnel fire.

In order to analyze the behavior of such improved ventilation system a real gallery (the "Caltanissetta Tunnel" in Italy) was considered as a case study for the simulations.

2. Methodology

In order to analyse the behaviour of the improved ventilation system, different scenarios were implemented, considering a real gallery as a case study for the simulations. In particular, the authors chose to model the active protection system of the tunnel by varying the ventilation capacity of the jet fans in the proximity of the fire occurrences.

The ventilation system chosen for this study consists in the use of piston effect fans, also called accelerators or jet fans, which induce an air flow inside the tunnel along its entire section and length (supporting the air flow already induced by vehicles). By using a forced ventilation of this type, vehicles stacked upstream of the fire are preserved by smoke, that is driven towards the parts of the tunnel from which vehicles have already been removed [18]. In particular reversible axial type stainless steel accelerators operating in pairs (and arranged along the tunnel at a pace of about 100 m) have been considered.

The simulations were carried out through the FDS (*Fire Dynamics Simulator*), that is indeed a CFD software created by the *National Institute of Standards and Technology* (NIST) during the investigation of the collapse of the World Trade Center. This program solves the Navier-Stokes equations according to a turbulence LES formulation, using finite volumes, and is particularly effective in predicting the dynamics of fire effluents (heat and smoke) once a suitable fire heat release rate (HRR) curve is prescribed.

2.1. The case study

The gallery considered for the simulations is the “Caltanissetta Tunnel” (in Sicily), part of the extension works of S.S.640 "Porto Empedocle" route, consisting in the adjustment of the existing S.S.640 (extra-urban single carriageway road, double direction) one.

The gallery is double-arched, with one-way traffic; it has a total length of about 8000 m and an average elevation a.s.l. of 519.50 m (with a maximum altitude a.s.l. of 550.81 m). The axes of the two arches are separated from each other by a minimum of 35 m (near the entrances) up to a maximum of 85 m.

The maximum coverage in the shell is 130 m. The road platform has a total width of 11.25 m, organized with 2 lanes (slow and overtaking), 3.75 m wide, flanked by a 0.50 m pier on the left and a 3.25 m pier on the right (emergency lane) along the entire length of the tunnel. Throughout the tunnel, free vertical francs ($\geq$ 5.00 m) are guaranteed on the carriageway [19].

The fire scenario considered in the case study involves a Heavy Goods Vehicle (HGV) and asphalt pyrolysis (to obtain a more realistic burning-fire model).

2.2. Simulation

Given the homogeneous characteristics of the whole road section, to simulate the fire scenario only a 100 meters section of the tunnel was implemented in the CFD software.

Considering that FDS is a computing code, it does not allow to draw curves; for this reason, they have been reproduced using a CAD model, approximating them with rectangles. For the sake of simplicity, the HGV used in the simulation was represented by a parallelepiped, whose characteristic dimensions are 2.5 m x 7 m x 3.5 m.

Concerning the size of the mesh, two types of cells were sat: moderate (of a size of 1 m) and fine (of a size of 0.5 m). By adopting the use of a fine mesh only in the fire proximity, in order to estimate the temperature of the vault more accurately, the calculation domain of 15 m x 100 m x 10 m was divided into 35250 cells distributed as follows:

- in the fire zone, with a domain of 15 m $\times$ 7 m $\times$ 10 m, 15600 cells of 0.5 m $\times$ 0.5 m $\times$ 0.5 m, of which 30 cells in the X (in the transverse direction to the section), 14 cells in the Y (longitudinal direction) and 20 cells in the Z (in height) directions, respectively;
- in the adjacent fire zone, for a domain of 15 m $\times$ 22 m $\times$ 10 m, 13200 cells of 0.5 m $\times$ 1.0 m $\times$ 1.0 m, of which 30 cells in the X (in the transverse direction to the section), 22 cells in the Y (longitudinal direction) and 10 cells in the Z (in height) directions, respectively;

- in the two zones, for a domain of 15 m × 21/22 m × 10 m, 6450 cells of 1 m × 1 m × 1 m, of which 15 cells in the X (in the transverse direction to the section), 21/22 cells in the Y (longitudinal direction) and 10 cells in the Z (in height) directions, respectively.

Fig. 1 shows the sketch of the tunnel used for simulations.

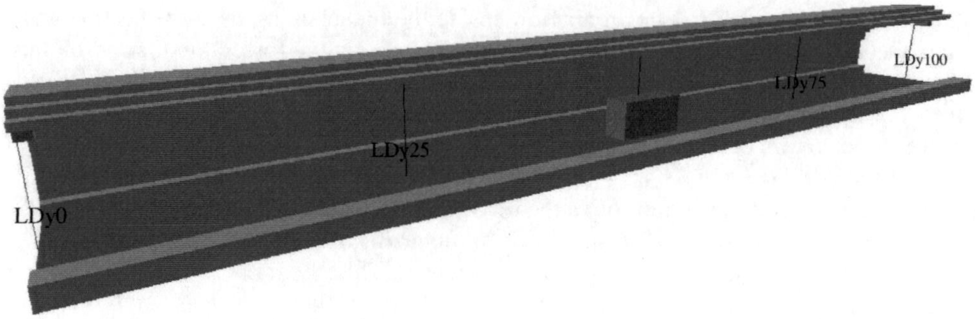

Figure 1. Sketch of the tunnel implemented into the FDS software (with outline of mesh domains).

Once the mesh size and the calculation domain have been defined, the two extremity portals must be defined in order to provide a continuity solution. To do this, the *open mesh boundaries* function was used to open the two portals, as if they were ventilation surfaces (i.e. VENTs).

The materials used in the modeling are concrete for the vault and asphalt (or bituminous conglomerate) for the road, which have been inserted in the FDS code by assigning specific heat, conductivity, density and emissivity values.

The duration of the simulations was set to 3000 seconds, given the high computational burden, leading to a calculation time of over 6 hours, using a *MacPro 6,1* equipped with a 12-Core *Intel Xeon E5* single processor (2.7 GHz, 30 MB Cache). Regarding the output quantities given by the software, two *boundary quantities* were required: temperature and smoke's height in different positions. In particular it was chosen to report in the present work the values of these parameters evaluated on the vault of the tunnel, in correspondence of the fire. Values of temperature and velocity of gasses have been also evaluated.

To estimate the temperature trend over time, virtual sensors have been used: they were positioned on the vault of the tunnel in correspondence of the fire point, as shown in Fig. 2.

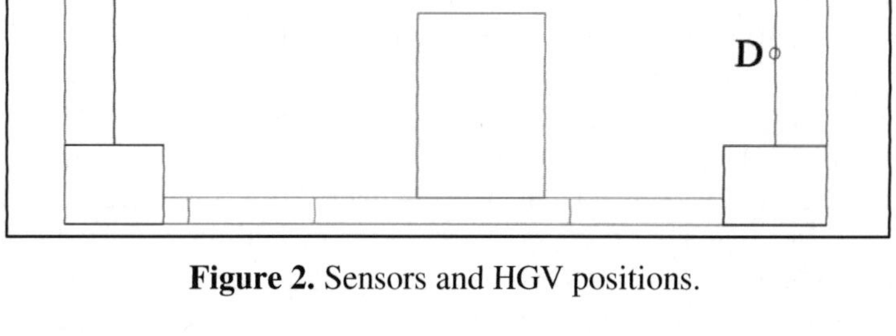

Figure 2. Sensors and HGV positions.

The study of the dynamics of the tunnel fire, here considered, starts from the evaluation of the Heat Release Rate (HRR) from a hypothetical flaming vehicle, which indicates the thermal energy emitted by a fire per unit of time. It can be obtained mathematically, with CFD simulations or by experimental tests in a controlled environment. In this work to simulate the combustion of the material it was used an HRR determined by Carvel et al. [20].

The predetermined HRR was assigned to the fire source by placing proper ventilation surfaces (VENTs) on its surface. The truck was therefore equipped with three VENTs, respectively on the top, right side and left side, to exactly reproduce the thermal release of the same truck that, after 1200 seconds, reaches a peak of about 200 MW (Fig. 3), according to literature data [21]. To do this, it has been hypothesized that all generic VENT cells emit an identical Heat Release Rate per Unit Area – HRRPUA (t) – given by:

$$HRRPUA\ (t) = \frac{HRR(t)}{A}$$

where A is the total area of the three VENTs, from which the thermal release comes, which in the case of the truck is equal to 61.6 m^2. Such HRRPUA (t) is thus identical for all cells and it is 3295.45 KW/m^2.

Thanks to these data it is possible to realize a more realistic scenario, by introducing a model of asphalt pyrolysis and one of forced longitudinal ventilation.

In this regard, as previously mentioned, for the present case study it was chosen to consider a reversible stainless-steel ventilation system, in which the fans operate in pairs and are arranged along the tunnel with a 100 m pace, having a flow rate of 24.0 m^3/s. In particular each end side of the tunnel has been equipped with two inlet fans that can be converted into suction fans by simply inverting the flow rate sign. In the FDS software, the fan was represented as a square with a side of 0.75 m.

Regarding the behavior of the bituminous conglomerate (and its ability to increase the load of fire), in order to create a model in FDS, the results of the BRE center for fire safety engineering at the University of Edinburgh were taken into account, concordant with the results of the PAVIREX project [22]. In particular, the following features were implemented in the FDS code:

- the asphalt combustion heat is equal to 35166 kJ/kg;
- the pyrolysis reaction heat is equal to 3000 kJ/kg;
- the reference temperature is equal to 450°C;
- a pyrolysis range of 60°C;
- a binder content of 6% in bulk.

3. Results

3.1. Base simulation (first scenario)

In order of confirming the value of HRR assumed in our study (that according to the method must be considered as a pre-determined parameter), we have conducted here a basic simulation of the fire phenomenon, in which standard forced longitudinal ventilation has been considered.

The obtained results confirm an HRR slightly exceeding 203 MW (Fig. 3), which ripple is due to the presence of oxygen from forced ventilation.

Fig. 4 show the gasses temperatures, where A, B, C and D are the sensors, indicated in Fig. 2, positioned in the fire occurrence point, namely at y = 50 m, while A_{-10} and A_{+10} represent the sensors positioned in A but at a y position of respectively 40 m and 60 m, that is at 10 m from the fire center (y = 50 m).

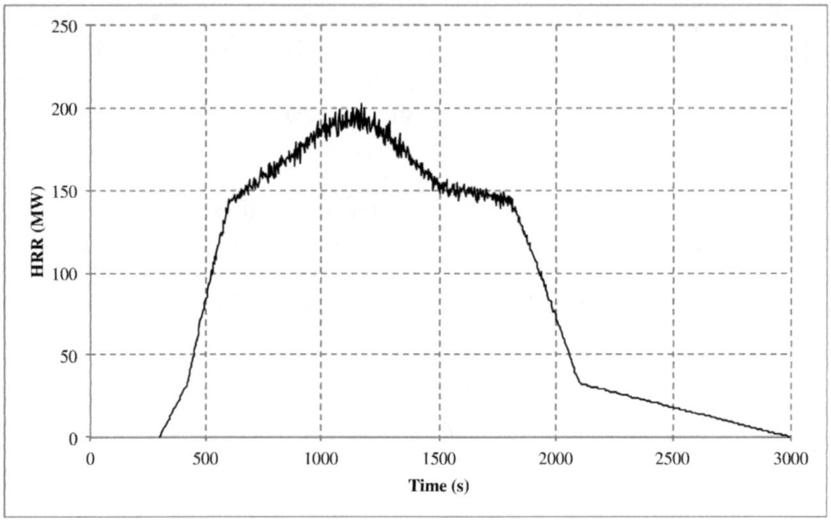

Figure 3. Heat Release Rate (first scenario).

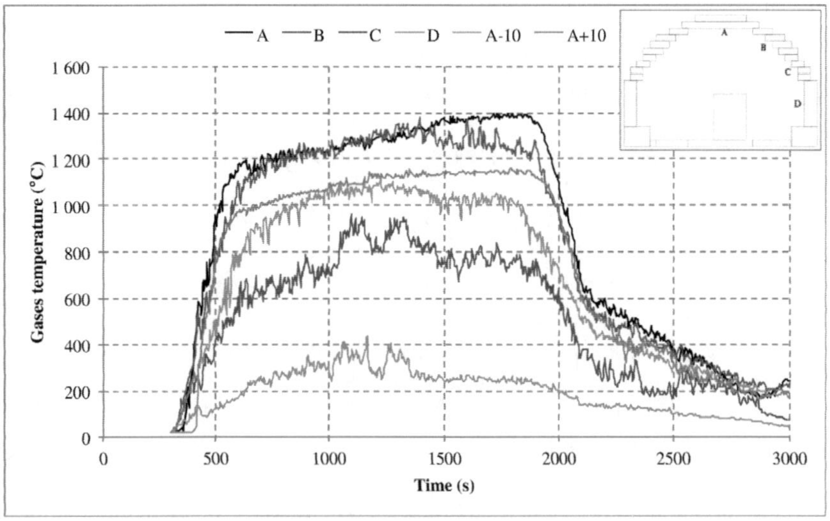

Figure 4. Gasses temperatures (first scenario).

In order to evaluate the effects of smoke on people, and their ability to react to the fire, the smoke stratification was also analyzed throughout the longitudinal section of the tunnel axis. For this purpose, layer devices have been placed, other than in the point where the fire occurs (HGV), also at 25 m (i.e. y = 25 m and y = 75 m) and at 50 m (y = 0 m and y = 100 m) from the HGV, respectively.

The obtained values (Fig. 5) allow to evaluate the smoke stratification along the longitudinal section of the tunnel.

Although their absolute values are questionable, these results therefore provide a comparison term for the subsequent simulations with improved ventilation.

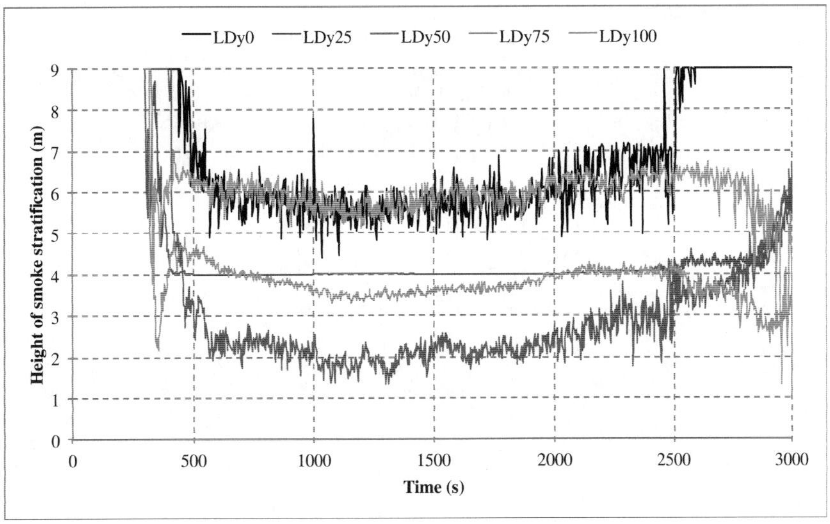

Figure 5. Smoke stratification (first scenario).

3.2. Simulations of alternative scenarios (second and third scenarios)

As previously reported, the improved ventilation system was modelled by varying the ventilation capacity and characteristics of the jet fans in the proximity of the fire occurrences; the axial fans used in the tunnel, in fact, give the possibility to vary the flow rate and they can also be switched off in conditions of complete evacuation of the tunnel.

This way, despite having the negative effect of providing more oxygen to the combustion reaction, allowing the benefit of removing smoke - which causes suffocation - more quickly, and of having a better visibility for the evacuation of the area. In particular, two different scenarios were used: in the first, an increase of twice the standard airflow was assumed; in the second, the same hypothesis of the base simulation were assumed, but considering all fans working in aspiration.

Since the standard airflow rate is 24.0 m^3/s for each fan placed on each portal, in the *second scenario* such flow rate was set at a value of 48.0 m^3/s. Each simulation lasted almost 7 hours.

The obtained results for the first scenario (Fig. 6) showed a delay of the back-layering effect of about 700 s (almost double respect to that of the base simulation) and a better smoke zoning, albeit slightly, going from a visibility of 1 m to a 2 m one.

The values of gasses' temperature (Fig. 7) have decreased compared to those of base simulation; anyway, it must be noted that the investigated cases didn't take into account a fire active protection system for controlling the gasses temperature's values.

Very variable data were obtained both for the smoke stratification and the gasses temperatures (comparison of Fig. 6 with 5 and Fig. 7 with 4, respectively), probably due to the fan speed acting on the flames. In return though, an increase in the evacuation time and a better smoke stratification have been obtained in the second part of the tunnel, thanks to the contrast between ventilation and back-laying speeds.

Overall, regarding people evacuation time, these results represent a worsening compared to the basic simulation (comparison Fig. 6 with 5); they do not therefore contribute to an increase in tunnel safety from this point of view.

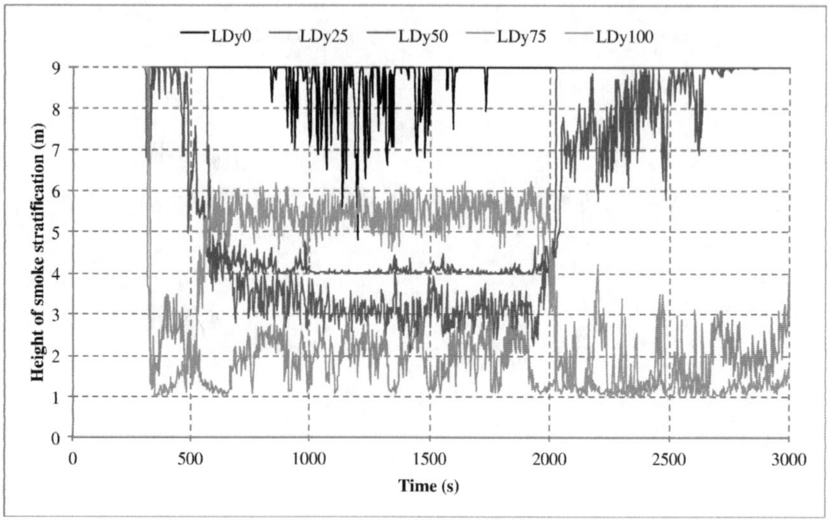

Figure 6. Smoke stratification with improved ventilation – second scenario.

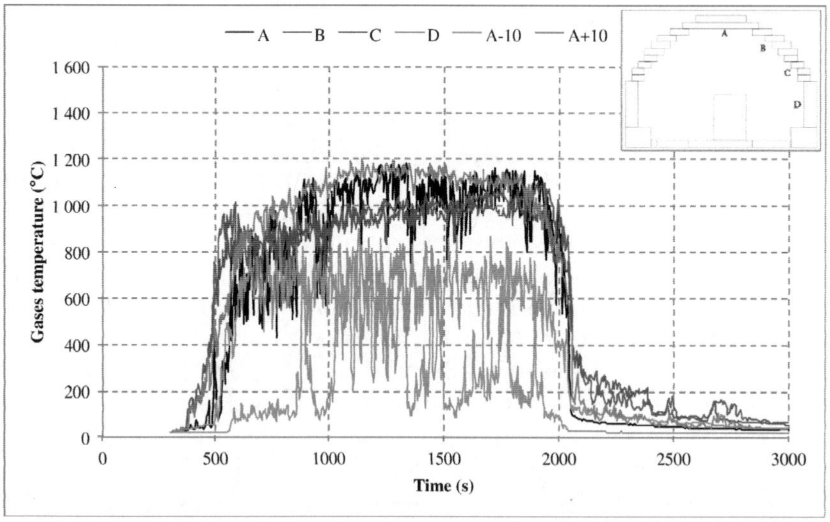

Figure 7. Gasses temperatures with improved ventilation – second scenario

In the *third scenario*, in an attempt to increase the sliding speed of the smoke, all fans were considered working in aspiration, by inverting the airflow direction of the fan positioned at y = 0.

The results regarding the smoke (Fig. 8) show an obvious increase in their stratification height, which allows a safer evacuation of the tunnel. While, regarding the gasses' temperatures (Fig. 9) it can be noted an increase of the values in the area near the tunnel ceiling in the point where the fire starts, and a decrease, instead, at lower heights.

Hence, acting on the fans' airflow rate direction, when a fire begins, actually enables to improve the control of smoke stratification inside a tunnel.

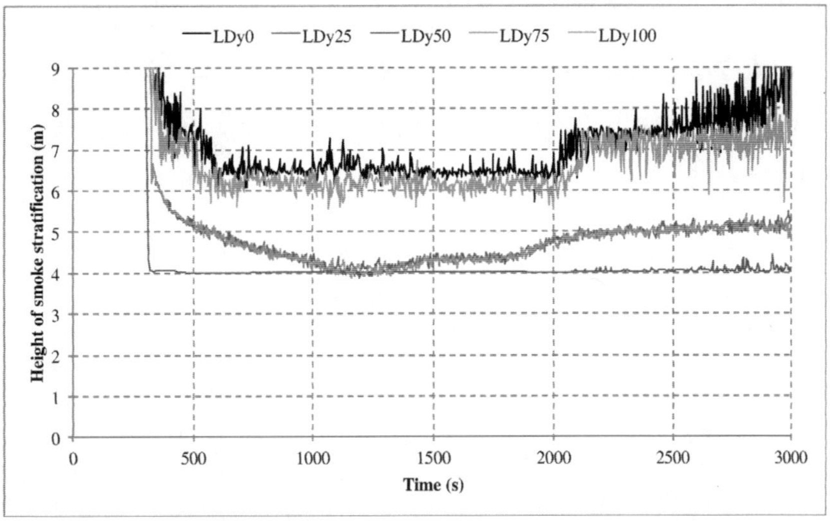

Figure 8. Smoke stratification with improved ventilation – third scenario.

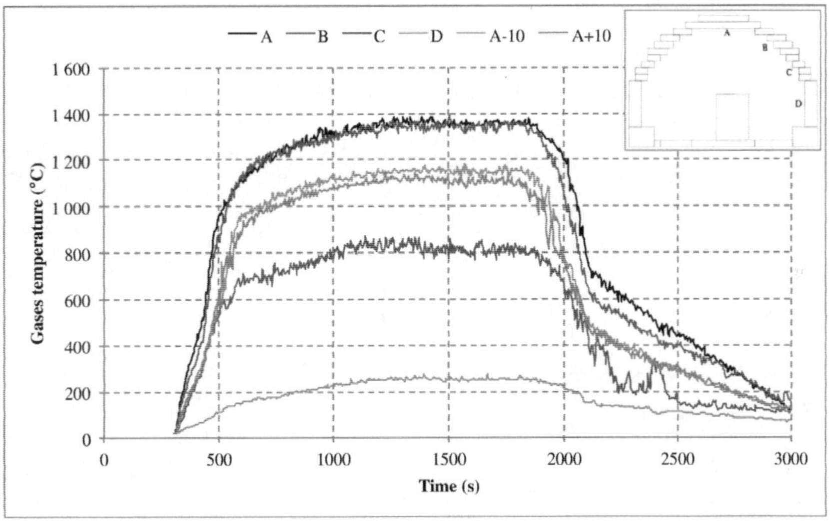

Figure 9. Gasses temperatures with improved ventilation – third scenario.

4. Discussions

Comparing the results of the first and second scenario, it can be noted how increasing the air flow rate in the second scenario will certainly lower the temperature of the gasses (Fig. 10), especially in the position, A_{-10}, foregoing the point where the fire occurs (considering that the direction of the airflow is assigned by the fans). Conversely, there is an increase in the smoke temperature at lower altitudes and, above all, the height of smoke stratification in the second scenario often remains below 2 m in positions following the point where the fire occurs (Fig. 11).

Regarding the third scenario, instead, the gasses temperatures are similar to those of the first scenario, but, in some cases, they are even slightly higher. In contrast, however, the height of the smoke stratification is always higher than 4 m.

Consequently, the third scenario would seem to be the most interesting one.

IOP Conf. Series: Journal of Physics: Conf. Series **1224** (2019) 012005 doi:10.1088/1742-6596/1224/1/012005

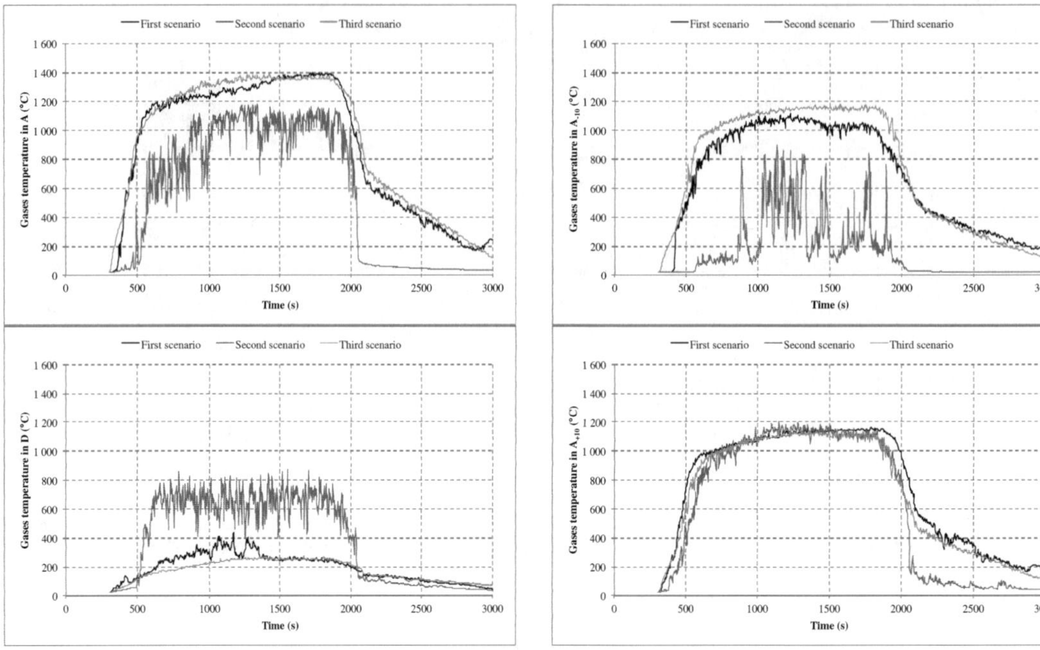

Figure 10. Comparison of gasses temperatures in the sensors A, A_{+10}, A_{-10} and D for the three scenarios.

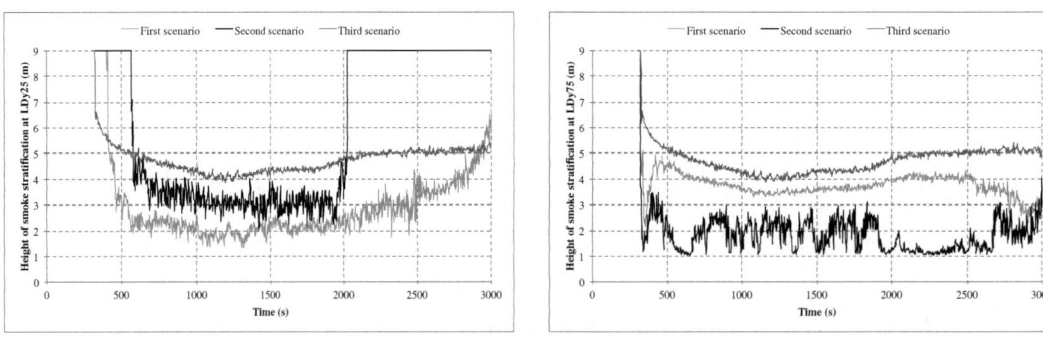

Figure 11. Comparison of smoke stratification in the layer devices positioned at y = 25 m and y = 75 m for the three scenarios.

The previous considerations lead to think that an improvement to the fire safety of tunnels, where there is no an active fire protection system, could be made through a fans direction management, by means of a control system that identifies the position of the fire through some sensors specifically installed along the gallery

Moreover, such a solution, that is a combination of proper sensors and a system for controlling and managing the rotating direction of the fans, should be a fairly simple and a relatively low-cost system to be implemented.

IOP Conf. Series: Journal of Physics: Conf. Series **1224** (2019) 012005 doi:10.1088/1742-6596/1224/1/012005

5. Conclusions

The present work started from considerations regarding the use of CFD to evaluate how the use of an active protection method, that is forced longitudinal ventilation, can have an impact on the heat extraction and people evacuation time during a tunnel fire.

The obtained results confirmed that CFD simulation do provide a powerful analysis tool for tunnel fires protection analysis; geometry, materials and ventilation can, in fact, be represented by taking advantage of a three-dimensional prediction. Thus, when applied to the design of a new and/or improved ventilation system, the CFD simulation provides a local understanding of the general behavior of fire (in terms of heat release rate), smokes stratification and people evacuation time from which an optimal configuration of the system can be developed.

In fact, the analysis of the effects of such active protection system in the tunnel, object of this case study, provided useful data to be taken into consideration in an eventual design phase.

It would be interesting, as further future development of this work, to elaborate the study by implementing a model for the entire tunnel which could also take into account, in addition to the longitudinal ventilation, the use of a behavioral evacuation model, a fire extinguishing system and a material behavior analysis, in order to have a complete representation of the interaction phenomena.

In conclusion, it has been shown that CFD simulations give an enough realistic indication about the physical fire's phenomenon and protection methods, and therefore provide a good starting point for a suitable design, recovery and restoration of artifacts.

References

[1] Beard A and Carvel R, The handbook of tunnel fire safety, 2005, London E14 4JD, 1 Heron Quay: Thomas Telford Publishing, *Thomas Telford Ltd,* pp. 127-177.

[2] Ingason H, Li Y Z, Lönnermark A, Runehamar Tunnel Fire Tests, January 2015, *Fire Safety Journal*, vol. **71**, pp. 134-149.

[3] Ji J, Guo F, Gao Z, Zhu J, Sun J, Numerical investigation on the effect of ambient pressure on smoke movement and temperature distribution in tunnel fires, 2017, *Applied Thermal Engineering*, vol. **118**, p. 663–669.

[4] Wang F, Wang M, Carvel R, Wang Y, Numerical study on fire smoke movement and control in curved road tunnels, 2017, *Tunnelling and Underground Space Technology*, vol. **67**, p. 1–7.

[5] Harish R, Venkatasubbaiah K, Effects of buoyancy induced roof ventilation systems for smoke removal in tunnel fires, 2014, *Tunnelling and Underground Space Technology*, vol. **42**, p. 195–205.

[6] Li Q, Fang Z, Yuan J, Tang Z, Numerical Simulation on Impacts of Longitudinal Ventilation on Tunnel Fire Detection, 2016, *Procedia Engineering*, vol. **135**, p. 275 – 280.

[7] Wang Q, Tang F, Li L, Zhang X, Fan C, Large eddy simulation on the effect of smoke exhaust openings arrangement on the smoke spread in tunnel fires, 2016, *Procedia Engineering*, vol. **135**, p. 309 – 315.

[8] Barbato L, Cascetta F, Musto M, Rotondo G, Fire safety investigation for road tunnel ventilation systems – An overview, 2014, *Tunnelling and Underground Space Technology*, vol. **43**, p. 253–265.

[9] Gannouni S, Maad R B, Numerical analysis of smoke dispersion against the wind in a tunnel fire, 2016, *Journal of Wind Engineering and Industrial Aerodynamics*, vol. **158**, p. 61–68.

[10] Heidarinejad G, Mapar M, Pasdarshahri H, A comprehensive study of two fire sources in a road tunnel: Considering different arrangement of obstacles, 2016, *Tunnelling and Underground Space Technology*, vol. **59**, p. 91–99.

[11] Danišovič P, Jančaříková E, Šrámek J, Zuziak M, Fire spread models and Tunnel Traffic & Operation Simulator, 2017, *Procedia Engineering*, vol. **192**, p. 92 – 95.

[12] Liang Q, Li Y, Li J, Xu H, Li K, Numerical studies on the smoke control by water mist screens with transverse ventilation in tunnel fires, 2017, *Tunnelling and Underground Space Technology*, vol. **64**, p. 177–183.

IOP Conf. Series: Journal of Physics: Conf. Series **1224** (2019) 012005 doi:10.1088/1742-6596/1224/1/012005

[13] Vermesi I, Rein G, Colella F, Valkvist M, Jomaas G, Reducing the computational requirements for simulating tunnel fires by combining multiscale modelling and multiple processor calculation, 2017, *Tunnelling and Underground Space Technology*, vol. **64**, p. 146–153.
[14] Du S, Zhang Y, Sun Q, Gong W, Geng J, Zhang K, Experimental study on color change and compression strength of concrete tunnel lining in a fire, 2018, *Tunnelling and Underground Space Technology*, vol. **71**, p. 106–114.
[15] Tang F, Mei F Z, Wang Q, He Z, Fan C G, Tao C F, Maximum temperature beneath the ceiling in tunnel fires with combination of ceiling mechanical smoke extraction and longitudinal ventilation, 2017, *Tunnelling and Underground Space Technology*, vol. **68**, p. 231–237.
[16] Wang F, Wang M, Huo J, The effects of the passive fire protection layer on the behavior of concrete tunnel linings: A field fire testing study, 2017, *Tunnelling and Underground Space Technology*, vol. **69**, p. 162–170.
[17] Król M, Król A, Koper P, Wrona P, Full scale measurements of the operation of fire ventilation in a road tunnel, 2017, *Tunnelling and Underground Space Technology*, vol. **70**, p. 204–213.
[18] Art Bendelius, Tunnel ventilation — state of the art, Parsons Brinckerhoff, USA, p. 128-140 (https://doi.org/10.1680/hotfs.31685.0007)
[19] Empedocle2 S.C.p.A., Galleria Caltanissetta SS n.640 di "Porto Empedocle", ANALISI DI RISCHIO.
[20] Carvel R, Beard A, Jowitt P W, How much do Tunnel Enhance the Heat Release Rate of Fires, 2001, Proc. 4th Int. Conf on Safety in Road and Rail Tunnels, Madrid.
[21] Ingarson H, Lonnemark A, Large Scale Fires tests in the Runehamar tunnel – Heat Release Rate, November 2003,nProceedings of the international seminar on catastrophic tunnel fires, Sweden.
[22] Pérez I et al., Asphalt mixture fire performance at full scale. PAVIREX project, June 2016, 6th Eurasphalt & Eurobitume Congress, Prague, Czech Republic.

IOP Conf. Series: Journal of Physics: Conf. Series **1224** (2019) 012006 doi:10.1088/1742-6596/1224/1/012006

Optimal Control of the Wilcox turbulence model with lifting functions for flow injection and boundary control

L Chirco[1], A Chierici[1], R Da Vià[1], V Giovacchini[1] and S Manservisi[1]

[1] Department of Industrial Engineering DIN
Laboratory of Montecuccolino
University of Bologna
Via dei Colli 16, Bologna 40136 (BO), Italy

E-mail: leonardo.chirco2@unibo.it

Abstract. This paper deals with boundary optimal control problems for the Navier-Stokes equations and Wilcox turbulence model. In this paper we study adjoint optimal control problems for Navier-Stokes equations to improve the advantages of using simulations where turbulence models play a significant role in designing engineering devices. We assess first distributed optimal control problems with the purpose to control the fluid behavior by injecting a flow on the boundary solid region to obtain a desired control over the fluid velocity and the kinetic turbulence energy in specific parts of the domain. Then, with the same purpose, we use lifting functions and boundary control. For this reason we reformulate the boundary optimal control problem into a distributed problem through a lifting function approach. The stronger regularity requirements which are imposed by standard boundary control approaches can then be avoided. The state, adjoint and control equations are derived and the optimality system solved for some simple cases with a finite element. Furthermore, we propose numerical strategies that allow to solve the coupled optimality system in a robust way for a large number of unknowns. The approach presented in this work is general and can be used to assess different objectives and types of control.

1. Introduction

In the last several years optimization techniques have gained attention in several research fields of fluid dynamics. Optimal control theory has been applied successfully to many fields ranging from heat transfer problems to fluid-structure interaction systems. Most of the works refers to the standard Navier-Stokes system [1, 2]. However, if a turbulence problem is considered then a direct simulation of the Navier-Stokes system is unpractical due to the computational effort required. In this work we consider the Reynolds averaged Navier-Stokes (RANS) system for the computation of the average pressure and velocity fields. The closure of the problem is guaranteed by the implementation of the Wilcox $k-\omega$ model. Its accuracy and stability allow the coupling between turbulence and the theory of optimal control. In particular, if appropriate limits are taken into account the $k-\omega$ model has regular solutions over large range of parameters [3].

Distributed control strategies, feedback control and adjoint mehods, have been successfully achieved in previous works both for DNS and RANS system while the implementation of a boundary control is not a trivial task [4, 5, 6]. The main difficulty concerns the mathematical

IOP Conf. Series: Journal of Physics: Conf. Series **1224** (2019) 012006 doi:10.1088/1742-6596/1224/1/012006

definition of the boundary control parameters. Boundary functions belong to the trace spaces restriction of the natural state volume spaces. It would be necessary to introduce a fractional norm to express the penalization of the cost functional associated to the control parameter. Due to the difficulty of this task usually stronger regularity requirements are imposed. This process leads to controls smoother than necessary due to over-penalization that can affect the accuracy of the solution. For this reasons in this work we aim to achieve a boundary control of the RANS system through a lifting function approach which consists of reformulation of the boundary control problem into a distributed one. By introducing the lifting function approach boundary controls are defined in the appropriate function boundary spaces restriction of the natural state volume spaces [7, 8].

In the present work we introduce an alternative formulation of the lifting function approach, where a distributed parameter **f** is defined over an external auxiliary domain that includes the controlled boundary. The optimal boundary control is then obtained by restricting the state solution corresponding to the optimal distributed control. In this way, we can express the cost functional by using standard norms on volume spaces. Furthermore, the information obtained with the distributed control can be used to design blowing or suction on the fluid from the boundary. The method presented here can replace a traditional optimization process based on a parametric analysis of the blowing section velocity profile. We use the adjoint method based on the Lagrange multipliers method [9]. The optimality system for the RANS equations closed with the $k-\omega$ turbulence model is derived. We consider two objectives. The first one is a velocity profile matching problem and the second one a turbulence enhancement case. Once that the type of control and the objective are chosen, the complete Lagrangian can be written considering the RANS system and the $k-\omega$ Wilcox model as constraints in order to derive the so-called optimality system. The numerical solution of the optimality system is then performed through a standard Galerkin method implemented in an in-house finite element C++ code with a steepest descent algorithm for the search of the local minimum of the functional.

2. Mathematical model

In this section we introduce the mathematical model describing our optimal control problem. We first report the conservation equations modeling the physical system and then define the objective functional through which we achieve the optimality system.

We use standard notation $H^s(\Omega)$ for the Sobolev spaces of square integrable functions with square integrable weak derivatives up to order s. We denote with $\langle\cdot,\cdot\rangle_s$ the scalar product in $H^s(\Omega)$ and the induced norm is $||\cdot||_m = \sqrt{(\cdot,\cdot)_m}$. Let $H_0^s(\Omega)$ be the space of all functions in $H^s(\Omega)$ that vanish of the boundary of the bounded open set Ω. We denote with $H^{-s}(\Omega)$ the dual space of $H_0^s(\Omega)$. The trace space for the functions in $H^1(\Omega)$ is denoted by $H^{1/2}(\Gamma)$ and its dual by $H^{-1/2}(\Gamma)$. Given any subset $\Gamma_s \subset \Gamma$, we denote by $H^1_{\Gamma_s}(\Omega)$ the subspace of $H^1(\Omega)$ containing functions with vanishing trace on Γ_s, i.e.

$$H^1_{\Gamma_s}(\Omega) = \Big\{u \in H^1(\Omega) | \gamma_{\Gamma_s} u = 0\Big\}, \tag{1}$$

where we use the trace restriction operator on Γ_s denoted with γ_{Γ_s} [10]. At last, we define the space of square integrable functions having zero mean over Ω as

$$L_0^2(\Omega) = \Big\{p \in L^2(\Omega) | \int_\Omega p \, dx = 0\Big\}. \tag{2}$$

2.1. Conservation equations

In this work we consider the Reynolds averaged Navier-Stokes system for the computation of the average velocity and pressure fields $(\mathbf{u}, p)$ coupled with Wilcox $k-\omega$ turbulence model. The

RANS system consists in the averaged conservation equation of mass and momentum for a fluid. We use Boussinesq's assumption to define the Reynolds' stress tensor introducing the turbulent or eddy viscosity ν_t

$$\nabla \cdot \mathbf{u} = 0 \,, \tag{3}$$

$$(\mathbf{u} \cdot \nabla)\mathbf{u} \;= -\nabla p + \nabla \cdot [(\nu + \nu_t)\mathbf{S}(\mathbf{u})] + \mathbf{f} \,, \tag{4}$$

where $\mathbf{f}$ is a force acting on the flow. The closure is obtained by modeling the eddy viscosity through two differential equation for the turbulent kinetic energy k and its specific dissipation rate ω

$$(\mathbf{u} \cdot \nabla)k = \nabla \cdot [(\nu + \sigma_k \nu_t)\nabla k] + P_k - \beta^* k\omega \,, \tag{5}$$

$$(\mathbf{u} \cdot \nabla)\omega = \nabla \cdot [(\nu + \sigma_\omega \nu_t)\nabla \omega] + \alpha \frac{\omega}{k} P_k - \beta\omega^2 + \frac{\sigma_d}{\omega} \nabla k \cdot \nabla \omega \,, \tag{6}$$

where $\mathbf{S}(\mathbf{u})$ is the deformation tensor and P_k is the production term of the turbulent kinetic energy

$$\mathbf{S}(\mathbf{u}) = \nabla \mathbf{u} + \nabla \mathbf{u}^T \qquad\qquad P_k = \frac{\nu_t}{2} \mathbf{S}(\mathbf{u}) : \mathbf{S}(\mathbf{u}) \,. \tag{7}$$

Furthermore, for the eddy viscosity ν_t of the turbulence model we assume

$$\nu_t = \min\left[\frac{k}{\omega}, \nu_{max}\right] \,. \tag{8}$$

The system (3-6) is completed with the appropriate boundary conditions reported in Table 1, where we denoted with Γ_i, Γ_w and Γ_o the fluid inlet, walls and outlet, respectively. We use a near-wall approach [11] for the solution of the turbulence problem, so the RANS equations are integrated throughout the viscous layer. We set the boundary conditions in the linear region at a distance y_d from the wall, so that the value of ω remains limited in the computational domain.

Variable	Γ_i	Γ_w	Γ_o
$\mathbf{u} \cdot \widehat{\mathbf{n}}$	$\mathbf{u}_i$	0	$\boldsymbol{\tau} \cdot \widehat{\mathbf{n}} = 0$
$\mathbf{u} \cdot \widehat{\mathbf{t}}$	0	$\boldsymbol{\tau} \cdot \widehat{\mathbf{n}} = \dfrac{\nu \mathbf{u} \cdot \widehat{\mathbf{t}}}{y_d}$	–
P	–	–	0
k	$k = k_i$	$\nabla k \cdot \widehat{\mathbf{n}} = \dfrac{2k}{y_d}$	$\nabla k \cdot \widehat{\mathbf{n}} = 0$
ω	$\omega = \omega_i$	$\omega = \dfrac{2\nu}{\beta^* y_d^2}$	$\nabla \omega \cdot \widehat{\mathbf{n}} = 0$

Table 1. State variables boundary conditions with near wall approach.

2.2. Boundary optimal control and lifting function

Let Γ_c be a subset of the boundary Γ of the domain $\Omega \subset \mathbb{R}^3$. The first step to obtain the optimality system is to choose an objective for our optimal control problem. The objective functional is denoted by $\mathcal{F}(\mathbf{u}, k)$ and is described by the following expression

$$\mathcal{F}(\mathbf{u}, k, \mathbf{f}) = \frac{a}{2} \int_\Omega ||\mathbf{u} - \mathbf{u}_d||^2 \, d\Omega + \frac{b}{2} \int_\Omega |k - k_d|^2 \, d\Omega \,, \tag{9}$$

where a and b are weight functions. If $b = 0$ we have a velocity matching problem and the objective is expressed as the distance in norm of the velocity from a target value $\mathbf{u}_d$. Similarly, if $a = 0$ the objective concerns the turbulent kinetic energy and its distance from the target k_d. We introduce the boundary condition

$$\mathbf{u} = \mathbf{g} \quad on\ \Gamma_c\,, \tag{10}$$

where $\mathbf{g} \in H^{1/2}(\Gamma_c)$ is the boundary control parameter. In standard boundary control approaches the objective functional is reformulated as

$$\mathcal{J}_b(\mathbf{u}, k, \mathbf{g}) = \frac{a}{2}\int_\Omega ||\mathbf{u} - \mathbf{u}_d||^2\, d\Omega + \frac{b}{2}\int_\Omega |k - k_d|^2\, d\Omega + \frac{\beta}{2}\int_{\Gamma_c} ||\mathbf{g}||^2\, d\Gamma + \frac{\gamma}{2}\int_{\Gamma_c} ||\boldsymbol{\nabla}\mathbf{g}||^2\, d\Gamma\,. \tag{11}$$

The last two terms are penalty regularization contributions introduced to limit the norm of the boundary control function $\mathbf{g}$ and of its gradient. However, by doing so some boundary control solutions becomes unacceptable. We then choose to determine an optimal distributed control from which any and all possible boundary controls can be determined by trace restriction. The cost functional changes so that the penalization terms involves the distributed control. To this end, we define the decomposition of the velocity field given by

$$\mathbf{u} = \mathbf{w} + \mathbf{u}' \quad \text{on } \Omega\,, \tag{12}$$

so that

$$\gamma_{\Gamma_c}\mathbf{u}' = \mathbf{g}\,, \quad \gamma_{\Gamma_c}\mathbf{w} = \mathbf{0}\,. \tag{13}$$

The functions $\mathbf{u}'$ is called lifting function and is the new control parameter, while the velocity $\mathbf{w}$ is the new state variable. Both of them are divergence-free fields

$$\boldsymbol{\nabla}\cdot\mathbf{u}' = 0\,, \quad \boldsymbol{\nabla}\cdot\mathbf{w} = 0 \quad \text{on } \Omega\,. \tag{14}$$

The velocity on Γ_c is determined by trace restriction of lifting function $\mathbf{u}'$ on that boundary, in this way $\mathbf{g} \in H^{1/2}(\Gamma_c)$, the natural space of definition. By substitution it is obtained the following form of state equation of Navier-Stokes

$$\mathbf{w}\cdot\boldsymbol{\nabla}\mathbf{w} + \boldsymbol{\nabla}p - \boldsymbol{\nabla}\cdot[(\nu + \nu_t)\mathbf{S}(\mathbf{w})] = \mathbf{f}(\mathbf{u}', \mathbf{w})\,, \tag{15}$$

where we have a force term depending on the lifting function and state velocity. We define the cost functional

$$\mathcal{J}(\mathbf{w}, \mathbf{u}') = \frac{a}{2}\int_\Omega ||\mathbf{w}+\mathbf{u}'-\mathbf{u}_d||^2\, d\Omega + \frac{b}{2}\int_\Omega |k - k_d|^2\, d\Omega + \frac{\beta}{2}\int_\Omega ||\mathbf{u}'||^2\, d\Omega + \frac{\gamma}{2}\int_\Omega ||\boldsymbol{\nabla}\mathbf{u}'||^2\, d\Omega\,. \tag{16}$$

We remark the difference between the regularization terms appearing in (11) and (16). The former involves the $H^1(\Gamma)$ norm whereas the latter involves the $H^1(\Omega)$ norm [12].

We now present a new lifting function strategy. Let Ω_{aux} be an extension of the domain Ω sharing with it the boundary Γ_c. We consider the Navier-Stokes equation for the velocity field $\mathbf{u} \in H^1(\Omega)$ without distributed forces

$$\begin{aligned} \mathbf{u}\cdot\boldsymbol{\nabla}\mathbf{u} + \boldsymbol{\nabla}p - \boldsymbol{\nabla}\cdot[(\nu + \nu_t)\mathbf{S}(\mathbf{u})] &= \mathbf{0} \quad \text{on } \Omega\,, \\ \mathbf{u} &= \mathbf{g} \quad \text{on } \Gamma_c\,. \end{aligned} \tag{17}$$

We define the lifting function on the auxiliary domain $\mathbf{u}' \in H^1(\Omega_{aux})$ so that $\mathbf{u}' = \mathbf{g}$ on Γ_c. It is then possible to formulate a Navier-Stokes equation also for the velocity field $\mathbf{u}'$ on the domain Ω_{aux}

$$\begin{aligned} \mathbf{u}'\cdot\boldsymbol{\nabla}\mathbf{u}' + \boldsymbol{\nabla}p - \boldsymbol{\nabla}\cdot[(\nu + \nu_t)\mathbf{S}(\mathbf{u}')] &= \mathbf{f} \quad \text{on } \Omega_{aux}\,, \\ \mathbf{u}' &= \mathbf{g} \quad \text{on } \Gamma_c\,. \end{aligned} \tag{18}$$

The force term $\mathbf{f}$ appears only in the right-hand-side of this equation since we can model the velocity in the auxiliary domain as we wish, as long as the boundary condition on Γ_c is respected. Due to this boundary condition there has to be continuity between the solutions of problem (17) and (18) so that we can write

$$\mathbf{u} \cdot \nabla \mathbf{u} + \nabla p - \nabla \cdot [(\nu + \nu_t)\mathbf{S}(\mathbf{u})] = \mathbf{f} \quad \text{on } \Omega \cup \Omega_{aux} \,. \tag{19}$$

Since the optimal distributed control $\mathbf{f}$ acts only on Ω_{aux} then the cost functional becomes

$$\mathcal{J}(\mathbf{u}, \mathbf{f}) = \frac{a}{2} \int_\Omega ||\mathbf{u} - \mathbf{u}_d||^2 \, d\Omega + \frac{b}{2} \int_\Omega |k - k_d|^2 \, d\Omega + \frac{\lambda}{2} \int_{\Omega_{aux}} ||\mathbf{f}||^2 \, d\Omega \,. \tag{20}$$

The last term represents the cost of the regularization of the control parameter $\mathbf{f}$ constraining $\mathbf{f} \in L^2(\Omega)$. The amount of regularization is controlled by the positive parameter λ, called regularization parameter and its value plays a key role in the solution of the control problem. If $\lambda = 1$ we have that $\mathbf{f}$ is a square integrable function but the control can be not effective. If $\lambda \to 0$ the control becomes irregular and this lack of regularization leads to convergence issues.

2.3. Optimality system

To obtain the optimality system we write the total Lagrangian of the optimal control problem, by considering the objective functional (20) and the constraints system (3-6). In order to include (8) in the Lagrangian we add the constraint

$$\left(\nu_t - \frac{k}{\omega}\right)(\nu_t - \nu_{max}) = 0 \,, \tag{21}$$

and introduce two new real variables γ_1 and γ_2 defined as

$$\gamma_1^2 = \nu_{max} - \nu_t \,, \qquad \gamma_2^2 = \frac{k}{\omega} - \nu_t \,. \tag{22}$$

The unknown state is defined by $(\mathbf{u}, p, \nu_t, k, \omega, \gamma_1, \gamma_2)$. The total Lagrangian is then given by

$$\begin{aligned}
&\mathcal{L}(\mathbf{u}, \mathbf{u}_a, p, p_a, k, k_a, \omega, \omega_a, \nu_t, \nu_a, \gamma_1, \gamma_{a1}, \gamma_2, \gamma_{a2}, \mathbf{f}) = \mathcal{J}(\mathbf{u}, k, \mathbf{f}) + \int_\Omega p_a \nabla \cdot \mathbf{u} \, d\Omega + \\
&+ \int_\Omega \mathbf{u}_a \cdot \{(\mathbf{u} \cdot \nabla)\mathbf{u} + \nabla p - \nabla \cdot [(\nu + \nu_t)\mathbf{S}(\mathbf{u})] - \mathbf{f}\} \, d\Omega + \\
&+ \int_\Omega k_a \{\mathbf{u} \cdot \nabla k - \nabla \cdot \Big[\Big(\nu + \frac{\nu_t}{\sigma_k}\Big)\nabla k\Big] - P_k + \beta^* k \omega\} \, d\Omega + \\
&+ \int_\Omega \omega_a \{\mathbf{u} \cdot \nabla \omega - \nabla \cdot \Big[\Big(\nu + \frac{\nu_t}{\sigma_\omega}\Big)\nabla \omega\Big] - \alpha \frac{\omega}{k} P_k + \beta \omega^2 - \frac{\sigma_d}{\omega} \nabla k \cdot \nabla \omega\} \, d\Omega + \\
&+ \int_\Omega \nu_a \Big(\nu_t - \frac{k}{\omega}\Big)(\nu_t - \nu_{max}) \, d\Omega + \int_\Omega \gamma_{a1}(\nu_{max} - \nu_t - \gamma_1^2) \, d\Omega + \\
&+ \int_\Omega \gamma_{2a} \Big(\frac{k}{\omega} - \nu_t - \gamma_2^2\Big) \, d\Omega \,.
\end{aligned} \tag{23}$$

To obtain the optimality system we impose the minimization conditions $\delta\mathcal{L} = 0$. Since the variations with respect to all the variables involved are independent then each variation has to be zero. In this way we obtain the weak form of the adjoint equations used in a finite element discretization (see more details in [13])

$$\int_\Omega \nabla \cdot \mathbf{u}_a \psi_a \, d\Omega = 0 \,, \quad \psi_a \in L^2(\Omega) \tag{24}$$

$$\int_\Omega \{a(\mathbf{u}-\mathbf{u}_d)\boldsymbol{\phi_a} + p_a\boldsymbol{\nabla}\cdot\boldsymbol{\phi_a} + \mathbf{u}_a\cdot(\boldsymbol{\phi_a}\cdot\boldsymbol{\nabla})\mathbf{u} + \mathbf{u}_a\cdot(\mathbf{u}\cdot\boldsymbol{\nabla})\boldsymbol{\phi_a} + \\ -\mathbf{u}_a\cdot\boldsymbol{\nabla}\cdot[(\nu+\nu_t)\mathbf{S}(\boldsymbol{\phi_a})] + +k_a\boldsymbol{\phi_a}\cdot\boldsymbol{\nabla}k - k_a\nu_t\mathbf{S}(\boldsymbol{\phi_a}):\mathbf{S}(\mathbf{u}) + \\ +\omega_a\boldsymbol{\phi_a}\cdot\boldsymbol{\nabla}\omega - \alpha\omega_a\mathbf{S}(\boldsymbol{\phi_a}):\mathbf{S}(\mathbf{u})\}\, d\Omega = 0\ , \qquad \forall\boldsymbol{\phi_a}\in\mathbf{H}^1_{\Gamma_{i,},\Gamma_{o,t},\Gamma_{w,n}}(\Omega) \tag{25}$$

$$\int_\Omega \Big\{b(k-k_d)\varphi_a + k_a\mathbf{u}\cdot\boldsymbol{\nabla}\varphi_a - k_a\boldsymbol{\nabla}\cdot\Big[\Big(\nu+\frac{\nu_t}{\sigma_k}\Big)\boldsymbol{\nabla}\varphi_a\Big] + k_a\beta^*\omega\varphi_a + \\ -\omega_a\frac{\sigma_d}{\omega}\boldsymbol{\nabla}\varphi_a\cdot\boldsymbol{\nabla}\omega + \frac{\gamma_{2a}-\nu_a(\nu_t-\nu_{max})}{\omega}\varphi_a\Big\}d\Omega = 0, \qquad \forall\varphi_a\in H^1_{\Gamma_i}(\Omega) \tag{26}$$

$$\int_\Omega \Big\{k_a\beta^*k\gamma + \omega_a\mathbf{u}\cdot\boldsymbol{\nabla}\gamma - \omega_a\boldsymbol{\nabla}\cdot\Big[\Big(\nu+\frac{\nu_t}{\sigma_\omega}\Big)\boldsymbol{\nabla}\gamma\Big] + 2\beta\omega\gamma\omega_a + \omega_a\frac{\sigma_d}{\omega^2}\gamma\boldsymbol{\nabla}k\cdot\boldsymbol{\nabla}\omega + \\ -\omega_a\frac{\sigma_d}{\omega}\boldsymbol{\nabla}k\cdot\boldsymbol{\nabla}\gamma + \frac{\nu_a(\nu_t-\nu_{max})-\gamma_{2a}}{\omega^2}k\gamma\Big\}d\Omega = 0, \qquad \forall\gamma\in H^1_{\Gamma_i,\Gamma_w}(\Omega) \tag{27}$$

The natural boundary conditions for the optimality system in strong form can be obtained by setting to zero the surface integrals that contain unknown terms or non-integrable functions. The boundary conditions of the adjoint variables for a near wall approach are reported in Table 2. Finally, we present the algebraic equation for the adjoint eddy viscosity and the control equation for the distributed force

$$\nu_a = \frac{k_a}{2}\mathbf{S}(\mathbf{u}):\mathbf{S}(\mathbf{u}) - \mathbf{S}(\mathbf{u}):\boldsymbol{\nabla}\mathbf{u}_a - \frac{1}{\sigma_k}\boldsymbol{\nabla}k\cdot\boldsymbol{\nabla}k_a - \frac{1}{\sigma_\omega}\boldsymbol{\nabla}\omega\cdot\boldsymbol{\nabla}\omega_a \tag{28}$$

$$\mathbf{f} = \frac{\mathbf{u}_a}{\lambda}\ . \tag{29}$$

The force $\mathbf{f}$ on the domain is equal to the adjoint velocity $\mathbf{u}_a$ scaled by the regularization parameter λ. In the case one might be interested in a finite volume discretization it is necessary

Variable	Γ_i	Γ_w	Γ_o
$\mathbf{u}_a\cdot\hat{\mathbf{n}}$	0	0	$\boldsymbol{\tau_a}\cdot\hat{\mathbf{n}} = -\mathbf{u}_a(\mathbf{u}\cdot\hat{\mathbf{n}})$
$\mathbf{u}_a\cdot\hat{\mathbf{t}}$	0	$\boldsymbol{\tau_a}\cdot\hat{\mathbf{n}} = \frac{\nu\mathbf{u}_a\cdot\hat{\mathbf{t}}}{y_d} + \frac{k_a\nu\mathbf{u}\cdot\hat{\mathbf{t}}}{y_d}$	0
p_a	–	–	0
k_a	0	$\boldsymbol{\nabla}k_a\cdot\hat{\mathbf{n}} = \frac{2(k_a+\sigma_d\omega_a)}{y_d}$	$\boldsymbol{\nabla}k_a\cdot\hat{\mathbf{n}} = 0$
ω_a	0	0	$(\nu+\frac{\nu_t}{\sigma_\omega})\boldsymbol{\nabla}\omega_a\cdot\hat{\mathbf{n}} = -(\mathbf{u}\cdot\hat{\mathbf{n}})\omega_a$

Table 2. Adjoint variables boundary conditions with near wall approach.

to recover the strong form of the adjoint system. We integrate by parts (24-27) and obtain the following adjoint system, which is completed by the boundary condition reported in Table 2

$$\boldsymbol{\nabla}\cdot\mathbf{u}_a = 0\ , \tag{30}$$

$$-(\boldsymbol{\nabla}\mathbf{u})^T + (\mathbf{u}\cdot\boldsymbol{\nabla})\mathbf{u}_a + \boldsymbol{\nabla}\cdot[(\nu+\nu_t)\mathbf{S}(\mathbf{u}_a)] + \boldsymbol{\nabla}p_a = a(\mathbf{u}-\mathbf{u}_d) + k_a\boldsymbol{\nabla}k + \\ +\omega_a\boldsymbol{\nabla}\omega + 2\boldsymbol{\nabla}\cdot[(k_a\nu_t+\alpha\omega_a)\mathbf{S}(\mathbf{u})]\ , \tag{31}$$

$$(\mathbf{u}\cdot\boldsymbol{\nabla})k_a + \boldsymbol{\nabla}\cdot\Big[\Big(\nu+\frac{\nu_t}{\sigma_k}\Big)\boldsymbol{\nabla}k_a\Big] - k_a\beta^*\omega = b(k-k_d) + \\ +\boldsymbol{\nabla}\cdot\Big(\sigma_d\frac{\omega_a}{\omega}\boldsymbol{\nabla}\omega\Big) + \frac{\gamma_{2a}-\nu_a(\nu_t-\nu_{max})}{\omega}\ , \tag{32}$$

$$\begin{aligned}\mathbf{u}\cdot\boldsymbol{\nabla}\omega_a + \boldsymbol{\nabla}\cdot\left[\left(\nu+\frac{\nu_t}{\sigma_\omega}\right)\boldsymbol{\nabla}\omega_a\right] - k_a\beta^* k - 2\beta\omega\omega_a = \boldsymbol{\nabla}\cdot\left(\sigma_d\frac{\omega_a}{\omega}\boldsymbol{\nabla}k\right)+ \\ +\frac{\omega_a\sigma_d}{\omega}\boldsymbol{\nabla}k\cdot\boldsymbol{\nabla}\omega + \omega_a\frac{\sigma_d}{\omega^2}\boldsymbol{\nabla}k\cdot\boldsymbol{\nabla}\omega - \frac{\nu_a(\nu_t-\nu_{max})-\gamma_{2a}}{\omega^2}k\ .\end{aligned} \tag{33}$$

2.4. Steepest descent algorithm

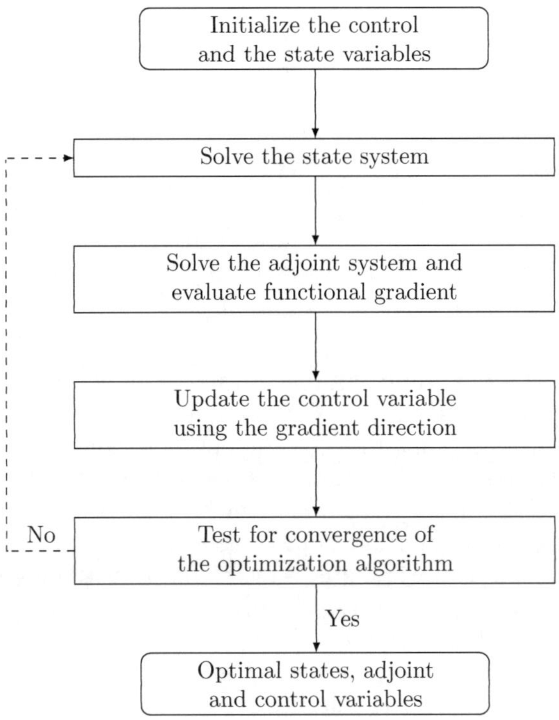

Figure 1. Schematic diagram of the use of steepest descent algorithm.

The system (24-29) is strongly non-linear so it is not possible to solve it with one-shot techniques. We then apply the steepest descent method that is a first-order iterative algorithm for finding the minimum of a functional. A schematic diagram of the general iterative process is shown in Figure 1. Once initialized the state and control variables, the adjoint system is iteratively solved. The adjoint variables are used to update the control parameter that acts on the state variables, until a functional local minimum is reached. The algorithm implemented in our in-house code is presented below.

1) Initialization:
   a) Set up an initial state $(\mathbf{u}^0, p^0, k^0, \omega^0, \nu_t^0)$ solution of the state system;
   b) Set up $\mathbf{f}^0 = 0, r^0 = 1$;
   c) Compute functional $\mathcal{J}^0(\mathbf{u}^0, k^0, \mathbf{f}^0)$;
2) External loop (i index):
   a) Solve the adjoint system and find adjoint variables $(\mathbf{u}_a^i, p_a^i, k_a^i, \omega_a^i, \nu_a^i)$;
3) Internal loop (j index):
   a) Compute control parameter $\mathbf{f}^i = \mathbf{f}^{i-1} + r^{i,j}\left(\mathbf{f}^{i-1} - \mathbf{u}_a^i/\lambda\right)$;
   b) Solve the state system $(\mathbf{u}^{i,j}, p^{i,j}, k^{i,j}, \omega^{i,j}, \nu_t^{i,j})$;

c) Compute functional $\mathcal{J}^{i,j}(\mathbf{u}^{i,j}, k^{i,j}, \mathbf{f}^i)$
  * if $\mathcal{J}^{i,j} - \mathcal{J}^{i,j-1} > 0$ set $r_{i,j+1} = 0.7 r_{i,j}$ and return to step (3a);
  * if $\mathcal{J}^{i,j} - \mathcal{J}^{i,j-1} < 0$ set $r_{i,j+1} = r^0$ and return to step (2a);
  * if $||\mathcal{J}^{i,j} - \mathcal{J}^{i,j-1}||/\mathcal{J}^{i,j-1} < toll$ convergence is reached.

3. Numerical results

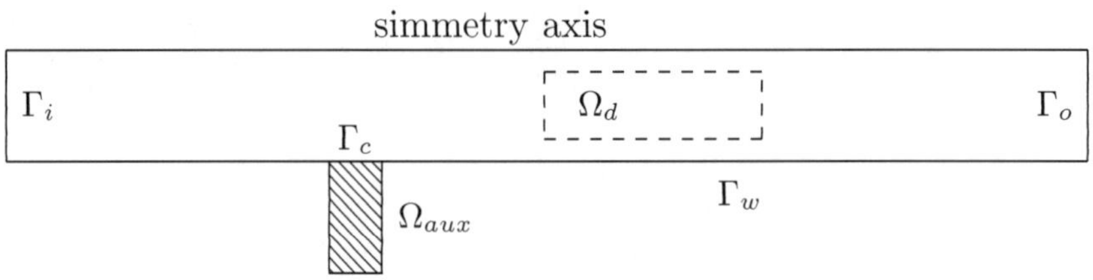

Figure 2. Plane channel geometry. Ω_{aux} is the lifting function auxiliary domain and Ω_d is the target region.

In this section we report some results obtained from the numerical solution of the optimality system described in the previous section. We use a segregated solver for the state and adjoint system and the steepest descent algorithm for the optimal control iterative solution. We use a finite element discretization of the optimality system written in weak form, replacing the infinite dimensional spaces by appropriate finite dimensional spaces. We use a *differentiate-then-discretize* approach since we apply the finite element approximation after computing the Fréchet differentials. We use quadratic finite elements for all variables except the pressure, for which we use linear finite elements to satisfy the *BBL inf-sup* condition needed for the stability of the discrete Navier-Stokes approximation. By doing so we obtain $\mathbf{u}_h$,$\mathbf{u}_{ah} \in \mathbf{X}_h^2(\Omega) \subset \mathbf{H}^1(\Omega)$, p_h,$p_{ah} \in S_h(\Omega) \subset L_0^2(\Omega)$, k_h,k_{ah},ω_h,$\omega_{ah} \in X_h^2(\Omega) \subset H^1(\Omega)$. The algorithm has been implemented in the in-house finite element C++ code FEMuS (available at [14]) which is parallelized with MPI libraries and uses PETSc to handle the algebraic solver of the systems.

We study the two-dimensional geometry shown in Figure 2. The half-channel width W is $0.1m$ and the length L is $1m$. The x-axis is set along the flow direction while the transverse one is y-axis. On the left boundary Γ_i we set an inlet Dirichlet condition $\mathbf{u} = (0.4, 0)\frac{m}{s}$. On the right surface Γ_o a standard outflow boundary condition is imposed. On the symmetry axis a homogeneous Neumann boundary condition is set for all the state and adjoint variables apart from u_y and u_{ay} that vanish on the symmetry axis. On the wall Γ_w we use the near-wall boundary conditions defined in Tables 1 and 2. Let $\Omega_{aux} = [0.3, 0.35] \times [-0.1, 0]$ be the auxiliary domain where the lifting function and control parameter are defined, while $\Omega_d = [0.5, 0.7] \times [0.02, 0.08]$ is the region where the aim to minimize the objective functional. The control surface is denoted with Γ_c. The kinematic viscosity is $\nu = 10^{-5}\frac{m^2}{s}$, so the Reynolds number $Re = UW/\nu$ is 8000 at the inlet of the channel.

3.1. Velocity matching case

In the velocity matching case we set $\mathbf{u}_d = (0.7, 0)\frac{m}{s}$ and solve the optimal control problem with $\lambda = 0.5$, 0.05 and 0.005. On the left of Figure 3 the x-velocity profile on a vertical line at $x = 0.6$ is reported for different values of the regularization λ. The dashed line shows the velocity target $u_{xd} = 0.7$. As λ decreases velocity profiles become more flat with a higher gradient at the wall. On the right of the same Figure the square difference $||u - u_d||^2$ on a vertical line at $x = 0.6$ is shown for $y \in (0.02, 0.08)$. The area under the curve represents $||u - u_d||_{L^2}^2$, i.e. the distance in

IOP Conf. Series: Journal of Physics: Conf. Series **1224** (2019) 012006 doi:10.1088/1742-6596/1224/1/012006

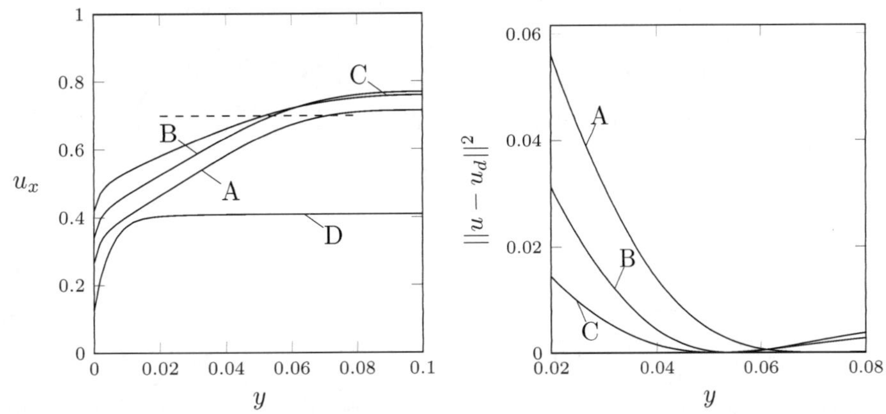

Figure 3. On the left axial velocity u_x, profile on a line at $x = 0.6$. On the right square difference $||u - u_d||^2$ for $y \in (0.02, 0.08)$ on the same line. Result (A) obtained with $\lambda = 0.5$, (B) with $\lambda = 0.05$ and (C) with $\lambda = 0.005$, (D) with no control. The dashed line is the target velocity $u_{xd} = 0.7$.

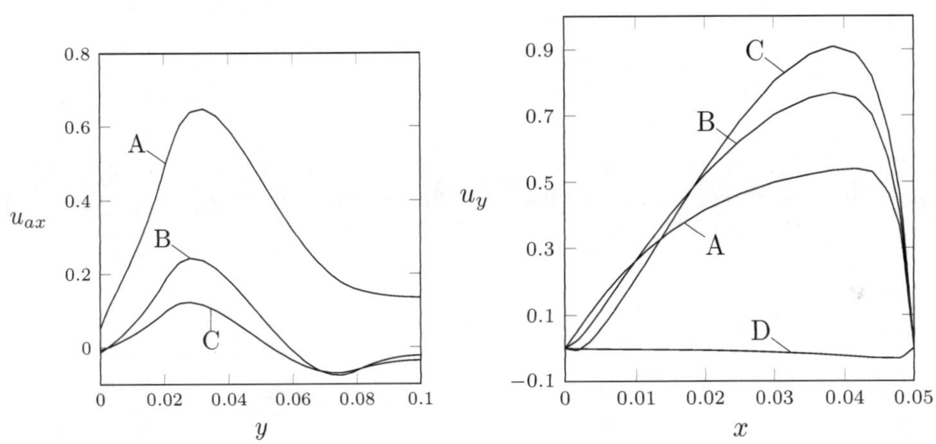

Figure 4. On the left adjoint axial velocity profile u_{ax}, profile on a line at $x = 0.6$. On the right transversal velocity u_y profile on a line at $y = 0$. Result (A) obtained with $\lambda = 0.5$, (B) with $\lambda = 0.05$, (C) with $\lambda = 0.005$ and (D) with no control.

L^2. As it is possible to observe that as λ decreases the L^2-distance from the objective becomes smaller.

In the left of Figure 4 the profile of the adjoint velocity u_{ax} is reported for three values of λ on a vertical line at $x = 0.6$. The adjoint velocity is negative where the fluid velocity has to decrease while shows a strong peak where the velocity has to increase significantly. When λ is smaller the control can act stronger, the velocity is closer to the desired profile and then the adjoint velocity profile is more flat. In the right of the same Figure the lifting function u_y profile on the control boundary Γ_c is reported for different λ values. As λ decreases, the velocity profile is more irregular and harder to be reproduced in the practice. Nevertheless, lower λ values provide a better matching with the desired profile. Finally, in Table 3, the objective functional is reported for the different regularization parameters.

λ	∞	0.5	0.05	0.005
$\mathcal{J}(\mathbf{u}) \cdot 10^5$	53.5874	7.68298	3.79369	1.8032

Table 3. Objective functionals computed with no control ($\lambda = \infty$) and different λ values in the velocity matching profile problem.

3.2. Turbulence enhancement case

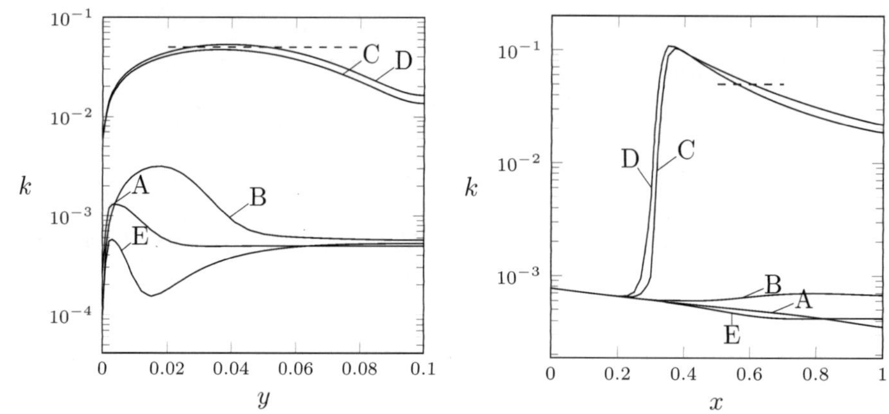

Figure 5. On the left turbulent kinetic energy k profile on a line at $x = 0.6$. On the right on a line at $y = 0.05$. Result (A) obtained with $\lambda = 0.01$, (B) with $\lambda = 0.001$, (C) with $\lambda = 0.0001$, (D) with $\lambda = 0.00001$, (E) with no control. The dashed line is the target value $k_d = 0.05$.

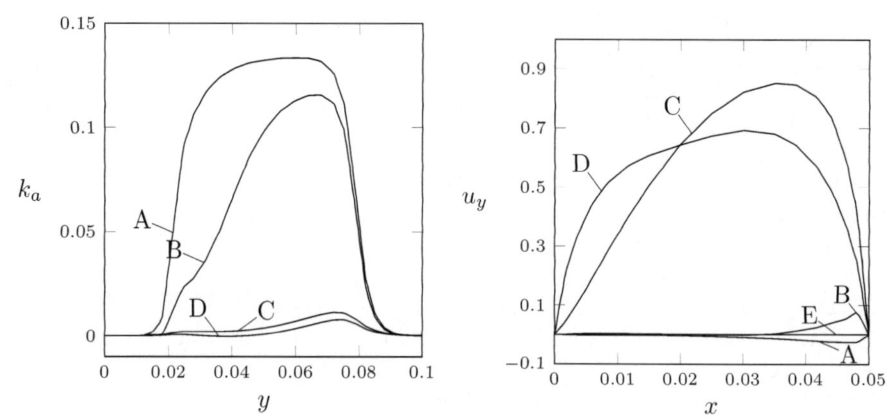

Figure 6. On the left adjoint turbulent kinetic energy k_a profile on a line at $x = 0.6$. On the right transversal velocity profile on a line at $y = 0$. Result (A) obtained with $\lambda = 0.01$, (B) with $\lambda = 0.001$, (C) with $\lambda = 0.0001$, (D) with $\lambda = 0.00001$, (E) with no control.

In this section we consider a turbulence enhancement problem which consists in increasing the turbulent kinetic energy k. We set $a = 0$, $b = 1$ and the target kinetic energy value $k_d = 0.05$, two orders of magnitude higher than that of the not controlled case. We solve the optimal control problem with $\lambda = 0.01$, 0.001, 0.0001 and 0.00001.

On the left of Figure 5 the turbulent kinetic energy k profiles on a line crossing the controlled region Ω_d at $x = 0.6$ are reported for different amount of regularization (A-D) and for the not controlled case (E) for comparison. We notice that by reducing λ the peak of the turbulent kinetic energy k grows and moves towards the center of the channel, where the objective functional is computed. On the right of the same Figure are reported k profiles on a line aligned with the main fluid flow at $y = 0.05$. Upstream of the auxiliary domain Ω_{aux} the flow regime is not altered by the control, while downstream, $x > 0.3$, the desired turbulence enhancement is obtained for cases C-D. The adjoint turbulent kinetic energy k_a profile is reported on a line at $x = 0.6$ on the left of Figure 6. When considering higher λ values we obtain that k is generally far from k_d then the source term $k - k_d$ in the k_a-equation is sizeable thus leading to adjoint turbulence peaks. The lifting function u_y profile over the control boundary Γ_c is reported for different λ values on the right of the same Figure. In order to enhance the turbulence in the middle of the channel more fluid has to be injected through the control boundary, with higher values of λ the control is too weak, while choosing $\lambda \leq 0.0001$ the optimal injection velocity profile is found. Finally, the objective functionals are reported in Table 4.

λ	∞	0.01	0.001	0.0001	0.00001
$\mathcal{J}(\mathbf{k}) \cdot 10^6$	15.4392	15.3898	15.0848	1.42287	1.07756

Table 4. Objective functionals computed with no control ($\lambda = \infty$) and different λ values in the velocity matching profile problem.

4. Conclusions

In this work we have studied a boundary optimal control problem for the Reynolds Averaged Navier-Stokes system coupled with a two-equation turbulence model in Wilcox $k-\omega$ formulation. We have used a *lifting function* approach that allows to reformulate a boundary control problem into a distributed one. In addition, the control parameter obtained belongs to its natural space. The distributed parameter is a force acting on an auxiliary domain that shares with the domain the boundary where the optimal boundary control is searched. Moreover, the force profile obtained can be used to design flow injection section. The optimality system has been derived with the minimization of the Lagrangian. We have used different objective functionals to study a velocity matching or a turbulence enhancement problem. We have obtained the adjoint equations and adjoint boundary conditions. The numerical simulations presented have shown the effectiveness of this approach in achieving the desired minimization and robustness.

References

[1] Gunzburger M D, Hou L and Svobodny T P 1992 *SIAM journal on control and optimization* **30** 167–181
[2] Gunzburger M and Manservisi S 2000 *SIAM J. Control Optim* **39**(2) 594–634
[3] Wilcox D 1994 *Turbulence Modeling for CFD* (DCW Industries, Incorporated)
[4] Bewley T R, Moin P and Temam R 2001 *Journal of Fluid Mechanics* **447** 179–225
[5] Manservisi S and Menghini F 2016 *Computers & Fluids* **125** 130–143
[6] Gunzburger M and Manservisi S 2000 *Comput Methods Appl Mech Eng* **189**(3) 803–823
[7] Bornia G 2013 *AIP Conference Proceedings* vol 1558 (AIP) pp 879–882
[8] Aulisa E, Bornia G and Manservisi S 2015 *Commun. Comput. Phys.* **189**(3) 621–649
[9] Gunzburger M D 2003 *Perspectives in flow control and optimization* vol 5 (Siam)
[10] Brenner S and Scott R 2007 *The mathematical theory of finite element methods* vol 15 (Springer Science)
[11] Bredberg J 2000 *Chalmers University of Technology, Internal Report 00/4. Goteborg*
[12] Bornia G, Gunzburger M and Manservisi S 2013 *Commun. Comput. Phys.* **14** 722–752
[13] Manservisi S and Menghini F 2016 *Computers & Mathematics with Applications* **71** 2389–2406
[14] Code FEMuS URL `https://github.com/FemusPlatform/femus`

Numerical analysis of temperature stratification in the CIRCE pool facility

Francesco EDEMETTI[1], Alessandro TASSONE[1], Vincenzo NARCISI[1], Fabio GIANNETTI[1], Luisa FERRONI[1], Mariano TARANTINO[2]

1 DIAEE - Nuclear Section, "Sapienza" University of Rome, Corso Vittorio Emanuele II, 244, 00186, Roma, Italy
2 Italian National Agency for New Technologies, Energy and Sustainable Economic Development, C.R. ENEA Brasimone, Italy

Corresponding author e-mail: francesco.edemetti@uniroma1.it

Abstract. In the framework of Heavy Liquid Metal (HLM) GEN IV Nuclear reactor development, the focus is in the combination of security and performance. Numerical simulations with Computational Fluid Dynamics (CFD) or system codes are useful tools to predict the main steady-state phenomena and how transitional accidents could unfold in GEN IV reactors. In this paper, to support the validation of CFD as a valid tool for the design, the capability of ANSYS CFX 15 to simulate and reproduce mixed natural convection and thermal stratification phenomena inside a pool is investigated. The 3D numerical model is based on the CIRCE facility, located in CR ENEA Brasimone. It is a pool facility, structured with all the components necessary to simulate the behavior of an HLM reactor, where LBE flows into the primary circuit. For the analysis, the LBE physical properties are implemented in CFX by using recent NEA equations [2]. Previously published RELAP5-3D© results [1] are employed to derive accurate boundary conditions for the simulation of the steady-state conditions in the pool and for CFX validation. The analysis focuses on the pool natural circulation with the presence of thermal structures in contact with LBE, considered as constant temperature sources. The development of thermal stratification in the pool is observed and evaluated with a mesh sensitivity analysis.

1. Introduction

In the Heavy Liquid Metal (HLM) - GEN IV Nuclear reactor development one of the main task deals with the thermal-hydraulics analysis of HLM-cooled fast reactor.

The attention is mainly focused on Lead and Lead-Bismuth Eutectic alloy (LBE) cooled reactors, due to their high capability to meet the safety requirements. Lead-cooled Fast Reactor (LFR) is one of the six concepts selected by Generation IV International Forum (GIF) and includes lead and LBE as primary coolant; both are chemically inert, and they offer other attractive characteristics in terms of interaction with structural materials and thermodynamic features.

Main interest for safety is the establishment of the natural circulation, an essential condition for a GEN IV reactor: these coolants guarantee the natural circulation cooling during the normal and transient operation. Key feature is the transient operation; this because, in case of an accident, the reactor is able to guarantee a long-term refrigeration of the core, thus preventing its meltdown. This aspect offers a considerable grace time in order to face an unprotected loss of flow accident and it permits to introduce

IOP Conf. Series: Journal of Physics: Conf. Series **1224** (2019) 012007 doi:10.1088/1742-6596/1224/1/012007

a combined mix of active and passive systems. Passive systems, as Decay Heat Removal system (DHR), assure optimal safety features over long time periods without needing human actions [3], as studied for ALFRED reactor [10], [11].

Nowadays, numerical simulations play an increasingly key role in the design of components and systems and new models and new software tools are developed to investigate thermal-hydraulics issues related to HLM technology. All these incoming tools must be validated in order to be used for safety analyses.

The activity aims to define, implement and carry out experimental results on the CIRCE (CIRColazione Eutettico) pool facility with the ICE (Integral Circulation Experiment) test section at ENEA Brasimone research center, in order to support the qualification of CFD as a valid tool for the design, control and analysis of complex three dimensional convection patterns in a large pool configuration, simulating the pool stratification phenomena in LBE and Lead cooled reactors and to investigate HLM mixing, which should induce thermo-mechanical stress on the structures. Another important scope is the validation of the models for the design of new HERO test section experimental campaign [9].

2. The CIRCE – ICE facility

The CIRCE – ICE facility, which aims to simulate the thermal-hydraulic behavior of the primary system in an HLM cooled pool reactor and the convection currents within the pool, is equipped with a Fuel Pin Simulator (FPS), which electrically simulates the core of the facility, the Heat Exchanger (HX) that simulates the steam generator and the DHR system, which simulates the security system for the transient operation.

CIRCE is a multipurpose pool facility designed to host different test sections, like ICE, welded to and hung from bolted vessel heads for the investigation of thermal-hydraulic aspects related to the HLM pool system.

The facility consists of a main vessel, earmarked for containing test section and filled with about 70 tons of molten LBE, two auxiliary tanks, dedicated to store LBE during maintenance phases and to transfer liquid metal during loading and drainage phases, and a data acquisition system [4].

The ICE test section is placed inside the main vessel, allowing to simulate the heat exchanger for primary cooling system of HLM pool-type reactor coupled with the secondary cooling system for the steam production. Main components and their positioning are showed in Figure 1 while the main flow path is highlighted in Figure 2. The feeding conduit is the inlet pipe of the test section and connects the lower plenum and the FPS. A Venturi-nozzle flow meter is installed into the feeding conduit in order to measure the primary flow rate through the FPS bundle, which consists of an electrical pin bundle with a nominal thermal power of 800 kW and an active length of 1000 mm. Molten LBE flows upward and increases his temperature passing through this part. The FPS is assembled with 37 electrical pins arranged in a wrapped hexagonal lattice and characterized by a pitch to diameter ratio equal to 1.8. Each pin has an outer diameter of 8.2 mm, a maximum thermal power of 25 kW and a heat flux at the pin wall of 1 MW/m^2. The hot fluid is introduced into the fitting volume and then flow upward into the riser which is a double wall pipe that connects the fitting volume with the separator. At the inlet section of the riser, a nozzle is installed allowing the injection of Argon that ensure the LBE assisted circulation into the primary circuit through the riser. The mixture flows upward and collects in the separator, where the gas is outstripped from the liquid metal: the former flows upward to the gas plenum, whereas the LBE is directed to the HX. The HX is made of 91 bayonet tubes with an active length of 3462 mm. The feed-water flows downward into the inner tube and then upward into the annular region between inner and middle tube, where the change of phase take place; outside the tubes, LBE flows downward decreasing the temperature. The volume between middle and outer tube is filled by pressurized helium to detect any leakage. The LBE, released by the HX, enters the pool, flowing downward, and reaching the lower plenum [4].

The DHR system consists of only one bayonet tube that has to remove the decay power by forced circulation of air; its height is 5012 mm, with an active length of 3205 mm. The bayonet is inside a

IOP Conf. Series: Journal of Physics: Conf. Series **1224** (2019) 012007 doi:10.1088/1742-6596/1224/1/012007

double wall shell with a thin air insulation gap to thermally decouple the DHR from the external LBE pool. During the transient operation, the hot LBE enters the DHR by the upper inlet section, it flows downward decreasing its temperature and exiting the component in the bottom half of the pool.

The ICE test section is installed in the pool; it has a radius of 585 mm and it is high 8490 mm, height where there is the closure of the vessel.

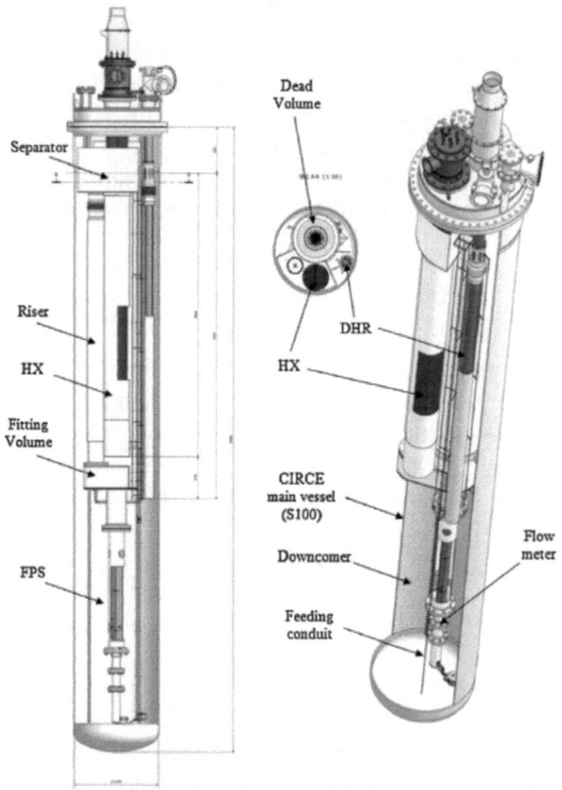

Figure 1. ICE test section [1]

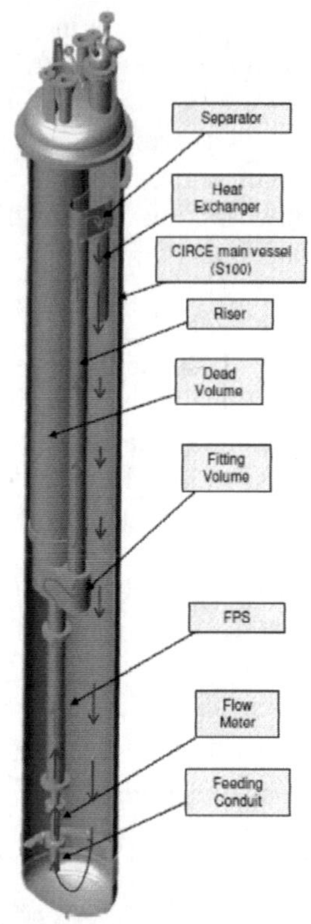

Figure 2. ICE test section: primary system main flow path

3. Problem formulation

This work is based on Ref. [5] to reproduce the geometry of CIRCE-ICE test facility and on Ref. [1] to initialize the simulation and to compare the obtained results.

The aim is to investigate the thermal stratification and convective phenomena of LBE. The primary cooling circuit of LBE is not investigated but it is assumed as data boundary condition for the model. To perform this analysis the CFD code ANSYS – CFX v15.0 is used.

3.1. ANSYS – CFX Geometry

A three-dimensional finite volume model, shown in the figures below (Figure 3, Figure 4 and Figure 5), has been built first by reproducing the structural components, and then subtracting them to the volume enclosed by the containment vessel.

Fluid domain has been sliced by fixed height multiple plane, creating 38 fluid volumes with heights equal to those volumes used in the RELAP5-3D© calculations presented by Narcisi et al. [1] to allow the use of these results as boundary conditions for the present simulation.

The geometry of CIRCE-ICE facility is not reproduced in its completeness, but only up to the entrance of the separator Mesh generation is a key part of the analysis and will be discuss further.

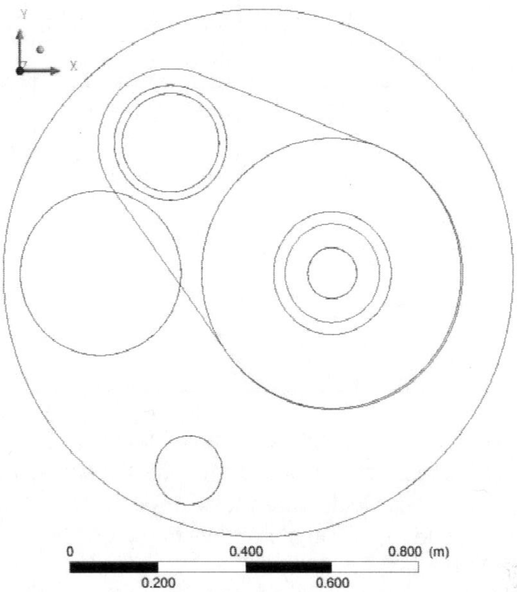

Figure 3. Numerical model CIRCE-ICE Z vertical view

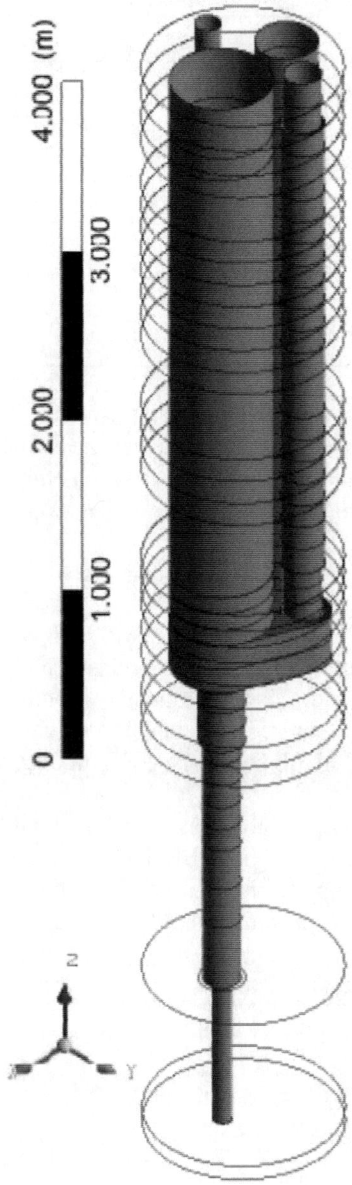

Figure 4. Numerical model CIRCE-ICE isometric view

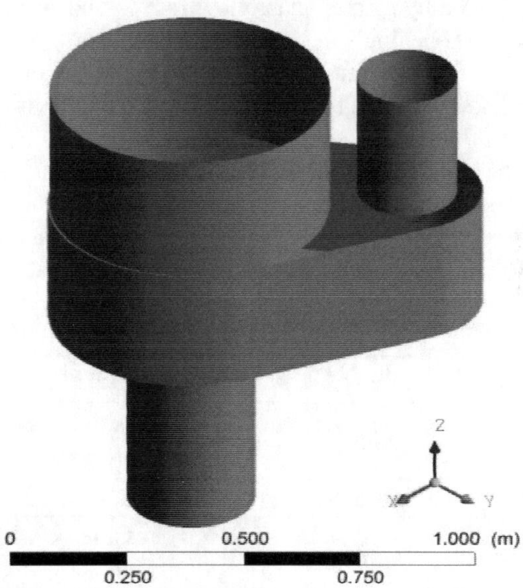

Figure 5. CIRCE-ICE fitting volume numerical model

3.2. Governing Equations

The simulation is a Reynolds Averaged Navier-Stokes (RANS) calculation, single-phase, using the k-ω Shear Stress Transport (SST) turbulence model, in order to account for the turbulence generated in the LBE domains, because the LBE is expected to pass from points where velocity is around 0.5 m/s to other points where the velocities decrease in order of magnitude of $10^{-2} \div 10^{-3}$.

The LBE was modelled with temperature-dependent thermophysical properties in order to obtain more accurate results; the equations are summarized in Table 1.

IOP Conf. Series: Journal of Physics: Conf. Series **1224** (2019) 012007 doi:10.1088/1742-6596/1224/1/012007

Table 1 Thermophysical properties of LBE [2] (Temperature in K)

Variable	Unit	Expression
Density	[kg m^{-3}]	$\rho = 11065 - 1.293 \cdot T$
Heat Capacity	[J kg^{-1} K^{-1}]	$c_p = 164.8 - 3.94 \times 10^{-2} \cdot T + 1.25 \times 10^{-5} \cdot T^2 - 4.56 \times 10^5 \cdot T^{-2}$
Viscosity	[Pa s]	$\mu = 4.94 \times 10^{-4} \cdot e^{\left(\frac{754.1}{T}\right)}$
Thermal Conductivity	[W m^{-1} K^{-1}]	$\kappa = 3.284 + 1.617 \times 10^{-2} \cdot T - 2.305 \times 10^{-6} \cdot T^2$
Thermal Expansion	[K^{-1}]	$\beta = \frac{1}{8558 - T}$

Since the density is represented as function of the temperature, the full Buoyancy model [12] is employed to calculate the buoyancy forces arising in the pool to provide better accuracy in the evaluation of the heat transfer and liquid metal velocity field. A turbulent-Prandtl number of 2 has been employed, as suggested in [6] [7], to hence the conduction heat transfer prevails on the convection (i.e. turbulence-driven) heat transfer.

3.3. Boundary Conditions

The inlet is set at the exit of the HX and outlet at the LBE ingress in the feeding conduit: both BCs are characterized by a fixed mass flow rate to and from the, not-modelled, test section internal hydraulic circuit. An "opening", to simulate the behavior of the argon gas plenum, is set as boundary condition at the top surface condition that is necessary to obtain the correct pressure profile along the height of the pool without the need to simulate a more complex multiphase flow involving the Ar-LBE mixture (BCs Table 2).

From the results obtained by the RELAP5-3D© analysis, [1], it is possible to set as boundary conditions the temperatures or the heat fluxes on the surfaces of structures in contact with LBE. Due to instabilities that the imposed heat flux may create, it was considered as appropriate choice to set the wall temperatures as a boundary condition. The thermal boundary conditions are reported in Table 3.

Table 2 CFX model momentum and mass boundary conditions

BC Name	BC Detail	Value
Inlet	Mass flow rate [kg s^{-1}]	57.737
	Temperature [K]	555
Outlet	Mass flow rate [kg s^{-1}]	57.737
Opening	Pressure [kPa]	107.5
	Temperature [K]	613

Table 3 CFX model heat transfer wall temperature boundary conditions among the CIRCE height

BC	H	FPS	Release Pipe	Fitting Vol	Dead Vol	Riser	HX	Pool	DHR
[#]	[mm]	[K]	[K]	[K]	[K]	[K]	[K]	[K]	[K]
1	170					638.6	620.1	613.8	
2	170					638.6	616.1	610.3	
3	150					616.0	610.1	603.7	
4	300					606.0	604.3	598.0	
5	150					604.4	600.7	596.2	
6	150					601.5	597.0	592.9	
7	150					599.9	593.6	591.2	
8	150					597.8	590.2	588.8	
9	150					596.2	586.9	587.0	
10	150					594.6	583.7	585.0	
11	150					593.0	580.5	583.1	
12	450					589.2	574.5	579.0	
13	150					582.7	568.7	574.1	
14	150					579.3	566.1	572.5	
15	150					576.5	563.7	571.8	
16	150					576.5	561.4	571.5	
17	150					576.1	559.2	571.6	
18	150					575.5	557.1	571.6	
19	170					575.5	557.9	571.6	
20	110				626.1	598.4	557.7	571.6	
21	110				628.0	634.6	557.6	570.8	
22	110				629.8	634.7		569.3	
23	160			619.6				563.6	
24	160			619.3				560.0	
25	132.5		622.5	610.6				557.5	
26	132.5		615.8					557.5	
27	132.5		617.1					556.5	
28	132.5		619.0					556.5	
29	100	620.6						556.5	
30	150	619.5						556.5	
31	150	618.1						556.5	
32	150	612.5						556.5	
33	150	603.8						556.5	
34	262.5	589.7						556.5	
35	262.5	576.6						556.5	
36	415	560.1						556.5	
37	970							556.5	
38	440							556.5	

IOP Conf. Series: Journal of Physics: Conf. Series **1224** (2019) 012007 doi:10.1088/1742-6596/1224/1/012007

3.4. Domain Discretization

To further evaluate the accuracy of the results, three different meshes have been generated. Each of them has an increased number of nodes and elements to perform a mesh sensitivity analysis. The resolution between the meshes is given by an increase of thinning in the axial divisions of the model. Meshes are characterized by hexahedral and tetrahedral elements and a near wall inflation layer with $y^+_1 = 20$: this is a good compromise between the accuracy and the computational cost, as solving the viscous sublayer is not an objective of the present analysis; statistics information about the three meshes are summarized in Table 4. All the boundary conditions are the same for each simulation as shown in Table 2 and Table 3.

Table 4 Mesh statistics

Mesh	Nodes	Elements	Ave. Orth. Quality	Ave. Skewness
#1	1965090	2044636	0.96856	0.10708
#2	3485837	3495472	0.97682	0.09512
#3	8961424	8637335	0.98981	0.07959

4. Result and discussion

The steady state CFD analyses have been carried out to evaluate the thermal stratification inside the CIRCE pool, comparing the calculated axial temperature profile with the experimental data; the convective phenomena, recirculation and stagnation zones within the pool which can cause hotspots that can produce strong thermal stress, have been also investigated

The value of mass flow rate at the entrance of the feeding conduit to be compared with the results from [1] and experimental data [8]: as expected, the result is the same of the RELAP5-3D© simulation due to the fixed boundary condition, returning a mass flow rate of 57.74 [kg s^{-1}]

The evaluated mass flow rate at the opening is close to 0 [kg s^{-1}], as expected, for all the meshes. Refining the mesh, the mass flow through the area increases, passing from a magnitude of 10^{-7} (mesh #1) to 10^{-5} (mesh #2) and 10^{-4} (mesh #3). Therefore, no relevant LBE flow from and to the pool is present at the top surface and no significant error is introduced by not modelling the Argon in the gas plenum.

The axial pressure profile along the height of the pool is shown in Figure 6. Pressure grows linearly along the height, resulting at the opening 107.5 [kPa] and at the entrance of feeding conduit: 829.06 [kPa] (mesh #1), 828.80 [kPa] (mesh #2) and 829.27 [kPa](mesh #3).

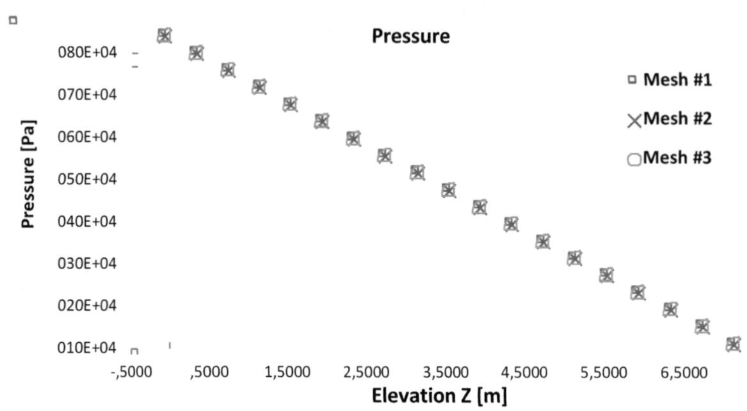

Figure 6 Axial pressure profile among the height

In order to investigate thermal stratification and mixing convection phenomena into the pool, CIRCE is equipped with several thermocouples (TCs), arranged in different vertical and circumferential positions, as showed in Figure 7 [8]. The 3D model used in CFX allows to compare the LBE temperature in the actual position of the TCs.

IOP Conf. Series: Journal of Physics: Conf. Series **1224** (2019) 012007 doi:10.1088/1742-6596/1224/1/012007

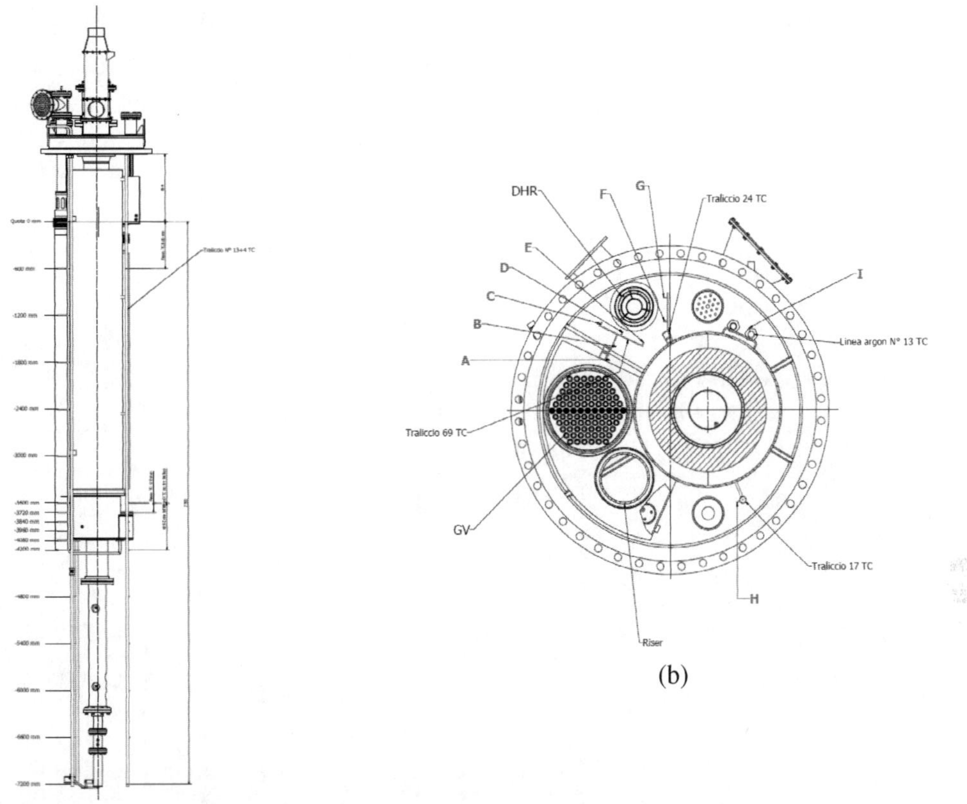

Figure 7 Thermocouples positions (a) (b)

The figure below shows a comparison between experimental and axial temperature profiles with the relative mesh in different circumferential positions. In Figure 8(a) and (b), the experimental values are obtained as the average temperature measured respectively by six and two TCs. Temperature profiles show a similar trend for each mesh in the different TCs positions. Some differences with the experimental data, as it is shown in 8(a) and 8(c) between 3000 mm and 4000 mm, have been obtained and there are some instabilities on the temperature trend for mesh#1 and mesh#2, therefore the experimental temperature plateau is not accurately reproduced in those calculations. Mesh #2 and #3 overestimates the temperature plateau in the same region but have a linear trend: these differences are probably due to the accuracy of the inflation layer that determines the y+ and to the axial mesh discretization of the model.

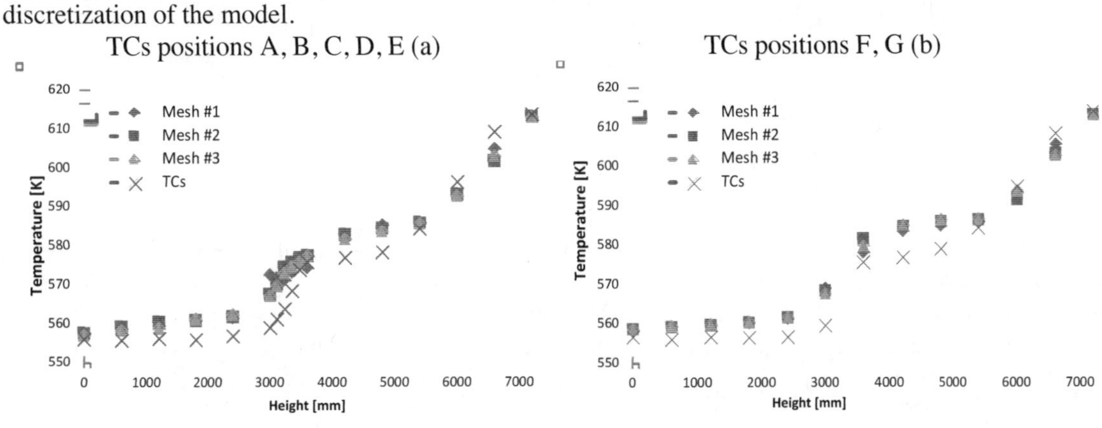

IOP Conf. Series: Journal of Physics: Conf. Series **1224** (2019) 012007 doi:10.1088/1742-6596/1224/1/012007

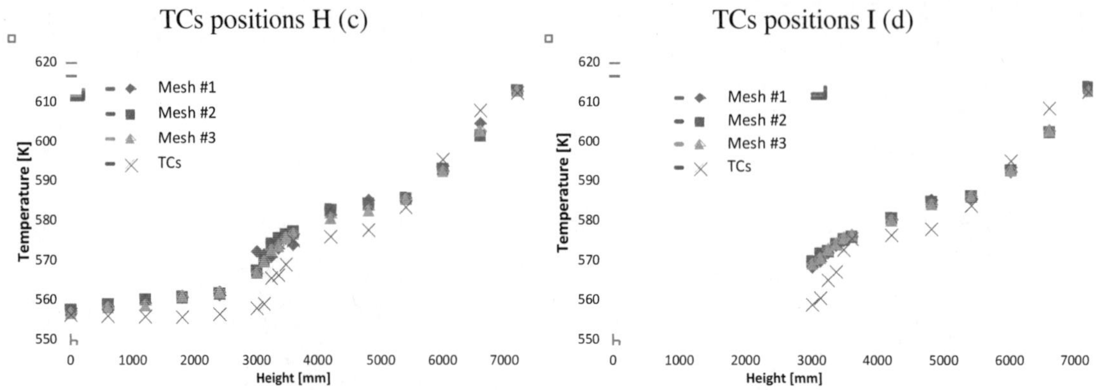

Figure 8 Thermal stratification along the axial profile

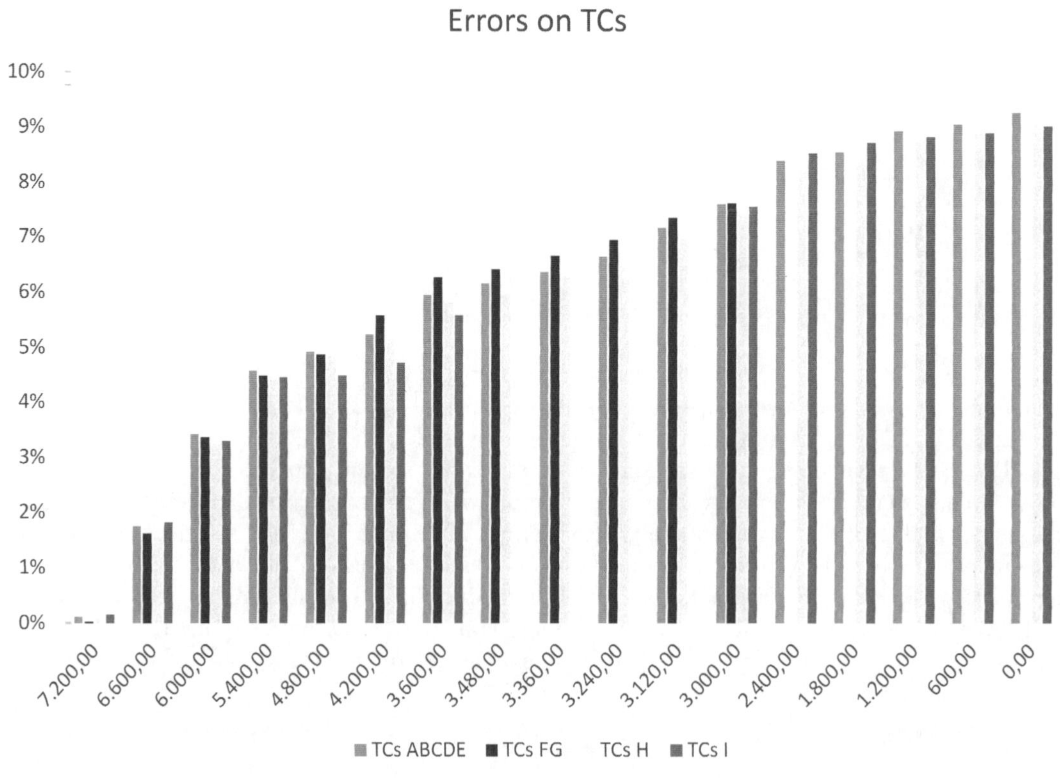

Figure 9: Errors bars regarding the TCs position of Mesh #3

The establishment of natural circulation is also confirmed in all simulations, thanks to the pressure drop and the temperature gradient inside the pool.

In Figure 10 (a) is represented the velocity field inside the pool of the mesh #3, the mixing phenomena is shown due to the vortex creation in the bottom part of the pool, the maximum velocity is reached at the entrance of the feeding conduit. The Figure 10 (b) shown the presence of vortex near the HX outlet and among the downward height, with recirculation points that also increase the mixing phenomena.

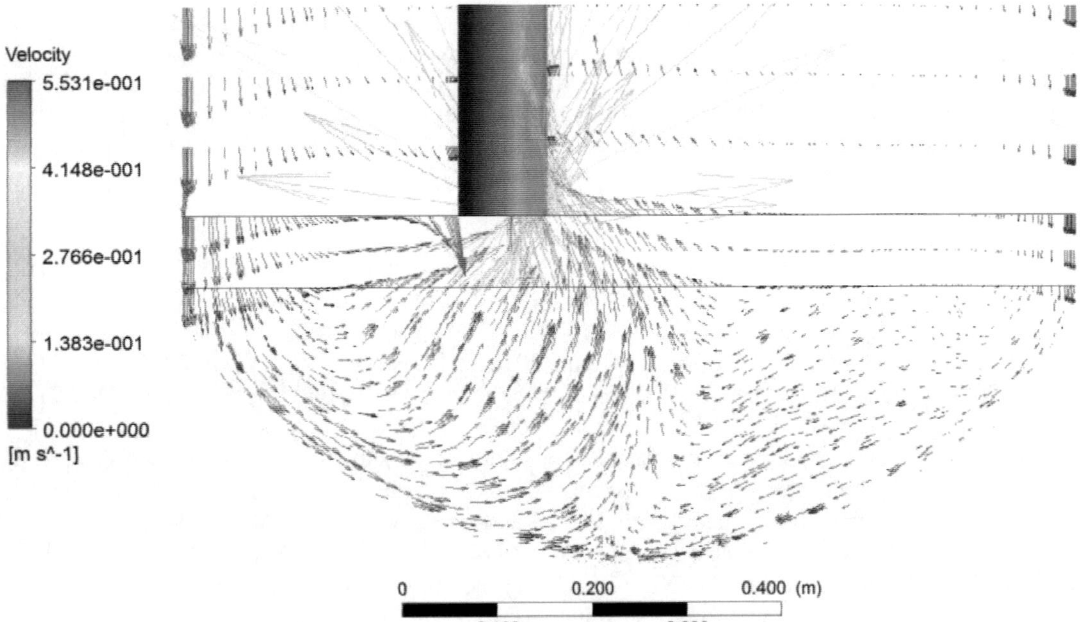

(a)Velocity vectors inside the pool of Mesh #3 at the bottom of the pool

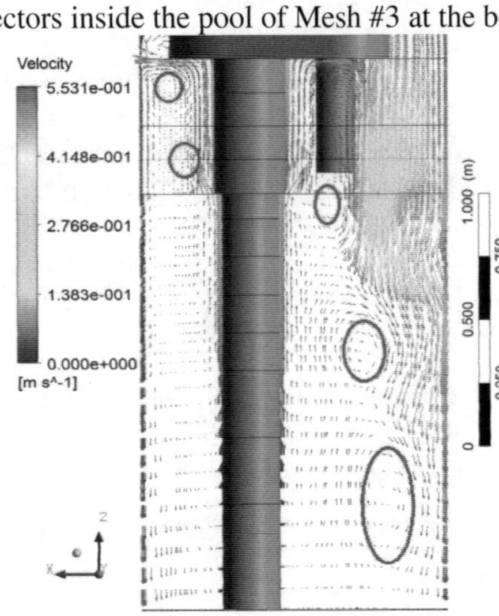

(b) Recirculation point among the height in Mesh #3

Figure 10: Mixing phenomena in Mesh #3 (a) (b)

5. Conclusion

The experimental campaign conducted on the CIRCE-ICE facility at ENEA-Brasimone offers additional data for the validation of CFD thermal-hydraulics code in the frame of HLM pool-type reactors.

The main objective of this work was to investigate the thermal stratification and mixing convection phenomena into the pool. The developed 3D model consists of a pool with the structures of the primary system immerged in the LBE to reproduce only the volume of the pool. The boundary conditions are given from the RELAP5-3D© analysis carried out in [1].

The thermal-hydraulic behavior of the natural convection inside the pool is well reproduced by the CFX code. In particular, the activation of the buoyancy forces allows to identify a large number of recirculation point with the consequence of intense mixing convection phenomena.

The LBE temperature is well reproduced for all the meshes, except for the mesh #1, where the simulations return instabilities that differ from the experimental value.

An important parameter to correctly evaluate the thermal stratification is the heat losses distribution.

Further investigation for the thermal stratification profile in this zone is needed, with greater mesh nodes number to solve the viscous sublayer, and a thicker axial mesh. A fully coupled model with RELAP5-3D© can return also a better solution to reduce the instabilities.

References

[1] V Narcisi, F Giannetti, M Tarantino, D Martelli, G Caruso "Pool temperature stratification analysis in CIRCE-ICE facility with RELAP5-3D© model and comparison with experimental tests" J. Phys. Conf. Ser. 923; doi:10.1088/1742-6596/923/1/012006
[2] P Balestra, F Giannetti, G Caruso, A Alfonsi 2016 *Sci. Technol. Nucl. Install.* **2016**
[3] OECD Nuclear Energy Agency 2014 Technology Roadmap Update for Generation IV Nuclear Energy Systems 1–66
[4] M Tarantino, P Agostini, G Benamati, G Coccoluto, P Gaggini, V Labanti, G Venturi, A Class, K Liftin, N Forgione and V Moreau, 2011, "Integral Circulation Experiment: Thermal-hydraulic simulator of a heavy liquid metal reactor" *J. Nucl. Mater.* **415** 433–48
[5] D Martelli "CIRCE Experimental facility" CI-R-162 ENEA Internal Report, 2015
[6] Fei Chen, Xiulan Huai, Jun Cai, Xunfeng Li, Ruixue Meng "Investigation on the applicability of turbulent-Prandtl-number models for liquid lead-bismuth eutectic" *Nucl. Eng. Des.* **257:** 128-133, April 2013
[7] F Vitillo et al 2015 *J. Phys.*: Conf. Ser. 655 012058
[8] Tarantino M, Martelli D, Barone G, Di Piazza I and Forgione N 2015 *Nucl. Eng. Des.* **286:** 261-77
[9] Narcisi V, Giannetti F, Del Nevo A, Tarantino M, Caruso G 2018 *Nucl. Eng. Des.* **333:** 181-195
[10] Caramello M, Gregorini M, Bertani C, De Salve M, Alemberti A, Panella B 2017 *Prog. Nucl. Energy* **99:** 127-139.
[11] Giannetti F, Vitale Di Maio D, Naviglio A, Caruso G 2016 *Nucl. Eng. Des.* **305:** 168-178
[12] ANSYS Documentation v 15.0, *CFX-Solver Modeling Guide*

IOP Conf. Series: Journal of Physics: Conf. Series **1224** (2019) 012008 doi:10.1088/1742-6596/1224/1/012008

Numerical simulation of oil-water two-phase flow in a horizontal duct with a Venturi flow meter

M Guilizzoni, G Salvi, G Sotgia, L P M Colombo

Politecnico di Milano, Department of Energy, Via Lambruschini 4, 20156 Milano, Italy

E-mail: manfredo.guilizzoni@polimi.it

Abstract. The progressive depletion of on-shore and light-oil reserves is forcing an increased use of transitional and heavy oils, which implies new challenges both during the extraction and the transportation. Focusing on the latter, a technique to reduce the pressure drop is water injection in the oil stream to create the so-called core annular flow (CAF), a flow regime with an oil core enveloped in a water annulus wetting the pipe wall, so that the apparent viscosity of the mixture is considerably reduced. Behaviour of CAF in ducts with non-uniform sections is still under research. This work is devoted to a CFD investigation about the pressure drop, pressure gradients, velocity profiles and *in situ* volume fractions in a duct including a Venturi flow meter. Unsteady RANS simulations were carried out using the Volume-Of-Fluid *interFoam* solver of OpenFOAM. Numerical results were experimentally validated for oil superficial velocities in the range 0.25–0.75 m/s and water superficial velocities in the range 0.44–1.10 m/s and comparisons between different approaches and sensitivity analyses were performed. Satisfactory agreement was found for the pressure drop and pressure gradients, and also for the *in situ* volume fraction with respect to the predictions of the Arney correlation.

1. Introduction

A large majority of the world energy demand is nowadays still satisfied by oil, and since the last decades progressive depletion of on-shore and light-oil reserves forced an increased use of transitional and unconventional/heavy oils. These are available in a larger total amount and in a higher number of countries, but also pose new challenges. In fact, despite different classifications, they share the main characteristic of having higher densities, viscosities, sulfur and impurities contents than the conventional oil, so that they that cannot be extracted, transported and processed using conventional techniques [1, 2]. In particular, pipeline transport of unconventional/heavy oils is problematic due to their high densities and particularly high viscosities. Among the possible solution to make it feasible, the creation of a peculiar two-phase flow pattern, in which the apparent viscosity is reduced, is at present under study and development [3]. Such a flow pattern is named Core Annular Flow (CAF): an oil core flows in the central part of the pipeline, while a “lubricating” water annulus wets the duct wall [4-6], thus reducing the wall shear stress and lowering the pressure drop. This favorable flow pattern only occurs for suitable velocities of the phases, so it is very important to measure and/or to predict the mass flow rates of the phases and of the total mixture. Moreover, the stability of Core Annular Flow in presence of variations in the pipeline (e.g. bends, changes in the cross-section, in-line measurement devices) is still under investigation.

This paper is therefore devoted to a Computational Fluid Dynamics (CFD) study aimed at characterizing the flow of an oil-water mixture in the Core Annular Flow regime within a pipeline section including a Venturi flow meter. Such a computational domain offers the opportunity to get information about the flow behavior both along straight ducts (upstream and downstream) and within a measurement device whose converging/diverging profile also represents a reduction and an expansion of the flow cross-section. Moreover, such a device can be of great practical interests for the oil and gas industry because it has no moving parts, thus greatly reducing the need for maintenance [7]; but an issue in its use is that it is necessary to rely on models to correlate the measured pressure difference to the mass flow rates. Thus, accurate characterization of the multiphase flow across it is needed.

2. Literature review

Even if the literature is not as vast as for gas-liquid flows, nowadays many papers can be found about liquid-liquid oil-water flows. Reference to many of the most important papers about this topic can be found in [8]. Focusing the attention on Core Annular Flow, pioneering studies were performed by Charles [9], followed (in a not exhaustive list) by Ooms et al [10], Oliemans et al. [11], Brauner [12], Arney et al. [13], Joseph et al. [6], Angeli and Hewitt [14], Sotgia et al. [6], Colombo et al. [15], Colombo et al. [16]. A review can be found in the book by Joseph and Renardy [17].

Concerning the numerical simulation of two-phase flows and particularly of oil-water flows, simplified (e.g. one-dimensional) models are implemented also in commercial software commonly used by companies in the oil and gas sector (e.g. OLGA Dynamic Multiphase Flow Simulator [18]). On the contrary, reliable three-dimensional simulations are still an open research problem and only a minor number of papers are available in the literature [19-21]. The work by Babakhani Dehkordi et al [21], focused on the flow within a Venturi meter, also includes a review about measurement devices used for two-phase and particularly oil-water flows.

3. Numerical model

3.1. Volume-Of-Fluid in interFoam

In the Core Annular Flow pattern, two immiscible phases are separated by an interface having a length scale comparable to the pipe diameter (particularly for the oil core, but also for the oil drops if present). Such interface is continuously varying in terms of shape and extension, due to the evolution of the flow structures, including also possible breakup and coalescence. The two phases can be both approximated to incompressible. One of the most suitable CFD approaches for similar scenarios is the Volume-Of-Fluid (VOF), introduced by Hirt and Nichols [22]. The latter was originally developed for different two-phase applications (e.g. drop impacts), but examples of their use for two-phase flow are already reported in the literature [19, 20, 23]. More complex methods could also be used (e.g. [24]), but no ready-to-use solver is available, so at present their use is practically restricted to academic studies. On the contrary, the aim of this paper is to investigate the performances of an open source software package when used in "field" conditions, i.e. with common hardware, an intermediate knowledge about CFD and requiring reasonably short simulation times. Therefore, VOF was selected for this study and simulations were performed on 3D domains using the *interFoam* solver of the OpenFOAM open source CFD toolbox. OpenFOAM was selected due to its free and open source nature and to the many favourable reviews and successful cases of use described in the literature (e.g. see references 13-17 in [25]). *interFoam* is an isothermal finite volume solver implementing a modified version of the VOF model, with the continuum surface force (CSF) model to include surface tension at the interface [26]. It solves a single set of equations, constituted by the usual mass and momentum conservation equations for incompressible flow, plus an additional equation for the *in situ* volume fraction γ (i.e. the ratio between the volume occupied by one phase and the total volume in a certain domain region) that is transported by the fluid velocity field *w*:

$$\partial\gamma / \partial\tau + \nabla\cdot(w\,\gamma) + \nabla\cdot[w_r\,\gamma\,(1-\gamma)] = 0 \qquad (1)$$

As it can be noticed from Eq. 1, *interFoam* introduces an additional term in the volume fraction equation with respect to the original VOF model. In this term, w_r is the "compression velocity" that is obtained by multiplying the maximum velocity magnitude in the transition region (i.e. the region where $0 < \gamma < 1$) by the normal to the interface vector and by an adjustable coefficient. Thanks to the additional term, which is active only in the interfacial region, the latter is compressed to very few cells. *interFoam* performs volume tracking and interface reconstruction, with no explicit interface tracking, so parasitic velocities due to spurious capillary effects [27] may arise. On macroscopic domains as those studied in this work, this is not an issue. Further details about *interFoam* can be found in [28].

3.2. Fluid properties and turbulence models
Both oil and water were assumed as Newtonian and incompressible fluids. Density and viscosity were set for oil at ρ_o = 890 kg/m^3 and μ_o = 0.830 Pa s respectively, and for water at ρ_w = 1000 kg/m^3 and μ_w = 1.026x10^{-3} Pa s respectively, corresponding to the values in the experiments used for validation.
At the investigated velocities of the phases, flow is laminar in the oil core and turbulent in the water annulus. As the VoF method models a single fictitious fluid, it is not possible to consider the two phases separately; so unsteady Reynolds-Averaged Navier-Stokes (RANS) simulations were performed for the whole fluid domain. The k-ε turbulence model, with standard wall functions and values for the involved constants, was selected as a first choice for the simulations. Additional simulations were also performed with the k-ω Shear Stress Transport model. Details about the two models can be found in Versteeg and Malalasekera [29].

3.3. Discretization schemes and CFL settings
The implicit Euler scheme (first-order accurate) was used to discretize the time derivative. The Gauss linear scheme was used for the gradient terms, while the Gauss linear corrected scheme for the Laplacian terms. For the divergence terms, Gauss upwind was used for the velocity and the turbulence variables, Gauss van Leer and Gauss linear or *interfaceCompression* (a peculiar scheme of OpenFOAM for the compression velocity) for the volume fraction.
The PISO (Pressure Implicit with Splitting of Operators) algorithm was used to deal with the pressure-velocity coupling. The linear solver PCG (preconditioned conjugate gradient) with a DIC (Diagonal Incomplete-Cholesky) preconditioner was selected for pressure, while the linear *smoothSolver* with a symmetric Gauss Seidel smoother for all the other variables. Some volume fraction sub-cycles were also performed. The maximum allowed Courant–Friedrichs-Lewy (CFL) number was set to 0.3 (sensitivity analysis was then performed raising it to 0.5): the solver is let free to adapt the time step to keep the CFL number under this desired limit (that is the requirement for the interface compression scheme used in *interFoam* for 3D cases)

3.4 Domain, mesh and boundary conditions
Figure 1 shows a sketch of the simulation domain and mesh (coarsened for better visualization) with an example of results in terms of shape of the oil core and of velocity field on the duct axial vertical plane. Boundary conditions for the different variables are also summarized. The insets show a cross-sectional view of the mesh and an example of the oil core shape observed in the Venturi flow meter during the experiments.
The simulated domain includes a section of a pipeline with 40 mm i.d., equipped with a Venturi flow meter (designed according to the ISO standards) having a 30 mm i.d. throat and a total length of 120 mm (35 mm before the throat and 85 mm after the same). Three domains, differing in the lengths of the upstream and downstream pipes, were tested:

- domain A: upstream duct 200 mm, downstream duct 150 mm;
- domain B: upstream duct 500 mm, downstream duct 150 mm;
- domain C: upstream duct 600 mm, downstream duct 200 mm;

IOP Conf. Series: Journal of Physics: Conf. Series **1224** (2019) 012008 doi:10.1088/1742-6596/1224/1/012008

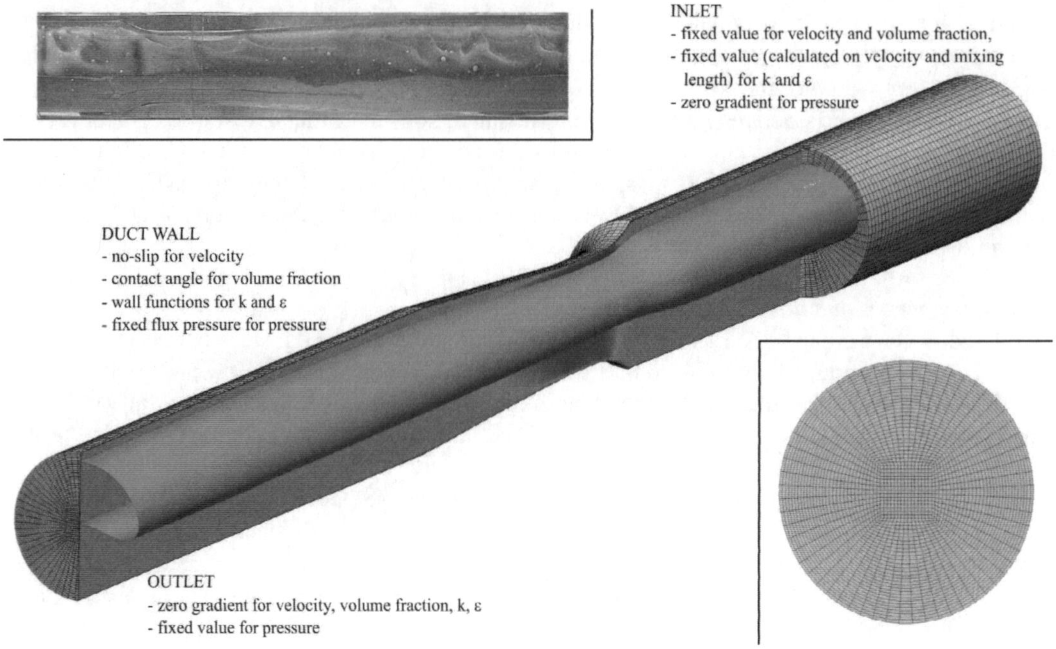

Figure 1. Sketch of the investigated domain and mesh (coarsened for better visualization) with an example of results in terms of shape of the oil core and velocity field along the duct axial vertical plane. In the insets: an example of the oil core shape observed in the Venturi flow meter during the experiments and a cross-sectional view of the used mesh.

Shorter lengths obviously affected negatively the determination of the distributed pressure drop, so the three domains were selected as a compromise between this aspect and the computational effort.
The meshes – in all cases structured – were created by the *blockMesh* OpenFOAM utility. Different mesh refinements were used to test mesh independence. The latter is always an issue when using the VOF approach and for the situation under study it was expected to be problematic due to the presence of the water annulus, which for some flow conditions is quite thin towards the top part of the duct. Strict mesh independence was not reached, but the following meshes were selected, again as a reasonable compromise between the mesh sensitivity and the computational effort:

- coarse mesh used for domain A: 143k cells;
- coarse mesh used for domain B: 233k cells;
- fine mesh used for domain C: 747k cells.

The values are relatively low, but they must be evaluated considering that transient simulations were performed and that the oil core – entering the domain at the inlet – had to move across the whole domain and then to reach a pseudo-steady configuration. With these domains and meshes the simulation time was in the range 15-30 hours (depending on the phase velocities) for cases A and B and between 80 and 140 hours for case C, on desktop/laptop computers equipped with Intel i5 and i7 series processors. A different choice might have been to initialize a certain shape of the oil core along the whole domain and to let it reach its equilibrium structure, possibly using a modified version of *interFoam* that is able to use local time stepping (*LTSInterFoam*) to make convergence towards pseudo steady-state conditions faster.
For all cases, mesh quality was checked in terms of aspect ratio, skewness and orthogonality.

A side-effect of using a mesh that is quite coarse near the wall is that for some cases the oil sticks to the wall for flow conditions in which experimentally this is not observed. As it will be discussed in Section 6, to solve this issue the duct wall was modelled as super-hydrophilic. This is why in Table 1 some cases are referred to as "standard wall" (zero-gradient boundary condition for the volume fraction, i.e. contact angle for a water drop in a surrounding oil ambient equal to 90°) and others as "hydrophilic wall" (contact angle for a water drop in a surrounding oil ambient equal to 3°).
Initial conditions of the two phases and the position and shape of the oil inlet were set using the *setFields* and *snappyHexMesh* OpenFOAM utilities.

4. Experimental set-up, procedures and results

The numerical results were validated against experimental data obtained during previous campaigns. Details about the experimental setup, procedure and results that were used for the validation can be found in [6, 30, 31].

5. Investigated flow conditions and setting of the inlet boundary conditions

Among the available experimental flow conditions, eight representative cases were selected for the CFD analysis, with superficial and actual velocities of oil and water as reported in Table 1. All of them show a Core Annular Flow pattern.
The Arney correlation [13] was used to estimate the *in situ* volume fractions at the duct inlet. Such fractions were in turn used to split the inlet section into two parts – cells belonging to the oil core and cells belonging to the water annulus – and to set the cross-sectional averaged effective phase velocities as boundary conditions for the velocity in the two regions.
Concerning the subdivision of the inlet between oil and water, in addition to the area also the shape of the two regions has to be decided. A basic choice is to use a circle centered on the axis for the oil and a circular crown for the water, but this neglects the effect of buoyancy. So also different choices in terms of shapes and position of the oil inlet were tested, as it will be discussed in Section 6.
In terms of implementation, the oil and water inlet regions were set using the *blockMesh* and *snappyHexMesh* OpenFOAM utilities.

6. Result and Discussion

6.1. Pressure drop and pressure gradients

Preliminary simulations about single-phase water flow within the domain showed excellent agreement with the predictions of the classic theory about pressure drop along the constant-section ducts and within the Venturi flow meter, both in laminar and turbulent conditions.
For the two-phase flow simulations, Table 1 summarizes the majority of the obtained numerical results and their comparison with the experimental data, in terms of:

- pressure drop in the Venturi flow meter (calculated as the difference between the pressure immediately before the device and the pressure at the throat);
- pressure gradient in the upstream duct;
- pressure gradient in the downstream duct.

Additional cases, whose results showed no significant difference from the ones reported, are omitted in the table and will only be commented hereinafter.
As already said, the investigated flow conditions are identified in terms of superficial velocities of the phases J and of cross-sectional averaged actual velocities at the inlet w.
In the first case (DA-CM-NW-GL), the upstream and downstream ducts proved to be too short for a reliable calculation of the pressure gradients, as their lengths are not sufficient for the flow to develop after the inlet region and re-develop after the Venturi flow meter. Nevertheless, the pressure drop in the latter is already captured with discrete accuracy: on the eight investigated cases, the mean absolute percentage error (MAPE) is around 16 %.

Table 1. Investigated flow conditions and comparison between the numerical and experimental results for: a) Venturi pressure drop, b) pressure gradient in the upstream duct, c) pressure gradient in the dowstream duct.

DA-CM-NW-GL Domain A, coarse mesh, standard wall, "Gauss linear" scheme for γ
DB-CM-NW-GL Domain B, coarse mesh, standard wall, "Gauss linear" scheme for γ
DB-CM-HW-GL Domain B, coarse mesh, hydrophilic wall, "Gauss linear" scheme for γ
DB-CM-HW-IC Domain B, coarse mesh, hydrophilic wall, "Interface compression" scheme for γ
DC-FM-HW-GL Domain C, fine mesh, hydrophilic wall, "Gauss linear" scheme for γ

Case	J [m/s]		w [m/s]		Venturi pressure drop [kPa]						PE [-]				
	Oil	Water	Oil	Water	EXP	DA-CM-NW-GL	DB-CM-NW-GL	DB-CM-HW-GL	DB-CM-HW-IC	DC-FM-HW-GL	DA-CM-NW-GL	DB-CM-NW-GL	DB-CM-HW-GL	DB-CM-HW-IC	DC-FM-HW-GL
1	0.25	0.44	0.89	0.62	0.56	0.52	0.52	0.52	0.52	0.52	-7.1	-7.1	-7.1	-7.1	-7.1
2	0.46	0.44	1.09	0.77	1.04	0.86	0.88	0.86	0.88	0.87	-17.3	-15.4	-17.3	-15.4	-16.3
3	0.52	0.67	1.47	1.03	2.11	1.46	1.47	1.47	1.48	1.48	-30.8	-30.3	-30.3	-29.9	-29.9
4	0.64	0.67	1.60	1.12	2.06	1.78	1.80	1.80	1.81	1.82	-13.6	-12.6	-12.6	-12.1	-11.7
5	0.64	0.89	1.92	1.33	2.74	2.45	2.43	2.43	2.43	2.47	-10.6	-11.3	-11.3	-11.3	-9.9
6	0.64	1.10	2.24	1.55	3.50		3.19	3.19	3.18	3.22		-8.9	-8.9	-9.1	-8.0
7	0.75	0.67	1.70	1.20	2.19		2.10	2.09	2.10	2.12		-4.1	-4.6	-4.1	-3.2
8	0.75	0.89	2.02	1.41	2.93		2.79	2.79	2.78	2.82		-4.8	-4.8	-5.1	-3.8
										MAPE [-]	15.9	11.8	12.1	11.8	11.2

Case	J [m/s]		w [m/s]		Upstream pressure gradient [kPa/m]						PE [-]				
	Oil	Water	Oil	Water	EXP	DA-CM-NW-GL	DB-CM-NW-GL	DB-CM-HW-GL	DB-CM-HW-IC	DC-FM-HW-GL	DA-CM-NW-GL	DB-CM-NW-GL	DB-CM-HW-GL	DB-CM-HW-IC	DC-FM-HW-GL
1	0.25	0.44	0.89	0.62	0.05										
2	0.46	0.44	1.09	0.77	0.14										
3	0.52	0.67	1.47	1.03	0.31		0.38	0.38	0.38	0.39		22.4	22.4	22.1	26.0
4	0.64	0.67	1.60	1.12	0.38		0.45	0.45	0.45	0.46		18.4	18.4	18.0	19.9
5	0.64	0.89	1.92	1.33	0.52		0.59	0.59	0.58	0.59		12.5	12.5	10.4	12.2
6	0.64	1.10	2.24	1.55	0.67		0.74	0.74	0.73	0.73		9.2	9.2	8.1	8.8
7	0.75	0.67	1.70	1.20	0.44		0.52	0.52	0.51	0.52		18.0	17.9	16.2	18.1
8	0.75	0.89	2.02	1.41	0.61		0.66	0.66	0.64	0.66		7.9	7.9	5.5	8.8
										MAPE [-]		14.7	14.7	13.4	15.6

Case	J [m/s]		w [m/s]		Downstream pressure gradient [kPa/m]						PE [-]				
	Oil	Water	Oil	Water	EXP	DA-CM-NW-GL	DB-CM-NW-GL	DB-CM-HW-GL	DB-CM-HW-IC	DC-FM-HW-GL	DA-CM-NW-GL	DB-CM-NW-GL	DB-CM-HW-GL	DB-CM-HW-IC	DC-FM-HW-GL
1	0.25	0.44	0.89	0.62	0.07										
2	0.46	0.44	1.09	0.77	0.17										
3	0.52	0.67	1.47	1.03	0.31		0.74	0.26	0.36	0.31		141.0	-16.8	17.8	-0.1
4	0.64	0.67	1.60	1.12	0.39		0.79	0.33	0.55	0.37		102.7	-16.5	40.0	-4.1
5	0.64	0.89	1.92	1.33	0.50		0.39	0.37	0.53	0.45		-23.0	-27.2	4.9	-10.1
6	0.64	1.10	2.24	1.55	0.64		0.43	0.43	0.44	0.53		-33.9	-33.9	-31.3	-17.0
7	0.75	0.67	1.70	1.20	0.51		0.43	0.37	0.50	0.43		-16.1	-28.5	-3.5	-17.4
8	0.75	0.89	2.02	1.41	0.62		0.42	0.42	0.45	0.50		-31.9	-32.1	-27.5	-18.4
										MAPE [-]		58.1	25.8	20.8	11.2

The accuracy in the prediction of the concentrated pressure drop in the Venturi flow meter is improved when using the longer domains, with a reduction of the MAPE to around 12 % for domain B and to around 11 % for domain C. Given the domain, different settings (hydrophilic wall, Gauss linear or *interfaceCompression* schemes for the volume fraction, different shapes for the oil inlet, variations in the inlet boundary conditions for the main variables) cause practically no difference in the results. So it can be concluded that the prediction of the concentrated pressure drop in the Venturi meter is quite robust.

Concerning the pressure gradients, for the upstream duct the different investigated domains (B and C), mesh types and wall wettability (standard or hydrophilic wall) cause practically no difference. Slight improvement is offered by the use of the *interfaceCompression* scheme instead of the Gauss linear scheme for the volume fraction. The most significant observation is that the accuracy of the simulation is quite different between the investigated flow conditions. The numerical simulations always overestimate the experimental values, but no significant trend (e.g., improving/worsening of the prediction for increasing/decreasing phase velocities) could be identified.

Overall, the MAPE is around 14-15 % which in any case can be considered as satisfactory. For comparison, the uncertainties in the estimation of the two-phase pressure gradient when using experimental models and correlations usually reach much more than 20 %.

For the downstream duct, in two cases (case 3 and case 4 in Table 1) the oil touched the duct wall, while in the corresponding experimental tests Core Annular Flow without oil fouling was observed. This is due to the fact that for those cases, at low phase velocities, the buoyancy effect is more significant and the top part of the water annulus is very thin, so that the mesh is too coarse to keep oil away from the wall. To overcome this problem without being forced to refine the mesh, the wettability of the duct surface was modified in the simulations, changing the value of the contact angle in the volume fraction boundary condition from 90° to 3°, i.e. a super-hydrophilic surface was artificially set. This effectively prevents the oil from sticking to the wall and the results become more reliable. As the domain dimensions are such that capillary effects are negligible, this unrealistic boundary condition should have no influence on the simulations for all the cases in which water is correctly in contact with the duct wall. Comparison of the DB-CM-NW-GL and DB-CM-HW-GL columns in Table 1 a, b, c confirms this hypothesis: results for the Venturi pressure drop and upstream pressure gradient are practically identical, and only the downstream cases in which the oil with the DB-CM-NW-GL settings was touching the duct wall show a changed value.

Still focusing the attention on the downstream duct, it can be seen that in this case the choice of the discretization scheme for the volume fraction and the use of domains B and C imply a significant difference in the results.

Overall, the use of the *interfaceCompression* scheme offers an improvement, even if an even better improvement is given by the use of the longer domain C with a finer mesh.

With the latter, the MAPE is as low as 11 %, which once again can be considered as a very satisfactory value, particularly considering that for case DC-FM-HW-GL the results show a more consistent trend: they always underestimate the experimental results, with an error that increases for higher velocities of the phases (but still remaining under 20 %).

Further tests, not reported in Table 1, were also performed to investigate the effects of:

- a variation in the maximum admitted Courant number. The latter was increased to 0.5 with no effect on the results;
- variations in the values of the initial velocity field in the duct (still or already flowing water), of the turbulent kinetic energy and its dissipation rate, of the turbulent viscosity, of the oil viscosity (that was reduced of 20 %). All these factors were varied independently, leading to changes in the results lower than 1 %. In the case of the oil viscosity this may seem surprising, but it is physically correct because in a Core Annular Flow only water is in contact with the duct wall and thus experiences the vast majority of the friction;
- a change in the shape of the oil and water region at the duct inlet. Oil cores shaped as a circle centered on the duct axis, as an eccentric half-circle, ellipse or half-moon region were tested.

No effect was observed on the Venturi pressure drop, while the pressure gradients showed minor differences, but with no indication that a certain shape is the best choice;

- the use of the *k-ω* SST turbulence model instead of the *k-ε* model. Motivation for this additional test is the fact that the *k-ε* model was used with the standard wall functions, but in some cases the *y+* values are partially out from the correct range for use of the wall functions themselves. The average *y+* value for domain B with coarse mesh is around 28, and when using the fine mesh the value obviously becomes lower. Using coarser meshes near the duct wall would worsen the description of the water annulus and increase the risk of spurious oil fouling. So case 7 of Table 1 was tested with the DB-CM-HW-GL settings and the *k-ω* SST turbulence model. Contrarily to what was expected, results were worse than with the *k-ε* model, with errors equal to 13.2 % upstream and 52.2 % downstream. Further investigation about this aspect is therefore needed, also using further refined meshes that can be more suitable for the *k-ω* model.

6.2. In situ volume fraction

In addition to analysing the pressure evolution along the flow, the velocity profiles and the *in situ* volume fractions of the phases can be extracted from the simulation results to study the flow structure. Of particular interest for comparison with simplified 1D models and experimental correlations is the analysis of the cross-sectional averaged *in situ* volume fractions *H*, that can be also related to the slip ratio between the phase velocities. Figure 2 shows the results for the *in situ* volume fractions of water H_w for three different positions along the duct axis, compared to the predictions of the homogeneous model and of the Arney correlation [13]:

$$H_w = \varepsilon_w [1 + 0.35 (1 - \varepsilon_w)] \qquad (2)$$

where ε_w is the water volume ratio, i.e. the ratio between the water volume flow rate and the total volume flow rate at the duct inlet.

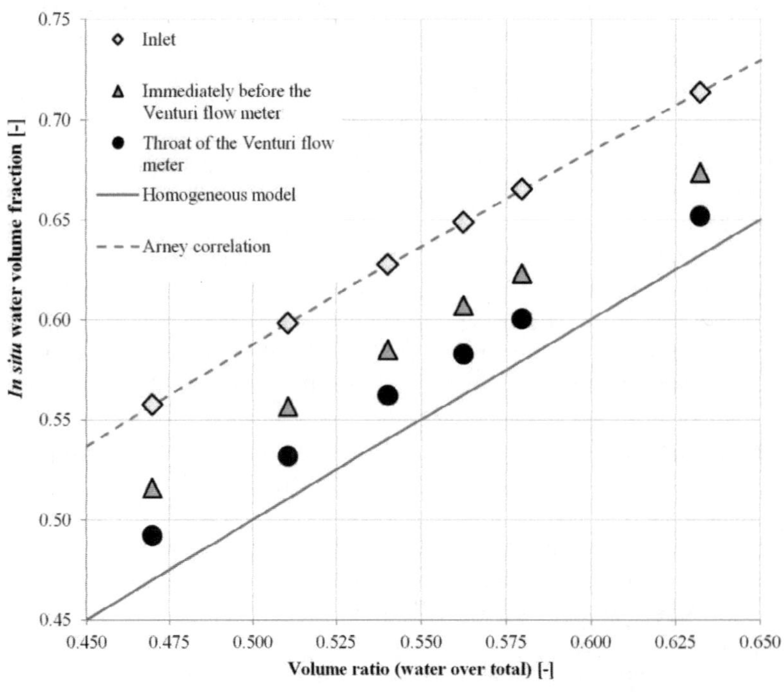

Figure 2. *In situ* water fraction calculated from the numerical simulations, compared to the predictions of the homogeneous model and the Arney correlation.

At the inlet the values are obviously superposed to the latter, as it was such a correlation to be used to set the oil and water regions. Then an evolution occurs and at the Venturi flow meter throat the values are much nearer to the predictions of the homogeneous model, confirming previous results both experimental [31] and numerical [21]. As already reported in [21] this should be related to the fact that in the Venturi water accelerates more than oil (due to the higher viscosity of the latter), so that the difference between the phase velocities is reduced. This also allows using the Venturi flow meter for the determination of the flow rates of the phases [31][32].

7. Conclusions

The performed CFD simulations evidenced that the *interFoam* solver of OpenFOAM allows predicting the pressure drop of oil-water two-phase flows along straight pipes and within a Venturi flow meter with good accuracy. With the used domains and settings, the deviations with respect to experimental data, in terms of MAPE, are within 11-12 % for the concentrated pressure drop, around 14-15 % for the pressure gradient upstream of the Venturi flow meter and around 11 % downstream of the same. Simulations can be performed on a common personal computer in a reasonably short time. Results are quite robust with respect to the settings of the inlet conditions, in terms both of cross-sectional shape of the inlet regions for the two phases and of boundary conditions for the turbulence variables, and the oil physical properties. Prediction of the pressure drop within the Venturi flow meter is also very robust with respect to the length of the investigated domain; pressure gradients along the upstream and downstream ducts are on the contrary more sensitive to the latter variable. A very good agreement was additionally found between the calculated *in situ* volume fractions and the predictions of the Arney correlation. As some investigated cases showed much larger errors (up to 30%) and the use of the k-ω SST model with respect to the k-ε model implied significantly different results, further studies are needed to investigate the modeling options more thoroughly.

8. Nomenclature

H	*in situ* volume fraction (-)
J	superficial velocity (m/s)
k	turbulent kinetic energy (J/kg)
MAPE	mean absolute percentage error (%)
PE	percentage error (%)
w	actual velocity (m/s)

Greek Symbols

ε	volume ratio (-)
ε	turbulent dissipation (J/kg-s)
γ	volume fraction for the VoF method (-)
μ	dynamic viscosity (Pa s)
ρ	density (kg/m^3)
τ	time (s)
ω	turbulence specific dissipation rate (1/s)

Subscripts and superscripts

o	oil
w	water

References

[1] IEA International Energy Agency 2013 *Resources to Reserves 2013* (available online at: http://www.iea.org/publications/freepublications/publication/Resources2013.pdf)
[2] Gordon D 2012. Understanding Unconventional Oil *The Carnegie papers, Energy and climate* (available online at: http://carnegieendowment.org/files/unconventional_oil.pdf)
[3] Saniere A, Hénaut I, Argillier J F 2004 *Oil & Gas Science and Technology – Rev. IFP* **59**(5) 455–66
[4] Oliemans R, Ooms G 1986 Core annular flow of oil and water through a pipeline. In: Hewitt G F, Delhaye J M, Zuber N (Eds.) *Multiphase Science and Technology* (Hemisphere Publishing Corp., Washington) **2**
[5] Joseph D, Bai R, Chen K P and Renardy R 1997 *Annu. Rev. Fluid Mech.* **29** 65–90
[6] Sotgia G, Tartarini P and Stalio E 2008 *Int. J. Multiphas. Flow* **34** 1161–74

IOP Conf. Series: Journal of Physics: Conf. Series **1224** (2019) 012008 doi:10.1088/1742-6596/1224/1/012008

[7] Atkinson I et al 2004/2005 A New Horizon in Multiphase Flow Measurement. (available online at: http://www.slb.com/resources/publications)
[8] Wang W, Gong J, Angeli P 2011 *Int. J. Multiphas. Flow* **37** 1156–64
[9] Charles M E, Govier G W and Hodgson G W 1961 *Can. J. Chem. Eng.* **39**(1) 27–36
[10] Ooms G, Segal A, Van Der Wees A J, Meerhoff R and Oliemans R V A 1984 *Int. J. Multiphas. Flow* 10 41–60
[11] Oliemans R V A, Ooms G, Wu H L and Duijvestijn A 1987 *Int. J. Multiphas. Flow* **13**(1) 23–31
[12] Brauner N 1991 *Int. J. Multiphas. Flow* **17**(1) 59–76
[13] Arney M S, Bai R, Guevara E, Joseph D D and Liu K 1993 *Int. J. Multiphas. Flow* **19**(6) 1061–76
[14] Angeli P, Hewitt G F 2000 *Int. J. Multiphas. Flow* 26 1117–40
[15] Colombo L P M, Guilizzoni M and Sotgia G M 2012 *Exp. Fluids* **53** 1617–25
[16] Colombo L P M, Guilizzoni M, Sotgia G M and Marzorati D 2015 *Int. J. Heat Fluid Fl.* **53** 91–97
[17] Joseph D, Renardy Y 1993 *Fundamentals of Two-Fluid Dynamics* (Springer, New York US)
[18] OLGA Dynamic Multiphase Flow Simulator (available online at: https://www.software.slb.com/products/olga)
[19] Ghosh S, Das G and Das P K 2010 *J. Chem. Eng. Process* **49** 1222–28
[20] Kaushik V V R, Ghosh S, Das G and Das P K 2012 *J. Pet. Sci. Eng.* **86–87** 153–64
[21] Babakhani Dehkordi P, Colombo L P M, Guilizzoni M and Sotgia G M 2017 *J. Pet. Sci. Eng.* **149** 540–52
[22] Hirt C W, Nichols B D 1981 J. Comp. Phys. **39**(1) 201–25
[23] Desamala A B, Vijayan V, Dasari A, Dasmahapatra A K and Mandal T K 2016 *J. Hydrodyn. Ser. B* 28 (4) 658–68
[24] Menard T, Tanguy S and Berlemont A 2007 *Int. J. Multiphas. Flow* **33**(5) 510–24
[25] Guilizzoni M, Santini M, Lorenzi M, Knisel V, Fest-Santini S 2014 *J. Phys.: Conf. Ser.* **547** 012028
[26] Brackbill J U, Khote D B and Zemach C 1992 *J. Comput. Phys.* **100**(2) 335–354
[27] Harvie D J E, Davidson M R and Rudman M 2006 *Appl. Math. Model.* **30** 1056–66
[28] Deshpande S S, Trujillo M F, Wu X and Chahine G 2012 *Int. J. Heat Fluid Flow* **34**(1) 1–14
[29] Versteeg H K, Malalasekera W 2007 *An Introduction to Computational Fluid Dynamics: The Finite Volume Method* (Pearson)
[30] Colombo L P M, Guilizzoni M, Sotgia G M, Babakhani Dehkordi P and Lucchini A 2017 *J. Phys.: Conf. Ser.* **923** 012012
[31] Colombo L P M, Guilizzoni M, Sotgia G M 2017 *Proc. 14th Int. Conf. on Multiphase Flow in Industrial Plant*, Desenzano del Garda (Brescia), Italy, Sept. 13 – 15, 2017
[32] Patent request Nr 102017000025126, “Procedimento e dispositivo per la misura delle portate dei componenti non miscibili di un flusso bifase liquido-liquido, Ministero dello Sviluppo Economico, 2017/03/07.

Modeling heat conduction in open-cell metal foams by means of the Three-Dimensional Thermal Fin theory

M Iasiello[1*], C Savarese[1], P J Damian[2], N Bianco[1], A Andreozzi[1], W K S Chiu[2] and V Naso[1]

[1] Dipartimento di Ingegneria Industriale, Università degli studi di Napoli Federico II, P.le Tecchio 80, 80125, Napoli, Italy
[2] Department of Mechanical Engineering, University of Connecticut, 191 Auditorium Road, 06269, Storrs (CT), USA

[*] E-mail: marcello.iasiello@unina.it

Abstract. Their relatively high thermal conductivity makes metal open-cell foams promising heat transfer enhancers for lightweight applications. Conduction occurs through both solid and fluid phases, depending on the material and the microstructure. The effective conductivity of a foam is predicted by means of analytical and numerical methods, based on both idealized and realistic foam geometries, as well as by empirical correlations based on experiments. In this paper, a new model for the prediction of the effective thermal conductivity of foams is proposed. The proposed Three-Dimensional Thermal Fin (TTF) theory is based on the Electrochemical Fin (ECF) theory previously applied to the analysis of ion and electron transport through nanoporous materials. It has computational time two – three times shorter than those of current numerical techniques, such as finite-element methods. Starting from a tomography-reconstructed metallic open cell foam, the 3-D structure is modeled as a network. Conventional fin heat transfer scaled equations, applied to each part of the network, allow for the evaluation of the temperature distribution and heat rates. The effective thermal conductivity of the metal foam is then derived. The proposed model is validated by comparison with experimental results as well as with predictive models from the open literature and a finite-element based model.

Nomenclature

A	area, m^2	R_{ct}	unit charge transfer resistance, Ω m^2
B, c	coefficients in equation (3.3)	SQ	sintering quality
C	coefficient in equation (3.4)	T	temperature, K
$1/\bar{h}$	unit surface thermal resistance, m^2 K W^{-1}	V	volume, m^3
i	electrical current, A	x	strut axial coordinate, m
k	thermal conductivity, W m^{-1} K^{-1}	*Greek symbols*	
l	fin length, m	ε	porosity
L	length of the computational domain, m	ρ	resistivity, Ω m^{-1}
L^*	sample scaled size	σ_{io}	ionic conductivity, S m^{-1}
$\dot{Q}$	heat rate, W	φ	voltage, V

Subscript		*max*	maximum
b	fin base	*min*	minimum
c	local cross section	*s*	solid
ct	charge transfer	*surf*	fin surface
eff	effective	*Superscript*	
f	fluid	*	non-dimensional
io	ionic		

1.Introduction

A foam is a cellular structure made up by a network of solid struts, which are the edges and faces of cells containing a large volume fraction of fluid. The foam structure is generally represented as a three-dimensional polyhedral structure whose pores either are sealed or form an interconnected network. The shape of the cross-section of this finite-dimension network depends on the manufacturing technique, that can also make the cross section of the strut variable.

With reference to their structure, metal foams are classified as open-cell foams or closed-cell foams. Open-cell foams consist of interconnected cells that allow a fluid to pass through, whereas in a closed-cell foam the cells are single enclosures within the material. In other words a cell is the unit that repeats through the space; a pore is the window between adjacent cells. The selection of the foam's solid material is generally based on the end-application. Because of their mechanical characteristics [1], closed-cell foams are mostly used in structural, load-bearing applications. Open-cell foams have good thermal properties [1, 2] and are currently used in many applications, such as automotive and biomedical applications or compact heat exchangers and solar air receivers [2 - 4].

Heat conduction in foams is currently investigated experimentally, analytically and numerically. The effective thermal conductivity of a foam is defined as the equivalent thermal conductivity of the two-phase porous medium. Different experimental techniques were employed. Bhattacarya et al.[5] carried out experiments by brazing metal foams between two aluminum plates at different uniform temperatures, measured by four thermocouples, attached to each plate. Analytical models were developed, based on idealized foam geometries. Boomsma and Poulikakos [6] and Yang et al. [7] presented analytical predictive models based on a tetrakaidekahedron geometry. Predictions of both studies were in good agreement with experiments for porosities higher than 0.90. The numerical solution of governing equations allows for the prediction of the effective thermal conductivity of more complex geometries [8, 9].

The Electrochemical Fin (ECF) theory is an analytical method used in the microstructural analysis of $Sr_2Fe_{1.5}Mo_{0.5}O_{6-\delta}$ (SFM), a redox stable solid oxide fuel cell (SOFC) electrode [10]. By using digital image processing tools, the microstructure is separated into an underlying network, made up of individual segments, that behave analogously to connected thermal fins. The fins are characterized to determine the parameters to be used in the ECF theory. The potential distribution along the network segments is obtained by the analytical solution of a system of one-dimensional ordinary differential equations developed from charge balances. With ECF theory, the electrode structure is developed as a complex combination of electrical resistances with different geometries. Compared to traditional techniques, such as the finite-element based ones, the above described approach leads to significant savings in computational time and cost, say two or three times [11].

In this work, starting from the ECF theory, a new tool for the prediction of the effective thermal conductivity of metal foams, the Three-Dimensional Thermal Fin (TTF) theory, is first presented. The electrical analogy for heat conduction is employed to extend the ECF theory to conduction heat transfer problems. The geometry of a 0.94 nominal porosity and 40 PPI open-cell aluminum foam is reconstructed by using x-ray computed tomography. The TTF theory is applied on the skeletonized tomographic image. The code is validated comparing its predictions with those given by models and experiments taken from the open literature as well as with predictions from COMSOL Multiphysics code. The comparisons validated the reliability of the TTF model and demonstrated it to be a promising tool for the prediction of conduction heat transfer through foam materials.

2. Mathematical model

2.1. Tomographic scans of the sample

The 0.94 nominal porosity and 40 PPI open-cell aluminium foam sample analyzed in the present work was manufactured by ERG Aerospace. The sample is a 10 mm long cylinder with a 10 mm diameter. The sample geometry was described by x-ray tomography carried out with an Xradia MicroXCT-400 scanner. A 17.8 µm (1x lens) resolution was chosen in order to highlight the foam strut shape. Ambrosio et al [12] showed that higher resolutions are not needed since they provide with similar CT-data morphological features. A filtered back-projection software provided by Xradia was employed to perform the 3-D object reconstruction [13].

2.2. The ECF and TTF theories

Under the assumption that the current flow continuously passes through the struts, that allows to neglect the effects of the space charge regions on the electrode performance as well as the assumption of one-dimensional Ohmic conduction in the solid phase (conduction occurs only along the fin axial coordinate, x), the governing equation for the ECF theory is [11]

$$\frac{d^2\varphi}{dx^{*2}}+\frac{1}{A_c}\left(\frac{dA_c}{dx^*}\right)\left(\frac{d\varphi}{dx}\right)-\frac{1}{\sigma_{io}\,R_{ct}\,A_c}\left(\frac{dA_{surf}}{dx}\right)\varphi=0 \tag{2.1}$$

The comparison of predictions from the ECF theory with those given by finite element methods showed that the Ohmic conduction could reliably be assumed to be one-dimensional [14].

A key benefit of the above discussed analytical approach is the possibility to scale equations in order to account for the morphology and performance of SOFC electrode microstructures. The interplay between microstructural morphology, material properties and operating conditions is described by a set of basic non-dimensional parameters: a non-dimensional length, $x^* = x/l$, a non-dimensional voltage, $\varphi^* = \varphi/\varphi_b$, two non-dimensional areas, $A_c^* = A_c/A_{c,b}$ and $A_{surf}^* = A_{surf}/A_{c,b}$.

A non-dimensional form of equation (2.1) is now derived

$$\frac{d^2\varphi^*}{dx^{*2}}+\frac{1}{A_c^*}\left(\frac{dA_c^*}{dx^*}\right)\left(\frac{d\varphi^*}{dx^*}\right)-\frac{1}{\sigma_{io}\,R_{ct}\,A_c^*}\left(\frac{dA_{surf}^*}{dx^*}\right)\varphi^*=0 \tag{2.2}$$

The number of variables in equation (2.2) is further reduced with scaling techniques. To this aim, reference is made to some significant parameters describing the electrode morphology and charge transfer characteristics.

The first parameter is the ionic electrical resistivity ratio

$$\frac{\rho_{io}}{\rho_{ct}}=\frac{l^2}{\sigma_{io}\,R_{ct}}\left(\frac{1}{A_c}\right)\left(\frac{dA_{surf}}{dx}\right)\approx\frac{l^2}{\sigma_{io}\,R_{ct}}\left(\frac{A_{surf}}{V}\right) \tag{2.3}$$

that is the ratio of the ionic electrical resistivity of the electrode to the charge transfer electrical resistivity. It depends on the electrode thickness, on the variable fin cross-sectional area and fin surface area. As an operator on the normalized potential term, the ionic electrical resistivity ratio represents a non-dimensional surface charge transfer coefficient that quantifies the balance between the rate of conductive charge transfer through the solid structure to the chemically active surface and

IOP Conf. Series: Journal of Physics: Conf. Series **1224** (2019) 012009 doi:10.1088/1742-6596/1224/1/012009

the rate of charge transfer from the chemically active surface. It can be recognized as the coefficient of the third term on the left-hand side of equation (2.2).

The second parameter, the sintering quality, can be derived from the coefficient of the second term on the left-hand side of equation (2.2)

$$SQ = \frac{1}{\left(\frac{l}{A_c}\right)\left(\frac{\mathrm{d}A_c}{\mathrm{d}x}\right)} = \left(\ln\frac{A_{c,\min}}{A_{c,\max}}\right)^{-1} \tag{2.4}$$

that accounts for the variation in the cross-sectional area along the length of an electrode. The partitioning of the solid phase and the consequent necessity of morphological and topological information for the analytical modelling approach imply to make reference to the average values of the sintering quality and resistivity ratio for each segment of the structure. As to the sintering quality, this implies that the profiles which can be rigorously treated are only the exponential variations of the cross-section. The above profiles require to measure only the intra-phase interfacial surfaces $A_{c,min}$ and $A_{c,max}$. As to the resistivity ratio, the geometric factor in parentheses is approximated using the measured surface area to volume ratio (see the second term in equation (2.3)).

The final form of the scaled equation is

$$\frac{d^2\varphi^*}{dx^{*2}} + \frac{1}{SQ}\left(\frac{d\varphi^*}{dx^*}\right) - \frac{\rho_{io}}{\rho_{ct}}\varphi^* = 0 \tag{2.5}$$

It is worth underlining that the assumption of an exponential variation of the strut cross section along its length fits well for open-cell foams, as it is shown in figure 1, where the strut magnification of a 0.94 nominal porosity 40 PPI aluminum-foam is reported and the strut surface is highlighted in yellow.

Starting from the ECF theory, the analogy between thermal and electrical transport allows one to extend the ECF theory to the study of heat conduction in metal foams. The thermoelectric analogy between Fourier's and Ohm's laws is summarized in table 1. The non-dimensional governing equation of the ECF theory, equation (2.2), also holds for heat transfer. Continuous temperature and heat flux across each junction are assumed. Temperature differences as small as to make the dependence of the thermal conductivity on the temperature negligible are imposed.

2.3. Foam skeletonization

The generation of the skeleton starting from the tomography of the of 0.94 nominal porosity 40 PPI

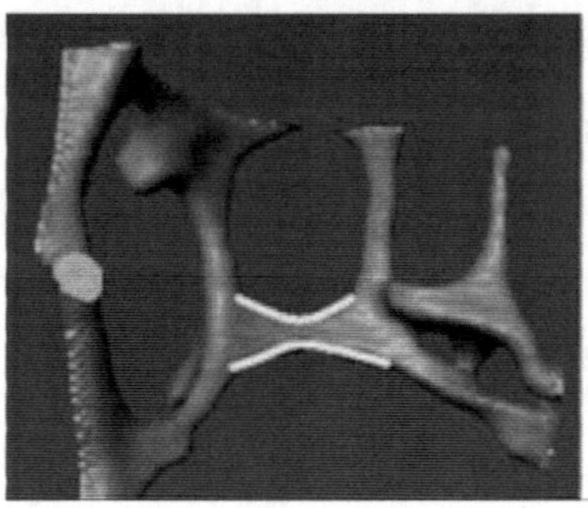

Figure 1. Strut magnification of a 0.94 nominal porosity 40 PPI aluminum-foam.

IOP Conf. Series: Journal of Physics: Conf. Series **1224** (2019) 012009 doi:10.1088/1742-6596/1224/1/012009

Table 1. Thermoelectric analogy.

Ohm's law (Electrical) $i = \sigma_{io}\ A\ \Delta\varphi / l$		Fourier's law (Thermal) $Q = k\ A\ \Delta T / l$	
i	Electrical current, A	$\dot{Q}$	Heat rate, W
$\Delta\varphi$	Voltage difference, V	ΔT	Temperature difference, K
σ_{io}	Ionic conductivity, S m^{-1}	k	Thermal conductivity, W m^{-1} K^{-1}
R_{ct}	Unit charge transfer resistance, Ω m^2	$1/\bar{h}$	Unit surface thermal resistance, m^2 K W^{-1}

metal foam is presented in figure 2. The skeletonization process is performed by means of postprocessing tools like Avizo and MATLAB. The process starts with the reconstruction of the X-ray microtomography data to generate a grayscale tomogram (figure 2a), and adjusted by using a Gaussian blur filter to improve the contrast in the image. A thresholding of a region of interest (ROI) within the grayscale image (figure 2b) is performed using the Otsu method. Erosion/dilation operations are also performed to clean the surface. Finally, a skeletonization algorithm generates a microstructural skeleton (figure 2c), which is further partitioned into a network. The particles in the network are characterized to yield fin morphological parameters and the network performance is subsequently assessed (figure 2d).

3. Results and discussion

3.1. Data reduction

Under the previously described appropriate boundary conditions, a COMSOL finite element code generates output data like heat flux and heat rate. Based on the boundary conditions, the effective

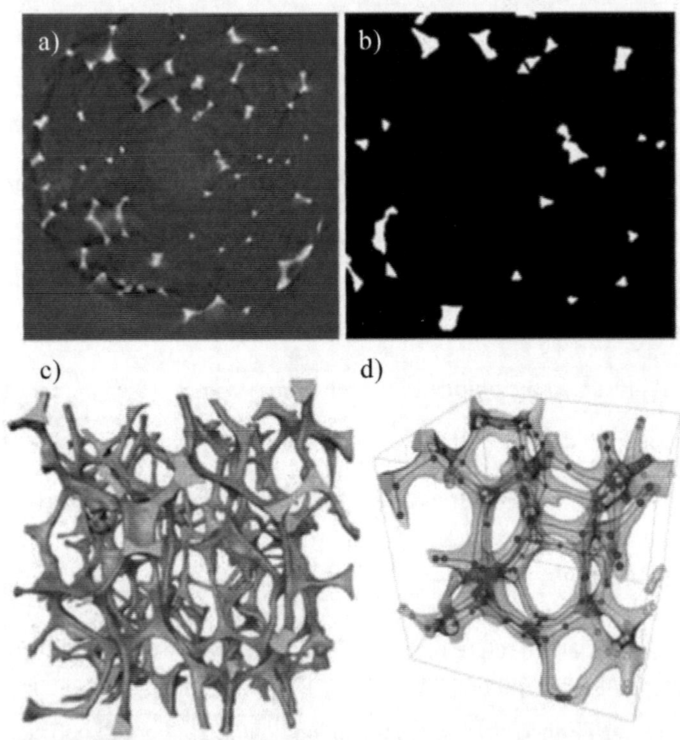

Figure 2. Skeleton work-flow: a) grey-scale image; b) thresholded image; c) reconstructed skeleton; d) network.

thermal conductivity can be determined by assuming heat transfer along each of the three principal directions in the foam domain. Convection heat transfer is neglected, thus only conduction through the foam is accounted for. Under the above assumptions, the Fourier law allows to derive the effective thermal conductivity of each phase

$$k_{eff} = k_{eff,s} + k_{eff,f} = \frac{QL}{A_s \Delta T} + \varepsilon k_f = \frac{QL}{(1-\varepsilon) A \Delta T} + \varepsilon k_f \quad (3.1)$$

where the heat rate is given as a code output and the porosity, as well as the cross-section area, are taken from imaging tools. With reference to equations (2.3) and (2.5), it is worth remarking that, the effective thermal conductivity obtained with the proposed Three-Dimensional Thermal Fin theory accounts also for convective heat transfer effects, since the model includes the unit charge transfer resistance, R_{ct}. Convection heat transfer was assumed to be negligible, in order to predict overall heat transfer in the struts without carrying out the expensive simulation of the foam fluid phase.

3.2. Comparisons with literature data

3.2.1. Experiments. The average values of the effective thermal conductivities in the three directions from the TTF theory and the experimental data presented in [5, 15], for a 0.94 porosity and 40 PPI foam, are reported in table 2. Since few experimental data for 0.94 porosity foams were found in the literature, also data for 0.94 ± 1% porosity have been reported in the table. It is important to remark that experimental data do not provide information on the anisotropy of the analyzed samples. The importance of anisotropy will be pointed out in Subsection 3.3. The good agreement with experimental data validates the feasibility of the herein proposed TTF theory.

3.2.2. Predictive models. Predictions from the TTF theory are now compared with those from models available in the open literature and a finite element code. Lemlich [16] derived analytically the correlation

$$k_{eff} = \left(\frac{1-\varepsilon}{3}\right) k_s + \varepsilon k_f \quad (3.2)$$

Calmidi and Mahajan [17], on the basis of experimental results, derived the correlation

$$k_{eff} = k_f \left[\varepsilon + B(1-\varepsilon)^c k_s / k_f \right] \quad (3.3)$$

with the coefficients B = 0.181 and c = 0.763 for air, with 6.9% maximum and 3.7% average deviations.

Bhattacarya et al. [5] presented the empirical correlation

$$k_{eff} = C\left[\varepsilon k_f + (1-\varepsilon) k_s \right] + \frac{1-C}{\varepsilon / k_f + \left[(1-\varepsilon)/k_s \right]} \quad (3.4)$$

with the best fit for C = 0.35, with R^2 = 0.97.

Table 2. Average value of the effective thermal conductivities in the three directions from the TTF theory and experimental data presented in [5, 15] for a 0.94 porosity and 40 PPI foam.

Average value from the TTF theory k_{eff} (W m^{-1} K^{-1})	Experimental data [5] k_{eff} (W m^{-1} K^{-1})	Experimental data [15] k_{eff} (W m^{-1} K^{-1})
3.9	4.5	3.8

IOP Conf. Series: Journal of Physics: Conf. Series **1224** (2019) 012009 doi:10.1088/1742-6596/1224/1/012009

Numerical simulations on the same scanned foam were also carried out. The binary data, sketched in figure 2b, was imported into MATLAB and a tetrahedral mesh for the solid phase was built up by using the open-source tool iso2mesh. The heat conduction equation was solved with the finite-element commercial code COMSOL Multiphysics, by assuming the solid/fluid internal walls and four boundaries to be adiabatic; the remaining two boundaries were assumed to be at uniform temperatures, 20.5 °C and 20.0 °C, respectively. The above boundary conditions make the problem one-dimensional and the effective thermal conductivity is obtained from the calculated heat rate. The reason of the adiabatic solid/fluid internal walls assumption is the four order of magnitude difference between aluminum and air thermal conductivities, which makes negligible the heat transferred through the fluid phase.

The average of the effective thermal conductivities in the three directions from the TTF theory and the percent differences among it and those predicted by equations (3.2 - 3.4) for a 0.94 porosity 40 PPI foam are reported in table 3. The higher percent differences between TTF theory predictions and predictions from [5] and [16] is likely due to different values of the effective thermal conductivity when evaluated along the stretch direction, due to strut elongation during the manufacturing process. This anisotropic effect will be further investigated in the next subsection.

3.3. Anisotropy analysis

The effective thermal conductivity was observed to depend on the directions on which it is computed [8]. The effective thermal conductivities in the three directions, their average value and the value predicted by the analytical model [16] as a function of the scaled sizes of cubic samples, are presented in figure 3. The scaled size of a cubic sample, L^*, is the ratio of the sample size to the pore size taken from [12]. Figure 3 shows that, above a certain cubic sample scaled size, the effective thermal conductivity can be considered to be nearly independent of the scaled size of the sample.

A value of $L^* = 4.8$ can, therefore, be assumed as the scaled size of cubic sample that guarantees the invariance of the foam effective thermal conductivity [18]; the maximum difference among the values predicted in the three directions is 33%. The figure also exhibits a more marked anisotropy along the direction 3, which, as it is shown in figure 4, is the foam stretching direction.

The comparison of predictions from TTF theory along the three directions with values predicted by [16] shows that direction 3 provides values closer to those given by [16]. The effective thermal conductivity in the direction 3 is affected by the viscous and gravity forces during manufacture. One can remark that, for $L^* \geq 4.8$, the minimum differences are those between predictions from [16] and those along direction 3; differences with average TTF theory predictions are larger while the maximum differences are exhibited in comparison with directions 1 and 2. This means that the average value might be a good compromise in predicting the effective thermal conductivity.

Visual evidence of foam stretch is shown by the tomographic scan of a 0.94 porosity and 40 PPI aluminum foam cell reported in figure 4. The yellow-highlighted cells show that the foam is stretched along the direction 3. The cell elongation in direction 3 is suspected to be due to viscous and gravity forces during the foaming process.

Table 3. Averaged effective thermal conductivities in the three directions from the TTF theory and percent differences among it and those from other models for a 0.94 porosity 40 PPI foam.

Averaged effective thermal conductivities in three directions from TTF theory k_{eff} (W m^{-1} K^{-1})	Percent differences in the average effective thermal conductivity from TTF theory and those from other models		
	Bhattacarya et al. [5]	Lemlich [16]	COMSOL
3.7	23.5	19.1	4.4

IOP Conf. Series: Journal of Physics: Conf. Series **1224** (2019) 012009 doi:10.1088/1742-6596/1224/1/012009

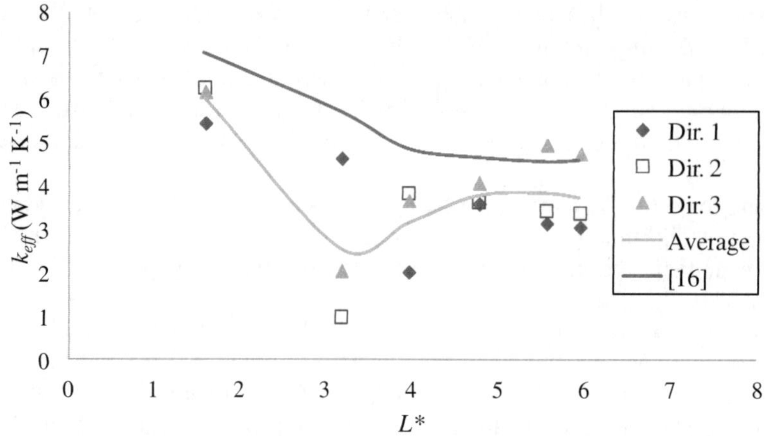

Figure 3. The effective thermal conductivities predicted by TTF in the three Cartesian directions, their average value, and values predicted by [16] *vs*. the scaled sizes of cubic samples for a 0.94 and 40 PPI porosity foam.

4. Conclusions

A new tool for the prediction of the effective thermal conductivity of metal foams, the Three-Dimensional Thermal Fin (TTF) theory has been presented. It extends the Electrochemical Fin (ECF) theory to conduction heat transfer problems. Its computational efficiency is attractive when the effective thermal conductivity of foams must be iteratively evaluated.

The temperature distribution and heat rates in a 0.94 nominal porosity 40 PPI open-cell aluminium foam sample have been predicted employing the TTF theory. Then the effective thermal conductivity of the foam has been predicted. The TTF theory has been validated by comparison to published experimental data and predictions from models taken from the open literature as well as with predictions from COMSOL Multiphysics finite element simulations.

Differences between the average value of the effective thermal conductivities along the three orthogonal directions predicted by TTF theory and experimental data were about 20%, similar to those compared with predictions from models taken from the literature. The differences could likely depend on the foam anisotropy, caused by viscous and gravity forces during the foaming process. Far smaller differences were found in the comparison of TTF with COMSOL predictions.

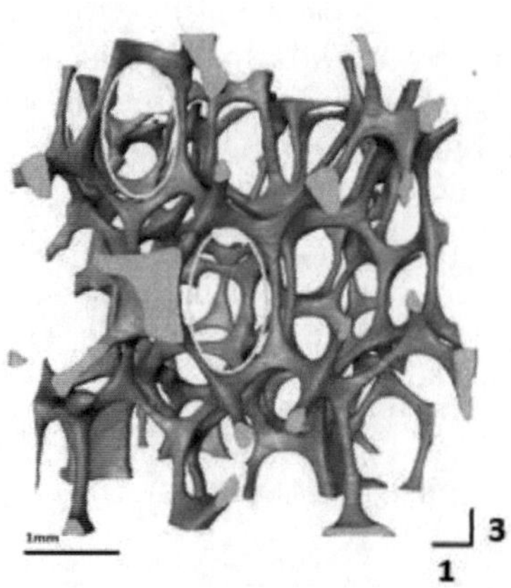

Figure 4. Tomographic scan of a 0.94 porosity aluminum foam cell in the 1 - 3 plane (2 is orthogonal to the 1 - 3 plane).

IOP Conf. Series: Journal of Physics: Conf. Series **1224** (2019) 012009 doi:10.1088/1742-6596/1224/1/012009

The foam anisotropy has also been analysed and it is found to be larger along the foam stretch direction. The best agreement was found between the TTF theory predictions along the stretching direction and predictions from [16]. Since the agreement between the average value of TTF theory predictions along three orthogonal directions is also rather satisfactory, we can conclude that the average value might be a good compromise in predicting the effective thermal conductivity of the foam.

In the future, authors will also perform simulations both at equal computational cost and equal time, in order to quantify savings, when TTF is used instead of a finite element method.

References

[1] Dukhan N ed 2013 *Metal foams: fundamentals and applications* DEStech Publications, Inc
[2] Odabaee M and Hooman K 2012 Metal foam heat exchangers for heat transfer augmentation from a tube bank, *Appl. Thermal Eng.* **36** 456-63
[3] Andreozzi A, Bianco N, Iasiello M and Naso V 2017 Numerical study of metal foam heat sinks under uniform impinging flow J. *Physics: Conference Series* **796(1)** 012002
[4] Fend T, Reutter O, Bauer J and Hoffschmidt B 2004 Two novel high-porosity materials as volumetric receivers for concentrated solar radiation *Solar energy materials and solar cells* **84(1-4)** 291-304
[5] Bhattacharya A, Calmidi V V and Mahajan R L 2002 Thermophysical properties of high porosity metal foams *Int. J. Heat Mass Transfer* 45 1017–31
[6] Boomsma K and Poulikakos D 2001 On the effective thermal conductivity of a three-dimensionally structured fluid-saturated metal foam *Int. J. Heat Mass Transfer* **44(4)** 827-36
[7] Yang X H, Bai J X, Yan H B, Kuang J J, Lu T J and Kim T 2014 An analytical unit cell model for the effective thermal conductivity of high porosity open-cell metal foams *Trans. porous media* **102(3)** 403-26
[8] Bodla K K Murthy J Y and Garimella S V 2010 Microtomography-based simulation of transport through open-cell metal foams *Num. Heat Transfer, Part A: Applications* **58(7)** 527-44
[9] Amani Y, Takahashi A, Chantrenne P, Maruyama S, Dancette S, and Maire E 2018 Thermal conductivity of highly porous metal foams: Experimental and image based finite element analysis *Int. J. Heat Mass Transfer* **122** 1-10
[10] Nelson G J, Peracchio A A, and Chiu W K S 2011 Analytical investigations of varying Cross section microstructures on charge transfer in solid oxide fuel cell electrodes *J. Power Sources* **196** 4695–704.
[11] Nelson G J, Nakajo A, Cassenti B N, DeGostin M B, Bagshaw K R, Peracchio A A and Chiu W K 2014 A rapid analytical assessment tool for three dimensional electrode microstructural networks with geometric sensitivity *J. Power Sources* **246** 322-34
[12] Ambrosio G, Bianco N, Chiu, W K, Iasiello M, Naso V and Oliviero M 2016 The effect of open-cell metal foams strut shape on convection heat transfer and pressure drop *Appl. Thermal Eng.* **103** 333-43
[13] Izzo J R, Joshi A S, Grew K N, Chiu W K, Tkachuk A, Wang S H and Yun W 2008 Nondestructive reconstruction and analysis of SOFC anodes using X-ray computed tomography at sub-50 nm resolution *J. Electrochemical Soc* **155(5)** B504-8
[14] Nelson G J, Cassenti B N, Peracchio A A and Chiu K S 2012 *J. Power Sources* **205** 48-56
[15] Bianchi E, Heidig, T, Visconti C, Groppi G, Freund H and Tronconi E 2012 An appraisal of the heat transfer properties of metallic open-cell foams for strongly exo-/endothermic catalytic processes in tubular reactors *Chem. Eng. J.* **198–199** 512–28
[16] Lemlich R 1978 A theory for the limiting conductivity of polyhedral foam at low density J. colloid and Interface Science **64(1)** 107-10

[17] Calmidi V and Mahajan R 1999 The effective thermal conductivity of high porosity fibrous metal foams *J. Heat Transfer* **121** 466–71

[18] Harris W M and Chiu W K S 2015 Determining the representative volume element size for three-dimensional microstructural material characterization, Part 1: Predictive models *J. Power Sources* **282**, 552-61

IOP Conf. Series: Journal of Physics: Conf. Series **1224** (2019) 012010 doi:10.1088/1742-6596/1224/1/012010

Numerical modelling of flashing flow phase change in convergent-divergent nozzle: A sensitivity analysis

Quang DANG LE, Riccardo MEREU, Giorgio BESAGNI, Vincenzo DOSSENA and Fabio INZOLI

Department of Energy, Politecnico di Milano, via Lambruschini 4, 20156 Milan, Italy

riccardo.mereu@polimi.it

Abstract. A sensitivity analysis of a computational fluid dynamics (CFD) model of flashing flow phenomenon including the thermal non-equilibrium effect is proposed in the present paper. The model uses a two-phase mixture approach and the phase-change process is based on the difference between the vaporization pressure and the vapour partial pressure. Thermal non-equilibrium effect is included in the model by a sub-routine for the boiling delay. A two-dimensional axisymmetric convergent-divergent nozzle is used as benchmark for the proposed study. Such geometry is representative of well-known applications in nuclear, refrigeration and energy engineering (e.g., primary flow in the motive nozzle of ejectors). The results are compared to experimental data and previous numerical results available in literature. Present results show good agreement with global and local experimental fluid dynamic quantities used to validate the model. The paper includes in the first part a brief description of physical phenomenon of flashing flow, the experimental benchmark geometry and operating conditions. In the second part, the physical model and the numerical modeling approach are reported, showing the validation and the results related to the sensitivity analyses of the artificial coefficients characterizing the modelling approach.

1. Introduction

The flashing flow, a phase change phenomenon, is encountered in accelerated flow inside nozzles with the boiling delay phenomenon (thermal non-equilibrium effect) [1, 2]. In general to model flashing flow, CFD approaches must take into account the two aspects including: (i) the phase change phenomenon during flashing flow and (ii) the two-phase interaction during the evaporation process between the liquid and the vapour phases. For the phase change during vaporization process, there are two main categories including: (i) the nucleation process is considered in models providing a better agreement to behavior of the real fluid-flow, but experimental data must be supported to control the source term of the nucleation equation (Table 1); (ii) models assume constant bubble size or number density and neglect the nucleation process (Table 2). For the interaction between two phases during the vaporization process, there are two approaches including: (i) the Eulerian-Eulerian model and (ii) the mixture model. Within the Eulerian-Eulerian model, the conservation equations are solved for each phase and then interfacial terms are considered for coupling two phases. Conversely, the mixture approach assumes the same velocity of individual phases in flow over short length scales (viz. homogeneous flow). In the mixture model, the average velocity field of the mixture flow is derived by a single momentum. Volume fraction is, thus, extracted by solving the individual equation for the second phase. In some cases such as high-speed flows, relative motion between two phases becomes more significant and must be taken into

account. Several models for velocity differences between two phases are suggested in literature, including: the drift-flux model [36], the mixture model [37], the algebraic-slip model [38], the suspension approach [39], the diffusion model [37, 40] and the local-equilibrium model [41].
This paper performs an improved CFD model for flashing flows, which includes the thermal non-equilibrium effect. In the proposed CFD model, the mixture two-phase approach is considered with the phase change process, assuming constant bubble size distribution, based on the difference between the vaporization pressure and the vapour partial pressure.
In this paper, the first part is devoted to the physical phenomenon of flashing flow and the experimental benchmark of a circular convergent-divergent nozzle is presented along with operating conditions. The second part introduces the physical model and the numerical modelling approach. Validation of the numerical approach and a sensitivity analysis of artificial coefficients are reported in the last part.

Table 1. *Flashing Flow Simulation with Nucleation Models: a literature survey*

References and code	Numerical model	Benchmark for validation
[3] CFX 4.2	- Eulerian model with Scalar transport equation for bubble number density using wall nucleation as source term. - Nucleation rate is determined by model of Jones [4-7].	- Flashing flows in pipes and nozzles [8].
[9] FLUENT 6.2.16	- Eulerian model. - Effects of bubble nucleation and interfacial heat transfer to phase change process, momentum and energy conservation are considered	- Flashing phenomenon is accounted in depressurization process of a piping system at Rusal Aughinish.
[10] NEPTUNE-CFD	- Eulerian model - unphysical heat transfer coefficient - A modified version of Jones' model [4-7] is used for nucleation effects to vapour generation rate, momentum and energy transfer.	- Critical flow in nozzle [11] with initially subcooled inlet and flashing outlet. - Experimental test from [12] with cavitation developments downstream an orifice.
[13] CFX 14.5	- Eulerian model. - Interphase momentum transfer with drag, lift, lubrication, virtual mass and turbulent dispersion forces. - Interphase mass transfer is derived by interfacial heat transfer - Nucleation models of [7] RPI model [14-16], Riznic model [17, 18], Rohatgi model [19] are used to bubble number transport equation.	- Vertical circular convergent-divergent nozzle from [20].
[21] STAR-CD	- Mixture model with barotropic phase change - The bubble growth is modeled by Rayleigh-Plesset equation.	- A single-hole diesel injector with different needle's lifts.

Table 2. *Flashing Flow Simulation Neglecting Nucleation Process*

References and code	Numerical model and assumptions	Experimental benchmark and remarks
Laurien and his colleagues [22-24] CFX 4.2	- Eulerian model - Assuming bubble number density is used for growth of bubble size by.	- Cavitation flow in pipes.
[25] CFX	- Eulerian model - Bubble diameter is is fixed to 1mm	- Edward blowdown test
[26] CFX 14.0	- Eulerian model - Nucleation process is neglected by assuming constant bubble number density.	- Transient pressure release in vertical pipe with flashing flow of [27].
[28] FLUENT 12.0	- Mixture model with slip phenomenon between two phases - phase change phenomenon is driven by pressure [29] and temperature - Rate of phase change is based on sonic velocity equation [30].	- Convergent-divergent motive nozzle of [31] inside a two-phase ejector.
[32] CFX 14.5	- Same model as in [26]	- Flashing of initial sub-cooled water flow in vertical convergent-divergent nozzle from [8].
[33] CFX	- Same model as in [26] uses different models for bubble number density (mono-disperse assumption and poly-disperse method).	- An overview of previous works with assumptions of both mono-disperse approach and poly-disperse approach.
[34] In-house code	- Mixture model. - Mass transfer models consider (i) pressure phase change model from [35] and (ii) thermal phase change model. - Modelling cavitating flow on a cylindrical head form and a Clark-Y hydrofoil geometry to confirm empirical coefficients of phase change model.	- Circular convergent-divergent nozzle geometry with operating conditions in Abuaf et al. [8] is used as experimental benchmark.

2. The phenomenon of flashing flow

Figure 1a presents the thermodynamic diagram of boiling process of water. This process happens at constant pressure in a saturation region under thermal equilibrium conditions (further details in [42]). However, in high-speed flows, as ejector-based systems, the phase change phenomenon inside nozzles shows thermal non-equilibrium behaviours. Particularly, the depressurization inside the nozzle leads to have superheated water and, consequently, a phase change occurs. The bubble nucleation growth in this case is related to the non-equilibrium effects [43]. In general, the flashing flow, phase change process driven by sudden depressurization, is separated in 3 stages, as presented in Figure 1b and summarized in the following:

1. **Stage 1**. After the pressure of flow drops below saturation line, the nuclei existing in cavities of wall and in the flow start growing. The surface tension forces dominate the behaviour of the bubbles at this stage , in which growth of the nuclei is limited (this stage is named delay period in [44] or idle period in[45]).
2. **Stage 2.** When the diameter of the bubbles reaches a critical radius (reference in [43]), the growth of the bubble is dominated by the pressure difference between the bubble surface (and the surrounding water. Rayleigh-Plesset equation as in [46] can be used to approximate the bubble growth rate in this stage.

3. **Stage 3.** Heat transfer around the bubble interface, combined with turbulence fluctuations and relative motion between phases, will have a big influence to boiling process in this stage.

a) Thermodynamic diagram of phase change of water with equlibrium assumption

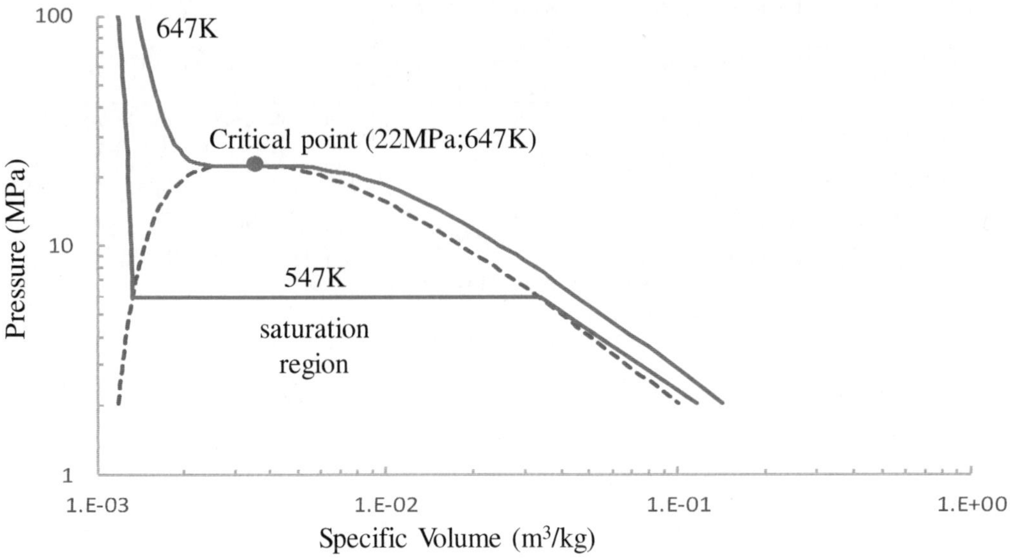

b) Start of vaporization

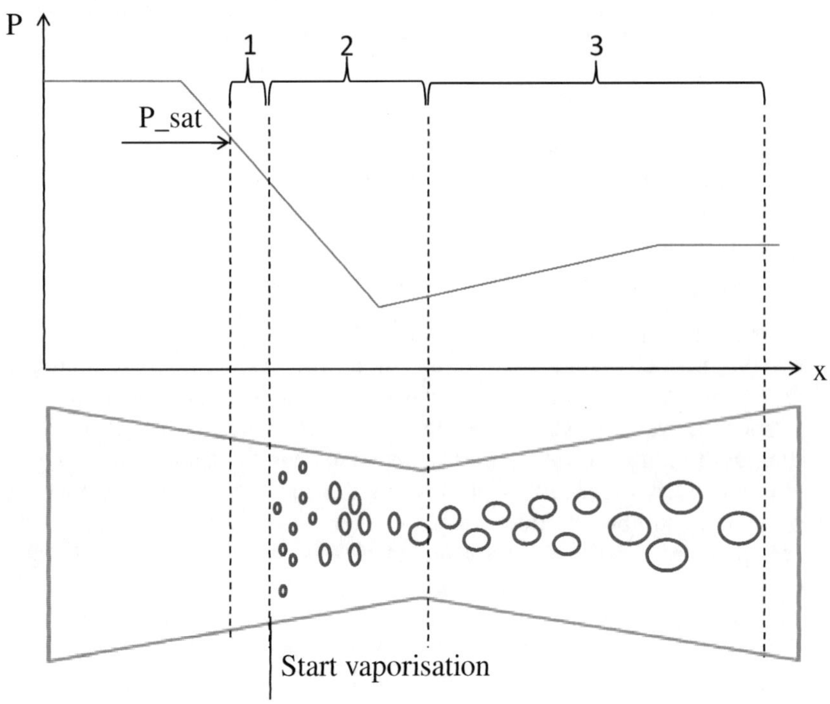

Figure 1. *Physical behaviour of flashing flow*

3. Experimental benchmark

The experimental measurements proposed by [8] have been used to validate the numerical results. In this benchmark, net vapour generation rates with non-equilibrium conditions are measured. Particularly, this benchmark includes a circular convergent-divergent nozzle, with sub-cooled water at the nozzle inlet and flashing flow at the outlet (Figure 2 presents the detailed dimensions of the nozzle). Abuaf has designed the converging-diverging nozzle with 49 pressure taps and two windows for observation at the exit point. Experimental data used to validate include (a) distributions of the pressure along nozzle, (b) the average vapour fraction, (c) inception point of the flashing and (d) radial void fraction profiles at axial positions. Table 3 describes the operating conditions of the experiments. For more information the experimental conditions, reader may refer to [8].

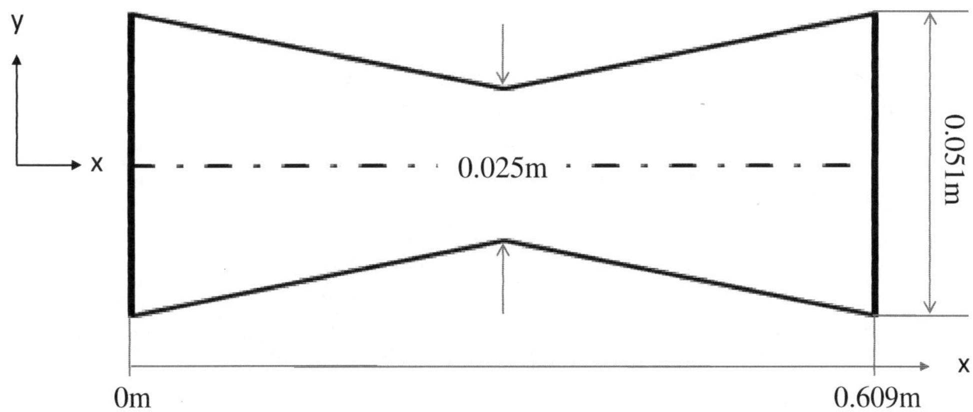

Figure 2. Vertical circular convergent-divergent nozzle in Abuaf et al. [8]

Table 3. Operating conditions and case code names in [8]

Case code name	Upstream pressure [Pa]	Inlet temperature [K]	Outlet pressure [Pa]	Flashing inception pressure [Pa]	Saturation pressure [Pa]	Mass flow rate [kg/s]
BNL284	530000	422.35	456000	404700	466000	7.3
BNL309	555900	422.25	402500	393500	464800	8.8
BNL273	573500	421.85	442100	419200	459800	8.7
BNL268	575200	422.05	443000	405700	462300	8.7
BNL304	577700	422.15	441000	399700	463500	8.8
BNL278	688600	421.95	434100	425700	461000	11.7
BNL296	764900	421.95	432600	417000	461000	13.1

4. Physical model

Model of flow

In this study, the mixture model with slip effect in ANSYS-FLUENT 16.0 is used with assumptions of pseudo single-phase flow. With this assumption, the same governing equations (mass, momentum and energy conservation) are used for both phases. For more information about this model, [47] is referred.

Phase change model

The transport equation for the phase change is given by:

$$\frac{\partial(\alpha_v \rho_v)}{\partial t} + \nabla \cdot (\alpha_v \rho_v \vec{v}_m) = \dot{m} \tag{1}$$

The source term $\dot{m}$ to consider the evaporation-condensation flux at the interface is defined:

$$\dot{m}\left[\frac{kg}{m^3 s}\right] = FA_i = A_i \beta \sqrt{\frac{M}{2\pi R T_{sat}}}(P_v - P^*) \tag{2}$$

Where β is the accommodation coefficient, M the molar mass [kg/kmol], R the universal gas constant [kJ/kmol K]. The turbulence effects to flashing process is taken into account via vaporization pressure in (3)

$$P_v = P_{sat} + 0.195\rho k \tag{3}$$

In Eq. (2), P^* is the value of the pressure inside saturation region and should be close to saturation pressure P_{sat} and k the turbulence kinetic energy [m^2s^{-2}]. Interfacial area density A_i, presented in (4)

$$A_i[m^{-1}] = (6\alpha_v)^{2/3}(\pi N_b)^{1/3} \tag{4}$$

Numerical set-up and boundary conditions

The computational domain is represented by 2D-axisymmetric approach and a structured quadrilateral elements grid has been set up. **Figure 3** and Table 4 present mesh and boundary conditions used in the simulations. In the benchmark cases, there is no reversed flow observed at the outlet and only outlet static pressure can be imposed to generate the depressurization inside the nozzle. Please note that a grid independency study has been performed to ensure accuracy. In this study, the $k-\omega$ SST with Standard option for the near wall treatment is chosen in combination with flashing model.

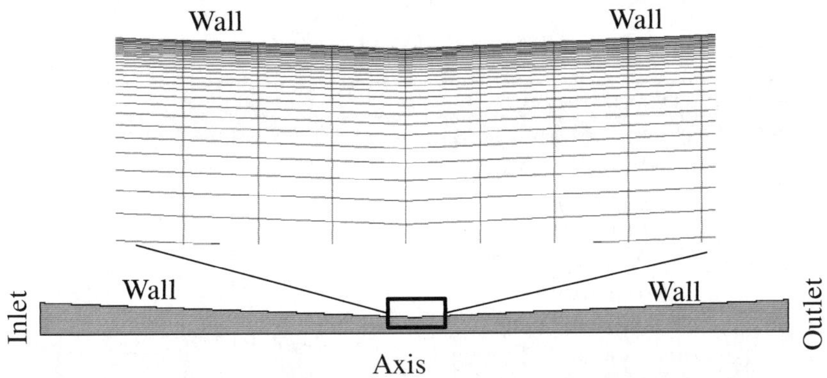

Figure 3. Mesh and boundary condition of convergent-divergent nozzle

Table 4. Boundary conditions

Boundary	Flow boundary	Turbulence boundary
Inlet	Total pressure	Turbulence intensity and hydraulic diameter
Outlet	Static pressure	
Wall	Adiabatic wall	Wall function
Axis	Axisymmetric	Axisymmetric

The proposed approach performs PISO scheme for Pressure-Velocity Coupling algorithm and Second Order Implicit scheme for the transient term. The spatial discretization of different quantities is presented in Table 5. In this simulation, the time step is 10^{-5}s with convergence criteria: (i) relative error of mass flow rate between inlet and outlet <2% and (ii) the maximum residual below 10^{-6} for all equations is imposed. Artificial coefficients such as bubble number density and maximum of $dp = P_{sat} - P^*$ are set at the values of $4 \cdot 10^8$ and 75Pa, respectively. The mass flow rate error can be explained by the use of artificial coefficients and the mesh approximation, as reported below. Further investigation on this aspect can be done to improve the accuracy of the numerical results. The phase change process is trigged by a vapour fraction of 10^{-5} imposed at the inlet. An analysis of mesh convergence has been performed for the BNL309 case with four different meshes including: (i) 21350, (ii) 53184, (iii) 109728 and (iv) 214650 elements, with a ratio of the elements $\frac{N_{i+1}}{N_i} \approx 2$ and the value of y^+ ranging from 6 to 16 for all tested meshes. Finally, mesh of 109728 elements is applied for further calculations.

Table 5. Numerical method

Discretization	Scheme
Transient formulation	Second Order Implicit
Pressure-Velocity Coupling	PISO
Gradient	Green-Gauss Cell Based
Pressure	PRESTO!
Density	Second Order Upwind
Momentum	
Volume Fraction	
Turbulence Kinetic Energy	
Specific Dissipation Rate	
Energy	

5. Numerical results

Validation

The numerical results, obtained with the present model, are validated with available measurements in [8]. The nucleation stage has been neglected in this model when imposing $N_b = 4 \cdot 10^8$ and introducing an accommodation coefficient to tune the boiling delay effect. In this part of paper, different results are compared and reported including mass flow rate, vapour volume fraction, pressure drop and vapour volume fraction profile.
Table 6 shows comparison between the computed mass flow rate and the experimental data for all the operating conditions. There is an agreement between numerical results and experiment when average and maximum relative errors equal to 4.7% and 6.8%, respectively. The accommodation coefficient β was imposed to a value close to unity for all cases.

Table 6. Comparison of Experimental Mass Flow Rate and CFD Mass Flow Rate

Cases	Accommodation coefficient β	Experimental Mass Flow Rate (kg/s)	CFD Mass Flow Rate (kg/s)	Relative error
1	0.8	7.3	7.7	+5.4%
BNL309	1.2	8.8	8.8	+0.2%
BNL273	1.05	8.7	9.3	+6.8%
BNL268	1.0	8.7	9.2	+5.7%
BNL304	1.11	8.8	9.1	+3.4%
BNL278	1.15	11.7	12.3	+5.1%
BNL296	1.18	13.1	13.9	+6.1%

Comparison of average vapour fractions and absolute pressures along the nozzle with experimental data is performed and shown in **Figure 4**. A general agreement of the present model with the measurements is observed for all tested cases. Average vapour starts from close to zero at the convergent part and increases near the nozzle throat region (x=0.3045m) up to the nozzle outlet. Besides, the present model shows a good prediction of the average vapour volume faction for tested cases BNL268, BNL273, BNL284, BNL304 and capability of model to correctly capture the flashing inception position is confirmed. There is an agreement with measured data of absolute pressure at convergent section, whereas a slight difference at the divergent section is observed. For all the analysed fluid dynamic quantities, the proposed CFD flashing model shows a good prediction with respect to measurements.

BNL268

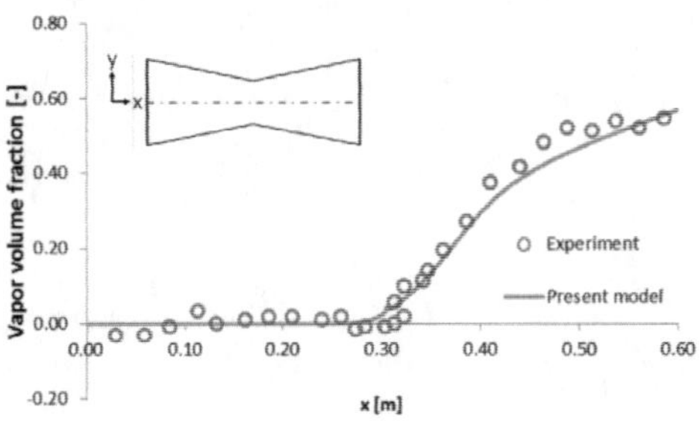

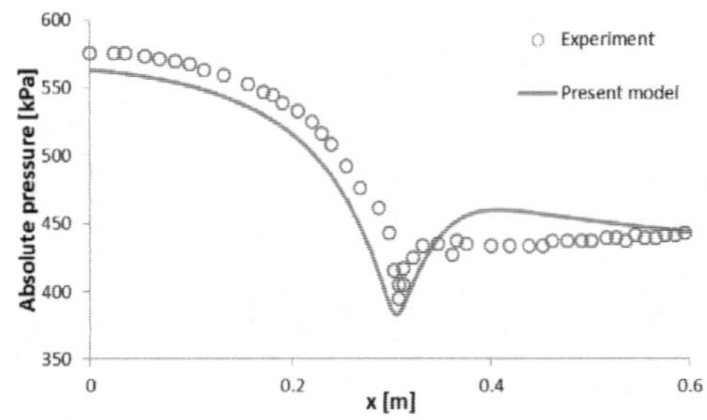

BNL273

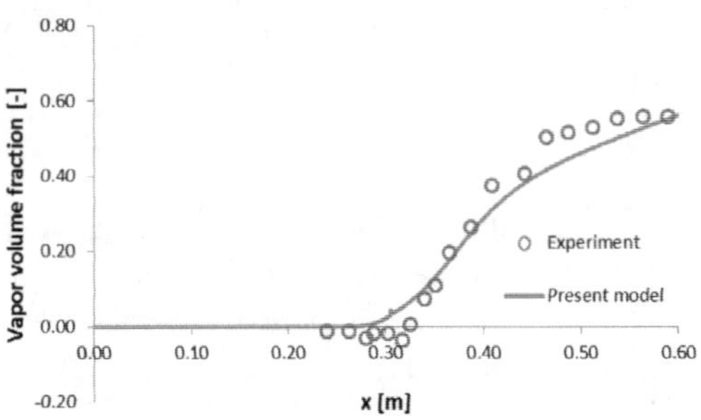

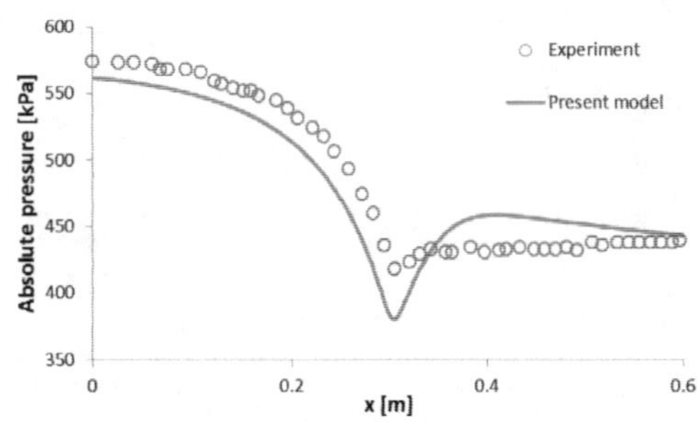

BNL278

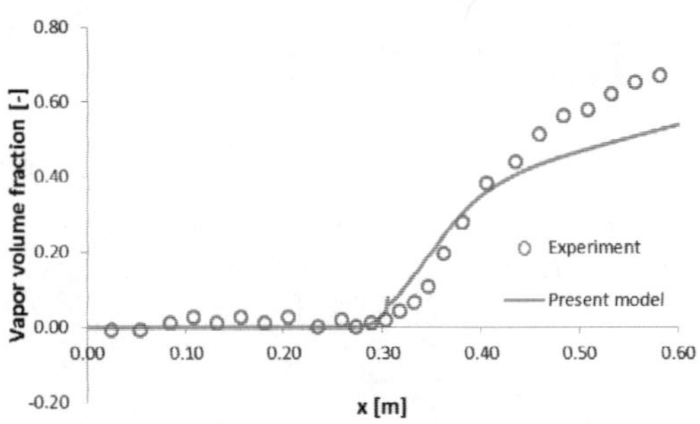

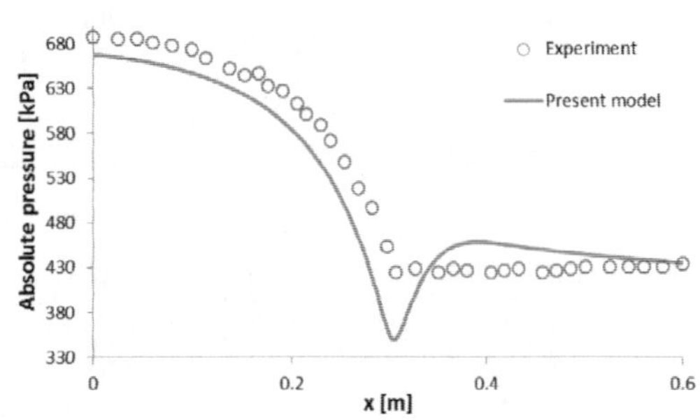

BNL284

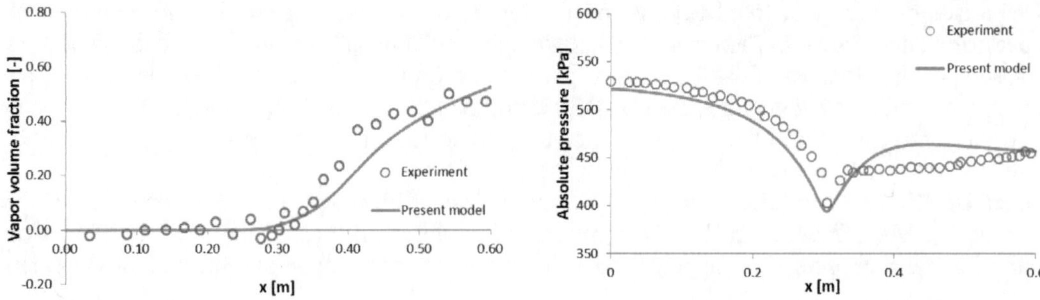

BNL296

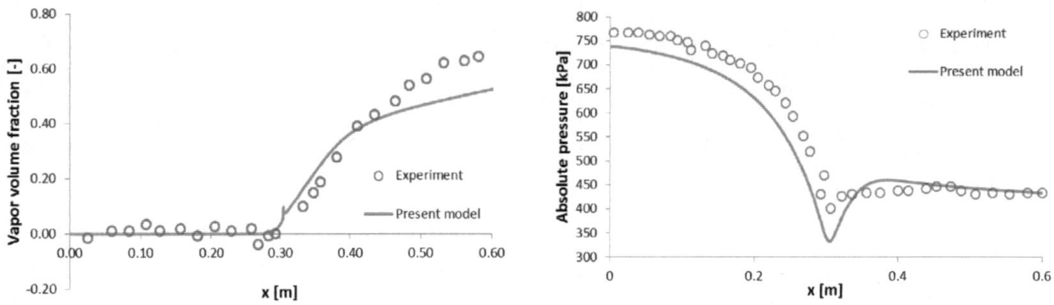

BNL304

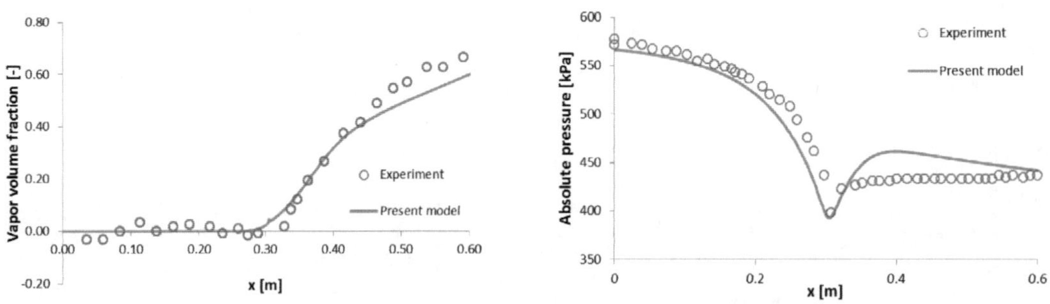

BNL309

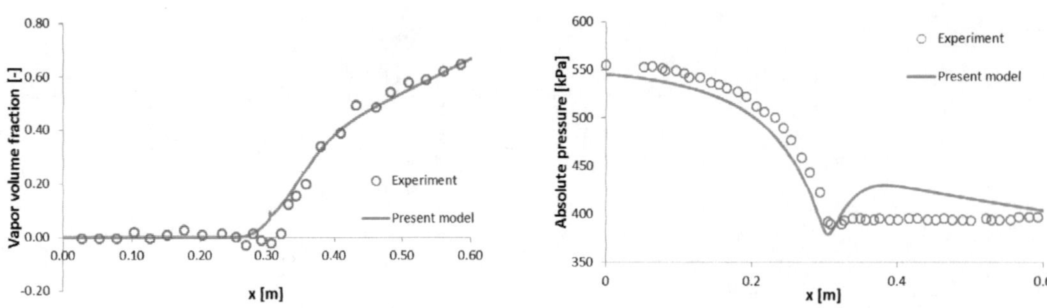

Figure 4. Averaged vapor fraction and absolute pressure along nozzle

Radial vapour distribution for case BNL309 is presented in **Figure 5** and Figure 6 from the near the nozzle throat at x=0.306m to near the nozzle outlet at x=0.577m. There is a comparison between the present model and experimental data at different sections across the nozzle. At x=0.306 (position A), the present flashing model shows a peak near the wall region, whereas, vapour volume fraction value of measurements equals to zero, implying a later flashing inception. In the present model, the turbulence effects near the wall to phase change process is taken into account leading to a good agreement with the measured data in sections from x=0.319m to x=0.577m at near wall region.

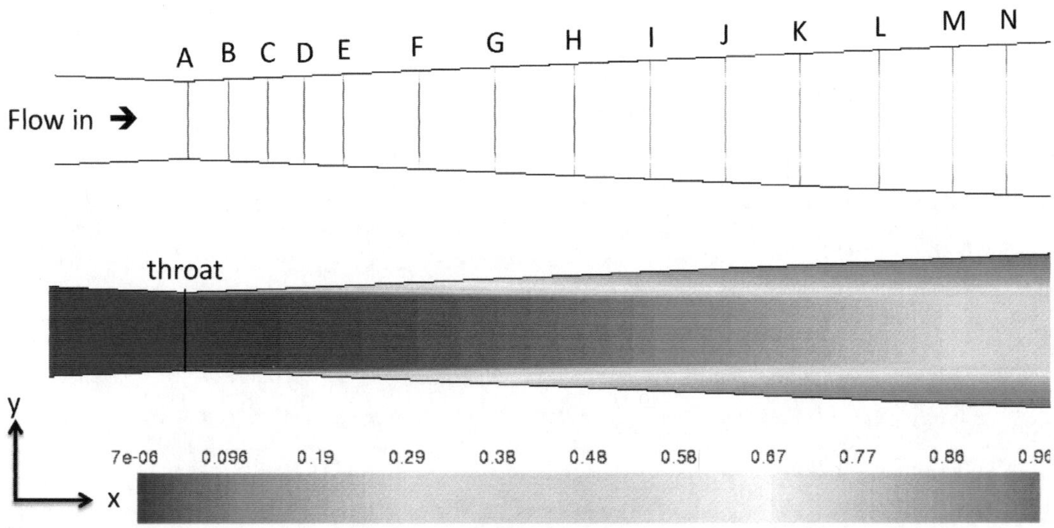

Figure 5. Radial void fraction profile at positions along divergent nozzle

Section A (x=0.306m)

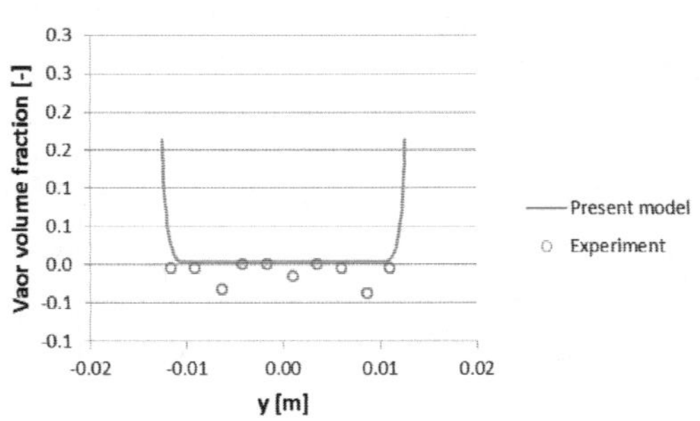

Section B (x=0.319m)

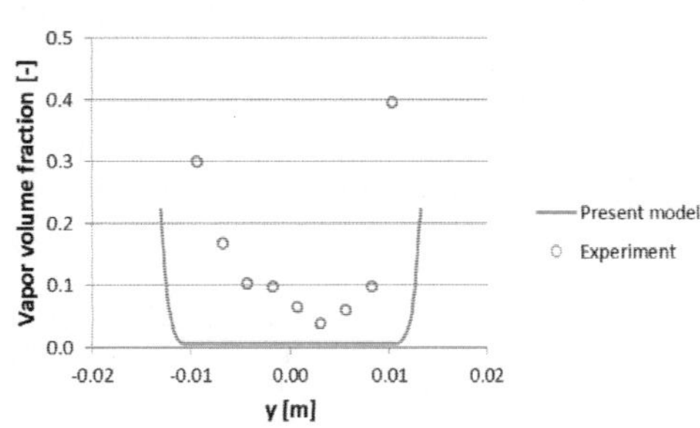

Section C (x=0.332m)

Section D (x=0.344m)

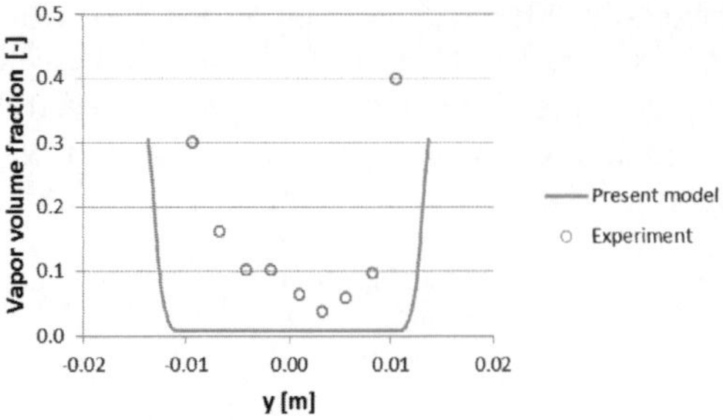

Section E (x=0.357m)

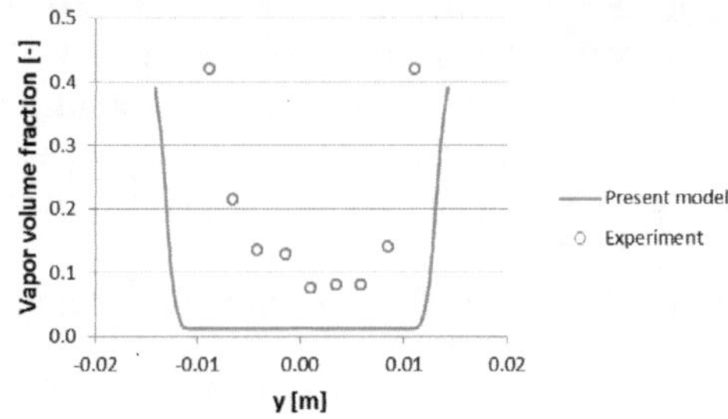

Section F (x=0.382m)

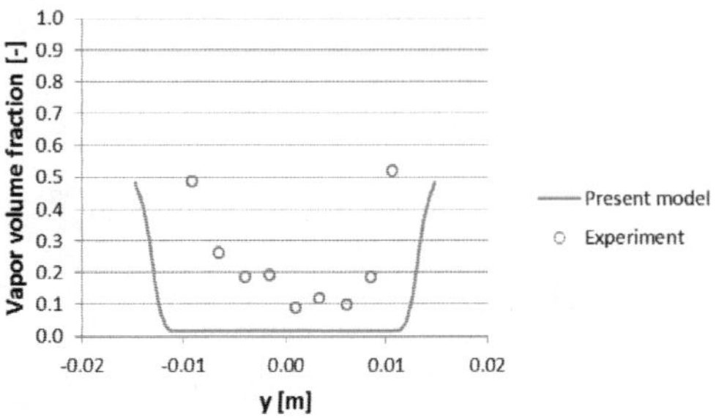

Section G (x=0.407m)

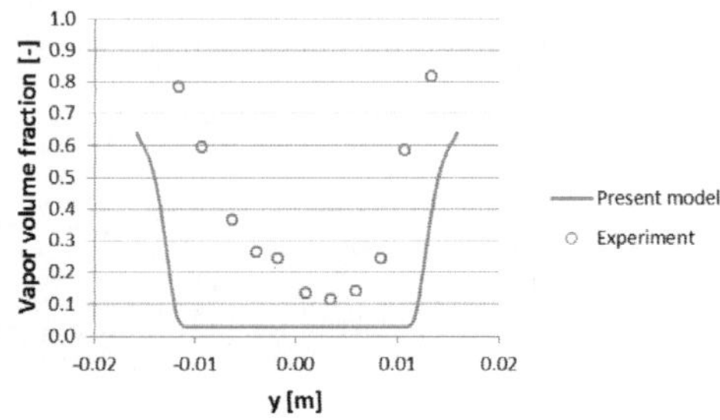

Section H (x=0.433m)

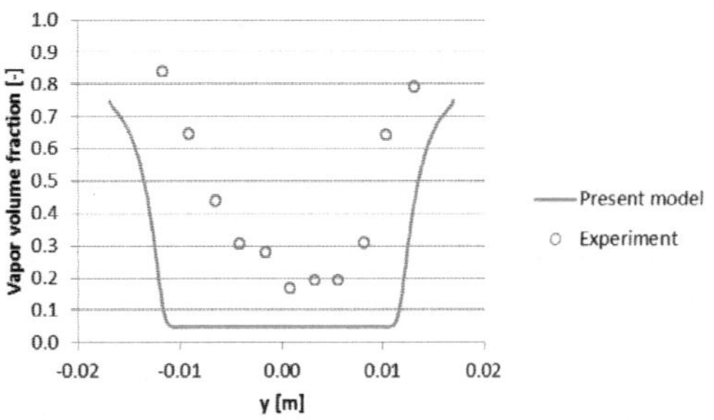

Section I (x=0.458m)

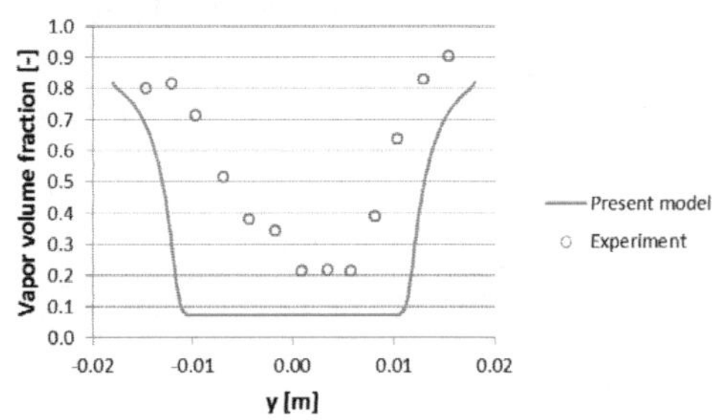

Section J (x=0.483m)

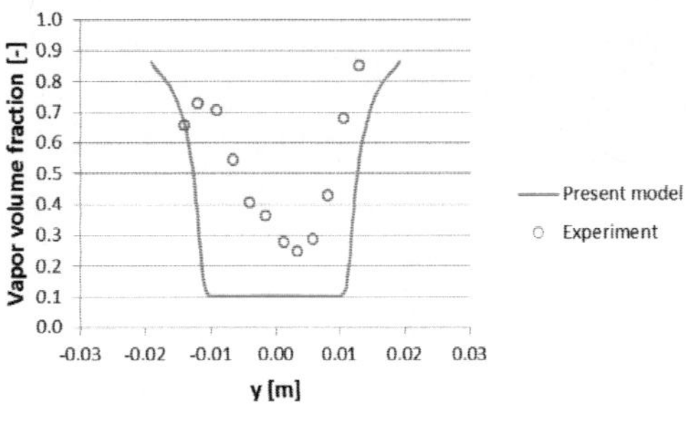

Section K (x=0.508m)

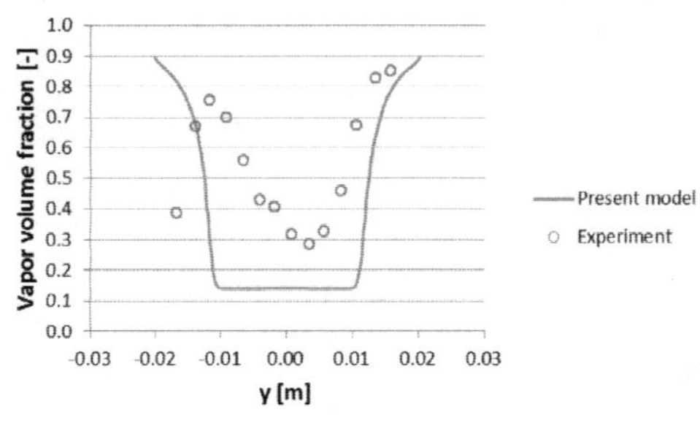

Section L (x=0.534m)

IOP Conf. Series: Journal of Physics: Conf. Series **1224** (2019) 012010 doi:10.1088/1742-6596/1224/1/012010

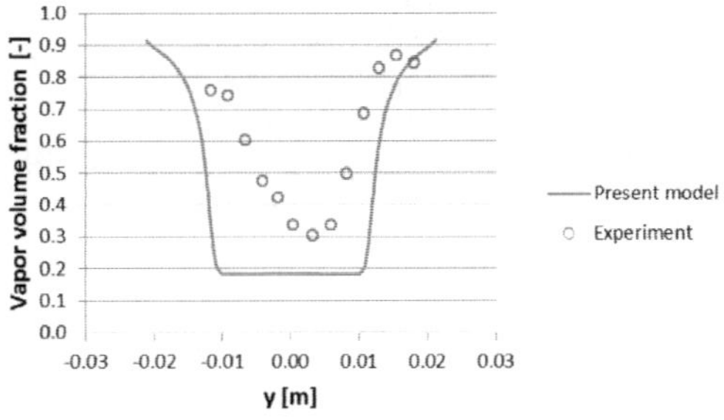

Section M (x=0.559m)

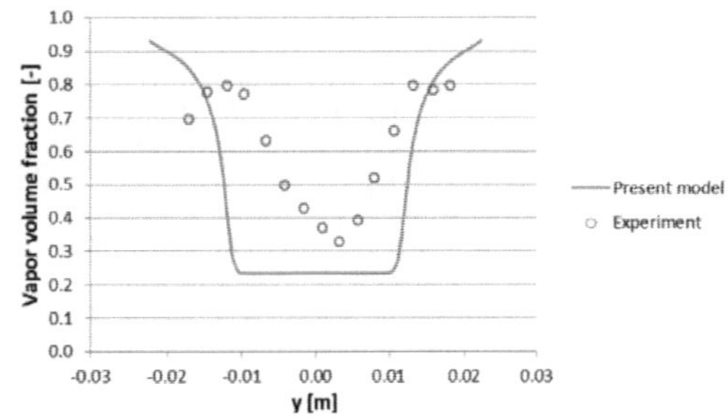

Section N (x=0.577m)

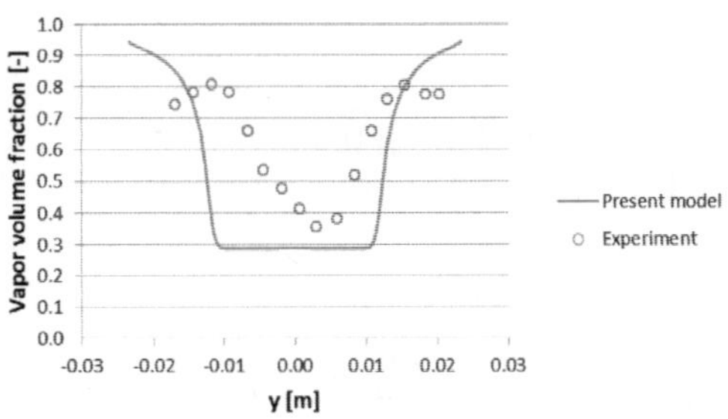

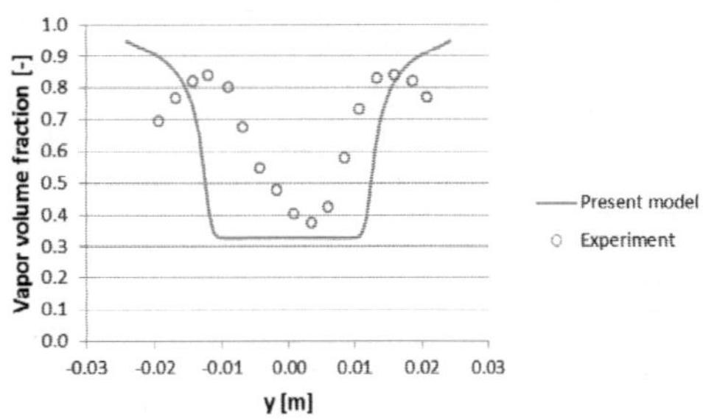

Figure 6. Radial vapour profile of BNL309 compared to experiment

Sensitivity analysis of artificial coefficients

A systematic procedure to evaluate effects of artificial coefficients including dp and the accommodation coefficient, β, is performed. In this approach, one coefficient will be fixed to perform sensitivity analysis on the other quantity. Generally, higher vapour volume fraction at divergent sections is observed when the either β or dp increases, as shown in Figure 4. Besides, the pressure at the throat region also increases corresponding to the increase of artificial coefficients. For static pressure variable along nozzle, it can be observed how the increase of artificial coefficients leads to an increase of pressure in correspondence of the throat region. This behaviour can be associated to an earlier vaporization process. Indeed, if water flow is incompressible from convergent part to throat, giving an explicit relation between static pressure and velocity, at throat level the flow becomes compressible and static pressure is not linked to velocity trend anymore and pressure does not keep decreasing further with the increase of velocity.
The influence of the artificial coefficients including the pressure drop dp and the accommodation coefficient, β, have been reported in Figure 7 for cases BNL268, BNL273, BNL284, BNL304, BNL309, respectively. A similar trend among different cases is observed when varying coefficients. The importance of accommodation coefficient β is highlighted by the sensitivity analysis presented here, showing a coherent behavior for different operating conditions.

BNL268

a) Vapour profile and static pressure along nozzle with fixed $dp = 75\text{Pa}$

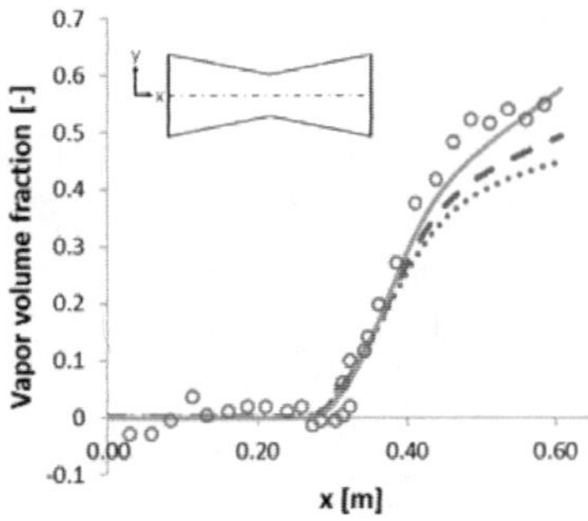

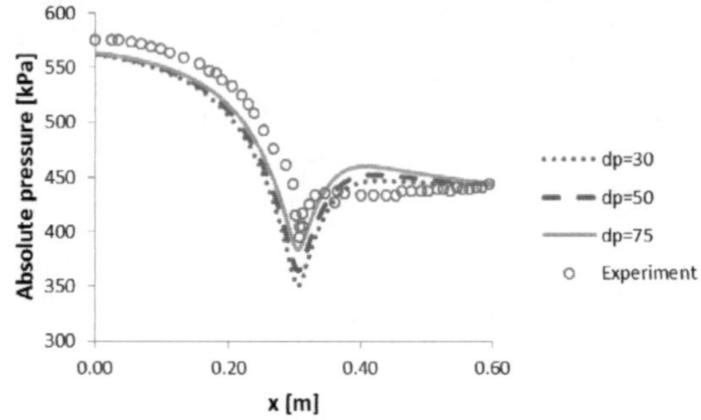

b) Vapour profile and static pressure along nozzle with fixed $\beta = 1.0$

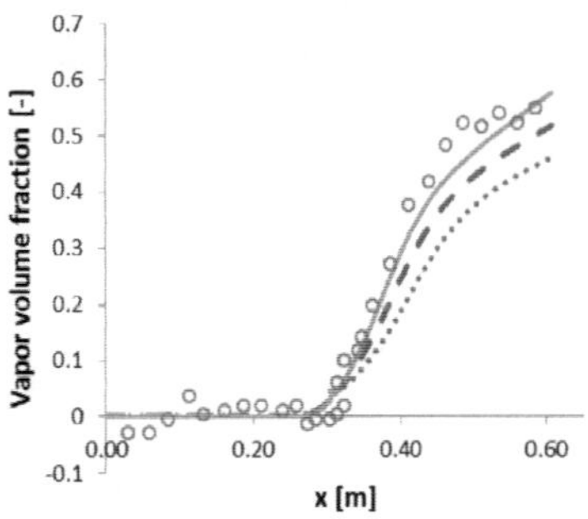

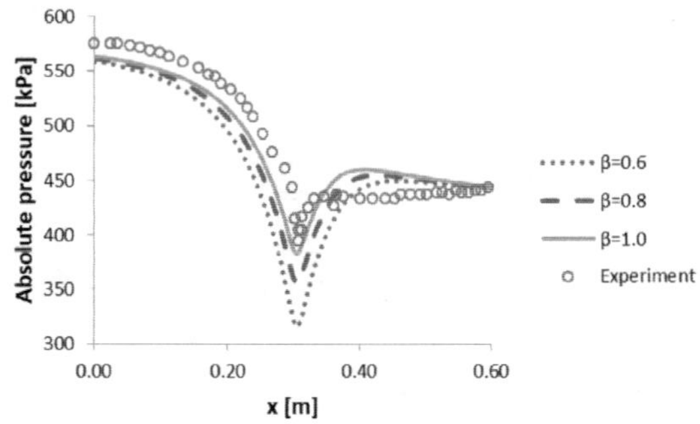

BNL273

a) Vapour profile and static pressure along nozzle with fixed $dp = 75\text{Pa}$

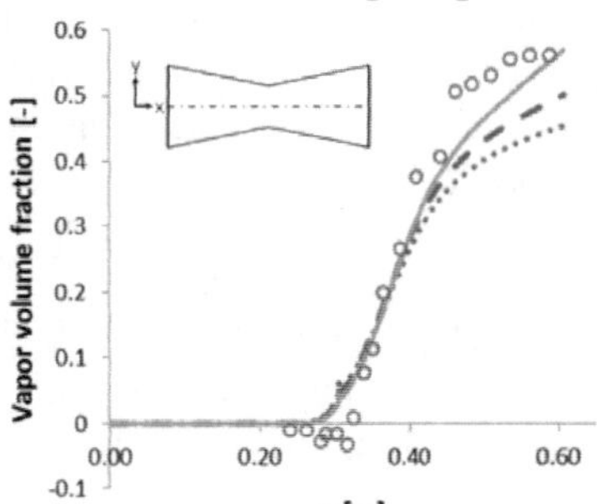

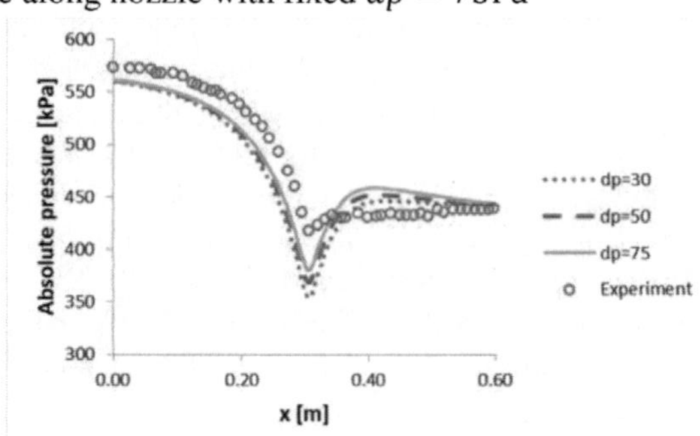

b) Vapour profile and static pressure along nozzle with fixed $\beta = 1.05$

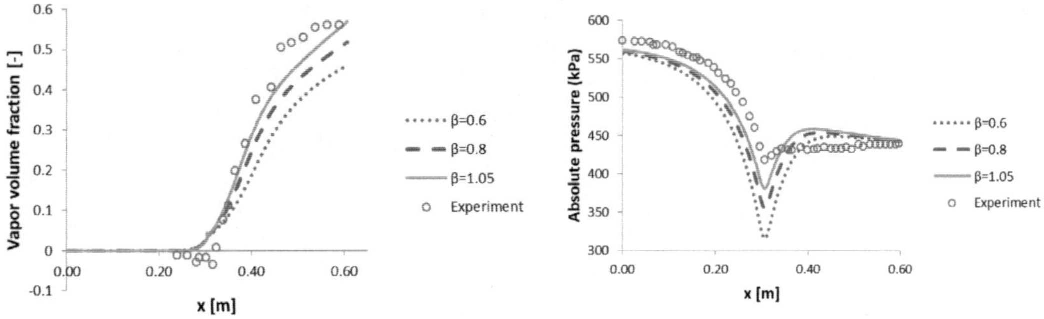

BNL284

a) Vapour profile and static pressure along nozzle with fixed $dp = 75$Pa

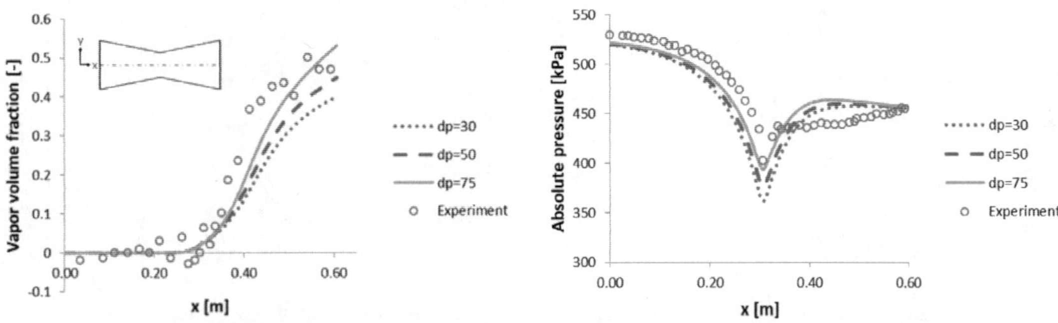

b) Vapour profile and static pressure along nozzle with fixed $\beta = 0.8$

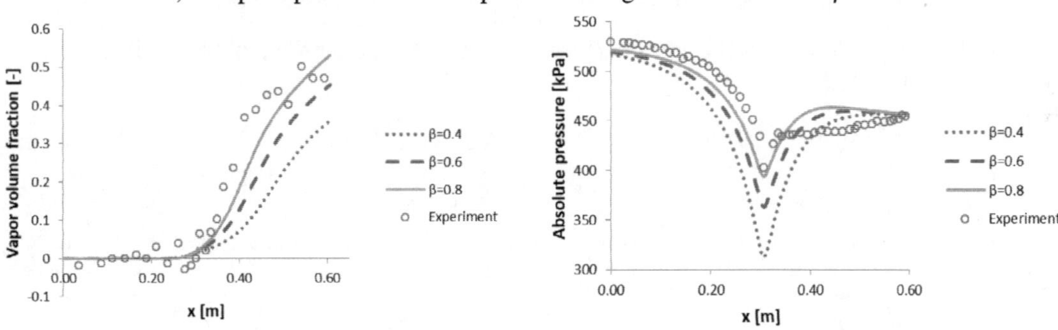

BNL304

a) Vapour profile and static pressure along nozzle with fixed $dp = 75\text{Pa}$

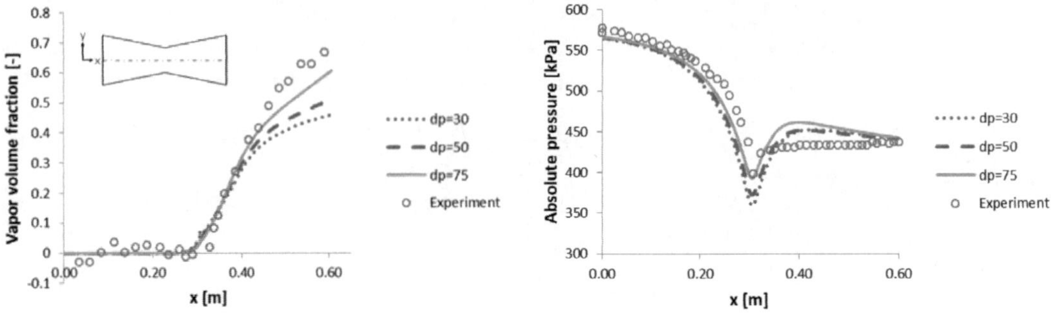

b) Vapour profile and static pressure along nozzle with fixed $\beta = 1.11$

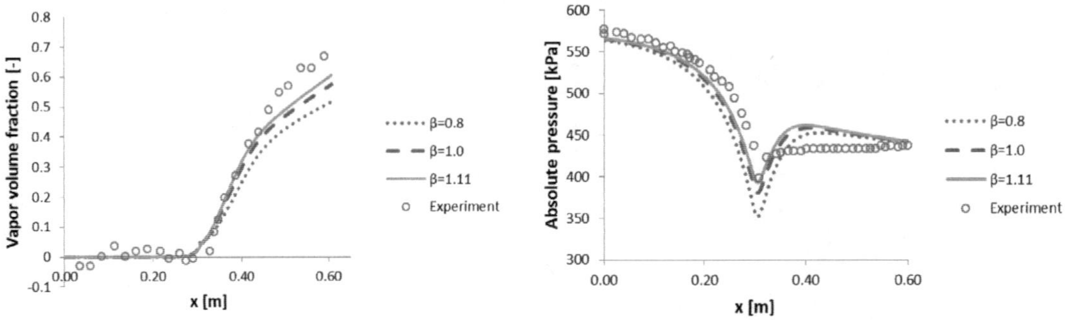

BNL309

a) Vapour profile and static pressure along nozzle with fixed $dp = 75\text{Pa}$

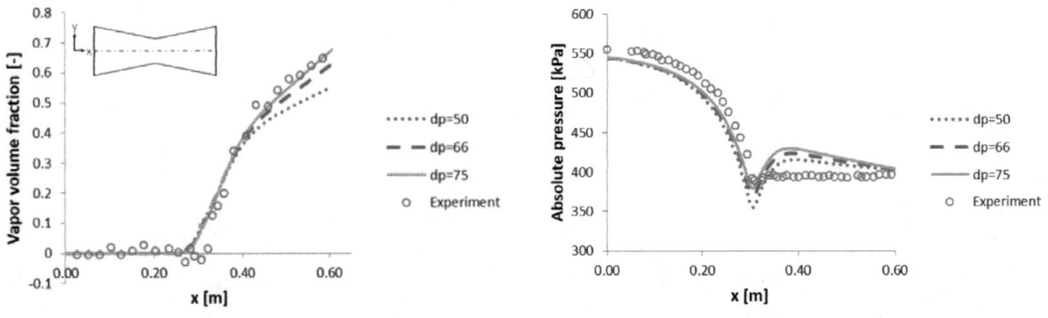

b) Vapour profile and static pressure along nozzle with fixed $\beta = 1.2$

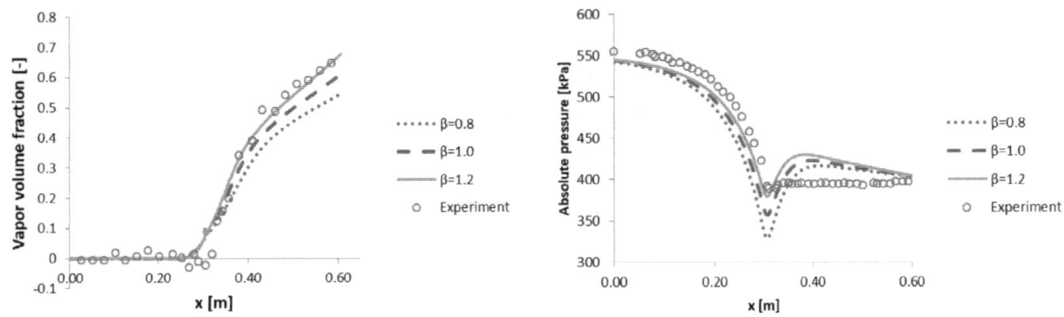

Figure 7. Artificial coefficients sensitivity in BNL268, BNL273, BNL284, BNL304 and BNL309

6. Conclusion

A CFD flashing model for flashing flows has been performed and validated. The model is based on the two-phase mixture model considering the slip effect and the thermal non-equilibrium effect (viz. boiling delay phenomenon). The model has been validated by using the available measurements of a circular convergent-divergent nozzle. Generally, there is an agreement in the comparison between numerical results and measured data including (a) mass flow rate (e.g., maximum relative error below 6%), (b) average vapour fraction and (c) static pressure along the nozzle and (d) radial vapour volume fraction profiles at different positions along nozzle. Besides, sensitivity analyses of artificial coefficients have been performed. With the accommodation coefficient and the difference between saturation pressure and vapour partial pressure at the interface, an earlier flash inception point is observed from numerical results when increasing the coefficients.

References

[1] Saha, P., 1978, *A Review of Two-phase Steam–Water Critical Flow Models with Emphasis on Thermal Non-equilibrium*, U.S. Nuclear Regulatory Commission, Washington, DC.

[2] Wallis, G. B., 1980, "Critical two-phase flow," Int. J. Multiphase Flow, **6**, pp. 97–112.

[3]Maksic, S., and Mewes, D., 2002, "CFD-calculation of the flashing flow in pipes and nozzles," In: Proc. ASME FEDSM'02, Montreal, Quebec, Canada, 14–18 July.

[4] Jones, O. C., and Zuber, N., 1978, "Bubble growth in variable pressure fields," J. Heat Transfer **100**, pp. 453–459.

[5] Shin, T. S., and Jones, O.C., 1986, "An active cavity model for flashing," Nucl. Eng. Des., **95**, pp. 185–196.

[6] Shin, T. S., and Jones, O.C., 1993, "Nucleation and flashing in nozzles – 1," Int. J. Multiphase Flow, **19**, pp. 943–964.

[7] Blinkov, V.N., Jones, O.C. and Nigmatulin, B. I., 1993, "Nucleation and flashing in nozzles – 2," Int. J. Multiphase Flow, **19**, pp. 965–986.

[8] Abuaf, N., 1981, *A Study of Nonequilibrium Flashing of Water in a Converging-diverging Nozzle: Volume 1*, Experimental. U.S. Nuclear Regulatory Commission, Washington, DC.

[9] Marsh, C.A., and O'Mahony, A.P., 2009, "Three-dimensional modelling of industrial flashing flows," In: Proc. CFD2008, Trondheim, Norway, 10–12 June.

[10] Mimouni, S., Boucker, M., Lavieville, J. and Bestion, D., 2008, "Modelling and computation of cavitation and boiling bubbly flows with the NEPTUNE CFD code," Nucl. Eng. Des., **238**, pp. 680–692.

[11] Robert, M., Farvacque, M., Parent, M., Faydide, B., 2003, "CATHARE 2 V2.5: a fully validated CATHARE version for various application," In: Proceedings of the 10th International Topical Meeting on Nuclear Reactor Thermal-Hydraulics (NURETH 10), Seoul, Korea.

IOP Conf. Series: Journal of Physics: Conf. Series **1224** (2019) 012010 doi:10.1088/1742-6596/1224/1/012010

[12] Archer, A., 2002. "A predictive model for cavitation erosion downstream orifices," In: Proceedings of ASME, Montreal.
[13] Janet, J. P., Liao, Y. X., and Lucas, D., 2015, "Heterogeneous nucleation in CFD simulation of flashing flows in converging-diverging nozzles," Int. J. Multiphase Flow, **74**, pp. 106–117.
[14] Lemmert, M., and Chawla, J., 1977, "Influence of flow velocity on surface boiling heat transfer coefficient," Heat Transfer Boil, **237**, pp. 247.
[15] Kurul, N., Podowski, M., et al., 1991, "On the modeling of multidimensional effects in boiling channels," In: ANS Proceedings of the 27th National Heat Transfer Conference. Springer-Verlag, Minneapolis, MN, pp. 28–31.
[16] Cole, R., 1967, "Bubble frequencies and departure volumes at subatmospheric pressures," AIChE J., **13**, pp. 779–783.
[17] Kocamustafaogullari, G., Ishii, M., 1983, "Interfacial area and nucleation site density in boiling systems," Int. J. Heat Mass Transfer, **26**.
[18] Riznic, J., and Ishii, M., 1989, "Bubble number density in vapor generation and flashing flow," Int. J. Heat Mass Transfer, **32**, pp. 1821–1833.
[19] Rohatgi, U., and Reshotko, E., 1975, "Non-equilibrium one-dimensional two-phase flow in variable area channels," In: Non-Equilibrium Two-Phase Flows; Proceedings of the Winter Annual Meeting, Houston, Tex., November 30–December 5, 1975, Meeting Sponsored by the American Society of Mechanical Engineers, New York, vol. 1. American Society of Mechanical Engineers, pp. 47–54.
[20] Wu, B., Saha, P., and Abuaf, N., 1981. "A Study of Nonequilibrium Flashing of Water in a Converging–Diverging Nozzle," Modeling, vol. 2. NUREG/CR.U.S.N.R.C., Commission, U.N.R.
[21] Pelletingeas, A., Dufresne, L., and Seers, P., 2016, "Characterization of Flow Structures in a Diesel Injector for Different Needle Lifts and a Fluctuating Injection Pressure," ASME J. Fluids Eng., **138**(8), pp. 081105.
[22] Giese, T., et al., 2002, "Experimental and numerical investigation of gravity-driven pipe flow with cavitation," In: Proc. ICONE10, Arlington, VA, 14–18 April.
[23] Laurien, E., and Giese, T., 2003, "Exploration of the two-fluid model of two-phase flow towards boiling, cavitation and stratification," In: Proc. 3rd ICCHMT, Banff, Canada, 26–30 May.
[24] Laurien, E., 2004, "Influence of the model bubble diameter on three-dimensional numerical simulations of thermal cavitation in pipe elbows," In: Proc. 3rd Inter-national Symposium on Two-Phase Modelling and Experimentation, Pisa, Italy, September 22–25.
[25] Frank, Th., 2007, "Simulation of flashing and steam condensation in subcooled liquid using ANSYS CFX," In: Proc. 5th FZD & ANSYS MPF Workshop, 25–27 April.
[26] Liao, Y., Lucas, D., Krepper, E., and Rzehak, Roland, 2013, "Flashing evaporation under different pressure level," Nucl. Eng. Des., **265**, pp. 801–813.
[27] Schaffrath, A., Krüssenberg, A.-K., Weiß, F.-P., Hicken, E.F., Beyer, M., Carl, H., Prasser, H.-M., Schuster, J., Schütz, P., and Tamme, M., 2001, "TOPFLOW—A new multipurpose thermal hydraulic test facility for the investigation of steady state and transient two-phase flow phenomena," Kerntechnik, **66**, pp. 209–212.
[28] Yazdani, M., Alahyari, A. A., and Dradcliff, T.D., 2013, "Numerical Modeling and Validation of Supersonic Two-Phase Flow of CO_2 in Converging-Diverging Nozzles," ASME J. Fluids Eng., **136**(1), p. 014503.
[29] Singhal, A. K., Athavale, M. M., Li, H., and Jiang, Y., 2002, "Mathematical Basis and Validation of the Full Cavitation Model," ASME J. Fluids Eng., **124**(3), pp. 617–624
[30] Brennen, C. E., 2005, *Fundamentals of Multiphase Flow*, California Institute of Technology.
[31] Nakagawa, M., Berana, M. S., and Kishine, A., 2009, "Supersonic Two-Phase Flow of CO2 Through Converging-Diverging Nozzles for the Ejector Refrigeration Cycles," Int. J. Refrig., **32**(6), pp. 1195–1202.
[32] Liao, Y., and Lucas, D., 2015, "3D CFD simulation of flashing flows in a converging-diverging nozzle," Nucl. Eng. Des., **292**, pp. 149–163.

[33] Liao, Y., and Lucas, D., 2017, "Possibilities and Limitations of CFD Simulation for Flashing Flow Scenarios in Nuclear Applications," Energies **10**(139), pp. 1-22.
[34] Jin, M. S., Ha, C. T. and Park, W. G., 2017, "Numerical study on heat transfer effects of cavitating and flashing flows based on homogeneous mixture model," Int. J. Heat and Mass Transfer **109**, pp. 1068-1083.
[35] Ha, C.T., Park, W. G. and Merkle, C. L., 2009, "Multiphase flow analysis of cylinder using a new cavitation model," in: Proceeding of the 7th International Symposium on Cavitation.
[36] Zuber, N. and Findlay, J. A., 1965, "Average Volumetric Concentration in Two-Phase Flow Systems," J. Heat Trans., **87**, pp. 453 - 468.
[37] Ishii, M., 1975, *Thermo-fluid Dynamics Theory of Two-phase Flow.* Eyrolles, Paris
[38] Pericleous, K. A. and Drake, S. N., 1986, *An Algebraic Slip Model of PHOENICS for Multi-phase Applications*, Numerical Simulation of Fluid Flow and Heat/Mass Transfer Processes, Springer-Verlag, Berlin, Germany.
[39] Verloop, W. C., 1995, "The inertial coupling force," Int J. Multiphase Flow., **21**, pp. 929 - 933.
[40] Ungarish, M., 1993, *Hydrodynamics of suspensions: Fundamentals of centrifugal and gravity separation*. Berlin: Springer.
[41] Johansen, S. T., Anderson, N. M. and De Silva, S. R., 1990, "A two-phase model for particle local equilibrium applied to air classification of powers," Power Technology, **63**, pp. 121–132.
[42] Morgan, M. J., and Shapiro, H. N., 2006, *Fundamentals of Engineering Thermodynamics*, John Wiley & Sons, Inc, England.
[43] Xi, X., Liu, H., Jia, M., Xie, M., and Yin, H., 2017, "A new flash boiling model for single droplet," Int. J. Heat Mass Transfer, **107**, pp. 1129-1137–.
[44] Miyatake, O., Tanaka, I., Lior, N., 1997, "A simple universal equation for bubble growth in pure liquids and binary solutions with a non-volatile solute," International Journal of Heat and Mass Transfer, **40**, pp. 1577–1584.
[45] Oza, R.D., and Sinnamon, J.F., 1983, "An experimental and analytical study of flashing boiling fuel injection," SAE Paper, p. 830590.
[46] Barbone, R., 1994, "Explosive Boiling of a Depressurized Volatile Liquid," M.Sc. Thesis, Department of Mechanical Engineering, McGill University, Montreal, Quebec, Canada.
[47] User's guide of ANSYS-FLUENT 16.2

IOP Conf. Series: Journal of Physics: Conf. Series **1224** (2019) 012011 doi:10.1088/1742-6596/1224/1/012011

Parametric analysis of input data on the CFD fire simulation

Andreozzi Assunta[1], Bianco Nicola[1], Musto Marilena[1*], Rotondo Giuseppe[1,2]

[1] DII - Università degli Studi di Napoli Federico II, Napoli, Italy
[2] ITT Marie Curie, Napoli. Italy

*Corresponding author: marilena.musto@unina.it

Abstract. During the last decades the CFD computer modelling of fire has increased significantly. The CFD fire simulator has been purposely developed and optimized for fire protection; nowadays they have a high reliability and ample perspectives of further development. The main drawbacks of CFD fire simulators are mainly due to the complexity of its setting. They are not user friendly and require a good competence on fires and fair competence in computing. To be able to estimate how the simulation outcome is influenced by model assumptions and simplifications, the user has to evaluate the dominating physical processes and involved empirical parameter that has an essential influence on the specific fire scenario. In this paper the Thermo-gravimetric (TGA) and Cone-calorimetric tests will be used to show in which ranges the input parameters such as reference temperature (i.e. the temperature of the released mass rate peak), thermal conductivity, pyrolysis range, specific heat, in a fire simulation with the "Fire Dynamics Simulator", will vary based on the measurement errors, measurement uncertainty, or misinterpretation of the user. Fire simulation validated on experimental tests are used to evaluate the impact of those variant input data on the simulation. A parametric analysis has been carried out and the results are evaluated and discussed.

1. Introduction

Over the years, thanks to the progress of Computational technique of Fluid Dynamics (CFD), the fire protection has been continuously evolving, shifting from a 'prescriptive-based' to a 'performance-based' design. The method based on prescriptive procedures even if simpler to use, is not free of design limitations. Instead, the performance-based design allows more flexibility than prescriptive regulations in order to satisfy the fire protection needs of modern buildings. Nevertheless, there are essential prerequisites to computer simulations to achieve representative realistic fire conditions of high quality and reliability i.e. the evaluation of the dominating physical processes and the involved empirical parameters that have an essential influence on the specific fire scenario, such as chemical and thermal material properties, ignition temperature, reaction rate, pyrolysis range, heating rate, etc. The latter aspect is to evaluate them accurately and the subsequently transfer them to the CFD code adequately. This task is crucial to obtain representative realistic fire evolution. In fact, very often the 'quality' of CFD simulations depends more on this task than on the capabilities of the CFD code itself [1-3]. In fact, FDS is used to study the influence of different parameters on the results of the simulation [4-6] in order to have useful indication, sometimes on specific aspects as the gasification performance [7], the auto-ignition [8], the carbon monoxide concentration distribution [9]; other times to mark the influence of the CFD models and settings on the ability of FDS to correctly simulate the real processes [8,10,11].

Usually, the input parameters are validated by modeling the laboratory scale experiments, like the cone calorimeter or thermogravimetric analysis. Afterwards the input parameters are transferred into the real scale scenario or user specific fire scenario. Different authors show the challenge of modeling [12-14] and reproducing a real experiment with FDS [15]. The approach to implement data from laboratory scale experiments requires the knowledge about the fire behavior of the used materials and about the transferability of the experiments to the considered fire scenario.

To be able to estimate how the simulation outcome is influenced by model assumptions and simplifications, the user has to evaluate the dominating physical processes and involved empirical parameters that have an essential influence on the specific fire scenario.

In this work the Thermo-gravimetric (TGA) and Cone-calorimetric tests will be used to show in which ranges the input parameters such as reference temperature (i.e. the temperature of the released mass rate peak), reaction rate, pyrolysis range, heating rate, in a fire simulation with the "Fire Dynamics Simulator", will vary based on the measurement errors, measurement uncertainty, or misinterpretation of the user. Fire simulation validated on experimental tests has been used to evaluate the impact of those variant input data on the simulation. A parametric analysis is carried out and the results are evaluated and discussed.

2. Experimental tests

To validate fire spreading with the pyrolysis model of FDS, thermogravimetric and cone-calorimetric analyses on chipboard were conducted.

2.1. TGA analysis

TGA measurements provided the pyrolysis rates and mechanisms and char yield in nitrogen from 30 to 900 °C; the latter corresponds to a heating rate of a material in the cone calorimeter and in real fires for some cases.

Thermogravimetric analysis was carried out on 4.5 mg of low-cost board sample shavings using a Perkin Elmer Pyres 1 TGA system. This analysis was useful in determining the amount of sample mass lost during heating and in assessing the percentage of glass fiber in the tested sample. The test was performed in the presence of an air flow of 150 ml/min, starting from 30 °C up to 900 °C with a heating ramp of 5 K/min. The test is reported in the figure 1, where two mass release peaks at two different temperatures at approximately 80 °C and 350 °C are shown.

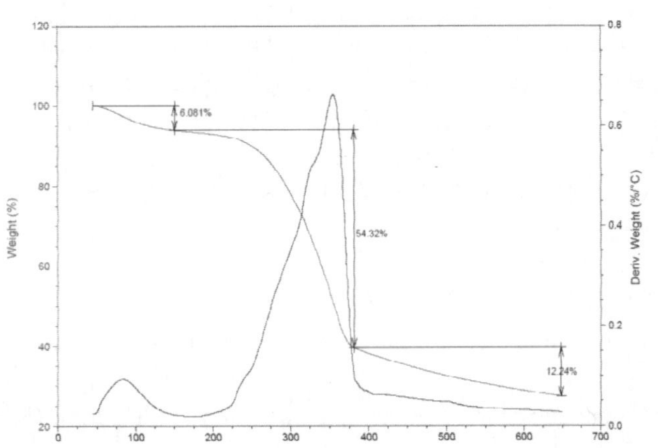

Figure 1. Thermogravimetric analysis

IOP Conf. Series: Journal of Physics: Conf. Series **1224** (2019) 012011 doi:10.1088/1742-6596/1224/1/012011

In particular, the pyrolysis interval value represents the range in which the specimen burns. As shown in the figure 2, the weight loss starts at 220 °C and ends at 320 °C. The mass loss, before the 220 °C, is due to the evaporation of humidity (peak at 100 °C) and other substances. The pyrolysis interval is one of most important parameters to set FDS pyrolysis and was reported in table 2.

2.2. Cone Calorimeter analysis

Most of the key flammability properties, as thermal properties of the original material (conductivity, density, specific heat), ignition temperature, heat of pyrolysis or gasification of the solid, thermal properties of char, extinction properties etc., to assess fire spread and fire growth of chipboard material can be obtained by experiments conducted in a cone calorimeter in according to standard.

The test specimens are 88.4 cm^2 and 2 cm thick, while their mass is m=0.130 kg. The Cone Calorimeter test was repeated 3 times with a heat flux of 50 kWm^{-2}. The results of the test are shown in the table 1, with the corresponding measurement extended uncertainties, and in the figure 2.

Table 1. Cone Calorimeter test results

		First test	Second test	Third test	Mean value	Uncertainty F.C. 90%
Test end time	s	1920	1920	1920	1920	
Ignition time	s	73	30	42	48	37.41
Average values for the first 180 s after ignition	kW/m^2	136.2	137.9	133.5	135.9	3.74
Average values for the first 300 s after ignition	kW/m^2	117.7	124.7	117.1	119.8	7.12
Heat released for unit of surface (HRUS)	MJ/m^2	143	159.8	148.1	150.3	14.52
Heat release rate peak	kW/m^2	294.9 (88 s)	247.6 (58 s)	276.2 (66 s)	272.9 (70 s)	40.16
Average Heat Release Rate	kW/m^2	93.9	102.9	95.1	97.2	8.24
Mass remaining after the test	g	34.6	34.4	35.6	34.9	1.08
Sample mass loss	g/sm^2	8.38	8.60	8.54	8.51	0.19

The results of the test at the Cone Calorimeter show that during the whole test the spaceman does not flame out until the end of the test (figure 2) and the time to ignition is on average 48 s with a heat release rate peak of about 272 kW/m^2 on average at 70 s (see table 1).

The measurement reported in table 1 was used to calculate the setting parameters for FDS reported in the table 2. From the Heat Release for unit of surface (HRUS) by specimen during Cone Calorimeter tests, it was been possible to evaluate the average Lower Heating Value (LHV) of the material, as:

Total Heat Release=HRUS·A= 150.3·88.4·10^{-4}=1328.65 kJ
where A is the specimen surface

LHV=THR/m= 1328.65/0.095=13971 kJ / kg
with 0.095 kg the mass of the specimen and where THR is the Total Heat Released

The relative vapor fuel vapor and char were calculated from the residual mass (RM) after the combustion process:

Char=Rm/m=0.035/0.130=0.3
Vapor Fuel = Bm/m=0.095/0.130=0.7
where Bm is the mass burned and Rm the residual mass.

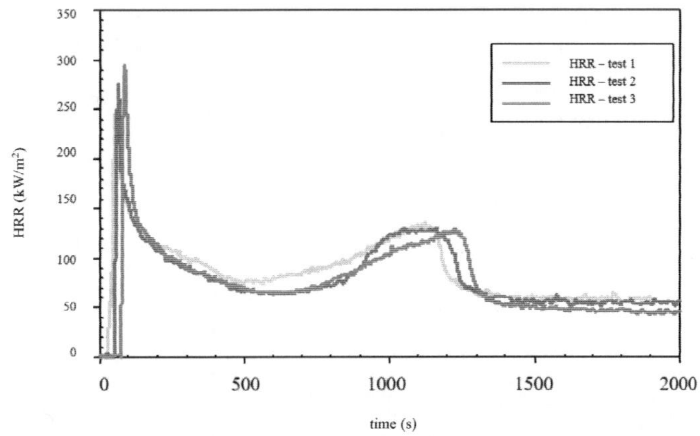

Figure 2. HRR curves from Cone Calorimetric tests

3. Numerical simulation

The Computational Fluid Dynamics technique is applied in order to solve a set of non-linear partial differential equations derived from basic laws of nature. They solve numerically the equation of continuity, the conservation of momentum and energy as a system of partial differential equations, relying on few sub models to take into consideration particular aspects. In the case of fire simulation, a combustion model is used to simulate the evolution of combustion, a turbulence model has to be included for the prediction of the buoyancy driven turbulent flow as well as a radiation model is needed to simulate the thermal radiation. Moreover, reaction kinetics, radiation transport and pyrolysis have to be considered [16]. The CFD code allows to simulate the entire development of fire [17]. For momentum conservation, FDS solves the Navier–Stokes equations using large eddy simulation (LES) to account for sub grid turbulence, and for the combustion reactions, it uses the mixture fraction model. Heat transfer to solid surfaces and convection within the fluid are taken into account. Moreover, the radiative transport equation for an absorbing/emitting and scattering medium is also solved.
Charring materials produce a barrier between the pyrolysis zone and the exposed surface. The process of decomposition is reduced depending on the properties of the material, the char and heat source. The user of FDS should implement the residue of the charring and needs to be aware, that breaking the char layer during the decomposition cannot be modeled; it is impossible with FDS because the user needs to characterize liquid properties (e.g. boiling temperature) for the sample, although the initial state is solid. Considering additionally layers for modeling the producing and breaking of char layers helps to take into account the effects of the combustion.

3.1. FDS simulation for the determination of chipboard combustion parameters

Thanks to the results of the tests to the cone calorimeter the thermal quantities and the pyrolysis parameters necessary for the thermo-fluid dynamics simulation have been obtained. In particular, the parameters obtained from the pyrolysis tests have been set directly as shown in table 2, while for the parameters of thermal conductivity and specific heat, various simulations have been carried out by varying these parameters in the intervals shown in table 3. Figure 3 shows the simulation model in FDS that reproduces the test at the cone calorimeter in order to evaluate the correctness of the combustion parameters and the properties of the material to be set in the code for the simulations of the fire scenarios. The boundary conditions that reproduce the test at the cone calorimeter were imposed; in particular, a thermal flow of 50 kW/m^2 was imposed on the chipboard surface. Comparison between the CFD and cone calorimeter test results is shown in the figure 4. The validated FDS simulation, on experimental test results, has been used to evaluate the impact of those variant input data on the simulation.

IOP Conf. Series: Journal of Physics: Conf. Series **1224** (2019) 012011 doi:10.1088/1742-6596/1224/1/012011

Table 2. Material properties of chipboard obtained from Cone Calorimetric and Thermogravimetric Tests

Temperature of maximum heat release rate	430	°C
Heating rate	5	K/min
Pyrolysis interval	100	°C
Heat of reaction	800	kJ/kg
LHV	13895	kJ/kg
Density	720	kg/m^3
Fuel vapor	0.7	-
Residual	0.3	-

Table 3. Material properties of chipboard obtained from literature

		Residual	chipboard
density	kg/m^3	36	720
specific heat	kJ/kgK	0.3÷0.50	2÷4
thermal conductivity	W/mK	0.10÷0.2	0.1÷0.2
emissivity	-	0.0	0.9
absorption thickness	1/m	50000	50000

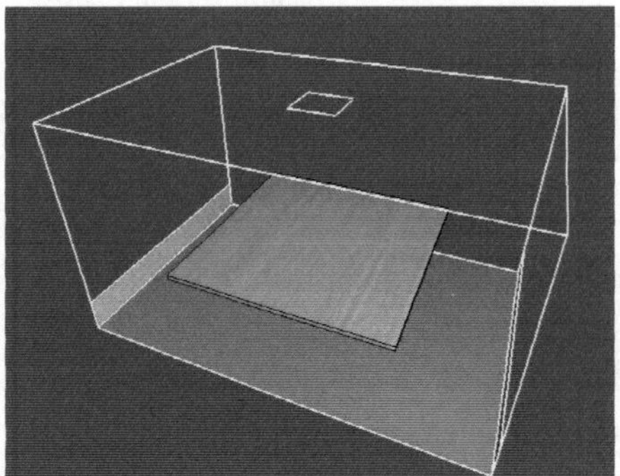

Figure 3. Cone Calorimeter test: FDS simulation model

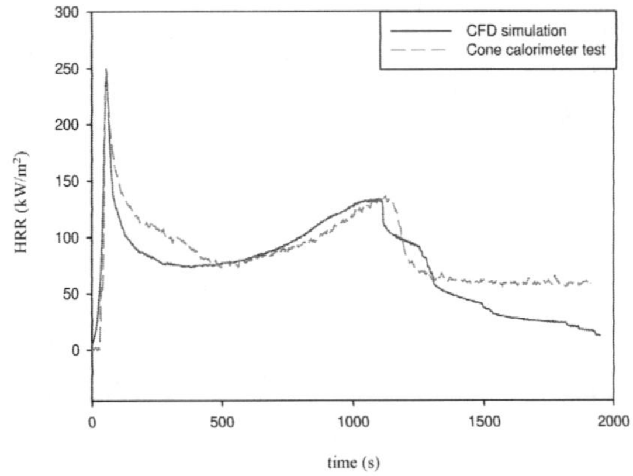

Figure 4. Comparison between the CFD and cone calorimeter test results

3.2. FDS results and parametric analysis

The figures from 5 to 11 show the parametric analysis carried out for different thermal and combustion parameters.

In particular, the figure 5 shows the influence of thermal conductivity on HRR-time curve, for fixed k=1.9 W/mK, for variation of the thermal conductivity values: – 20%, -10%, -5%, +5%, +10%, +20% with respect of its imposed value. It is noted that a -20% error on the correct value (k=1.5 W/mK) of the thermal conductivity, involves a decrease in the time of about 15 s with a first heat release rate peak of about 235 kW/m^2 (see figure 6).Vice versa, an error of + 20% (k=2.3 W/mK) involves a delay of about 12 s of the first heat release rate peak, with an HRR value equal to about 229 kW/m^2. Moreover, in figure 5, for a + 20% error, the time of maximum mass loss, compared to the base case, is reduced by about 100 s, while for a -20% error, the time increases by about 120 s. The value of the HRR, at the second peak, decreases as the thermal conductivity decreases from 157 kW/m^2 to about 110 kW/m^2.

With regards to the influence of heat specific capacity on HRR-time curve, as shown in the figures 7 and 8, considering the same errors used for the conductivity, it is possible to note that, for fixed c_p=2.8 kJ/kgK, an error equal to -20% (c_p=2.2 kJ/kgK) on the correct value of the heat specific capacity, involves a decrease in the time of 15 s at which the first peak of combustion is obtained with a value about 315 kW/m^2. An error equal to +20% (c_p=3.4 kJ/kgK) involves a delay of about 64 s of the first heat release rate peak with a value about 170 kW/m^2. Moreover, in figure 7, for a -20% error, the time of maximum mass loss, compared to the base case, is reduced at about 285 s, while for a +20% error, the time increases by about 300 s. The value of the HRR, at the second peak, decreases as heat specific capacity increases from 195 kW/m^2 to about 100 kW/m^2.

With regards to the influence of reference temperature on HRR-time curve, as shown in the figures 9 and 10, considering the same errors, it is possible to note that, for fixed T_{ref}=380 °C, an error equal to -20% (T_{ref}=304 °C) on the correct value of the reference temperature, involves a decrease in the time (35 s) at which the first peak of combustion is obtained with a value of HRR of about 420 kW/m^2. An error equal to +20% (T_{ref}=465 °C) involves a delay of about 70 s of the first heat release rate peak with a value of about 180 kW/m^2. Moreover, in figure 9, for a -20% error, the time of maximum mass loss, compared to the base case, is reduced by about 540 s, while for a +20% error, the time increases about 500 s. The value of the HRR, at the second peak, decreases as reference temperature increases from 308 kW/m^2 to about 55 kW/m^2.

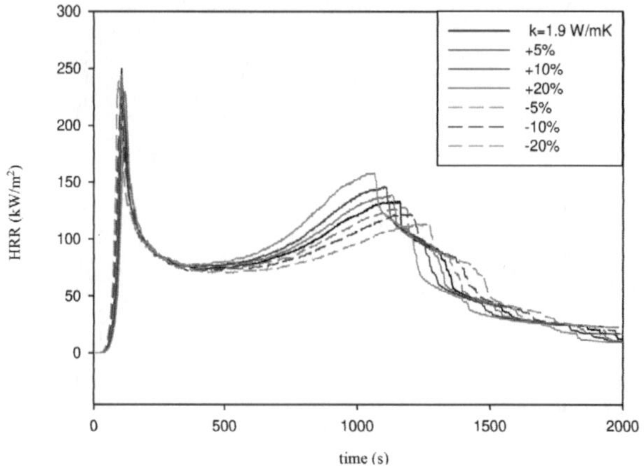

Figure 5. Influence of thermal conductivity on HRR-time curve

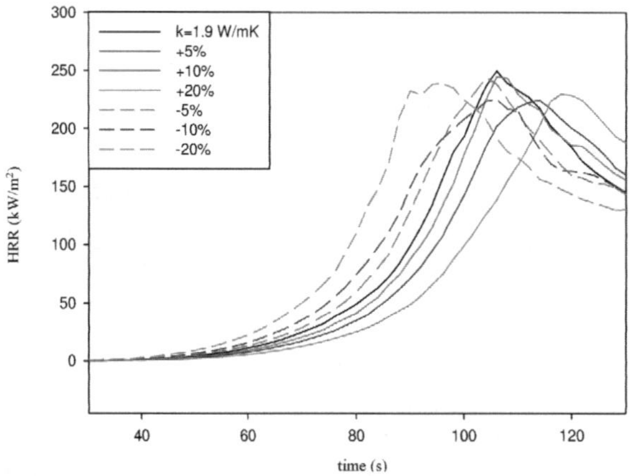

Figure 6. Particular of the influence of thermal conductivity on the first peak of HRR-time curve

With regards to the influence of pyrolysis range on HRR-time curve, as shown in the figures 11 and 12, considering the same errors, it is possible to note that, for fixed pyrolysis range =120 °C, an error equal to -20% (pyrolysis range =96 °C) on the correct value of the reference temperature, involves a decrease in the time (10 s) at which the first peak of combustion is obtained with a value of about 407 kW/m^2 and an error equal to +20% (pyrolysis range =144 °C) involves a delay of about 60 s of the first heat release rate peak with a value about 123 kW/m^2. Moreover, in figure 11, the influence due to the pyrolysis range variation is negligible for both time and HRR pick values.

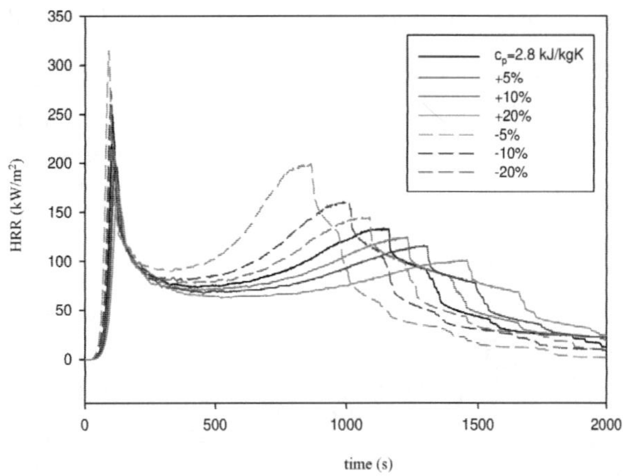

Figure 7. Influence of heat specific capacity on HRR-time curve

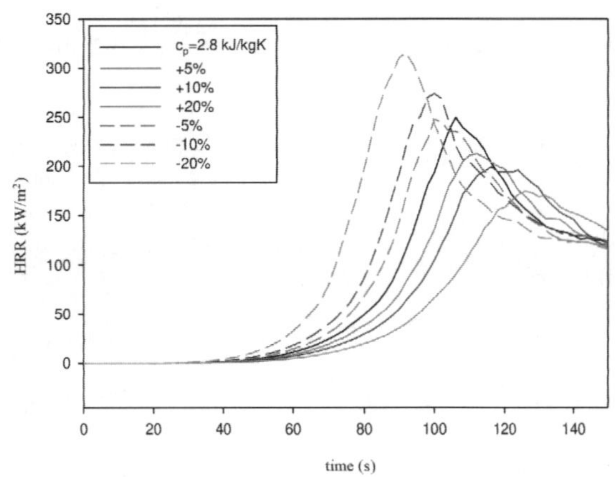

Figure 8. Particular of the influence of heat specific capacity on the first peak of HRR-time curve

4. CONCLUSION

The main drawbacks of CFD fire simulators are due to the not user friendly and require a good competence on fires and fair competence in computing. To estimate accurately the influence of the model assumptions and simplifications on the simulation outcome, it should know the dominating physical processes and involved empirical parameters that have an essential influence on the specific fire scenario.

In the present paper, a parametric analysis was carried out to evaluate the influence of the main thermal and combustion parameters on goodness of CFD simulation. The results showed that the variation of ±20% of these parameters, such as thermal conductivity, heat capacity, reference temperature and pyrolysis range, with respect to their nominal values, implies, in some cases, an important variation in terms of HRR peak and combustion time. For example, in the case of reference temperature, its variation of +20% with respect to the fixed base value, implies an advance of the combustion peak of about 8 min with a reduction of combustion time of about 10 min. This variation could be significantly important for fire risk analysis. Furthermore, the value of the HRR, at the second peak, decreases as reference temperature increases from 308 kW/m^2 to about 55 kW/m^2. Therefore, an accurate parametric a-priori analysis is necessary in order to obtain reliable results for a subsequent fire risk analysis. The sensitivity of the error in the setting of some parameters is very high and would lead to completely incorrect evaluations.

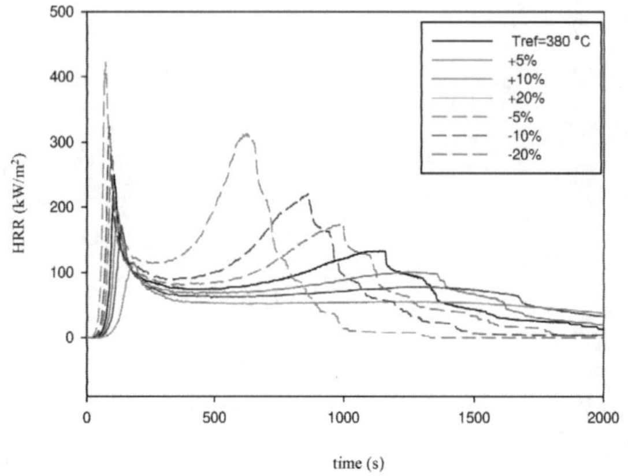

Figure 9. Influence of reference temperature on HRR-time curve, fixed T_{ref}=380 °C

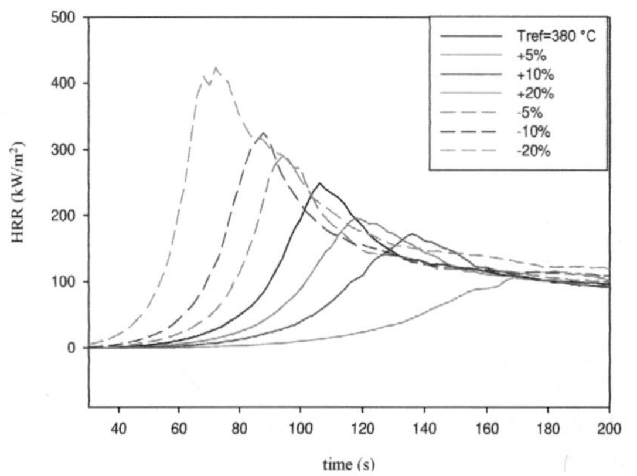

Figure 10. Particular of influence of reference temperature on the first peak of HRR-time curve

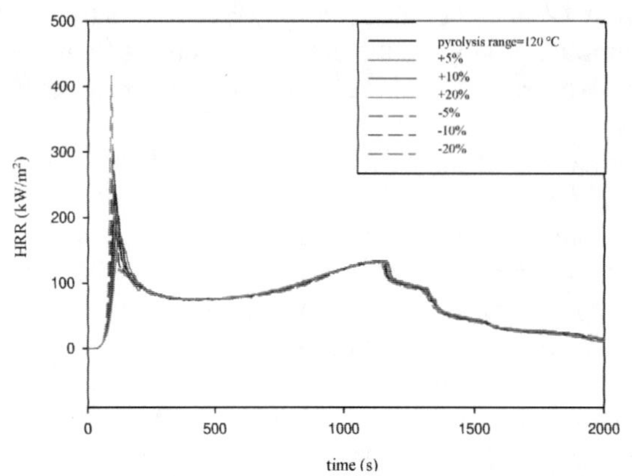

Figure 11. Influence of pyrolysis range on HRR-time curve, fixed pyrolysis range equal to 120°C

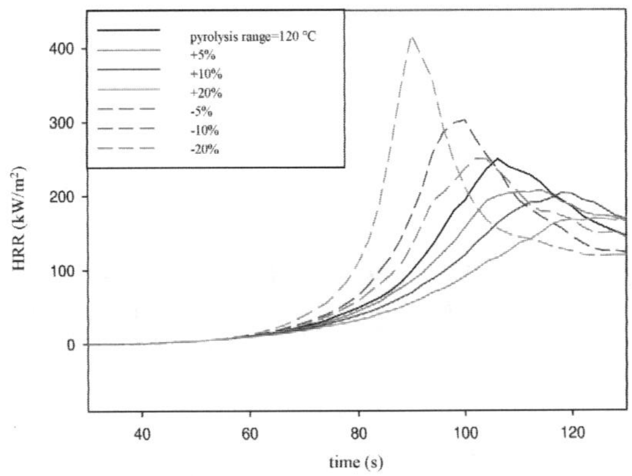

Figure 12. Particular of the influence of pyrolysis range on the first peak of HRR-time curve

References

[1] Johansson N and Ekholm M 2018 Variation in Results Due to User Effects in a Simulation with FDS *Fire Technology* **54 (1)** 97-116

[2] Steen-Hansen A 2007 Prediction of fire classification for wood based products multivariate statistical approach based on the cone calorimeter *Fire and Materials* **31** 207–223

[3] Trettin C, Hagemann P, Werrel M, Meinert M and Wittbecker FW 2014 Influence of input parameters on the fire simulation *Proceedings of Fire and Evacuation Modeling Technical Conference* 1-8

[4] Huang S S, Lu S X, Li C H and He Q Z 2015 Numerical investigation of fire safety of an indoor pedestrian street *Proceedings of 7th International Conference on Intelligent Computation Technology and Automation ICICTA 2014* 383-387

[5] Ayala P, Cantizano A, Gutierrez-Montes C and Rein G 2013 Influence of trium roof geometries on the numerical predictions of fire tests under natural ventilation conditions *Energy and Buildings* **65** 382-390

[6] Zhang P, Fu J and Wu T 2008 Numerical simulation of the Cooperation effect of high-pressure water mist fire protection and mechanical ventilation *Journal of Shenyang Jianzhu University (Natural Science)* 854-858.

[7] Wang M Y, Lin C P, Li Y T and Ma H K 2014 Utilization of fire dynamics simulator model to study rice husk gasification in fixed-bed gasifier *BioResources* **9(3)** 3792-3804

[8] Nasif M S and Al Waked R 2013 Using computational fluid dynamics simulation to perform investigation *Applied Mechanics and Materials* **393** 845-850

[9] Hu L H, Fong N K, Yang L Z, Chow W K, Li Y Z and Huo R 2007 Modeling fire-induced smoke spread and carbon monoxide transportation in a long channel: Fire Dynamics Simulator comparison with measures data *Journal of Harzadous Materials* **140(1-2)** 293-298

[10] Tharima A F, Yusoff M Z and Ramhan M M 2018 Evaluation of a large eddy simulation on thermal boundary condition in underground car park *Journal of Advanced Research in Fluid Mechanics and Thermal Sciences* **48(1)** 17-24

[11] Mijorski S and Stankov P 2012 CFD modeling of Dalmarnock uncrontrolled fire test *Central European Journal of Engineering* **2(2)** 279-288

[12] Camillo A 2013 Multi-scale investigation of fire behaviour of a seat and a wall panel from European railway transport system ENSMA.

IOP Conf. Series: Journal of Physics: Conf. Series **1224** (2019) 012011 doi:10.1088/1742-6596/1224/1/012011

[13] Jiang Y 2006 Decomposition, Ignition and Flame Spread on Furnishing Materials *Centre for Environment Safety and Risk Engineering* Victoria University, Australia.
[14] Meunders A, Schroeder B, Baker G, Arnold L, Pau Dand Spearpoint M 2014 Parameter Optimization and Sensitivity Analysis for Fire Spread Modelling with FDS *SFPE 10th International Conference on Performance-Based Codes and Fire Safety Design Methods* University of Queensland, Australia.
[15] McGrattan K, Hostikka S, McDermott R, Floyd J, Weinschenk C and Overholt K 2013 *Fire Dynamics Simulator (Version 6) Users Guide* National Institute of Standards and Technology, Building and Fire Research Laboratory
[16] Bilger R W 1994 Computational field models in fire research and engineering *Fire safety science-proceedings of the fourth international symposium* 95-110
[17] Lazaro M, Boehmer H, Alvear D, Capote J and Trouve A 2008 Numerical Simulation of Fire Growth, Transition to Flashover, and Post-Flashover Dynamics in the Dalmarnock Fire Test *Fire Safety Science–Proceedings of the ninth international symposium* **9** 1377-1388

IOP Conf. Series: Journal of Physics: Conf. Series **1224** (2019) 012012 doi:10.1088/1742-6596/1224/1/012012

Experimental validation of numerical model for evaluation of local heat transfer coefficient in coiled tubes

P Vocale[a], M Abbenante[b], F. Bozzoli[a,b], S Rainieri[a,b], G Pagliarini[a,b]
[a]Department of Engineering and Architecture, University of Parma, Parco Area delle Scienze 181/A, 43124 Parma – Italy
[b]CIDEA Interdepartmental Centre, University of Parma, Parco Area delle Scienze 181/A, 43124 Parma – Italy

E-mail: pamela.vocale@unipr.it

Abstract. The efficiency of the enhanced heat transfer equipment was widely investigated in literature by adopting both experimental and numerical approaches. Obviously, the numerical tools allow improving the knowledge of the fundamental heat transfer mechanisms in curved geometries, since they enable to evaluate punctually the velocity and temperature distributions. On the other hand, it has to be pointed out that the numerical results can be considered reliable only if they are validated throughout the comparison with either analytical results or experimental data. The validation of numerical models is usually carried out by considering average heat transfer performances, therefore this approach does not guarantee that these models enable to evaluate the local heat transfer phenomena with the same accuracy. In the present paper a promising numerical tool for evaluating local convective heat transfer in coiled-tubes, within the laminar regime, is presented. The Navier-Stokes and energy equations are solved by means of the open-source CFD software OpenFOAM®, with second-order accurate finite-volume schemes, within the framework of the SIMPLE algorithm. The robustness of the numerical model is assessed by comparing the numerical outcomes to the local experimental data available in literature.

1. Introduction

The performances of passive heat transfer enhancement techniques were widely investigated in literature by adopting both experimental and numerical approaches. Although the numerical methodologies enable to evaluate what happens within the fluid, thus providing a better understanding of the heat transfer enhancement mechanism, it has to be pointed out that numerical models can be considered relevant if they are validated against either analytical results or experimental evidences.

One of the simplest passive heat transfer enhancement technique, commonly used in several industrial applications, is based on the use of coiled tube because, for given operating conditions, in this kind of devices heat transfer rate is more effective than in straight pipes [1].

Due to the influence of the wall conformation on the thermal performance of heat transfer equipment, many research group developed numerical models to investigate laminar heat transfer in curved tubes [2-11]. In particular, Mori and Nakayama [2,3] investigated the effect of the secondary flow due to the curvature on the heat transfer coefficient both analytically and experimentally considering different thermal boundary conditions, namely constant wall heat flux and uniform wall temperature. By means of first and second order solutions of the boundary-layer approximation equations, they proposed simple practical correlations to evaluate the Nusselt number as a function of the main parameters that affect the fluid behaviour.

The limits of the boundary-layer approximation approach was highlighted by Mitsunobu and Cheng [4] who analysed fully developed laminar forced convection in curved pipes and presented a finite-difference solution obtained by means of a combination of line iterative method and boundary vorticity method. Their numerical outcomes showed the effect of Prandtl number on the convective heat transfer coefficient.

The influence of Dean number on the fluid flow behaviour was investigated by Patankar et al. [5] who presented a numerical analysis of local convective heat transfer coefficient for Prandtl number equal to 0.7.

Kozo and Aoyama [6] theoretically and numerically investigated the effect of the secondary flow on heat transfer in uniformly heated helically coiled tube, by accounting for both the centrifugal and buoyancy forces. By analysing several operating conditions, they proposed an approximate correlation that enables to evaluate the Nusselt number as a function of main parameters that influence the fluid behaviour. The effect of buoyancy in helical pipe was also studied by Zheng et al. [7] who accounted for the influence of thermal radiation in a participating medium, as well.

Yang et al. [8, 9] presented numerical investigations on the fully developed laminar convective heat transfer in a helicoidal pipe, with particular attention to the effects of Dean and Prandtl numbers and of torsion on the local convective heat-transfer coefficient. In particular, as concern the effect of torsion induced by the coil pitch [9], the authors reported the Nusselt number distribution for different values of the coil pitch, and they showed that, due to torsion, the local heat-transfer coefficient, compared to the case of an ideal torus, is increased on half of the tube wall while it is decreased on the other half.

Recently, Ko [10] numerically studied the entropy generation rate in helical pipes, by considering water as working fluid and several values of wall heat flux. For each operating condition analysed, his numerical results suggested the optimal Reynolds number to minimize the entropy generation.

However, it has to be pointed out that many of the works available in literature are focused only on the numerical analysis of the average performance of the coiled tubes. Since the flow velocity and temperature fields in coiled ducts are strongly asymmetrical over the cross-section of the tube, local effects have to be accounted for when a numerical model has to be developed. Moreover, local behaviour can affect the performances of many industrial processes. For instance, in food pasteurisation, the irregular temperature field induced by the wall curvature could reduce the effect of the bacteria heat-killing or could locally overheat the product. To the Authors' best knowledge, only few papers present numerical results in terms of local Nusselt number.

Furthermore, it has to be pointed out that the numerical models that were developed to analyse the local convective heat transfer coefficient, were validated only in terms of average performances. Therefore, it is not guaranteed that the so validated models enable to properly evaluate local heat transfer phenomena.

More recently, Bozzoli et al. [11] presented a preliminary numerical analysis of local convective heat transfer in coiled tubes within laminar flow regime. The adopted model was validated through the comparison of the numerical results with the experimental data in terms of convective heat flux distribution along the boundary of the duct. The comparison highlighted that the adopted numerical model did not allow to capture correctly the local heat transfer phenomena.

Therefore, the present work aims to extend the numerical analysis proposed in [11] by adopting a different numerical model that enables to accurately investigate the local heat transfer phenomena in helically coiled tubes for a fully developed laminar flow. To account for the local interaction between the solid walls and the fluid, the analysis is carried out by considering the conjugate heat transfer problem. The Navier-Stokes and energy equations are solved by using the open-source CFD software OpenFOAM®,

The validation of the here presented numerical model is carried out by comparing the numerical results with the experimental data in terms of local convective heat transfer coefficient.

2. Experimental data

The experimental data used to validate the numerical results were obtained by Bozzoli et al. [11,12] who evaluated the convective heat flux and heat transfer coefficient distributions along the boundary of a helically coiled tube. For sake of clarity it's important to recall here the main operating conditions of the experiments carried out in [12]. The tube was characterized by eight coils following a helical profile along the tube's axis, as shown in Figure 1. The duct was characterized by an internal diameter, a helix diameter and a pitch, 14 mm, 310 mm and 200 mm, respectively (the curvature ratio was about 22.14) while the wall thickness was 1 mm.

Ethylene glycol was used as working fluid while the Reynolds number was varied between 135 and 1006. The fluid was heated by Joule dissipation in the tube wall; indeed the inlet and outlet coiled test sections were equipped with stainless-steel fin electrodes and connected to a power supply, as shown in Figure 1.

The test section was located approximately 9 m downstream of the inlet section, in the region where the laminar boundary layers reached the asymptotic profiles. Therefore, the results obtained for that particular section were representative of the fully developed region.

To minimize the heat exchange with the environment, the heated section was insulated by using neoprene, as shown in Figure 1; on the other hand, a small portion of the external tube wall, near to the outlet section, was uninsulated to make it accessible to an infrared imaging camera. A thin film of opaque paint characterized by uniform and known emissivity coated this portion.

Figure 1. Sketch of test rig used for assessing the performance of the helically coiled tube.

The temperature distribution maps on the external surface of the wall were acquired by using a FLIR SC7000 infrared camera, while the inlet and the outlet fluid bulk temperatures were measured with type-T thermocouples. The acquired temperature distributions were employed as input data of the Inverse Heat Conduction Problem (IHCP) in the wall under a solution approach based on Tikhonov regularization method.

3. Numerical model

The local performance of helically coiled tube was numerically investigated by solving the conjugate heat transfer problem, as shown in Figure 2.

The tube under investigation had the same geometrical and physical properties of the duct tested in [11,12] and described in the previous section.

A Cartesian coordinate system *x*, *y*, *z* was introduced, whose origin was located as shown in Figure 2.

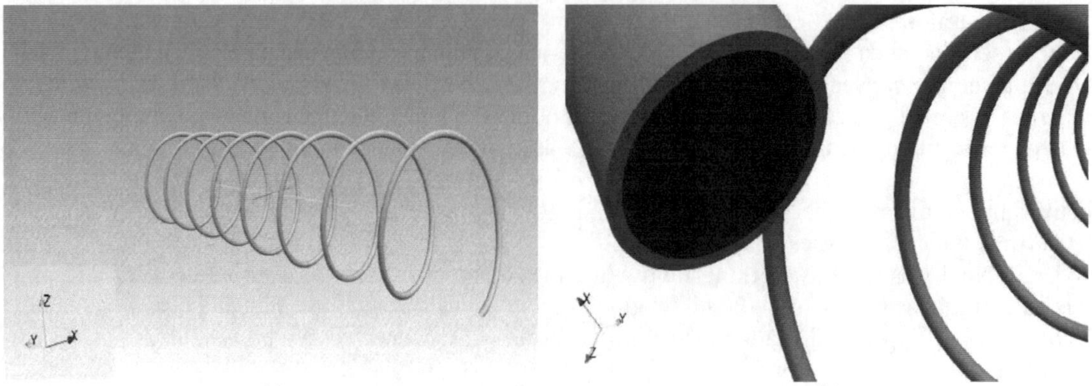

(a) (b)

Figure 2. Geometry under investigation. Sketch of the whole geometric domain (a); zoom on fluid and wall regions, in blue and in grey, respectively (b).

By considering a Newtonian fluid with constant physical properties and by assuming that the flow is laminar and steady-state, the governing equations in Cartesian coordinates are as follows:

$$\frac{\partial u}{\partial x}+\frac{\partial v}{\partial y}+\frac{\partial w}{\partial z}=0 \tag{1}$$

$$\rho\left(u\frac{\partial u}{\partial x}+v\frac{\partial u}{\partial y}+w\frac{\partial u}{\partial z}\right)=-\frac{\partial p}{\partial x}+\mu\left(\frac{\partial^2 u}{\partial x^2}+\frac{\partial^2 u}{\partial y^2}+\frac{\partial^2 u}{\partial z^2}\right)$$

$$\rho\left(u\frac{\partial v}{\partial x}+v\frac{\partial v}{\partial y}+w\frac{\partial v}{\partial z}\right)=-\frac{\partial p}{\partial y}+\mu\left(\frac{\partial^2 v}{\partial x^2}+\frac{\partial^2 v}{\partial y^2}+\frac{\partial^2 v}{\partial z^2}\right) \tag{2}$$

$$\rho\left(u\frac{\partial w}{\partial x}+v\frac{\partial w}{\partial y}+w\frac{\partial w}{\partial z}\right)=-\frac{\partial p}{\partial z}+\mu\left(\frac{\partial^2 w}{\partial x^2}+\frac{\partial^2 w}{\partial y^2}+\frac{\partial^2 w}{\partial z^2}\right)$$

$$\rho\left(u\frac{\partial i}{\partial x}+v\frac{\partial i}{\partial y}+w\frac{\partial i}{\partial z}\right)=\lambda_f\left(\frac{\partial^2 T}{\partial x^2}+\frac{\partial^2 T}{\partial y^2}+\frac{\partial^2 T}{\partial z^2}\right) \tag{3}$$

where u, v and w are the components of the velocity field, ρ, μ, and λ_f are the fluid density, dynamic viscosity and thermal conductivity, respectively, p is the pressure, i the fluid specific enthalpy and T the fluid temperature.
In the solid region, the energy conservation equation reads as follows:

$$\lambda_s\left(\frac{\partial^2 T}{\partial x^2}+\frac{\partial^2 T}{\partial y^2}+\frac{\partial^2 T}{\partial z^2}\right)+q_g=0 \tag{4}$$

being T the temperature of the solid domain, q_g the heat generated by Joule effect in the wall and λ_s the thermal conductivity of the wall.
The governing equations in the fluid region were solved by considering the following boundary conditions at x=0:

$$u=W,\qquad v=0,\qquad w=0,\qquad T=T_{in} \tag{5}$$

while pressure condition was imposed at the outlet section.

At the fluid-solid interface, no-slip boundary condition was considered for the fluid flow problem while continuity of temperature and heat flux was imposed for the thermal problem. The energy equation was simultaneously solved for enthalpy in solid and fluid regions. Consequently, the thermal condition at the fluid-solid interface was automatically met by the local energy balance. Other surfaces of the solid domain were considered to be adiabatic.
The local Nusselt number was evaluated as follows:

$$Nu = \frac{hD}{\lambda_f} = \frac{q_w D}{\lambda_f \left(T_w - T_b\right)} \tag{6}$$

where q_w is the wall heat flux, h is the convective heat transfer coefficient, D the tube internal diameter, T_w and T_b are the wall and fluid bulk temperature, respectively.
By considering the wall heat flux and the wall temperature averaged along the wall circumference, the circumferentially averaged Nusselt number was evaluated:

$$\overline{Nu} = \frac{\overline{q_w} D}{\lambda_f \left(\overline{T_w} - T_b\right)} \tag{7}$$

being $\overline{q_w}$ and $\overline{T_w}$ the circumferentially averaged wall heat flux and temperature, respectively.

The conjugate heat transfer problem (i.e. Eqs.(1) and (4)) together with its boundary conditions was investigated using the finite volume method. The simulations were carried out with the open-source CFD package OpenFOAM®, where the flow and energy equations were solved using unstructured and second-order accurate finite-volume schemes, within the framework of the SIMPLE (Semi-Implicit Method for Pressure-Linked Equations) algorithm [13]. For all governing equations, linear iterative solvers were used with an absolute tolerance equal to 1e-9. In particular, the pressure distribution within the fluid was solved using a Preconditioned Conjugate Gradient (PCG) solver, while an iterative solver that uses a smoother was used for determining both the velocity and temperature distributions in the fluid domain [14]. The solver with smoother was also used to solve the energy conservation equation in the solid region.

A limited linear or central convection scheme are used as divergence and gradient schemes, while a corrected linear scheme was used for the diffusive term of the governing equations. The computations were performed by means of an hybrid grid, consisting of hexahedral cells and created using the open-source software for mesh generation Gmsh [15].

4. Results

The model validation was carried out by considering the following operating conditions for which local experimental data were available in [11,12]: Pr=173 and Re=135.
To make sure that the solution was independent of the mesh resolution, several grids with different sizes were tested. Due to the interest of the present analysis in the hydrodynamically and thermally fully developed flow, for each considered grid, the local and the circumferentially averaged Nusselt number in the fully developed region were monitored.
To better evaluate the fluid-solid interface for all of the considered grids several thin layers were generated in both solid and fluid domains, as shown in Figure 3, where two representative outlet sections of the investigated grids are presented. It has to be pointed out that due to complexity of here considered geometry, several mesh parameters (i.e. the number of thin layers in both fluid and solid domains, the number of sections along the duct axis and the number of radial elements in the cross-section) can be varied to test the influence of the grid size on the numerical results. To check the

influence of the cells number on the fluid flow behaviour, all of the grid parameters were varied at the same time.

In Figure 4, the results of the mesh independence study are reported in terms of both circumferentially averaged and local Wall Shear Stress (WSS) for all of the here considered mesh sizes. In particular, in Figure 4a WSS averaged along the duct perimeter is depicted as a function of the dimensionless cell size (i.e. V^* was obtained by dividing the total cells volume by the minimum cells volume), while in Figure 4b local distribution of WSS is presented along the angular coordinate α, whose origin was taken at the extrados. It was observed that for a grid characterized by V^*=1.01e07 (i.e. about 4'400'000 cells) both the average and the local WSS could be assessed with a good accuracy.

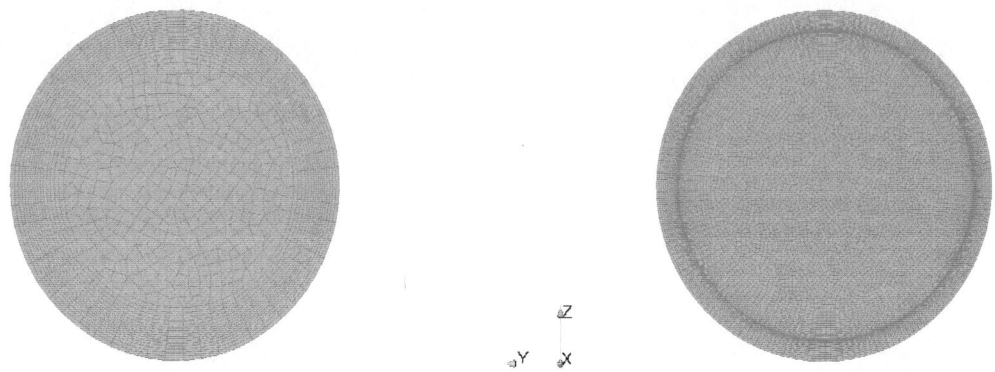

(a) (b)

Figure 3. Sketch of the investigated grids. Coarsest mesh (976'000 cells) (a) and finest mesh (12'500'000 cells) (b).

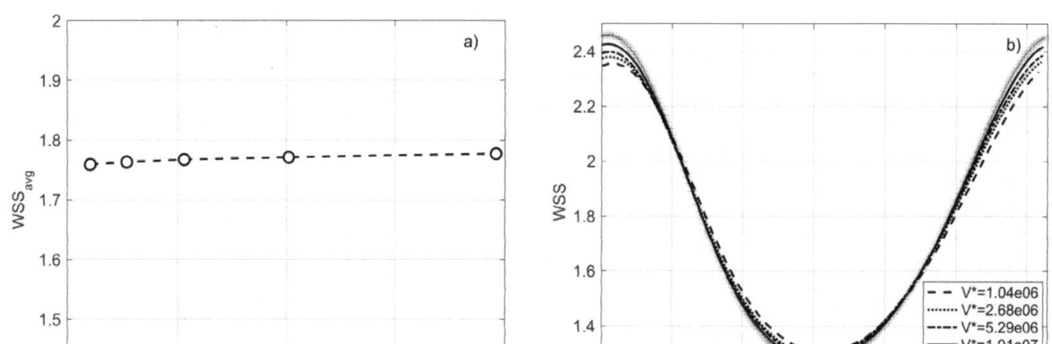

(a) (b)

Figure 4. Mesh convergence analysis. Average WSS (a) and local WSS (b).

Although the local fluid flow behaviour was correctly evaluated by using a mesh characterized by at least 4'385'000 cells, a further refinement was necessary to obtain a good accuracy in the evaluation of the fluid thermal behaviour, as shown in Figure 5, where the results of the grid convergence analysis are reported in terms of both circumferentially averaged and local Nusselt numbers. *Nu* averaged along the wall circumference as a function of the dimensionless cell size is depicted in Figure 5a. It can be noticed that with the grid refinement ranging from V^*=1.01e07 to V^*=1.96e07 the average Nusselt number only changes by less than 1.6%, which indicates that the results obtained by adopting the grid with about 7'400'000 can be considered accurate. To assess the influence of the grid size on the evaluation of the local heat transfer phenomena, in Figure 5b the Nusselt number distribution along the angular coordinate is reported for all of the here considered mesh sizes. The data in Figure 5b confirm that the results obtained by adopting a mesh characterized by at least 7'400'000 cells (i.e. V^*=1.01e07) allow evaluating the local heat transfer with a good accuracy.

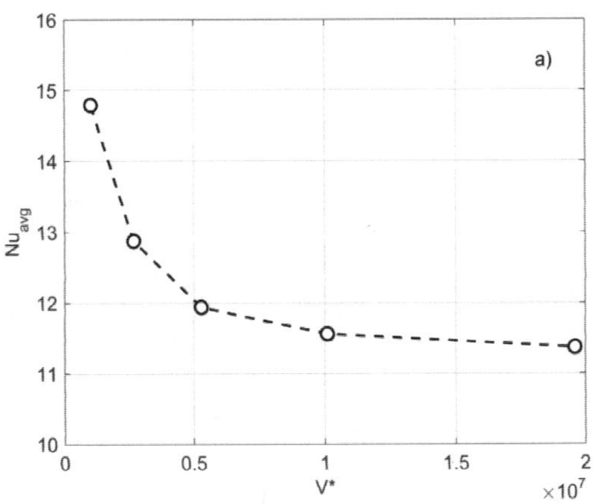

(a)

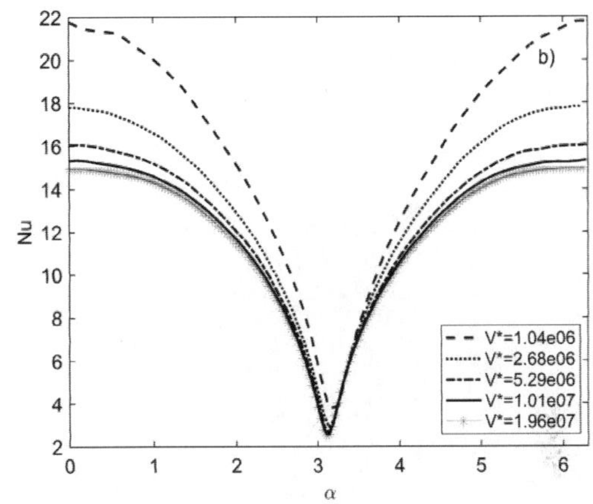

(b)

Figure 5. Mesh convergence analysis. Circumferentially averaged Nusselt number (a) and local Nusselt number (b).

To validate the numerical model here adopted, the results obtained by using the finest mesh were compared to the experimental data presented in [11,12] in terms of local Nusselt number. In Figure 6 the numerical outcomes in terms of normalized Nusselt number distribution (i.e. $Nu_{ratio}=Nu/Nu_{max}$) along the angular coordinate are compared to experimental results. The comparison reveals that the numerical model enables to evaluate the local distribution of the heat transfer coefficient with a good accuracy, since the numerical results and experimental data are in good agreement. Moreover, in the same figure the results obtained by adopting the numerical model presented in the preliminary analysis [11] are presented. It can be observed that the present numerical model is more accurate in the evaluation of the local distribution of the Nusselt number ratio.

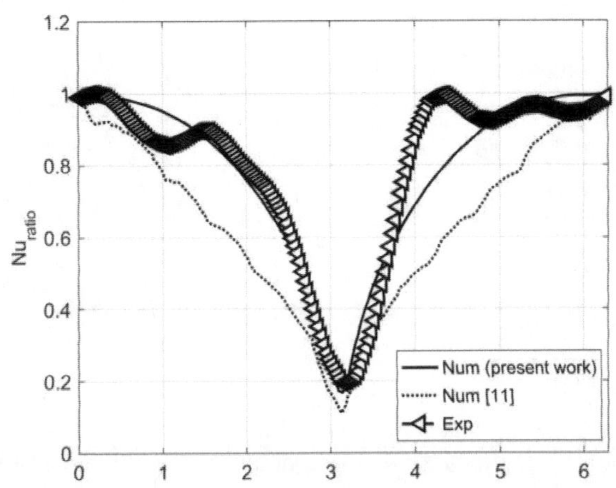

Figure 6. Comparison with experimental data and preliminary numerical results.

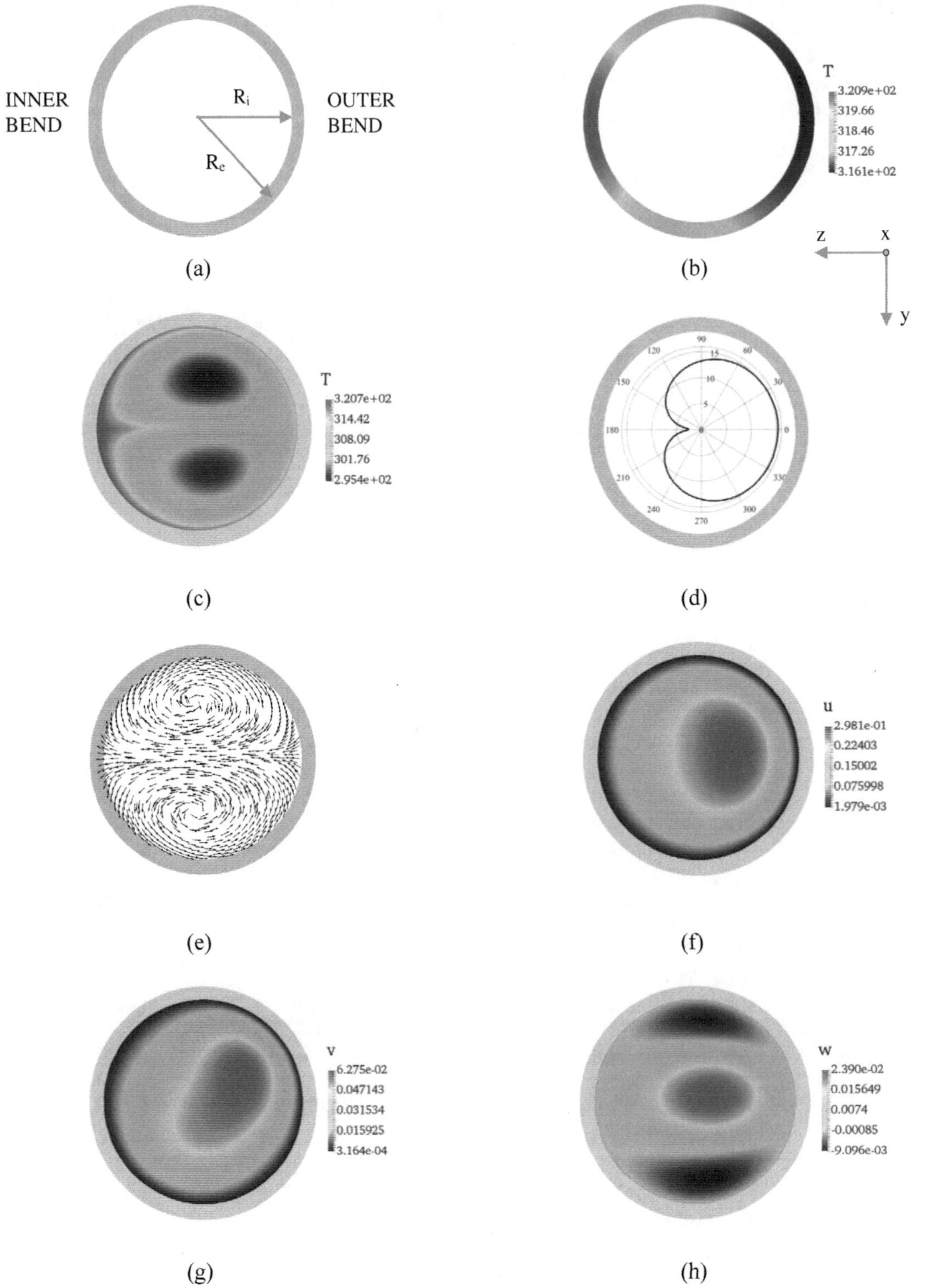

Figure 7. (a) Sketch of the outlet section; (b) wall temperature distribution [*K*]; (c) fluid temperature distribution [*K*]; (d) Nusselt number; (e) vortices; (f) *x*-component of the velocity vector [*m/s*]; (g) *y*-component of the velocity vector [*m/s*]; (h) *z*-component of the velocity vector [*m/s*].

Since the numerical model here adopted is robust and accurate, the numerical results can provide a good understanding of the physical phenomena, which take place in this kind of heat transfer device. In Figure 7 several representative results of the numerical runs are presented. The numerical outcomes clearly show that, as expected, the centrifugal force induces a secondary motion orthogonal to the main flow direction, which is characterized by the formation of a pair of vortices (i.e. the Dean vortices). These vortices presented in Figure 7e are responsible of the fluid temperature distribution shown in Figure 7c and of the Nusselt number distribution depicted in Figure 7d. In fact, as expected, the minimum of the convective heat transfer coefficient occurs at the extrados. To better evaluate the influence of the torsion induced by the coil pitch on the fluid velocity, the distributions of the velocity components u,v and w, which represent the components of the velocity vector along x, y and z, respectively, are also reported in Figure 7.

5. Conclusions

A numerical model able to capture the local heat transfer phenomena in laminar flow through helically coiled tubes was presented. To provide a good understanding of the local phenomena, the analysis was carried out by considering the conjugate heat transfer problem.

The flow and energy equations were solved by using the open-source CFD package OpenFOAM®, based on unstructured and second-order accurate finite-volume schemes, within the framework of the SIMPLE algorithm.

The experimental validation of the here adopted numerical model was carried out by comparing the experimental data obtained for ethylene glycol with the numerical results, in terms of both average and local distribution of the convective heat transfer coefficient. The good agreement between the predictions and the experimental evidences reveals that the numerical model offers a better insight in the heat transfer enhancement mechanism thus providing a useful tool to optimize the heat transfer devices.

Therefore, the here adopted numerical model will adopt to evaluate the influence of the main parameters that affect the fluid flow behavior, such as the Reynolds and Prandtl numbers, the Dean number, the thermal and geometrical properties of the wall (in order to evaluate the influence of the thermal boundary conditions as well). Moreover, the effect of the torsion induced by the coil pitch on the fluid flow behaviour will be investigated by considering several configurations.

Acknowledgments

The present work was partially supported by Regione Emilia Romagna, within POR-FESR 2014-2020, project HEGOS - nuove pompe di calore per l'Harvesting EnerGeticO in Smart buildings.
The numerical simulations were performed at CINECA High Performance Computing facilities, under the *HeTrE4TI* Project.

References

[1] Naphon P and Wongwises S 2004 A review of flow and heat transfer characteristics in curved tubes *Renew. Sust. Energ. Rev.* **10** 463–90
[2] Mori Y and Nakayama W 1967 Study on forced convective heat transfer in curved pipes (1st report, laminar region) *Int. J. Heat Mass Tran.* **8** 67–82
[3] Mori Y and Nakayama W 1967 Study on forced convective heat transfer in curved pipes (3rd report, theoretical analysis under the condition of uniform wall temperature and practical formulae) *Int. J. Heat Mass Tran.* **10** 681–95
[4] Mitsunobu A and Cheng KC 1971 Boundary vorticity method for laminar forced convection heat transfer in curved pipes *Int. J. Heat Mass Tran.* **14** 1659–75
[5] Patankar SV, Pratap VS and Spalding DB 1974 Prediction of laminar flow and heat transfer in helically coiled pipes *J. Fluid Mech.* **62** 539–51
[6] Kozo F and Aoyama Y 1988 Laminar heat transfer in a helically coiled tube *Int. J. Heat Mass*

Tran. **31** 387–96
[7] Zheng B, Lin CX and Ebadian MA 2000 Combined laminar forced convection and thermal radiation in a helical pipe *Int. J. Heat Mass Tran.* **43** 1067–78
[8] Yang G and Ebadian MA 1994 Perturbation solution for laminar convective heat transfer in a helix pipe *J. Heat Transf.* **116** 771–75
[9] Yang G, Dong ZF and Ebadian MA 1995 Laminar forced convection in a helicoidal pipe with finite pitch *Int. J. Heat Mass Tran.* **38** 853–62
[10] Ko TH 2006 Numerical investigation of laminar forced convection and entropy generation in a helical coil with constant wall heat flux *Numer. Heat Tr. A-Appl.* **49** 257–78
[11] Bozzoli F, Cattani L, Rainieri S and Zachár A 2015 Numerical and experimental study of local heat transfer enhancement in helically coiled pipes. Preliminary results *J. Phys. Conf. Ser.* **655** doi:10.1088/1742-6596/655/1/012047
[12] Bozzoli F, Rainieri S, Cattani L, and Pagliarini P 2013 Compound convective heat transfer enhancement in helically coiled wall corrugated tubes *Int. J. Heat Mass Tran.* **59** 353–62
[13] Ferziger, JH and Peric, M 2001 *Computational Methods for Fluid Dynamics* 3rd Ed. Springer: Berlin.
[14] Moukalled, F, Mangani, L and Darwish, M 2016 *The Finite Volume Method in Computational Fluid Dynamics* Springer International Publishing: Switzerland.
[15] Geuzaine C and Remacle JF 2013 A 3-D finite element mesh generator with built-in pre- and post-processing facilities *Int. J. Numer. Meth. Eng.* **79** 1309–31

Numerical analysis of advection-diffusion problems on 2D general-shaped domains by means of a RBF Collocation Meshless Method

Riccardo Zamolo[1] and Enrico Nobile[1]

[1] Dipartimento di Ingegneria e Architettura, Università degli Studi di Trieste, via A. Valerio 10, 34127 Trieste, Italy

E-mail: `riccardo.zamolo@phd.units.it`

Abstract. A Collocation Meshless Method based on local Radial Basis Function (RBF) interpolation is employed to solve two-dimensional advection-diffusion problems with particular reference to the incompressible Navier-Stokes equations in their transient form, i.e., unsteady flows, using primitive variables $(\mathbf{u}, p)$. A projection scheme is employed to decouple the continuity and momentum equations; particular attention is given to the choice of the required solvers. This approach is applied to the simulation of unsteady flows for two typical test cases, i.e., the lid-driven cavity problem and the flow past a circular cylinder between parallel walls. Numerical results compare very favorably with literature ones, confirming that this approach can be effectively employed in the numerical simulation of unsteady flows on practical geometries where complex node distributions and large number of nodes are required.

Keywords:
Navier-Stokes equations, Poisson equation, multigrid, node distribution, fluid flow

1. Introduction

The simulation of engineering problems with widely used numerical approaches such as Finite Element and Finite Volume methods requires a spatial discretization of the domain; this meshing phase can require a significant effort, especially when dealing with complex geometries of practical relevance where the mesh must fulfill particular properties.

In order to avoid the need of a classic mesh, in recent years several meshless approaches have been proposed [1]: the mesh is replaced by a simple distribution of nodes which easily adapts to complex geometries; then some formal interpolation on this set of nodes is required in order to approximate differential quantities from nodal values. Being this a very generic approach, it can be applied to a wide range of problems including heat transfer [2, 3], solid mechanics and fluid flow [4].

Meshless methods employing local Radial Basis Function (RBF) interpolation [5–8], also known as RBF generated Finite Differences methods (RBF-FD), are particularly attractive for practical fluid flow problems because of their geometrical flexibility and computational efficiency due to the local formulation, and will be used in this paper for such purpose; furthermore, the local formulation allows the use of very efficient solvers.

The numerical approach presented here will be applied to the simulation of two-dimensional unsteady flows for two well-known test cases: the lid-driven cavity at $Re = 1000$ and $Re = 5000$, and the flow past a circular cylinder between parallel walls at Re ranging from 100 to 1000. Very good agreement to literature results is found for each of these test cases, showing that the proposed approach can be an accurate and efficient tool in the numerical simulation of unsteady fluid flow problems over geometries of practical relevance.

2. Governing equations and boundary conditions

Let us consider an incompressible fluid with density ρ and constant kinematic viscosity ν. The resulting two-dimensional Navier-Stokes equations can be made dimensionless by taking H, u_0, H/u_0 and ρu_0^2 as reference quantities for length, velocity, time and pressure, respectively, obtaining the following nondimensional momentum and continuity equations:

$$\frac{\partial \mathbf{u}}{\partial t} + (\mathbf{u} \cdot \nabla)\mathbf{u} = -\nabla p + \frac{1}{Re}\nabla^2 \mathbf{u} \tag{1}$$

$$\nabla \cdot \mathbf{u} = 0 \tag{2}$$

where $Re = u_0 H/\nu$ is the flow Reynolds number. No-slip conditions, i.e., $\mathbf{u} = \mathbf{u}_W$, are imposed at all solid walls with velocity $\mathbf{u}_W$; in the case of the flow past a circular cylinder, velocity is imposed at the inlet while fully developed flow conditions, i.e., $\partial \mathbf{u}/\partial \mathbf{n} = \mathbf{0}$ with $\mathbf{n}$ the outward unit normal, are imposed at the outlet.

3. Numerical method

3.1. 2D node distributions

The two-dimensional node distributions, i.e., a set of N nodes $\mathbf{x}_i$, required by the RBF-FD meshless discretizations have been obtained through the modified Quadtree algorithm followed by repel refinement proposed in [9]; such algorithm is briefly summarized here.

Given a prescribed spacing function $s(\mathbf{x})$ which defines the local spacing between the nodes, an initial node distribution is obtained through the Quadtree algorithm modified with a dithering correction in order to reduce the nodal quantization error, i.e., the difference between the integer number of nodes that can be contained in any set Ω and the prescribed (non-integer) number $\int_\Omega s(\mathbf{x})^{-2}\, d\omega$. The resulting node distribution is then improved by the application of 50 refinement iterations based on radial repel forces between the nodes.

3.2. Local RBF interpolation

RBF interpolation approximates the generic field ξ around $\mathbf{x}$ through the following expansion:

$$\xi(\mathbf{x}) = \sum_{j \in J_q(\mathbf{x})} a_j \varphi(\|\mathbf{x} - \mathbf{x}_j\|) + \mathbf{b}^T \mathbf{x} + c \tag{3}$$

where $J_q(\mathbf{x})$ is the set of the indices j of the q nodes $\mathbf{x}_j$ closest to $\mathbf{x} = (x, y)^T$. $\xi(\mathbf{x})$ is a linear combination of q radial functions φ centered at the q nodes $\mathbf{x}_j$ plus a linear polynomial in $\mathbf{x}$. In this work $q = 7$ local nodes have been chosen because the employed multicloud solver [10] is specifically developed and optimized for this choice. Hardy's Multiquadric has been chosen as radial function since it seems to offer the better results if an appropriate shape factor ε is employed [11]:

$$\varphi(r) = \sqrt{1 + (\varepsilon r)^2} \tag{4}$$

where the shape factor has been rescaled as $\varepsilon = \bar{\varepsilon} s_M / d_n(\mathbf{x})$ with s_M the maximum spacing function over the domain and $d_q(\mathbf{x})$ the local subdomain size [3]. In this work we employed $\bar{\varepsilon} = 1$ since it is a reasonable trade-off between accuracy and numerical stability [12].

The coefficients a_j, $\mathbf{b}$ and c can be formally computed imposing the canonical interpolation conditions for the $m \le q$ local nodes $\mathbf{x}_i$, $i \in J_q(\mathbf{x})$ which do not lie on the portion of the boundary where Neumann conditions are imposed on ξ:

$$\xi(\mathbf{x}_i) = \xi_i \tag{5}$$

where ξ_i is the nodal value of ξ in $\mathbf{x}_i$.

The following additional relations are needed in order to ensure linear reproducibility of Equation (3):

$$\sum_{j \in J_q(\mathbf{x})} a_j = 0 \quad , \quad \sum_{j \in J_q(\mathbf{x})} x_j a_j = 0 \quad , \quad \sum_{j \in J_q(\mathbf{x})} y_j a_j = 0 \tag{6}$$

If any of the local nodes $\mathbf{x}_i$, $i \in J_q(\mathbf{x})$ lies on the Neumann boundary, for each of these $q - m$ Neumann nodes the corresponding boundary condition must be satisfied:

$$\left. \frac{\partial \xi}{\partial \mathbf{n}} \right|_{\mathbf{x}=\mathbf{x}_i} = 0 \tag{7}$$

where the normal derivative of ξ is obtained from Equation (3), and only homogeneous Neumann boundary conditions are required in the presented problems.

Collecting the n coefficients a_i and the m unknown values ξ_i in column vectors $\mathbf{a}$ and $\boldsymbol{\xi}$, respectively, the interpolation system is the following:

$$\mathbf{G} \begin{Bmatrix} \mathbf{a} \\ \mathbf{b} \\ c \end{Bmatrix} = \begin{Bmatrix} \boldsymbol{\xi} \\ \mathbf{0} \end{Bmatrix} \tag{8}$$

3.3. Collocation

A generic linear partial differential equation $L(\xi) = b$ with the local RBF approximation expressed by Equation (3) becomes:

$$\sum_{j \in J_n(\mathbf{x})} a_j L\big(\varphi(\|\mathbf{x} - \mathbf{x}_j\|)\big) + \mathbf{b}^T L(\mathbf{x}) + c L(1) = b(\mathbf{x}) \tag{9}$$

Writing the previous equation in compact form for a generic node $\mathbf{x}_k$ near $\mathbf{x}$ gives:

$$L(\{\boldsymbol{\varphi}^T \ \mathbf{x}^T \ 1\})|_{\mathbf{x}=\mathbf{x}_k} \begin{Bmatrix} \mathbf{a} \\ \mathbf{b} \\ c \end{Bmatrix} = b(\mathbf{x}_k) \tag{10}$$

where $\boldsymbol{\varphi}$ is the column vector of the q local RBF φ and L acts element-wise on vectors. Finally, recalling $\mathbf{a}$, $\mathbf{b}$ and c from Equation (8) we obtain:

$$L(\{\boldsymbol{\varphi}^T \ \mathbf{x}^T \ 1\})|_{\mathbf{x}=\mathbf{x}_k} \mathbf{G}^{-1} \begin{Bmatrix} \boldsymbol{\xi} \\ \mathbf{0} \end{Bmatrix} = b(\mathbf{x}_k) \tag{11}$$

Writing Equation (11) for each internal node $\mathbf{x}_k$, i.e., which does not lie on the boundary, gives the final linear system:

$$\mathbf{A}_{II} \boldsymbol{\xi}_I = \mathbf{b} - \mathbf{A}_{ID} \boldsymbol{\xi}_D \tag{12}$$

where the subscripts I and D refer to *Internal* and *Dirichlet* contributions, respectively; $\mathbf{A}_{II}$ is sparse and non-symmetric.

The presented RBF-FD discretization approach is proven to be second order accurate in space for second order elliptic problems [9, 10].

3.4. Solution procedure

At each time step, the computation of velocity and pressure through Equations (1) and (2) is decoupled using a traditional projection scheme with a three-level Gear scheme for the time discretization; first, a tentative velocity $\mathbf{u}^*$ is computed from the linearized momentum equation:

$$\frac{3\mathbf{u}^* - 4\mathbf{u}^n + \mathbf{u}^{n-1}}{2\Delta t} + \tilde{\mathbf{u}}^{n+1}\nabla\mathbf{u}^* = -\nabla p^n + \frac{1}{Re}\nabla^2\mathbf{u}^* \tag{13}$$

where $\tilde{\mathbf{u}}^{n+1} = 2\mathbf{u}^n - \mathbf{u}^{n-1}$ is the extrapolated velocity at timestep $n+1$ and Δt is the time step size. Then, the computed tentative velocity $\mathbf{u}^*$ is forced to satisfy the continuity equation (2) by means of an irrotational correction $\mathbf{u}^{n+1} = \mathbf{u}^* - \nabla\phi$ (Helmholtz-Hodge decomposition) resulting in a Poisson equation in the auxiliary variable ϕ with homogeneous Neumann boundary conditions:

$$\nabla^2\phi = \nabla\cdot\mathbf{u}^* \tag{14}$$

The pressure is then updated as $p^{n+1} = p^n + \phi/\Delta t$.

Equations (13) and (14) are discretized using the previously presented local RBF collocation technique, resulting in a scheme with second order accuracy in both time and space. The momentum equation (13), consisting in 2 scalar equations for $\mathbf{u} = (u, v)^T$, is solved using the BiCGSTAB [13] iterative method with an additive correction multicloud (ACMC) technique as preconditioner [10], using 10^{-9} as target normalized RMS norm for the residuals; such choice for the momentum equations solver is motivated by the fact that the coefficient matrices for these equations change at each time since the advective term is treated implicitly, therefore a fast and iterative solver must be used. For the present test cases, the following ACMC parameters have been used: SOR iteration multiplier $r_I = 2$, SOR relaxation factor $\omega \in [0.7, 1.3]$ and overcorrection factor $t_{oc} \in [0.7, 1.8]$; the smallest values for ω and t_{oc} have been employed for the high Re cases in order to ensure the convergence of each scalar equation in less than 180-200 Work Units (1 WU = 1 residual evaluation at the fine level).

The Poisson equation (14) is solved through a LU decomposition of the corresponding coefficient matrix since it is constant and therefore the factorization can be performed once at the beginning of the simulation and the Poisson solution phase at each time step is very fast; nonetheless, this approach is not efficient with large number of nodes, e.g., $N > 2.5 \div 5\cdot 10^5$, because of the excessive growth of memory and time requirements of the LU approach. In such cases an iterative approach similar to the one used for momentum equations must be employed.

The computational costs are comparable to the ones obtained with a FEM discretization with linear triangular elements on isotropic meshes where the stencil size for each node is comparable to the present choice $q = 7$.

In the case of the extension of the method to practical 3D problems, the following techniques can be employed for the efficient solution of the systems of equations arising from the RBF-FD discretization of both Poisson and momentum equations:

- BiCGSTAB method with incomplete LU factorization (ILU) [14];
- algebraic multigrid (AMG) [15];
- a 3D extension of the multicloud techniques presented in [10].

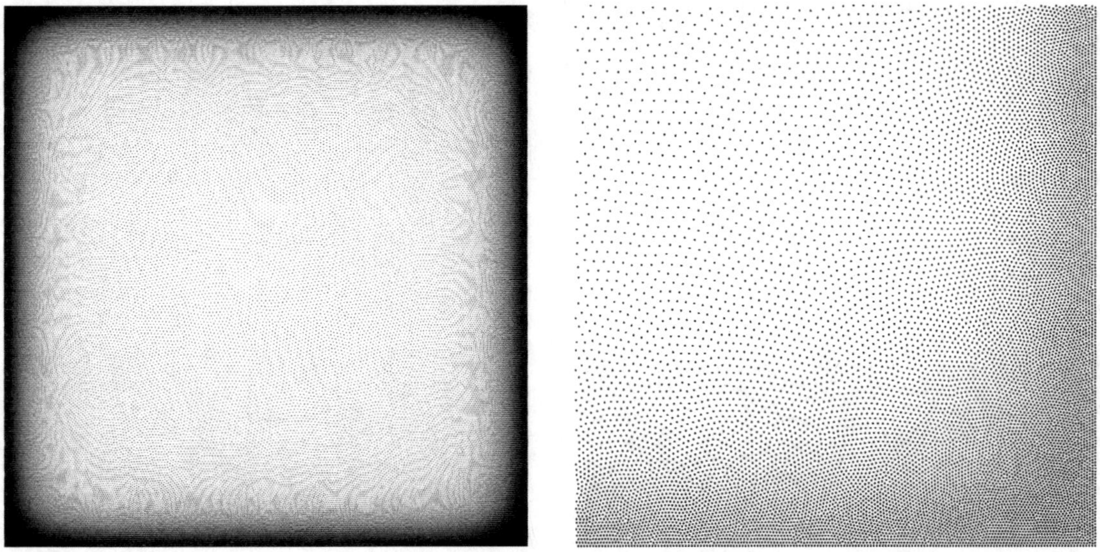

Figure 1. Node distribution for the lid-driven cavity problem with $N \approx 10^5$ nodes (left) and enlarged view of the bottom right corner (right).

3.5. Auxiliary computations

The streamfunction ψ is calculated solving the Poisson equation $\nabla^2\psi = -\nabla \times \mathbf{u}$ using the same node distributions and the same local RBF collocation technique employed for the discretization of the momentum and Poisson equations of the problem itself. Dirichlet boundary conditions are employed: $\psi = 0$ at all solid walls in the case of the lid-driven cavity problem, $\psi = \bar{\psi}$ in the case of the flow past a circular cylinder, where $\bar{\psi}$ is coherent with velocity boundary conditions, i.e., $\bar{\psi} = const$ along solid walls and $\partial\bar{\psi}/\partial y = u$ along the vertical inlet and outlet.

3.6. Implementation details

The presented procedures has been implemented in MATLAB® environment using MATLAB linked MEX functions which are compiled from C source code for the computational expensive tasks, on a modern laptop Intel® i7 2.6GHz with 4 cores and 8 threads. Multi-core parallelism is achieved using OpenMP® API for the C source code employed in the coefficients calculations and multicloud operations, i.e., SOR smoothing iterations, residual evaluation and restriction. Some of the remaining MATLAB operations, including BiCGSTAB algorithm, are natively parallelized on all available cores in MATLAB by default. The typical computing time is 0.9 s/time step with $N \approx 10^5$ nodes for each Re and Δt employed in this work.

4. Numerical results and discussion

4.1. Lid-driven cavity problem

The domain of the problem is the square cavity $[0, H]^2$ where the top wall moves with velocity $\mathbf{u}_W = (u_0, 0)^T$.

The spacing function $s(\mathbf{x})$ is defined as follows:

$$\frac{5s}{s_M} = 1 + \Big[1 + \cos\big(\pi(2\bar{x} - 1)\big)\Big]\Big[1 + \cos\big(\pi(2\bar{y} - 1)\big)\Big] \tag{15}$$

where $\bar{x} = x/H$ and $\bar{y} = y/H$ are the nondimensional coordinates. A node distribution with $N \approx 10^5$ nodes is depicted in Figure 1 together with an enlarged view of one corner; as it can

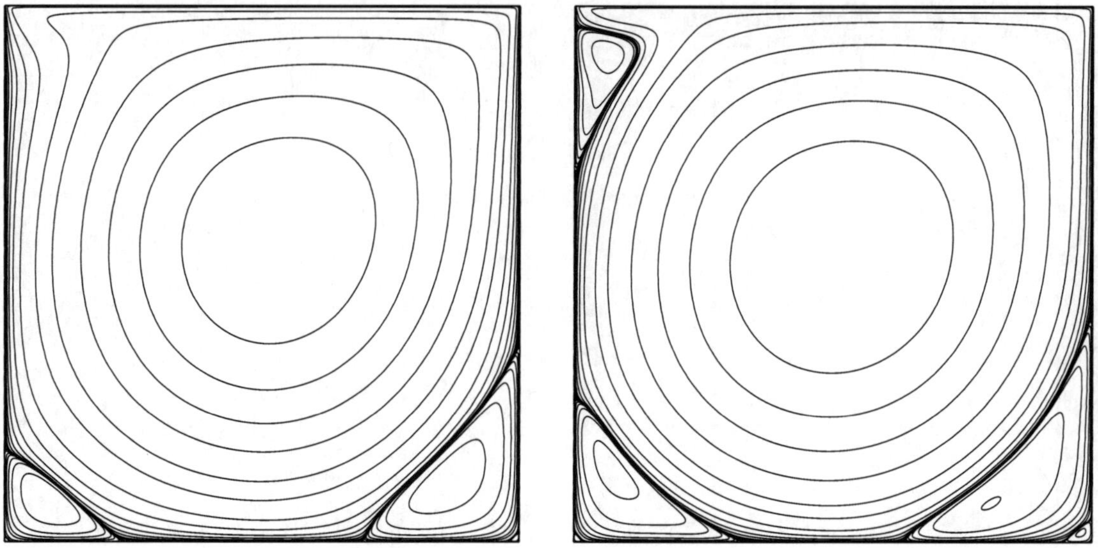

Figure 2. Streamfunction contours for the lid-driven cavity problem at Re = 1000 (left) and Re = 5000 (right).

Table 1. Streamfunction values for streamline plots.

-1×10^{-1}	-8×10^{-2}	-6×10^{-2}	-4×10^{-2}	-2×10^{-2}	-1×10^{-2}	-3×10^{-3}	-1×10^{-3}
-3×10^{-4}	-1×10^{-4}	-3×10^{-5}	-1×10^{-5}	-3×10^{-6}	-1×10^{-6}	-1×10^{-7}	-1×10^{-8}
-1×10^{-9}	-1×10^{-10}	-1×10^{-11}	-1×10^{-12}	-1×10^{-13}	-1×10^{-14}	0	1×10^{-14}
1×10^{-13}	1×10^{-12}	1×10^{-11}	1×10^{-10}	1×10^{-9}	1×10^{-8}	1×10^{-7}	1×10^{-6}
3×10^{-6}	1×10^{-5}	3×10^{-5}	1×10^{-4}	3×10^{-4}	1×10^{-3}	3×10^{-3}	1×10^{-2}

be seen, the previously defined spacing function decreases from the center of the cavity, where $s = s_M$, towards the walls, where $s = s_M/5$, in order to better resolve the boundary layers.

Starting from rest, a steady state solution for $Re = 1000$ is obtained after about 200 time units with $\Delta t = 0.05$ and $N \approx 10^5$ nodes. The case $Re = 5000$ is started from the steady state solution at $Re = 1000$ and a steady state solution is again found after about 200 time units with $\Delta t = 0.025$. The streamfunction contours for the values reported in Table 1 are depicted in Figure 2 for both $Re = 1000$ and $Re = 5000$ and show very good agreement, to graphical accuracy, with the results reported in [5, 16].

The values of the extrema of streamfunction and the corresponding locations for primary and secondary vortexes are reported in Table 2 where the results from other authors are also reported for comparison: these results show very good agreement with reference values in both cases $Re = 1000$ and $Re = 5000$, with less than 1% deviations for both strengths and locations of the streamfunction extrema in each case.

4.2. Flow past a circular cylinder between parallel walls

Table 2. Comparison of streamfunction extrema for $Re = 1000$ and $Re = 5000$.

	Primary vortex, ψ (location)	Secondary vortex BR, ψ (location)	Secondary vortex BL, ψ (location)
	$Re = 1000$		
Bayona [5]	-0.1189 (0.5308,0.5652)	1.730E-3 (0.8641,0.1118)	2.334E-4 (0.0832,0.0781)
Nobile [16]	-0.1175 (0.5313,0.5625)	1.770E-3 (0.8594,0.1094)	2.320E-4 (0.0859,0.0781)
Present res.	-0.1188 (0.5327,0.5674)	1.726E-3 (0.8623,0.1111)	2.321E-4 (0.0840,0.0781)
	$Re = 5000$		
Bayona [5]	-0.1223 (0.5151,0.5352)	3.077E-3 (0.8046,0.0727)	1.379E-3 (0.0728,0.1371)
Nobile [16]	-0.1214 (0.5156,0.5357)	3.096E-3 (0.8058,0.0737)	1.370E-3 (0.0714,0.1384)
Present res.	-0.1227 (0.5141,0.5362)	3.068E-3 (0.8032,0.0715)	1.366E-3 (0.0734,0.1380)

BR = bottom right; BL = bottom left.

Figure 3. Particular of the node distribution near the circular cylinder ($-0.5H < x < 2.5H$) with $N \approx 10^5$ nodes.

Table 3. Comparison of nondimensional period T and time-averaged drag coefficient C_D.

	Re	100	200	240	300	500	750	1000
Present results	T	–	–	0.85	0.80	0.73	0.69	0.66
	C_D	4.65	3.37	3.15	2.98	2.76	2.75	2.75
Zovatto and Pedrizzetti [17]	T	–	–	0.85	0.81	0.73		0.67
	C_D	4.68	3.40	3.19	2.97	2.71	2.68	2.64

The domain of the problem is a plane channel with width $11H$ and height H between the horizontal walls, containing a circular cylinder with diameter $d = 0.2H$ centered at the half of the channel height and $3H$ from the vertical left inlet; therefore the vertical right outlet is positioned $8H$ to the right of the cylinder center, which is also the origin of the coordinate system. A Poiseuille velocity profile $u = 6u_0[1/4 - (y/H)^2]$, which has an average velocity u_0, is imposed at the inlet, while fully developed flow conditions $\partial\mathbf{u}/\partial x = \mathbf{0}$ are imposed at the outlet; although a real fully developed flow is not obtained at the end of the channel, especially for high

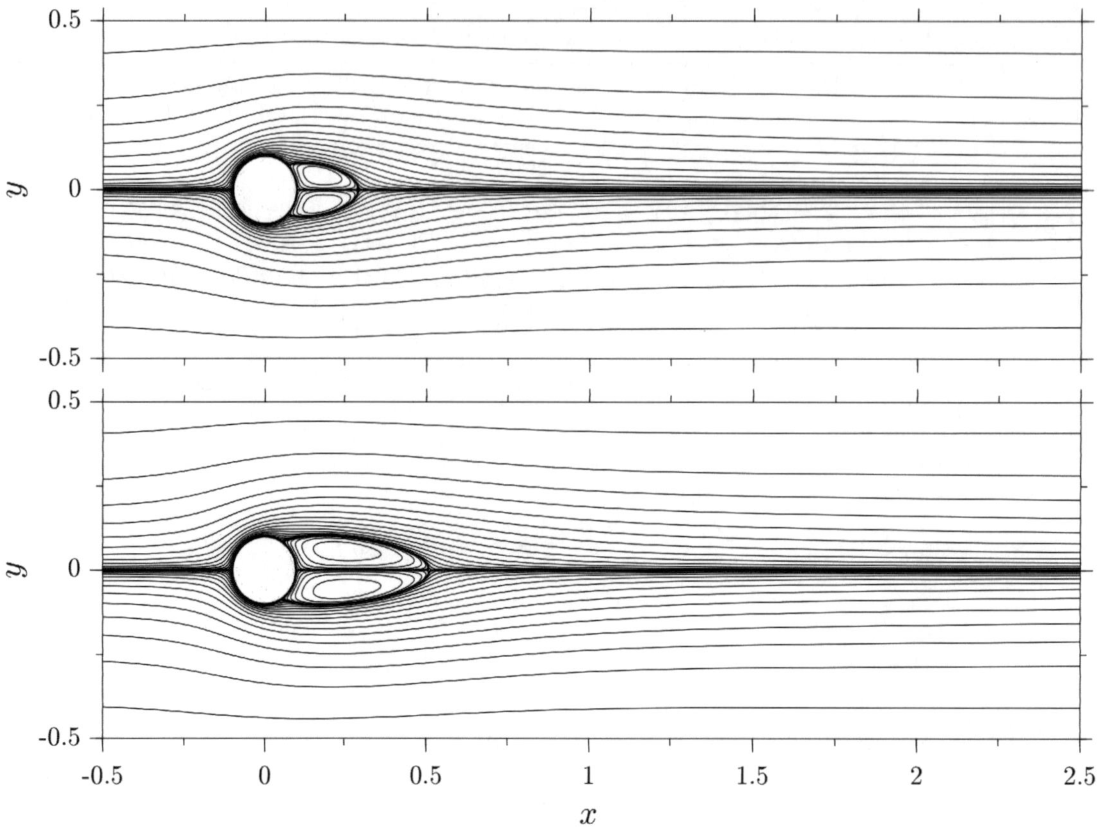

Figure 4. Streamlines for the steady flow past a circular cylinder with $Re = 100$ (top) and $Re = 200$ (bottom).

Re, the channel extension downstream of the cylinder, i.e., $8H = 40d$, is enough to avoid any appreciable influence of the outlet boundary conditions on the flow near the cylinder.

The spacing function is defined as follows:

$$\frac{s_M}{s} = 1 + k_W \left(\frac{2y}{H}\right)^{2e_W} + k_C \frac{d}{2r} \tag{16}$$

where $k_W = 3$, $e_W = 1.75$, $k_C = 24$ and r is the distance from the center of the cylinder; therefore the ratio between the nodal spacing at the horizontal walls and the maximum spacing is approximately $1/(k_W + 1) = 1/4$, while the ratio between the nodal spacing at the cylinder wall and the maximum spacing is approximately $1/(k_C + 1) = 1/25$. An enlarged view of the node distribution near the cylinder for $N \approx 10^5$ nodes is depicted in Figure 3 where it can be seen that the nodal density near the cylinder is very high in order to better resolve the boundary layers.

Simulations are carried out with the same node distribution with $N \approx 10^5$ nodes for $Re = 100$, 200, 240, 300, 500, 750 and 1000; each case is started with the last available solution at Re immediately below, except for the case $Re = 100$ which is started from rest.

Steady state solutions are obtained after about 20 time units for $Re = 100$ and $Re = 200$ with $\Delta t = 0.025$. The boundary conditions for streamfunction are set to obtain $\psi(x, 0) = 0$ at the half of the channel height; the streamfunction contours for these steady solutions are depicted in Figure 4 for values of ψ ranging from $\psi(x, -H/2)$ (streamfunction at bottom wall)

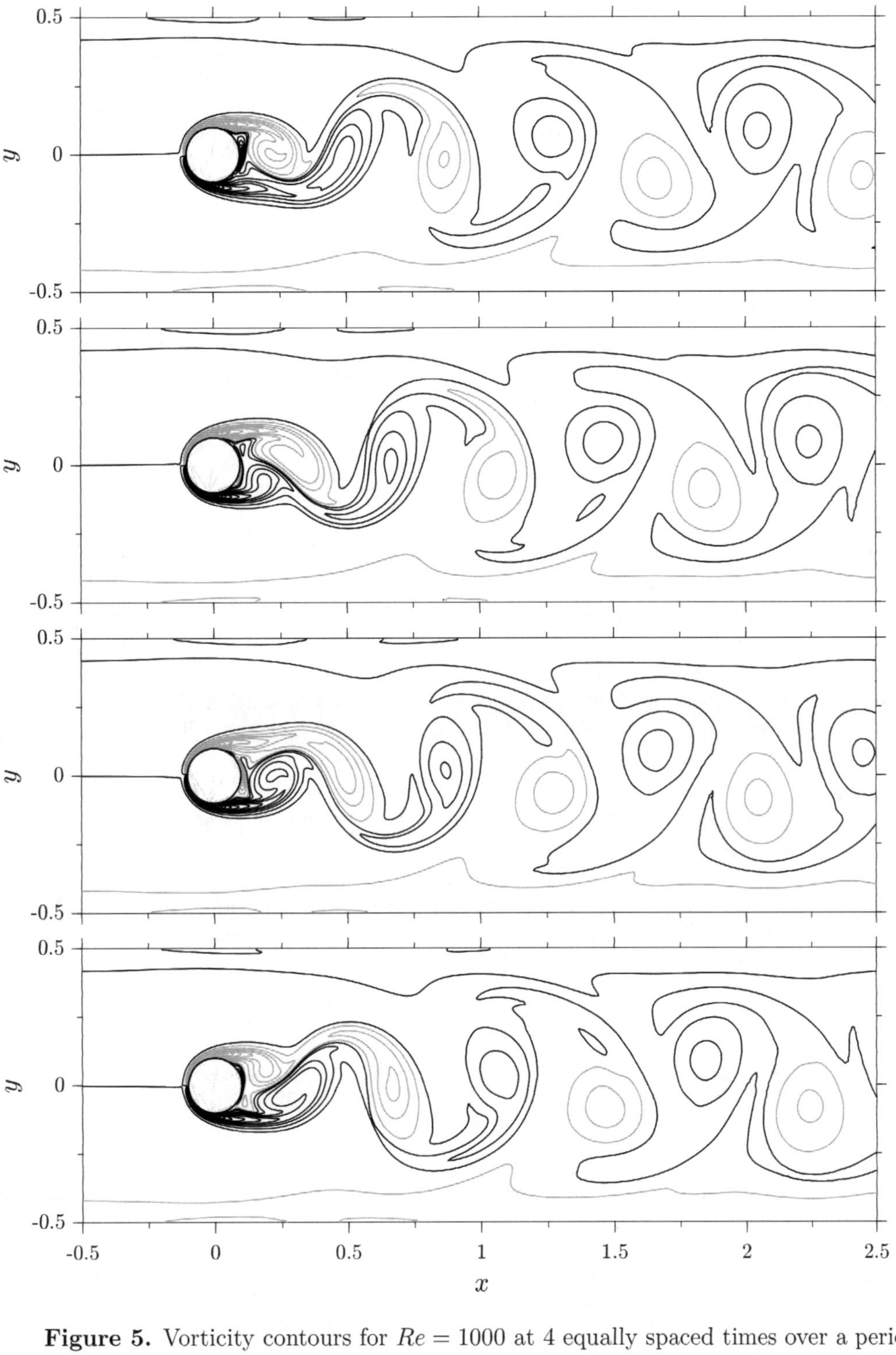

Figure 5. Vorticity contours for $Re = 1000$ at 4 equally spaced times over a period T.

to $\psi(x, H/2)$ (streamfunction at top wall) as the power ψ^5 in order to highlight the streamlines in the wakes. Very good agreement is found, to graphical accuracy, with the results reported in

[18].

The solution for $Re = 240$, as expected, is not stable and a periodical solution is found after about 90 time units with $\Delta t = 0.025$; this is in perfect accordance with the results of [19] which show that the critical Re is 231. Periodical solutions with period T are found for each of the higher Re considered, with periodical vortex shedding behind the cylinder. For $Re = 750$ and $Re = 1000$ the time step has been reduced from $\Delta t = 0.025$ to $\Delta t = 0.02$ in order to allow the convergence of momentum equations in less than 200 Work Units, at the price of more time steps per period T. Contour plots for vorticity in the case $Re = 1000$ are depicted in Figure 5 for vorticity values starting from ± 5 and increasing with constant steps ± 10; again, very good agreement is found, to graphical accuracy, with the results reported in [17].

A comparison with literature results is reported in Table 3 for the period T and the drag coefficient $C_D = 2F_x/\rho u_0^2 Hd$, where F_x is the x-component of the force exerted by the fluid on the cylinder, which is time-averaged over a period T in the case of periodical flows. Numerical results compare very favorably with reference values for each of the considered Re, with less than 5% deviations for C_D and less than 2% deviations for T.

5. Conclusions

In this work a collocation meshless approach, based on local RBF interpolation, is used for the numerical simulation of unsteady fluid flow problems using primitive variables, i.e., velocity and pressure, on 2D domains with complex node distributions of practical relevance. A projection scheme is employed to decouple the momentum and continuity equation; particular attention is given to the choice of the solvers in order to obtain an efficient solution procedure. The lid-driven cavity problem and the flow past a circular cylinder have been chosen as test cases to highlight the properties of the presented numerical scheme; several computations for different Reynolds numbers are carried out in order to validate the proposed meshless approach at different flow regimes.

Numerical results for these test problems agree very favorably with previous numerical studies for each of the considered cases; along with the computational efficiency, such results suggest that the proposed meshless approach can be extremely effective in the numerical simulation of unsteady fluid flow problems on general-shaped 2D domains with node distributions of practical engineering relevance; this approach can also be extended to 3D problems with remarkable geometrical advantages over standard mesh-based methods.

References

[1] Li H and Mulay S 2013 *Meshless Methods and Their Numerical Properties* (CRC Press) ISBN 978-1-4665-1747-9
[2] Šarler B and Vertnik R 2006 *COMPUT MATH APPL* **51**(8) 1269–1282
[3] Mavric B and Šarler B 2015 *INT J NUMER METHOD H* **25**(6) 1488–1510
[4] Kosec G and Šarler B 2008 *INT J NUMER METHOD H* **18**(7/8) 868–882
[5] Bayona V, Flyer N, Fornberg B and Barnett G 2017 *J COMPUT PHYS* **332** 257 – 273
[6] Fornberg B and Flyer N 2015 *ACTA NUMER* **24** 215–258
[7] Sarra S and Kansa E 2009 *Multiquadric Radial Basis Function Approximation Methods for the Numerical Solution of Partial Differential Equations* (*Advances in Computational Mechanics* vol 2) ed Atluri S N (Tech Science Press)
[8] Waters J and Pepper D 2015 *NUMER HEAT TR B-FUND* **68**(3) 185–203
[9] Zamolo R and Nobile E 2018 *COMPUT MATH APPL* **75**(1)
[10] Zamolo R and Nobile E 2018 *Multicloud techniques for convergence acceleration with local RBF meshless discretizations* (Submitted)
[11] Franke R 1980 A critical comparison of some methods for interpolation of scattered data Tech. rep. NAVAL POSTGRADUATE SCHOOL, Monterey, California
[12] Zamolo R and Nobile E 2017 *J PHYS CONF SER* **923**
[13] van der Vorst H 1992 *SIAM J SCI STAT COMP* **13**(2) 631–644

[14] Saad Y 2003 *Iterative Methods for Sparse Linear Systems* (Philadelphia, Pennsylvania: SIAM) chap 10: Preconditioning Techniques 2nd ed
[15] Stüben K 2001 *J COMPUT ANAL APPL* **128**(1-2) 281–309
[16] Nobile E 1996 *NUMER HEAT TR B-FUND* **30**(3) 351–370
[17] Zovatto L and Pedrizzetti G 2001 *J FLUID MECH* **440** 1–25
[18] Singha S and Sinhamahapatra K 2010 *OCEAN ENG* **37** 757–769
[19] Chen J H, Pritchard W G and Tavener S J 1995 *J FLUID MECH* **284** 23–41

IOP Conf. Series: Journal of Physics: Conf. Series **1224** (2019) 012014 doi:10.1088/1742-6596/1224/1/012014

Comparison between cooling strategies for power electronic devices: fractal mini-channels and arrays of impinging submerged jets

N Baraldi[1], A Fregni[1], M Sabato[1], E Stalio[1], F Brusiani[2], M Tranchero[2] and T Baritaud[2]

[1] Dipartimento di Ingegneria "Enzo Ferrari", Università di Modena e Reggio Emilia
[2] Ferrari S.p.A.

E-mail: enrico.stalio@unimore.it

Abstract. Power electronic devices like Insulated Gate Bipolar Transistors (IGBTs) and diodes are often characterized by power densities and dimensions that could result in very high heat flux densities. In order to guarantee the expected performance and lifetime for these components, dedicated active cooling devices are usually adopted. In the present paper, the comparison between two different cooling strategies for power electronics is presented: fractal-channel design and submerged impinging jets. Each cooling strategy is tested on two different geometrical configurations. Water is used as coolant in all cases. Assessment of the considered cooling methods is done through application of the selected configuration in a simplified system composed by a rectangular chip (heat source) separated from the coolant by a solid block.

Three-dimensional conjugated heat transfer simulations are performed by using RANS solver implemented in OpenFOAM and two-equations turbulence models, resolving also the viscous sublayer. Numerical results allow to compare the cooling strategies in terms of maximum chip temperature, overall chip-to-coolant thermal resistance, and pumping power required. In summary, the fractal-channel design shows limitations in guaranteeing low chip temperatures at an affordable pumping power. The submerged impinging jets approach shows very high local heat transfer coefficient by which it is possible to tailor the cooling effect on specific hot spots.

Nomenclature

Greek Symbols

δ	Nozzle height $[m]$
η , ψ	Numerical coefficients for fractal-channel geometries
Γ	Ratio between height and width of a fractal-channel branch
γ	Ratio between hydraulic diameters of two consecutive branches
ρ	Fluid density $[kg/m^3]$
θ	Ratio between lengths of two consecutive branches
ζ, α, β	Dimensionless coefficients in Nusselt number correlation

Roman Symbols

A	Chip plan area $[m^2]$
AR	Aspect ratio
c	Specific heat $[J/(kgK)]$
c_D	Discharge coefficient
C_{k+1}	Ratio of heat transfer rate between two consecutive branches in fractal channels
D	Nozzle diameter $[m]$
d	Dimensionless hydraulic diameter of a fractal-channel branch
H	Dimensionless height of a fractal-channel branch
h	Heat transfer coefficient $[W/(m^2K)]$
L	Dimensionless length of a fractal-channel branch
l	Length $[m]$
M	Wet surface per unit length of a fractal-channel branch $[m]$
m	Mass $[kg]$
N	Number of nozzle
n	Number of bifurcations in fractal channels
Nu	Nusselt number
P	Power $[W]$
p	Static pressure $[Pa]$
P_g	Heat generation rate of the chip $[W]$
P_p	Pumping power $[W]$
Pr	Prandtl number
Q	Heat $[J]$
R_{th}	Thermal resistance $[K/W]$
Re	Reynolds number
S	Inlet-section area $[m^2]$
T	Temperature $[K]$
u	Velocity $[m/s]$
V	Fluid volume $[m^3]$
W	Dimensionless width of a fractal-channel branch
y^+	Dimensionless wall distance
Z	Jet-to-target spacing $[m]$

IOP Conf. Series: Journal of Physics: Conf. Series **1224** (2019) 012014 doi:10.1088/1742-6596/1224/1/012014

Subscripts, superscripts and other symbols

0	Relative to the first fractal-channel branch
1, 2	Sides of the chip in plan view
$\bar{}$	Surface averaged
$\dot{}$	Per unit time
b	Bulk condition
id	Ideal
in	Inlet
max	Maximum
out	Outlet
r	Real
ref	Reference condition
w	At liquid-solid interface

1. Introduction

Heat removal from power electronic devices is a technical problem common to many industrial fields, including the recent application of power electronics to hybrid traction systems. This study investigates and compares two single-phase liquid cooling devices for a single chip: the first one employs a network of fractal channels while the second one uses an array of submerged impinging jets. The efficiency of each cooling strategy is investigated on two test benches: a baseline configuration and an improved configuration designed to improve the baseline thermal and fluid-dynamics performance.

Application of fractal channels is considered a promising solution for the cooling of power electronics thanks to the large heat transfer surface involved. In Ref. [1], it is demonstrated that a tree-shaped structure increases its heat transfer capabilities and decreases its pumping power request for larger number of fractal levels. In Ref. [2], fractal channels are compared against serpentine channel. Results show the improvements of tree-shaped structures in heat transfer efficiency and pressure drop reduction. In Ref. [3], tree-shaped nets on a square chip are investigated and compared to straight and serpentine networks also. Results show that at the cost of a moderate increment in pressure drop, the temperature of the chip decreases for large numbers of branching levels. Also Escher and coworkers [4] compare fractal channels against a parallel channel cooler. In their study, parallel channels having a constant cross section show better cooling efficiency than fractal ones. From this brief literature survey, it appears clearly that tree-shaped channels have a good potential but their performance can depend strongly on the specific application. In the present paper the crucial importance of the ratio between wet surface and heat transfer area in tree-shaped channels is discussed.

A different technique for the cooling of power electronics is the use of two-phase or single-phase arrays of impinging jets. By this technique, it is possible to guarantee a high cooling efficiency on specific hot spots at an affordable pumping power request. Jörg *et al.* [5] clearly showed as the individual cooling of semiconductors, for example by micro-jets, offers new perspectives in the design of power electronic modules. As showed in [6], in general, with respect to the single jet the use of jet arrays is a preferred cooling configuration because of the higher area-averaged heat transfer coefficient reached for a specific coolant flow rate. As for the fractal cooling approach, also for the impinging jet method there are possible drawbacks due to the larger non uniformity in chip surface temperature and cooling efficiency sensitiveness to the manufacturing tolerances [7].

In the present paper, the application of the tree-shaped fractal channels and of the impinging jet arrays in cooling electronic devices have been evaluated by performing 3D-CFD conjugated simulations. The adopted numerical procedure is validated against experimental data and results are compared and discussed from both heat transfer coefficients and pumping power requirement point of views. Heat generation rate of the chip and inlet temperature of the coolant are the same for all the simulated configurations, although simulations on the fractal channels are carried out for imposed volumetric flow rate of the coolant while submerged jets are computed under a given pressure drop, this is done in view of numerical considerations and the two cases represent the same practical problem, very close to electric traction applications.

2. Method of analysis

Results reported in the present paper are obtained by simulation and using the open source code OpenFOAM (www.openfoam.com). The specific solver employed is the *chtMultiRegionSimpleFoam*, a steady-state solver which imposes mass conservation through the SIMPLE algorithm and allows for the inclusion of conduction heat transfer in solid walls, thus being able to calculate conjugate heat transfer cases. As the Reynolds number in the heat transfer passages investigated is *transitional* in most of the cases considered here, a RANS method is employed, using a $k - \omega$ SST turbulence model [8] for the submerged impinging jets case and a $k - \varepsilon$ model for the fractal channels, see for example [9]. For low turbulence levels these models calculate a very small eddy diffusivity and provide a total viscosity term which is very close to the coolant molecular viscosity. Spatial discretisation is done by a second order central scheme for the diffusive term and by a first order upwind scheme for advection.

Given the linear temperature field behaviour, the mesh in the solid region requires in general a smaller number of cells with respect to the fluid regions. As a consequence of that, a non-conformal mesh interface is present at solid-fluid interface in the fractal channels simulations, while for the impinging jet cases a conformal mesh is adopted. The mesh is in general very fine both for the fractal channels and the submerged impinging jets, with a maximum distance from the wall (y^+) always lower than 4. The average y^+ value is below 1 in each simulation, therefore a low-Reynolds approach is followed and the flow field is directly calculated also in wall proximity. Further hypotheses considered in the present study are: viscous dissipation is negligible, the Knudsen number is very small, and buoyancy effects are negligible. All of them have been checked by a preliminary analysis.

Performance of the different cooling strategies have been assessed using the following criteria

- Maximum chip temperature T_{max},
- overall heat transfer coefficient,
- thermal resistance,
- pumping power.

The overall heat transfer coefficient $\overline{h}$ is defined as:

$$\overline{h} \equiv \frac{\dot{Q}}{A_w \left(\overline{T}_w - T_{b,in}\right)} \tag{1}$$

where $\dot{Q}$ is the heat transferred from the chip to the liquid coolant in the time unit, A_w is the liquid-solid interface areas, $\overline{T}_w$ is the average temperature of this interface and $T_{b,in}$ is the bulk temperature at the inlet.

The thermal resistance R_{th} between chip and coolant is defined similarly

$$R_{th} = \frac{T_{max} - T_{b,in}}{\dot{Q}} \tag{2}$$

see also Ref. [10].

The mechanical energy per time unit for maintaining an assigned flow rate is calculated as

$$P_p = \dot{V}\Delta p \tag{3}$$

where $\dot{V}$ is the volumetric flow rate and Δp is the static pressure drop between the inlet and the outlet sections, $\Delta p = p_{in} - p_{out}$.

To preserve confidentiality of configurations and conditions, data are presented in a non-dimensional form using reference quantities defined as follows. The reference length is

$$l_{\text{ref}} = \sqrt{l_1 l_2} \tag{4}$$

where $l_1 l_2 = A$ is the chip plan area. The reference velocity definition is based upon the bulk velocity at the inlet or the imposed pressure drop, depending on the specific case considered:

- for imposed pressure drop (impinging jet cases)

$$u_{\text{ref}} = \sqrt{\frac{2\,\Delta p}{\rho}} \tag{5}$$

- for given volumetric flow rate (fractal channels)

$$u_{\text{ref}} = u_b = \frac{\dot{V}}{S} \tag{6}$$

where $\dot{V}$ is the volumetric flow rate and S is the inlet-section area. The reference temperature is

$$T_{\text{ref}} = \frac{P_g}{\rho u_{\text{ref}}\, c\, S} \tag{7}$$

where P_g is the heat generation rate of the chip, $\rho = \rho_{\text{ref}} = 978$ kg/m^3 and $c = c_{\text{ref}} = 4187$ J/(kg K) are respectively density and specific heat, which are kept constant in this study.

3. Cooling devices and computational domains

3.1. Fractal-channels

The first configuration of fractal channel considered in this study is the classical fractal-shaped network reported in figure 1 (Case A). Geometry is fully described by the number of bifurcations n, length L_0 and hydraulic diameter d_0 of the first branch, and the ratio between lengths and hydraulic diameters of two consecutive branches, respectively $\theta = L_{k+1}/L_k = N^{-1/\eta}$ and $\gamma = d_{k+1}/d_k = N^{-1/\psi}$. Geometric dimensions introduced are illustrated in figure 2. Channels of rectangular cross-section with uniform depth of $0.501\, l_{\text{ref}}$ are selected for manufacturing reasons, see for example refs. [1], [11] and [12]. Values of θ and γ are related to the parameters η and ψ. in Ref. [13] and [14] values $\eta = 3$ and $\psi = 3$ are recommended for reducing pressure drop. In this study it has been possible to set $\psi = 3$ while, due to the size of the chip selected $\eta = 3$ would cause channels to overlap. The channel lengths have been instead selected in order to fill the chip plan area with flow passages. Table 1 summarizes the geometric parameters of the first configuration studied.

The computational domain, depicted in figure 3, includes a single chip, which acts as the volumetric heat source, an aluminium plate, where the coolant passages are manufactured, the heat exchanger passages and the fluid flowing within the passages. An imposed mass flow rate of $0.010\, \rho_{\text{ref}}\, u_{\text{ref}}\, l^2_{\text{ref}}$ enters the domain from the side of the aluminium region at the constant temperature of $26.5\, T_{\text{ref}}$; it passes through the channels and at the highest fractal branch, it is

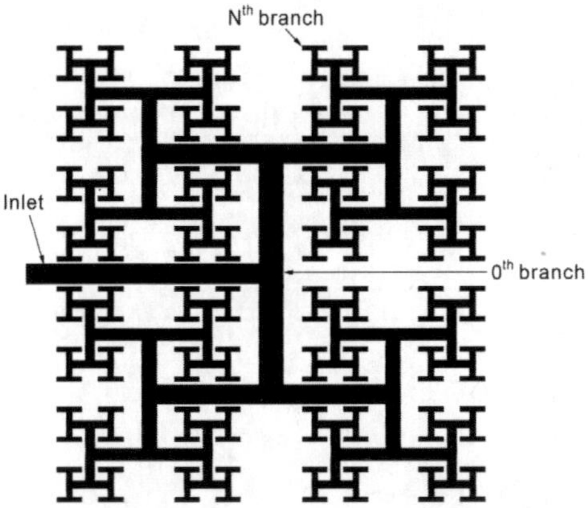

Figure 1. Fractal network.

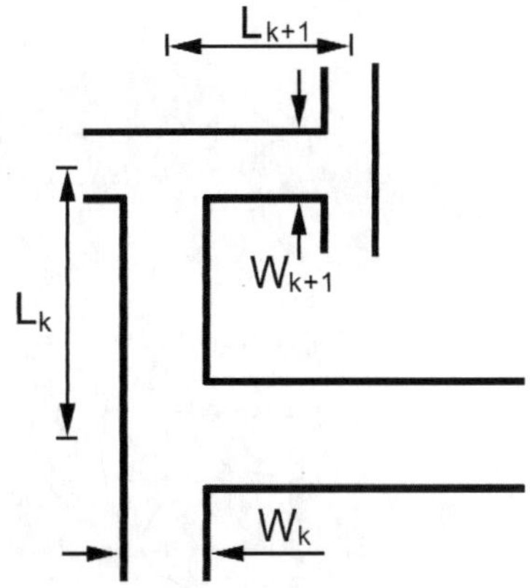

Figure 2. Fractal characteristic dimensions.

Table 1. Fractal-channel dimensions of the k-th level, Case A. Lengths are made non-dimensional using $l_{\rm ref}$, defined as in equation (4).

k	H_k	W_k	L_k	d_k	L_k/d_k
Inlet	0.501	0.040	0.488	0.076	6.4
0	0.501	0.050	0.244	0.094	2.6
1	0.501	0.039	0.244	0.075	3.3
2	0.501	0.031	0.122	0.059	2.1
3	0.501	0.024	0.122	0.047	2.6
4	0.501	0.019	0.061	0.037	1.6
5	0.501	0.015	0.043	0.030	1.5
6	0.501	0.012	0.031	0.023	1.3
7	0.501	0.009	0.026	0.019	1.4

discharged into a plenum through rectangular holes. The fluid exits from the domain at the sides of this plenum at a given pressure.

At the top of the computational domain, which is the upper boundary of the discharge plenum, symmetry conditions are applied in order to set the first-order normal derivatives to zero and to reduce the influence of boundaries to the solution. The outlet pressure is fixed and homogeneous Neumann conditions are specified for each velocity component and temperature fields.

For the thermal field, lateral domain boundaries are adiabatic and heat is therefore extracted from the domain only through the liquid coolant. Since the geometry and fluid dynamics conditions are symmetric about the longitudinal middle plane of the inlet channel, only half of the domain described is considered in the simulations, see figure 3.

The computational grid used in the fractal-channel simulations involves about 3.9 millions hexahedral cells and is topologically structured, both in the solid and fluid regions. At the fluid-solid interface the AMI interpolation implemented in OpenFOAM is used in order to reduce the number of control volumes in the aluminium region. The maximum interpolation error on the wall heat flux is of 0.14% . The average and maximum distances between wall and first computational point distance are $y^+ = 0.55$ and $y^+ = 4.4$ wall units. Accordingly, a Low-Reynolds approach is applied.

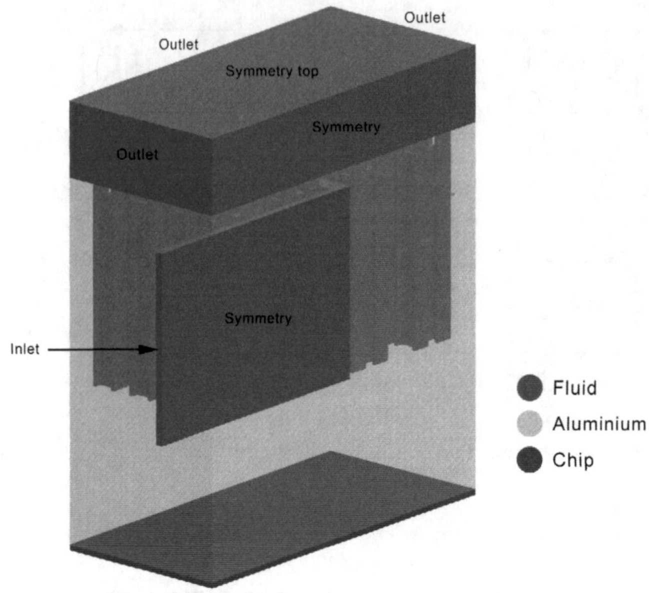

Figure 3. Computational domain and boundary conditions of the first fractal-channels configuration.

3.2. Submerged impinging jets

A second set of simulations has been performed on arrays of submerged impinging jets. Prior to the study of jet arrays, the accuracy of the numerical procedure has been checked by comparing results of two simulations of a single axisymmetric jet against experimental and numerical results, respectively published in Ref. [15] and [16]. Simulations are conducted resorting to an axisymmetric geometrical model. The jet diameter is 3.18 mm and a uniform heat flux of 25 W/cm^2 is imposed on the impingement plate; further details on the domain dimensions and boundary conditions can be found in Ref. [16]. Two turbulence models have been tested: the k-ε and the k-ω SST. In both cases the Low Reynolds approach is used for the near wall treatment, average and maximum distance of the first computational volumes from the wall are of $y^+ = 0.9$ and $y^+ = 2.9$ respectively. The comparison between present results and data from Ref. [15] and [16] is summarized in table 2. The radial profile of local heat transfer coefficient is reported in figure 4, where it appears that the k-ω SST model reproduces the experimental local heat transfer coefficient with a smaller error with respect to the k-ε model. Table 2 shows that also mass flow rate, reattachment point and discharge coefficient are predicted by the k-ω SST turbulence model more accurately. The discharge coefficient is computed as

$$c_D = \frac{\dot{m}_r}{\dot{m}_{id}} = \frac{4\,\dot{m}_r}{N\pi D^2\sqrt{2\,\rho\Delta p}} \tag{8}$$

where pressure losses are assumed lumped in the nozzle. In equation (8) $\dot{m}_r$ is the real mass flow rate, N and D are the number of jets and their diameter, Δp is the imposed pressure drop between inlet and outlet and ρ is the fluid density. As a consequence of this validation procedure the k-ω SST model is used in all the simulations of submerged impinging jets presented here.

The study of the jet array cooling strategy has been conducted performing several simulations varying the number of jets, their diameter and the aspect ratio(AR $= D/\delta$, where δ and D are the nozzle height and diameter), also by taking into account results reported in refs. [17], [18], [15],

Table 2. Comparison between present results, using k-ε and k-ω SST turbulence model, and data from papers [15] and [16]. In the table D is the nozzle diameter.

	k-ε	k-ω SST	Experiment [15]	Simulation [16]
Pressure drop (Pa)	7116	7116	7116	7250
Mass flow rate (kg/s)	0.0288	0.0310	0.0326	0.0326
Discharge coefficient	0.713	0.768	0.806	-
Reattachment length	-	0.92D	-	0.80D

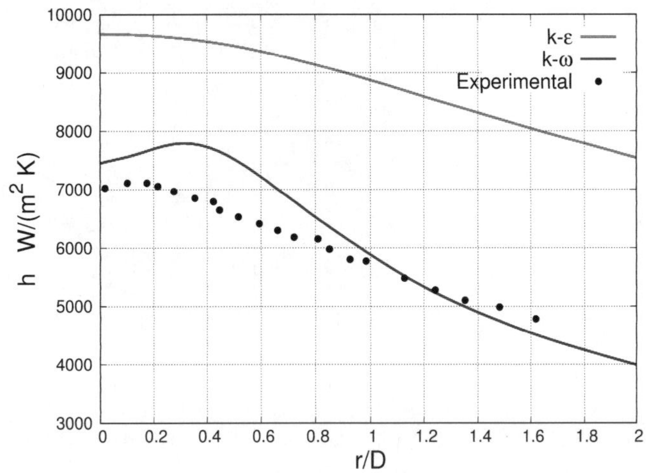

Figure 4. Radial profiles of local heat transfer coefficient in the single axisymmetric jet simulation: comparison between present results and experimental data [15].

[19], [20] and [21]. Among all the simulations performed, two configurations arise for thermal and fluid dynamics performance: arrays of 6×6 and 7×7 jets, with jets of diameter $0.050\,l_{\mathrm{ref}}$, unit aspect ratio (AR $= 1$) and a distance between the nozzle plate and the impingement plate of $Z/D = 3$. In the following text, only these two cases are presented. Since the configuration is symmetric with respect to two orthogonal directions, the computational domain considers only a quarter of the full system. Configurations considered are shown in figure 5. The simulated domain is depicted in figure 6 and it involves a stack of different regions: a chip (which is the heat source), two layers of copper and ceramic, an aluminium plate and the coolant region, which include an inlet plenum, the nozzles and a discharge zone. The copper, ceramic and aluminium layers introduce a conductive thermal resistance for the heat flux per unit area of $60.2\,T_{\mathrm{ref}}\,l_{\mathrm{ref}}^2/P_{\mathrm{ref}}$, $261.7\,T_{\mathrm{ref}}\,l_{\mathrm{ref}}^2/P_{\mathrm{ref}}$ and $1105.9\,T_{\mathrm{ref}}\,l_{\mathrm{ref}}^2/P_{\mathrm{ref}}$ respectively.

The coolant enters the domain by the lateral boundaries of the inlet plenum at a temperature of $60.9\,T_{\mathrm{ref}}$, passes through the nozzles, exchanges heat from the aluminium plate and exits through the lateral boundaries in the discharge zone, see figure 6. The energy required for the fluid flow is provided by imposing a relative pressure of Δp_{ref} at the inlet and ambient pressure at the outlet. Symmetric boundary condition is imposed both at the top of the inlet plenum (for the same reasons provided for fractal channels, see section 3.1) and at the symmetry planes, while the no-slip condition is imposed at the other boundaries of the fluid region. The thermal input of $3.952 \times 10^{-3}\,P_{\mathrm{ref}}$ relative to the entire chip is provided by setting a volumetric heat

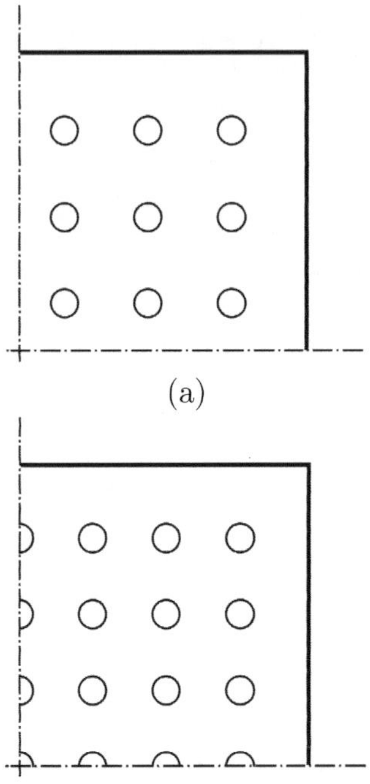

Figure 5. Nozzles arrangement: (a) 6 × 6 array, (b) 7 × 7 array.

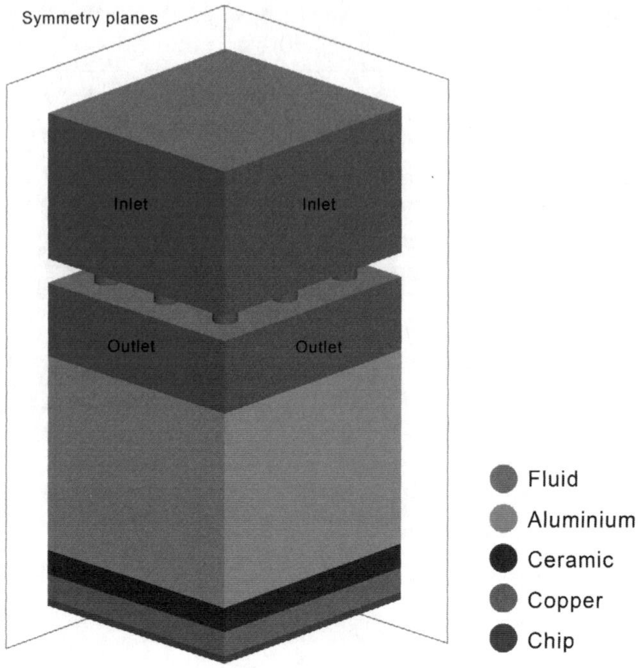

Figure 6. Computational domain of the 6 × 6 array configuration.

source in the chip, then heat is transferred in the vertical direction through the solid regions, and finally to the coolant. Lateral boundaries of the solid regions are set to adiabatic or symmetric, which at second order accuracy corresponds to the adiabatic condition.

The computational domain is discretised with about 2 million hexahedral cells and is topologically structured in both configurations. At the solid-fluid interface the mesh is conformal, thus no interpolation is applied. Since the computational grid allows to achieve y^+ average values of 0.92 and 0.88 for the 6 × 6 and 7 × 7 configuration respectively, the low Reynolds near wall treatment is imposed on the impingement plate in both cases.

4. Results

4.1. Fractal-channels

Figure 7(a) shows the temperature field on a section parallel to the chip, at the mid channel depth for case A. As expected, the temperature of the coolant increases but far enough from the inlet the fluid temperature is higher than the solid wall. The extent of this phenomenon is made clear in figure 7(b), in which the black line represents the perimeter of the solid-liquid interface where the solid walls are cooled, the red line indicates the portion of interface where the solid walls are heated. At the mid channel depth, the three last fractal levels are mostly characterized by inverse heat transfer and give place to an idle heat transfer loop.

Figures 8(a) and 8(b) reveal that the largest pressure drop is in the fluid discharge to the outlet plenum. No cooling effect is brought by this highly dissipative jet flow and mechanical energy is wasted in the discharge. A second region of remarkable pressure drop is in the first, largest branch.

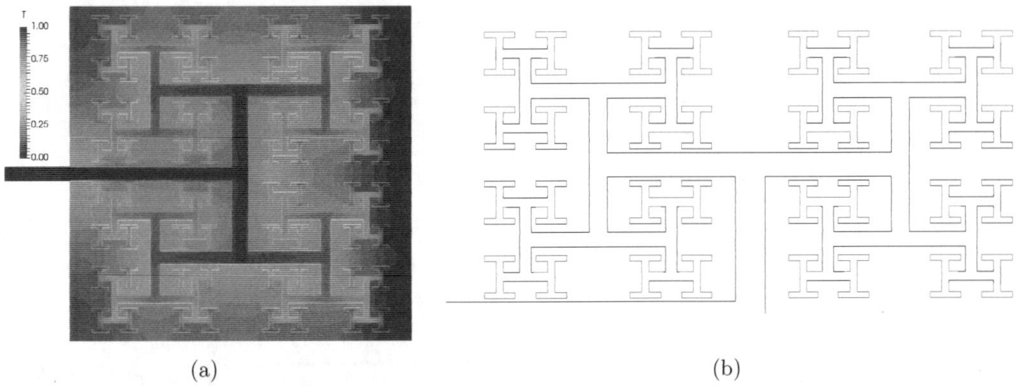

Figure 7. (a) Temperature field on a section parallel to the chip at mid channel depth, in both aluminium and fluid regions, and (b) perimeter of the solid-fluid interface coloured according to the direction of heat transfer: black from solid to fluid, red from fluid to solid.

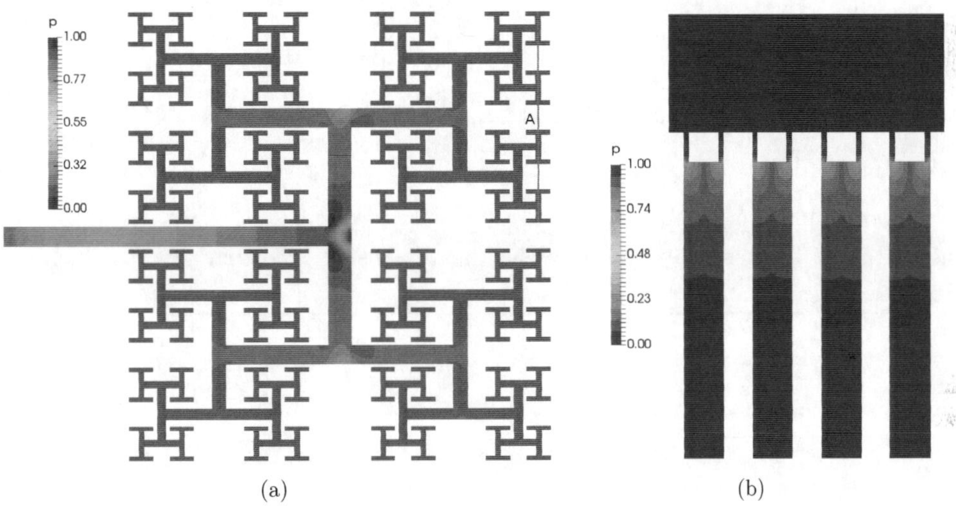

Figure 8. Pressure field on a section parallel to the chip at mid channel depth (a), and on the section indicated by the red line (b).

Global results are summarized in table 3. The main disadvantage of this solution is in the large portion of wet area which exhibits negative heat transfer ($A_{w,h<0}$): it includes almost half the total wet area in the conditions set for this study. The amount of heat per unit time transferred back from the coolant to the aluminium region, $\dot{Q}_{\mathrm{inv}}$ is more than one fourth of the heat input rate, and an idle heat transfer loop is formed. In summary this fractal-channel configuration allows to achieve a relatively low maximum temperature of the chip but the cost in terms of pumping power is high.

Results obtained in Case A suggested a set of geometrical modifications and the design of a potentially more efficient fractal-channel configuration (Case B). Geometrical modifications are listed below:

- removal of the two highest fractal levels;
- increased hydraulic diameter of the first branch;
- increased hydraulic diameter of the last branches;

Table 3. Overall performance of fractal-channels, Case A. Thermal resistance is made dimensionless by $T_{\text{ref}}/P_{\text{ref}}$, heat transfer coefficient by $P_{\text{ref}}/\left(l_{\text{ref}}^2 T_{\text{ref}}\right)$ and power by P_{ref}.

R_{th}	$\overline{h}$	P_{p}	$A_{w,h<0}$ (%)	$\dot{Q}_{\text{inv}}$ (%)
28401	$1.23{\cdot}10^{-5}$	$1.61{\cdot}10^{-8}$	49.7	27.3

Table 4. Fractal-channels dimensions in the enhanced configuration (Case B). Lengths are made non-dimensional using l_{ref}, defined as in equation (4).

k	H_k	W_k	L_k	d_k	L_k/d_k
Inlet	0.501	0.080	0.488	0.145	3.4
0	0.501	0.062	0.244	0.114	2.1
1	0.501	0.048	0.244	0.090	2.7
2	0.501	0.037	0.134	0.071	1.9
3	0.501	0.029	0.127	0.056	2.3
4	0.501	0.023	0.064	0.044	1.4
5	0.501	0.036	0.064	0.068	0.9

Table 5. Overall performance of the enhanced fractal-channels configuration. Thermal resistance is made dimensionless by $T_{\text{ref}}/P_{\text{ref}}$, heat transfer coefficient by $P_{\text{ref}}/\left(l_{\text{ref}}^2 T_{\text{ref}}\right)$ and power by P_{ref}.

R_{th}	$\overline{h}$	P_{p}	$A_{w,h<0}$ (%)	Q_{inv} (%)
3528	$1.75{\cdot}10^{-4}$	$3.06{\cdot}10^{-8}$	29.3	11.8

- reduction of the dimensions of channels close to the entrance duct.

Geometrical modifications applied are motivated also by the temperature field of figure 7(a), where a cold region close to the entrance duct is observed due to its proximity to the cold fluid.

The dimensions of Case B configuration are summarized in table 4. As reported in table 5, Case B configuration requires a much smaller pumping power, while the maximum temperature of the chip is not very different from Case A. Also in this second configuration about one third of the wet surface is characterized by an inverse heat flux. To overcome the inverse heat flux problem while keeping the same thermo-physical properties of the coolant, the volumetric flow rate should be substantially increased. Since the pressure drop required by fractal channel is already very high this solution is to be disregarded. The fractal channel configuration selected (Case A) and also its geometrically improved version (Case B) need to undergo a optimization procedure before being conveniently employed for the electronic cooling.

4.2. Submerged impinging jets

Contours of heat transfer coefficient on the aluminium plate are reported in figure 9. Note that this images are obtained mirroring the computational region about the symmetry planes. It is observed that jets interact, as shown by the different shapes of the high heat transfer coefficient

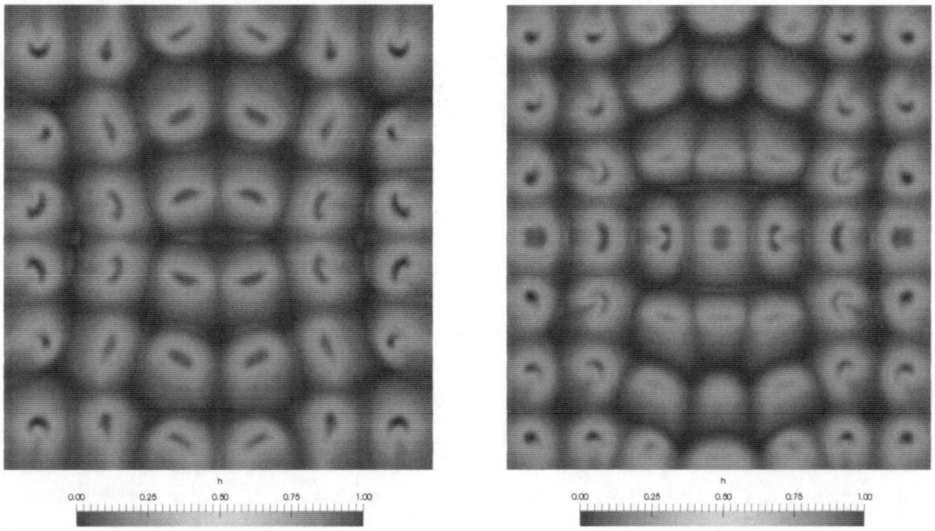

(a) (b)

Figure 9. Normalized heat transfer coefficient on the impingement plate: (a) 6×6 array, (b) 7×7 array.

Table 6. Comparison between overall results of the 6×6 and 7×7 array configurations.

Parameter	6×6	7×7
R_{th} $(\cdot T_{\mathrm{ref}}/P_{\mathrm{ref}})$	$3.73 \cdot 10^{3}$	$4.94 \cdot 10^{3}$
P_{p} $(\cdot P_{\mathrm{ref}})$	$8.87 \cdot 10^{-8}$	$1.09 \cdot 10^{-7}$
$\overline{h}$ $(\cdot P_{\mathrm{ref}}/(l_{\mathrm{ref}}^{2} T_{\mathrm{ref}}))$	$4.49 \cdot 10^{-4}$	$3.45 \cdot 10^{-4}$
c_{D}	0.63	0.57

spots corresponding to each jet. The horizontal flow toward outlet deviates the jets, preventing the undisturbed impingement of every jets.

The overall performance of these configurations are summarized in table 6. Comparing solutions at constant pressure drop, the 7×7 array keeps the chip at almost 2 ℃ cooler than the 6×6 array, but the cost in terms of pumping power is 20% higher. The choice between these configurations depends upon the cooling system characteristics.

5. Discussion

The four cases considered are characterized by

- equal inlet temperature
- equal heat transfer rate

and following the reasoning in "Performance evaluation criteria for use of enhanced heat transfer surfaces in heat exchanger design" by Webb [22] the efficiency of each of the four configurations investigated can be assessed by comparison of the pumping power. It appears in table 7 where the ratios between dimensional pumping power required by fractal channels and by the two jet

Table 7. Comparison of pumping power required, thermal resistance and space averaged heat transfer coefficient. Numbers are obtained by division of the dimensional results by the smallest value. Subscripts A, B and refers to the case A and B respectively.

Parameter	Case A	Case B	6×6	7×7
$R_{\rm th}/R_{\rm th,B}$	1.00	1	1.68	1.64
$\overline{h}/\overline{h}_{\rm A}$	1	1.77	2.86	2.99
$P_{\rm p}/P_{\rm p,6x6}$	5.31	1.26	1	1.23

configurations are reported, that the 6×6 jet configurations exchange the same heat at the lowest cost in terms of mechanical power required.

On the other side, from the point of view of maximum temperature of the chip, fractal channels ensure a lower temperature, despite the occurrence of negative heat fluxes. This is clearly shown in table 7, where the ratios between dimensional thermal resistance of impinging jets and fractal channels are reported. Given the definition of thermal resistance of a heat exchanger, see equation (2), and the circumstance that heat transfer rate and inlet temperature of the coolant is equal in all cases $R_{\rm th}$ provides a measure of the maximum temperature of the cooled electronic device.

Fractal channels For fractal channels exchanging heat through only a small part of the wet surface, it can be demonstrated that aside from the pumping power required, heat transfer enhancement which is expected to derive from the large heat transfer surface of the highest fractal levels, is at best very small. Let us consider the ratio between heat transfer rate per unit length of two channels of diameter d_{k+1} (smaller diameter) with respect to a single channel of hydraulic diameter d_k (larger diameter) carrying the same volumetric flow rate of the coolant. This ratio represents heat transfer enhancement to be ascribed to heat transfer surface increase.

Considering the canonical form of the Nusselt number dependence upon Re which applies to fully developed flow in pipes,

$$\mathrm{Nu} = \zeta\, \mathrm{Re}^{\alpha}\, \mathrm{Pr}^{\beta} \tag{9}$$

and Dittus-Boelter correlation which is reported to be valid for liquids in turbulent flow regime $\mathrm{Re} > 10^4$ prescribes $\alpha = 4/5$ while for a given fluid in laminar conditions $\alpha = 0$

$$\frac{\dot{Q}_{k+1}}{\dot{Q}_k} = \left(\frac{\mathrm{Re}_{k+1}}{\mathrm{Re}_k}\right)^{\alpha} \frac{d_k}{d_{k+1}} \frac{2\, M_{k+1}}{M_k} \frac{\Delta T_{k+1}}{\Delta T_k} \tag{10}$$

where M_k indicates wet surface at fractal level k, and leads to

$$\frac{\dot{Q}_{k+1}}{\dot{Q}_k} = \left(\frac{2M_{k+1}}{M_k}\right)^{(1-\alpha)} \frac{d_k}{d_{k+1}} \frac{\Delta T_{k+1}}{\Delta T_k} \tag{11}$$

For self-similar cross sections where $M_k/d_k = M_{k+1}/d_{k+1}$

$$\frac{\dot{Q}_{k+1}}{\dot{Q}_k} = 2^{(1-\alpha)} \left(\frac{M_k}{M_{k+1}}\right)^{\alpha} \frac{\Delta T_{k+1}}{\Delta T_k} \tag{12}$$

and for laminar flow ($\alpha = 0$)

$$\frac{\dot{Q}_{k+1}}{\dot{Q}_k} = 2\, \frac{\Delta T_{k+1}}{\Delta T_k} \tag{13}$$

raising fractal level can increase heat transfer rate, provided temperature differences at subsequent time levels do not decrease too fast.

Let us consider coefficient C_{k+1}, defined as

$$C_{k+1} \equiv \frac{\dot{Q}_{k+1}}{\dot{Q}_k} \frac{\Delta T_k}{\Delta T_{k+1}} \tag{14}$$

so that for $C_{k+1} > 1$ heat transfer augmentation is expected. In the present case, a series of channels of rectangular cross section is considered of sides W_k and H, where $W_k < H$. In case $W_{k+1} = W_k/2$ and $H = \Gamma_{k+1} H_{k+1}$

$$C_{\mathrm{k+1}} = \left(\frac{2 + 2\Gamma_{k+1}}{2 + \Gamma_{k+1}} \right)^{(2-\alpha)} \tag{15}$$

which for $\Gamma_{k+1} > 0$ is a increasing function of Γ_{k+1} tending to $2^{(2-\alpha)}$ for elongated rectangles (large Γ_{k+1}).

In fractal channel applications it might occur that not all the wet surface is involved in the heat transfer. In case just one of the small sides W_k contributes to heat transfer the heat augmentation coefficient in turbulent flow conditions is given by

$$C_{\mathrm{k+1}} = \left(\frac{2 + 2\Gamma_{k+1}}{2 + \Gamma_{k+1}} \right)^{(1-\alpha)} \tag{16}$$

Coefficient C_{k+1} increases with Γ_{k+1} tends to $2^{(1-\alpha)}$ for elongated channels which equals $2^{(1-\alpha)} \approx 1.15$ in turbulent conditions. In laminar conditions

$$C_{\mathrm{k+1}} = \frac{2 + 2\Gamma_{k+1}}{2 + \Gamma_{k+1}} \tag{17}$$

tends to 2 for elongated channels. We have shown that heat transfer augmentation of fractal channels, whose favourable heat transfer features are expected to descend from the large heat transfer surface of the highest fractal levels, is only marginal.

6. Conclusions

Simulation results show that before fractal channels can be employed for electronics cooling applications their geometry need to undergo a multi-objective optimization procedure, which would probably modify their canonical shape and geometry.

Results reported suggest that the space averaged temperature difference between solid wall and fluid is too small, the difference becomes negative over a considerable portion of the fluid-solid interface, giving place to an idle heat transfer loop. In view of improving performances while keeping the same thermo-physical characteristics of the liquid, an increase in the coolant mass flow rate might be tried but the pumping power required for circulation is already too large, for example with respect to the submerged impinging jet solution.

Aside from the pumping power required, it is demonstrated in the text that when fractal channels are used for exchanging heat, these can be conveniently applied only in the case of heat transferred trough the whole channel wet surface.

As soon as a small part of the wet surface is used for heat transfer, a raise in fractal level is demonstrably characterized by a heat transfer augmentation which is at best very limited because the main heat transfer enhancement mechanism of fractal channels is almost neutralized.

Considering also the temperature difference between fluid and solid, it can be concluded that fractal channels are not conveniently applied in the cooling of flat, solid plates, *e.g.* for electronics cooling.

On the other side, heat transfer studies on arrays of submerged impinging jets under very similar conditions as the fractal channels show that these jets can exchange heat at a low cost in terms of mechanical pumping power. Their space-averaged overall heat transfer coefficient is very large. The few drawbacks affecting this solution include the temperature of the chips, which is in our case slightly higher than for fractal channels and the observation that, given the small nozzle diameters, impinging jets might be prone to fouling and obstruction and very sensitive to construction tolerances. Impinging jets are instead very flexible in terms of geometry, because the distribution of orifices can be specifically designed to specifically cool regions generating heat at high rates.

References

[1] Chen Y and Cheng P 2002 *Int. J. Heat and Mass Transf.* **45** 2643–48
[2] Senn S M and Poulikakos D 2004 *J. Power Sources* **130** 178–191
[3] Wang X-Q, Mujumdar A S and Yap C 2006 *Int. J. Thermal Sciences* **45** 1103–1112
[4] Escher W, Michel B and Poulikakos D 2009 *Int. J. Heat and Mass Transf.* **52** 1421–1430
[5] Jörg J, Taraborrelli S, Sarriegui S, De Doncker R W, Kneer R and Rohlfs W 2018 *IEEE Trans. Power Electron.* **33** 4224–4237
[6] Rau M J and Garimella S V 2013 *Int. J. Heat and Mass Transf.* **24** 715–726
[7] Penumadu P S and Rao A G 2017 *Appl. Therm. Eng.* **110** 1511–1524
[8] Menter F R, 1993 *AIAA Journal,* **32** 1598-1605.
[9] Ferziger J H and Perić M, 2002 *Computational Methods for Fluid Dynamics.*
[10] Tuckerman D B and Pease R F W 1981 *IEEE Elect. dev. letters* **2** 126–9.
[11] Hong F J, Cheng P, Ge H and Joo G T 2007 *Int. J. Heat and Mass Transf.* **50** 4986–98
[12] Chen Y, Zhang C, Shi M and Yang Y 2010 *AIChE J.* **56** 2018–29
[13] Bejan A and Errera M R 1997 *Fractals* **5** 685–95
[14] Murray C D 1926 *Proc. National Academy of Sciences* **12** 207–14
[15] Garimella S V and Rice R A 1995 *J. Heat Transfer* **117** 871–7
[16] Morris G K and Garimella S V 1998 *J. Elect. Packaging* **120** 68–72
[17] Garimella S V and Nenaydykh B 1995 *ASME-PUBLICATIONS-HTD* **319** 49–58
[18] Garimella S V and Nenaydykh B 1996 *Int. J. Heat and Mass Transf.* **39** 2915–23
[19] Li C and Garimella S V 2001 *Int. J. Heat and Mass Transf.* **44** 3471–80
[20] Robinson A J and Schnitzler E 2007 *Exp. Thermal and Fluid Sc.* **32** 1–13
[21] Whelan B P and Robinson A J 2009 *Appl. Therm. Eng.* **29** 11–2
[22] Webb R L 1981 *Int. J. Heat and Mass Transf.* **24** 715–726

IOP Conf. Series: Journal of Physics: Conf. Series **1224** (2019) 012015 doi:10.1088/1742-6596/1224/1/012015

Finite heating duration based-sensitivity analysis in transient heat conduction

G D'Alessandro[1] **and F de Monte**[1]

[1] Department of Industrial and Information Engineering and Economics, University of L'Aquila, via G. Gronchi 18, 67100 L'Aquila, Italy

giampaolo.dalessandro@graduate.univaq.it

Abstract. In the parameter estimation the sensitivity analysis of the temperature to the unknown parameter (such as thermal conductivity, heat capacity) plays a fundamental role. In the experimental setup for thermal properties measurements of solid materials, the heater in contact with the specimen can be modelled through an high conductivity thin layer. In such a case the one dimensional finite rectangular body, representing the sample, is subject to a boundary condition of the 4th kind at the heated boundary. Also, a constant heat flux applied for a finite period of time is considered in the analysis and the sample backside is assumed insulated. The temperature field within the finite body is obtained by using the superposition principle. Then, the scaled sensitivity coefficients are computed analytically for two different locations: at the interface between the heater and the sample, and at the sample backside. The results show that the sensitivity coefficients with respect to the thermal conductivity and to the volumetric heat capacity of the sample are uncorrelated.

1. Introduction

Sensitivity coefficients have many applications such as parameter estimation, optimal experiment design [1] and uncertainty or error analysis [2]. In parameter estimation, valuable insight is provided by carrying out a sensitivity analysis of the temperature to the unknown parameters, such as thermal conductivity, volumetric heat capacity or thermal diffusivity. For instance, in an apparatus for which the sensitivity coefficients had not been investigated it is possible that additional materials in the experimental configuration (such as the thin heater giving up heat to the sample) can have a large impact on the temperature than the material of interest (sample), with direct consequences on the quality of the results [1].

Also, on the one hand the sensitivity coefficients make possible a preliminary evaluation of the goodness of the experimental results (at least from a qualitative point of view), on the other hand they are directly involved in the estimation of the parameters when minimizing the ordinary least square norm [3-5].

As described in [6], the parameter estimation technique requires measured temperature values. It is essential that the measured response be sensitive to the parameters of interest: the more sensitive the temperature (or large the sensitivity coefficient) is, the more valuable the temperature measurements are [1]. Also, in order to gain much insight and information from the results as possible, it must be that the sensitivity coefficients with respect to the parameters of interest are uncorrelated and large in magnitudes [7].

The focus of the current paper is to provide sensitivity coefficients in the event that the effect of the thin heater in contact with the specimen is simulated through an high-conductivity thin layer to which a constant heat flux is applied for a finite period of time. Sensitivity coefficients for a similar case, but involving a constant heat flux applied for a unlimited period of time, are provided in a previous work of the same authors [7], where they were computed numerically using a two-points central difference approximation for large times. The addressed heat conduction problem concerns a one dimensional finite rectangular body, representing the sample, subject to a boundary condition of the fourth kind [8, Chap. 2] at the heated boundary $x=0$, and insulated at the backside ($x=L$). As the problem is linear, the thermal field is obtained by using the superposition principle. Then, the so-called "scaled" sensitivity coefficients are computed analytically by performing the partial derivatives of the temperature with respect to the parameters of interest. The coefficients are presented in a graphical form and they are calculated for two different location: 1) at the interface between the heater and the sample, and 2) at the sample backside. The results of the analysis show that the sensitivity coefficients with respect the thermal properties of the sample are uncorrelated and, therefore, the thermal conductivity and the heat capacity can be estimated simultaneously.

2. Mathematical formulation

The schematic of an experimental apparatus for the thermal properties measurements of solid materials is shown in figure 1a. In particular, a thin layer heater is located at the interface of two samples of the same material and thickness and gives heat up at surface of both samples. Also, all the external surfaces are insulated and the uniform volumetric heat generation within the heater, $g(t)$, is applied for a finite period of time (from t=0 to t_h) as follows.

$$g(t)=g_0 H(t)-g_0 H(t-t_h) \quad (\tilde{t}>0) \tag{1}$$

where g_0 is the constant volumetric heat generation, while $H(\cdot)$ is the Heaviside function.

For the sake of thermal symmetry, the three-layer configuration (specimen-heater-specimen) reduces to a simplified two-layer configuration depicted in figure 1b. In particular, it is assumed that 1) the heater is in perfect thermal contact with the sample; 2) the heater can be modelled as a lumped body, and 3) the thermal properties of the sample (thermal conductivity k and volumetric heat capacity C) are temperature-independent as well as the volumetric heat capacity of the heater (C_f). Note that due to the second assumption the volumetric heat generation can be considered as a surface heat flux, $q''_{f,0}=g_0 L_f$, applied for a finite period of time as shown by figure 1b, where L_f is a half of the heater thickness. Also, the sample and the heater are initially at the same uniform temperature T_{in}.

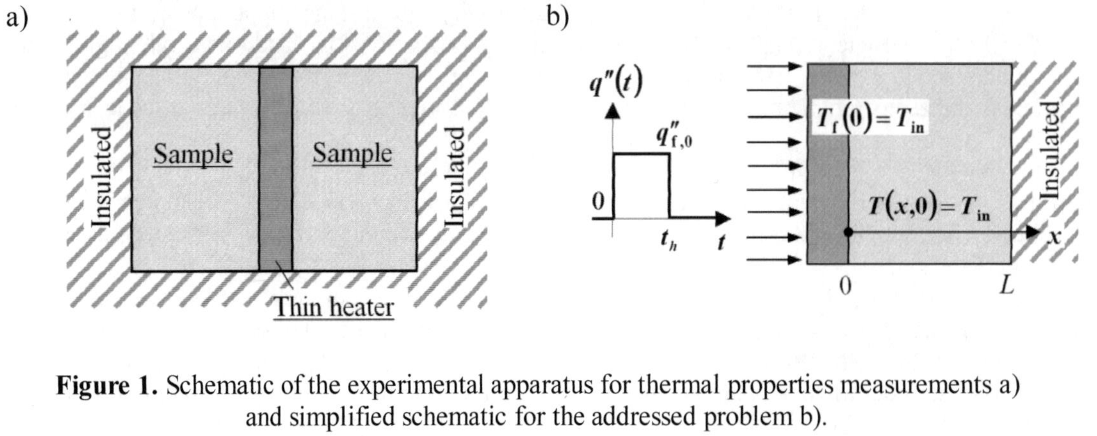

Figure 1. Schematic of the experimental apparatus for thermal properties measurements a) and simplified schematic for the addressed problem b).

The mathematical formulation of this transient, linear, rectangular heat conduction problem is defined in dimensionless form as follows.

$$\frac{\partial^2\tilde{T}}{\partial\tilde{x}^2}=\frac{\partial\tilde{T}}{\partial\tilde{t}} \qquad (0<\tilde{x}<1;\tilde{t}>0) \tag{2a}$$

$$-\left(\frac{\partial\tilde{T}}{\partial\tilde{x}}\right)_{\tilde{x}=0}+P\frac{\partial\tilde{T}_f}{\partial\tilde{t}}=\underbrace{H(\tilde{t})-H(\tilde{t}-\tilde{t}_h)}_{q''_f(\tilde{t})}; \quad \tilde{T}_f(\tilde{t})=\tilde{T}(0,\tilde{t}) \qquad (\tilde{t}>0) \tag{2b}$$

$$-\left(\frac{\partial\tilde{T}}{\partial\tilde{x}}\right)_{\tilde{x}=1}=0 \qquad (\tilde{t}>0) \tag{2c}$$

$$\tilde{T}_f(0)=0; \qquad \tilde{T}(\tilde{x},0)=0 \qquad (0<\tilde{x}<1) \tag{2d}$$

where equation (2b) represents the boundary condition of the fourth kind at $\tilde{x}=0$ [8] and the dimensionless variables appearing in the above equations are defined as:

$$\tilde{T}=\frac{T-T_{in}}{q''_{f,0}L/k},\ \tilde{T}_f=\frac{T_f-T_{in}}{q''_{f,0}L/k},\ \tilde{q}''=\frac{q''}{q''_{f,0}},\ \tilde{q}''_f=\frac{q''_f}{q''_{f,0}},\ \tilde{x}=\frac{x}{L},\ \tilde{t}=\frac{kt}{CL^2},\ \tilde{t}_h=\frac{kt_h}{CL^2},\ P=\frac{C_fL_f}{CL}$$

In detail the P parameter appearing in equation (2) denotes the ratio between the heat capacity of the heater and the sample.

According to the numbering system devised in [9] this linear problem may be denoted by X42B50T00. In detail, this number denotes a transient heat conduction problem concerning a 1D rectangular finite body (by the "X"), subject to a boundary condition of the fourth kind at the surface x=0 (by the "4" in X42) where a finite duration heat flux is applied (by the B5), and having an insulated boundary at x=L (heat flux boundary condition by the "2" in X42, which is homogeneous by the "0" in B50); also, T00 denotes a zero initial temperature for both layers.

3. Temperature solution

The solution to the current problem, equations (2a)-(2d), has to be calculated both when the heating is "on" ($0<\tilde{t}\le\tilde{t}_h$) and when it is "off" ($\tilde{t}>\tilde{t}_h$). By using the superposition principle it results in

$$\tilde{T}=\begin{cases}\tilde{T}_{X42}(\tilde{x},\tilde{t}) & (0\le\tilde{t}\le\tilde{t}_h)\\ \tilde{T}_{X42}(\tilde{x},\tilde{t})-\tilde{T}_{X42}(\tilde{x},\tilde{t}-\tilde{t}_h) & (\tilde{t}>\tilde{t}_h)\end{cases} \tag{3}$$

where $\tilde{T}_{X42}$ is the response to a constant heat flux applied at $\tilde{t}=0$ for an infinite duration. For such a heat flux the problem is denoted by X42B10T00 whose solution is available in references [10, p. 128, # 5] and [11] (note that the "1" in B10 denotes a constant heat flux applied for any $\tilde{t}>0$). It is

$$\tilde{T}_{X42}(\tilde{x},\tilde{t})=\underbrace{\frac{\tilde{t}}{\tilde{N}_0}+\frac{\tilde{x}^2}{2\tilde{N}_0}-\frac{\tilde{x}}{\tilde{N}_0}+\frac{1}{3\tilde{N}_0^2}}_{\tilde{T}_{X42,qs}(\tilde{x},\tilde{t})}\underbrace{-\sum_{m=1}^{\infty}\frac{\cos(\beta_m\tilde{x})-P\beta_m\sin(\beta_m\tilde{x})}{\tilde{N}_m\beta_m^2}e^{-\beta_m^2\tilde{t}}}_{\tilde{T}_{X42,ct}(\tilde{x},\tilde{t})} \tag{4a}$$

where the dimensionless norm $\tilde{N}_m$ and the eigenvalues β_m are defined as, respectively,

$$\tilde{N}_m=\begin{cases}P+1 & m=0\\ \dfrac{(P\beta_m)^2+P+1}{2} & m=1,2,\ldots\end{cases} \tag{4b}$$

$$\beta_m \cot(\beta_m) = -P^{-1} \qquad m = 0,1,2,\ldots \tag{4c}$$

The $\beta_0 = 0$ eigenvalue is responsible of the term $\tilde{t}/\tilde{N}_0$ appearing in equation (4a). The computation of the remaining β_m eigenvalues as well as the number of terms m_{max} are discussed in the next subsections.

3.1. *Computation of the eigenvalues*

The eigenvalues appearing in equation (4a) are computed as roots of the eigencondition (transcendental equation) given by equation (4c).
Its roots may be computed by using explicit approximate relations based on the second-order modified Newton method [12]. These relations provide an approximate value of the exact eigenvalue with high accuracy (8-decimal place after one iteration, and 15-decimal place after two iterations) for the P range $[0, \infty)$.

3.2. *Computational solution*

The infinite series appearing in the temperature solution, equation (4a), exhibits an exponential convergence, that is, very fast. However, as we cannot take into account infinite terms a convergence criterion for this series is needed. Following the procedure given in [13], the maximum number of required terms $m_{\max}$ to get a truncation error of 10^{-A} (A=2,3,..15) in the series solution may be taken as:

$$m_{\max} = \text{ceil}\left\{\frac{1}{2} + \frac{1}{\pi}\left[\frac{A\ln(10)}{\tilde{t}}\right]^{1/2}\right\} \tag{5}$$

where $(m-1/2)\pi$ has been used as a conservative estimate for β_m and "ceil(z)" is a Matlab function that rounds the number z to the nearest integer greater than or equal to z.

3.2.1. *Early time solution.*

In addition, according to equation (5) a large number of terms is required for early times. However, for times less than the so-called deviation time $\tilde{t}_{\mathrm{d}}$, defined as [9, 14]

$$\tilde{t}_{\mathrm{d}} = \frac{(2-\tilde{x})^2}{10A} \qquad (A = 2,3,\ldots,15), \tag{6}$$

the temperature $\tilde{T}_{X42}$ appearing in equation (3) can be replaced by the solution of the semi-infinite transient X40B1T00 problem with an error less than 10^{-A}.
However, as $\tilde{t}_{\mathrm{d}}$ can be less or greater than $\tilde{t}_h$, two cases has to be considered:

- $\tilde{t}_{\mathrm{d}} \le \tilde{t}_h$. In this case, the temperature solution is simply given by

$$\tilde{T}(\tilde{x},\tilde{t}) \cong \tilde{T}_{X40}(\tilde{x},\tilde{t}) \qquad \left(0 \le \tilde{t} \le \tilde{t}_{\mathrm{d}} \le \tilde{t}_h\right) \tag{7a}$$

- $\tilde{t}_{\mathrm{d}} > \tilde{t}_h$. In such a case, equation (3) holds the two parts and may be rewritten as

$$\tilde{T}(\tilde{x},\tilde{t}) \cong \begin{cases} \tilde{T}_{X40}(\tilde{x},\tilde{t}) & \left(0 \le \tilde{t} \le \tilde{t}_h < \tilde{t}_{\mathrm{d}}\right) \\ \tilde{T}_{X40}(\tilde{x},\tilde{t}) - \tilde{T}_{X40}(\tilde{x},\tilde{t}-\tilde{t}_h) & \left(\tilde{t}_h < \tilde{t} \le \tilde{t}_{\mathrm{d}}\right) \end{cases} \tag{7b}$$

where $\tilde{T}_{X40}(\tilde{x},\tilde{t})$ denotes the solution of the X40B1T00 problem. It is available in the heat conduction literature [10, p 306, #12] and it is also discussed in [15]. In dimensionless form it results in

$$\tilde{T}_{X40}(\tilde{x},\tilde{t})=\sqrt{\frac{4\tilde{t}}{\pi}}\exp\left(-\frac{\tilde{x}^2}{4\tilde{t}}\right)-\tilde{x}\,\mathrm{erfc}\left(\frac{\tilde{x}}{\sqrt{4\tilde{t}}}\right)-P\left[\mathrm{erfc}\left(\frac{\tilde{x}}{\sqrt{4\tilde{t}}}\right)-\mathrm{Am}(\tilde{x},\tilde{t},P)\right]\quad\left(0\le\tilde{t}\le\tilde{t}_{\mathrm{d}}\right)\tag{8a}$$

where $\mathrm{Am}(\tilde{x},\tilde{t},P)$ is the so-called Amos function defined as

$$\mathrm{Am}(\tilde{x},\tilde{t},P)=\exp\left(\frac{\tilde{x}}{P}+\frac{\tilde{t}}{P^2}\right)\mathrm{erfc}\left(\frac{\tilde{x}}{\sqrt{4\tilde{t}}}+\frac{\sqrt{\tilde{t}}}{P}\right)=\exp\left(-\frac{\tilde{x}^2}{4\tilde{t}}\right)\mathrm{erfcx}\left(\frac{\tilde{x}}{\sqrt{4\tilde{t}}}+\frac{\sqrt{\tilde{t}}}{P}\right)\tag{8b}$$

Moreover, $\tilde{T}_{X40}(\tilde{x},\tilde{t}-\tilde{t}_h)$ may be derived from equation (8a) by simply replacing $\tilde{t}$ with $\tilde{t}-\tilde{t}_h$.

A verified computer code in Matlab ambient for calculating the solution of the X42B10T00 problem is provided in [11]. Also, for the sake of completeness, a plot of the dimensionless temperature as a function of time with $\tilde{t}_h$ as a parameter is given in figure 2.

4. Sensitivity coefficients

The scaled sensitivity coefficients are defined as partial derivatives of the temperature with respect to the model parameter p of interest (e.g. k, C, C_f) multiplied by the parameter itself:

$$X_p=p\frac{\partial T}{\partial p}\quad\text{with}\quad p=k,C,C_{\mathrm{f}}\tag{9}$$

Note that the scaled sensitivity coefficients have units of °C and their calculation requires the use of the chain rule [16].

When the parameter of interest is the thermal conductivity, the corresponding sensitivity coefficient can be expressed as follows

$$X_k=k\frac{\partial T}{\partial k}=k\frac{\partial}{\partial k}\left(\tilde{T}\frac{q''_{\mathrm{f}0}L}{k}\right)=\frac{q''_{\mathrm{f}0}L}{k}\left[k\frac{\partial\tilde{T}}{\partial k}-\tilde{T}\right]\tag{10}$$

where $\tilde{T}=\tilde{T}[\tilde{x},\tilde{t},P,\beta_m(P),\tilde{t}-\tilde{t}_h]$, with $\tilde{t}(k,C)$, $P(C,C_{\mathrm{f}})$, $\tilde{t}-\tilde{t}_h(k,C)$. Therefore, by using the chain

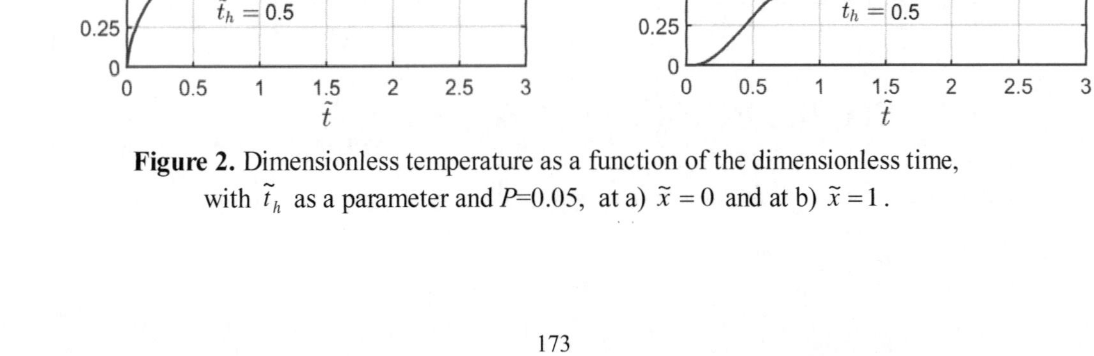

Figure 2. Dimensionless temperature as a function of the dimensionless time, with $\tilde{t}_h$ as a parameter and P=0.05, at a) $\tilde{x}=0$ and at b) $\tilde{x}=1$.

rule it is found that

$$\frac{\partial \tilde{T}}{\partial k} = \frac{\partial \tilde{T}}{\partial \tilde{t}}\frac{\partial \tilde{t}}{\partial k} + \frac{\partial \tilde{T}}{\partial (\tilde{t} - \tilde{t}_h)}\frac{\partial (\tilde{t} - \tilde{t}_h)}{\partial k} \tag{11}$$

By substituting equation (11) in equation (10), one can obtain the scaled sensitivity coefficient with respect to the thermal conductivity. In dimensionless form it results in:

$$\tilde{X}_k = \tilde{t}\frac{\partial \tilde{T}}{\partial \tilde{t}} + (\tilde{t} - \tilde{t}_h)\frac{\partial \tilde{T}}{\partial (\tilde{t} - \tilde{t}_h)} - \tilde{T} \tag{12}$$

where $\tilde{X}_k = X_k / (q''_{f0} L / k)$.

By following the same procedure described above, the dimensionless scaled sensitivity coefficient with respect to the volumetric heat capacity C is obtained.

$$\tilde{X}_C = C\frac{\partial \tilde{T}}{\partial C} = -\tilde{t}\frac{\partial \tilde{T}}{\partial \tilde{t}} - P\frac{\partial \tilde{T}}{\partial P} - (\tilde{t} - \tilde{t}_h)\frac{\partial \tilde{T}}{\partial (\tilde{t} - \tilde{t}_h)} \tag{13}$$

In similar manner the scaled sensitivity coefficient with respect to the volumetric heat capacity of the heater C_f, in dimensionless form, results in:

$$\tilde{X}_{C_f} = C_f\frac{\partial \tilde{T}}{\partial C_f} = P\frac{\partial \tilde{T}}{\partial P} \tag{14}$$

Also, the following relationship involving the scaled sensitivity coefficients listed above is valid for all values of time and position.

$$\tilde{X}_k + \tilde{X}_C + \tilde{X}_{C_f} = -\tilde{T} \tag{15}$$

4.1. Computation of the scaled sensitivity coefficients $\tilde{X}_k$, $\tilde{X}_C$ and $\tilde{X}_{C_f}$

The sensitivity coefficients discussed above may be computed both when the heating is "on" ($0 < \tilde{t} \le \tilde{t}_h$) and when it is "off" ($\tilde{t} > \tilde{t}_h$). By using the superposition principle they may be given as

$$\tilde{X}_i = \begin{cases} \tilde{X}_{i,X42}(\tilde{x},\tilde{t},P) & (0 \le \tilde{t} \le \tilde{t}_h) \\ \tilde{X}_{i,X42}(\tilde{x},\tilde{t},P) - \tilde{X}_{i,X42}(\tilde{x},\tilde{t}-\tilde{t}_h,P) & (\tilde{t} > \tilde{t}_h) \end{cases} \quad \text{with } i = k, C, C_f \tag{16}$$

where $\tilde{X}_{i,X42}$ (with *i=k, C, C_f*) denotes the sensitivity coefficients of the X42B10T00 case which, bearing in mind equations (12)-(14), can be evaluated using the following relations

$$\tilde{X}_k(\tilde{x},\tilde{t},P) = \tilde{t}\frac{\partial \tilde{T}}{\partial \tilde{t}} - \tilde{T}; \quad \tilde{X}_k(\tilde{x},\tilde{t}-\tilde{t}_h,P) = (\tilde{t} - \tilde{t}_h)\frac{\partial \tilde{T}}{\partial (\tilde{t} - \tilde{t}_h)} - \tilde{T} \tag{17a}$$

$$\tilde{X}_C(\tilde{x},\tilde{t},P) = -\tilde{t}\frac{\partial \tilde{T}}{\partial \tilde{t}} - P\frac{\partial \tilde{T}}{\partial P}; \quad \tilde{X}_C(\tilde{x},\tilde{t}-\tilde{t}_h,P) = -(\tilde{t} - \tilde{t}_h)\frac{\partial \tilde{T}}{\partial (\tilde{t} - \tilde{t}_h)} - P\frac{\partial \tilde{T}}{\partial P} \tag{17b}$$

$$\tilde{X}_{C_f}(\tilde{x},\tilde{t},P) = P\frac{\partial \tilde{T}}{\partial P}; \quad \tilde{X}_{C_f}(\tilde{x},\tilde{t}-\tilde{t}_h,P) = P\frac{\partial \tilde{T}}{\partial P} \tag{17c}$$

Then, by applying the above equations to equation (4a) the sensitivities $\tilde{X}_{i,X42}(\tilde{x},\tilde{t},P)$ (with *i=k, C,*

C_f) result in

$$\tilde{X}_{k,X42} = \frac{\tilde{x}(2-\tilde{x})}{2\tilde{N}_0} - \frac{1}{3\tilde{N}_0^2} + \sum_{m=1}^{\infty} \frac{\cos(\beta_m \tilde{x}) - P\beta_m \sin(\beta_m \tilde{x})}{\tilde{N}_m \beta_m^2} \left(1 + \beta_m^2 \tilde{t}\right) e^{-\beta_m^2 \tilde{t}} \quad (18a)$$

$$\tilde{X}_{C,X42} = -\frac{\tilde{t}}{\tilde{N}_0} + \frac{P[2\tilde{t} + \tilde{x}(\tilde{x}-2)]}{2\tilde{N}_0^2} + \frac{2P}{3\tilde{N}_0^3} + \sum_{m=1}^{\infty} \frac{D_m \cos(\beta_m \tilde{x}) - E_m \sin(\beta_m \tilde{x})}{\tilde{N}_m^2 \beta_m^4} e^{-\beta_m^2 \tilde{t}} \quad (18b)$$

$$\tilde{X}_{C_f,X42} = -\frac{P[2\tilde{t} + \tilde{x}(\tilde{x}-2)]}{2\tilde{N}_0^2} - \frac{2P}{3\tilde{N}_0^3} + P\sum_{m=1}^{\infty} \frac{F_m \sin(\beta_m \tilde{x}) + G_m \cos(\beta_m \tilde{x})}{\tilde{N}_m^2 \beta_m^4} e^{-\beta_m^2 \tilde{t}} \quad (18c)$$

where

$$D_m = \tilde{N}_m \beta_m^4 \tilde{t} - PG_m, \qquad E_m = P\left(\tilde{N}_m \beta_m^5 \tilde{t} + F_m\right) \quad (18d)$$

Also,

$$F_m = -P\beta_m^2 \frac{d\beta_m}{dP}\left(2P^2\beta_m^2 + \tilde{N}_0\right) + \tilde{N}_m \beta_m^2 \left[\frac{d\beta_m}{dP}\left(\tilde{x} - 2P\beta_m^2 \tilde{t} - P\right) + \beta_m\right] - P\beta_m^3 \left(P\beta_m^2 + \frac{1}{2}\right) \quad (18e)$$

$$G_m = \beta_m \frac{d\beta_m}{dP}\left[2P^2\beta_m^2 + \tilde{N}_m \beta_m^2 \left(P\tilde{x} + 2\tilde{t}\right) + \tilde{N}_0\right] + \beta_m^2 \left(P\beta_m^2 + \frac{1}{2}\right) \quad (18f)$$

It is worth noting that the sensitivity coefficients given by equations (18a)-(18c) can be computed analytically as the derivative $d\beta_m/dP$ appearing in equations (18e) and (18f) may be given in an algebraic form. In fact, the eigencondition defined by equation (4c) can be rewritten as

$$P = -\frac{\tan(\beta_m)}{\beta_m} \quad (19)$$

Then, by differentiating equation (19) with respect to the eigenvalue β_m and by taking the corresponding reciprocals, it is found that

$$\frac{d\beta_m}{dP} = \frac{\tan(\beta_m)[1 - \beta_m \tan(\beta_m)] - \beta_m}{\beta_m^2} \quad (20)$$

Moreover, the sensitivity coefficients $\tilde{X}_{i,X42}(\tilde{x}, \tilde{t} - \tilde{t}_h, P)$ may be derived from equations (18a)-(18f) by simply replacing $\tilde{t}$ with $\tilde{t} - \tilde{t}_h$.

For early times (less than the deviation time $\tilde{t}_d$) the sensitivity coefficients may be evaluated analytically by using equations (7a) and (7b).

5. Results

Figures 3-5 show the dimensionless sensitivity coefficients $\tilde{X}_k$, $\tilde{X}_C$ and $\tilde{X}_{C_f}$ as a function of time for different P values. They are computed for a heating duration of $\tilde{t}_h = 1$ and for two temperature sensor locations: 1) at the interface between the heater and the sample ($\tilde{x} = 0$), and 2) at the sample backside ($\tilde{x} = 1$). In addition, in these figures a comparison with the sensitivities for P=0 (when the heater is completely neglected - X22B50T0 case [6, Chap. 8]) is also provided.

As shown by figure 3 the sensitivity to thermal conductivity of the sample $\tilde{X}_k$ increases in absolute

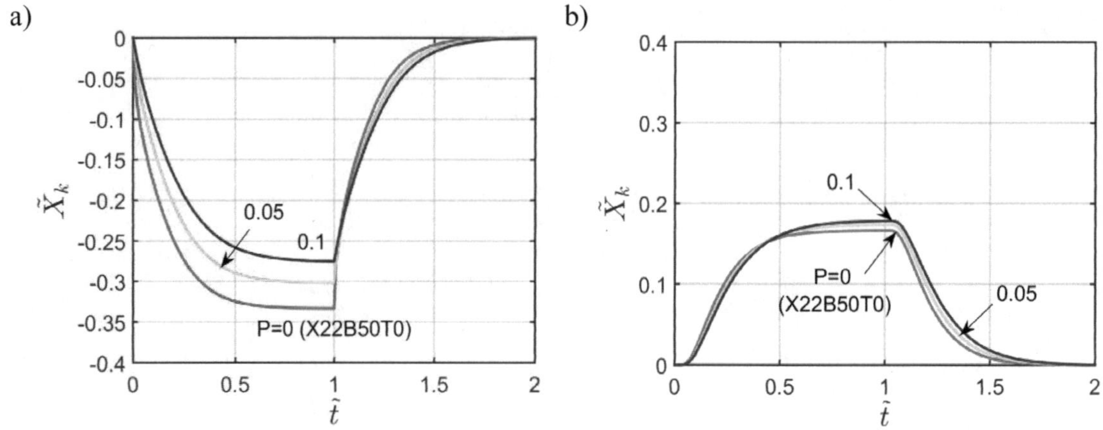

Figure 3. Dimensionless scaled sensitivity coefficient to the thermal conductivity as a function of the dimensionless time, for different P values, at a) $\tilde{x}=0$ and at b) $\tilde{x}=1$.

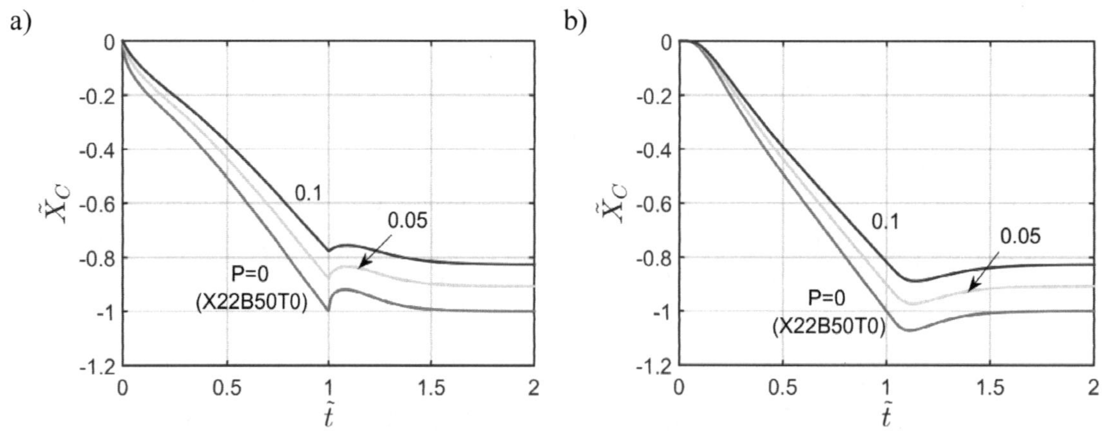

Figure 4. Dimensionless scaled sensitivity coefficient to the heat capacity of the sample as a function of the dimensionless time, for different P values, at a) $\tilde{x}=0$ and at b) $\tilde{x}=1$.

value with time during the heating period; then, when the heater is turn off ($\tilde{t}_h=1$) it decreases approaching zero for large times. In particular, figure 4a shows that during the heating the sensitivity to k at $\tilde{x}=0$ is lower in absolute value when P is higher, while the opposite occurs when the heating is off. At the sample backside the behaviour is quite different (see figure 3b). In fact, after a certain time (of about 0.4), $\tilde{X}_k$ increases when P increases even during the heating period. Note, also, that the greatest sensitivity to k occurs at the interface heater-sample ($\tilde{x}=0$).

The sensitivity to the volumetric heat capacity of the sample $\tilde{X}_C$ is shown by figure 4. In particular, in figure 4a the sensitivity to C at $\tilde{x}=0$ increases in absolute value with time during the heating period, while it exhibits a slight fall when the heater is turn off ($\tilde{t}_h=1$); then it continues to increase until a constant value is approached for large times. On the contrary, the sensitivity at $\tilde{x}=1$ continues in absolute value to rise even during the cooling period and, then, it decreases to reach the same constant value occurring at $\tilde{x}=0$, as shown by figure 4b. In addition, the sensitivity to C at $\tilde{x}=1$ is larger than that at $\tilde{x}=0$ for some times. Unlike the sensitivity to k, the coefficient $\tilde{X}_C$ is lower in absolute

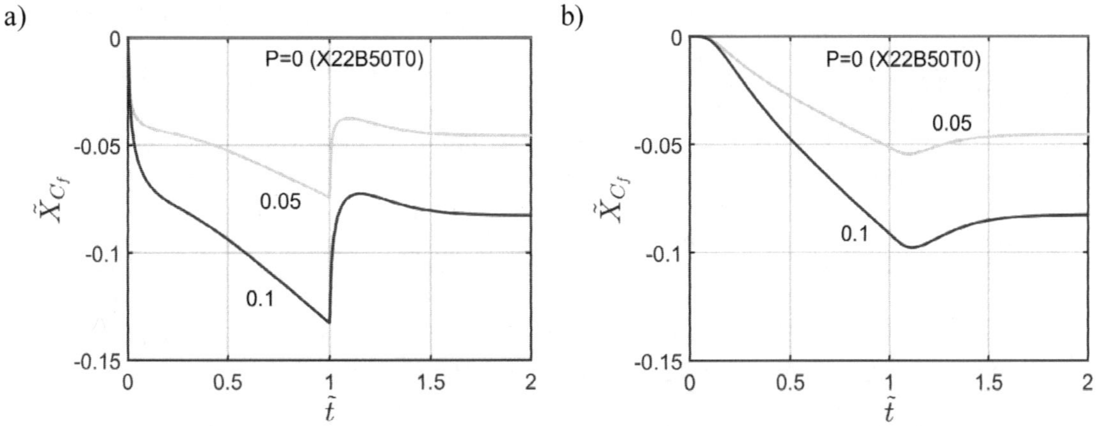

Figure 5. Dimensionless scaled sensitivity coefficient to the heat capacity of the heater as a function of the dimensionless time, for different P values, at a) $\tilde{x}=0$ and at b) $\tilde{x}=1$.

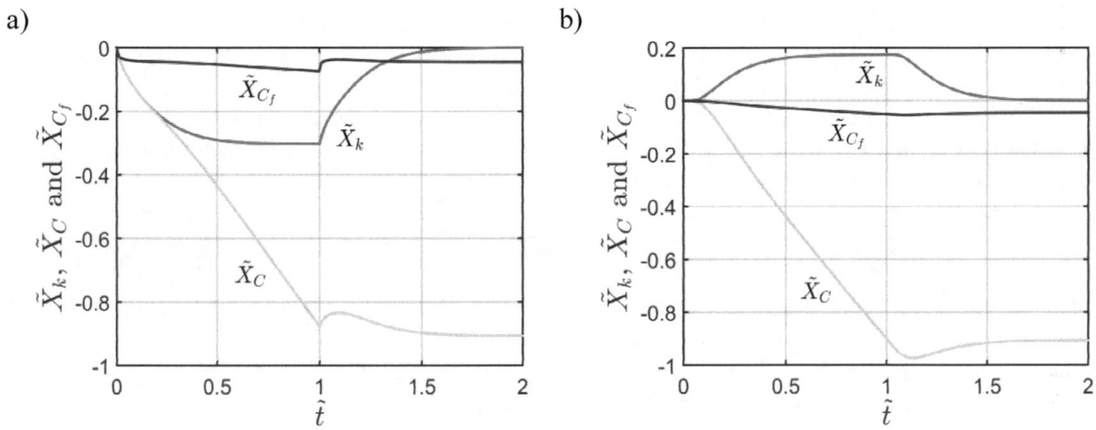

Figure 6. Comparison among dimensionless scaled sensitivity coefficients for $P=0.05$ and $\tilde{t}_h=1$ at a) $\tilde{x}=0$ and at b) $\tilde{x}=1$.

value when P is larger for all the experiment duration.

Figure 5 shows the sensitivity to the volumetric heat capacity of the heater $\tilde{X}_{C_f}$. Similarly to $\tilde{X}_C$, the sensitivity to C_f at $\tilde{x}=0$ drops steeply as soon as the heater is turn off (see figure 5a), whereas the sensitivity at $\tilde{x}=1$ continue to rise as shown by figure 5b. However, at both locations the same constant value is approached for large times. During the heating period the largest sensitivity occurs at $\tilde{x}=0$, while for $\tilde{t}>1$ it is larger at the sample backside.

Moreover, a comparison among the sensitivity coefficients computed at both $\tilde{x}=0$ and $\tilde{x}=1$ for P=0.05 is given by Fig. 6. It suggests that, at early times, the sensitivities $\tilde{X}_k$ and $\tilde{X}_C$ are nearly equal, that is, correlated while, for larger times, they behave quite differently and, hence, they are uncorrelated. In addition, when the steady state is reached the sensitivity to the thermal conductivity k approaches zero, while the sensitivity to the heat capacity of the sample C approaches a not null steady value. It is also evident that the sensitivity function $\tilde{X}_{C_f}$ is very small at both $\tilde{x}=0$ and $\tilde{x}=1$. In other words the temperature of the sample is more sensitive to its heat capacity than to the heat capacity of the heater.

A comparison with the sensitivity obtained when a constant heat flux is applied for a unlimited period of time [7] is now discussed. As shown by figures 3-6 the sensitivity coefficients changes in shape when the heating is off. These changes in shape do not occur when a constant heat flux is applied for a unlimited period. In fact, in such a case the sensitivity to the thermal conductivity would approach a constant value, while the sensitivities to the volumetric heat capacity of the heater and the sample would continue to increases monotonically with time. Also, this is valid at both locations ($\tilde{x} = 0$ and $\tilde{x} = 1$). However, sensitivity coefficients having changes in shape or in sign are useful to reduce the degree of correlation among parameters and also to maximize the determinant of the $\boldsymbol{X}^T\boldsymbol{X}$ matrix (where $\boldsymbol{X}$ is the sensitivity matrix) involved in the parameter estimation procedure. For these reasons a finite heating period plays a key role on sensitivity coefficients.

6. Conclusions

The temperature solution for the X42B50T00 problem has been derived by using the superposition principle. Then, the sensitivity coefficients with respect to the thermal properties of the sample (thermal conductivity k and volumetric heat capacity C) and the volumetric heat capacity of the heater have been investigated in order to establish the best location for the temperature sensor.

The analysis shows that the sensitivity coefficients with respect to k and to C are uncorrelated and, therefore, these thermal properties can be estimated simultaneously. However, the maximum sensitivity to the thermal conductivity occurs the interface between the heater and the sample, while the sensitivity to the volumetric heat capacity of the specimen exhibits its maximum at the sample backside. For this reason, in order to estimate k and C simultaneously it should be more effective using two sensors (one located at $\tilde{x} = 0$ and the other at $\tilde{x} = 1$) rather than using only one sensor.

References

[1] Dowding K J, Blackwell B F and Cochran R J 1999 Application of sensitivity coefficients for heat conduction problems *Numerical Heat Transfer Part* B **36** 33-55

[2] Blackwell B F and Beck J V 2010 A technique for uncertaninty analysis for inverse heat conduction problems, *Int. J. Heat Mass Transfer* **53** 753-9

[3] Aviles-Ramos C and Haji-Sheikh A 2001 Estimation of thermophysical properties of composites using multi-parameter estimation and zeroth-order regularization *Inverse Problems Engineering* **9** 507-36

[4] Sakly A, Jemni A, Lagonotte P and Petit D 2011 Estimation of the effective thermal properties in a matallic medium by an inverse technique using temperature measurements *J. Appl. Fluid Mechanics* **4** 23-9

[5] Cui M, Gao X and Zhang J 2012 A new approach for the estimation of temperature-dependent thermal properties by solving transient inverse heat conduction problems *Int. J. Thermal Sciences* **58** 113-9

[6] Beck J V and Arnold K J 1977 *Parameter Estimation in Engineering and Science* (New York: John Wiley and Sons)

[7] D'Alessandro G and de Monte F 2017 Sensitivity coefficients for thermal property measurements using a boundary condition of the 4th kind *J. Phys. Conf. Series* **796** 012011

[8] Cole K D, Beck J V, Haji-Sheikh A and Litkouhi B 2011 *Heat Conduction Using Green's Function 2nd Edition* (Boca Raton FL: CRC Press Taylor&Francis)

[9] Cole K D, Beck J V, Woodbury A and de Monte F 2014 Intrinsic verification and a heat conduction database *Int. J. Thermal Sciences* **78** 36-47

[10] Carslaw H S and Jaeger J C 1959 *Conduction of Heat in Solids 2nd edition* (New York: Oxford University Press)

[11] D'Alessandro G and de Monte F 2016 X42B10T00 *Exact Analytical Conduction Toolbox*, exact.unl.edu

[12] Haji-Sheikh A and Beck J V 2000 An efficient method of computing eigenvalues in heat conduction *Numerical Heat Transfer Part* B **38** 133-56

[13] de Monte F, Beck J V and Amos D E 2011 A heat-flux based building block approach for solving heat conduction problems *Int. J. Heat Mass Transfer* **54** 2789-800

[14] de Monte F, Beck J V and Amos D E 2012 Solving two-dimensional Cartesian unsteady heat conduction problems for small values of the time *Int. J. Thermal Sciences* **60** 106-13

[15] de Monte F and Beck J V 2013 X40B10T00 *Exact Analytical Conduction Toolbox*, exact.unl.edu

[16] Blackwell B F, Cochran R J and Dowding K J 1999 Development and implementation of sensitivity coefficient equations for heat conduction problem *Numerical Heat Transfer Part* B **36** 15-32

IOP Conf. Series: Journal of Physics: Conf. Series **1224** (2019) 012016 doi:10.1088/1742-6596/1224/1/012016

Occupancy modelling of buildings based on CO_2 concentration measurements: an experimental analysis

Alessandro Franco, Francesco Leccese, Lorenzo Marchi

Department of Energy, Systems, Territory and Constructions Engineering (DESTEC), University of Pisa, Largo Lucio Lazzarino, 56126 PISA (ITALY),

email address: alessandro.franco@ing.unipi.it

Abstract. The problem of real-time estimation of occupancy of buildings (number of people in various zones at every time instant) is relevant to a number of emerging applications that achieve high energy efficiency through feedback control. The measurement of CO_2 concentration can be considered an important indicator that allows to define the occupation of closed and crowded spaces. Interesting cases can be school buildings and other buildings used in civil and residential (shopping centers, hospitals, etc.). This paper starting from an experimental analysis conducted of a University campus in different operating conditions, with different model for occupancy simulation in multi-zone multi-occupant buildings and its preliminary validation from experimental data acquired in different cases. The acquired data are used to present some graphical models and perform occupancy estimates. Starting from an accurate analysis of the data, attempts are made to define a preliminary estimation method through the development of a mathematical models of occupancy dynamics in a building, which show promising results.

1. Introduction

During the last decades, the issue of energy saving is increasingly attracting given the goal of decreasing carbon emissions and energy consumption set by the EU through the European 2020 standard.

The buildings sector is the largest energy consuming sector, accounting for over than one-third of final energy consumption globally and an equally important source of carbon dioxide (CO_2) emissions.

High energy consumption in buildings can be caused by several factors including a reduced insulation determined by the thermal envelope, a bad management of the thermal and electrical system, improper use of energy equipment and energy waste of any kind. Moreover, given the final use of the building, it may be difficult to identify a clear path of utilization and for that reason more energy consumption can occur [1].

Among the various buildings for public activities, University Campuses and Schools in general are interesting structures made up of different types of buildings, with different end-use (sometimes with different end-use within the same envelope), different thermal envelopes and different occupancy profiles. An energy characterization of such kind of buildings may reveal a key factor to identify possible actions or refurbishments with the purpose of reducing consumptions. Several papers in the literature have considered these aspects, analyzing the problem with a general perspective, for example [2] and [3], or with reference to specific cases, for example some interesting papers, such as [4-6], or Master Thesis, [7].

Assessing university building with several end-use, different problems emerge. Within the same envelope there could be laboratories, offices, lectures room, classes, open spaces, halls, gyms which have different energy needs. For that reason, a building cannot be characterized by its consumptions only, but in relation with its end use. Moreover, university has different usage profile during the year: during lectures period, classes will be relatively full and many internal gains will occur so in order not to waste thermal energy the heating system should consider all those factors. Furthermore, during lecture periods while high occupancy is registered a high plug-in load might occur, so high consumptions during those periods should not be considered waste all the same. Moreover, during examination period, lower occupancy might be registered in classes but, somewhere else like cafeteria, library, or laboratories there might be a higher occupancy, without considering the high occupancy during written or oral tests. Also, both in lecture and exams periods during

weekends, pre-holidays and holidays very reduced occupancy is expected. In order to maintain the comfort within certain parameters anyway, specific profiles should be considered for the thermal control system. Anyway, a relevant element of analysis is the occupancy of the various structures.
Occupant presence release latent and sensible heat that changes thermal conditions warranting increased in air-conditioning. They also release pollutants such as carbon dioxide (CO_2) which is also an indicator for ventilation adequacy. Monitoring of presence and crowding is important both for the purpose of maintaining controlled thermo-hygrometric comfort conditions and also for the purpose of forecasting and optimizing energy consumption. In particular, forecasting of the occupation of the buildings is very important to have a better control of the operation of the thermal control systems.
At present, the energy certification of buildings is done on a basis exclusively linked to the facility and the structure of the building, without taking into account in any way the real use of the structures.
In fact, one of the elements that defines the difference between design conditions (ideals) and operating conditions (real) in all the structures using thermal control plants, is the number of people inside the building. This determines significant deviations from the design conditions (for example, in a crowded place the internal temperature will be higher than the one that would be in the same place with less overcrowding) both concerning the use of thermal energy as well as of electricity. Furthermore, the current legislation does not consider for example what are the electrical uses related to the overcrowding of places. In spaces such as universities, this is manifested by the presence of numerous devices (notebooks, smartphone chargers, various educational equipment) that significantly increase, and sometimes not easily predictable, the consumption of electricity.
High consumption in buildings could have been caused by several factors including a bad behavior of the thermal envelope, a bad management of the thermal and electrical system, improper use of energy equipment and energy waste of any kind. Moreover, given the final use of the building, it may be difficult to identify a clear path of utilization and for that reason more energy consumption can occur.
University Campuses are made up of different types of buildings, with different end-use (sometimes with different end-use within the same envelope), different thermal envelopes and different occupancy profiles.
An energy characterization of such kind of buildings may reveal a key factor to identify possible actions or refurbishments with the purpose of reducing consumptions. The study of the building energy demand has become a topic of great importance, because of the significant increase of interest in energy sustainability.
University campuses represent specific groups of diverse buildings, with significant energy consumption. They consist of many different buildings, representing small-scale town for itself. Therefore, they provide an interesting testbed to characterize and understand energy consumption of groups of "mixed-use" buildings.
Moving from some recent trend in the research, the paper try to analyze the perspectives of CO_2 based prediction and evaluation of occupancy of building, as an effective mean to obtain operational efficiency increase and of the quality of the environment. The method has been recently proposed in the literature, as clearly evident analyzing, among the other, recent papers such as [8-10]. The step presented concern an attempt of obtaining a correlation between the CO_2 concentration and some quantitative indicator starting from an experimental analysis carried out in a University building.

2. Energy efficiency, environmental quality and the effective use of building: the importance of occupancy estimation and the connection with CO_2 concentration

In the literature there are several studies that highlight the most important aspects to be considered for carry out an analysis of the energy efficiency of a public building. An energy characterization of the buildings may reveal a key factor to identify possible actions or refurbishments with the purpose of reducing consumptions.
An overview of the various energy performance assessment models for common buildings in literature has been made in [11].
The first datum that characterizes a building from an energy point of view is its primary energy consumption (gas and/or electric energy). In general, there are three strategies with which the energy consumption of a building can be estimated, namely: the analysis of bills and total energy consumption, direct consumption measurement and modelling.
Energy bill is a type of high quality measurement data. The method based on the analysis of the energy bills is, among the three, the simplest method, economic but the less accurate since it is not possible to obtain consumption with a frequency other than the monthly one not even isolate the energy consumption of a certain area if there is only one counter. Disaggregating energy bills into end-use provides a better understanding on energy use and results in a better assessment for systems and equipment.
Direct measurement, on the other hand, is a method of energetic characterization of the building much more accurate and that allows you to get much more information. This strategy is implemented through the use of

specific meters, carefully positioned inside the building can provide data with variable sampling frequency (hourly, daily, monthly consumption etc.) and related to particular areas.

The third strategy, largely diffused in the recent literature is to provide a mathematical model of the building and to perform a dynamic simulation of the behaviour. The construction of the dynamic model of a building requires inputs that should be firstly collected and then be fed into a so-called simulation engine to describe detailed mathematic models. Typical inputs for a dynamic simulation tool of this kind may include four groups of parameters, as weather conditions, building description and component description.

There are many influencing factors in energy efficiency benchmarking, such as end-use characterization, building typology (old, refurbished or new structures) and occupancy is one of those.

Energy consumption and indoor environment quality (IEQ) of buildings appears to be strongly linked to human occupants. Predicting the number of occupants in a space is essential for the effective management of various building operation functions as well as improve energy efficiency.

Ever since occupancy detection has been of primarily importance in energy efficiency of public buildings, many techniques have been proposed, each one with its advantages and disadvantages.

Currently, the occupancy sensor market is dominated by infrared and ultrasonic detectors. These relatively simple devices provide fine-grained information on peoples' presence and location. Their output is of binary type, it detects the state "occupied" vs. "not occupied". In addition, these sensors are often unreliable in capturing immobile occupants. Hence, infrared and ultrasonic sensors can be only detectors of presence and do not provide information on count. This fact determines their applicability, which is limited to lighting control systems. The simplest way to obtain information regarding the level of room occupancy is to employ more advanced, especially dedicated equipment, such as vision sensors for example video cameras and RFID (Radio Frequency IDentification) tags.

Video cameras are a reliable and accurate way to determine not only if a space is occupied, but also the number of occupants. However, counting people directly may be challenging due to partial/complete object occlusion. Video images are either observed by human operators or analysed by computer tools. Main drawbacks of this technology are: the need of large data storage capacity, the fact that commercially available software-based image analysis for application to building system control is still under development and, above all, this solution interferes with privacy concerns. That is why several works using low-resolution cameras or even developing "reduced" sensor from camera, with a very different appearance from conventional video camera, have been proposed to obtain enough information to detect person's position and movement status, while reducing the psychological resistance against having a picture taken.

Electromagnetic signal detection systems, radio frequency identification have the capacity to provide occupancy information on location, number of occupants, their identities and track.

RFID is an effective technology for indoor localization that can provide adequate accuracy and fine-grained occupancy information. It is cost-efficient, does not require line of sight conditions. In addition, it has on-board data storage capacity that can be used for another purpose such as building asset management. The disadvantages of this technology result from the privacy (psychological resistance of the tag users), technical problems and some inconvenience. Occupants must possess the RFID tag. Therefore, users feel they are constantly being monitored. Frequently, occupants use more than one device generating EM (electromagnetic) signals, which can be detected by the radio frequency identification system. It results in false registration of presence and count as well as incorrect location. The terminal device held by the user is usually battery powered and not sustainable for long-term data acquisition. A complex and advanced signal processing station is often required. The RFID tag must be attached to the occupant. It causes some inconveniences for the user performing everyday activities.

The drawbacks of the direct determination of occupant number make this approach unsuitable in certain situations. Therefore, indirect inference of occupancy levels in rooms or buildings is considered. This alternative approach exploits signals originating from different buildings systems and indoor environment.

Considering the correlation between the use of structures and the CO_2 concentration in indoor environments, it seems very interesting to elaborate models that, starting from an evaluation on indirect experimental bases of the occupation, can also define a correlation of this with the energy consumption. This could be really interesting in order to determine a prediction of energy consumption referred to the occupants and to produce a feedback on the operation of the thermal control system in order to obtain beneficial effects both in term of air quality and environmental comfort and energy saving.

Recently, it was proposed to estimate the number of indoor occupants from environmental parameters, in particular such as temperature, humidity, and CO_2 concentration, see for example [12]. Amongst these parameters, CO_2 concentration is the one that most correlates with the number of occupants. As mentioned above, humans naturally exhale this gas depending on their metabolic rate. Hence the analysis of the dynamics of CO_2 concentration may be employed in determining the number of occupants inside room.

Moreover, the measurement of CO_2 is useful not only for the occupancy detection but, since CO_2 is a gas which can cause undesired effects on the occupants after long time exposure, also for comfort purposes.
In order to understand the qualitative value of CO_2 concentration that can be measured a preliminary quantitative definition is important.
Carbon dioxide typical exposure limits for outdoors and for different public spaces have been established by a number of governmental health and industrial safety groups. Each of these standards of air concentrations is expressed in parts per million (ppm). Carbon dioxide concentration in the range between 250-600 ppm is the typical level for ambient air in outside spaces: the upper levels indicate a very crowded town, while the lower levels can be measured in countryside. Typical values can be 350-400 ppm. The reference outdoor value is 400 ppm atmosphere, considering 300-320 ppm a level typical of pure air and up to 700 ppm the polluted atmosphere [13]. The acceptable level of carbon dioxide is related to building category and is defined in European Standard 13779 [14]. The standard states that the typical ranges of CO_2 concentration above the outdoor level are for building category I lower than 400 ppm, category II, 400-600 ppm, category III, 600-1000 ppm and category IV over than 1000 ppm.
For confined spaces, several sources indicate that indoor air problems are significantly reduced at 600 ppm or less of carbon dioxide. In general, a guideline of 600-1000 ppm or less is preferred in offices and schools due to the fact that the majority of occupants are young and considered to be a more sensitive population in the evaluation of environmental health status, [15]. The level of 5000 ppm is considered a Permissible Exposure Limit/Threshold Limit Value meaning an upper limit for which no acute (short term) or chronic (long-term) health effects. But a level above 2000 ppm is certainly deleterious for long time exposure.
Considering confined spaces, a concentration over the value of 1000 ppm can be connected with a relevant occupation of the spaces and/or an indicator of ventilation adequacy for the confined spaces so that all the standards for the air quality control recommends adjustment of the building's ventilation system. In recent years it is indicated that the level of 800 ppm is the one for which further investigation is necessary. But the case of Universities and University campus is a particular one due to the fact that different activities with different levels of occupation are possible in the same day, making very difficult to control the air quality without a direct feedback on the control system. Implementing CO_2 sensors into the air handling unit could bring to a double benefit, that is occupancy definition and comfort assessment. Considering the fact that CO_2 sensors are little and do not represent a disturbance of any kind to the unfolding of activities within the rooms in which they are installed, the developing of techniques of occupancy definition using CO_2 concentration represent a valid and effective way.

3. CO_2 concentration measurement activity: selection of appropriate instrumentation and rooms

From the analysis exposed in the previous section, it clearly emerges the fact that CO_2 concentration monitoring within university building is a not well explored field even if it may be a very promising one. For these reasons it has been set up an analysis procedure that comprehend different kind of sites and environmental conditions. All the monitoring activities have been performed within the School of Engineering at the University of Pisa. The School of Engineering area (a satellite view is reported in figure 1) is a mixed stock of new and old buildings, characterized by different building materials and inner room size. The School of Engineering is composed by 4 main buildings.
In order to evaluate the features that influence the comfort, and consequently energy savings that could arise, within highly crowded classrooms in University buildings, several campaigns have been performed. Said campaigns consists of the utilization of sensors in properly crowded classrooms during exams and lessons.

3.1 Sensors used for the measurement

The sensors utilized are of commercial type, model Chauvin Arnoux CA1510. These sensors are very small and do not emit any kind of sound or substance that can pollute the surrounding are. For that reason, they can be easily placed in crowded situation without being even noticed. The CA1510 is an instrument for measuring physical quantities that combines the following functions:

Measurement of Carbon Dioxide concentration in the air (CO_2);
Measurement of the ambient temperature;
Measurement of the relative humidity (RH).

The particular instrument works out air quality criteria based either on the level of CO_2 or on a combination of the three physical quantities measured. The characteristics of the sensors used for the experimental investigation are the following:

- for the measurement of CO_2 concentration, the instrument is provided with a Dual-beam infrared cell, using the Non-Dispersive Infrared technology. The measurement is from 0 to 5000 ppm with an intrinsic uncertainty of ±3% at 25°C and atmospheric pressure. The resolution of the instrument is 1ppm. The temperature error is 1ppm/°C in the range from −10 to 45°C. The influence of pressure can be obtained with the following equation: CO_2 real=CO_2 measured multiplied for (1+(1013-P) x 0,0014) (P is the pressure expressed in mbar);
- for the measurement of temperature, the instrument is provided with a CMOS sensor that can furnish a measure in the range between -10 and 60°C with uncertainty of ±0.5°C at 50% RH;
- for the measurement of relative humidity (RH), the instrument is provided with a capacitive sensor that range from 5 to 95 %RH the intrinsic uncertainty is ±2% at RH from 10 to 90%. But this measurement is not considered for the purposes of the present paper.

3.2 Locations for the measurement
The monitoring activity has been carried out in various classrooms in four different buildings where the measurements have been performed have different characteristics.

• Building A: This is identified by the green circle in figure 1 and is the historical location of Engineering in Pisa, its structure is composed by heavy masonry walls and the classroom have high ceilings. Moreover, there are various kind of classroom such as, drawing room, restrooms, computer labs, corridors, scientific labs, cellular office, lecture room and open spaces. The measurements have been done in a classroom, that is well representative of the various, that has been identified with number 1.

• Building B: This is identified with the yellow circle in figure 1. It was built at the end of the 60s of the last century. Its walls are preassembled light weight walls and the rooms have lower ceilings. There are study rooms, restrooms, corridors, classrooms, computer labs and a cafeteria. The structure presents a geometric articulation articulated by three fronts, one with respect to the other, with a rigid partition due to the repetition of the modules of the opaque covering and the fixtures. The inner walls are fine and characterized by poor materials and the insulation is provided by glass wool and wood fibers. Within its envelope there are two big-sized classrooms up to 1426 m^3 and 370 seats available, some little and medium sized classrooms, computer labs, restrooms, a study room and a cafeteria. The classrooms are, except for the two big classrooms previously mentioned, characterized by low ceilings and thus by a small volume/surface ratio. In this building two classrooms have been selected for the monitoring: a computer lab, identified with the name CL1 and a classroom, identified with the number 2.

• Building C: This is identified with the blue circle in figure 1. The structure has been constructed in the early nineties with classrooms only. The building outer design is simple and plain, being made of bricks in a very geometrical fashion. The building is articulated on 4 store and a central stairwell that constitutes a big inner open space. The purpose of the building is merely educational being provided with only classrooms of various sizes, spacing from a minimum of 18 seats to a maximum of 150. The classrooms are characterized by low ceiling as was the trend during the period the building was built. In this building, that has been identified with number 3, only one classroom has been monitored, the room is located at the 4th floor and has a small volume.

• Building F: This is identified with the red circle in figure 1. It was previously a barn and it have been refurbished with educational purposes, and was operating in 2006. The classroom inside are characterized by high ceilings (most of the rooms have "sheds") and the walls thickness is halfway between Building A and Building B. Those triangular structures, visible in figure 1, provided the rooms with a good light penetration but also with a great volume and thus a great volume to surface ratio. In this structure seven classrooms have been selected for monitoring activity, identified with the numbers 4, 5, 6, 7, 8, 9 and 10.

It is possible to observe that the highest volume to surface ratio is, as expected, reached in some classrooms of Buildings F and A, given the fact that these buildings design is characterized by high ceilings.
CL1 and classroom 3, meanwhile, reach the lowest value whether classroom 2 (although its big volume) and 10 are halfway. The highest air volume per student ratio in maximum occupancy conditions is generally reached by the building F classrooms, and in the Computer Lab (CL1).
In this case it is determined by the low number of seats available. Not being previously aware of the classrooms' features that most influence air composition, and CO_2 concentration in particular, it has been necessary to monitor a wide range of classrooms. Fortunately, the study site is a heterogeneous mix of

various classroom types and it has been possible to program an effective monitoring activity. The main characteristics that distinguish the different classrooms tested are:

- volume;
- surface;
- maximum occupancy:
- tightness;
- air conditioning system;
- end-use;
- structure of the walls.

In Table 1 an accurate description of the buildings and the classrooms is given. As already mentioned, the monitoring activity thus has been performed in 11 classrooms in different periods of the year 2018.
The monitoring campaigns have been performed during both the examination period (January and February 2018) and the lecture period (March and April 2018), in various classrooms situated in the different buildings of the School. The classrooms are different in typology (typical didactic use only or informatics rooms) volume, walls composition, air exchange rate, occupancy, daytime and external conditions.
For each classroom a minimum of three measurement points and a maximum of four measurement points have been considered. The data have been subsequently analyzed with a double objective:

• To identify the features of the room that most influence CO_2 concentration. Factors like volume, air change ration, surface are undoubtedly bound to the air pollution and acknowledging their influence can help future building in reducing the risk of high CO_2 concentration without spending too much energy through the air handling unit.

• To define a possible correlation between CO_2 concentration, the occupancy of a room and the time. The recognition of certain path in the CO_2 concentration during certain occupancy levels could bring to the possibility to set the cooling system in order to provide the needed comfort reducing the energy consumption. Moreover, such expression might be useful to the identification of certain occupancy patterns that could be used in appropriate computer software that forecast energy consumption and thus, the ability of the energy manager to minimize it.

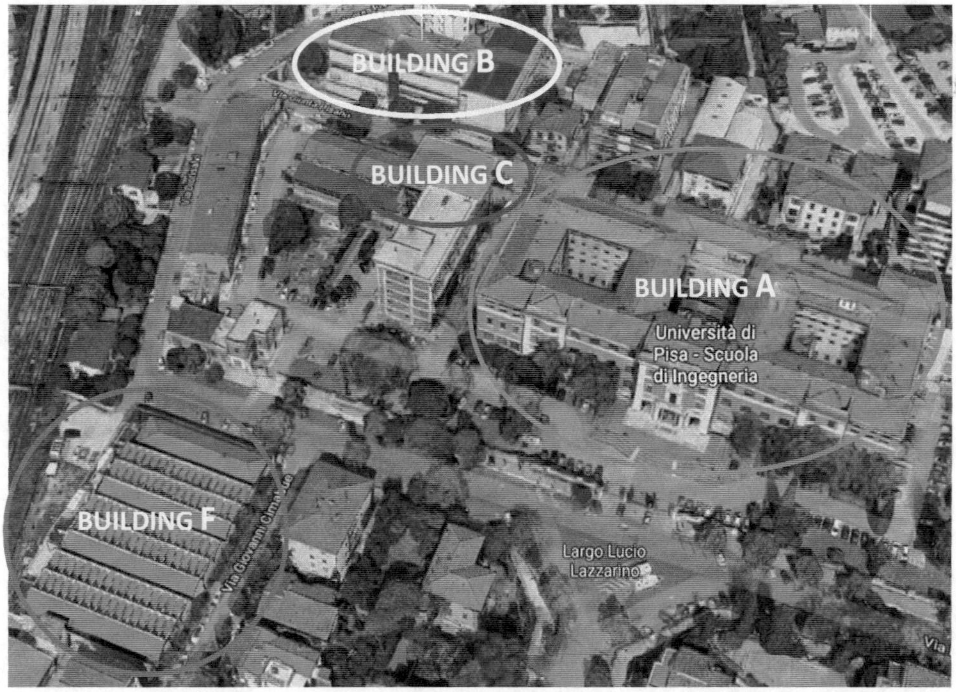

Figure 1. A complete view of the buildings of the School of Engineering (University of Pisa)

Table 1. Characteristics of the rooms object of monitoring

Room	Building	Floor	Maximum occupancy (n)	Floor Surface (m^2)	Volume (m^3)	Ratio Vol./surface (m^3/m^2)	Minimum vol. for student (m^3)
1	A	1st	108	88	481	5.47	4.45
CL1	B	1st	116	216	583	2.70	5.02
2	B	1st/2nd	370	336	1426	4.24	3.85
3	C	4th	74	73	212	2.90	2.86
4	F	1st	305	286	1587	5.55	5.20
5	F	1st	212	216	1220	5.65	5.75
6	F	1st	109	130	721	5.54	6.61
7	F	1st	198	197	1093	5.54	5.52
8	F	1st	104	129	716	5.55	6.88
9	F	1st	109	128	710	5.55	6.50
10	F	1st	140	131	438	3.34	3.12

4. CO_2 air concentration campaigns report and analysis of the results

The experimental analysis consists in the utilization of sensors in properly crowded classroom during exams and lessons. The sampling period is 4 data per minute. During the experiments the activities within the room under analysis are performed without particular caution. The operator places the instruments in proper positions (generally at the two opposite side of the rooms) and records the occupancy schedule, observing whether any particular event occurred (for example the opening of a door, the sudden change in the students' activity etc.). Figure 2 shows the results obtained in the measurements in a specific room during a lesson.

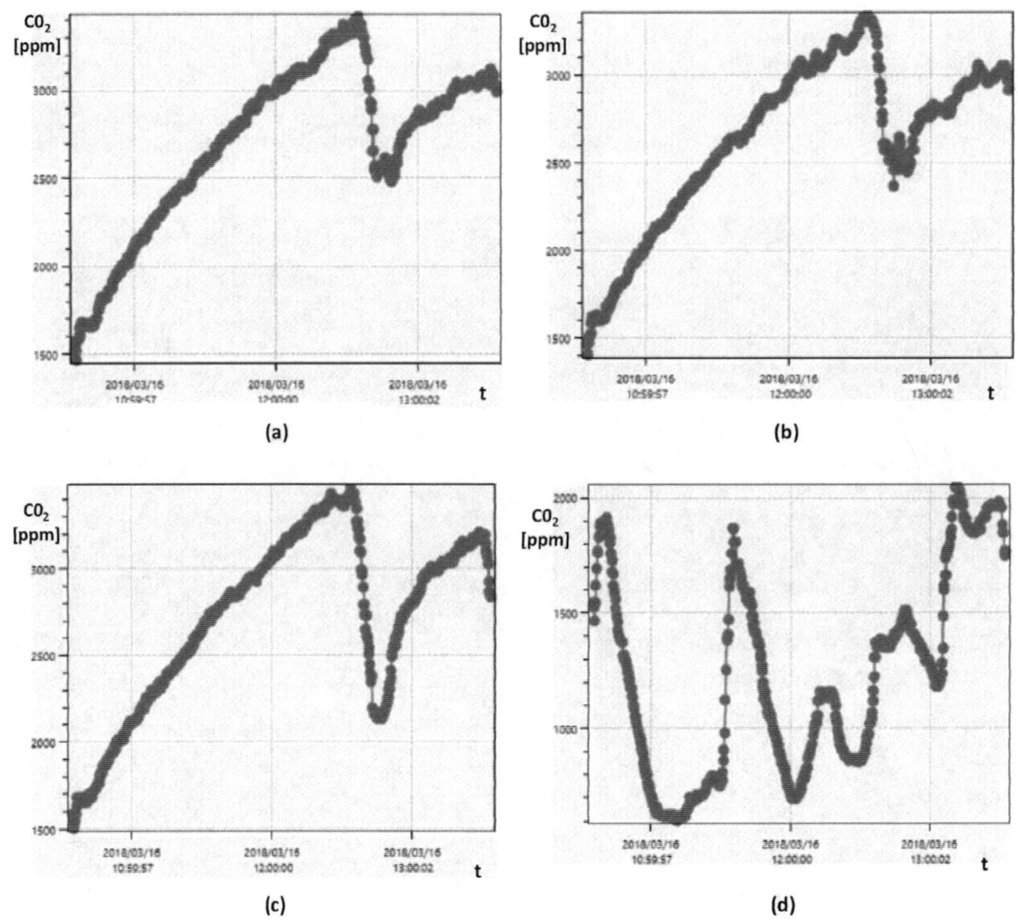

Figure 2. CO_2 concentration (ppm) as a function of time (t) in three different zones of the classroom 9 (a), (b) and (c) and in the outside corridor (d) during a lesson of three hours.

IOP Conf. Series: Journal of Physics: Conf. Series **1224** (2019) 012016 doi:10.1088/1742-6596/1224/1/012016

In the y-axis the CO_2 concentration, expressed in part per million (ppm) is represented while in x-axis time t is reported. Considering the results of figure 2, that represent a typical report of the various measurements, it is evident how the quite fast increase of the CO_2 concentration during the first part of the lesson when the doors are closed. A sensible decrease, that can be quantified in the order of magnitude of 1000 ppm can be obtained when the lesson is suspended for some minutes. After that the lesson starts again and consequently a new increase of the CO_2 concentration can be evidenced.

Another interesting result, that can be evidenced from the analysis of figure 3 is the connection between air change ratio and CO_2 concentration. This represent the typical CO_2 and temperature increase trends during a lesson. In this case it is evident the beneficial effect of the air conditioning system and of the air change, that permits to reduce both the temperature in the room and the CO_2 concentration. The connection between the air change ratio (ACR) and the measured CO_2 level has been already evidenced in [6]. A summary of the results obtained in the various experiments in the different rooms is provided in Table 2. In the left column it is possible to read the classroom and the sequential number of the experiment. The double number as #2.1 and #2.2 identifies two experiences obtained in the same classroom and in the same day but in different time (for example morning and afternoon).

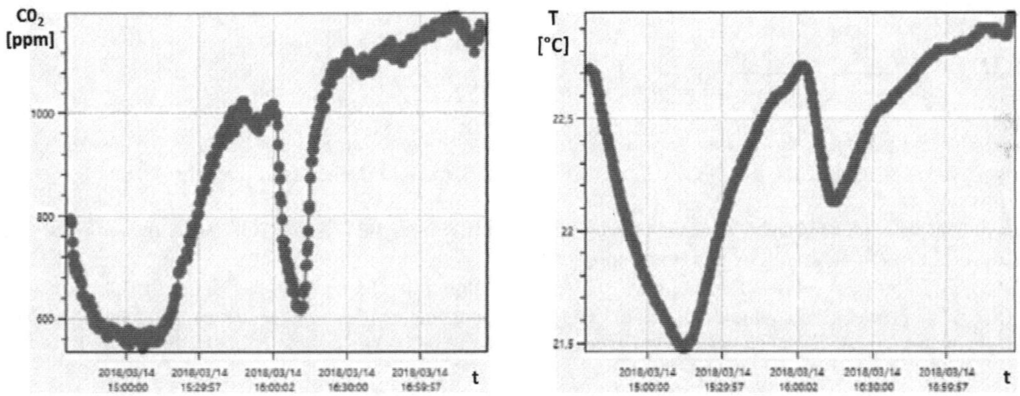

Figure 3. CO_2 concentration and temperature variation in the computer lab (CL1)

Table 2. Results of monitoring activity

Room/ Experience	Volume (m^3)	Occupation (n)	VAPS (m^3/n)	dCO_2/dt (ppm/s)	CO_2 (t=0) (ppm)	CO_2 (t=t*) (ppm)	t* (min)
7 #2	1093	181	6.03	0.46	2581	3885	47
1 #2.2	481	75	6.41	0.75	1056	3719	59
1 #2.1	481	75	6.41	0.67	900	3102	55
5 #1	1220	167	7.31	0.52	1138	4914	120
7 #1	1093	146	7.48	0.40	791	3296	105
10 #1.2	438	56	7.71	0.35	2077	3640	74
10 #1.1	438	56	7.71	0.32	1733	3579	95
10 #2.2	438	50	8.64	0.44	697	2409	65
10 #2.1	438	50	8.64	0.40	695	2763	87
5 #3	1220	106	11.51	0.25	1100	3408	152
9 #2.1	710	59	12.03	0.26	1647	3389	51
9 #2.2	710	59	12.03	0.28	2386	3115	42
9 #1.2	710	54	13.15	0.44	1495	2810	50
9 #1.1	710	54	13.15	0.43	1511	2333	32
9 #1.3	710	54	13.15	0.36	1861	3246	64
8 #1	716	54	13.26	0.37	1205	2382	53
4 #3	1587	93	17.06	0.17	596	2684	206
8 #2	716	39	18.35	0.28	1358	2124	45
4 #2	1587	86	18.45	0.21	792	1409	48
5 #2.1	1220	62	19.68	0.13	2165	2994	110
4 #1	1587	58	27.36	0.11	678	1740	155

Considering the various experiments, the CO_2 concentration (expressed in ppm) with time (t), $[CO_2](t)$ is a function of different variables, like the volume of the room, V, the number of occupants, n, and others:

$$[CO_2](t) = f(V, n, t, [CO_2](t=0), VAPS, ACR) \tag{1}$$

where VAPS is the air volume per each student in the full occupancy conditions, T is the total duration of the experiments; $[CO_2](t=0)$ is the CO_2 concentration at the starting time t=0 (i.e. the initial concentration) and ACR, the air change ratio that in general can be obtained only with a mechanical system. In figure 4 an attempt is made to establish a correlation between the increase of CO_2 concentration with time, represented by the derivative of CO_2 concentration and the volume available in each room for the single occupant: this last value is representative of the occupancy level. Though if an interesting correlation among the two variables can be evidenced, some discrepancies with the linear trend can be observed.
Considering the data of figure 4, it is possible to put in evidence the predominant role of some variables, like the specific volume available for each person (VAPS). This appears to be close to the linearity, so that it is possible to state that the first derivative of the CO2 concentration can be considered mainly dependent on the volume available for each student:

$$\frac{d[CO_2]}{dt} = \frac{[CO_2](t)-[CO_2](0)}{t} \sim f\left(\frac{V}{n}\right) \tag{2}$$

Considering the results of the various experimental analyses, some qualitative and quantitative elements can be highlighted concerning the link between the CO_2 concentration and the occupation of the room:

- the volume per student is an influencing factor: the higher this ratio the less the CO_2 concentration rise;
- the initial CO_2 concentration level is important;
- the CO_2 concentration follows the occupancy pattern: the higher the occupation, the higher the CO_2 levels;
- a time delay between occupancy pattern and CO_2 concentration increase can be observed, due to the fact that it takes some time for the CO_2 to spread all over the room, the bigger the room, the higher the delay.

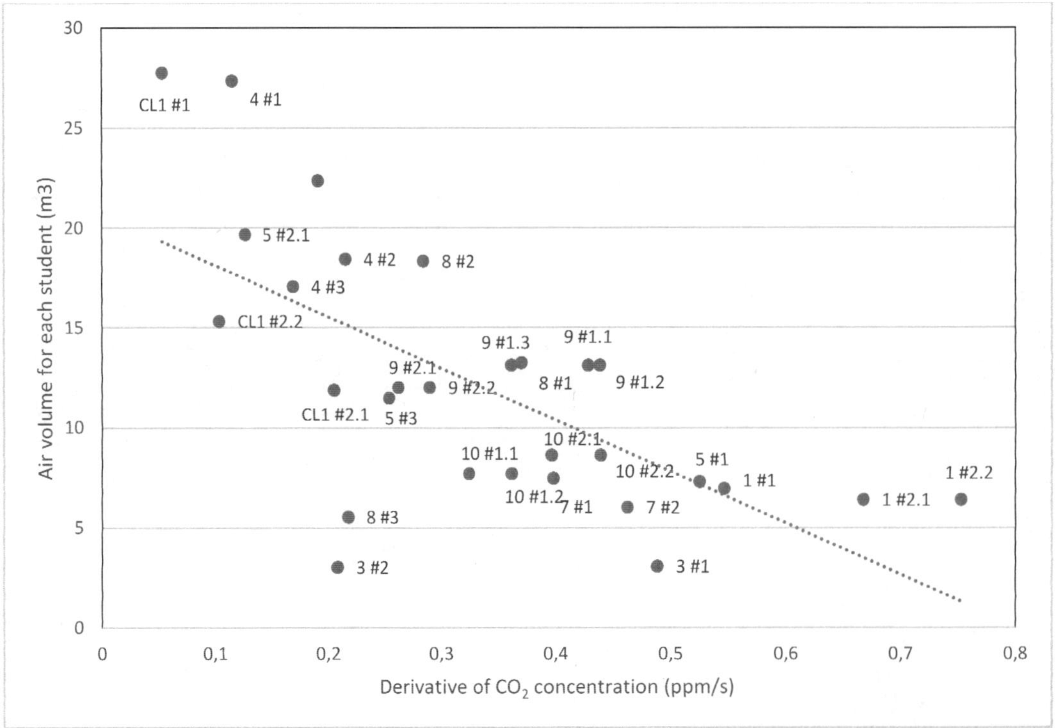

Figure 4. CO_2 concentration correlation with the available volume: a possible correlation

5. Conclusions

Energy consumption and indoor environment quality (IEQ) of buildings are clearly linked to the occupation. Predicting the number of occupants in a space is essential for the effective management of various building operation functions as well as to improve energy efficiency and environmental quality. One of the methods to estimate the number of indoor occupants is the analysis of CO_2 concentration and of CO_2 concentration variation. CO_2 measurement is useful not only for occupancy detection but for comfort purposes too.
This study analyzes the correlation between CO_2 measured values and actual occupant numbers. Analysis is performed in different classroom of the School of Engineering at the University of Pisa, characterized by different characteristics and volumes, during a period of 4 months. The objective has been to test the correlation between CO_2 concentration and occupancy in various condition giving quantitative elements.
Some data can be remarked: the CO_2 concentration increase with time is a very dispersed value ranging from 0.1 to 0.8 ppm for seconds; this means that for each minute an average increase of 20-25 ppm for each minute can be observed. The higher values of the concentration increase are directly correlated with the reduction of the volume available for each student. Values over 0.40 ppm/s can be observed if the volume for student is less than 10 m^3. The attempt to establish a correlation between the increase of CO_2 concentration with time, represented by the derivative of CO_2 concentration and the volume available in each room shows that CO_2 can be really an important qualitative indicator to establish the occupancy level of a space.
At the end of the analysis it is surely possible to conclude that the CO_2 concentration follows surely the occupancy pattern. But a lot of secondary factors like the opening of windows and doors, and the delay time between occupancy pattern and CO_2 concentration, due to the fact that it takes some time for the CO_2 to spread all over the room in dependence on the total volume of the classroom may have a strong influence on the CO_2 concentration and their effect is hardly forecastable. For this motivation, the use of this indicator for quantitative analysis is not so simple excluding the case of "fully closed" volumes.

References

[1] Yang L, Yan H and Lam JC 2014 Thermal comfort and building energy consumption implications – A review *Applied Energy* **115** 164-173.
[2] Magrini A, Gobbi L and D'Ambrosio FR 2016 Energy audit of public buildings: the energy consumption of a University with modern and historical buildings. *Energy Procedia* **101** 169-175
[3] Corgnati SP, Corrado V and Filippi M 2008 A method for heating consumption assessment in existing buildings: A field survey concerning 120 Italian schools *Energy and Buildings* **40** 801–809.
[4] Altan H, Douglas JS and Kim YK 2014 Energy Performance analysis of University Buildings: case studies at Sheffield University *J Archit Eng Tech* **3** 129.
[5] Fathi S and Srinivasan RS 2015 Analysis of energy performance of University campus buildings using statistical and energy modelling approaches. Proceedings of the 2015 Winter Simulation Conference.
[6] Allab Y, Pellegrino M, Guo X, Nefzaoui E and Kindinis A 2017 Energy and comfort assessment in educational building: case study in a French university campus *Energy and Buildings* **143** 202-219.
[7] Sretenovic A 2013 Analysis of energy use at university campus, M. Sc. Thesis, Norwegian University of Science and Technology, Department of Energy and Process Engineering.
[8] Tagliabue CL, Manfren M, Ciribini ALC and De Angelis E 2016 Probabilistic behavioral modeling in building performance simulation - The Brescia eLUX lab *Energy and Buildings* **128** 119–131.
[9] Menezes AC, Cripps A, Bouchlaghem D and Buswell R 2012 Predicted vs. actual energy performance of non-domestic buildings: Using post-occupancy evaluation data to reduce the performance gap *Applied Energy* **97** 355–364.
[10] Zuraimi MS, Pantazaras A, Chaturvedi KA, Yang JJ, Tham KW and Lee SE 2017 Predicting occupancy counts using physical and statistical CO_2-based modeling methodologies *Building and Environment* **123** 517-528.
[11] Wang S, Yan C and Xiao F 2012 Quantitative energy performance assessment methods for existing buildings *Energy and Buildings* **55** 873–888.
[12] Candanedo L M, Feldheim, V 2016 Accurate occupancy detection of an office room from light, temperature, humidity and CO_2 measurements using statistical learning models, *Energy and Buildings* **112** 28–39.
[13] Schnelle, KB Dunn RF and Ternes ME 2016 Air Pollution Control Technology Handbook, CRC Press:
[14] Persily A. 2015, Challenges in developing ventilation and indoor air quality standards: The story of ASHRAE Standard 62, *Building and Environment* **91** 61-69.
[15] Ramalho O, Wyart G, Mandin C, Blondeau P, Cabanes PA, Leclerc N, Mullot JU, Boulanger G and Redaelli M, 2015, Association of carbon dioxide with indoor air pollutants and exceedance of health guideline values, *Building and Environment* **93** 115-124.

IOP Conf. Series: Journal of Physics: Conf. Series **1224** (2019) 012017 doi:10.1088/1742-6596/1224/1/012017

Simplified numerical modelling of infrared radiation effects in tomato dry peeling

G Cuccurullo and L Giordano
Department of Industrial Engineering, University of Salerno,
Via Giovanni Paolo II 132, 84084 Fisciano (SA), Italy

E-mail: gcuccurullo@unisa.it

Abstract. In view of the increasing interest in developing sustainable systems for tomato peeling, the unsteady thermal field in a rotating tomato exposed to an array of infrared emitters was studied. First, a predictive 3D numerical model was developed to simulate heat transfer in a rotating tomato undergoing infrared heating. Since simulations required great computational efforts, it was considered appropriate to introduce a simplified model. To this purpose, the net radiative heat flux on the tomato surface was approximated by a suitable periodic function: the related angular velocity accounted for tomato rotation, while its amplitude was linked up with view factors between the tomato and the heating source. On such basis, a 1D axisymmetric model involving radiative heating from the source regarded as a periodic II-type boundary condition was set up. It was found that temperatures predicted by the simplified model agreed well with the 3D model output. Finally, a parametric analysis was performed to investigate the effect of the position and angular velocity of the tomato samples on the rate and uniformity of IR heating. Results were pursued both by the 3D and 1D model, in the latter case obtaining faster yet accurate responses.

1. Introduction

The main critical concerns about the sustainability of tomato processing industry regard energy consumptions and wastewater disposal problems associated with lye and steam peeling methods, [1-3]. The steam peeling does not use chemicals, yet, it results in higher peeling loss than lye peeling. Thus, tomato processors are seeking alternatives to replace traditional methods. In this connection, the development of sustainable tomato peeling technology seems to converge toward infrared (IR) dry-peeling: this process avoids using lye and water, meanwhile requiring almost the same time to heat samples. Moreover, it produces thinner thickness of peeled-off skin and slightly firmer texture of peeled, while a similar ease of peeling is achieved. Other attractive properties are claimed such as increased Young's Modulus of tomato peels and reduced peel adhesiveness. These findings demonstrated the effectiveness of the novel IR dry-peeling process for tomatoes [4]. However, some critical aspects of peeling tomatoes using infrared radiation heating were outlined by [5].

Wishing to perform successful infrared peeling requires to realize both rapid and uniform heating on the tomato surface. A typical configuration for industrial peeling is realized by means of a plane matrix of infrared emitters, looking at tomatoes in relative motion; in fact, tomatoes are carried on by the conveyor belt which induces their rotation in order to improve heating uniformity. In order to achieve an optimal design of infrared heating systems, the irregular shape of tomatoes and their different expositions to the heating source when transported along by the conveyor belt are to be considered. Due to the complexity of the geometry at hand, a numerical approach is required to retrieve tomato thermal response to infrared

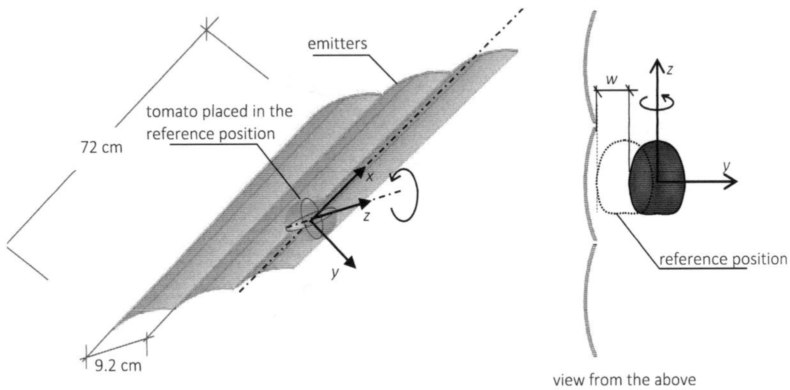

Figure 1. Geometrical configuration tomato/emitters

heating. However, a preliminary description based on an analytical model was previously attempted by the authors: by considering that the proper time scale of the process at hand is small, they considered a semi-infinite body subjected to a suitably chosen pulsating heat source [7;8]. With regard to the numerical approach, computational simulation of IR heating in peeling process has not yet been satisfyingly explored, probably because the IR dry-peeling technology is too recent. Pan et al. [9;10] proposed a numerical model with the tomato exposed to double sided infrared heating, yet no rotation was accounted for. Some authors considered volumetric absorption of infrared energy at food surface; consequently, they dealt with semi-empirical exponentially decaying models with a finite penetration depth [11-13]. In any case, tomato rotation was not considered although it is intuitively a critical operating parameter, since it strongly influences heating uniformity.

In this paper a three-dimensional geometric model for the unsteady temperature field in rotating tomatoes exposed to a matrix of infrared emitters was developed. The heating process duration was assumed to end when a specified temperature level was attained which turned out to be useful for peeling purposes. Having in mind to achieve the best heating uniformity, rotation speed and relative position with respect to the source are key parameters to feature; therefore, different geometrical configurations were considered when solving the problem by the 3D-numerical model. Finally, a careful characterization of the heating source intensity in terms of the view factors between the tomato and the source enabled an alternative 1D-model, independent of source geometrical details. Thus, a simplified equivalent model for the heat source was allowed which led to satisfyingly recover the description of the surface temperature field, whilst ensuring a sensible reduction of computational efforts.

2. The three-dimensional numerical model

2.1 Geometrical configuration

A three-dimensional FEM model was developed by employing the commercial code COMSOL Multiphysics [14]. The unsteady temperature field for the rotating tomato subjected to the irradiation coming from an array of IR emitters was obtained. A schematic representation of the geometry under study is reported in figure 1: three IR plane emitters, 72 cm x 9.2 cm, were supposed to heat the tomato; a reference position was chosen so that the tomato was placed in front of the central emitter, midway along the emitter longitudinal axis, at a distance $w = 0$. Of course, the reference position was such as the tomato experienced the maximum heating intensity. The tomato sample was placed orthogonally with respect to emitters longitudinal axis.

The computational domain of the tomato was generated based on the two-dimensional axisymmetric parametric profile introduced by Li et al., [15]; the related dimensionless shape coefficients were sought as: $a = 31$ mm, $b = 24$ mm, $c_1 = 0.24$, $c_2 = 0.38$, that identifying a realistic shape for medium size tomato. The tomato computational domain was discretized using unstructured tetrahedral grid elements. A mesh convergence study was formerly performed until results turned out to be independent of the mesh size during the warming up period, [7]. In practice, the warming up period is assumed to end when the

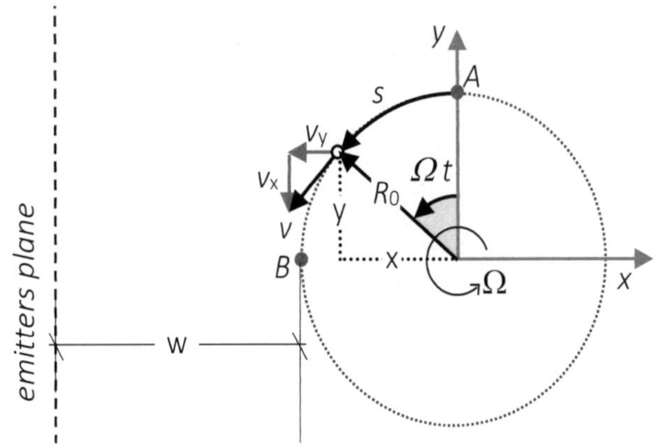

Figure 2. Decomposition of the local velocity

surface average temperature recovers a threshold value of 90°C; the latter temperature is known to be useful for peeling purposes.

The net radiative heat flux on the tomato surface was numerically evaluated for several distances (w in figure 1) between the tomato and the emitters. The distance w varied between 0 and 2 cm, this range allowing heat rates to be as high as to ensure heating was fast enough for the addressed purposes.

2.2 Basic equations

The energy balance equation applied to the tomato undergoing transient conduction heat transfer is expressed by the fundamental equation of conduction, which, assuming constant properties, turns out to be:

$$\rho c\,\frac{\partial T}{\partial t} = \underline{\nabla}\cdot\left(k\underline{\nabla}T\right) \tag{1}$$

where T is the tomato temperature, t is the time, ρ, c and k are the density, the specific heat and the thermal conductivity of tomato, respectively.

On the tomato surface, the total heat flux is the sum of the radiative and convective heat transfer contributions. Hence, the first boundary condition is given by:

$$\underline{\mathrm{n}}\cdot\left(-k\underline{\nabla}T\right) = \varepsilon\left(E_b\left(T\right)-G\right)+h\left(T-T_{ext}\right) \tag{2}$$

where $\underline{\mathrm{n}}$ is the outward normal vector of the boundary, G [W/m^2] is the surface irradiation, E_b is the blackbody emissive power, ε is the tomato emissivity, h is the average surface heat transfer coefficient, T_{ext} is the air temperature. In particular, the view factor is automatically determined by the numerical code for each of the facets of the discretized geometry using a "hemicube" or "direct area integration" approach. The surface irradiation is mainly due to the infrared emitters. The average surface temperature of the emitters was assumed to be 650 °C, meanwhile the surrounding temperature was set to 20 °C. The tomato sample was assumed to be a grey body with a surface emissivity equal to 0.95, while air was assumed as a non-participating medium.

At the tomato centre, the symmetry boundary condition was imposed:

$$\underline{\mathrm{n}}\cdot\left(k\underline{\nabla}T\right) = 0 \tag{3}$$

The initial temperature of the tomato was assumed to be uniform:

$$T(x,y,z,t=0)=T_i \tag{4}$$

where T_i represents the initial temperature of tomato.

The 3D model accounts for tomato rotation around its main axis by assigning the angular velocity Ω and then calculating the local velocity components as in figure 2:

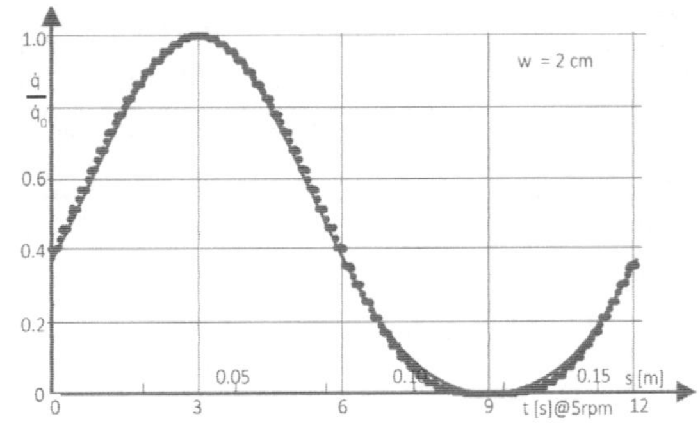

Figure 3. Analytical curve vs numerical data related to the maximum net radiative heat flux

$$v_x = -\Omega\, R \sin(\Omega\, t\,) = -\Omega\, x \;; \qquad v_y = \Omega\, R \cos(\Omega\, t\,) = \Omega\, y \tag{5}$$

Note that the angle Ωt can be related to the path $s = \Omega t\, R_0$ described by a point moving along the circumference with peripheral velocity $v = \Omega R_0$.

2.3 Modelling the pulsating heat source

Numerical simulations allowed to determine the net radiative heat flux crossing the widest tomato circumference. The addressed circumference is the more critical because it experiences the highest temperature increases. The points A and B of figure 2 belong to such a circumference. The former point is the first to experience heating from the rising source, while the latter is the one experiencing the maximum heating along the selected circumference. Due to the marked difference of temperature between the emitters and the tomato, the flux crossing the selected circumference essentially coincides with the irradiation due to the emitters; it was determined for several configurations, i.e. for selected distances w, namely 0, 0.5, 1 and 2 cm. The following normalized function was found to satisfyingly describe the pulsating heat source:

$$\frac{\dot{q}(\phi,t)}{\dot{q}_B} = \frac{1}{2}\left(\sin(\Omega\, t) + \sin(\Omega\, t)^2 + \phi\cos(2\,\Omega\, t) + \phi\right) \tag{6}$$

where $\dot{q}_B = \dot{q}(\phi, t = \pi/2/\Omega)$ is the radiative heat flux experienced by point B, and ϕ is a parameter which is determined by imposing that, for each distance w, the average value along the circumference for the function at hand recovers the corresponding one given by the numerical 3D model. Since $\dot{q}_A = \dot{q}(\phi, t = 0)$, the parameter ϕ can be regarded as the ratio of the minimum to maximum net radiative heat flux: $\phi = \dot{q}_A / \dot{q}_B$. As an example, in figure 3 the net radiative heat flux along the selected circumference as calculated by the numerical model and the one obtained through eq. 6 are compared for an arbitrarily selected distance. Apart from this specific case, the goodness of the fit is shown in Table 1 as a function of the distance w.

Table 1. RMSE and ϕ as a function of w

w [cm]	$\phi = \dot{q}_A / \dot{q}_B$	RMSE	$F_{12}^{\perp} / F_{12}^{\parallel}$
0	0.42	0.019	0.430
0.5	0.41	0.018	0.418
1	0.39	0.017	0.403
2	0.37	0.015	0.381

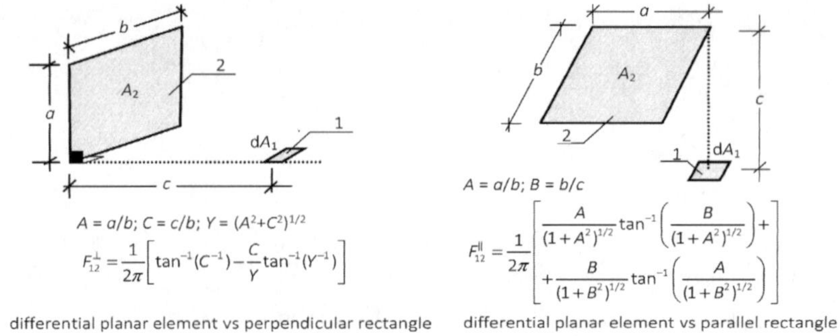

Figure 4. View factors between a differential planar element and a finite area

In order to figure out the relationship between the distance w and the parameter ϕ, consider that the latter can be further interpreted as the view factor between the point A and the emitters, normalized with respect to the view factor related to the point B, see figure 4. Therefore:

$$\frac{\dot{q}_A}{\dot{q}_B} = \frac{F_{12}^{\perp}}{F_{12}^{\parallel}} \tag{7}$$

By invoking the view factors definitions shown in figure 4, the right-hand side of the previous equation turns out to be a known function of the distance w, [16, 17]. The last column in table 1, shows the ratio of the view factors appearing in eq. 7 evaluated by means of the corresponding analytical expressions whereas the left-hand side is given up in the second column of the same table in accordance with eq. 6. A relative error of about 2.5% is achieved. Hence, equation (7) establishes a relationship of the kind $\phi = \phi(w)$, accounting for the shape of the heat flux along the heated circumference as depending on the distance w.

The knowledge of the view factor $F_{12}^{\parallel}$ allows to determine the radiative flow that crosses the point B when the distance w changes. In fact, if the maximum heat flux in the reference configuration is assumed to be known, the definition of the view factor between the emitter and the point B yields:

$$\dot{q}_B(w) = \dot{q}_B(0)\frac{F_{12}^{\parallel}(w)}{F_{12}^{\parallel}(0)} \tag{8}$$

Thus, determining the maximum heat flux as a function of the distance w requires a single simulation for featuring the maximum heat flux at the reference distance, i.e. $\dot{q}_B(0)$; the latter calculation can be conveniently carried out in the stationary regime in order to minimize time for computation, since it only depends on the geometry. Of course, such a simulation was performed by imposing a convective coefficient sufficiently large, in order to keep the tomato surface temperature lower than the peeling threshold: in this way, the net radiative heat flux turned out to be almost constant and independent of the tomato temperature.

3. One-dimensional axisymmetric numerical model

In order to predict the thermal response of the tomato subjected to radiative heat flux with lower computational efforts, a numerical 1D axisymmetric model was implemented in Comsol Multiphysics. Therefore, the rotating tomato was described as an infinite cylinder, whose radius R_0 was assumed to be equal to the segment connecting the points A and B of figure 2. The cylinder was subjected to periodic heat flux according to eq. 6. A combined II-type and III-type boundary condition was considered providing for both radiative infrared heating and convective cooling to an ambient at fixed temperature; the latter assumption is justified by considering that surface temperatures of infrared heaters are much higher than the surrounding ones during the warming-up period. This strong simplification allows to avoid the three-dimensional description of the emitters.

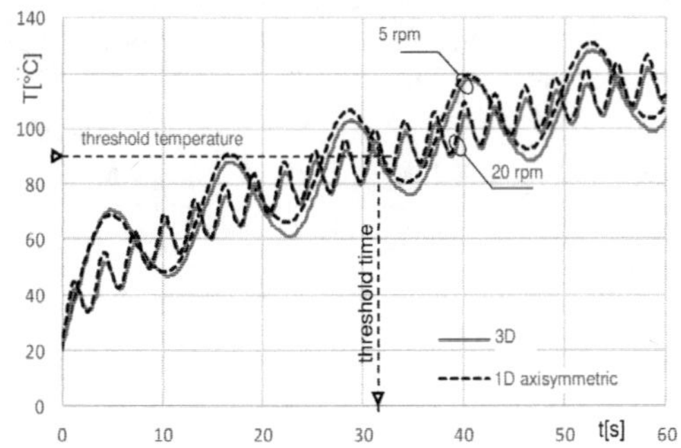

Figure 5. 3D and 1D axisymmetric temperature fluctuations, when w = 0 cm

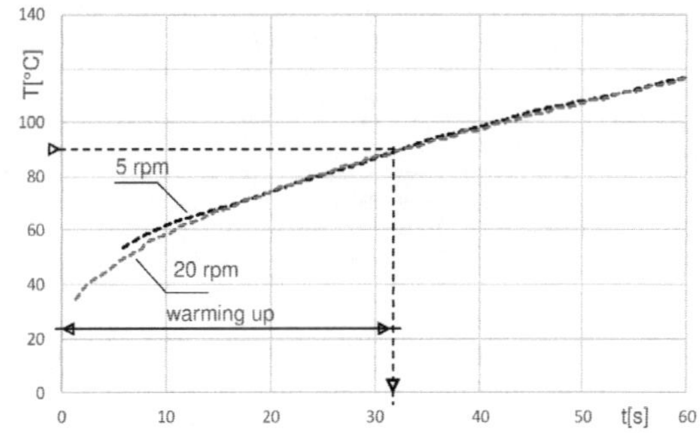

Figure 6. Average surface temperature for 5 and 20 rpm

Under such assumptions, the dimensionless energy balance and the related boundary conditions led to the following linear problem:

$$\rho c_p \frac{\partial T}{\partial t} = k \frac{1}{R} \frac{\partial}{\partial R}\left(R \frac{\partial T}{\partial R} \right) \tag{9}$$

$$-k \left. \frac{\partial T}{\partial R} \right|_{R=R_0,t} = \dot{q}(\phi,t) - h \cdot (T - T_{\text{ext}}) \tag{10}$$

$$k \left. \frac{\partial T}{\partial R} \right|_{R=0,t} = 0 \tag{11}$$

$$T(R,t=0) = T_{\text{i}} \tag{12}$$

This approximate description allowed to numerically predict the thermal response of the tomato points lying on the more exposed circumference.

4. Results

4.1 Comparison between 3D and 1D numerical models

As expected, the 1D model was attractive because of its execution speed: only few seconds were required, whereas the 3D model required almost one hour in order to describe the tomato thermal transient until the threshold-time. On the other hand, a subsequent reduction in the related accuracy could be expected when using the simplified model. A measure of the achieved approximation is given in figure 5, where temperature evolution for a point moving along the circumference is shown for both 3D and 1D calculations. To two typical angular velocities, i.e. 5 and 20 rpm, are considered. For the sake of simplicity, calculations were performed by neglecting convective heat transfer. Results inherent the 1D model satisfyingly approach the corresponding 3D model calculations: the relative percentage error during the warming up period resulted to be contained within about 3.5% and 5.2% for the two rotational velocities. It is interesting to observe that the time interval needed to complete the warming up turns out to be independent of the angular velocity as shown in figure 6, where the average surface temperatures are represented.

4.2 Parametric analysis of thermal uniformity

Since heating uniformity all over the tomato surface is mandatory in order to realize good peeling performance while ensuring high quality of peeled tomatoes and energy saving, [18,19,20], a surface temperature uniformity analysis was carried out by the 3D model. In order to quantify the surface

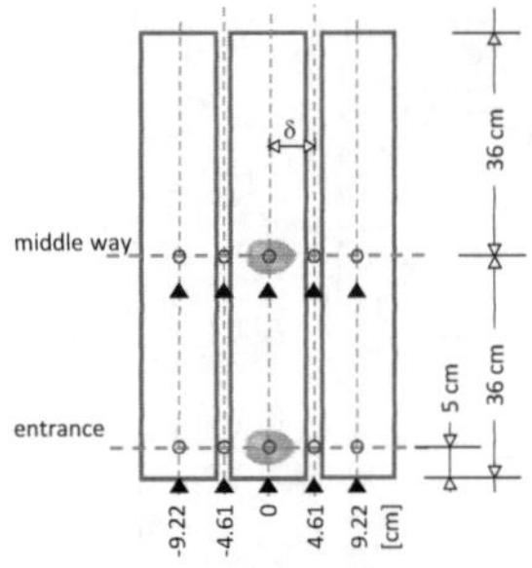

Figure 7. Positions selected for the parametric analysis of thermal uniformity

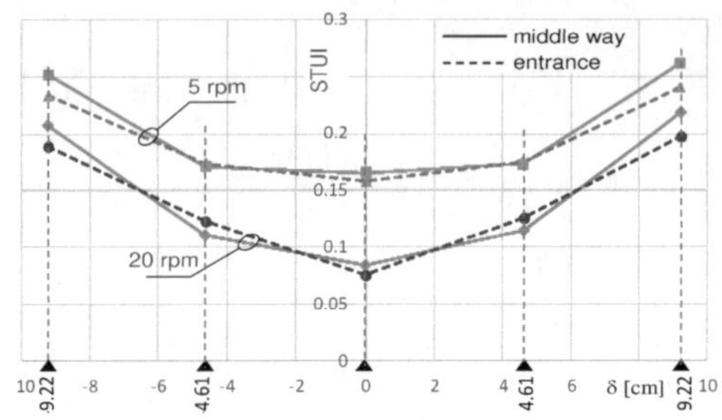

Figure 8. STUI as a function of the displacement δ and of the rotational velocity

temperature uniformity while varying the angular velocity of the sample under test and its position with respect to the matrix of emitters, the surface temperature uniformity index (STUI) was introduced [7]:

$$STUI = \frac{1}{T_{objective} - T_f} \cdot \sqrt{\frac{\int_A \left(T - T_{average}\right)^2 dA}{A}} \tag{13}$$

the area A representing the whole tomato surface.

The STUI was evaluated at the end of the warming-up period for different tomato positions. In particular, for a selected distance between the tomato and the emitters, namely $w = 0$, ten different tomato positions were considered, see figure 7: positions were identified by selecting five displacements with respect to the emitters longitudinal axis (δ) and two distances, 5 and 36 cm, measured far away from the emitters leading edge, labelled "entrance" and "middle way" in the figure. Results for the STUI are represented in figure 8: for a fixed angular speed, the STUI exhibits a minimum on the symmetry longitudinal axis while higher values are attained in correspondence of the exterior emitters ($\delta = -9.22$ cm e $\delta = 9.22$ cm): unfavourable positions corresponding to the highest lateral displacements worsen noticeably temperature uniformity on the tomato surface. On the other hand, the STUI appears scarcely sensitive to the distance from the leading edge. Finally, with increasing the angular speed, the STUI significantly reduces.

5. Conclusions

A 3D numerical model was developed in order to predict the temperature field of a rotating tomato subjected to radiative heating for peeling purpose. The resulting net radiative heat flux on the tomato surface was modelled by a suitable periodic function, depending on the view factors between the tomato and the heating source. This function was used to introduce a II-type boundary condition for a 1D asymmetric numerical model, thus allowing a simplified description of the problem at hand since 3D description of the emitters was avoided. The simplified model was able to recover the thermal response of the tomato with considerably reduced computational efforts; yet, it was able to approximate the 3D model results with satisfying accuracy within the framework of interest.

Finally, in order to provide useful information to support innovative dry-peeling process, 3D simulation results showed how rotational velocity and position of the tomato influence the surface temperature uniformity.

6. References

[1] Cheryl R, Wade Y, Renee G-S, Hao F 2012 Conventional and Alternative Methods for Tomato Peeling, *Food Eng Rev* **4** 1–15

[2] Shi J, LeMaguer M 2000 Lycopene in tomatoes: chemical and physical properties affected by food processing. *Critic Tev Food Sci Nutr* **40**(1) 1-2

[3] Garcia E, Barrett DM 2005 Peelability and yield of processing tomatoes by steam or lye, *J. Food Process Preserv* **30**(1) 3–14

[4] Li X, Pan Z, Atungulu G G., Zheng X, Wood D, Delwiche M, McHugh T H 2014 Peeling of tomatoes using novel infrared radiation heating technology, *Innovative Food Science & Emerging Technologies*, **21** 123-130

[5] Li X, Pan Z, Atungulu G G., Wood D , McHugh T H 2014 Peeling mechanism of tomato under infrared heating: Peel loosening and cracking, *Journal of Food Engineer*ing **128** 79–87

[6] Li X *A Study of Infrared Heating Technology for Tomato Peeling: Process Characterization and Modeling* 2012, ISBN - 9781267968708

[7] G. Cuccurullo, L. Giordano, Temperature field for radiative tomato peeling, *Journal of Physics: Conference Series*, **796** 2017

[8] G. Cuccurullo, L. Giordano 2017 *Analytical solutions for tomato peeling with combined heat flux and convective boundary conditions*, Journal of Physics: Conference Series, **923** (1)

[9] Li X, Pan Z 2014 Dry peeling of Tomato by Infrared Radiative Heating: Part I. Model Development, *Food Bioprocess Technol* **7** 1996–2004.

[10] Li X, Pan Z 2014 Dry peeling of tomato by infrared radiative heating: Part II. Model validation and sensitivity analysis, *Food and Bioprocess Technology* **7** 2005–2013.

[11] Datta A, Ni H 2002 Infrared and hot-air-assisted microwave heating of foods for control of surface moisture, *Journal of Food Engineering*, **51** 355–364.

[12] Prakash B 2012, *Mathematical modeling of moisture movement within a rice kernel during convective and infrared drying*, Ph.D. Dissertation (Department of Biological and Agricultural Engineering, University of California at Davis, Davis, CA, USA).

[13] Trivittayasil V, Tanaka F, Uchino T 2011 Investigation of deactivation of mold conidia by infrared heating in a model-based approach, *Journal of Food Engineering*, **104** 565–570.

[14] COMSOL Multiphysics Version 4.3a User Guide, October 2012.

[15] Li X, Pan Z, Upadhyaya S K, Atungulu G G & Delwiche M 2011 Three-dimensional geometric modeling of processing tomatoes, *Transactions of the ASABE* **54** 2287–2296.

[16] Hamilton D C and Morgan W R 1952 Radiant-interchange configuration factors, NASA TN 2836

[17] Hottel H C 1954, Radiant heat transmission, in William H. McAdams (ed.), Heat Transmission, 3rd ed., McGraw-Hill Book Co., New York., 55-125,

[18] Pan Z, Li X, Khir R, El-Mashad H M, Atungulu G G, McHugh T H., Delwiche M 2015 A pilot scale electrical infrared dry-peeling system for tomatoes: Design and performance evaluation, *Biosystems engineering* **137**, 1-8.

[19] Geedipalli S, Rakesh V, Datta A 2007 Modeling the heating uniformity contributed by a rotating turntable in microwave ovens, *Journal of Food Engineering*, **82** 359–368.

[20] Tanaka F, Verboven P, Scheerlinck N, Morita K, Iwasaki K, Nicola B 2007 Investigation of far infrared radiation heating as an alternative technique for surface decontamination of strawberry, *Journal of Food Engineering*, **79** 445–452.

IOP Conf. Series: Journal of Physics: Conf. Series **1224** (2019) 012018 doi:10.1088/1742-6596/1224/1/012018

Experimental study of PM emissions from wood pellet stoves with an innovative burning pot

L F Polonini[1], D Petrocelli[2], S P Parmigiani[2] and A M Lezzi[1]

[1] Dipartimento di Ingegneria Meccanica ed Industriale, Università degli Studi di Brescia, via Branze 38, 25123, Brescia, Italy

[2] AICO S.p.A., Via Kupfer 31, 25036, Palazzolo sull'Oglio (BS), Italy

E-mail: adriano.lezzi@unibs.it

Abstract.

In this study, results of an extensive set of experiments on a wood pellet stove model equipped with innovative burning pots are presented. Tests have been performed to investigate the relations between design data and operation parameters of the stove and its emissions. In particular, carbon monoxide (CO) and particulate matter (PM) emissions are correlated to the burning-pot depth.

The burning pot A, adopted by AICO S.p.A, allows to obtain near-to-zero CO emissions and low PM emissions: however, PM emissions reduction with respect to standard burning pots is not as enhanced as CO emission reduction. Here it is shown that a further reduction of PM emissions (down to 6–16 mg/Nm^3) can be attained increasing the burning pot depth.

Experiments show also an increment of PM emissions in subsequent tests in the same day: there is an apparent dependence of PM release on the time from ignition. It is likely that this effect is due to ashes accumulation on the bottom of the burning pot and to their interaction with the primary air stream that flows over them.

1. Introduction

Nowadays concern about the greenhouse effect, caused also by the intense exploitation of fossil fuels, is shifting interest towards biomass fuels which have nominally zero-net emissions of greenhouse gases: generation of heat and power from biomass sources is becoming an important element of the energy scenario that will characterize the next years. Unfortunately, a poor exploitation of biomass can cause a significant generation of pollutant and particle emissions. It is generally acknowledged that aerosols are the cause of many diseases and premature death [1, 2], hence the problem of analyzing and reducing aerosols produced by biomass fueled appliances is receiving increasing attention. In particular, it is becoming more and more important to identify which factors control and influence the emission of pollutants and particulate matters and how to reduce them.

Limits of emissions in international and national regulations are becoming more restrictive. Within the framework set by the EU Ecodesign Directive [3], for example, limit of PM emissions for pellet stoves has been fixed to 20.0 mg/Nm^3. This value will be enforced starting on January 1, 2022. In Italy, the public subsidies given to substitute an old heating appliance with a new, efficient one are defined by "Conto Termico 2.0" [4] and are based on a classification that considers PM emissions: for example, the upper limit for the best class is 15.0 mg/Nm^3 for pellet stoves and 10.0 mg/Nm^3 for pellet boilers.

On one hand, it is clear that the need for high efficiency and non-polluting devices that use renewable energy sources is becoming more important, both for the electricity generation and for the direct heating of spaces. On the other hand, manufacturers need clear and long term indications to plan the development of low emission appliances.

An overview on emissions due to biomass combustion can be found in several papers [5, 6, 7, 8, 9]. The flue gases downstream of an ideal combustion are nitrogen (N_2), oxygen (O_2), carbon dioxide (CO_2) and water vapor (H_2O), but in real combustion conditions several other components are present: carbon monoxide (CO), nitrogen oxides (NO and NO_2), hydrogen (H_2), hydrocarbons (HC), poly-cyclic aromatic hydrocarbons (PAHs), besides particulate matter (PM). In [10] the effect of air excess on wood pellet combustion is analyzed: taking into consideration CO emissions, it is shown that for each stove exists an optimal O_2 concentration in the flue gas – hence an equivalent air excess – that minimizes CO generation.

In addition to the O_2 content effect, also the effect of the ratio between primary and secondary air is analyzed [10, 11]: in order to reduce NO_x emissions primary air has to be reduced as much as possible, even under stoichiometric air ratio; at the same time its residence time in the combustion chamber has to grow as much as possible; secondary air, instead, increases turbulence of gases in the combustion chamber promoting oxidation of CO and HC and tertiary air completes all the reactions in the combustion chamber feeding directly the flame.

Aerosol formation processes are investigated in several papers. For example in [5, 12, 13] theoretical models to explain how particles nucleate and how they change as a function of temperature is presented. Besides, information are given about chemical composition of PM in the flue gas. Different studies demonstrate that particle size from wood pellet combustion is mostly submicron with concentration peaks around 0.1 μm and 0.9 μm [7, 9, 12]. They are composed by inorganic compounds of potassium, chlorine, nitrogen, calcium, magnesium and phosphorus – ash particles – and by soot from incomplete combustion [14].

Other works [15, 16, 17] analyzed chimney temperatures related to fire risks of adjacent structures. Flue gas temperature, besides representing safety issues, is an important parameter of combustion efficiency. During the experiments presented in this paper, flue gas temperature, at stove outlet ranges between 450 K and 510 K.

This paper reports results of a collaboration between AICO S.p.A and the University of Brescia aimed to reduce PM emissions of pellet stoves and boilers manufactured by AICO, by improving the combustion conditions and maintaining CO emissions low.

During the first step of the project, attention was concentrated on CO and PM reduction upon optimization of the burning pot and primary air inlet. In standard burning pots, adopted by most of pellet stoves manufacturers, primary air flows through an array of holes on the bottom of the pot. In 2016, AICO developed and patented an innovative burning pot [18], shown in Figure 1, left. Extensive comparison tests showed that AICO's innovative burning pot allowed a strong CO emission reduction, together with PM reduction as well, although not as significant as the one of CO: results of these tests are described at length in [19]. The shape of primary air inlets in AICO's burning pot produce very good combustion conditions and quasi-zero CO emissions. As a consequence the organic part of PM is removed, decreasing total amount of particulate.

In the second step of the project which is described here and which aims to further improve PM reduction, a literature survey on primary measures to decrease wood pellet stove emissions has allowed to identify a few high performance reactors, as those described in [12, 20]. These reactors are high cylinders with several stages of secondary and tertiary air flows that feed the combustion oxidizing all the combustible gases. Considering the main features of these reactors, the burning pot patented by AICO has been modified: in particular, its depth, as well as the distance between primary and secondary air inlets, has been increased. The findings of an experimental campaign to compare the original of AICO's burning pot, with its modified version, are reported here. A definite improvement of PM emissions was recorded.

2. Materials and methods

2.1. Burning pot models and combustion air

The experimental data reported in this paper have been collected at AICO Testing Lab during the development of new pellet stove models. In particular, two burning pot models will be considered here, quite similar in design and dimension. The two burning pots tested will be denoted A and B and they have a nominal burning power of 8.5 kW: burning pot A is developed and patented by AICO S.p.A [18], while burning pot B is a modification of the former, characterized by a larger depth.

Burning pot A, shown in Figure 1 on the left, consists in a square solid base (85 mm per side) with a depth of 50 mm: primary air is supplied through a meandering slit between the base and the sidewall. This air flow does not pass through the bed of embers, but along/over its side and upper surface. Secondary air is supplied at the top of the burning pot and impinges directly on the flame: it passes through several holes with 7 mm of diameter. Burning pot B, shown in Figure 1 on the right, differs from burning pot A only for depth which is 120 mm. Primary and secondary air inlets are equal to those of burning pot A.

The stove is equipped with a tertiary air supply that keeps the glass door clean and feed the flame directly in the combustion chamber from the top. This inlet does not depend on the burning pot used.

Primary, secondary and tertiary air flow rates are not directly controlled. In Figure 2 a scheme of the stove with all air inlets is shown.

In all tests the same batch of pellet was used: its quality is ENPLUS A1 [21]. Elemental concentrations and heating value are specified in Table 1.

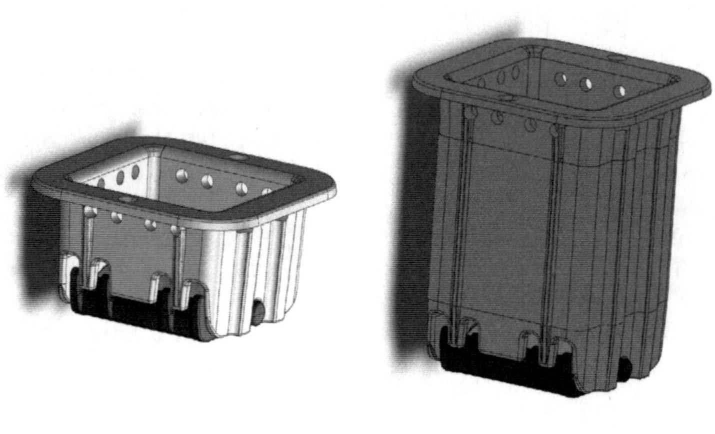

Figure 1. On the left: burning pot adopted in the stove A; on the right: burning pot adopted in the stove B

IOP Conf. Series: Journal of Physics: Conf. Series **1224** (2019) 012018 doi:10.1088/1742-6596/1224/1/012018

Table 1. Characteristics of pellet adopted during tests: values on wet basis

Element	Concentration (%)
Carbon (C)	48.12
Hydrogen (H)	5.70
Oxygen (O)	39.80
Nitrogen (N)	0.11
Water (W)	6.00
Ash	0.18
Low heating value (LHV)	17371 kJ/kg

2.2. Experimental setup and test session

For all tests reported here, we used the experimental setup and procedure proposed by the standards [22, 23, 24] and used by other authors [5, 10, 12, 14]. The whole stove, including the hopper, is put on a platform scale in order to measure the burning rate. The composition of flue gas is measured at the flue gas socket.

Each stove required some days of testing (that will be denoted as "test sessions"). A regular test session starts in the morning with operation parameters setting and the stove ignition: the burning pot, in particular, is emptied of all the residuals from the previous session. After about one hour from ignition combustion reaches quasi-steady conditions, detected upon monitoring the flue gas temperature. At this point a sequence of tests is performed. Each test lasts 30 minutes and several tests are performed during a test session.

During a test CO_2, O_2, CO and NO are measured and recorded every 5 seconds. At the end, average values over the 30 minutes are computed. At the same time, PM is also collected by a filter–gravimetric method: the sampling time being 30 minutes, i.e. the test duration. Time lag between two tests is necessary to replace the filter and calibrate the instrument and usually ranges between 15 and 30 minutes.

The gas analyzers used is ABB EL3020, equipped with cells to measure the concentration (volume fraction) of O_2 and CO in dry gas, i.e. the flue gas without H_2O. Measurement accuracy are $\pm 0.2\%$ for O_2 measures and $\pm 3\,\mathrm{mg/Nm^3}$ for CO measures.

The quartz fibre filter adopted has a retention efficiency of 99.998% of the particulate with $0.3\,\mu$m diameter. Measurement accuracy of the balance adopted for measuring PM weight deposited on the filter is ± 0.08 mg. The pump sucks flue gases at constant flow of 25 Nl/min and it is equipped with a flow meter with an accuracy of ± 0.1 Nl/min. Measures of PM emissions are normalised at 13% of O_2. A schematic sketch of PM and flue gas sampling system used in this study is shown in Figure 2 and reproduces the method described by the normative [23]. All measurements were performed at the end of thermal transient.

In order to make a proper comparison of results, CO and PM are normalized at the same concentration of O_2, using the relation $X_{ref} = X(21 - O_{2,ref})/(21 - O_2)$ (see [24]). Where X is the actual (averaged or instantaneous) concentration of CO or PM in $\mathrm{mg/Nm^3}$, O_2 the actual (averaged or instantaneous) concentration of O_2 in %. $O_{2,ref}$, the reference concentration of oxygen, here 13%, and X_{ref}, the concentration of CO or PM at reference conditions, in $\mathrm{mg/Nm^3}$. All data shown in Figures 3 and 4 and Table 2 are normalized.

3. Results and discussion

3.1. Comparison of emissions from different burning pots

Comparison of combustion tests performed with burning pot A and B shows different performances in terms of particulate emissions. Average results, obtained with the two different burning pots, are shown in Table 2. As pointed out before, the two groups of data (A and B) differ for the burning pot depth only. The stove has, in all tests, the same primary, secondary and tertiary air inlets, it works at the same nominal burning power (8.5 kW) and at the same nominal air excess ratio obtained upon imposing the flue gas extractor velocity. As matter of fact air excess does not depend only on flue gas extractor, but also by flue draught, air inlets surface, instant amount of burnt pellet and flue gas temperature. In particular, instant amount of burnt pellet and flue gas temperature vary continuously during the combustion due to the fact that pellet is a solid fuel and each pellet load can vary in number. Therefore it is not possible to have an exact air excess ratio, then an exact O_2 concentration in the flue gas with this stove technology. Concentrations of O_2 fluctate in a range during the combustion.

In Table 2, the average values during 30 minutes test of O_2, CO, and PM emissions are reported.

Apparently CO emissions were not affected by the burning pot depth: value for each test range between 5 and 115 mg/Nm^3, with average value and standard deviation of all tests of 20 mg/Nm^3 and 27 mg/Nm^3, with burning pot A; while they range between 8 and 65 mg/Nm^3, with average value and standard deviation of all tests of 28 mg/Nm^3 and 15 mg/Nm^3, with burning pot B. All these values are referred at 13% of O_2. PM emissions were instead reduced from value for each test ranging between 10.7 and 23.3 mg/Nm^3, with average value and standard deviation of all tests of 15.0 mg/Nm^3 and 3.0 mg/Nm^3, with burning pot A; to average values ranging between 6.5 and 16.4 mg/Nm^3, with average value and standard deviation of all tests of 11.0 mg/Nm^3 and 3.1 mg/Nm^3, with the burning pot B. All these values are referred at 13% of O_2. These conclusions can be seen at glance from Figure 3.

Possible internal factors that are at the base of the PM decrement obtained with burning pot B were analyzed. This reduction is probably connected with more straight flows that take place inside the burning pot B, in particular primary air feeds pellet without secondary air interferences. Pellet at the top of burning pot A is situated close to secondary air inlets and to combustion chamber entrance, therefore that portion of pellet is more exposed to primary and secondary air interferences than pellet to the top of burning pot B where entrainment effect is probably less. Secondary air feeds the flame and burns all pyrolysis gases generated from pellet heated up in air lacking conditions. In particular, with burning pot B, secondary air does not invest the pellet on the top as can probably occur with burning pot A.

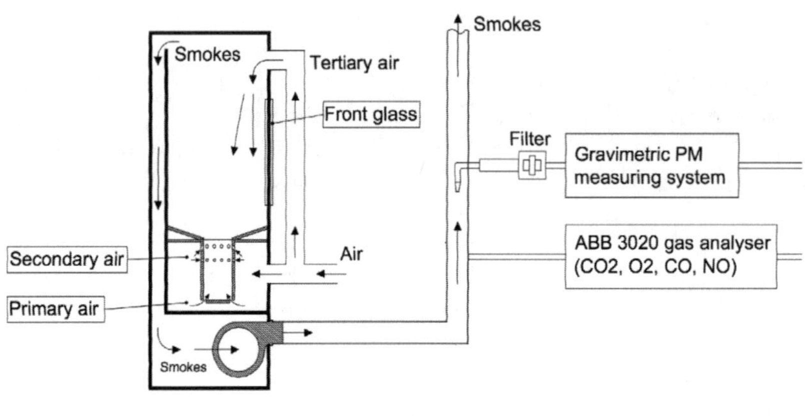

Figure 2. Scheme of the stove with primary, secondary and tertiary air inlets and sketch of PM and flue gas sampling system

Table 2. Emissions from burning pots A and B. CO and PM values are referred at 13% of O_2. Data represent average values over a 30 minutes test. At the bottom of the table average values and standard deviations of all tests with burning pot A and B respectively.

Burning pot A				**Burning pot B**			
Test number	O_2	CO	PM	**Test number**	O_2	CO	PM
	%	mg/Nm^3 – 13% O_2	mg/Nm^3 – 13% O_2		%	mg/Nm^3 – 13% O_2	mg/Nm^3 – 13% O_2
day 1.1	9.3	8	23.3	**day 1.1**	11.1	12	6.5
day 2.1	9.3	8	13.0	**day 1.2**	10.0	31	15.1
day 2.2	8.3	7	14.2	**day 2.1**	8.0	46	7.7
day 2.3	7.8	5	17.4	**day 2.2**	9.2	19	7.0
day 2.4	7.8	7	18.5	**day 2.3**	7.0	14	7.5
day 3.1	9.6	15	11.4	**day 3.1**	9.3	8	7.5
day 3.2	8.3	7	16.0	**day 4.1**	9.6	9	6.5
day 4.1	6.6	7	13.5	**day 4.2**	7.9	19	12.1
day 4.2	5.5	19	15.7	**day 4.3**	8.5	22	14.1
day 4.3	5.2	40	15.2	**day 5.1**	8.6	23	8.1
day 4.4	4.9	115	19.4	**day 5.2**	7.9	26	12.2
day 5.1	7.3	8	10.7	**day 5.3**	9.6	24	10.6
day 5.2	7.1	10	11.8	**day 5.4**	6.6	65	11.7
day 5.3	6.6	8	14.8	**day 5.5**	7.2	23	11.7
day 5.4	6.8	10	14.5	**day 6.1**	8.0	26	12.4
day 5.5	5.7	34	13.3	**day 6.2**	7.7	33	12.6
day 6.1	7.0	11	14.7	**day 6.3**	8.3	34	13.7
day 7.1	9.7	59	14.6	**day 6.4**	7.6	40	14.2
day 7.2	8.3	12	13.4	**day 6.5**	8.1	53	16.0
				day 6.6	7.7	43	16.4
				day 7.1	9.1	44	9.8
				day 7.2	7.9	27	8.7
				day 7.3	8.4	14	11.7
Average	7.4	20	15.0	**Average**	8.4	28	11.0
St.dev	1.5	27	3.0	**St.dev**	1.0	15	3.1

3.2. Time dependence of PM emissions

Consecutive tests within the same test session exhibit an increment of PM emissions. At the beginning, the combustion generally releases a minimum quantity of PM, then PM emissions continue to increase with time. We explain this phenomenon as caused by a change in combustion conditions due to ash accumulation in the burning pots: as the ash levels in the burning pots rise above a threshold, it interacts with the primary air, enhancing entrainment.

In Figure 4 on the left, we plot average PM measured during five test sessions using burning pot A, while in Figure 4 on the right, we plot average PM measured during four test sessions using burning pot B. Each test session lasts one day. In each day, there is an apparent trend: PM emissions tend to increase from the first to the last test of the session.

With the burning pot A, for the first test of each day, that is usually performed after one hour from ignition, measured PM falls between 10.7 and 14.6 mg/Nm^3, at 13% of O_2, while for the last test of the day, that is usually performed after six hours from ignition, PM falls between 13.3 and 23.3 mg/Nm^3, at 13% of O_2. With burning pot B, for the first test of the day, that is usually performed after one hour from ignition, measured PM falls between 6.5 and 12.4 mg/Nm^3 while for the last test of the day, that is usually performed after six hours from

IOP Conf. Series: Journal of Physics: Conf. Series **1224** (2019) 012018 doi:10.1088/1742-6596/1224/1/012018

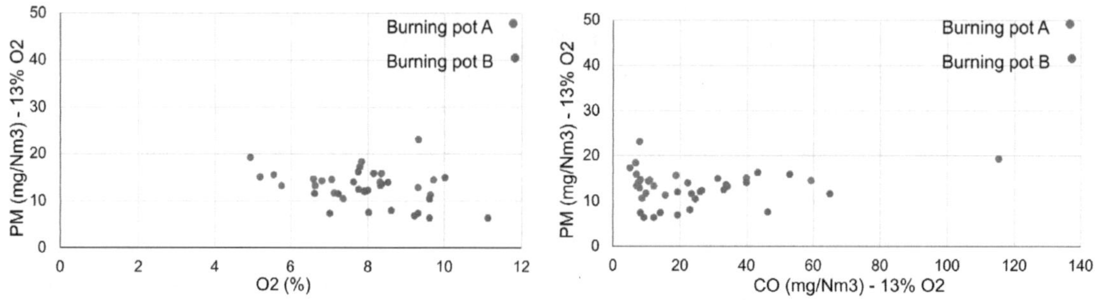

Figure 3. Average PM–O_2 values, on the left, and average PM–CO values, on the right, for test performed with burning pot A and B. CO and PM values are referred at 13% of O_2

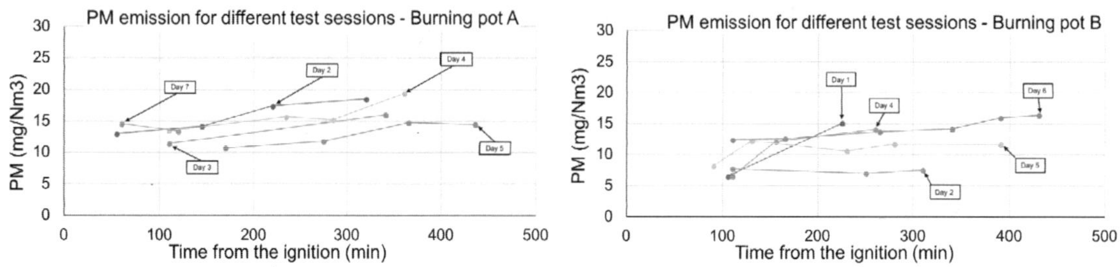

Figure 4. Average PM versus time from ignition for different test sessions with burning pot A, on the left, and with burning pot B, on the right. CO and PM values are referred at 13% of O_2

ignition, PM falls between 7.5 and 16.4 mg/Nm^3.

This increment is not vanished with burning pot B, but it is much less apparent and it converges at lower PM level than in case with burning pot A. Furthermore, with burning pot B, this effect is not present in all test sessions.

4. Conclusions

This paper provides new elements in the understanding of PM generation during the combustion in wood pellet stoves. Combustion depends heavily on design of the combustion chamber, in particular of the burning pot. In this study, two different patented burning pots developed at AICO S.p.A were analyzed: burning pot A, with primary air coming from a slit between the side wall and the bottom of the burning pot, and a modified version of A, much deeper (burning pot B). The stove operated for both burning pots, at the same nominal combustion conditions and burning power (8.5 kW).

The experimental results exhibit different PM levels in the flue gases depending on the burning pot tested. When burning pot A was mounted, the amount of PM released by the stove was larger than when burning pot B was mounted. We consider these results an evidence that burning pot depth is an important factor that influence PM emissions. It is probably determined by the flow structure in the burning pot B that is straighter, minimizing entrainment effect. Furthermore, burning pot B is characterized by a larger distance between pellet and secondary air, minimizing interference between secondary air and embers that is a likely cause of PM entrainment and direct combustion.

Another experimental evidence collected during tests on both burning pots is that PM emissions tend to increase with time passed from ignition. At the beginning, when burning conditions can be considered steady (about one hour after ignition) the quantity of PM is minimum, as time goes by, however, PM production grows. This trend is more apparent with burning pot A. We explain this behaviour as caused by the accumulation of ashes in the burning pot and by their interaction with the primary air streams.

Performance of these burning pots are due to a semi-pyrolysis process that take place in the bottom-centre of the burning pot where pellet heats up in air-absence conditions. This can happen because primary air flows from the bottom-side of the burning pot and does not touch the pellet in the bottom-centre.

References

[1] WHO 2016 Ambient Air Pollution: a global assessment of exposure and burden of disease. World Health Organization.
[2] Pruss-Ustun A, Wolf J, Corvalan C, Bos R and Neira M 2016 Preventing disease through healthy environments. Towards an estimate of the environmental burden of disease. World Health Organization.
[3] Directive 2012/27/EU 2012 Energy Efficiency Directive 2012/27/EU.
[4] DM 16/02/2016 (Italy) 2016 Aggiornamento della disciplina per l'incentivazione di interventi di piccole dimensioni per l'incremento dell'efficienza energetica e per la produzione di energia termica da fonti rinnovabili.
[5] Brunner T, Fluch J, Obernberger I and Warnecke R 2013 Investigations of aerosol formation pathways during MSW combustion based on high temperature impactor measurements, *Fuel Processing Technology* **105** 154
[6] Biswas A K and Umeki K 2015 Simplification of devolatilization models for thermally-thick particles: Differences between wood logs and pellets *Chemical Engineering Journal* **274** 181
[7] Zellagui S, Trouve G, Schonnenbeck C, Zouaoui-Mahzoul N and Brilhac J 2017 Parametric study on the particulate matter emissions during solid fuel combustion in a drop tube furnace *Fuel* **189** 358
[8] Fachinger F, Drewnick F, Giere R and Borrmann S 2017 How the user can influence particulate emissions from residential wood and pellet stoves: Emission factors for different fuels and burning conditions *Atmospheric Environment* **158** 216
[9] El May Y, Dorge S, Jeguirim M, Trouve G and Said R 2012 Measurement of Gaseous and Particulate Pollutants during Combustion of Date Palm Wastes for Energy Recovery *Aerosol and Air Quality Research* **12** 814
[10] Petrocelli D and Lezzi A M 2014 CO and NO emissions from pellet stoves: an experimental study *J. Phys.: Conf. Series* **501** 012036.
[11] Brunner T, Obernberger I and Scharler R 2009 Primary measures for low-emission residential wood combustion - comparision of old with optimised modern systems *Proc. 17th European Biomass Conference & Exhibition.*
[12] Wiinikka H, Gebart R, Boman C, Bostrom D, Nordin A and Ohman M 2006 High-temperature aerosol formation in wood pellets flames: Spatially resolved measurements *Combustion and Flame* **147** 278
[13] Wiinikka H and Gebart R 2004 Critical Parameters for Particle Emissions in Small-Scale Fixed-Bed Combustion of Wood Pellets *Energy & Fuels* **18** 897
[14] Fernandes U and Costa M 2013 Formation of Fine Particulate Matter in a Domestic Pellet-Fired Boiler *Energy & Fuels* **27** 1081
[15] Leppnen P, Neri M, Luscetti D, Bani S, Pentti M and Pilotelli M 2017 Comparison between European chimney test results and actual installations *Journal of Fire Sciences* **35** 62
[16] Neri M, Leppnen P, Bani S, Pentti M and Pilotelli M Experimental and Computational Study of the Temperatures Field Around a Chimney Roof Penetration *Fire Technology* **52** 1799
[17] Neri M, Luscietti D, Fiorentino A and Pilotelli M 2018 Statistical Approach to Estimate the Temperature in Chimney Roof Penetration *Fire Technology* **54** 395
[18] AICO S.p.A. 2016 Braciere autopulente per apparecchio da riscaldamento.
[19] Polonini L F, Petrocelli D, Parmigiani S P and Lezzi A M 2018 Experimental study of PM emissions from wood pellet stoves with innovative burning pots *Proc. of 26th European Biomass Conf. and Exhibition* 1525
[20] Obernberger I, Mandl C and Brandt J 2016 Demonstration of a new ultra-low emission pellet and wood chip small-scale boiler technology *Proc. of European Biomass Conf. and Exhibition* 367
[21] EN 17225-1:2014 Solid biofuels – Fuel specifications and classes – Part 1: General requirements
[22] EN 13284-1:2001 Determination of low mass concentration of dust - Part 1: Manual gravimetric method
[23] EN 13240+A2:2004 Roomheaters fired by solid fuel. Requirements and test methods
[24] EN 14785:2006 Residential space heating appliances fired by wood pellets. Requirements and test methods

IOP Conf. Series: Journal of Physics: Conf. Series **1224** (2019) 012019 doi:10.1088/1742-6596/1224/1/012019

Low-order dynamic model of a domestic electric oven Part I: Experimental characterization of the main heating functions

Michael Lucchi*, **Marco Lorenzini***, **Vincenzo Di Paola***

*Alma Mater Studiorum - Universitá di Bologna, Forlí, Via Fontanelle 40, Italy

E-mail: marco.lorenzini@unibo.it

Abstract.

Household electric ovens can be considered as one of the most common domestic appliances and they usually represent a low-efficiency category, whose value ranges between 10% and 12%. Usually, oven energy class is determined by means of a test procedure regulated by the EN 60350-1 European standard, which moreover requires a proper control of the oven centre temperature. However, during normal operating conditions, because of the presence of the food at the centre, the temperature control within the cavity occurs on the basis of the temperature measured by a probe usually placed in the proximity of one of the cavity corners. Any analysis aimed at decreasing energy consumption must therefore be able to model in sufficient detail the thermal behaviour of the appliance, while still retaining a degree of simplicity. Lumped parameter models have proved themselves a suitable choice, but, especially for gray-box and black-box types, their parameters need to be estimated and validated through experiments. In this work, a common household oven (Electrolux) was instrumented in order to investigate the temperature distribution within the appliance as a consequence of the operating conditions adopted, with the subsequent aim to model the dynamic behaviour of the oven.

The oven was set into a cabinet normally used during the energy consumption test, in order to capture the proper thermal dynamics of the system, and the two main heating functions (forced and natural convection) were also investigated in the range 160-240 $^{\circ}C$. In addition to the temperature measured by the probe (a Pt500 sensor), several thermocouples were placed within the cavity in order to measure the oven centre temperature and to obtain the mean temperatures of walls and heating elements, while one thermocouple of the same type monitored the temperature of the surrounding environment. The number of thermocouples was chosen by trial and error on the basis of modelling needs. Also, four power meters were used to record the electric power consumption of each heating element and the overall electric power absorbed. Since the ultimate goal of this research activity is to create a dynamic model of the oven for control design purposes, the experimental data obtained in the campaign have subsequently been used for this purpose.

1. Introduction

The increasing awareness of the finiteness of energy sources has spurred scientific research into identifying the main areas of energy consumption and into devising ways to reduce the energy demands at all levels. Among the pre-eminent energy consumers, buildings play a key role, and household appliances contribute almost 25% of the overall demand, [1]. Household ovens belong to this category, and their efficiency, between 10% and 12%, ranks them among the devices with

the largest potential for improvement, [2, 3]. One way to increase the oven performance is to devise control systems and strategies which allow proper control of the oven centre temperature. This is also a compelling requirement for the energy consumption test, which is carried out according to the EN 60350-1 European standard, [4] for electric ovens and determines in which energy category (from A+++ to D) the device will fall. The cheapest way of testing a wide spectrum of control strategies at the feasibility stage is through a mathematical model. The model must be able to replicate in a sufficiently accurate and simple way the dynamic behaviour of the oven in the range of temperatures normally associated with energy consumption tests ($160 - 200\,^{\circ}C$ for ventilated operation and $160 - 240\,^{\circ}C$ for static, i.e. non-ventilated, operation [4]). The model is used as a predictive tool for virtual tests and thus spares time-consuming and expensive experiments, but needs to be calibrated with experimental data and validated, again using experimental data as benchmark.
Part I of this paper details the experimental data acquisition procedure and results of tests carried out on a typical household electric oven for different sets of operational conditions. Some of the data were used for parameter estimation in the lumped-parameter model used to characterize the oven, as detailed in Part II [5], and the rest were subsequently employed to validate the model. The choice of a graybox model was mandated by the problem, which is not amenable to the analytical treatment such as for whitebox models, owing in particular to the difficulties in estimating the actual conductance of the oven's cavity walls. This parameter would be influenced by the actual contact resistance of the rock wool used as insulator, which is hard to estimate, and varies from oven to oven due to manual assembly. It would therefore be difficult to use the data to set up an analytical/semi-analitical model; on the contrary, temperature measurements on the different faces could be employed to estimate the heat flux through each wall and door using an inverse methodology as described in [6, 7].

2. The oven

The oven considered is a commercial model ("Electrolux -EOB6850BOX") with a 72 dm^3 volume of its inner cavity (where food is placed for cooking). The oven is set into a wooden cabinet of the kind usually employed for the energy consumption test. The oven has three heating elements, which can be powered alone or as a combination in the different cooking programmes. The top heater (TH) can supply a maximum heating power of 2300 W, is located in closed proximity to the top wall and is the only one actually inside the cavity. Close to the bottom wall, but on its outer side, is the bottom heater (BH), with a maximum power of 1000 W. The third heater is on the cavity's rear wall, and is ring-like, as it encircles the outer perimeter of the fan sucking air from the ambient and cavity and conveying it over the ring heater (RH), which has a maximum power of 1900 W, and again to the cavity itself.
Depending on the operational conditions, one or more heaters are active; in particular, the heating modes investigated are reported in Table 1. For each operation mode the active elements are reported in column 1, whilst columns 2 to 4 give the minimum, maximum temperature set-point and temperature interval between two consecutive tests, as prescribed by the European standard, [4].
In order to describe the system in a simple yet effective way, a lumped parameter approach was chosen, as detailed in Part II of this paper. The model entails a number of parameters (equivalent to conductances and capacitances in an electric network) which need to be determined through experiments. In order to do so, several tests were run, and some of the results were employed to determine the model parameters, whilst the remaining data were used for validation purposes. Since the model aims at describing, among others, the time evoultion of the temperatures of the air at the oven centre and of the oven temperature transducer (a Pt500 resistance temperature transducer, RTD), a single parameter describing the cavity, as in [8, 9] would not suffice, and a degree of discretisation was needed. Besides the air at the oven centre (OC), it was chosen to

assign representative temperatures to each of the cavity walls: rear (PW), bottom (BW), top (TW), right (RW), left (LW) and glass door (D), and to each heater: TH, BH, RH. The readings of the Pt500 were also needed, as was the ambient air temperature.
Prior to running experiments for the energy consumption tests, the temperature and weight of the wet brick were recorded and checked against the allowable values. During the tests themselves, as prescribed by EN 60350-1, two more temperatures from the thermocouples placed inside the brick were recorded. Under these conditions the temperature at the oven centre could not be measured, so this was estimated from the values of cavity air temperature at four locations 2 cm away from the side walls of the brick.

Table 1: Operation characteristics

Power On	$T_{min}\,°C$	$T_{max}\,°C$	$\Delta T\,°C$
F+RH	160	200	20
TH+BH	160	240	40
F+TH	160	200	20

The power instantly supplied to the heaters was measured independently by means of three LogiLight EM0002^{TM} power meters, Fig.1, and another device of the same type was employed to meter the total power absorbed. After calibration with a reference power meter (Yokogawa WT210^{TM}) the measurement uncertainty was found to be lower than $\pm 0.5\%$, whereas the standard for energy consumption requires a maximum uncertainty of $\pm 1\%$. The power meter were operated through an *ad hoc* Arduino^{TM} platform.

Figure 1: Reference instrument for calibration (left) and power meter assembly (right).

The raw temperature measurements from the built-in Pt500 were recorded through an Advanced Measurement Infrastructure (AMI) device and sent to a LabView application, which also recorded the temperature measurements from the various thermocouples employed, which were collected over a FieldPoint^{TM} module. All thermocouples were type K and were calibrated in the $1 - 500\,°C$ range. After calibration, the estimated uncertainty of temperature measurements, including the measurement chain, was $\pm 1.0\,°C$, up to the maximum set-point temperature for the test, i.e. $240\,°C$, and a maximum of $\pm 4.5\,°C$ in the whole range of temperatures recorded. Total uncertainties have been calculated according to customary practice [10, 11]. The experimental set-up also allowed control laws to be implemented in order to power the three heaters independently. This option was not used while acquiring data for model calibration and validation purposes. A schematic of the measurement and control loop is shown in Fig.2.

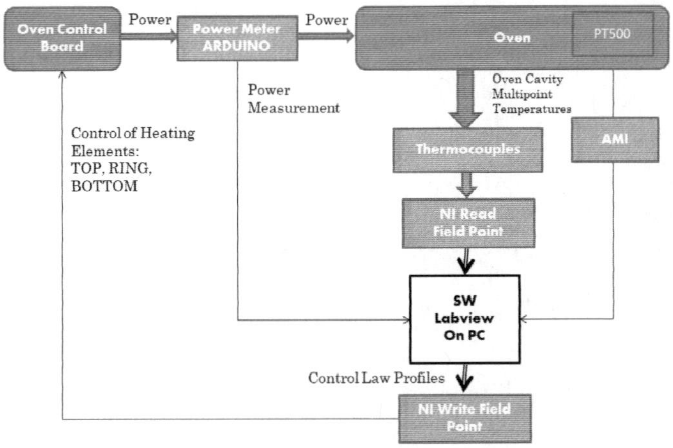

Figure 2: Schematic of the measurement and control loop.

One important issue was related to the number of temperature measurement points needed to obtain a representative description of the model components. In fact, the temperature field in the oven cavity and at its boundaries is far from uniform, owing to both the different composition of the walls and the ventilation system.
In particular, the door is composed of several uninsulated, selective glass panes with channels between them through which air drawn from the ambient circulates for cooling purposes, as shown in Fig.3: two fans, one behind the rear wall , the other above the top wall, draw ambient air in. The air entering from the rear is either blown into the oven cavity from its sides whence it exits from a grid in the rear wall, sucked in by the same, or directed towards the area above the top wall, where the electronic components of the oven are located. The other fan draws air through the channels in the oven door and from vanes at the upper backside of the oven through the electronics and out again. The mother board is a critical component and its temperature must be kept low through the use of high-performance heat sinks such as those described in [12]. A single measurement point would not suffice in some cases, as is illustrated in Fig.4, which refers to the top wall's temperature evolution during a transient in the static (i.e. non-ventilated) mode at a set-point of $200°C$. Three temperature sets were measured at points along one diagonal of the oven: the top-right corner, T_{right}, the middle, T_{centre}, and the bottom-left corner, T_{left}. The evolution of the temperature obtained from the average of the three recordings, T_{mean}, is also plotted: it is clear that the latter value is closer to that recorded at the top-right corner than to that of the thermocouple in the middle. During the calibration phase it was also verified that using the value of T_{centre} would yield unsatisfactory values for the parameter, which meant poor model performance in the validation tests, whilst good results were obtained when using T_{mean}. The same situation was encountered for several other locations; for this reason, three measurement points were adopted for each of the cavity's inner walls, according to the sketch of Fig.5.

The heating elements were also suitably instrumented with thermocouples fastened at different locations on their surface. Contact between the temperature transducers and the surfaces was ensured by heat resistant tape (Kapton^{TM}); in the case of the electric heater, where local temperature could exceed the application range of Kapton^{TM}, the thermocouples were also held

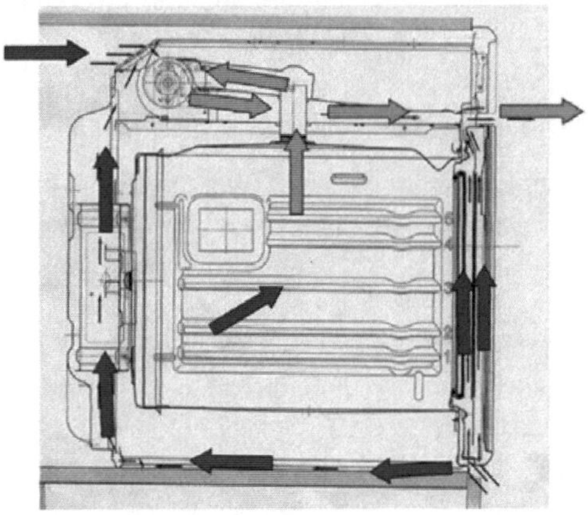

Figure 3: Airflow in the oven.

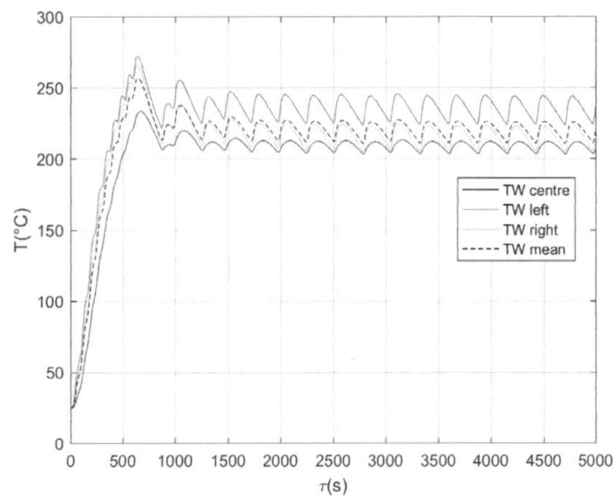

Figure 4: Temperature plots for three locations on the top wall and mean temperature. Static mode at 200°C

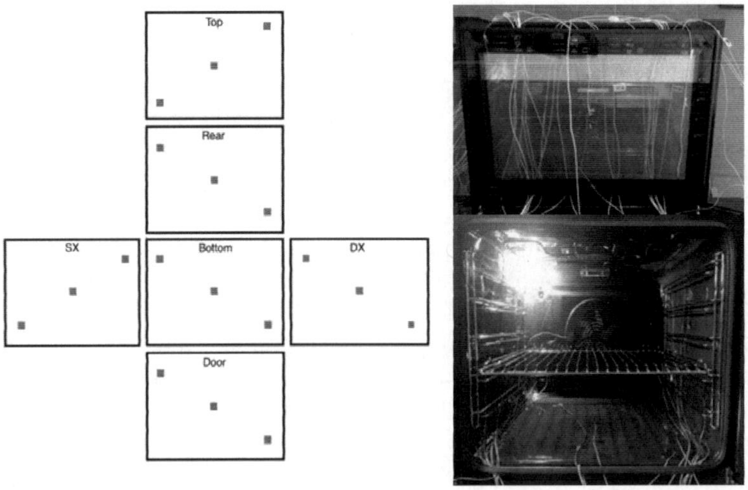

Figure 5: Layout of the thermocouples on the cavity walls (left) and pictures of the thermocouples in place (right).

in place by steel wires. Figure 6 shows the temperature evolution of three thermocouples on the TH during a test with the fan powered off (static mode) at a set-point of 200°C: the excellent agreement of the two thermocouples which are symmetric to the heater's supply's contacts is $\pm 3.7\,°C$, within the uncertainty bonds for a measured temperature of $413\,°C$.
When running energy consumption tests, a brick prepared according to the prescription of EN 60350-1 was laid onto a tray inside the cavity. The standard also requires that its temperature be determined by measurement at two points inside the brick [4].

In order to obtain a satisfactory description of the brick in the lumped-parameter model, a temperature representative of the brick surface was needed. One thermocouple was therefore embedded at surface level into each of the brick's faces, as shown left in Fig. 7. Also, in order to

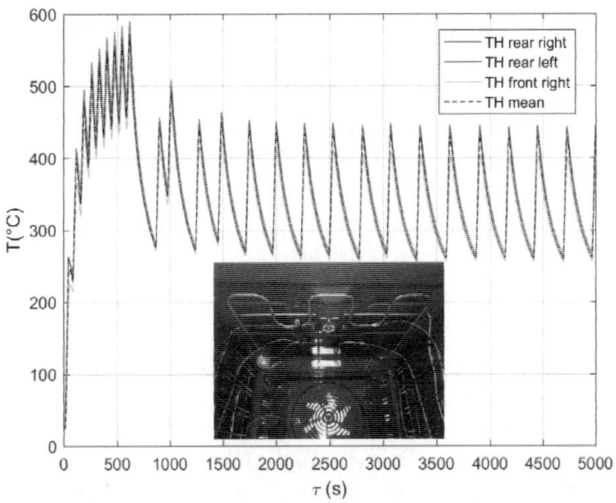

Figure 6: Temperature evolution of three thermocouples on the top heater during a test with the fan powered off (static mode) at a set-point of $200°C$.

Figure 7: Layout of the thermocouples on the brick walls (left) and onto the grid (right); the red rectangle marks the position of the brick during tests.

obtain a suitable piece of information to compute the temperature of the air at the oven centre, four thermocouples were placed perpendicular to the grid onto which the brick is laid during the EC test about $2\,cm$ from its side, see Fig. 7 right.

2.1. Overall uncertainty

The overall uncertainty is calculated straightforwardly, without any inolved sensitivity coefficients as for other cases, [13], since nearly all the quantities of interest to estimate the parameters of the graybox model of the oven presented in part II of this paper, [5], are either measured directly (e.g. the power supplied to the heaters) or are an average of measurements at different locations, such as the temperatures representative of the cavity's faces (in this case the value used is the mean of the measured temperatures at three locations). To adopt a conservative approach it can therefore be assumed that the uncertainties of the derived quantities coincide with those of the measured quantities. In the example above, the uncertainty on the surface temperature of one face is the same as that of the single temperature measurement, i.e. $\pm 1.0\,°C$, when application of the procedure described in [11] would yield a value of $\pm 1.0/\sqrt{3}\,°C$.

3. Results

Several tests were run to collect data for parameter estimation of the model and its validation, under the experimental conditions listed in Table 1. Some of the tests were aimed at determining the most satisfactory layout of the temperature sensors to yield results which could be meaningfully employed in the model, an example of which has already been shown in Fig.4. Some results are reported in the following to highlight peculiarities of the process.

Figure 8 shows both the power up and temperature history of the top heater during a static (i.e. without ventilation) test at $200\,°C$. It is clear how the temperature follows the power up cycles, but it can also be noticed that, when the heater is in operation, the power supply is not uniform, and decreases, if slowly, in time. This is due to both fluctuations in the network supply and changes in the value of the heater's resistance, which depends on temperature.

Figure 9 shows the temperature versus time during a stationary test at $200\,°C$ of set-point for the oven centre, of the Pt500 as obtained through the AMI, and that of a thermocouple fastened onto the surface of the RTD, as detailed in the picture. Both the ramp during the transient phase and the wobbling behaviour once steady state conditions are reached can be easily discerned. It is also interesting to notice that the OC temperature shows no overshoot, whilst both the Pt500 and the thermocouple do: this is due to the lack of ventilation and to the location of the Pt500, close to the top heater, which is used in this mode. As a consequence, the Pt500 heats up faster than the oven air, and the thermocouple even more so. It can also be noticed that the Pt500 lags slightly behind the thermocouple, which is exposed more directly to the heater and, above all, has a smaller thermal inertia. It would seem puzzling that the readings of the Pt500 show a temperature that oscillates around the set-point and consistent with that of the oven centre once the transient is over, whereas the recordings of the thermocouple onto the RTD sensor are higher. In fact, the Pt500 readings are automatically corrected by the manufacturer to account for the shift between actual temperature at the location of the RTD and that at the oven centre. The AMI device reads the compensated signal, which is consistent with the set-point.

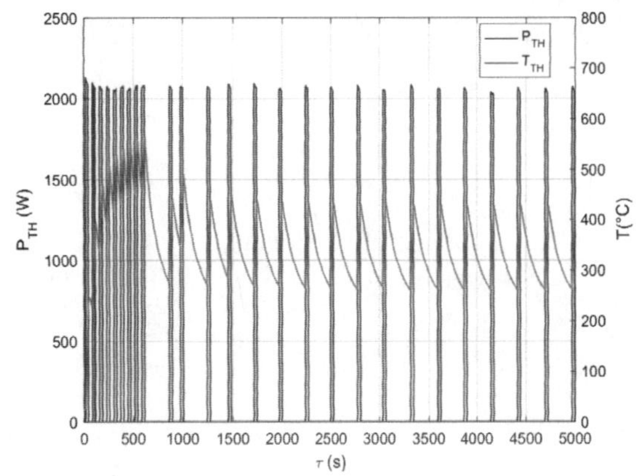

Figure 8: Power and temperature evolution for the top heater.

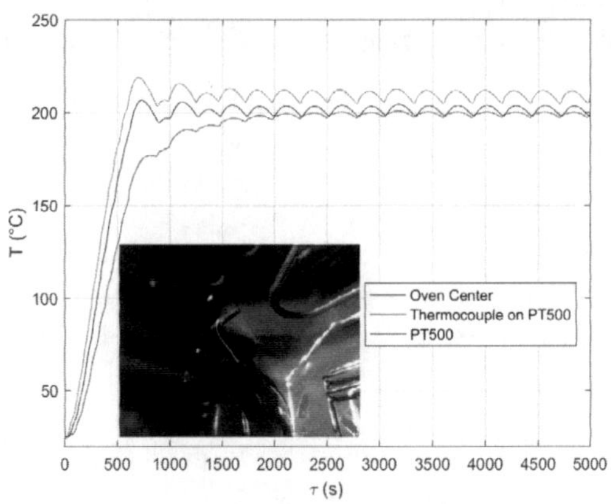

Figure 9: Temperature of the oven centre, Pt500 and thermocouple on the RTD for static mode at $200\,°C$.

The temperature evolution for one ventilated (top) and one static (bottom) test at $200\,°C$ set-point is shown in Fig.10. Whilst temperatures at the outer surface and in the channel between the glass panes are only slightly different, the effect of forced convection can be appreciated at the inner surface of the door, where the temperature is almost thirty degrees cooler than for the static case. This also explains why manufacturers carry out the energy consumption test with the fan on.

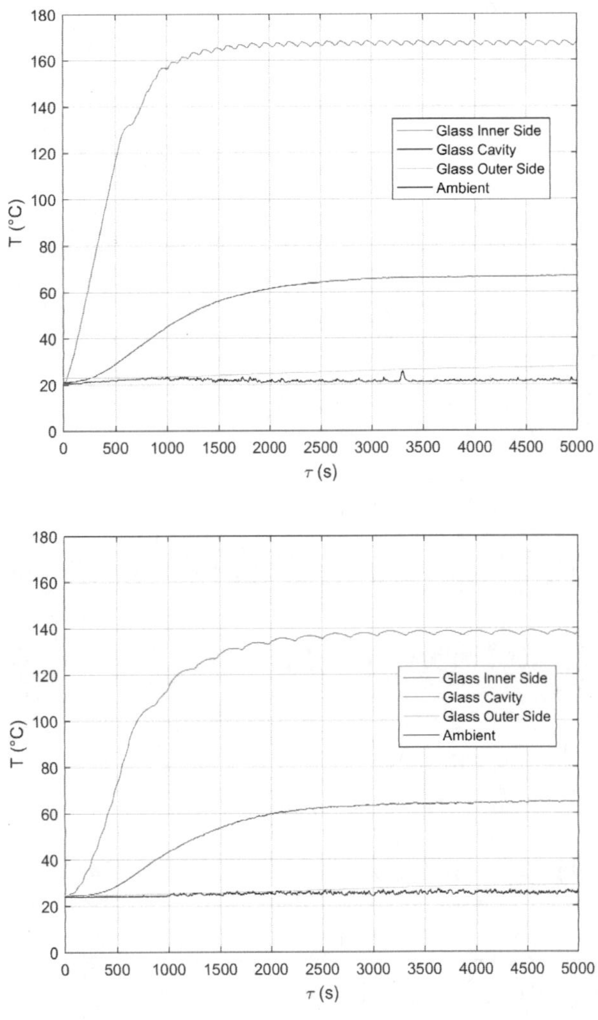

Figure 10: Temperature across the oven door in the ventilated (top) and static (bottom) cases, for an operational set-point of $200\,°C$.

The temperature at the six locations on the brick surface is shown in Fig. 11 for a set-point of $200\,°C$. In both cases the lowest temperatures are associated with the bottom surface; this is because the brick has been left to soak at least eight hours as per the requirements of the EC test, and the cold water seeps through the porous structure and drips from the bottom. The highest temperature during ventilation is that associated with the brick's rear: this is because in this heating mode the fan, which faces the rear of the brick, is powered and draws air from the cavity, thus accelerating surface evaporation and decreasing the local thermal capacity. Also, in the ventilated mode temperatures at the different locations, with the exception of the rear, are much closer to one another and higher than in the static case. The forced flow of air, which enters from the sides of the cavity and exits through the fan, acts so as to make conditions more homogeneous and to increase heat transfer from the cold brick to the hotter cavity air.

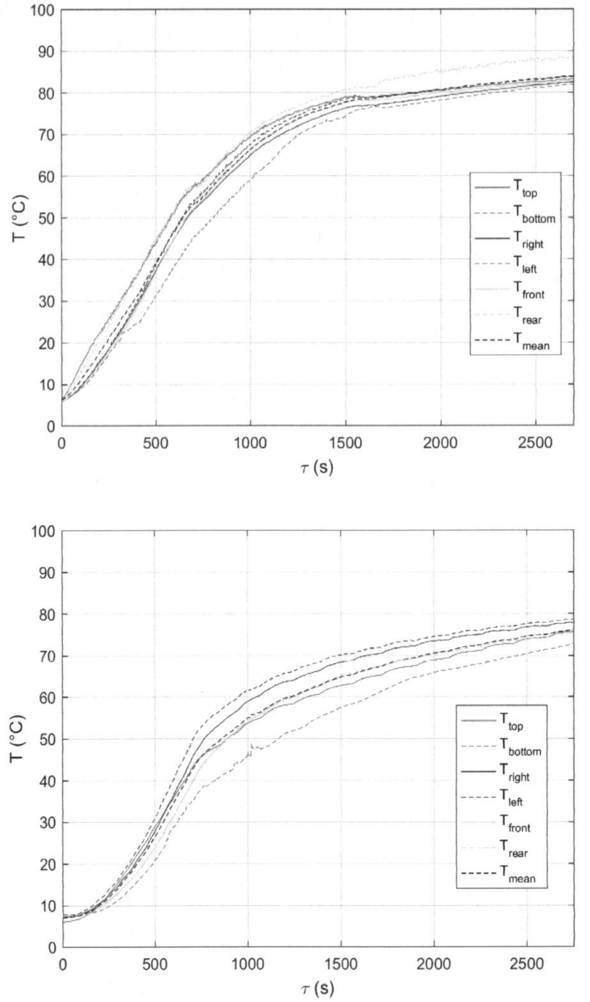

Figure 11: Temperature for the brick faces in the ventilated (top) and static (bottom) cases, for an operational set-point of 200 °C.

4. Conclusions

In this work the experimental set-up of a domestic oven has been described. Temperature and power measured at suitable locations are used to estimate the parameters of a lumped model which aims at describing the dynamic behaviour of the oven both empty and in the conditions of the energy consumption test. The data collected and not employed in the estimation procedure have been used to validate the model, with results illustrated in Part II of this paper in the case of non-ventilated operation with an empty oven.

Other forms of the model, for ventilated operation with an empty oven and for the EC test have also been validated with the data collected in the experimental campaign, and these results will be published at a later stage.

Acknowledgements
The Authors wish to acknowledge Electrolux Italia S.p.a. for supplying the test oven and for funding this research.

References
[1] Hoxha E and Jusselme T 2017 *Science of the Total Environment,* **596–597** pp. 405–416
[2] Amienyo D, Doyle J, Gerola D, Santacatterina G and Azapagic A 2016 *Sustainable Production and Consumption,* **6** 67–76
[3] Bansal P, Vineyard E and Abdelaziz O 2011 *Applied Thermal Engineering,* **31** 3748–3760
[4] CENELEC 2013 En 60350-1 household electric cooking appliances part 1: Ranges, ovens, steam ovens and grills methods for measuring performance (latest version including all amendments).
[5] Lucchi M, Lorenzini M and Roberti G 2018 *Proceedings of 36th UIT Heat Transfer Conference*
[6] Beck J, Blackwell B and StClair C 1985 *Inverse Heat Conduction: Ill-Posed Problems* (New York: Wiley Interscience)
[7] Beck J and Arnold K 1977 *Parameter Estimation in Engineering and Science* (New York: Wiley Interscience)
[8] Ramirez-Laboreo E, Sagues C and Llorente S 2014 *22nd Mediterranean Conference on Control and Automation, MED* pp 505–510
[9] Ramirez-Laboreo E, Sagues C and Llorente S 2016 *Applied Thermal Engineering,* **93** 683–691
[10] Lorenzini M, Morini G, Henning T and Brandner J 2009 *International Journal of Thermal Sciences* **48** 282–289
[11] Moffat R 1988 *Experimental Thermal and Fluid Science* **1** 3–17
[12] Lorenzini M, Fabbri G and Salvigni S 2007 *Applied Thermal Engineering* **27** 969–975
[13] Yang Y, Chalabi H, Lorenzini M and Morini G 2014 *Heat Transfer Engineering* **35** 159–170

IOP Conf. Series: Journal of Physics: Conf. Series **1224** (2019) 012020 doi:10.1088/1742-6596/1224/1/012020

Low-order dynamic model of a domestic electric oven Part II: parameter identification and model validation for the static heating mode

Michael Lucchi*, **Marco Lorenzini***, **Giovanni Roberti***
*Alma Mater Studiorum - Universitá di Bologna, Forlí, Via Fontanelle 40, Italy
E-mail: michael.lucchi2@unibo.it

Abstract. Appliances are responsible for almost the 25% of the overall environmental impact of buildings, and, among these, electric domestic ovens represent a low-efficiency category when considering flats, with an energy efficiency between 10% and 12%, thus stimulating the development of more efficient technologies. The energy performance of such an appliance is determined through a test according to the EN 60350-1 European standard, which, among others, requires suitable control of the oven centre temperature. Since the development of reliable dynamic models represents a high-potential tool to design better control strategies, the aim of this work is to create a model simulating the transient thermal behaviour during the static, i.e. non-ventilated, heating mode of an electric domestic oven within the cabinet adopted for energy consumption tests; the model must be able to predict the main temperatures within the cavity, with particular attention to the oven centre. A lumped parameter approach based on the thermoelectric analogy was chosen, in order to obtain a low-order dynamic model suitable for control design purposes. Since the model strives to reproduce the transient behaviour of the oven centre temperature, a single lumped parameter simulating the thermal behaviour of the whole oven cavity was unsuitable and a certain level of macroscopic discretization had therefore to be introduced. Thus, the oven cavity was modelled with eight parameters, representing its main parts like the walls, the glass door and the air in the cavity; also, the temperature probe, a Pt500 as is customary for such devices, was included in the model, since this is the only temperature feedback during operation and its presence in the model could lead to an easier implementation of efficient control logics. Three lumped parameters simulating the thermal behaviour of the heaters were also considered, leading to an 11th order model.

1. Introduction

Household appliances contribute to almost 25% of the overall environmental impact of buildings, as reported in [1]; among them, electric ovens represent significant sources of energy consumption when considering apartments, because of their very low energy efficiency (between 10% and 12%), which stimulates the development of new techniques, [2]-[3]. The EN 60350-1 European standard, [4], which regulates the energy consumption characterization of domestic electric ovens, requires a proper control of the oven center temperature, thus making the design of suitable control strategies an important field of investigation. Dynamic models can help designers to obtain control laws suitable for different purposes (e.g. minimizing the energy consumption, increasing the control precision), and this work aims at developing a model to simulate the transient thermal behaviour of an electric domestic oven, reproducing the main temperatures

within the cavity and, in particular, the oven center temperature.
Several works can be found in the literature about models of domestic ovens in transient regime, ranging from low to high levels of complexity. As to the former, the work of Abraham and Sparrow, [5], can be mentioned, who developed a simple model based on the first law of thermodynamics for closed systems, in order to predict the temperature of a thermal load placed inside an electric oven, but without giving information about the oven center temperature. Good examples of more complex models are the works where CFD techniques were applied to study the transient behaviour of ovens. Among these works are those of Mistry et al., who investigated the transient natural convection during the broiling and baking cycles in electric ovens, considering the radiation from the heating elements too, [6], and the heat transfer and fluid flow inside a residential gas oven, representing oven burners as wall temperature boundary conditions and trying to find a satisfactory compromise between computational time and accuracy, [7].
When considering models for control design purposes, high-complexity approaches like those adopting CFD are unsuitable: for several thermal systems, like e.g. vapour compression refrigerating machines [8], the best choice is usually represented by low-order dynamic models, often set up through a lumped parameters approach, since it represents a good trade-off between simplicity and accuracy.
On the basis of what was previously presented in [9], Laboreo et al. realized a lumped-parameters dynamic model of a domestic oven with a thermal load (taking into account the water evaporation rate too) using the thermoelectric analogy and investigating both the forced convective and the radiative coking modes, [10]. However, a single lumped parameter was used to represent the whole inner cavity, including metal sheets and air, thus making it impossible to estimate the oven center temperature per se, whose control is required during the energy consumption test. In [11], the dynamic behaviour of a professional electric oven was investigated through the lumped capacitance method, also considering the radiative heat exchange between surfaces and heaters without linearization, approximating the latter with planar surfaces to estimate their view factors. Yet, the validation shown in the paper was carried out at a single temperature set-point, tuning the average convection coefficients between air and walls and the thermal transmittances between air and heaters and between air and thermal masses so as to obtain a good agreement between numerical results and experimental data.
In this work, a control-oriented low-order lumped parameters model of an electric domestic oven without thermal load has been developed in MATLAB/Simulink®, and its transient behaviour in natural convective (i.e. static) heating mode was investigated. Different elements of the cavity (eg. walls, door, heaters) were included in the model, but particular attention was paid to the model ability to reproduce the transients of the oven center temperature. Also the radiative heat exchanges within the cavity were taken into account, without adopting linearization procedures, as a consequence of the benefits highlighted by the same authors in [12]. Moreover, the transient behaviour of the Pt500 sensor was included within the model, since it represents the only temperature feedback available for the control system during normal operating conditions; in this way, the actuation of simulated control logics in a real oven can be facilitated. Model parameters were identified via optimization from experimental data at a given temperature set-point and model validation was then carried out considering data acquired at different set-points.

2. The oven

The oven considered is a commercial model with a 72 dm^3 inner cavity volume placed within a typical cabinet for energy consumption tests. The oven is equipped with three heating elements: a 1900 W ring heater (RH), a 2300 W top heater (TH) and a 1000 W bottom heater (BH). In the static heating mode considered here, TH and BH are the active heaters. The details about the measurements setup and the experimental campaign are reported in Part I of this paper, [13].

IOP Conf. Series: Journal of Physics: Conf. Series **1224** (2019) 012020 doi:10.1088/1742-6596/1224/1/012020

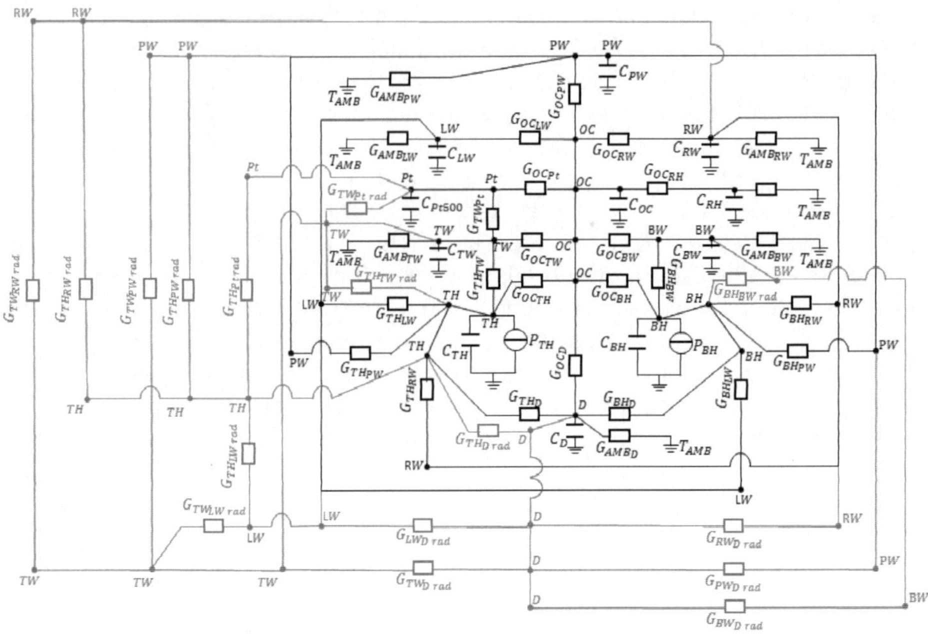

Figure 1: Thermoelectric scheme of the non-linear model.

3. Model

The oven model was developed on the basis of the thermoelectric analogy, thus interconnecting a set of capacitors by means of thermal conductances. While CFD models give a more detailed physical insight of the problem usually requiring long simulation times, the lumped-parameter approach based on the thermoelectric analogy allows to significantly reduce the computational cost, which has a fundamental importance when developing control-oriented models and must be as low as possible, and at the same time it offers a good level of accuracy when applied to transient heating and cooling problems, [10, 14]. Since the main aim was to reproduce the transients of the oven center and the temperatures of the Pt500 , a minimum level of macroscopic discretization was necessary. Thus, the oven cavity has been divided into seven main parts, representing five walls (bottom (BW), right (RW), left (LW), rear (PW), top (TW)), the glass door (D) and the air cavity (OC) and three parameters simulating the thermal behaviour of the heaters were considered (bottom (BH), top (TH) and, although not activated, ring heater (RH), in order to be able to investigate in the future also the dynamics of the forced convection mode). Considering the lumped capacitance introduced for the Pt500 sensor, a 11th order model was obtained. A scheme of the thermal network is shown in Fig. 1. In the static heating mode, since the cavity walls see high temperature elements like the top heater and because of the relatively low temperature characterizing the door, significant radiative heat fluxes can occur and influence the thermal phenomena of the whole cavity, especially during the transients. The influence of radiative terms in the thermal dynamics of the oven has been considered through fourth-power terms, as explained just below.

Equations (1)-(11) describe the energy balances for the 11 parameters constituting the model. $\dot{Q}_{i\,j}$ are linear terms representing the heat transfer between two generic nodes i and j, expressed as $G_{i\,j} \cdot (T_i - T_j)$, where $G_{i\,j}$ is the thermal conductance between nodes i and j, while T_i and T_j are the temperatures of the corresponding nodes. The radiative heat transfer is expressed through the terms $\dot{Q}^{rad}_{i\,j}$, calculated as $G^{rad}_{i\,j} \cdot \sigma \cdot (T_i^4 - T_j^4)$, where $G^{rad}_{i\,j}$ is the radiative thermal conductance between nodes i and j, σ is the Stefan-Boltzman constant and T_i and T_j are the node temperatures expressed in kelvins. Both $G_{i\ j}$ and $G^{rad}_{i\ j}$ obey a reciprocity law, such that $G_{i\ j} = G_{j\ i}$ and $G^{rad}_{i\ j} = G^{rad}_{j\ i}$. P_{TH} and P_{BH} represent the electric power absorbed by the top and bottom heaters respectively and are model inputs, whilst the ambient temperature T_{amb}, which influences energy dispersions, is kept constant at $23°C$ and is treated as a perturbation to the model. The ambient temperature is at $23°C$, which can be considered reasonable, since it lays between the temperature set-points usually considered for domestic environments, namely $20°C$ in the winter and $26°C$ in the summer, [15].

$$C_{OC}\frac{dT_{OC}}{d\tau} = \dot{Q}_{RW\ OC} + \dot{Q}_{LW\ OC} + \dot{Q}_{BW\ OC} + \dot{Q}_{TW\ OC} + \dot{Q}_{PW\ OC} + \dot{Q}_{D\ OC} + \dot{Q}_{Pt\ OC} + \dot{Q}_{TH\ OC} + \dot{Q}_{BH\ OC} + \dot{Q}_{RH\ OC} \quad (1)$$

$$C_{TH}\frac{dT_{TH}}{d\tau} = \dot{Q}_{OC\ TH} + \dot{Q}_{TW\ TH} + \dot{Q}^{rad}_{TW\ TH} + \dot{Q}_{PW\ TH} + \dot{Q}^{rad}_{PW\ TH} + \dot{Q}_{RW\ TH} + \dot{Q}^{rad}_{RW\ TH} + \dot{Q}_{LW\ TH} + \dot{Q}^{rad}_{LW\ TH} + \dot{Q}_{D\ TH} + \dot{Q}^{rad}_{D\ TH} + \dot{Q}^{rad}_{Pt\ TH} + P_{TH} \quad (2)$$

$$C_{BH}\frac{dT_{BH}}{d\tau} = \dot{Q}_{OC\ BH} + \dot{Q}_{BW\ BH} + \dot{Q}^{rad}_{BW\ BH} + \dot{Q}_{PW\ BH} + \dot{Q}_{RW\ BH} + \dot{Q}_{LW\ BH} + \dot{Q}_{D\ BH} + P_{BH} \quad (3)$$

$$C_{RH}\frac{dT_{RH}}{d\tau} = \dot{Q}_{OC\ RH} + \dot{Q}_{AMB\ RH} \quad (4)$$

$$C_{TW}\frac{dT_{TW}}{d\tau} = \dot{Q}_{OC\ TW} + \dot{Q}_{TH\ TW} + \dot{Q}^{rad}_{TH\ TW} + \dot{Q}_{AMB\ TW} + \dot{Q}_{Pt\ TW} + \dot{Q}^{rad}_{Pt\ TW} + \dot{Q}^{rad}_{PW\ TW} + \dot{Q}^{rad}_{LW\ TW} + \dot{Q}^{rad}_{RW\ TW} + \dot{Q}^{rad}_{D\ TW} \quad (5)$$

$$C_{BW}\frac{dT_{BW}}{d\tau} = \dot{Q}_{OC\ BW} + \dot{Q}_{BH\ BW} + \dot{Q}^{rad}_{BH\ BW} + \dot{Q}^{rad}_{D\ BW} + \dot{Q}^{rad}_{LW\ BW} + \dot{Q}^{rad}_{RW\ BW} + \dot{Q}^{rad}_{PW\ BW} + \dot{Q}_{AMB\ BW} \quad (6)$$

$$C_{RW}\frac{dT_{RW}}{d\tau} = \dot{Q}_{OC\ RW} + \dot{Q}_{TH\ RW} + \dot{Q}^{rad}_{TH\ RW} + \dot{Q}_{BH\ RW} + \dot{Q}^{rad}_{BW\ RW} + \dot{Q}^{rad}_{TW\ RW} + \dot{Q}^{rad}_{D\ RW} + \dot{Q}_{AMB\ RW} \quad (7)$$

$$C_{LW}\frac{dT_{LW}}{d\tau} = \dot{Q}_{OC\ LW} + \dot{Q}_{TH\ LW} + \dot{Q}^{rad}_{TH\ LW} + \dot{Q}_{BH\ LW} + \dot{Q}^{rad}_{BW\ LW} + \dot{Q}^{rad}_{TW\ LW} + \dot{Q}^{rad}_{D\ LW} + \dot{Q}_{AMB\ LW} \quad (8)$$

$$C_{PW}\frac{dT_{PW}}{d\tau} = \dot{Q}_{OC\ PW} + \dot{Q}_{TH\ PW} + \dot{Q}^{rad}_{TH\ PW} + \dot{Q}_{BH\ PW} + \dot{Q}^{rad}_{BW\ PW} + \dot{Q}^{rad}_{TW\ PW} + \dot{Q}^{rad}_{D\ PW} + \dot{Q}_{AMB\ PW} \tag{9}$$

$$C_{D}\frac{dT_{D}}{d\tau} = \dot{Q}_{OC\ D} + \dot{Q}_{TH\ D} + \dot{Q}^{rad}_{TH\ D} + \dot{Q}_{BH\ D} + \dot{Q}^{rad}_{BW\ D} + \dot{Q}^{rad}_{TW\ D} + \dot{Q}^{rad}_{LW\ D} + \dot{Q}^{rad}_{RW\ D} + \dot{Q}^{rad}_{PW\ D} + \dot{Q}_{AMB\ D} \tag{10}$$

$$C_{Pt}\frac{dT_{Pt}}{d\tau} = \dot{Q}_{OC\ Pt} + \dot{Q}_{TW\ Pt} + \dot{Q}^{rad}_{TW\ Pt} + +\dot{Q}^{rad}_{TH\ Pt} \tag{11}$$

As shown in Eq.(1), the oven center exchanges thermal power with each element of the grid. Equations (2)-(3) describe the energy balance of the top and bottom heaters, which are connected to the side walls (RW,LW,PW) and the door (D), whilst direct connections between bottom heater and top wall, and top heater and bottom wall were neglected. Moreover, radiative terms between the bottom heater and the bottom wall, and between the top heater and each wall in the cavity are considered. Equation (4) shows the energy balance for the ring heater, which receives no electric power in the static mode and exchanges heat with the oven center and the ambient only. Equations (5)-(10) describe the energy balance for the cavity walls and the door, where dispersions towards the ambient are included. As shown in Eqs. (5)-(6), the top and bottom walls are coupled with the other walls and the door through radiative terms, because they take higher temperatures due to heater proximity. Fourth-power terms coupling the door node with those associated to other walls are considered, because of its relatively low temperature, see Eq. (10). The energy balance of the Pt500 sensor is described by Eq.(11): in order to consider the temperature stratification typical of the natural convective heating mode, a linear term coupling the sensor node to the top wall node was introduced in addition to the heat exchange with the oven center. In addition, two non-linear terms coupling the Pt500 sensor to the top wall and the top heater are present to estimate the radiative heat exchange. Finally, the radiative heat exchange between the bottom wall and the top elements (wall and heater) was neglected, since several optimization attempts in the parameters identification process highlighted over-estimations of T_{BW} when these non-linear terms were taken into account, leading to less satisfactory results in the estimation of the oven center temperature.

4. Parameters identification

On the basis of experimental data (temperatures and electric power absorbed by the heaters) obtained in a three-hours test at $200°C$ (see Part I of this article), which represents the central value among the three temperature set-points at which to carry out the energy consumption test of the oven in non-ventilated mode, as prescribed by the European standard, [4], an optimum set of capacities and conductances was found through an optimization procedure conducted with the Parameter Estimation toolbox available in MATLAB/Simulink®. The Non-Linear-Least-Squares method with the Trust-Region-Reflective algorithm was applied, and the cost function minimized was:

$$f = \sum_{i=1}^{n} w_i \cdot \delta_i = \sum_{i=1}^{n} w_i \cdot \frac{\sum_{k=1}^{N}(T_i^{model}(k) - T_i^{exp}(k))^2}{N} \tag{12}$$

The weight w_i for the mean square error δ_i associated to the i-th lumped parameter, calculated between the temperature evaluated by the model $T_i^{model}(k)$ and the experimental data $T_i^{exp}(k)$ at each of the N time-steps of the simulation, was assumed equal to unity for each of the eleven nodes of the thermal grid, with the aim of reproducing equally well almost all the temperature trends within the cavity. The initial estimation of thermal capacities of the walls and the glass door was carried out considering the series of metal sheets, insulation layers and wooden walls capacities for the former and the series of four glass layers and air cavities for the latter. In particular, the relationship $1/C_i = \sum_{l=1}^{L} 1/(m_l \cdot c_l)$ was applied, where C_i is the initial capacity of the i-th parameter and m_l and c_l are the mass and the specific heat capacity of the l-th layer. As for the initial estimation of Pt500 and heaters capacities, the product between their mass and the specific heat capacity of AISI 430 steel was assumed, while the sum of thermal capacities of cavity air C_{air} and metallic grid C_{grid} was considered as initial estimation of oven center capacity. Neglecting contact resistances, a series of thermal resistances which took both the insulation layers and the furniture into account was used to calculate the initial values of conductances between the i-th lumped wall and the environment; as for the heat exchanges between cavity air and walls, the first estimation of the conductances was carried out through the correlations for natural convection on a vertical or horizontal surface available in [14]. A manual tuning procedure was carried out to estimate the initial values of conductances between the oven center and the heaters, the glass door and the environment and between the heaters and the side walls, because of the large amount of uncertainty which characterizes these quantities. Since the main aim of the model was to correctly predict the transients of the main temperatures within the cavity in response to a particular input electric power forcing function, neglecting the quantification of the heat exchanges between the grid nodes, a large field of variability was allowed for the model parameters in the optimization procedure, in order to ensure the best possible predictive capacity for the model. A fixed-step Runge-Kutta method (Matlab ode4) with a time-step of 1 s was chosen as solver for the integration in the time domain for both parameter identification and validation.

In order to investigate the predictive capability of the model, the electric power absorbed by the heaters in tests carried out at temperature set-points different from that used for the parameters identification ($200°C$) was used as a model input and the predicted temperatures were then compared with experimental data. Several simulations of about-three-hours heating cycles were performed: the validations reported here refer to temperature set-points T_{set} of $180°C$ and $220°C$.

5. Results

As anticipated in Sec. 3, the electric power absorbed by the heaters is an input of the model; an example of such an input is shown in Fig. 2, where the temporal profile of the electric power absorbed by the top and bottom heaters during a test at $200°C$ is plotted.

Figure 3 compares the values of the oven center temperature predicted by the model with the experimental data. In the optimization test-case at $200°C$, see Fig. 3-(a), the percentage displacement Δ remains under 2% for almost the whole duration of the simulation, with a peak at the beginning: this spike in the relative error appears for all the nodes of the grid, but can be considered insignificant, as it occurs when temperatures are very low. The trend of T_{OC} predicted by the model in the validation tests at $180°C$ and $220°C$ is shown in Figs. 3 (b)-(c): the model shows good predictive potentiality, since the percentage displacement is confined below 5%, with the mean temperature reached at the stabilized regime which is slightly underestimated at $T_{set} = 180°$ and overestimated at $T_{set} = 220°$. Although it can be considered negligible for control design purposes, a time-delay in achieving the stabilized regime has been highlighted for the test at $180°C$, probably due to an overestimation of thermal capacities.

The model response is good also when considering the temperature trend predicted for the Pt500

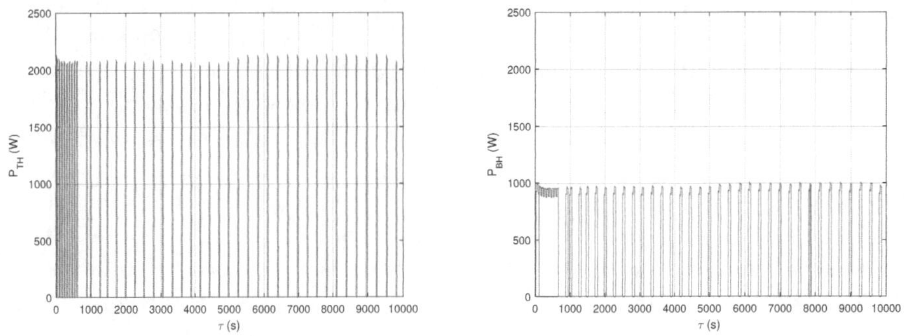

(a) (b)

Figure 2: Model power inputs. Static Mode 200°C. (a) Top heater. (b) Bottom heater.

probe, both in the optimization case, Fig. 4 (a), and in the validation procedure, Fig. 4 (b)-(c). A time-delay in reaching the stabilized regime and a slight overestimation of the mean stabilized temperature appear also for the Pt500 in the validation test-cases at 180°C and 220°C, as shown in Fig. 4 (b)-(c): this can be attributed to the oven center dynamics, which directly influences the transient behaviour of the probe. The temperature percentage displacement remains always below 5% in the stabilized regime in both validation tests, a maximum overestimation of the highest temperature sensed of about 5.5% occurring at 180°C set-point.

Figures 5-6 show the results obtained for the top heater and the glass door in the optimization test case (Figs. 5 (a) and 6 (a)) and in the two validation tests (Figs. 5 (b)-(c) and 6 (b)-(c)). The good predictive capability of the model is confirmed: for the top heater, the maximum percentage displacement Δ appears during the heating phase (about 500 s long) and ranges from 8.7% ($T_{set} = 200°$) to 11% ($T_{set} = 220°$), while a maximum error in the stabilized regime slightly above 4% is obtained for the door in the validation test at $T_{set} = 220°$. The underestimation and overestimation highlighted in the predicted temperatures of the oven center and the Pt500 in the validation cases appear also for the door node: this behaviour may be caused by the assumption of temperature independent conductances, whilst natural convection, which together with irradiation represents the main heat transfer mechanism within the cavity when the oven is in non-ventilated mode, is strongly temperature-dependent.

6. Conclusions

The transient thermal behaviour of a domestic electric oven in natural convective mode has been investigated through a low-order lumped parameter model based on the thermoelectric analogy and suitable for control design purposes. Fourth-power terms were introduced to take into account the radiative heat exchange between the elements of the cavity. In order to facilitate the development of new control laws, the dynamics of the Pt500 probe was also considered in the model, since it represents the only temperature feedback during normal oven operation. The model is of 11th order and includes cavity walls, the glass door, the heaters, the cavity air and the Pt500 probe, and takes the electric power absorbed by the heaters as inputs and the surrounding temperature as disturbance.
The set of capacities and conductances which allow to best predict the temperature's temporal trends of the elements in the heated cavity was identified through optimization based on experimental data obtained at a temperature set-point of 200° and on the Non-Linear-Least-Squares method. Validations were then carried out at different temperature set-points.
The model showed good predictive capabilities, with percentage errors with respect to

IOP Conf. Series: Journal of Physics: Conf. Series **1224** (2019) 012020 doi:10.1088/1742-6596/1224/1/012020

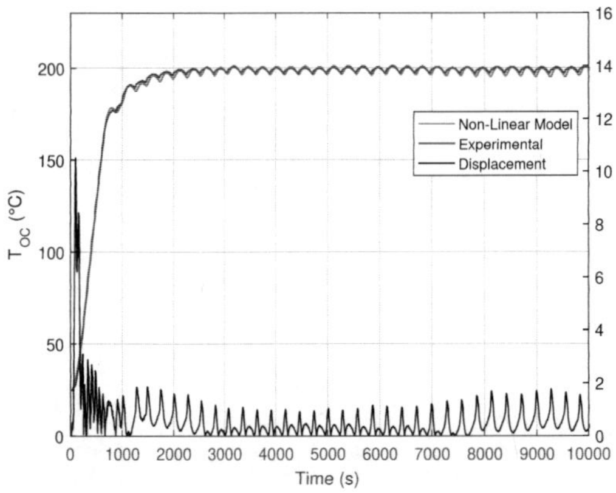

(a)

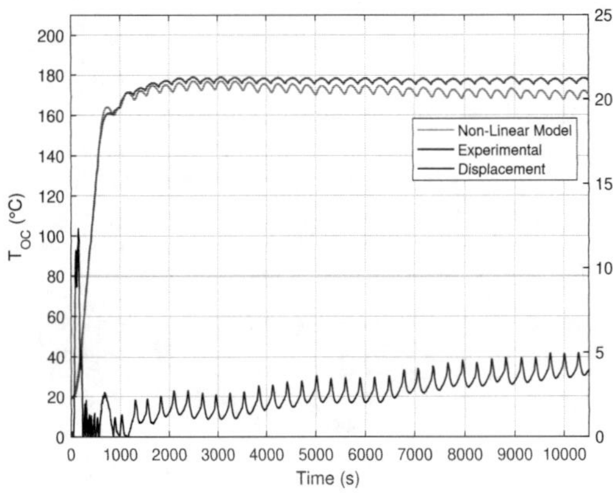

(b)

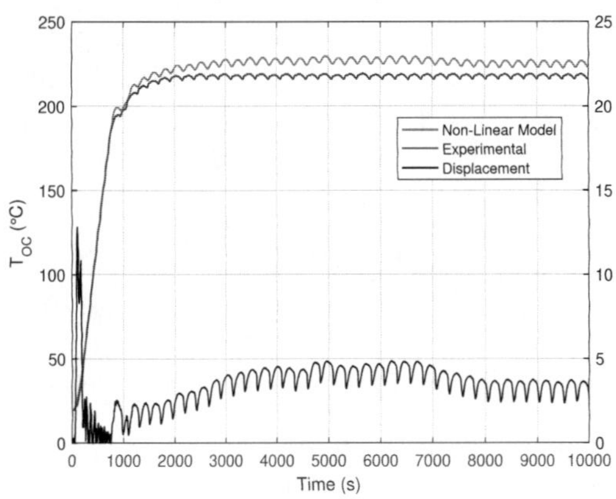

(c)

Figure 3: Oven center temperature prediction in the static heating mode. (a) Optimization at $T_{set} = 200°C$. (b) Validation at $T_{set} = 180°C$. (c) Validation at $T_{set} = 220°C$.

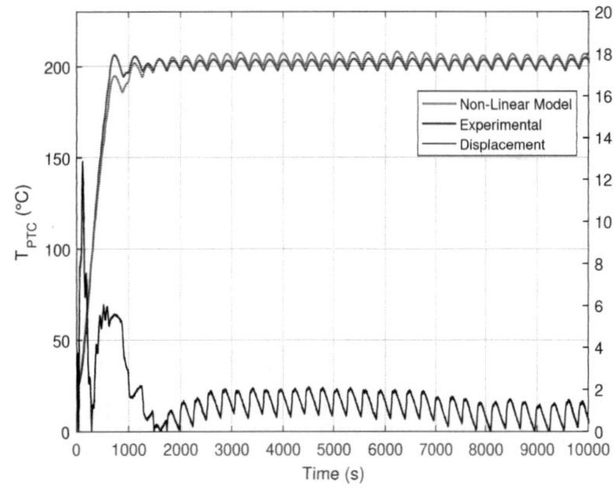

(a)

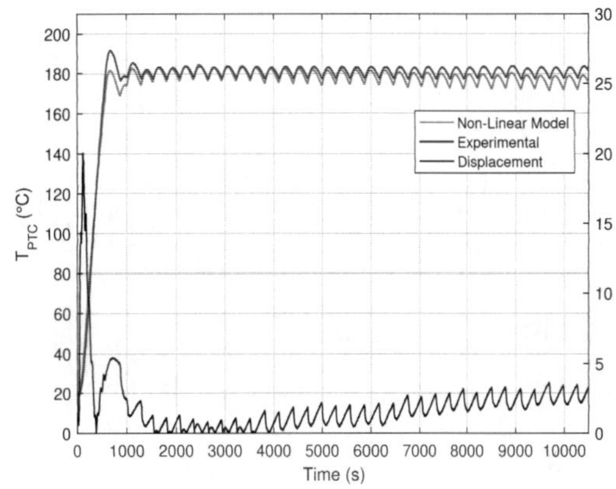

(b)

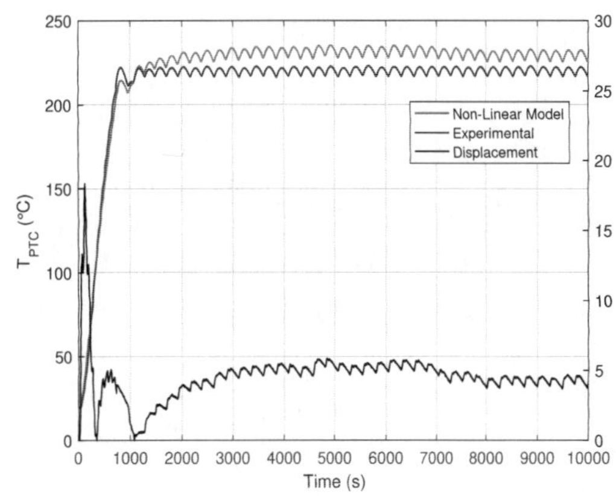

(c)

Figure 4: Pt500 temperature prediction in the static heating mode. (a) Optimization at $T_{set} = 200°C$. (b) Validation at $T_{set} = 180°C$. (c) Validation at $T_{set} = 220°C$.

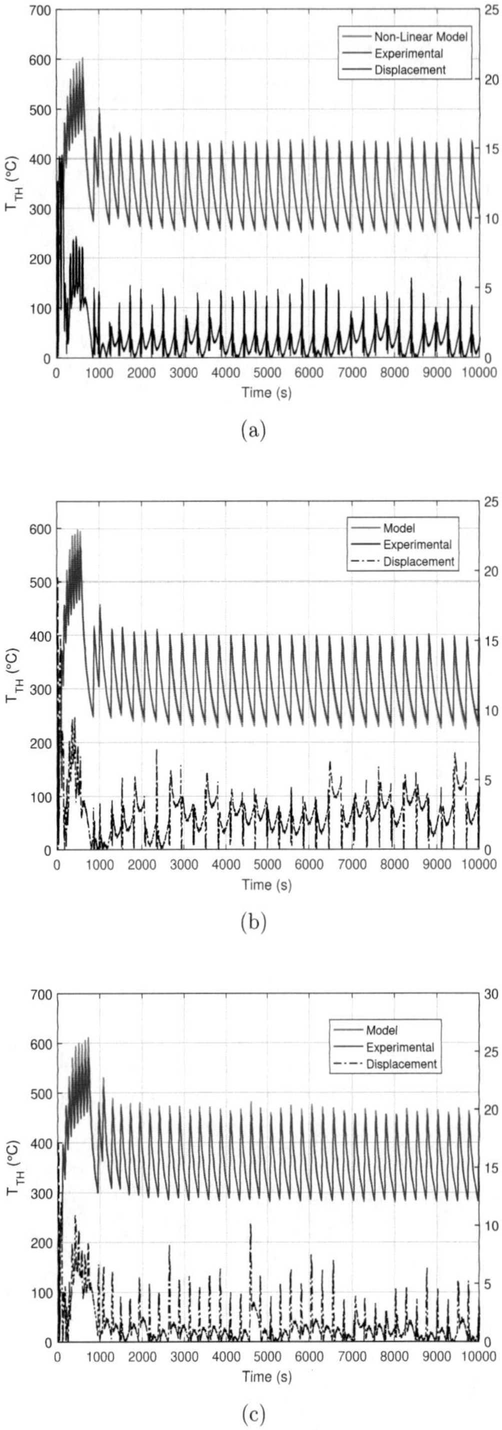

Figure 5: Top heater temperature prediction in the static heating mode. (a) Optimization at $T_{set} = 200°C$. (b) Validation at $T_{set} = 180°C$. (c) Validation at $T_{set} = 220°C$.

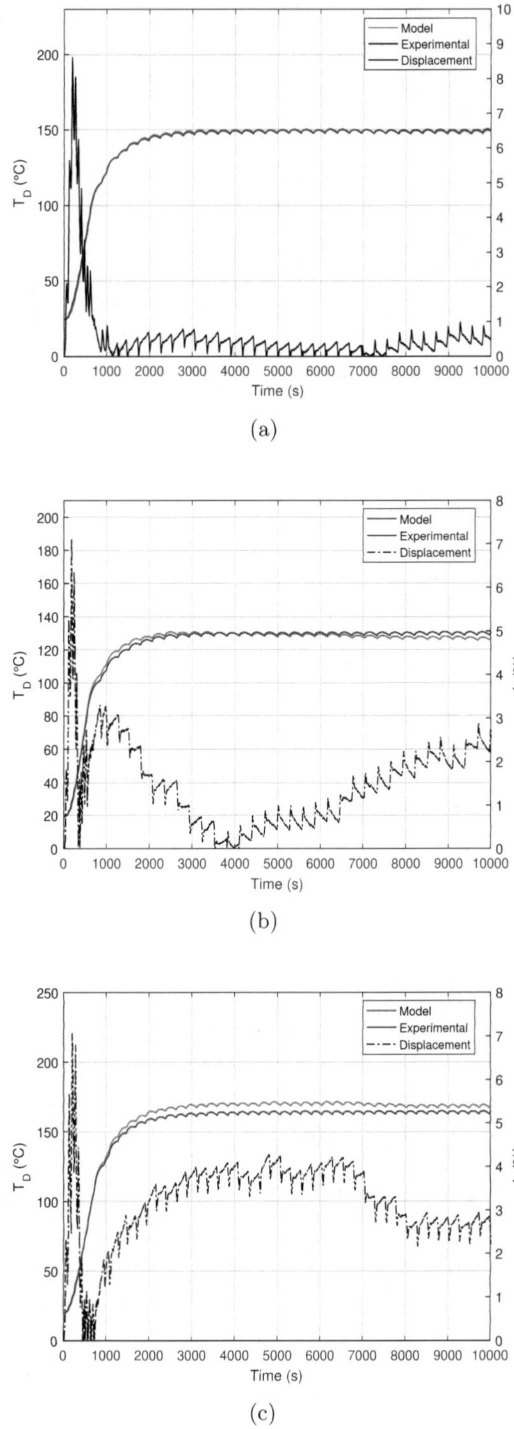

Figure 6: Door temperature prediction in the static heating mode. (a) Optimization at $T_{set} = 200°C$. (b) Validation at $T_{set} = 180°C$. (c) Validation at $T_{set} = 220°C$.

experimental data always below 5% both for the oven center and Pt500 temperatures: since one of the requirements of EN 60350-1 is to correctly control the oven center temperature, the usefulness of the model is confirmed. The model also exhibited good reliability in predicting the temperature of the other components of the oven.
Other forms of the model has been developed with the same numerical approach, in order to carry out transient analysis also for the forced convective heating mode, with and without thermal load: the main results of those analyses will be published at a later stage. Further developments include the investigation of the predictive capability in case of the temperature-dependent thermal conductances, thus leading to a model with non-linear parameters.

Acknowledgements
Funding of this research by Electrolux Italia S.p.a. is gratefully acknowledged.

References
[1] Hoxha E and Jusselme T 2017 *Science of the Total Environment,* **596–597** pp. 405–416
[2] Amienyo D, Doyle J, Gerola D, Santacatterina G and Azapagic A 2016 *Sustainable Production and Consumption,* **6** 67–76
[3] Bansal P, Vineyard E and Abdelaziz O 2011 *Applied Thermal Engineering,* **31** 3748–3760
[4] CENELEC 2013 En 60350-1 household electric cooking appliances part 1: Ranges, ovens, steam ovens and grills methods for measuring performance (latest version including all amendments).
[5] Abraham J and Sparrow E 2004 *Journal of Food Engineering,* **62** 409–415
[6] Mistry H, Ganapathisubbu S, Dey S, Bishnoi P and Castillo J 2006 *Applied Thermal Engineering,* **26** 2448–2456
[7] Mistry H, Ganapathisubbu S, Dey S, Bishnoi P and Castillo J 2011 *Applied Thermal Engineering,* **31** 103–111
[8] McKinley T and Alleyne A 2008 *International Journal of Refrigeration* **31** 1253–1264
[9] Ramirez-Laboreo E, Sagues C and Llorente S 2014 *2014 22nd Mediterranean Conference on Control and Automation, MED 2014* pp 505–510
[10] Ramirez-Laboreo E, Sagues C and Llorente S 2016 *Applied Thermal Engineering,* **93** 683–691
[11] Burlon F, Tiberi E, Micheli D, Furlanetto R and Simonato M 2017 *Energy Procedia,* **126** 2–9
[12] Lucchi M and Lorenzini M 2019 *Applied Thermal Engineering* **147** 438–449
[13] Lucchi M, Lorenzini M and Di Paola V 2018 *Proceedings of 36th UIT Heat Transfer Conference*
[14] Bergman T, Lavine A, Incropera F and Dewitt D 2011 *Fundamentals of Heat and Mass Transfer* (Hoboken: John Wiley & Sons)
[15] Lucchi M, MLorenzini and Valdiserri P 2017 *Journal of Physics: Conference Series,* **796**

IOP Conf. Series: Journal of Physics: Conf. Series **1224** (2019) 012021 doi:10.1088/1742-6596/1224/1/012021

Design and commissioning of an experimental facility for performance evaluation of pure and mixed refrigerants

Luca Molinaroli, Andrea Lucchini, Luigi Pietro Maria Colombo

Dipartimento di Energia, Politecnico di Milano, via Raffaele Lambruschini 4/A, 20156, Milano (MI), Italy

E-mail: `luca.molinaroli@polimi.it`

Abstract. The recent regulations about fluorinated greenhouse gases introduce a rapid phase-down of traditional, environmentally harmful refrigerants. Consequently, new working fluids are continuously appearing on the market, leading to the need of testing them to analyse their performance before making an ultimate choice. In the present paper, a test rig built to carry out experimental studies on different refrigerant in a drop-in application is described and first results for the set-up commissioning are commented. The experimental campaign was carried out in typical water-to-water heat pump operating conditions, varying the cold water temperature at evaporator inlet from 10 °C to 20 °C, with a step of 2.5 °C, and hot water temperature at condenser inlet from 30 °C to 50 °C, with a step of 10 °C and the compressor shaft rotational frequency from 30 Hz to 80 Hz, with a step of 10 Hz. The first findings demonstrate that the experimental results of some key performance parameters, namely heating capacity, COP and compressor volumetric efficiency, agree well with open literature trends, allowing to state that the experimental facility was correctly designed and can be used to deep the analysis on performance assessment of new refrigerants in a drop-in application.

1. Introduction

The recent EU 517/2014 regulation [1] and the Kigali amendment [2], force the air conditioning and refrigeration industry to find new refrigerants able to cope with more and more severe limits on Global Warning Potential (GWP). As a consequence, in the last years new refrigerants have been continuously introduced and the need of testing them in vapour compression systems to analyse their performance arises.

For example, the substitution of the largely used working fluid in refrigeration applications, R404A, was studied both from a theoretical point of view [3] and from an experimental point of view [4] with low GWP refrigerants and, similarly, the replacement of refrigerants traditionally used in air conditioning and heat pump applications, namely R410A for small to medium capacity systems and R134a for medium to large capacity systems, with new, low GWP alternatives was analysed by several authors considering both natural [5] and synthetic refrigerants [6, 7, 8]. The general findings of these wide screening of several alternatives were that the energy efficiency of system using the new refrigerants could achieve the same value of the traditional refrigerant-based based systems, but an adjustment of expansion device, charge and, above all, compressor swept volume are required.

In this framework, the experimental analysis of the performance of systems that use new refrigerants is of primary importance since it could provide not only data about the system operation with a new working fluid but also an useful feed-back for system adjustment or re-design. Therefore, the goal of the present paper is to describe a test rig recently built to carry out experimental studies on different refrigerants in a drop-in application. The experimental set-up mimics a water-to-water system equipped with variable speed compressor and fully instrumented to measure and acquire all the relevant parameters.

2. Experimental methodology

2.1. Experimental set-up

The layout of the experimental set-up used to assess the refrigerant performance is shown in Figure 1. The test rig mimics a water-to-water system and consists of three different loops:

(i) The first loop is the refrigerant loop, whose main components are a semi-hermetic, variable speed, reciprocating compressor, two stainless steel plate heat exchangers (i.e. the condenser and the evaporator) and two electronic expansion valves (in parallel). This loop was used to reach the desired refrigerant mass flow rate and superheating at evaporator outlet acting on the compressor shaft rotational frequency and on the expansion valve set-point respectively.

(ii) The second loop is the cold water loop and mainly consists of a variable speed pump, a thermal storage and a three-way valve. This loop was used to set the water mass flow rate and temperature at evaporator inlet to the desired values whereas the thermal storage allows to reduce the water temperature fluctuations and to reach stable testing conditions.

(iii) The third loop is the hot water loop and is built considering the same components found in the previous one and adding an auxiliary chiller. This loop was used to set the water mass flow rate and temperature at condenser inlet to the desired values and to reach stable operating conditions thanks to the thermal storage.

The main characteristics of the test rig components are provided in Table 1

Table 1. Main characteristics of the experimental set-up components.

Component	Parameter	Range
Compressor	Swept volume @ 50 Hz	$13.15\ m^3 \cdot h^{-1}$
	Shaft frequency	30 Hz - 87 Hz
Condenser	Height x Width x Depth	289 mm x 119 mm x 93.6 mm
	No of plates	40
Evaporator	Height x Width x Depth	376 mm x 119 mm x 71.2 mm
	No of plates	30
Expansion valve	Capacity range	1200 W - 12000 W
	Capacity range	1690 W - 16900 W
Liquid receiver	Volume	$2.8 \cdot 10^{-3}\ m^3$
Suction accumulator	Volume	$2.33 \cdot 10^{-3}\ m^3$
Oil separator	Volume	$2.8 \cdot 10^{-3}\ m^3$
Recuperator	Height x Width x Depth	193 mm x 76 mm x 71.2 mm
	No of plates	30

The water-to-water unit was charged under Group 1 reference condition (see Table 3) considering a superheating at evaporator outlet equal to 5 K and a subcooling at condenser outlet equal to 3 K as target values. The final charge of R134a was 4 kg.

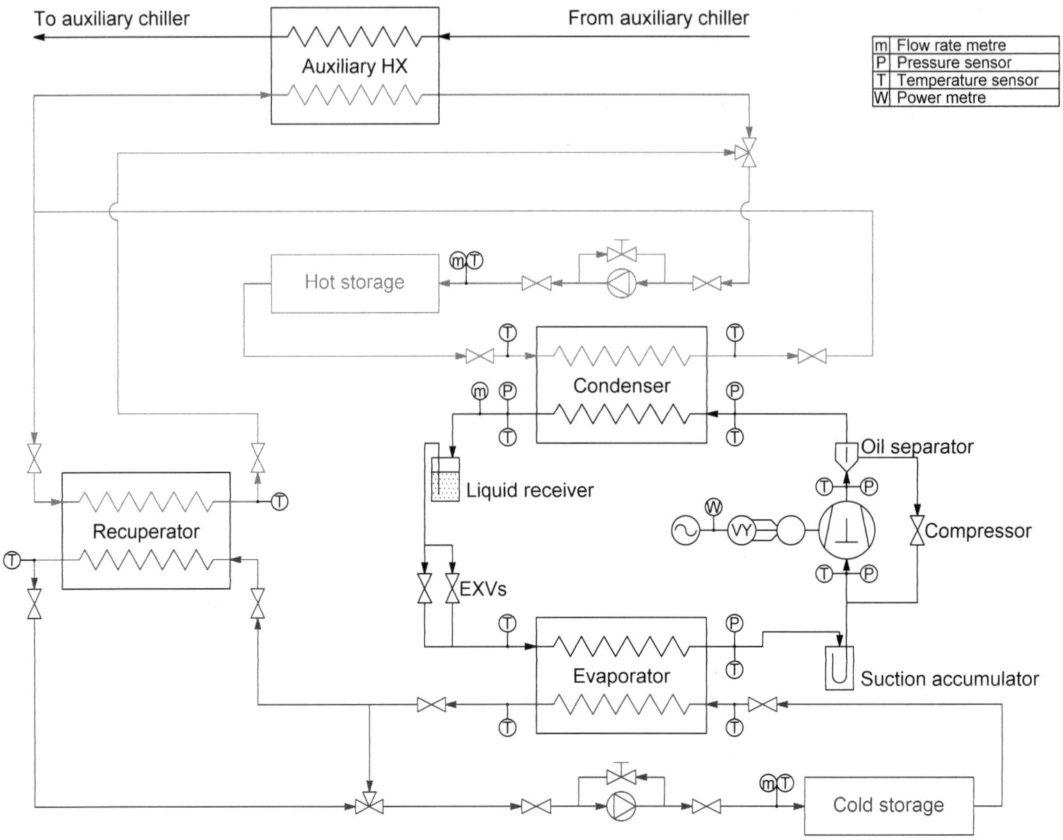

Figure 1. Schematic of the experimental set-up.

The test rig was equipped with instrumentations that allow for the data acquisition of the main parameters such as pressures, temperatures, flow rates and power. The positions of all the measurement devices are shown in Figure 1 while their main characteristics are collected in Table 2 where, for the sake of completeness, it is worth specifying that the water mass flow metre is equipped also with a RTD Pt 1000 temperature probe whose accuracy is equal to 0.1 K.

2.2. Experimental procedure
The following experimental procedure was adopted to run each test:

(i) The cold and hot water pumps are switched on and the RPM are set to the values needed to guarantee the water mass flow rates required by the test.
(ii) The compressor is switched on and the shaft rotational frequency is set to the value required by the test.
(iii) The cold and hot water temperatures required by the test are set at the inlet of the two tanks acting on the 3-way valves. Consequently, the water temperature inside both tanks slowly changes as a result of the mixing process and the same happens to the water temperature at evaporator and condenser inlets.

Table 2. Measurement instrumentation and accuracy.

Parameter	Instrument	Range	Accuracy
Refrigerant mass flow rate	Coriolis mass flow metre	0 kg·h^{-1} - 300 kg·h^{-1}	±0.15% r.v.
Refrigerant pressure	Pressure transducer	0 kPa - 4000 kPa	±0.3% f.s.
Refrigerant temperature	RTD Pt 100	243.15 K - 343.15 K	±0.1 K
Water mass flow rate	Vortex flow metre	0.21 m^3·h^{-1} - 3 m^3·h^{-1}	±2% r.v.
Water temperature	RTD Pt 100	253.15 K - 353.15 K	±0.1 K
Compressor power	Power transducer	0 W - 4000 W	±0.2% f.s.

(iv) Once the water temperature at evaporator and condenser inlet has reached the value required by the test, the data acquisition starts with a sample rate equal to 1 s. For each of the measured pressures and temperatures (see Fig. 1), the simple moving average is calculated considering the last 900 samples.

(v) If the deviation of each measured pressure and temperature lies within ± 10 kPa and ± 0.2 K respectively, the test is considered in steady-state condition and further 900 samples are recorded for data analysis.

(vi) Finally, at the end of the test, the evaporator and condenser capacities are calculated based on both the water temperatures and refrigerant enthalpies. If the values agree within ± 4% the experimental test is considered completed otherwise it is repeated.

2.3. Data reduction

The water-to-water system was operated as if it was a heat pump, therefore the condenser capacity is the useful effect and the performance index is the *COP*.

The condenser capacity, $\dot{Q}_{COND}$, was determined as the average between the refrigerant-side value and the water-side value according to the following equation:

$$\dot{Q}_{COND} = \frac{1}{2}\left[\dot{m}_R\left(h_{R,IN,COND} - h_{R,OUT,COND}\right) + \dot{m}_W c_{P,W}\left(T_{W,OUT,COND} - T_{W,IN,COND}\right)\right] \tag{1}$$

REFPROP [9] software was used to calculate the water isobaric heating capacity, as function of water temperature, and the refrigerant enthalpies, as function of refrigerant pressure and temperature.

The coefficient of performance, *COP*, was calculated neglecting the power consumption of the two pumps and the power consumption of the EXV controller as follows:

$$COP = \frac{\dot{Q}_{COND}}{\dot{W}_{COMP}} \tag{2}$$

where, it is worth specifying, the compressor power, $\dot{W}_{COMP}$, accounts also for inverter losses.

Finally, the volumetric efficiency, η_{VOL}, of the compressor is defined as the ratio between the actual, $\dot{v}_{ACT}$, and the maximum theoretical, $\dot{v}_{THE}$, volume flow rate at compressor suction:

$$\eta_{VOL} = \frac{\dot{v}_{ACT}}{\dot{v}_{THE}} = \frac{\left(\dot{m}_R/\rho_{R,IN,COMP}\right)}{V_{COMP} f_{ROT,COMP}} \tag{3}$$

where the density of the refrigerant at suction pressure and temperature, $\rho_{R,IN,COMP}$, is again calculated using REFPROP [9] software.

2.4. Uncertainty analysis
The uncertainty of all the quantities, both directly measured and calculated, was estimated according to Moffat [10]. More in detail, the experimental uncertainty u_x of each directly measured variable x was calculated as follows:

$$u_x = \pm\sqrt{u^2_{x_{INST}} + (t_{95}\sigma_{\overline{x}})^2} \tag{4}$$

where $u_{x_{INST}}$ is the instrumental uncertainty of the directly measured variable, t_{95} is the Student test multiplier at 95% confidence level and $\sigma_{\overline{x}}$ is the standard deviation of the mean value of the collected samples.

Similarly, the uncertainty u_y of the generic calculated quantity y was estimated using the combined standard uncertainty under the uncorrelated input quantities assumption:

$$u_y = \pm\sqrt{\sum_{i=1}^{N}\left(\frac{\partial y}{\partial x_i}u_{x_{INST}}\right)^2 + t^2_{95}\sum_{i=1}^{N}\left(\frac{\partial y}{\partial x_i}\sigma_{\overline{x_i}}\right)^2} \tag{5}$$

being N the number of uncorrelated input quantities y depends on.

3. Results
The commissioning of the experimental set-up was carried out through a set of experimental tests. Tests were organized in 4 groups that were chosen to assess the influence of the water temperature at evaporator inlet, of the water temperature at condenser inlet and of the rotational frequency of compressor shaft on system performance. The test conditions used in the experimental campaign are reported in Table 3. In the same table, the bold values shown for each group represent the values of the experimental conditions used to find the evaporator and condenser water flow rates that lead to a temperature difference across them equal to 5 K.

Table 3. Testing conditions (bold values are the reference values).

Group No	$f_{ROT,COMP}$ [Hz]	$T_{W,IN,EVAP}$ [°C]	$T_{W,IN,COND}$ [°C]
1	50	10, 12.5, **15**, 17.5, 20	30
2	50	10, 12.5, **15**, 17.5, 20	40
3	50	10, 12.5, **15**, 17.5, 20	50
4	30, 40, **50**, 60, 70	10	30

All tests where run acting on the EXV so that the target superheating was 5 K in stable working conditions.

The condenser heating capacity and the system COP found under groups No 1, 2 and 3 testing conditions (see Table 3) are shown in Fig. 2 and in Fig. 3 respectively. For each water temperature at condenser inlet, the heating capacity shows an increasing trend with the water temperature at evaporator inlet while the heating capacity reduces as the water temperature at condenser inlet increases with the same water temperature at evaporator inlet. This trend is justified considering that the higher the water temperature at evaporator inlet, the higher the evaporating temperature and pressure which, in turn, leads to a higher refrigerant density at compressor inlet. When the suction density gets higher, the refrigerant mass flow rate increases and, consequently, the capacity increases.

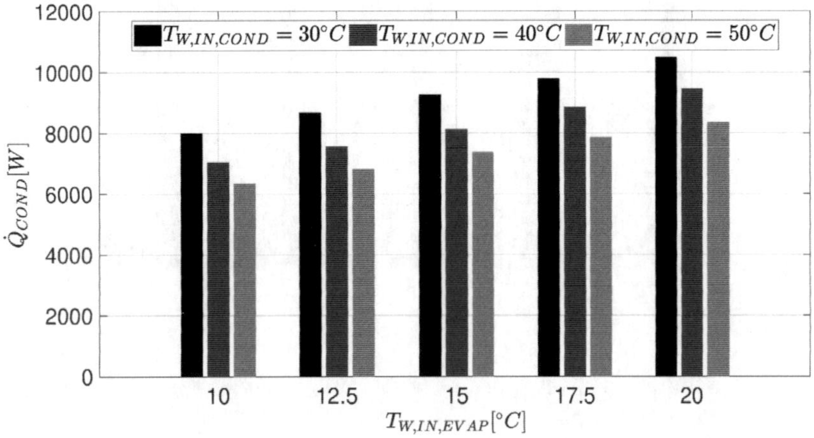

Figure 2. Condenser heating capacity as a function of evaporator water inlet temperature and for three different condenser water inlet temperature.

A similar trend is found for the COP, since it grows when the water temperature at evaporator inlet is increased (same condenser water inlet temperature) and it gets lower when the water temperature at condenser inlet increases (same evaporator water inlet temperature). This behaviour is explained remembering that the performance of any reverse cycle, the basis of a vapour compression system, is enhanced as the evaporating and the condensing temperatures get closer. As for the condenser heating capacity, when the water temperature at evaporator inlet gets higher, the evaporating temperature undergoes a strong increase, the condensing temperature a negligible one and the system performance improves. In the same way, the higher the water temperature at condenser inlet, the higher the condensing temperature, with slightly higher evaporating temperature, which leads to lower COP.

The influence of the rotational frequency of compressor shaft on condenser heating capacity and COP (group 4 in Table 3) is depicted in Fig. 4 and Fig 5 respectively. The trend of the condenser heating capacity comes from the increase of the refrigerant mass flow rate with the compressor shaft rotational frequency increase. On the other side, the COP reduces with the compressor shaft rotational frequency since the capacity increase, both at condenser and at the evaporator, under the same water inlet temperatures and flow rates leads to a lower evaporating temperature and a higher condensing temperature since the heat transfer area is constant. This, in turn, lowers the COP.

Finally, Fig. 6 shows the testing conditions inside compressor operating diagram whereas Fig. 7 reports the compressor volumetric efficiency experimentally found. From the analysis of these figures it is possible to state that the compressor was tested in a narrow range of evaporating temperatures, which is typical of water-to-water heat pump operating conditions, and that the compressor volumetric efficiency depends on the pressure ratio in a linear way, as widely discussed in technical literature [11].

4. Conclusions

In the present paper, an experimental facility for performance evaluation of pure and mixed refrigerant was presented. The facility allows to measure the most important parameters

IOP Conf. Series: Journal of Physics: Conf. Series **1224** (2019) 012021 doi:10.1088/1742-6596/1224/1/012021

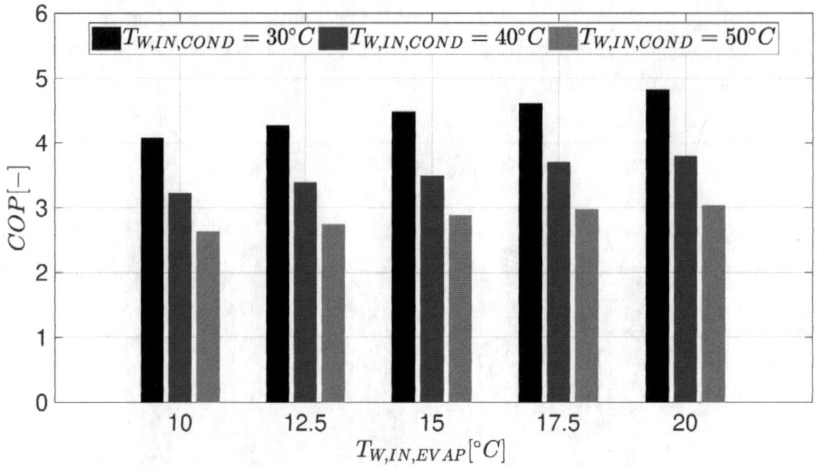

Figure 3. COP as a function of evaporator water inlet temperature and for three different condenser water inlet temperature.

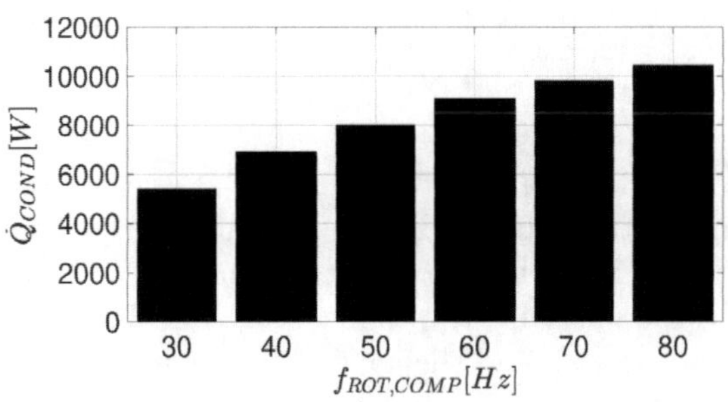

Figure 4. Condenser heating capacity vs compressor shaft rotational frequency.

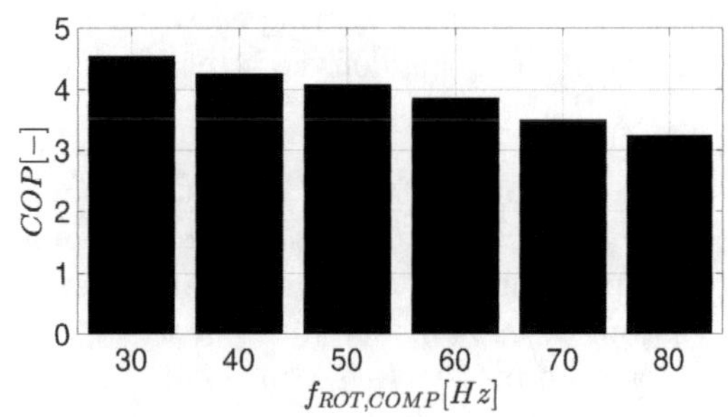

Figure 5. COP vs compressor shaft rotational frequency.

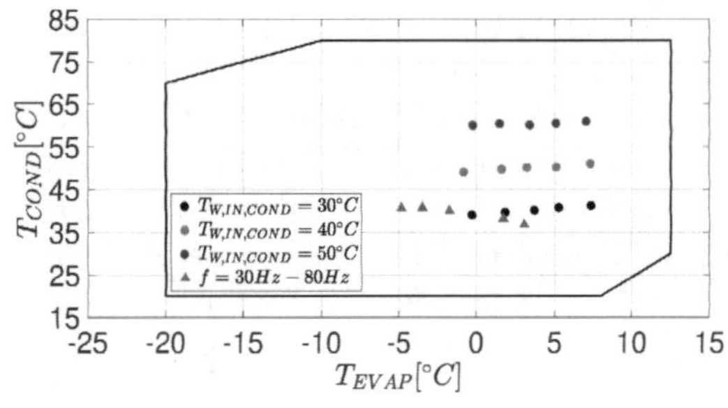

Figure 6. Representation of the testing point inside compressor operating diagram.

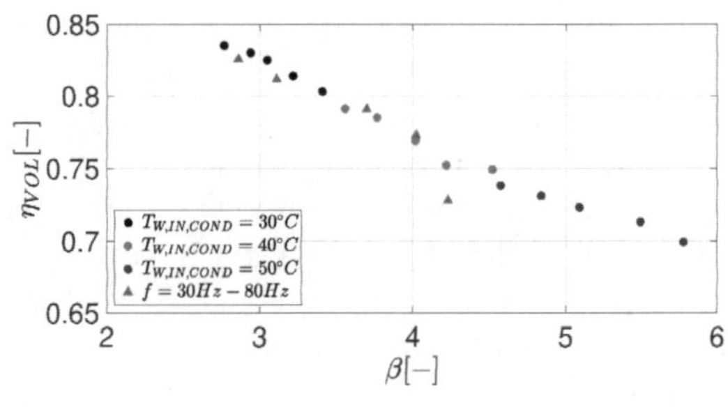

Figure 7. Compressor volumetric efficiency as a function of pressure ratio.

(pressures, temperatures, flow rates, power) for the performance characterisation of a vapour compression system under the range of operating conditions usually found in water-to-water heat pump or chillers. A variable speed reciprocating compressor is installed in the system to increase further the range of operating conditions.

First experimental campaign was aimed at introducing the most correct experimental procedure, reviewing data reduction and checking the accuracy of the results. First experimental findings show that the trends of condenser heating capacity, COP and compressor volumetric efficiency reliably agree with scientific and technical literature results.

Further studies will involve the analysis of system operation under different shaft rotational frequencies and the use of low GWP pure or mixed refrigerants.

Acknowledgments

The financial support of MIUR through the program PRIN 2015 (Grant Number 2015M8S2PA) is greatly acknowledged. Frascold company is acknowledged for providing the compressor used in the experimental campaign.

Nomenclature

c_P	Isobaric heating capacity [kJ·s^{-1}·K^{-1}]
f	Frequency [Hz]
h	Enthalpy [kJ·kg^{-1}]
$\dot{m}$	Mass flow rate [kg·s^{-1}]
$\dot{Q}$	Heat flow rate [W]
V	Volume [m^3]
$\dot{v}$	Volumetric flow rate [m^3·s^{-1}]
$\dot{W}$	Power [W]

Greek symbols

β	Pressure ratio [-]
η	Efficiency [-]
ρ	Density [kg·m^{-3}]

Subscripts

ACT	Actual
$COMP$	Compressor
$COND$	Condenser
IN	Inlet
OUT	Outlet
R	Refrigerant
THE	Theoretical
VOL	Volumetric
W	Water

References

[1] European Union, 2014 Regulation (EU) No 517/2014 of the European Parliament and the Council of 16 April 2014 on fluorinated greenhouse gases and repealing Regulation (EC) No 842/2006. *Off. J. Eur. Union* **150** 195-230.

[2] United Nations Environment Programme, 2016 Further Amendment of the Montreal Protocol.

[3] Mota-Babiloni A., Navarro-Esbri J., Barragán Á., Molés F., Peris B. 2014 Theoretical comparison of low GWP alternatives for different refrigeration configurations taking R404A as baseline *Int. J. Refrigeration* **44** 81-90.

[4] Makhnatch P, Mota-Babiloni A., Rogstam J., Khodabandeh R. 2017 Retrofit of lower GWP alternative R449A into an existing R404A indirect supermarket refrigeration system *Int. J. Refrigeration* **76** 184-192.

[5] Pitarch M., Navarro-Peris E., Gonzálvez-Maciá J., Corberán J.M. 2017 Experimental study of a subcritical heat pump booster for sanitary hot water production using a subcooler in order to enhance the efficiency of the system with a natural refrigerant (R290) *Int. J. Refrigeration* **73** 226-234.

[6] Sánchez D., Cabello R., Llopis R., Arauzo I., Catalán-Gil J., Torrella E., 2017 Energy performance evaluation of R1234yf, R1234ze(E), R600a, R290 and R152a as low-GWP R134a alternatives *Int. J. Refrigeration* **74** 269-282.

[7] Mota-Babiloni A., Makhnatch P., Khodabandeh R., Navarro-Esbri J., 2017 Experimental assessment of R134a and its lower GWP alternative R513A *Int. J. Refrigeration* **74** 682-688.

[8] Makhnatch P., Mota-Babiloni A., Khodabandeh R., 2017 Experimental study of R450A drop-in performance in an R134a small capacity refrigeration unit *Int. J. Refrigeration* **74** 682-688.

[9] Lemmon E.W., Huber M.L. and McLinden M.O. 2013 NIST Standard Reference Database 23: Reference Fluid Thermodynamic and Transport Properties-REFPROP, Version 9.1 *National Institute of Standards and Technology, Standard Reference Data Program* Gaithersburg, MD, USA.

[10] Moffat R.J. 1988 Describing the uncertainties in experimental results *Exp. Therm. Fluid Sci.* **1** 3-17.

[11] Wang S.K. 2000 Handbook of air conditioning and refrigeration *McGraw-Hill.*

IOP Conf. Series: Journal of Physics: Conf. Series **1224** (2019) 012022 doi:10.1088/1742-6596/1224/1/012022

A Mesoscale-Microscale approach for the energy analysis of buildings

S Montelpare[1], V D'Alessandro[2], C Lops[1], E Costanzo[2] and R Ricci[2]

[1] University "G. d'Annunzio" of Chieti-Pescara, Engineering and Geology Department (INGEO)
Viale Pindaro 42, 65127 Pescara (Italy)
[2] Marche Polytechnic University, Industrial Engineering and Mathematical Sciences Department (DIISM)
Via Brecce Bianche 1, 60131 Ancona (Italy)

E-mail: s.montelpare@unich.it

Abstract. This work is aimed to suggest a numerical approach able to select the most useful building orientation with respect to the local wind in complex urban areas. It is showed a mesoscale-microscale numerical approach able to predict local flow patterns for building designers. The city of Ancona was selected to analyse wind patterns over complex orography in presence of buildings. The analysed area is characterized by a densely built hilly promontory flanked by a poorly urbanized valley. Two buildings were chosen to carry out energy analyses. At first, the MM5 weather prediction model is used to assess the wind patterns and the wind occurrence distribution. The 2010 year was simulated and compared with real data from a met-mast. After, computational fluid dynamic (CFD) analyses are conducted with and without the surrounding buildings to appreciate speed and inflow effects. CFD results showed great changes in the examined building area. The final step was the energy simulation, by means of EnergyPlus, of two multistoried buildings equipped with a double skin façade; results show a great impact on the energy consumption by comparing a wrong orientation with respect to the best solution derived from the mesoscale-microscale approach.

1. Introduction

Nowadays, "sustainable" is the new keyword of progress and the low environmental impact of buildings is a must. The basic principles of sustainable design integrate the traditional criteria of functionality, cost containment and aesthetic results, with environmental factors, health and well-being of building occupants. In this background, surveys aimed at defining the typical characteristics of the climate in a building site and the study of the interaction between the wind resources and the urban landscape assumed an increasing interest. Geographic position, terrain orography, meteorological parameters, play an important role for the wind resource of a site, but a direct survey is not always feasible. More and more frequently, the techniques to obtain this information involve a numerical approach and the use of scale models in wind tunnel; in both this techniques there is a common fundamental data, how to recreate the ABL and the ASL. Toparlar et al. [1] provides a review of research reported in journal publications on this topic till the end of 2015, by categorizing them on the morphology of the urban area (generic versus real) and on the methodology (with or without a validation procedure). They observe that CFD studies for generic urban areas typically involve simple building shapes, such as cubes or rectangular prisms, and do not consider validation; when present, this is normally carried out with data from wind-tunnel measurements. CFD studies on real urban areas are normally realized without validation and they are comparative studies of several urban configurations and design parameters; almost all these studies are conducted with steady Reynolds Averaged Navier-Stokes (RANS) approaches. Blocken [2] reviews the importance of urban physics related to the societal challenges and provides some views on CFD: possibilities and limitations are discussed, and tips and tricks for accurate and reliable CFD simulations are illustrated. He underlines that the flow in an atmospheric boundary layer (ABL) is inherently unsteady and therefore, excluding (LES) because of the high computational efforts, an unsteady RANS approach should be required to resolve the unsteady mean-flow structures, while modelling the turbulence. Several studies [3, 4] analyse the impact of computational parameters on the numerical outcome for wind flow pattern. Other studies [5] analyse the coupling of mesoscale meteorological models (MMM) with computational wind engineering (CWE) models to produce realistic simulations; they focus the attention on up-scaling or downscaling methodologies [6] and on the effects related to energy buildings issues [7, 8, 9, 10, 11, 12, 13]. Some authors have focused their attention to the convective heat transfer coefficients in urban neighborhoods [14, 15, 16, 17], others on the air temperature in complex urban areas [18, 19] or on the Urban Heat Island (UHI) [20, 21].

This paper is aimed to evaluate the feasibility and the results of a holistic approach, based on a meso/micro scale numerical approach, which starts from a meteorological hindcast and finishes in a building energy simulation (BES) analysis by employing a CFD instrument to connect the different geometric scales. First of all, the paper checks the validity of the meteorological hindcasting simulation, then evaluate the effect of the local urbanization and finally quantifies the energy demand differences between a generic orientation and an orientation optimized by the proposed approach. In particular, the paper is divided into the following sections. Section 2 describes the case study. Section 3 illustrates the mesoscale approach, the adopted model explaining the settings and presenting the results obtained. Section 4 describes the microscale CFD approach, the computational domain, the numerical model and its outputs. Section 5 illustrates the energy analysis approach and the corresponding results. Finally, Section 6 summarizes the main achievements of the paper.

2. The Case Study

The methodology proposed in this paper is applied to a couple of buildings located in the city of Ancona and precisely in an urban expansion area characterized by a densely built hilly promontory flanked by a poorly urbanized valley (Fig.1). Both Buildings present a multy storey (Fig.2) double skin facade (DSF) whose performance is deeply influenced by the wind speed and direction and, for this reason, they are particularly suitable for the present work.

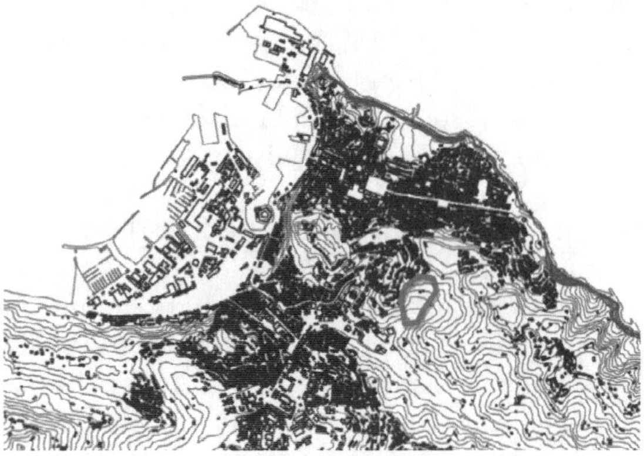

Figure 1: The analyzed area.

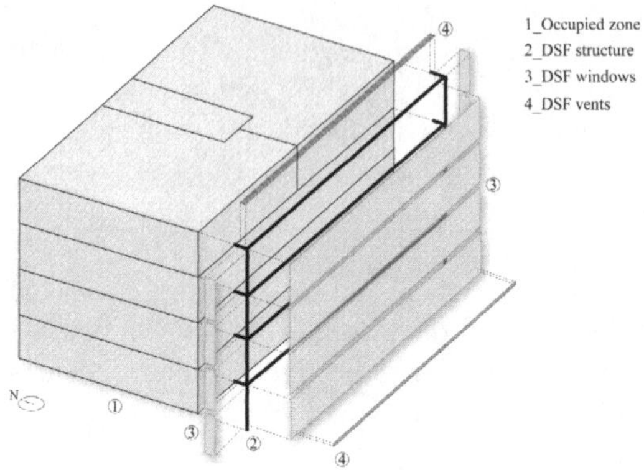

Figure 2: Simulated building.

3. The Mesoscale Approach

A mesoscale analysis was carried out in order to describe the climatology of the area of interest and in particular the wind assessment. A hindcasting procedure over a complete year (01/01/2010 - 31/12/2010) was performed adopting the version of a largely used meteorological code, properly modified by our research group.

3.1. The Meteorological Model

MM5v3, the PSU/NCAR Mesoscale Model Fifth Generation (MM5), is *a limited-area, nonhydrostatic, terrain-following sigma-coordinate model designed to simulate or predict mesoscale atmospheric circulation.* The MM5v3 official version is able to ingest digital terrain models and terrain land uses with a resolution of 30sec; the version used in this paper is a modified in order to be able to work also with the NASA Shuttle Radar Topography Mission (SRTM) 3sec Digital Terrain Elevation Model (DTEM) and with the high definition Corine European Land Cover database. These new skills were developed by our research group by modifying the FORTRAN code of the TERRAIN module; in this way it is possible to carry out simulations with a horizontal grid spacing down to 200 [m], describing with more accuracy complex orographies like the area of *Monte Conero* in Ancona (the object of this work).

3.2. The Numerical Set-up

The adopted operative procedure involves the use of a 2-way nesting approach with 5 domains, having the same center (longitude 13.519827°E, latitude 43.605847°N) and the same number of grid points (961), but with an increasing grid spacing; in particular, a growth ratio of 3 and an Arakawa-Lamb B-staggering (Fig.3b). Each domain extends in the vertical direction by 27 pressure sigma levels (Fig.3a), starting from 10 $[m]$ up to 10532 $[m]$ above surface level. Details about the domain settings are reported in Table 1:

Table 1: Domain settings table.

Name	Grid Spacing	Grid Points	Domain Extension
Domain 1	0.2 [km]	31x31	6x6 $[km^2]$
Domain 2	0.6 [km]	31x31	18x18 $[km^2]$
Domain 3	1.8 [km]	31x31	54x54 $[km^2]$
Domain 4	5.4 [km]	31x31	162x162 $[km^2]$
Domain 5	16.2 [km]	31x31	486x486 $[km^2]$

The use of a relatively high number of nested domain allows to take into account meteorological phenomena ranging from the synoptic scale down to the local one. The domain 1 (Fig.5) covers a large part of the Italian topography and also the Adriatic basin with the Croatian coasts; at this synoptic scale it is possible to evaluate the main perturbation phenomena that occur with weekly or monthly frequency (Fig.4). Domain 5 (Fig.6) covers, instead, the region nearby the area of interest and describes the terrain slopes with more accuracy; in this way it is possible to better describe local convective phenomena like sea or hill breezes that occur with 12h or 24h frequencies.

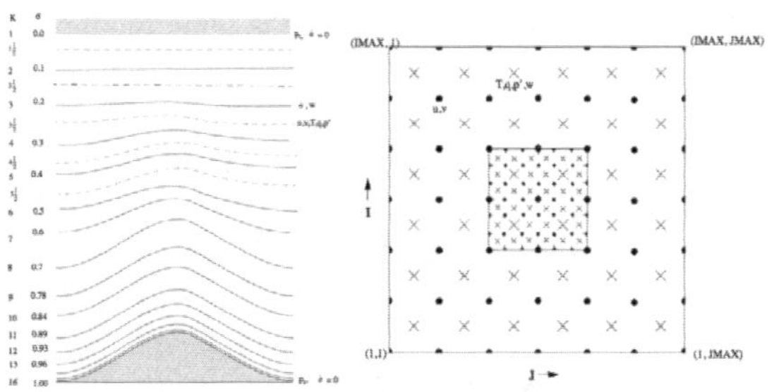

Figure 3: MM5 Pressure Sigma Levels (a) and Arakawa B Nesting Schema (b).

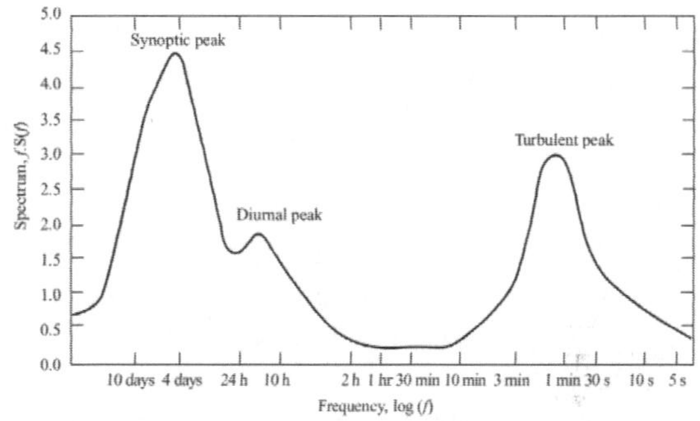

Figure 4: Wind Spectrum Farm Brookhaven based on work by Van der Hoven (1957).

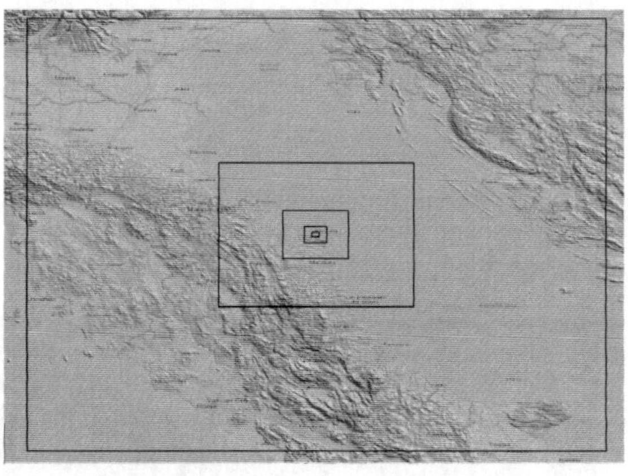

Figure 5: Domain nesting.

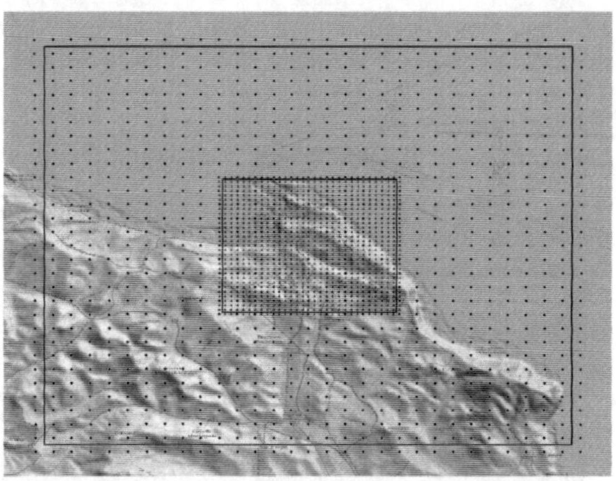

Figure 6: Domain 4 vs Domain 5.

The adopted meteorological input is the NCEP ds083.2 datasets, belonging to the FNL Operational Model Global Tropospheric Analyses; these data are available on 1-degree by 1-degree grids with a 6hs-time interval and they are from the Global Data Assimilation System (GDAS), which continuously collects observational data from the Global Telecommunications System (GTS) and other sources (as meteorological stations, met-balloons, met-tower, ships etc..) from all around the world. These inputs are interpolated on the 31x31 outer domain horizontal grid by the REGRID module and on 27 pressure full sigma levels by the INTERPF module. Finally, the MM5 module resolves a complete set of equations (pressure, momentum, thermodynamics, moisture etc.) to output a four dimensional database of climatic variables (air pressure, air and soil temperature, wind speed and direction, humidity, radiation etc.) with a 4 mins-time step and a 200 $[m]$ horizontal resolution.

The physics options adopted to solve the simulation domains are:

Table 2: MM5 Physics Options.

Physics Option	Domain 1	Domain 2	Domain 3	Domain 4	Domain 5
$IMPHYS\_{MoistureScheme}$	Simple Ice	Simple Ice	Simple Ice	Simple Ice	Simple Ice
$ICUPA\_{CumulusScheme}$	KF2	KF2	None	None	None
$IBLTYP\_{PlanetBound.Layer}$	MRF	MRF	MRF	MRF	MRF
$FRAD\_{Atm.Radiation}$	RRTM	RRTM	RRTM	RRTM	RRTM
$ISOIL\_{SoilTemp.Model}$	Noah LSM	Noah LSM	Noah LSM	Noah LSM	Noah LSM

The MM5 outputs are post-processed to obtain wind and solar maps, windroses and time histories of the main climatic parameters: i.e. planet boundary layer height, vapour mixing ratio, wind speed components, wind direction, ground temperature, air temperature, surface downward shortwave radiation, surface downward longwave radiation, pressure, humidity.

Time histories are rearranged in form of virtual meteorological masts having a measurement plane for every sigma level; for the purpose of the present research, only the lowest 8 heights were extracted, corresponding to $10, 30, 51, 71, 92, 122, 163, 204, 255, 321, 407$ and $509[m]$.

3.3. The Mesoscale Results

Figures 8 and 7 report some windroses extracted at the same positions but different heights above the surface layer: i.e. $10[m]$ and $509[m]$. Observing the figure 8 it is possible to note that the wind intensities distribution is the same while the occurrences are slightly different only for the western sectors. Figure 7 shows, instead, marked differences both in terms of speed intensities and prevalent directions: the points over the sea have lower velocities, as expected for their lower height, but the same behaviour, while others located around the city exhibit stronger velocity decrease and also marked inflows due to the terrain conformation and the land use. The presence of a steep hills in the southern part of the city influences the wind distribution by aligning the flow along the main axis.

Figure 7: Annual wind roses @ 10[m] asl.

Figure 8: Annual wind roses @ 509[m] asl.

These results corroborate the paper initial hypothesis that a meso-micro approach could better support the energy analysis in complex terrain. In fact, the use of the equation 1 in

BESs, as suggested by Ashrae 2001, can not completely describe the local modification of the meteorological mast. This equation converts the meteorological wind speed from the EnergyPlus Weather (EPW) file to the local wind speed by using the α and δ values in correspondence of the meteorological station and building location. These values, reported in table 3, take into account the effects that the land use induces on the height of an atmospheric surface layer and on the wind speed profile, but they do not consider orographic speed-up and inflow effects.

$$U_{loc} = U_{met} \left(\frac{\delta_{met}}{z_{met}} \right)^{\alpha_{met}} \left(\frac{z}{\delta} \right)^{\alpha} \tag{1}$$

Table 3: Terrain dependent coefficients (ASHRAE 2001).

Terrain	Description	Exponent α	Layer Thickness $\delta[m]$
1	Flat, open country	0.14	270
2	Rough, wooded country	0.22	370
3	Towns and cities	0.33	460
4	Ocean	0.10	210
5	Urban, industrial, forest	0.22	370

The equation 1 derives from equation 2 considering 2 different assumptions: the gradient wind, the wind speed at the atmospheric boundary layer (ABL) top, is the same over the meteorological station and over the local building, the boundary layer height assumes the values reported in the table 3 without transition mixing zones. The former assumption is quite acceptable while the latter is more difficult to satisfy in case of different land uses in relatively small areas.

$$U_1 = U_2 \cdot \left(\frac{z_2}{z_1} \right)^{\alpha} \tag{2}$$

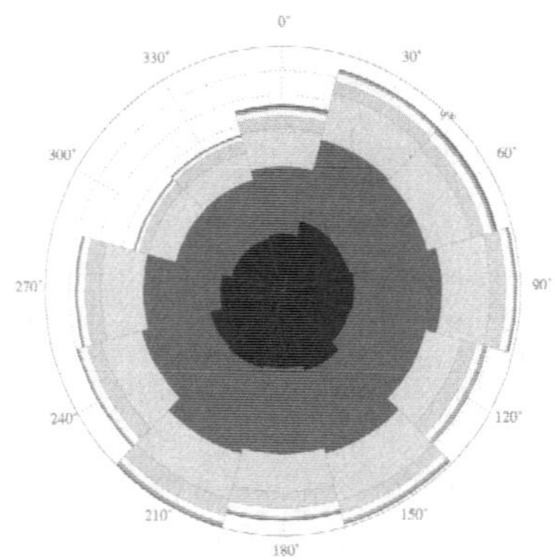

Figure 9: EnergyPlus Ancona EPW windrose

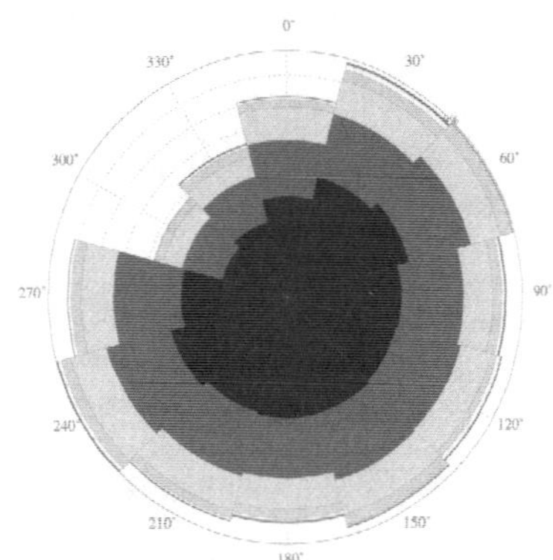

Figure 10: EnergyPlus Ancona-Falconara EPW windrose

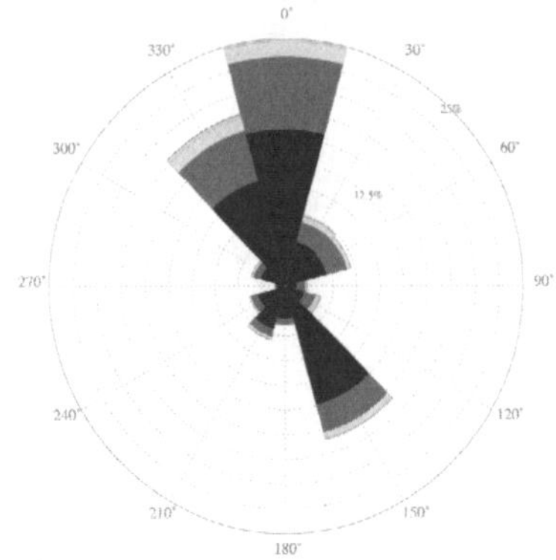

Figure 11: SIRMIP Ancona 2010 windrose

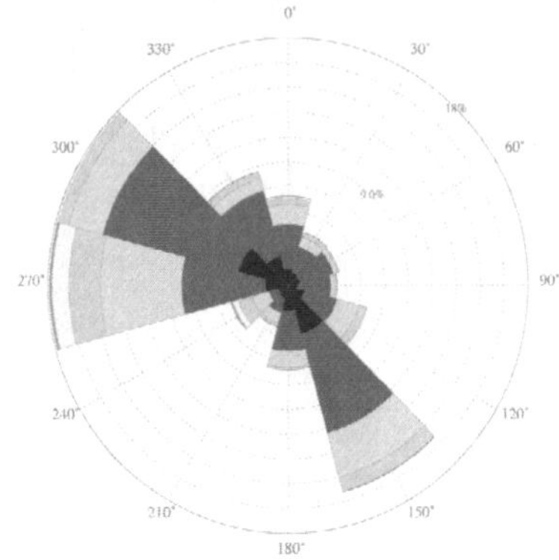

Figure 12: microMM5 Ancona 2010 windrose

Figures 9-12 show the annual windroses for: the available EnergyPlus Weather files (i.e. Ancona and Ancona-Falconara), the SIRMIP meteorological mast (installed on the roof of the building of Marche Region) and a microMM5 virtual mast extracted in the same coordinates of the real one. The SIRMIP met-mast is the closest to the area of study. Both the EnergyPlus windroses show very similar directional occurrences but exhibit different wind speeds: globally it is possible to note a large spread of direction occurrences. Contrarily, the SIRMIP met-mast and the microMM5 virtual mast exhibit more defined directions and show a different rotation of 60° only for the northern occurrences: this difference could be related to the SIRMIP positioning, because it is installed on the top of a large and tall building that locally modifies the flow pattern and the microMM5 code is not able to describe this local phenomenon. However, it is possible to say that in general the mesoscale approach is able to better describe the annual measurements of the closest real anemometer.

4. The CFD Approach

CFD analysis was carried out on a large computational domain to include the speed-up and inflow effects generated by the nearby complex orography. In addition, the presence of buildings in the surrounding districts was evaluated in order to estimate their influence on the observed area.

4.1. The Computational Domain

The first step of the microscale approach was the realization of the computational domain and the choice of the mesh sizes. The primary aspect to be considered is the digital terrain model because the coastal area of Ancona is characterized by high cliffs and the only use of a detailed cartography (1:200 CTR Charts) can correctly reproduce this complexity. A custom Matlab procedure was developed to extract the intersection points among each contour line and the radial lines starting from the domain center and spanning all the circumference at every 2°; in this way, a cloud of about 110k points was obtained to generate the mesh. Figure 13 reports these points from isocontour at $10[m]$ and the high density of points of the coastline is clearly visible. The cloud of points was then shifted to have the lower elevation corresponding to a $0[m]$ value and, subsequently, a smoothing procedure, using a bell function, was performed to obtain a flat terrain all around the external part of the simulation domain (Fig.14); these changes have been made to avoid discontinuity in the incoming inlet wind velocity profile. The obtained computational domain is a cylindrical volume having a circular plant with a radius of 6 $[km]$; it is subdivided into three concentric annular zones centered on the building position analysed [2]. The outer annulus, deformed by the bell function, is considered as smooth, the central one is described with an equivalent roughness, corresponding to the local land use, and the inner circle takes in account the real buildings. The height of the cylinder ($750[m]$) was selected in order to have a blockage ratio lower than or equal to 3% for the most disadvantaged section [2, 22]; i.e. the cross plane having the highest terrain profile. The cylinder external surface was subdivided into 12 sectors to describe all the wind directions reported in the windroses previously showed. Considering that the aim of this paper was to evaluate the proposed multiscale computational approach, the only results referred to the western winds are here reported.

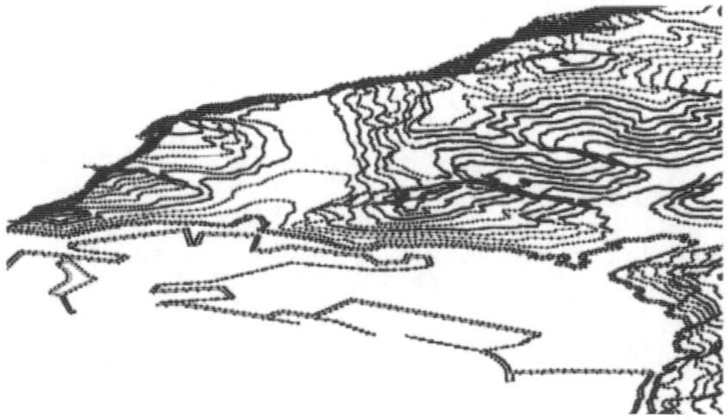

Figure 13: Isocontour Ancona.

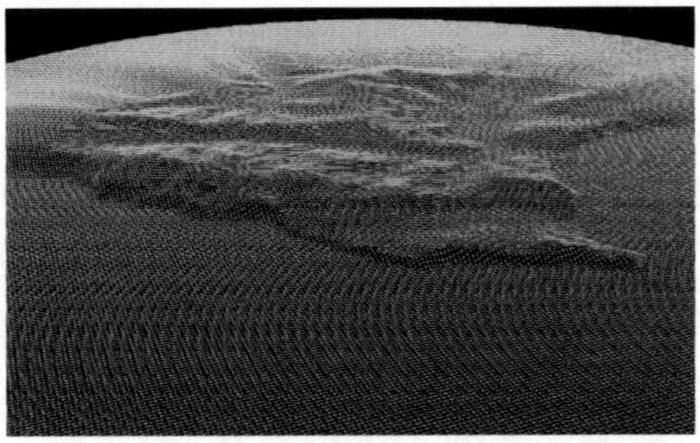

Figure 14: CFD terrain mesh.

Two grids were created: the former, with only the test case buildings, uses a refinement volume around the area of interest and has 2118000 cells (Fig.15). The latter, with all the other buildings present in the surrounding districts, has 1458811 cells and involves the use of two different roughness length to change the height of the first cell at wall (Fig.16). This double model was used to evaluate the flow perturbations caused by the existing surrounding built.

Figure 15: CFD computational domain.

Figure 16: Mesh detail around buildings.

4.2. The Boundary Conditions

The inlet boundary condition was set as an atmospheric flow, following the exponential wind profile described in the equation 2 with an α value of 0.2. This condition, by fixing the $x - y$ direction of the incoming wind, was applied for 6 sectors of the cylinder external surface, while in the other 6 a pressure inlet outlet condition was settled. The domain upper surface was defined with a no-slip condition, while the base of the outer annulus was settled with a slip condition on the smooth surface. The intermediate annulus and the central area were settled as rough surfaces. The numerical model does not use the aerodynamic roughness z_0 typically implemented in the von Karman law (eq.3) but the equivalent Nikuradse roughness K_s. The wall law used in the numerical model is reported in equation 4 where the E value is about 9 and the ΔB value is a constant depending on the dimensionless equivalent roughness K_s^+ (eq.5). The

IOP Conf. Series: Journal of Physics: Conf. Series **1224** (2019) 012022 doi:10.1088/1742-6596/1224/1/012022

ΔB value represents a vertical translation of the logarithmic wind profile due to the presence of the roughness. In the present study, a fully rough regime having K_s^+ greater than 90 could be assumed; with the present assumption, the Nikuradse equivalent roughness could be evaluated with the equation 6, where the constant $C_s = 0.5$ and $E \approx 9$.

$$U(z) = \frac{u^*}{k} \cdot \ln\left(\frac{z}{z_0}\right) \tag{3}$$

$$u_p = \frac{u^*}{k} \ln\left(E\frac{\rho u^* z_p}{\mu}\right) - \Delta B \tag{4}$$

$$K_s^+ = \frac{\rho K_s u^*}{\mu} \tag{5}$$

$$K_s = \frac{z_0 E}{C_s} \tag{6}$$

Finally, the values of the Nikuradse equivalent roughness assumed for the computational domain are reported in table 4:

Table 4: Domain settings table.

Domain Surface	K_s
Sea	0.19586
Test area terrain	0.2
Intermediate annulus terrain	3.52

The least boundary condition is referred to the building surfaces, that are considered smooth and with a slip condition.

4.3. The Numerical Model

Unsteady simulations were carried out by using a Reynolds Stress (RSM) turbulence model; preliminary analyses on 2D (Almeida Hill) and 3D (Askervein Hill) test cases showed that two equation models like $k - \epsilon$ and $k - \omega$ are not able to accurately describe flow separations. The seven equation Reynolds Stress Model has instead demonstrated to be able to simulate complex flows by using transport equations for the six independent Reynolds stresses in addition to the equation for the turbulent dissipation; the counterpart is a higher computational effort.

4.4. The CFD Results

CFD simulations showed a significant effect of the nearby buildings on the wind speed distribution in the proximity of the test buildings and also a great effect on the wind directions with marked deviations of the inflow angle. Figure 17 shows the points where a vertical profile of the mentioned variables were extracted. Results are presented both in terms of vertical profiles and as visual maps.

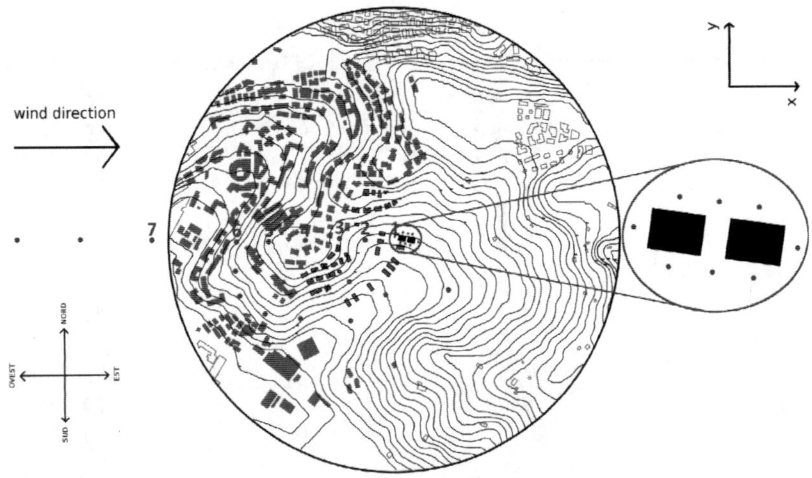

Figure 17: CFD extraction points.

As figures 18 and 19 show, the local terrain conformation and the presence of nearby buildings, even if at distance greater than $100[m]$, are able to strongly influence the wind speed distribution up to about $80[m]$ above the surface layer. It should be considered that the buildings located inside the area of study are typically three or five stories and with height below $15[m]$.

By observing the figure 17 it is possible to note that the point corresponding to the terrain maximum height is the number 4 and that the western wind encounters the test building downstream a hill, in a zone of pressure recovery.

The absence of nearby buildings better explains the effects directly related to the terrain local conformation: it is possible to see a marked speed-up from the point 7 up to the top of the hill (points 4 and 3), while the pressure recovery downstream the summit justifies the velocity decrease. The graph of the wind inflow deviation shows a significant deviation for the points upstream the hill, while the points near the area of study do not exhibit meaningful flow deflections. Globally, it is possible to say that the orographic effects induce, in the test buildings area and for western wind, appreciable effects on the wind speed but not on the direction.

Figure 19 shows very different behaviour with respect to the previous case: excluding the point 7, that is the region where buildings are simulated as an equivalent roughness, and the point 5, that is in an area with few buildings, all the other points exhibit irregular wind profiles with points 1 and 6 that have a constant velocity for the lower height, underlying the presence of flow separation areas. The point just upstream the test buildings (point 1) presents a velocity decrease, at a height of $12[m]$, of about 50%. The graph of the inflow deviation shows also great effects due to the presence of nearby buildings. The point 6 and the test building point (n^o1) have very opposite behaviour with angles of deviation up to $25°$ both positive and negative. Globally, it is possible to say that the presence of nearby buildings is a main parameter to take into account for evaluating the incoming wind speed profile to a specific area of interest: both the wind velocity and direction are strongly modified.

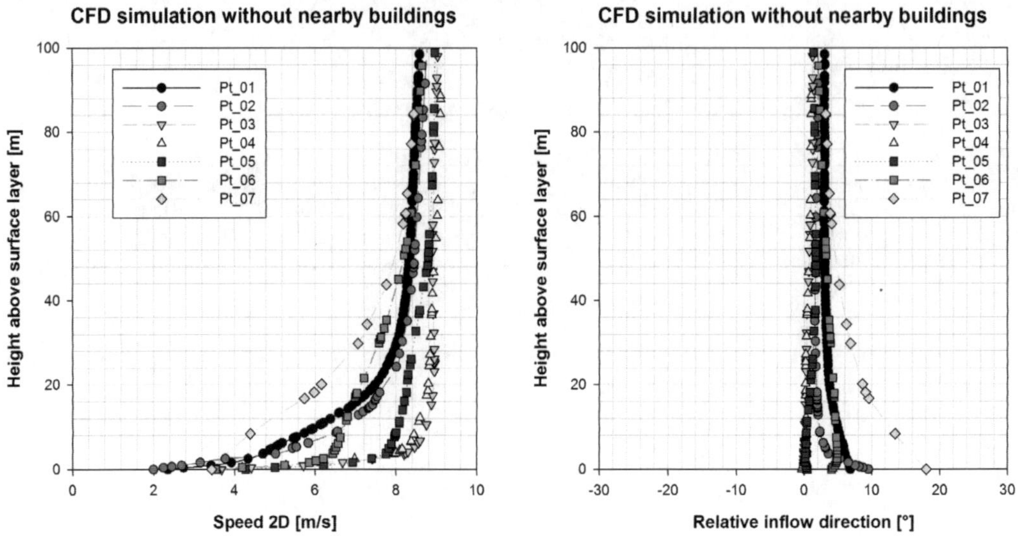

Figure 18: Wind Speed and Inflow deviation without nearby buildings.

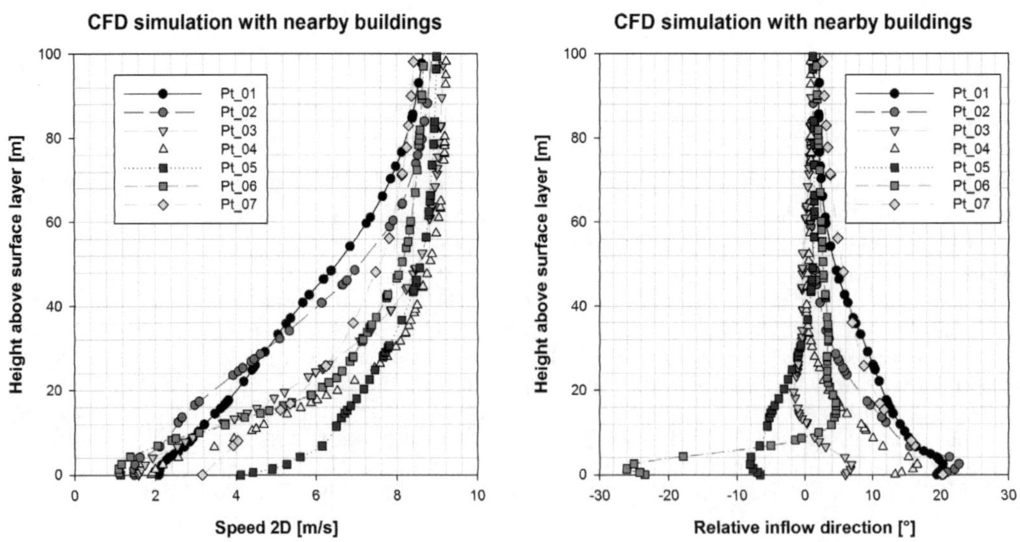

Figure 19: Wind Speed and Inflow deviation with nearby buildings.

5. The Building Energy Analysis

Dynamic energetic simulations have been carried out in order to estimate the reduction in terms of building cooling requirement and define the variation of the DSF performance, considering the main wind directions.

5.1. The Model Set-up

The case study is a residential building, dating from the second Italian postwar, and made of reinforced concrete. It has a rectangular shape with sizes equal to 22 x 11.5 metres, with four levels above ground for a total high of 12.2 metres. Its main axis direction is perfectly towards the East-West directrix. Table 5 summarizes the transmittance (K) values of the building envelope.

Table 5: K values for building components.

Element	$s[m]$	$K[\frac{W}{m^2K}]$
External wall	0.30	1.13
Floor	0.08	3.00
Roof	0.25	1.40
Partition	0.15	1.60
Window	0.01	3.16

On the South elevation, a naturally ventilated and multi storey double skin façade is present. It means that the cavity between the inner and the outer layers of the DSF has inlet and outlet vents that allow the natural ventilation inside the enclosed space. The cavity is 1 $[m]$ depth and extends over the total height of the building, without any separation. The external skin is made of single glazed windows with K value equal to 5.80 and, in order to avoid the overheating of the air volume in the cavity, it has an integrated shading system composed of external blinds with medium reflectivity slats and solar control type fixed at $120[W \cdot m^{-2}]$. The inlet and the outlet vents are located respectively on the bottom and the top of the double skin façade. Internal grills are placed on the inner layer of the DSF, ensuring an adequate ventilation of the occupied zone. These grills, both the internal and external ones, allow continually the air to enter the cavity and the whole building during the hottest period and they are closed in the winter time when the cavity becomes sealed and its ventilation is not allowed in order to decrease the transmittance of the façade and reduce the heat losses. The infiltration rate of the building is $0.7[m^3 \cdot s^{-1}]$ and the control mode for the calculated natural ventilation, both for the cavity and the building, is constant. This means that whenever an opening's operation schedule allows venting, all of the zone's openable windows and doors are open. The air temperature distribution model for each zone is mixed and it is considered uniform through the zone. The DSF cavity, at the zone level, is set as unoccupied by loading, heating/cooling and lighting template data. The plant system of the building involves the presence of natural gas boilers with an efficiency around 85%, for the heating need, and split system with a coefficient of performance equal to 1.8 for the summer period. The case study has been analysed through dynamic analysis in Energy Plus, considering as simulation period the typical summer week (19 – 25 August) that is identified by the weather data translator as being typical of the summer.

5.2. The Building Energy Simulation Results

The building energy simulations are performed in Energy Plus, using its graphical interface DesignBuilder. Table 6 and Figures 20 and 21 show respectively the impact that different wind directions have on the performance of a building with a double skin façade in terms of cooling requirement and air flow rate in the DSF cavity. The simulations consider as main wind directions two cases: the first case analyses the possibility of a wind which impacts directly on the DSF, coming from the South (180°), and the second one studies the opposite condition, with a Northern wind (0°). According to the energetic simulations, the Southern wind, impacting directly on the DSF, can reduce the building cooling requirement of the 10% as it is shown in Table 6.

Table 6: Building cooling requirement with different wind directions.

Wind direction	Cooling requirement $[kWh[$
South Wind	605
North Wind	662

In addition, the air flux that enters and goes out from the vents, located respectively at the bottom and the top of the cavity of the double skin façade, is higher with a 180°wind direction and the difference between the two configurations reaches $2[m^3 \cdot s^{-1}]$ during specific daily hours.

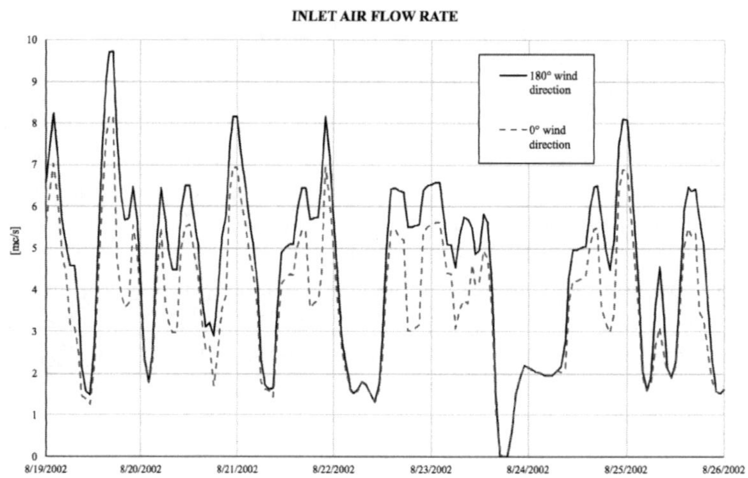

Figure 20: Inlet air flow rate.

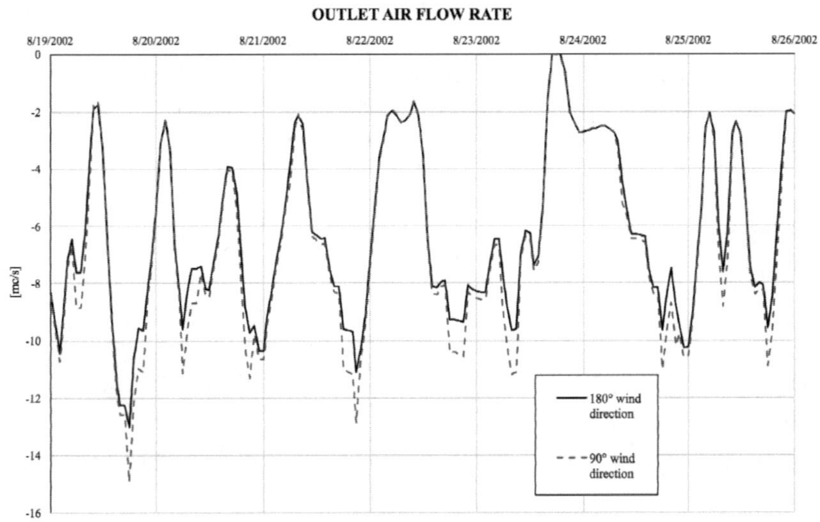

Figure 21: Outlet air flow rate.

The dynamic energy simulations highlight how the most efficient condition for a building with a double skin façade is reached when the wind impacts directly on the grills of the DSF cavity, ensuring the highest amount of natural ventilation, both in the cavity and in the whole building, and involving a consequent reduction in terms of energy consumption for the cooling need of the occupied zones. For a more accurate estimation of the natural ventilation inside the DSF cavity related to different wind directions, a coupling approach between energy simulation and computational fluid dynamic analysis should be adopted. As suggested in [23] and [24], in fact, the energy simulation programs assume that the air in an indoor space is uniform and the convective heat transfer coefficients available for the analyses are usually empirical and not accurate. Therefore, coupling the energy simulation with the CFD analysis is a good option to overcome these limits in order to obtain more realistic and reliable results.

6. Conclusions

This paper was aimed to evaluate the feasibility and the results of a holistic approach able to improve the energy analysis of buildings in complex urban environments. The city of Ancona was selected as the area of study because it shows the combination of a complex orography with high density urban areas; in particular, two buildings, equipped with a double skin façade, were identified as the object of the energy analyses. The first step of the proposed approach is a hindcasting meteorological analysis by means of the MM5v3 NWP model; this code was modified to ingest a high definition digital terrain model (the SRTM 3s DTM) and also a high resolution land use database (the European CORINE). A nesting procedure, with five domains, allowed to obtain information from synoptic scales down to local ones. Results show a strong influence of the local orography with marked differences among the extracted numerical windroses. The comparison between the MM5 virtual meteorological mast and the real one installed on the top of the "Marche Region" building show a very good agreement underlining the reliability of the proposed hindcasting procedure for local climates. The second step was the simulation, with a CFD code, of the urban context nearby the analysed buildings. A computational domain with a blockage lesser than 3% was defined and it was divided into three concentric subdomains having different roughness descriptions: the inner is centred on the two analysed buildings and considers also the nearby district buildings. The results of the CFD analysis allowed to obtain both the speed-up values and the inflow deviations of the wind approaching the test buildings. There are significant differences among the extracted points and between the simulations with and without the surrounding buildings: this underlines the necessity to correctly simulate the nearby urbanization in order to take into account local modification of the wind resource. The third step was an energy evaluation, by the DesignBuilder software, of the effects that a correct wind-building interaction may induce. A difference of about 10% on the cooling demand was obtained for the typical summer week, underlining the importance of a correct orientation and of a right local wind assessment. Synthetically the proposed procedure could be summarized in the subsequent step:

- A hindcasting simulation with a modified NWP code to obtain numerical met masts.
- A CFD simulation of the nearby urban area to estimate local speed-up and inflow values able to locally modify the numerical met mast.
- A BES analysis to evaluate the effect of natural ventilation on the cooling requirements.

Nomenclature

k	Von Karman constant
K_s^+	Dimensionless equivalent roughness
α	Wind power law exponent
μ	Dynamic viscosity $[\frac{m^2}{s}]$
ρ	Fluid density $[\frac{kg}{m^3 s}]$
K	Thermal transmittance $[\frac{W}{m^2 K}]$
K_s	Equivalent roughness $[m]$
U	Wind speed $[m/s]$
u^*	Friction velocity $[m/s]$
u_p	Wind speed at the center of the first wall cell $[m/s]$
z	Height above ground $[m]$
z_0	Roughness lenght $[m]$
z_p	Height of the center of the first wall cell $[m/s]$

References

[1] Toparlar Y, Blocken B, Maiheu B and van Heijst G 2017 *Renewable Sustainable Energy Rev.* **80** 1613–1640 ISSN 13640321
[2] Blocken B 2015 *Build. Environ.* **91** 219–245 ISSN 03601323
[3] Dhunny A, Samkhaniani N, Lollchund M and Rughooputh S 2018 *Urban Clim.* **24** 185–204 ISSN 22120955
[4] Liu S, Pan W, Zhang H, Cheng X, Long Z and Chen Q 2017 *Build. Environ.* **117** 11–23 ISSN 03601323
[5] Yamada T and Koike K 2011 *J. Wind Eng. Ind. Aerodyn.* **99** 199–216 ISSN 01676105
[6] Tewari M, Kusaka H, Chen F, Coirier W J, Kim S, Wyszogrodzki A A and Warner T T 2010 *Atmos. Res.* **96** 656–664 ISSN 01698095
[7] Srebric J, Heidarinejad M and Liu J 2015 *Build. Environ.* **91** 246–262 ISSN 03601323
[8] Allegrini J, Dorer V and Carmeliet J 2015 *Sustainable Cities and Society* **19** 385–394 ISSN 22106707
[9] Gracik S, Heidarinejad M, Liu J and Srebric J 2015 *Build. Environ.* **90** 15–29 ISSN 03601323
[10] Khoshdel Nikkho S, Heidarinejad M, Liu J and Srebric J 2017 *Appl. Therm. Eng.* **116** 850–865 ISSN 13594311
[11] Mauree D, Perera A and Scartezzini J L 2017 *Energy Procedia* **142** 2648–2654 ISSN 18766102
[12] Eleftheriadou A, Sfetsos A and Gounaris N 2016 *Energy Build.* **111** 176–183 ISSN 03787788
[13] Huang K T and Li Y J 2017 *Energy Build.* **154** 448–464 ISSN 03787788
[14] Liu J, Heidarinejad M, Gracik S and Srebric J 2015 *Energy Build.* **86** 449–463 ISSN 03787788
[15] Defraeye T, Blocken B and Carmeliet J 2011 *Energy Convers. Manage.* **52** 512–522 ISSN 01968904
[16] Montazeri H and Blocken B 2017 *Build. Environ.* **119** 153–168 ISSN 03601323
[17] Liu J, Srebric J and Yu N 2013 *Int. J. Heat Mass Transfer* **61** 209–225 ISSN 00179310
[18] Toparlar Y, Blocken B, Vos P, van Heijst G, Janssen W, van Hooff T, Montazeri H and Timmermans H 2015 *Build. Environ.* **83** 79–90 ISSN 03601323
[19] Allegrini J and Carmeliet J 2017 *Urban Clim.* **21** 278–305 ISSN 22120955
[20] Göndöcs J, Breuer H, Pongrácz R and Bartholy J 2017 *Urban Clim.* **21** 66–86 ISSN 22120955

[21] Allegrini J and Carmeliet J 2018 *Urban Clim.* **24** 340–359 ISSN 22120955
[22] Franke J, Hellsten A, Schlünzen H and Carissimo B 2007 *COST Action 732 - Best practice guideline for the cfd simulation of flows in the urban environment* ISBN 3-00-018312-4
[23] Zhai Z, Chen Q, Haves P and Klems J H 2002 *Build. Environ.* **37** 857–864
[24] Manz H and Frank T 2005 *Energy Build.* **37** 1114–1121 ISSN 03787788

IOP Conf. Series: Journal of Physics: Conf. Series **1224** (2019) 012023 doi:10.1088/1742-6596/1224/1/012023

Sustainability assessment and performance evaluation of a Ground Coupled Heat Pump system: coupling a model based on COMSOL Multiphysics and a MATLAB heat pump model.

C Alimonti[1], E Soldo[1], S Minniti[2] and MFM Speetjens[3]

[1]Sapienza University of Rome, DICMA, Via Eudossiana 18 00184 Rome
[2]Eindhoven University of Technology, Electrical Engineering Dept. 5612 AZ Eindhoven
[3]Eindhoven University of Technology, Mechanical Engineering Dept. 5612 AZ Eindhoven

claudio.alimonti@uniroma1.it

Abstract. The present study investigates the sustainable use of a ground coupled heat pump (GCHP). In order to assess the performance of this type of installation, a computer model composed by two parts has been developed. The Borehole Heat Exchanger (BHE) model is developed in COMSOL Multiphysics, based on numerical methods. Part of the results are fed to the heat pump energy model, developed in MATLAB. A real case study has been used to validate the model: the Faculty of Engineering of La Sapienza University in Latina, undertaking a renewal project for an abandoned part of the building. After the renovation, the building will host a research center on the topic of low-enthalpy geothermal systems. The analysis have demonstrated that the modelled GCHP system can supply a significant share of the energy required from the future research center. This amount of energy can be provided keeping almost stable the thermal balance of the surrounding region in the subsoil, operating in a sustainable way. The range of variation of the ground temperature with respect to the average value is within the limit of 5°C, which is the cap set by the international legislation.

1. Introduction

Our primary challenge in the energy transition is the research of solid alternatives for the global energy system; alternatives that could be sustainable over a long time on our planet. The current energy supply system, mainly based on fossil fuels must gradually be replaced by another system, based on new innovative principles. The sustainability should be considered as important as the renewability in the evaluation of "green" energy sources.

Geothermal energy is one of the main opportunities for this purpose and it consists of the heat contained within the Earth's crust and mantle. This resource is characterized by an enormous potential both for electricity production and direct use applications. The use of Ground Source Heat Pumps (GSHP) for building heating do not produce CO2 emissions and do not require specific characteristics for the subsoil. Therefore, the GSHPs can be strategic in order to reach the environmental goals confirmed by the Conference of Parties in Paris (COP21). The growth in installed capacity of direct-use applications reflects the increasing international interest in this type of technology: according to

[1] the number of installed units is increased of a 52% respect to the data of 2010 and three times respect to the data of 2005.
The aim of this work is to build an overall model in order to carry out a first sustainability analysis of a ground source heat pump system. The analysis has two targets:

- understand in what extent the surrounding ground is influenced by the GSHP installation over a long-term period (10 years);
- on the other side, evaluate the energy performance of the system depending on the coupling between heat exchanger and surrounding ground.

As the sustainability of a geothermal source depends on a large number of variables, often interdependent and transient, an analytical approach is quite difficult. In this work, a two-dimensional (2D) transient model of a GSHP system is developed.
The general structure of the model is divided in two parts:

- the Borehole Heat Exchanger (BHE) model built in COMSOL Multiphysics, which simulates the interaction between the system and the ground;
- the heat pump (HP) model developed in MATLAB, which analyses the energy performance of the system.

There is a one-way coupling between the two models: part of the results obtained from the BHE model is fed into the HP energy model to calculate and assess the performance indicators of the system. The case-study provides the relevant information regarding the stratigraphy of the subsoil as well as the climatic characteristics of the location.
A case study is used to validate the model in a real situation, the Faculty of Engineering of La Sapienza University, located in Latina, which is undertaking a renewal project for an abandoned part of the building. After the renovation works the abandoned wing will host a research center on the topic of low-enthalpy geothermal systems.

2. Methodology

The schematic reported in figure 1 depicts the structure of the model. The general problem was divided in two models: the BHE model and the HP energy model.

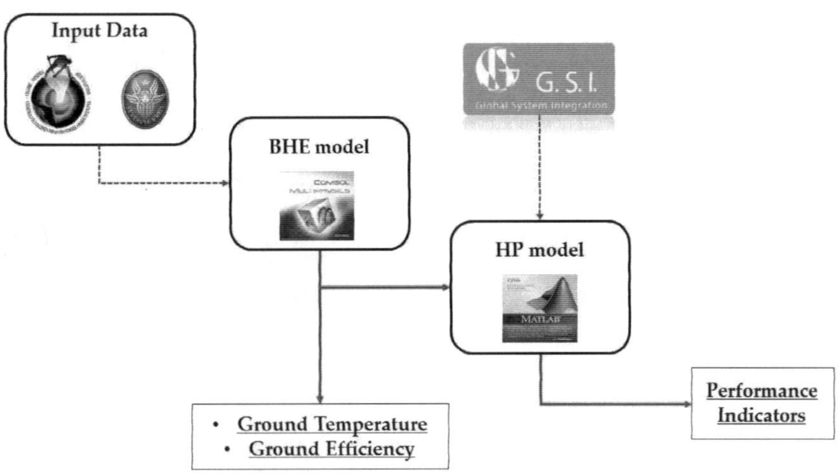

Figure 1. Structure of the model

The BHE model was developed in COMSOL Multiphysics, a modelling software based on numerical methods. A time simulation of 10 years has been used to study the effect of the borehole heat exchanger on the surrounding ground. The heat pump energy model was developed in MATLAB, starting from a catalogue data of a specific heat pump. The model used was the AQUA MAGIS (mod.110) made by the company Global System Integration. Some results were obtained directly from the BHE model such as the temperature field of the surrounding ground. The coupling between the BHE model and the heat pump model in MATLAB is a one-way coupling: the Leaving Fluid Temperature obtained from the BHE model was used as input data for the heat pump model, obtaining all the performance indicators for the specific case-study.

2.1. BHE model

The BHE model has been built with COMSOL Multiphysics®, general-purpose software based on advanced numerical methods for modelling and simulating physics-based problems. COMSOL is based on different modules, which model different physical problems such as the Structural Mechanics Module and the Computational Fluid Dynamics (CFD) Module. Another feature of the software is the presence of several Multiphysics Modules, built to couple different modules and different physics.

The computational power available was not enough to implement and simulate over a long period a 3-D model. Therefore, the proposed model was built in 2D, even considering that a high detail level was not relevant to determine with sufficient accuracy the performance of the heat pump. The long-term effect on the ground can be considered almost symmetric in every vertical plane passing through the center of the borehole.

The implemented geometry is a section made with a vertical plane passing through the axis of the two legs of the U-tube. A single U-tube configuration was considered for simplicity. In figure 2 the dimensions of the U-tube pipe inside the borehole are shown further in detail. The size of the hole is an average of typical values found in literature [2,3]. A pipe with a nominal diameter (and actual OD) of 40 mm is chosen and for the dimension ratio (DR) has been selected a DR11 (32,7 mm) which is the most common with nominal diameters lower than 63 mm.

The positioning of the pipe inside the borehole can be variable depending on the characteristic of the subsoil and the thermal properties of the grout. In the section implemented in the model, there is a 30-mm distance between the two legs of the U-tube pipe. The same distance is present between each leg and the borehole wall.

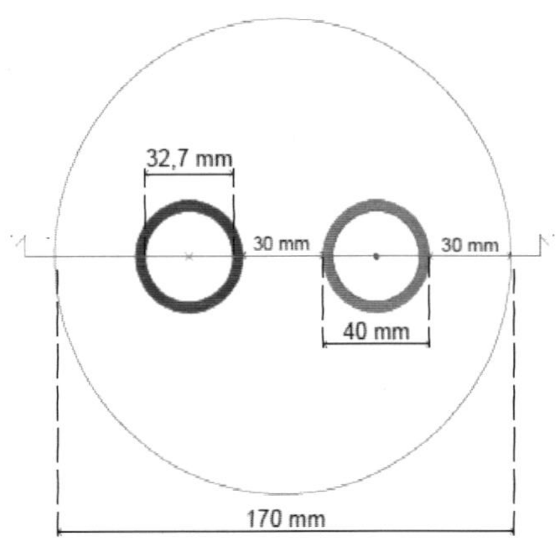

Figure 2. Horizontal section of the BHE model

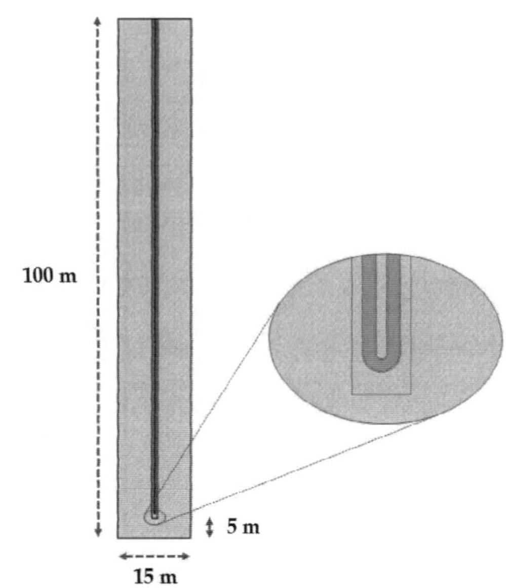

Figure 3. Dimensions of the BHE model implemented in Comsol

The overall dimensions of the model are shown in figure 3 from the Graphical User Interface (GUI) in COMSOL Multiphysics. The dimension of the lateral surrounding zone was selected considering the typical influence zone of this application.
The table 1 shows the materials implemented in the model.

Table 1. Materials of the model and their position

Material	Position in the model
Antifreeze Solution	Circulating in the pipe
Grout	Filling Material
HDPE	U-tube Pipe
Ground	Surrounding Region

The antifreeze solution is a mixture of water and Mono Ethylene Glycol (MEG), very common in this type of applications. Data sheets from the company MEGlobal [4] are used to evaluate temperature-varying properties such as viscosity, thermal conductivity and heat transfer coefficient at constant pressure. Regarding the share between MEG and water, a mixture with 30% of MEG (in weight, which means 27,7% of volume share) is chosen for the BHE model.
The High-Density Polyethylene (HDPE) has been selected for pipe material. The table 2 summarizes the average thermal properties of grout [5, 6] and of the HDPE [7, 8].

Table 2. Grout and HDPE thermal properties

Material	Density	Thermal conductivity	Heat capacity at constant pressure
Grout	1700 kg m^{-3}	1.5 W m^{-1} K^{-1}	2000 J kg^{-1} K^{-1}
HDPE	9500 kg m^{-3}	0.5 m^{-1} K^{-1}	1900 J kg^{-1} K^{-1}

The thermal properties of the subsoil are values that cannot be easily measured and stored. The values set in the model (figure 4) are an average of different references found in the literature [2, 9].

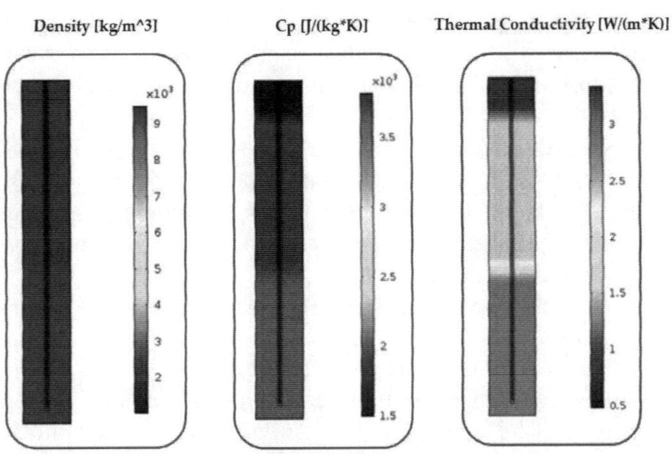

Figure 4. Thermal properties of the surrounding rock

It is easy to notice the change in thermal conductivity between different layers, passing from the value of 3,3 $Wm^{-1}K^{-1}$ to 1,2 $Wm^{-1}K^{-1}$ at deeper depths. The values in density are slightly different for

different layers, keeping the values around 2000 kg/m^3. The heat capacity at constant pressure decreases from deeper to shallow depths, since the deeper is the depth the greater is the density and capacity to store heat limiting its movement throughout different layers.
The figure 5 describes the modules structure used to build the BHE model. The **Non-Isothermal Flow** is the used Multiphysics Interface to simulate a moving fluid exchanging heat with the surrounding materials. The **Heat Transfer Module** uses the heat equation to calculate the temperature field in the whole model: the **Heat Transfer in Solids** evaluates the temperature of the ground, the HDPE and the filling material of the BHE; the **Heat Transfer in Fluids** evaluates the temperature of the antifreeze solution. Inside the U-tube, the **Turbulent Flow** module uses the Reynolds-Averaged Navier-Stokes equations (RANS) equations for the conservation of momentum and the continuity equation for the conservation of mass.

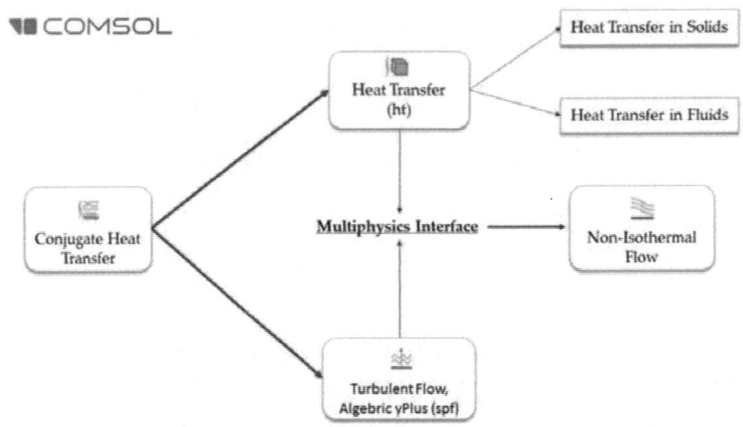

Figure 5. Modules framework in COMSOL multiphysics

Regarding the boundary conditions, a constant temperature is set on the external boundaries of the BHE model. The average monthly temperature of the case study's location is set on the upper boundary and it varies every month. The yearly average temperature of the subsoil is set on the lateral and on the lower boundaries of the model and it is constant for the entire simulation period. On the outlet boundary of the pipe a constant pressure is set, while in the inlet boundary a constant inlet velocity is set. The fluid has no slip along the walls because the vector field is equal to zero.
The output data obtained from the simulation are the ground temperature, the average linear heat flux and the fluid temperature.
The ground temperature data obtained from the BHE model are stored at 5, 40, 80 meters. Two values for every depth have been selected: at 2 meters left and 2 meters right from the BHE.
The average linear heat flux along the two legs of the U-tube pipe is representative of the heat exchange between the antifreeze solution and the ground. The average temperature of the heat transfer fluid is calculated along the outlet boundary of the pipe.

2.2. Heat pump energy model
The general structure of the heat pump energy model is reported in figure 6. The two inputs are the EFT derived from the BHE model, previously described, and the Fluid Temperature required on the building-side that depends from the type of distribution system within the building.
The heat pump is AQUA MAGIS unit, an high-efficiency water-to-water heat pump selected from the catalogue of Global System Integration (GSI).
Catalogue data are used as starting point to assess the energy performance of the heat pump. The Entering Fluid Temperature (EFT) is the independent variable, in function of which all the relevant parameters. The catalogue reports the evaluation only for 4-5 different values of EFT. A MATLAB

code was developed to determine the polynomial interpolation functions for each parameter in order to perform the calculation for every possible value of EFT.

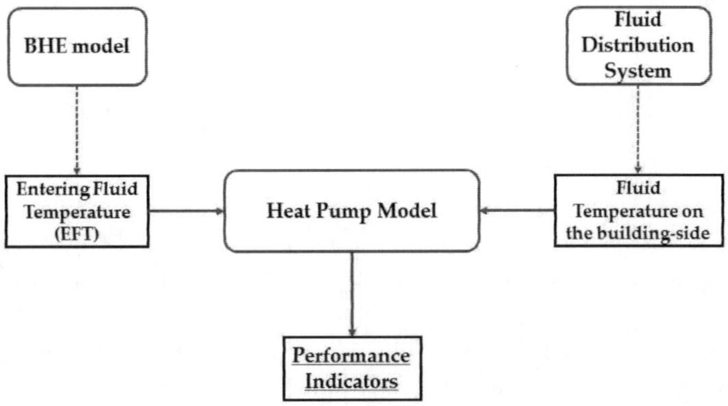

Figure 6. Modules framework in MATLAB

Regarding the building side, the code was developed considering the two possible distribution systems: the fan Coil units' system that requires temperature around 45°C in heating mode and around 7°C in cooling mode; the underfloor heating that requires temperature around 35°C in heating mode and around 15°C in cooling mode.

The catalogue data of the heat pump shows that the temperature difference (ΔT) of the antifreeze solution after the heat exchange with the refrigerant of the heat pump is nearly stable independently from the EFT. This justifies the use of a constant ΔT for the entire season.

The output data of the HP model are the power supplied to the building (output power), the Coefficient Of Performance (COP) while operating in the heating mode and the Energy Efficiency Ratio (EER) while operating in the cooling mode.

3. The case study: Faculty of Industrial Engineering in Latina

The project concerns the renovation works on one wing of the Faculty of Engineering in Latina, detached location of La Sapienza University.

The available wing (figure 7) is completely abandoned. A first step will be the renewal of the building to host a research center on the low-enthalpy geothermal systems (the first one in the central part of Italy). Secondly, a geothermal plant will be installed, to supply heating and cooling and to produce electricity to the research center. The additional value will be constituted by the installation of monitoring equipment to evaluate the performance of the system.

The space on the ground floor will be used as "Technical Room-Laboratory", where the internal unit of the geothermal plant and the Thermal Response Test equipment (TRT) will be hosted. The six rooms of the upper floor will be used for different purposes: a big conference room, which could be used also as lecture room; one room to control the performance parameters of the geothermal installation; a room for measurements and control for the ground parameters; one room for the researchers and the technical responsible; other two rooms will be used for design-consulting, support to design companies, software developments, geothermal databases, and technoeconomic feasibility studies.

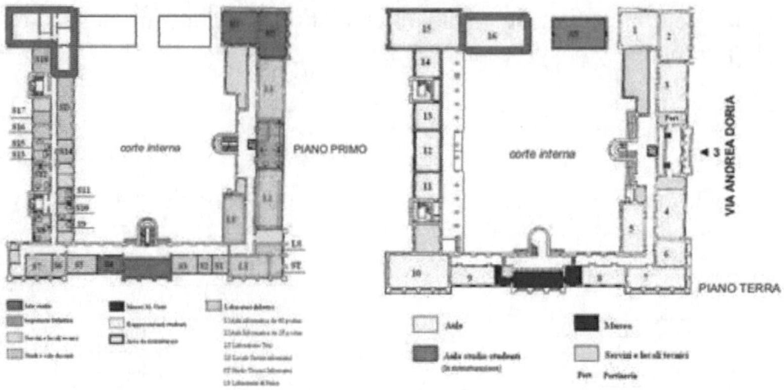

Figure 7. Plants of the two floor. In red is highlighted the abandoned wing

An energy analysis was conducted to assess the energy efficiency of the abandoned wing and to identify some improving measures: substitution of windows fixtures, installation of a thermal envelope to the external walls, insulation layers on the roof. These measures will reduce the heat dispersion during winter of a 48,5% and the external gains during summer of a 19,2%.
The table 3 reports the peak load of the energy requirement for each room of the two floors.
The design target is 17 kW for the cooling peak load and 10,5 kW for the heating peak load.

Table 3. Peak loads of the case study

Room	Heating Mode	Cooling Mode
Conference Room 1	1592	2841
Conference Room 2	4145	6218,3
Lecture Room 1	794,1	1253,1
Lecture Room 2	662,9	1216,5
Lecture Room 3	907,7	1918,9
Lecture Room 4	1455	2370
Corridor 1	395,7	493,6
Corridor 2	579,9	1081,3
Total	**10532,3**	**17392,7**

The building of the BHE model was strictly linked with some characteristics of the chosen location.
In table 4, meteorological data were stored from the meteorological station of Latina Airport in the 30-years period between 1971 and 2000. The yearly average temperature is 15,1°C and this is assumed as the temperature of the undisturbed ground. This is the first value of the location that will be set on the boundary conditions of the model.

Table 4. Monthly temperature of the case study

	Jan	Feb	Mar	Apr	May	Jun	Jul	Aug	Sep	Oct	Nov
Average Max. Temp. [°C]	13	14	15,8	18,1	23	26,8	29,9	30,5	26,8	22,4	17,4
Average Min. Temp. [°C]	3,5	3,6	5,2	7,3	11,2	14,8	17,5	18,1	15,5	12	7,8
Average Temp. [°C]	8,4	8,7	10,5	12,7	17,1	20,8	23,7	24,3	21,15	17,2	12,6

The used stratigraphy is an average of different subsoils near Latina and it is presented in table 5. The subsoil is simulated until 100m depth because this is also the total length of the BHE in the model. The thermal conductivity of the subsoil is expected to decrease going down with the depth.

Table 5. Stratigraphy of the case study

Depth [m]		Soil
0	10	Sand
10	60	Clayey sand
60	100	Clay

3.1. Simulation strategy

The simulation strategy was set considering 5 days (working days) of 24h operation and 2 days (weekend) of stop, according to the typical schedule of Italian universities, where no lectures are planned during the weekend. The two-days stop permits a recovery of the ground temperature surrounding the BHE and it allows to the heat transfer fluid to reach the external average temperature, improving the efficiency during operative days.

On the seasonal mode of operation, the simulation strategy was set considering 10 months of operations (5 in heating mode and 5 in cooling mode) and 2 months of stop (1 month for each operating period). This share was chosen considering the typical thermal load in Italy and considering also the results of the energy analysis of the site. This two months stop will also allow to the ground to spread the thermal load acquired/lost due to the heat injection/extraction of the BHE in a better way compared to the slight recovery in the weekends.

4. Results

4.1. Validation of the model

The BHE model was validate to verify the influence of the boundary effect on the BHE model. Therefore the total width of the lateral boundaries has been changed from 15 m to 30 m and a greater ΔT has been applied in the periodic boundary condition that links inlet and outlet of the U-tube pipe (4,6 °C during the heating mode and -5°C during the cooling season).

The figure 8 shows the effect of the modifications on the ground temperature at 5 km from the center of the BHE.

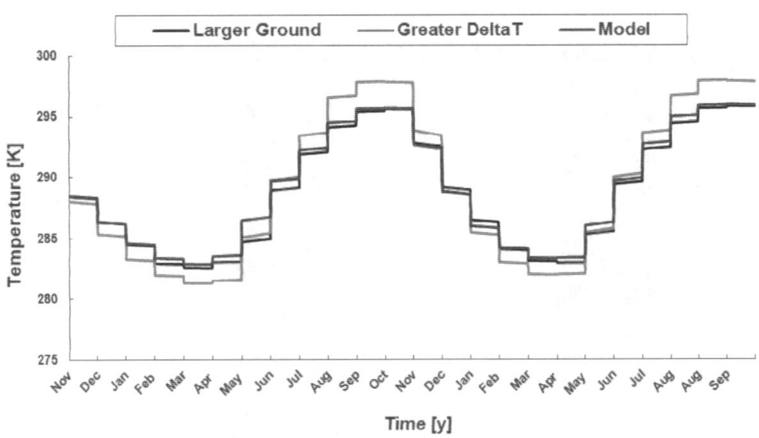

Figure 8. Ground temperature at 5m depth: comparison of the modified COMSOL models

Considering the model with larger lateral boundaries the temperature change of the ground is nearly equal to the normal model. This is enough to confirm that the boundary effect is not relevant for the evaluation of the ground temperature. The greater ΔT in the Periodic Boundary Condition between the inlet and the outlet of the pipe influences the change in temperature within the subsoil. A

larger range of variation is induced during the year. The influence of the geothermal installation is preponderant compared to the influence of the monthly fluctuations of the external temperature.
The influence-zone of the installation is evaluated for the two deeper types of subsoil, the clayey sand (figure 9 - a) and the clay (figure 9 - b). The simulation time was set to 3 years. The average temperature has been calculated for lines parallel to the main dimension of the borehole at the end of each operational season (two values per year). The initial alteration of the subsoil is significant but after that, there is a stabilization: the influence zone at the end of the installation's lifetime is just under 8,5 m for clayey sand and about 7 m for clay.

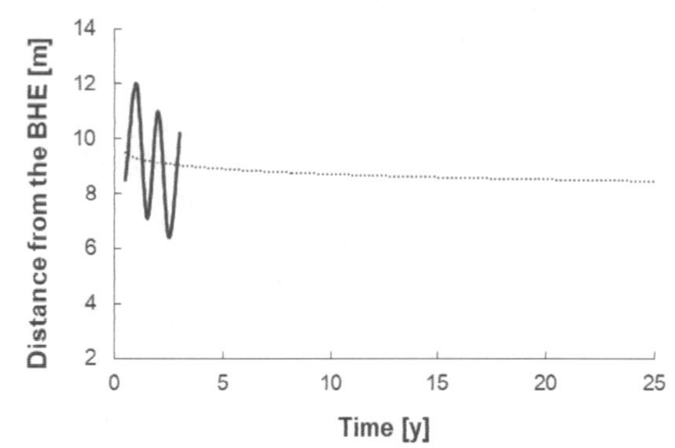

Figure 9 - a. Clayey subsoil

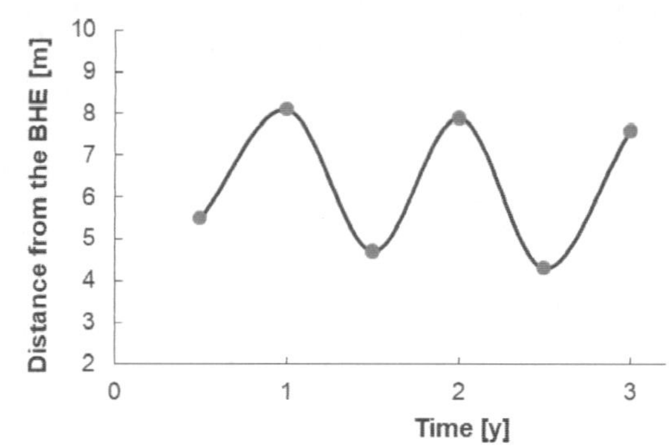

Figure 9- b. Clay subsoil

Figure 9. Influence zone of the GCHP installation

The figure 10 above shows the temperature distribution in the BHE model at the end of the Heating/Cooling season of the 3rd year. This is the larger model used to evaluate the influence of the boundary effect.

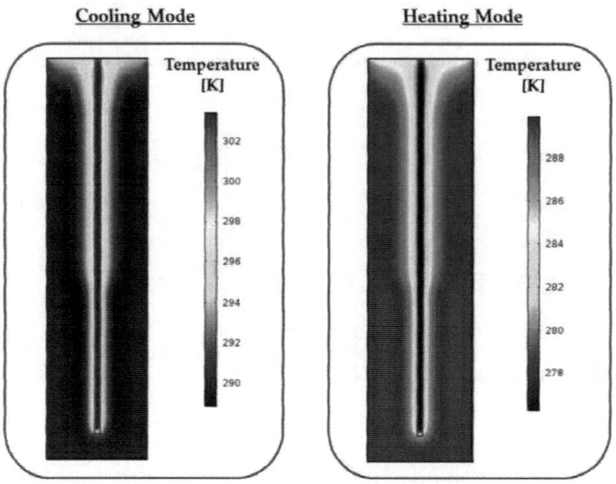

Figure 10. Temperature distribution for the larger model in the 3rd year of simulation

4.2. BHE model results
The figure 11 illustrates the fluctuation of the ground temperature in time at different depths. At the 10th year of operation, the yearly average temperature at 5 m of depth is 16,9°C while at the end of the

lifetime (25 years) of the installation the yearly average temperature is 290,6 K (17,45°C). The overall increase amounts to 2,3°C.
At 40m depth, the range of variation of the temperature is between 285 K (12°C) and 294,5 (21,6 °C). The yearly average temperature has an overall increase of 2,2°C at the end of the 25 years of operation with a final average temperature of 290,49°C (18,55°C). the overall trend has a larger increase in the first two years due to the modification of the undisturbed ground temperature. After that the trend is quite stable with a slight yearly increase.
Considerably different is the result at 80m depth where the type of subsoil is completely different (clay) and with lower thermal conductivity. The range of variation of the temperature is much tighter: between 287 K (14°C) and 292 K (19°C). Here, there is a larger increase in the first year (yearly rate of 0,24°C/y) and a lower increase in the next years. The temperature reaches the value of 290,05 K (16,9°C) after 25 years of operation.

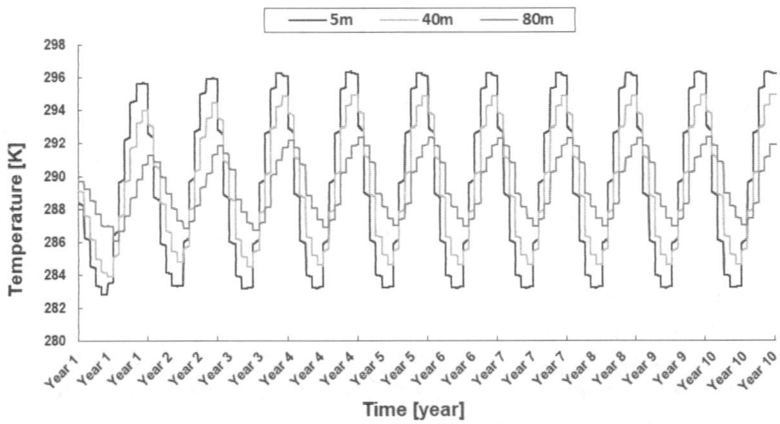

Figure 11. Comparison of ground temperature at different depths

In figure 12 the ground efficiency in the inlet leg (negative during the heating mode and positive during the cooling mode) is showed. The results show that 4 years are needed for the stabilization of the heat transfer process between the probe and the ground. The efficiency in the first heating season is lower because the ground is undisturbed. Afterwards, there is a stabilization because of the alternation of cooling and heating seasons that results in efficiency improving. The asymptotic value in the heating season is over 44 W/m, for the cooling season, the asymptotic value (around 47,5 W/m) is reached through more fluctuations. Furthermore, the amount of the heat transferred from the heat pump to the sink during the cooling season is higher than the one extracted during the heating season.

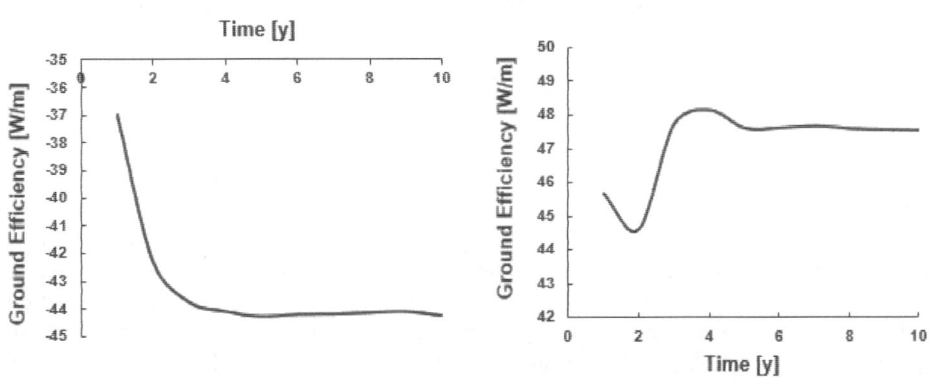

Figure 12. Average ground efficiency along the inlet leg. Left: heating mode - right: cooling mode

IOP Conf. Series: Journal of Physics: Conf. Series **1224** (2019) 012023 doi:10.1088/1742-6596/1224/1/012023

The situation is similar in the outlet leg (figure 13) because the system is almost symmetric with respect to a vertical line passing through the centre of the model. The only difference is the ΔT between the antifreeze solution and the subsoil. In the second part of U-tube pipe the fluid has already gained almost half of the total heat and the temperature has changed consequently. For this reason, the difference of temperature is lower and the ground efficiency is reduced. After the first 4 years, the value stabilises for the heating mode near 34 W/m, while for the cooling mode just under 37 W/m.

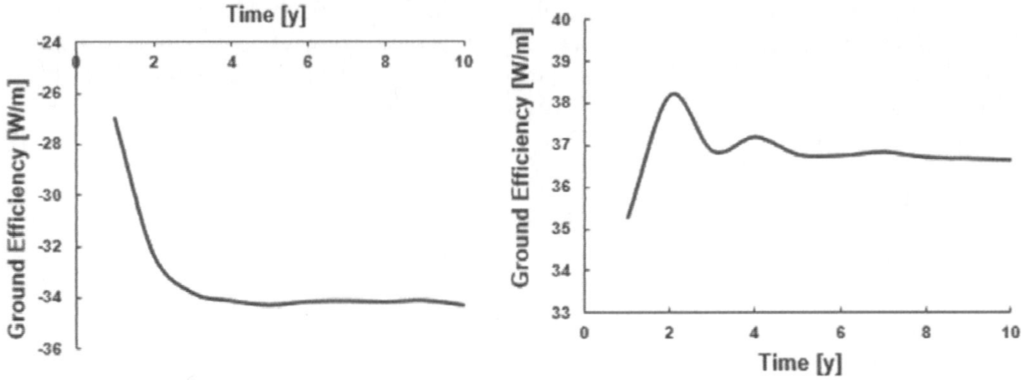

Figure 13. Average ground efficiency along the outlet leg. Left: heating mode - right: cooling mode

The entering fluid temperature (EFT) in the heat pump is the average of the temperature of the fluid along the upper boundary of the outlet leg of the U-tube pipe.
The figure 14 reports the trend of the EFT of the antifreeze solution during the 10 months of operation, for a total time of 10 years. After the first 2 years of stabilization, the variation range of the temperature is nearly stable: the range of variation is 2-12 °C for the heating mode, 18-33 °C for the cooling mode.

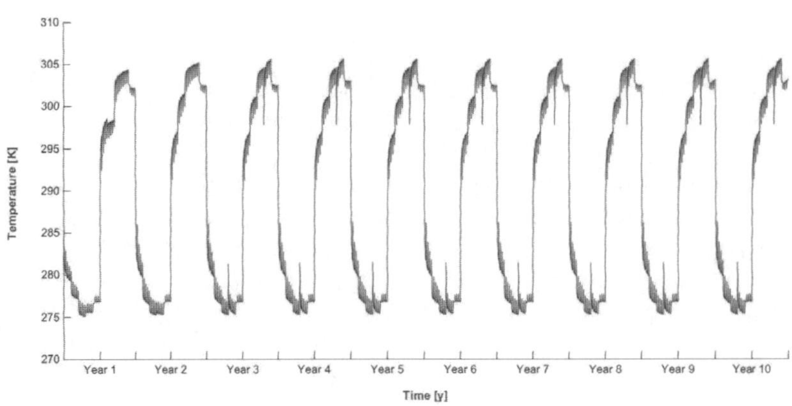

Figure 14. Entering fluid temperature for the 10 years of simulation

The figure 15 shows the yearly trend of EFT. During the 5 days of 24hours operation, the temperature varies logarithmically (it decreases in heating mode, it increases in cooling mode) but it does not reach the complete stabilization. Then, there are 2 days of stop where the heat transfer fluid is idle, and it reaches again the outside temperature. So the temperature in every month is slightly different to the monthly average temperature of the external environment, resulting in an efficient heat exchange.

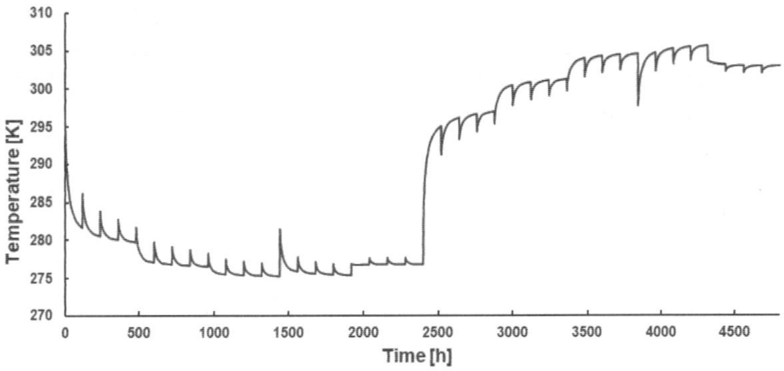

Figure 15. Entering fluid temperature for the 4th year

4.3. Heat Pump performance

Two types of distribution system are investigated because they are possible solution for the case-study: fan-coil units' system (figure 16) and underfloor heating system (figure 17).

Both of the solution can satisfy the peak heating request of the building (10.5 kW), but none of them is able to supply the peak load of cooling mode (17,32 kW). The average output power in the cooling mode is 10,98 kw for fan-coil units' system and 13,92 for underfloor heating system.

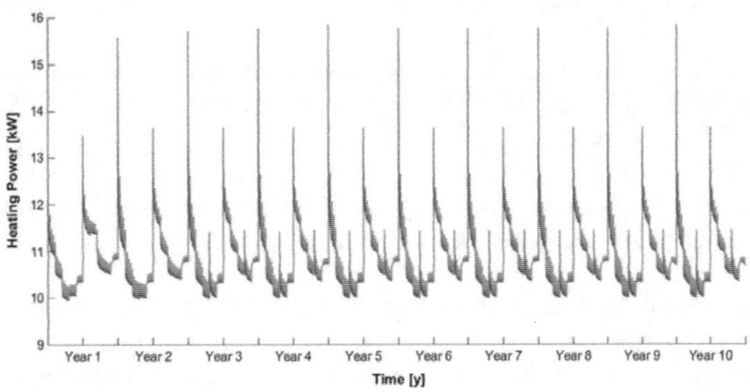

Figure 16. Output power with the fan-coil units' system

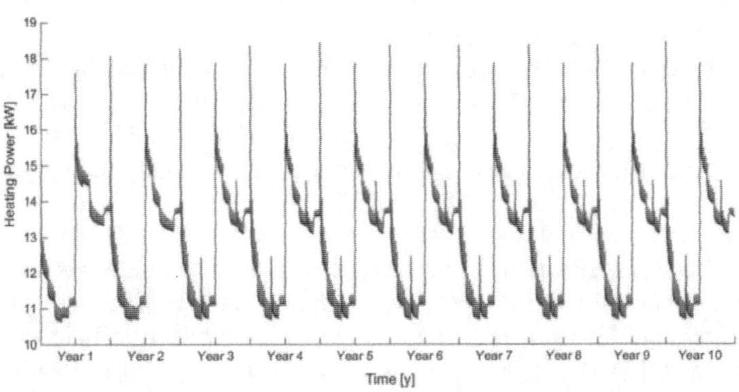

Figure 17. Output power with the underfloor heating system

Higher performance indicators are calculated for the underfloor heating system (figure 18) respect to the fan-coil distribution system. The stabilization of the COP is faster than the EER. The stabilized value of the COP is around 3,53 for fan coil system and 4,8 for underfloor heating system. The EER reaches the value of 5,3 for fan coil system and 6,44 for underfloor heating system.

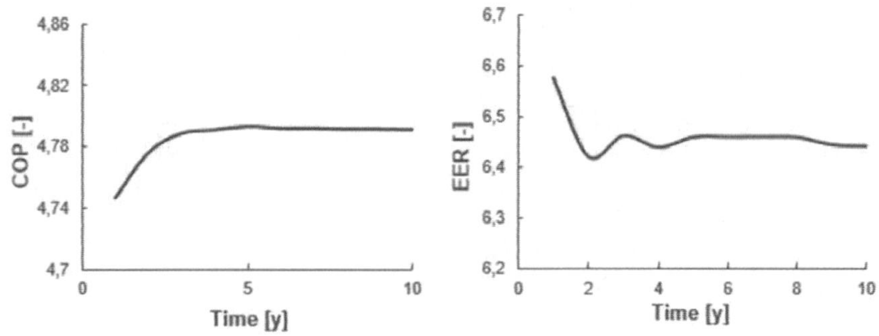

Figure 18. Seasonal performance indicators for the underfloor heating system

5. Conclusions

The present study investigates the sustainable use of a ground source heat pump applied to the Faculty of Industrial Engineering of La Sapienza University, located in Latina, which is undertaking a renewal project for an abandoned part of the building.
The analysis has been conducted with two main targets: understand in what extent the surrounding ground is influenced by the installation; evaluate the performance of the system during a long-term period of operation (at least 10 years).

Due to the number of coupled processes and variables involved The model has been divided in two parts: A 2-D Borehole Heat Exchanger (BHE) model is built on COMSOL to simulate the coupling between the system and the ground, and a heat pump energy model is developed on MATLAB using catalogue data as starting point.

The results of the simulations show that during 10 years of operation, the temperature change at a distance of 2 m from the center of the BHE, is around 1,5°C, reaching an increase of 1,9°C at the end of the 25 years of operation. This is the most relevant change, noticed at 5m depth from the surface. Going down in depth, there are different types of subsoil with lower thermal properties, which limit the temperature change. This is even lower at 40m and 80m depth. The thermal balance is almost maintained during the 10 years of operation because the plant is operating for the same time (5 months) both in the cooling and the heating season.

The stratigraphy of the subsoil in the location of Latina seems to be suitable for the application of a Ground Source Heat Pump system. The energy analysis illustrates that is not possible to meet the required peak load of the Faculty with just one single BHE, even with a length of 100m. The heating peak load can be reached with both types of distribution system but the output power is not enough during the cooling season. Highest output power and performance indicators are reached with the Underfloor heating system. The COP reaches an average value of 4,8 while the EER amounts to 6,4 in the cooling season. Those outputs confirm that geothermal heat pumps work more efficiently with distribution systems that do not require high temperature on the building side. By Lowering the lift between the two temperature levels, it results an increase of the overall efficiency of the system. On the other side, the Underfloor heating system is certainly the less cost-effective solution for the renewal of the faculty. A cost-benefit analysis should be performed to assess what is the best solution for supplying the energy demand of the research center.

Since the simulated BHE was not enough to satisfy the entire energy demand of the future research center, one option could be the installation of another BHE beside to the previous one. Considering this possibility, the influence-zone in the BHE model was evaluated, in order to avoid thermal interferences between different BHEs providing energy to the same building. The influence-zone amounts to 8,5 m from the center of the BHE. This means that the installation will modify the temperature field in the subsoil for 17 m in total.
Thus, in summary, the modelled GCHP system can supply a significant share of the energy required from the future research center. This amount of energy can be provided keeping almost stable the thermal balance of the surrounding region in the subsoil, operating in a sustainable way.

References

[1] Lund J W and Boyd TL 2016 Direct utilization of geothermal energy 2015 worldwide review *Geoth.* **60** 66–93
[2] Kavanaugh S and Rafferty K 2014 *Geothermal Heating and Cooling: Design of Ground-Source Heat Pump Systems* (Atlanta : ASHRAE)
[3] Yang H, Cui P and Fang Z 2010 Vertical-borehole ground-coupled heat pumps: A review of models and systems *Appl. En.* **87** 16-27
[4] MEGlobal. Ethylene Glycol - Product Guide . [Online] 2008. http://www.meglobal.biz/
[5] GeoPro Inc. The importance of thermal grout conductivity . [Online] www.geoproinc.com.
[6] CETCO. Geothermal Grout: Enhanced Thermally Conductive Grout. [Online] December 2013. http://www.cetco.com/DesktopModules/Bring2mind/DMX/Download.aspx?Command=Core_Download&PortalId=0&EntryId=4344
[7] INEOS Olefins and Polymers USA. Typical Engineering Properties of High Density Polyethylene. [Online] http://www.ineos.com/globalassets/ineos-group/businesses/ineos-olefins-and-polymers-usa/products/technical-information--patents/ineos-typical-engineering-properties-of-hdpe.pdf.
[8] Professional Plastics. Thermal Properties of Plastic Materials. [Online] http://www.professionalplastics.com/professionalplastics/ThermalPropertiesofPlasticMaterials.pdf.
[9] Hamdan I and Clarke B 2010 Determination of Conductivity of Coarse and Fine Sand Soils Proc. World Geoth. Congr. 2010 Bali, Indonesia

Performance maps for an air-cooled air conditioning system as a preliminary instrument for the diagnosis of evaporator fouling

P. Catrini[1], A. Piacentino[1], J. E. Braun[2], A. Patil[2], A. L. Hjortland[2]

[1] Dpt. of Energy, ICT and Mathematical Models (DEIM), University of Palermo, Viale delle Scienze, 90128, Palermo, Italy
[2] Ray W. Herrick Laboratories, School of Mechanical Engineering, Purdue University, 177 S. Russell Street, West Lafayette, IN 47907-2099, USA

E-mail:catrinipietro@gmail.com

Abstract. During the past few decades, there has been increased interest in the development of automated approaches for detecting and diagnosing faults in air conditioning systems. Among them, thermoeconomic diagnosis is a technique which involves the use of exergy analysis. As a first step towards improving its performance, this work is focused on the thermodynamic and exergy analysis of a direct expansion air conditioning system used in small commercial building applications. The analysis was carried out by means of experimental activities on a 17.5 kW rooftop unit installed at the Herrick Laboratories, Purdue University, Indiana (USA). The system under investigation was equipped with a variable-speed compressor and variable-speed fans, thus allowing the unit to meet different loads. A detailed mapping of the performance of this system was carried out by considering the effect of the outdoor temperature and the cooling load. In addition, the effect of evaporator fouling was investigated. Results showed that poor exergy performance were achieved due to the exergy destruction occurring in the evaporator. Also, when testing evaporator fouling, results showed that this fault contributed to increase the consumption of the mechanical exergy of the air, thus allowing for an easy detection of this fault.

1. Introduction

The increase in the global energy demand has been pushing research towards the exploitation of renewable energy sources and towards a more rational use of energy. To this aim, due to high energy consumption in air conditioning applications, an optimal design and operation of HVAC systems could lead to cost-effective energy savings and environmental benefits. Among HVAC systems, packaged rooftop unit (RTU) are widely used for air conditioning in commercial application. In order to improve the energy performance of these systems, more sophisticated control logics have been implemented, as the variable speed on compressor and fans which allows also for a better control of the indoor comfort conditions. Another important aspect is the energy consumption during the operation of these systems, which generally increases over time due to poor maintenance [1]. To this aim, some fault detection and diagnosis techniques (FDD) have been proposed and investigated in the literature [2],[3] and [4]. Among them, Thermoeconomic diagnosis is an innovative exergy-based FDD for detecting "faults" in energy systems [4]. This technique has been proven to be promising but still efforts are needed to improve its performance [5][6][7] [8] . To this aim, preliminary details into the exergy behavior of each component and the effects of the control system on the thermoeconomic model adopted are necessary [9].

This works aims at mapping the exergy performance of a packaged rooftop unit equipped with a variable-speed compressor and variable-speed fans by means of experimental data collected from a real unit. The performance of the system was evaluated by considering: (i) two values of outdoor temperatures and (ii) four levels of the RTU cooling capacity. The analysis was carried out on the basis of an ad-hoc thermoeconomic model, which is duly presented in this paper. After mapping the performance of the unit in the fault-free scenario, two levels of evaporator fouling were investigated. Assessment and comparison of the main exergy performance indicators for the fault-free and the faulty scenario will be presented.

2. Experimental set-up and testing procedures

A rooftop unit (RTU) was considered in this study having a rated capacity of 17.5 kW (5 ton) and a SEER rating of 20.0 (figure 1). The unit is equipped with a scroll compressor and a thermal expansion valve (TXV). The refrigerant used is R410a. The unit can be operated in a two-staged mode or in a variable speed mode. In a variable speed mode, both the compressor and the indoor fan speed are controlled in order to satisfy different cooling loads and avoiding the cycling of the unit. More specifically, the compressor speed is varied according to the supply air temperature setpoint (SAT_{sp}). Conversely, the indoor fan speed is controlled in order to meet the temperature setpoint for the indoor environment. A schematic of the test facilities is shown in figure 2. Two psychrometric chambers (each one equipped with its own reconditioning system) allowed to simulate the outdoor and indoor environment. A summary of the instrumentation installed on the RTU is presented in table 1. A real-time controller from National Instruments was used for data acquisition and control of the RTU.

Figure 1. Packaged rooftop air conditioning unit installed at Herrick Laboratories

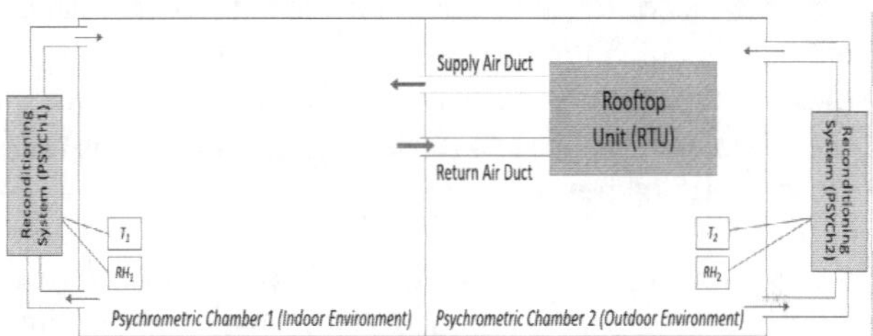

Figure 2. Packaged rooftop air conditioning unit installed at Herrick Laboratories

Table 1. Description of RTU instrumentation

	Sensors Type	Sensors Location	
Refrigerant Temperature	*T-type Thermocouples* (Accuracy ±0.5°C)	Compressor Suction Compressor Discharge Condenser Outlet	TXV inlet Evaporator Inlet Evaporator Outlet
Refrigerant Pressure	Pressure Transducers with voltage output (Accuracy ±0.5% of the full-scale reading)	Compressor Suction Compressor Discharge Liquid line outlet	
Refrigerant Mass flowrate	*Coriolis Mass flowmeter* (Accuracy ±0.5% of the full-scale reading)	On the liquid line	
Dry Bulb Air Temperature	*T-type Thermocouples* (Accuracy ±0.5°C)	<u>Return Air</u>: two-by-two grid was spaced equally in a rectangular grid placed before the evaporator <u>Supply Air:</u> two-by-two grid spaced equally in a rectangular grid placed after the supply fan <u>Condenser Inlet</u>: 8 thermocouples mounted in a four-by-two grid on the face of the condenser <u>Condenser Outlet</u>: 4 thermocouples mounted radially on the condenser outlet	

Dew Point Air Temperature	*Dew point hygrometer with chilled mirror probe* (Accuracy ±0.15°C)	Air samples were taken before the evaporator inlet and at the beginning of the supply air duct
Power Consumption	*Watt Transducers* (Accuracy of ±0.5 % of the full-scale reading)	Indoor Fan Supply Air Compressor Power Condenser Fan

2.1. Description of testing procedures

In order to evaluate the performance of the unit at different load level, a set of different air mass flowrate across the evaporator were tested by varying the indoor fan torque. More specifically, the torque was varied from 100% of its maximum value (in this case, the unit provides the maximum cooling capacity) down to 30% (minimum RTU cooling capacity) while keeping the supply air temperature setpoint (SAT_{sp}) equal to 15°C. For all tested conditions, the temperature and relative humidity of the air returning to the unit from the indoor environment were kept constant and equal to 26.5°C and 50%.

As concerns the effect of the boundary conditions, two values of the outdoor temperature (here indicated as Toutdoor) were investigated: 30°C and 35°C. The operation of the unit at each set of driving conditions was first recorded with no faults imposed. Then, the evaporator fouling was introduced and all tests were repeated. Measurements were taken every second for all of the input and output variables. The steady state of the system was detected by comparing data of a moving averaging window of 20 minutes.

Evaporator fouling was tested by acting on the opening of an iris damper placed on the supply air duct of the unit. Different levels were selected by measuring the air pressure drop across the RTU while running the indoor fan at its maximum speed (around 1250 rpm). Considering a 150 Pa air pressure drop (about 0.6 inch of water) across the RTU for the fault-free scenario (i.e. when the damper is fully open), two levels of fouling were tested: (i) a "light" fault which corresponds to a 280 Pa air pressure drop (about 1.0 inch of water) and (ii) a "heavy" evaporator fouling which corresponds to 380 Pa air pressure drop (about 1.5 inches of water).

3. Exergy analysis and thermoeconomic modelling of the examined system

Exergy analysis of the examined system was carried out on the basis of the thermoeconomic model adopted for this system. More specifically, in Thermoeconomics exergy flows entering and exiting a given component are defined on the basis of its *productive scope* [7]. For instance, an evaporator coil for cooling and dehumidifying an air stream aims at increasing the thermal and chemical exergy of the air (which is then identified as its productive scope) by exploiting the thermal and mechanical exergy of the refrigerant. The exergy flow produced is generally identified as a "Product", and the amount of exergy used to this aim is usually identified as "Fuel". Due to the need to identify a productive scope for components, exergy flow related to a stream of matter (for instance, refrigerant and air in cooling application) is splitted into its "thermal", "mechanical" and chemical" fractions, which are respectively related to the temperature, pressure and composition disequilibria between its thermodynamic state and the ambient dead state. The procedures followed for calculating these fractions are here omitted, thus details can be found in [10]. In figure 3, the thermoeconomic model of the investigated RTU is presented.

It is worth observing that:

1. The compressor increases both the thermal and the mechanical exergy of the refrigerant, i.e. $\Delta\dot{B}_1^{\mathrm{T}}$ and $\Delta\dot{B}_1^{\mathrm{M}}$ by using a mechanical exergy flow provided by an electric motor. In this case, since both the compressor and the electric motor were treated as a unique component, the electric power is the exergy flow directly consumed by the compressor.
2. The condenser aims at discharging the thermal exergy content of the refrigerant, i.e. $\left|\Delta\dot{B}_2^{\mathrm{T}}\right|$ into an air flowrate coming from the outdoor environment, by consuming: (i) the electric power to drive the fan, i.e. $\dot{W}_{e,2}$ (in this case the condenser and its fan are considered as a unique

component) and (ii) the mechanical exergy of the refrigerant $\left|\Delta\dot{B}_2^{\mathrm{M}}\right|$ due to pressure losses on refrigerant side.

3. The thermal expansion valve uses the mechanical exergy flow $\left|\Delta\dot{B}_3^{\mathrm{M}}\right|$ produced by the compressor to increase the thermal exergy of the refrigerant (i.e. $\Delta\dot{B}_3^{\mathrm{T}}$), then supplied to the evaporator.
4. The evaporator increases the thermal and chemical exergy of the air by means of the following exergy flows: (i) the mechanical exergy $\left|\Delta\dot{B}_4^{\mathrm{M}}\right|$ of the refrigerant dissipated by the pressure drop on the refrigerant side, (ii) the thermal exergy of the refrigerant $\left|\Delta\dot{B}_4^{\mathrm{T}}\right|$ and (iii) the mechanical exergy of the air $\left|\Delta\dot{B}_{\mathrm{air}}^{\mathrm{M}}\right|$ which is dissipated by pressure drop on air side.
5. the indoor fan increases the mechanical exergy of the air in order to balance the pressure drop across the evaporator, by using the mechanical power provided by an electric motor. Here, both the indoor fan and its electric motor are considered as a unique component, thus the electric power is the exergy flow consumed by this component.

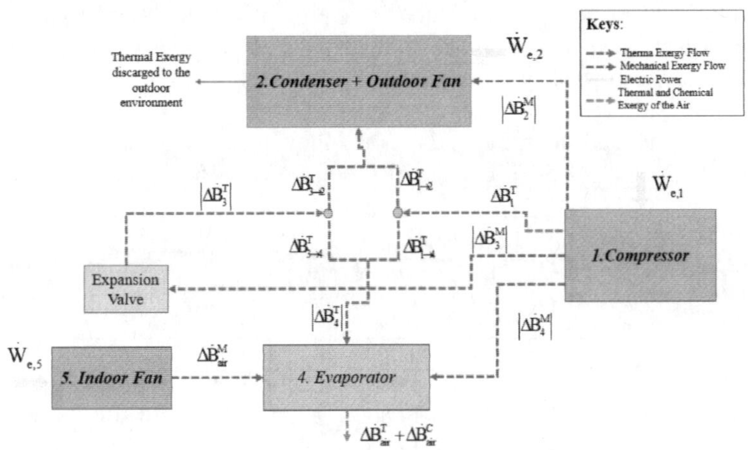

Figure 3. Thermoeconomic model of the rooftop unit

In order to assess the exergy performance of each component, the *unit exergy consumption of the i-th component* k_i was used in this work. This indicator is the ratio between the exergy used by the i-th component (here indicated as F_i) to produce the unit exergy of its product P_i, i.e. $k_i = F_i / P_i$. For instance, the unit exergy consumption of the evaporator k_4, here shown in equation 1, is the ratio between the following exergy quantities: (i) the reduction in the thermal and mechanical exergy of refrigerant plus the mechanical exergy of the air and (ii) the increase in thermal and chemical exergy of the air cooled and dehumidified across the evaporator.

$$k_4 = \frac{\left|\Delta B_4^{\mathrm{T}}\right| + \left|\Delta B_4^{\mathrm{M}}\right| + \Delta B_{\mathrm{air}}^{\mathrm{M}}}{\Delta B_{\mathrm{air}}^{\mathrm{C}} + \Delta B_{\mathrm{air}}^{\mathrm{T}}} \tag{1}$$

In order to quantify the percentage of the exergy destroyed in the thermostatic valve, the quantity χ_3 was defined as in equation 2. Particularly, ΔB_3^{T} is the increase in the thermal exergy of the refrigerant across the valve and $\left|\Delta B_3^{\mathrm{M}}\right|$ is the mechanical exergy of refrigerant consumed to this aim.

$$\chi_3 = \frac{\Delta B_3^{\mathrm{T}}}{\left|\Delta B_3^{\mathrm{M}}\right|} \tag{2}$$

4. Results and discussion

In this section, the following results are shown: (i) energy and exergy performance of the RTU in the free-fault scenario (ii) energy and exergy performance of the unit when evaporator fouling is imposed.

4.1. Energy and Exergy Performance in the fault-free scenario

In table 2, thermodynamic data measured during the experimental activities are shown. Only results for $T_{outdoor}$= 30°C are shown, being the same trends verified for $T_{outdoor}$=35°C. The indoor fan torque was set equal to 100%, 80%, 50% and 30%. The minimum cooling capacity, i.e. 10.3 kW, is achieved by setting the fan torque to 30%, and it is equal to the 55% of the maximum cooling capacity (i.e. 18.7 kW). Also, as concerns the sensible load fraction of the cooling capacity, here expressed by the sensible to heat ratio values SHR (which is the ratio between the sensible load and the cooling load), it is approximately constant for all load levels considered. Condensing pressure decreases when reducing the cooling capacity, conversely, the evaporating pressure is approximately constant. The supply dry bulb temperature is approximately equal to the setpoint (i.e. 15°C) and the electrical power consumed by the compressor decreases as the control system reduced the compressor speed.

Table 2. Thermodynamic Data measured for the following test conditions: T_{indoor}=26.5°C, RH= 50% $T_{outdoor}$=30°C and SAT_{sp}= 15°C

Indoor Fan Torque Percentage	*Air Mass Flowrate*	*Return Air*		*Supply Air*		*Indoor Fan Power*	*Outdoor Fan Power*	*Sensible Capacity*	*Cooling Capacity*	*SHR*
		Dry Bulb Temp.	*Dew Point Temp.*	*Dry Bulb Temp.*	*Dew Point Temp.*					
	[kg/s]	[°C]	[°C]	[°C]	[°C]	[W]	[W]	[W]	[W]	[W/W]
100%	1.273	26.5	11.98	12.07	15.20	1125	293	14700	18774	0.783
80%	1.157	26.5	12.21	11.97	14.80	809	292	14189	18322	0.774
50%	0.901	26.5	12.20	12.05	14.78	389	216	11013	14394	0.765
30%	0.678	26.5	12.30	12.26	14.90	173	162	8102	10362	0.782

Indoor Fan Torque Percentage	*Refrigerant Quality*	*Suction Pressure*	*Suction Temp.*	*Suction Superheat*	*Discharge Pressure*	*Subcooling*	*Refrigerant Mass Flowrate*	*Compressor Power*
	[-]	[kPa]	[°C]	[°C]	[kPa]	[°C]	[kg/s]	[W]
100%	0.16	1087.6	15.2	5.76	2500.1	8.93	0.101	3323
80%	0.16	1073.1	14.4	5.56	2495.9	8.82	0.099	3253
50%	0.17	1083.8	15.7	5.20	2387.6	8.11	0.089	2334
30%	0.17	1099.3	14.9	4.91	2283.8	6.59	0.066	1594

In figure 3, the energy efficiency ratio (EER) values, which is the ratio between the RTU cooling capacity and the sum of electrical consumption of the indoor fan, outdoor fan and the compressor, are shown for the two tested outdoor temperatures (i.e. 30°C and 35°C) and for different RTU cooling capacities. Regardless the temperature of the outdoor environment, the EER increases when reducing the cooling capacity of the unit due to the reduction in the condensing pressure. Also, for a fixed cooling capacity, a reduction of the EER is observed when increasing the outdoor temperature.

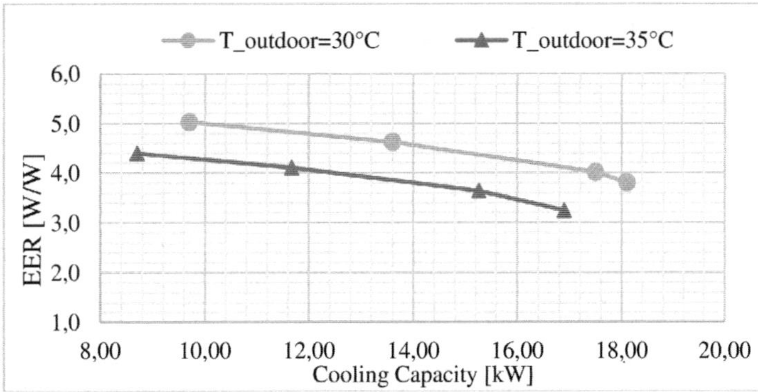

Figure 3. Energy Efficiency Ratio (EER) values for the fault-free scenario

In table 3, exergy results are presented on the basis of the model shown Figure 1. This analysis was carried out by means of Engineering Equation Solver [11]. It is worth observing that:

- the unit exergy consumption of the compressor (i.e. k_1) slightly increases when the capacity of the unit is reduced. This trend is related to the isoentropic efficiency of the compressor, which decreases in the examined range.
- regardless the considered cooling capacity, the mechanical exergy produced by the compressor is largely used by the TXV, being negligible the ones consumed respectively by the evaporator and the condenser. Also, both the mechanical and the thermal exergy provided by the compressor decreases when reducing the cooling capacity due to reduction in the refrigerant mass flowrate and in the pressure ratio.
- the percentage of exergy destruction in the TXV is approximately constant for all load level.
- high values of the evaporator unit exergy consumption (i.e. k_4) are observed.
- the mechanical exergy of the air decreases when reducing the RTU capacity, due to decreasing in both air mass flowrate and pressure drop across the evaporator. Also, the unit exergy consumption of the indoor fan is not sensitive to the load level.

Table 3. Exergy flows for each component for the following test conditions: T_{indoor}=26.5°C, RH= 50% $T_{outdoor}$=30°C and SAT_{sp}= 15°C

			Indoor Fan Torque Percentage			
			100%	**80%**	**50%**	**30%**
1.Compressor	$\dot{W}_{e,1}$	$[kW_{ex}]$	3.323	3.253	2.334	1.594
	ΔB_1^T	$[kW_{ex}]$	0.782	0.746	0.496	0.281
	ΔB_1^M	$[kW_{ex}]$	1.650	1.611	1.194	0.836
	k_1	$[kW_{ex}/kW_{ex}]$	1.366	1.380	1.381	1.428
2. Condenser	$\dot{W}_{e,2}$	$[kW_{ex}]$	0.293	0.292	0.216	0.162
	$\lvert\Delta B_2^T\rvert$	$[kW_{ex}]$	0.82	0.83	0.72	0.58
	$\lvert\Delta B_2^M\rvert$	$[kW_{ex}]$	0.012	0.010	0.005	0.003
3.TXV	ΔB_3^T	$[kW_{ex}]$	1.391	1.357	1.011	0.709
	$\lvert\Delta B_3^M\rvert$	$[kW_{ex}]$	1.636	1.598	1.188	0.832
	χ_3	$[kW_{ex}/kW_{ex}]$	0.150	0.149	0.151	0.152

	$\lvert\Delta B_4^{\mathrm{T}}\rvert$	[kW$_{ex}$]	1.346	1.374	1.354	1.306
	$\lvert\Delta B_4^{\mathrm{M}}\rvert$	[kW$_{ex}$]	0.010	0.009	0.011	0.012
4.Evaporator	$\lvert\Delta B_{\mathrm{air}}^{\mathrm{T}}\rvert$	[kW$_{ex}$]	0.443	0.427	0.329	0.242
	$\lvert\Delta B_{\mathrm{air}}^{\mathrm{C}}\rvert$	[kW$_{ex}$]	0.069	0.071	0.057	0.036
	k_4	[kW$_{ex}$/ kW$_{ex}$]	4.836	4.236	3.545	3.085
	$\dot{W}_{\mathrm{e},5}$	[kW$_{ex}$]	1.12	0.81	0.39	0.17
5. Indoor Fan	$\lvert\Delta B_{\mathrm{air}}^{\mathrm{M}}\rvert$	[kW$_{ex}$]	0.221	0.153	0.060	0.030
	k_{fan}	[kW$_{ex}$/ kW$_{ex}$]	5.07	5.27	6.45	5.79

The overall exergy performance is calculated according to equation 3.

$$\eta_{\mathrm{ex,RTU}} = \frac{\Delta\dot{B}_{\mathrm{air}}^{\mathrm{C}} + \Delta\dot{B}_{\mathrm{air}}^{\mathrm{T}}}{\dot{W}_{e,1} + \dot{W}_{e,5} + \dot{W}_{e,2}} \tag{3}$$

Results for $\eta_{\mathrm{ex,RTU}}$ are shown in figure 4. The overall exergy performance is very poor (i.e. under 15%). Most of the exergy destruction is located in the evaporator as testified by the values of k_4 in Table 3. The exergy performance of the RTU increases when the cooling capacity is reduced and when lower outdoor temperatures are considered.

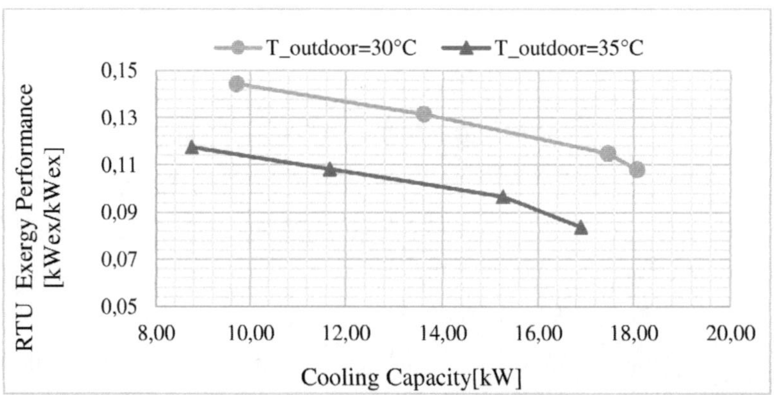

Figure 4. RTU exergy performance for different outdoor temperatures and different cooling capacity in the fault free scenario

4.2 Effect of evaporator fouling on the energy and exergy performance of the RTU.

Experimental results for the evaporator fouling are shown in figure 5. First of all, the main effect of this fault is the reduction in the air mass flowrate across the evaporator. When running the indoor fan at its maximum speed (i.e. 1250 rpm), the following percentage reductions in air mass flowrate were observed: (i) 10% for the light fouling scenario and (ii) 25% for the heavy one. In figure 5, the sensible capacity is shown on the x-axis instead of the total cooling capacity. In fact, since the unit is controlled on the basis of temperature measurement, it is reasonable to compare fault-free and faulty scenarios where the RTU provides the same sensible cooling capacity. It is worth noting that:

1. when the indoor fan is running at the maximum speed, only 12 kW of sensible capacity are provided with respect to 14.5 kW for a fault-free scenario.
2. a slight decrease in both condensing and evaporating pressure is observed;
3. for a fixed sensible capacity, the power consumed by the compressor increases due to the higher refrigerant mass flowrate.
4. a decrease in the EER values is observed. For instance, considering the maximum sensible capacity provided in the heavy fouling scenario, i.e. 12 kW, an EER=3.8 W/W is achieved with respect to 4.2 W/W in the fault-free one.

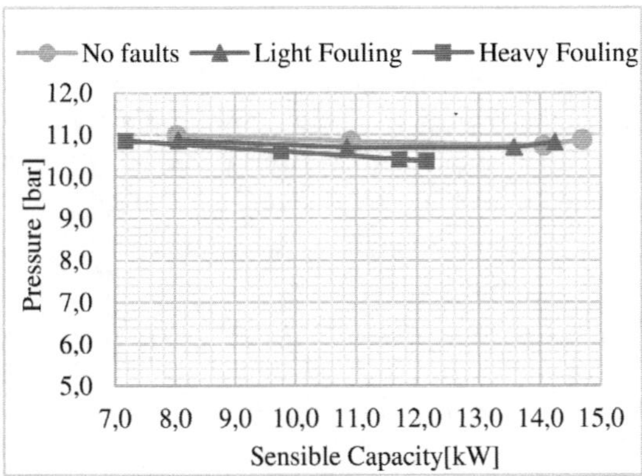

(a)

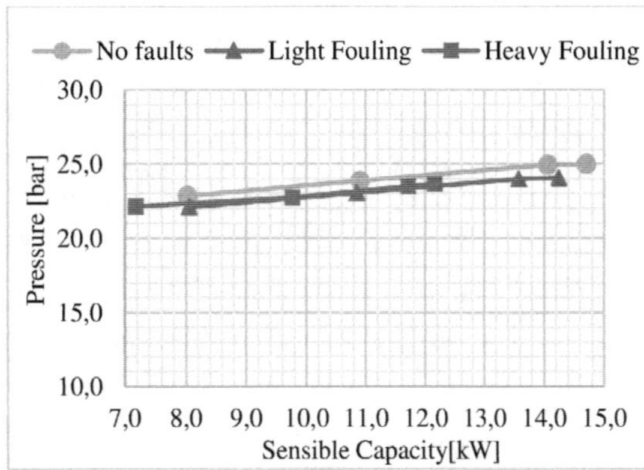

(b)

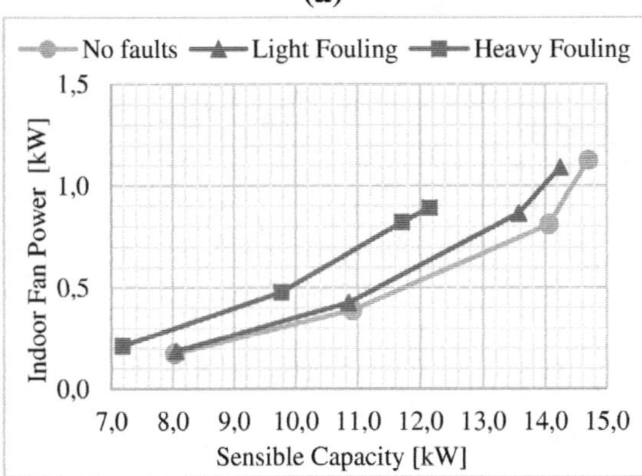

(c)

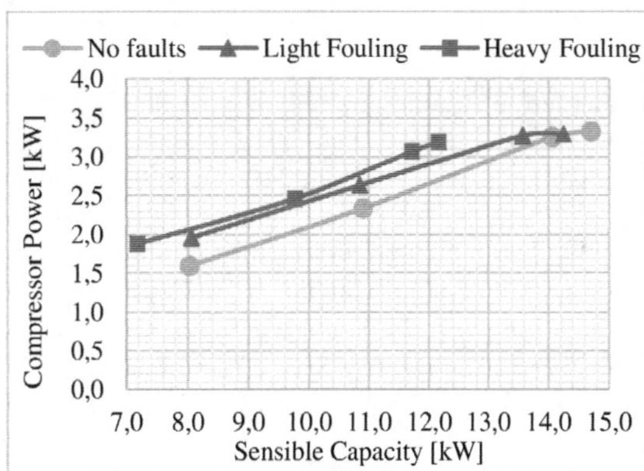

(d)

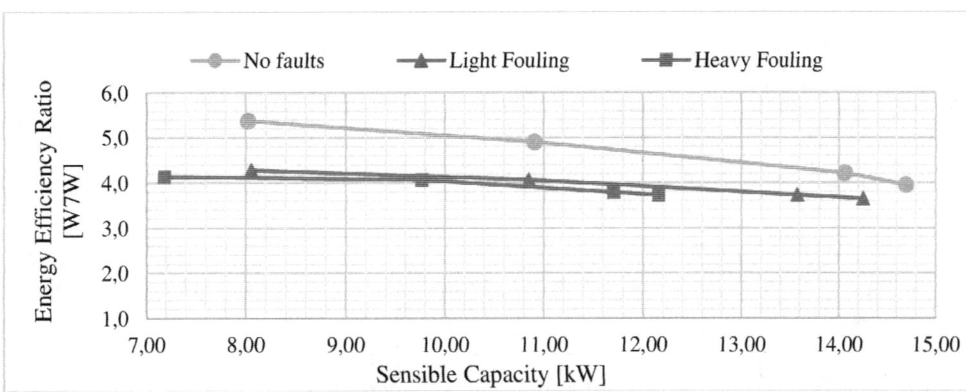

(e)

Figure 5: Thermodynamic results for evaporator fouling: **(a)** evaporating pressure **(b)** condensing pressure **(c)** indoor fan power **(d)** compressor power **(e)** EER values

For a sake of clarity, in Figure 6 only results for $T_{outdoor}$=30°C are shown. The same trends are observed for $T_{outdoor}$=35°C. The unit exergy consumption of the compressor, i.e. k_1, is plotted in Figure 6a. Results showed that for a fixed sensible load, the compressor unit exergy consumption is approximately constant regardless the level of fault imposed. The exergy destruction percentage in the TXV, i.e. χ_3 in Figure 6b show also that for a fixed sensible constant value regardless the level of fault imposed on the evaporator.

Results for the evaporator exergy unit consumption are shown in Figure 6c and 6d. More specifically, the consumption of air mechanical exergy per unit exergy product of the evaporator, i.e. $k_{4,5}$ increases when passing from a fault-free to a faulty scenario due to the increased air pressure drop across the evaporator. Conversely, the unit exergy consumption of refrigerant thermal exergy, i.e. $k_{4,ref}$ is not sensitive to the fault level. These results suggest that detection of evaporator fouling can be easily achieved only by checking variation in the $k_{4,5}$ values

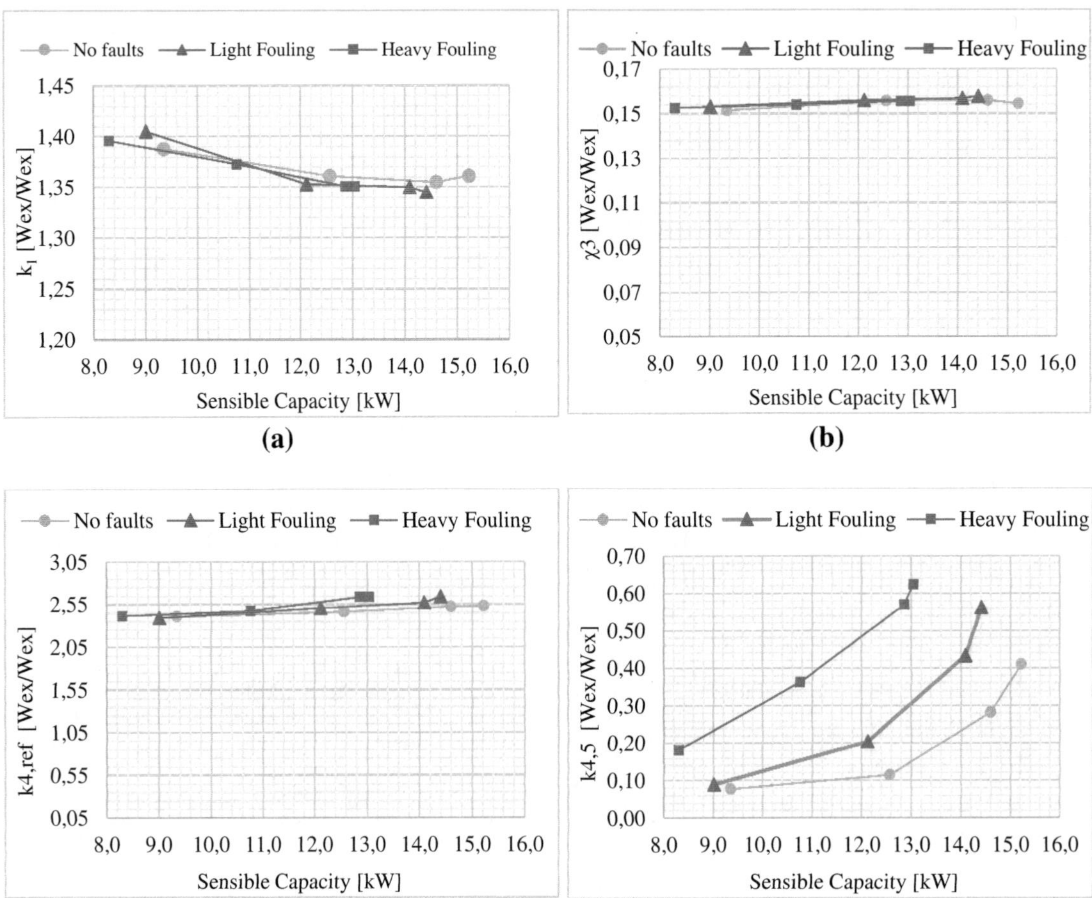

Figure 6: **(a)** Unit exergy consumption of the compressor k_1 **(b)** exergy destruction percentage χ_3 in TXV **(c)** evaporator unit exergy consumption of refrigerant thermal exergy **(d)** unit exergy consumption of mechanical exergy of the air in the evaporator

5. Conclusion

This work aims at mapping the exergy performance of a packaged air conditioning unit equipped with a variable speed-fans and variable speed-compressor. The results showed that energy performance ratio of the unit increases when the cooling load decreases as a consequence of the lower pressure ratios. The overall exergy performance of the unit is poor mainly due to exergy destruction occurring at the evaporator. Some improvements were observed when the RTU capacity is reduced. When investigating evaporator fouling, results suggested that the fault affected mainly consumption of the mechanical air exergy, thus allowing for an easy detection of the fault

References

[1] Breuker M G and Braun J E 1998 Common Faults and Their Impacts for Rooftop Air Conditioners *HVAC&R Research* **4** 303-18
[2] Rossi T M and Braun J E 1997 A statistical rule-based fault detection and diagnostic method for vapor compression air conditioners *HVAC&R Research* **3** 19-37
[3] Li H and Braun J E 2007 Decoupling features and virtual sensors for diagnosis of faults in vapor compression air conditioners *Int. J. Refrigeration* **30** 546–64
[4] Uson S and Valero A 2010 Thermoeconomic Diagnosis of Energy Systems, *Prensa Universidad de Zaragoza* Zaragoza Spain.
[5] Piacentino A. and Talamo M 2013 Innovative thermoeconomic diagnosis of multiple faults in air conditioning units: methodological improvements and increased reliability of results, *Int. J. Refrigeration* **36** 2343-65
[6] Torres C, Valero A, Serra L, Royo J 2002 Structural theory and thermoeconomic diagnosis: Part I. On malfunction and dysfunction analysis *Energy Conversion and Management* **43** 1503-18.
[7] Torres C 2004 Symbolic thermoeconomic analysis of energy systems. In: Frangopoulos CA, editor. *Exergy, energy system analysis and optimization, from encyclopedia of life support system (EOLSS)*. Oxford: EOLSS Publishers. Available at: (hhttp://www.eolss.net).
[8] Valero A, Correas L, Zaleta A, Lazzaretto A, Verda V, Reini M, Rangel V, 2004 On the thermoeconomic approach to the diagnosis of energy system malfunctions: Part 2. Malfunction definitions and assessment. *Energy*, **29** 1889-1907
[9] Verda V 2006 Accuracy level in thermoeconomic diagnosis of energy systems *Energy* **31** 3248-260
[10] Szargut J 2005 Exergy Method: Technical and Ecological Applications *WITpress*
[11] Klein S.A. 2010 Engineering equation solver V9.810. Middleton, WI: Solar Energy Laboratory, *F-Chart Software*

Comparison of two hygroscopic materials for a solar-assisted desiccant-based air handling unit

C Roselli[1], M Sasso[1] and F Tariello[2,*]

[1] Department of Enginereeng, Università degli Studi del Sannio, Piazza Roma 21, 82100 Benevento, Italy
[2] Department of Medicine and Health Sciences "Vincenzo Tiberio", Università del Molise, Campobasso, Italy
* Corresponding author e-mail: francesco.tariello@unimol.it

Abstract. Solar-driven air-conditioning systems are an interesting solution to reduce greenhouse gas emissions as they are essentially powered with a renewable energy source. Among the solar heating and cooling systems, desiccant-based air handling units (AHUs) are a very promising alternative as they require low temperature thermal energy for operation in cooling mode. In this paper the experimental plant of University of Sannio (Benevento, Southern Italy), consisting of an AHU with a desiccant wheel using as hygroscopic material silica gel is coupled with solar thermal collectors. This AHU has been also simulated with a different material for the desiccant wheel, the MIL101@GO-6 (MILGO). It is a composite material, consisting of graphite oxide dispersed in the MIL101 metal organic framework network structure, that shows a better dehumidification effectiveness. The energy and environmental performance of a desiccant-based solar heating and cooling system with this new material have been evaluated with respect to the conventional AHU based on cooling dehumidification. Yearly based dynamic simulations of the plant serving a university classroom located in Benevento have been performed. Furthermore, a comparison with the air-conditioning system equipped the currently adopted material (silica gel) has been carried out too. The optimal plant configuration varying collector type (flat plate and evacuated tube), tilt angle and surface has been found.

1. Introduction

The replacement of electric-driven air-conditioning systems with solar activated ones could make an important contributions in terms of buildings primary energy demand saving, GHG emissions reduction and summer peak electric load cut. Desiccant-based air handling units operate with a thermally-driven cycles as they require heat to regenerate the desiccant material. This thermal energy can be supplied by various conventional means like waste heat, gas burner or electric heater. However, the best coupling is with solar energy technologies because solar energy availability and cooling load are approximately in phase.

During 2016, in European Union up to 40% of the total energy consumption derives by the building sector, and at the same time also 40% of the total greenhouse gas emissions are released in this ambit [1]. The energy demand for air-conditioning in the residential sector is estimated to be about the 65% of the total energy consumption [2]. In addition to this, it could be observed that cooling energy demand is showing a growing trend both in developing and developed Countries in the last two decades.

In such a scenario, it seems interesting to investigate the benefits deriving by solar heating and cooling systems, in particular by those solutions based on AHU equipped with desiccant wheel (DW).

In [3] a solar-driven AHU equipped with a silica gel DW and a heat wheel was investigated. A Maisotsenko cycle cooler is also integrated in the AHU to reduce the process air temperature and so to meet the sensible load of the building. The solar field consists of 7.2 m^2 flat plate solar collectors and 4.7 m^2 evacuated tube collectors. On the basis of the experimental results the Authors verified a cooling capacity in the range 3.4 – 4.28 kW and a COP variable between 0.65 and 1.17. The dehumidification effectiveness was significantly affected by the inlet air temperature (reduction) and by the inlet air humidity (increase), varying from 29% to 49%. The solar fraction reaches the maximum value of 70%.

The Authors of [4] analysed from an energy, exergy and economic point of view seven different configurations of desiccant-based AHU using TRNSYS software. The air handling units are diversified by:

- the adoption or not of a heat recovery wheel, which allows to transfer thermal energy between the warm exhausted regeneration air and the same regeneration air (pre heating) before it is heated with solar energy;
- the number of dehumidification stages (five solutions provide one stage while two are with two stages);
- the amount of recirculated process air.

The best COP and exergy efficiency take place in Dunckle configuration and ventilation mode. They are 0.6 and 35%, respectively. The heat recovery systems improve the performance of AHUs. From an economic point of view there is 50% in electric energy saving of the Uckan and Dunckle configurations compared to air conditioning heat pumps, leading to a payback period of about 4 years.

In [5], a mathematical model of a desiccant-based AHU is developed. It is used to predict the plant behaviour considering the typical climatic conditions of the Algerian coast and the effect of operating parameters. Temperature and humidity ratio of the process air supplied to the room decrease when the DW rotation speed passes from 20 rpm to 10 rpm, ensuring better performance.

The most widely used material for DW is the silica gel, but many alternative substances have been analysed too. In [6], a synthesized metal silicate DW is adopted in a AHU together with a sensible heat wheel and an electric heat pump. The COP varying air stream flow rates, regeneration temperature and ambient conditions have been experimentally evaluated. The results are used to a TRNSYS model that demonstrates to efficiently simulate the cooling system.

In [7], two alternative desiccant materials (Silica-Gel and Titanium Dioxide) have been compared. A solar-desiccant cooling system is numerically investigated through a validated TRNSYS model in three East Asian climatic conditions (temperate, subtropical and tropical). Titanium Dioxide has proven to be a good alternative material as it can reach lower indoor temperature and humidity ratio with higher cooling performance than the Silica-Gel, considering the same specification of the solar thermal system and desiccant cooling system. The system coefficient of performance (SCOP) is within range 1.5–3 while the solar fraction is between 65% and 90%.

In this work an AHU equipped with a DW filled with an alternative hygroscopic material, named MILGO, is considered and compared with a conventional HVAC system for the air-conditioning service of an university classroom located in Benevento. Flat plate and evacuated tube solar thermal collectors are considered as heat source. The aperture area of the solar field is varied as well as the tilt angle. Furthermore the possibility to recover different percentage of the solar thermal energy surplus for further low temperature uses is assessed too. Finally, the energy, environmental and economic results obtained by the innovative AHU, with respect to the conventional air-conditioning system, are compared with the same results achieved by the plant incorporating a silica gel DW.

2. Plant description

The alternative plant analysed in this paper is a HVAC system mainly driven by solar thermal energy. Such thermal energy is used:

- to regenerate, during the cooling operation (Figure 1), the desiccant wheel (DW) of the air handling unit;
- to realize the pre and post heating processes in the heating period (Figure 2).

More in detail, the alternative system (AS) consists of a solar subsystem and a desiccant-based air handling unit. A solar field, with different collectors configurations, is connected to a 1000 l thermal energy storage tank (TS); to prevent the solar circuit from high temperature level (>100°C) a heat exchanger that dissipates thermal energy surplus or produces hot water for sanitary use or for other purposes (HW-HX) is installed. For the solar field the analyses developed later will consider two solar thermal collectors (SC) types (flat plate and evacuated tube) and three aperture areas: about 20, 27 and 34 m^2. The solar field layouts are described in Table 1 [8].

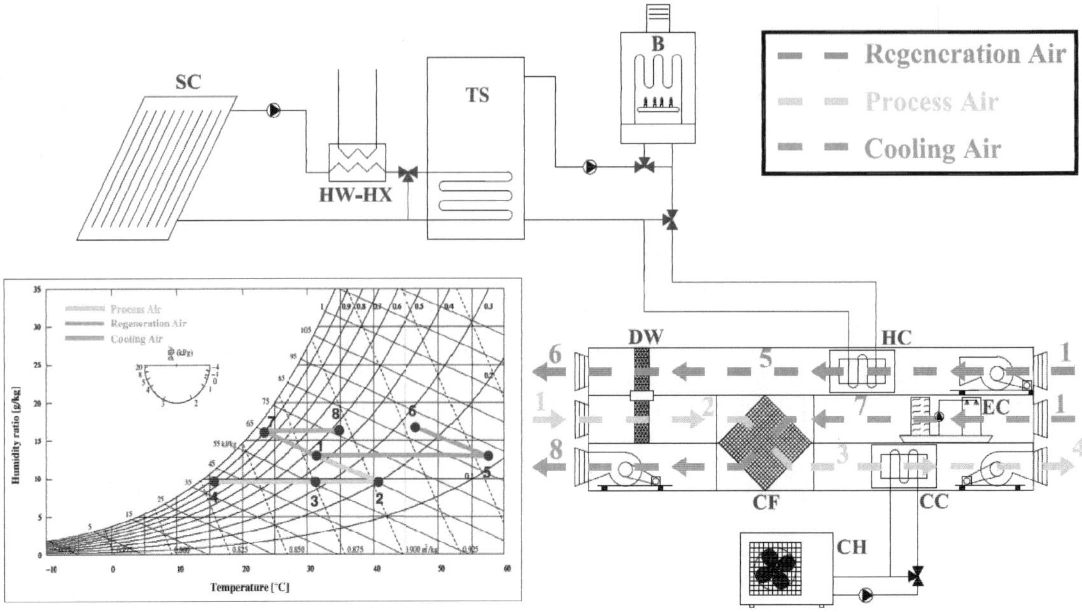

Figure 1. Alternative system layout (cooling mode).

The proposed AHU is arranged as the experimental system installed at the University of Sannio laboratory [9]. It is a hybrid desiccant-based air handling unit because air cooling is also controlled with an electric chiller (CH). It has three air channels: one for process air, that is the air supplied to the conditioned space after dehumidification (1-2) and cooling (2-3-4); one for the cooling air, that is outdoor air cooled down by humidification (1-7); the last one for regeneration air, that is outdoor air heated up (1-5) by solar thermal energy or if necessary by the heat supplied by a natural gas fired boiler (B) to regenerate (5-6) the hygroscopic material of the desiccant wheel. The adsorption process realized on the surface of the lower part of the desiccant wheel allows a nearly isenthalpic dehumidification process, (see Figure 1). The DW actually installed has a matrix composed of alternating layers of smooth and corrugated silica gel and metal silicates sheets, chemically incorporated into a support of inorganic fibers. This honeycomb structure maximizes the contact surface with air, reduces the pressure drop and the weight, and increases the structural strength. The wheel has a weight of 50 kg and its dimensions are 700 and 200 mm (diameter and thickness). The frontal area of the wheel exposed to process and regeneration air flows is characterized by a diameter of about 600 mm, because a circular crown of the total area is obstructed by the metallic frame of the wheel cassette. The rated rotation speed is 12 revolutions per hour. Sixty percent of the cross section of the DW is crossed by process air while the remaining 40% by regeneration air.

Hereinafter, it will be considered to replace the silica gel with an alternative hygroscopic material, the MIL101@GO-6 (MILGO), numerically analyzed in a previous paper [10]. It is a composite material, consisting of 6% w/w graphite oxide dispersed in the MIL101 metal organic framework network structure. On the basis of previous evaluations, the material shows a dehumidification efficiency 30% greater than silica gel, so it is a promising substitute of the conventional desiccant material in the solar driven system.

The performances of the AS, equipped with a DW filled of this new material, are compared with those of a conventional system (CS). In addition, the energy, environmental and economic indices are compared with the results reported in [8] for Benevento.

In the CS the AHU realizes dehumidification by cooling down the air flow below the dew point (cooling energy is removed by an electric chiller), subsequently it is heated up to supply to the conditioned space air not too cold.

In heating mode the innovative AHU is configured as in Figure 2, it is arranged as in the standard configuration that provides pre-heating (1-2-3), humidification (3-4) and post-heating (4-5), but thermal energy is supplied by the solar subsystem and if not sufficient by the boiler. In the conventional plant the boiler is the only heat source during the winter period. More detailed information about the components rated characteristics are described in the following Table 2.

Table 1. Solar collectors configurations [8].

Collector Types	Aperture Area [m^2]	Arrangements	Solar Loop Pump Power [W]
	9x2.25=20.3	1 row of 4 collectors + 1 row of 5 collectors	200
Flat-Plate	12x2.25=27.0	3 rows of 4 collectors	300
	15x2.25=33.8	3 rows of 5 collectors	375
	6x3.43=20.6	2 rows of 3 collectors	200
Evacuated Tube	8x3.43=27.4	4 rows of 2 collectors	300
	10x3.43=34.3	2 rows of 3 collectors +1 row of 4 collectors	330

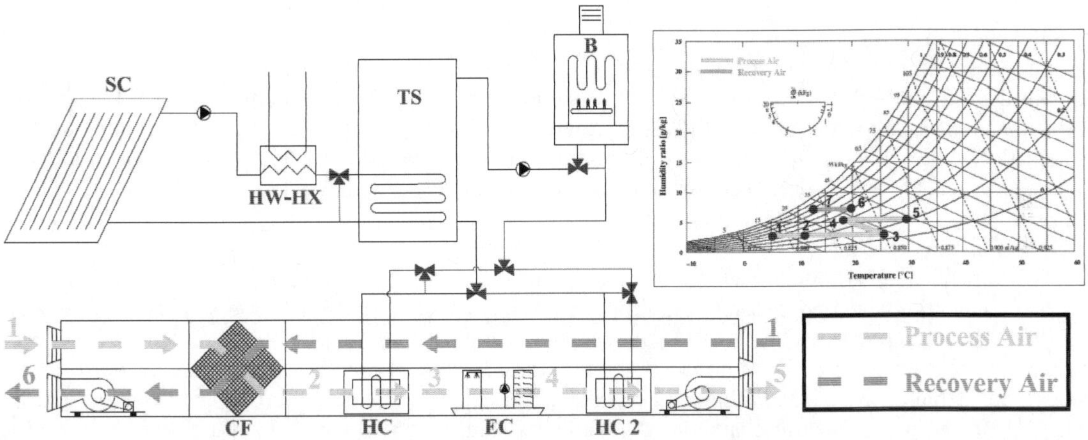

Figure 2. Alternative system layout (heating mode).

The operation of the AHU is determined by the user demand. It is a university classroom located in Benevento (Southern Italy) with a floor surface of 63.5 m^2 and 30 seats occupied in the weekdays from 9:00 to 18:00. The air conditioning system is switched on at 08:30 in the morning, half an hour before the opening of the classroom, and it is turned off at 18:00 in the afternoon, when the classroom is closed. The indoor air set-point temperatures in winter and summer operation are 20 °C and 26 °C, respectively, while the relative humidity is constantly maintained at 50%. A dead band of 0.5 °C and 10% is assumed in the control strategy respectively for temperature and humidity.

The circulation pump in the solar circuit starts working when the fluid temperature in the solar collectors is higher than that in the TS; thermal energy is taken from the tank to feed the heating coils (HC, HC2) when the AHU is ON, the boiler operates as a back-up system if the temperature of the hot water is not high enough. The cooling coil is fed by the electric chiller during the cooling operation if the process air has not been sufficiently cooled by the cooling air in the cross flow heat exchanger (CF).

Table 2. AS main components rated characteristics.

Devices	Rated main characteristics	Values
AHU fans	Volumetric flow rate (m^3/h)	800
	Power (W)	300
TS	Net volume (l)	971
	Tank loss coefficient (W/(m^2·K))	1.37
B	Heating capacity (kW)	24.0
	Efficiency (%)	90.2
CH	Cooling capacity (kW)	8.50
	COP	2.98

The main characteristics of the building envelope are listed in Table 3.

Table 3. Classroom envelope characteristics.

	Opaque Components				Transparent Components		
	Roof	External walls (N/S)	External walls (E/W)	On the ground floor	North	South	East/West
U (W/m^2K)	2.30	1.11	1.11	0.297	2.83	2.83	2.83
Area (m^2)	63.5	36	15.87	63.5	8.53	9.40	0.976
g (-)	-	-	-	-	0.755	0.755	0.755

3. Models and analyses

The software TRNSYS 17 [11] integrated with the "TESS" libraries [12] has been used to dynamically simulate the AS and CS. Simulations with a time-step of 1.5 min have been performed. In TRNSYS the system components, the controls and the building are reproduced by types, each of which implements a mathematical model. In order to simplify the discussion and to concentrate the attention on the DW, only this mathematical models will be described, the others have been already reported in [8], [9] and [13].

3.1 Desiccant Wheel model

The DW with silica gel adsorbent material is modelled with the type 1716 of TESS library. Its mathematical model calculates the performance using the simplified Maclaine-Cross and Banks method [14], that models the dehumidification process, a combination of mass and heat transfer, similarly to the simple thermal energy transfer in a heat exchanger. The coupled equations, that describe the two processes, are reduced to two uncoupled differential equations with two independent variables, called characteristic potentials F_1 and F_2, [15], [16]. In the psychrometric chart, the isopotential F_1 lines approximately represent constant specific enthalpy lines while the isopotential F_2 lines approximately describe constant relative humidity curves. The potential functions depend on the thermohygrometric properties of the air and on the thermophysical properties of the adsorbent material [17]. These relations have been expressed for the pair silica gel-air by Jurinak [16], and they are:

$$F_{1,j} = \frac{-2865}{(t_j+273.15)^{1.49}} + 4.344(\omega_j/1000)^{0.8624} \tag{1}$$

$$F_{2,j} = \frac{(t_j+273.15)^{1.49}}{6360} - 1.127(\omega_j/1000)^{0.07969} \tag{2}$$

where the subscript "*j*" refers to the generic thermo-hygrometric condition of the air at which the two potentials are evaluated, whereas t and ω are the air temperature (°C) and the humidity ratio (g/kg), respectively. The result of the intersection of the isopotential lines are the output conditions of the process air in the ideal case, when both the adsorption and the desorption processes are isoenthalpic. The Jurinak's model provides that the conditions of real output are estimated using two efficiency indices of the wheel, η_{F1} and η_{F2}, calculated in analogy with the efficiency of a heat exchanger as:

$$\eta_{F1} = (F_{1,2} - F_{1,1})/(F_{1,5} - F_{1,1}) \tag{3}$$

$$\eta_{F2} = (F_{2,2} - F_{2,1})/(F_{2,5} - F_{2,1}) \tag{4}$$

where potentials F_1 and F_2 must be evaluated in the states 1, 2 and 5 of Figure 1.
The parameter η_{F1} and η_{F2} for the silica gel DW of the experimental plant were validated and calibrate in [18] and their values are 0.207 and 0.717 respectively. Instead, for MILGO the two effectiveness parameters are

calculated as the average values obtained by 48 simulated conditions and they are equal to 0.029 and 0.904, respectively [10].

3.2 Energy, environmental and economic analysis

The results of dynamic simulation have been elaborated in order to perform a comparison between the AS and CS. The analyses consist in the evaluation of the:

- primary energy saving (PES) achieved by AS with respect to CS;
- equivalent CO_2 emissions avoided by AS with respect to CS;
- simple pay back period (SPB) related to the investment in the new plant.

In addition, these results obtained by the proposed plat with MILGO DW are compared with the same indices assessed for the AS with silica gel based DW and reported in [8].

From the energy point of view the comparison of AS an CS is carried out considering the primary energy requirements related to fossil fuels, in fact the Primary Energy Saving of no-renewable energy sources is:

$$PES = \left(1 - E_p^{AS}/E_p^{CS}\right) \times 100 \tag{5}$$

where:

$$E_p^{AS/CS} = \left(E_{el,CH}^{AS/CS} + E_{el,aux}^{AS/CS}\right)/\eta_{EG} + E_{th,B}^{AS/CS}/\eta_B \tag{6}$$

the primary energy of the alternative and conventional system ($E_p^{AS/CS}$) is evaluated taking into account that the energy efficiency of the Italian national electric system (η_{EG}), including transmission and distribution losses, is 42% [8] and using the boiler efficiency reported in Table 2, 90.2%. In addition, it is assumed that there is no primary energy associated to solar energy because it is a renewable energy source.

To assess the positive effects on the environment of the AS installation, equivalent CO_2 emissions of the two systems have been calculated and the equivalent CO_2 avoided emissions have been derived:

$$\Delta CO_2 = \left(1 - ECO_2^{AS}/CO_2^{CS}\right) \times 100 \tag{7}$$

where:

$$CO_2^{AS/CS} = \left(E_{el,CH}^{AS/CS} + E_{el,aux}^{AS/CS}\right) \cdot \alpha + E_{th,B}^{AS/CS} \cdot \beta/\eta_B \tag{8}$$

β is the specific emission factor of primary energy related to natural gas combustion, equal to 0.207 $kgCO_2/kWh_{Ep}$, [8] and α is the specific emission factor of electricity drawn from the Italian grid, equal to 0.573 $kgCO_2/kWh_{el}$ [8].

Concerning the economic analysis, the feasibility of the AS can be estimated by means of the Simple Pay Back (SPB) period:

$$SPB = EC/\textstyle\sum_{k=1}^{N} F_k \tag{9}$$

where N is the number of years to payback the investment, i.e. the number of years for which the equation is verified, EC is the extra cost of the AS (desiccant-based AHU, storage tank and collectors) with respect to the CS, F_k is the cash flow for the generic year k:

$$F_k = OC^{CS} - OC_k^{AS} \tag{10}$$

with OC_k^{AS} and OC^{CS} are indicated the operating costs of the AS and CS; the former is given by:

$$OC_k^{AS} = \textstyle\sum_r V_{NG,r} c_{NG,r} + \left(E_{el,CH}^{AS} + E_{el,aux}^{AS}\right) \cdot c_{el} + MC_B + MC_{CH}^{AS} + MC_{SC} - I_{a,tot} \tag{11}$$

where the following assumptions, according to Italian conditions and specific components used, were considered:

- Lower Heating Value (LHV) of natural gas equal to 9.52 kWh/Sm^3;
- the total volume of natural gas $V_{NG,tot} = \sum_r V_{NG,r} = E_{th,B}/(\eta_B \cdot LHV)$ should be divided according to the brackets reported in [8], to which different unitary costs ($c_{NG,r}$) are associated;
- unitary cost of electricity (c_{el}) equal to 0.211 €/kWh;

- extra cost of desiccant-based AHU with respect to the conventional one equal to 10,000 €, it also includes maintenance costs, as provided by the purchase contracts;
- investment cost of storage tank equal to 3,000 €;
- investment cost of chiller: 3,000 € for the AS, 6,000 € for the CS;
- specific cost of collectors (c_{SC}): 360 €/m^2 for flat-plate collectors; 600 €/m^2 for evacuated collectors;
- maintenance cost of the boiler (MC_B) equal to 80 €/y;
- maintenance cost of the chiller in the AS (MC_{CH}^{AS}) equal to 150 €/y [9];
- maintenance cost of the solar collectors ($MC_{SC}= mc_{SC}\cdot S\cdot c_{SC}$) with specific maintenance cost mc_{col} equal to 2% and S gross solar collectors area [19];
- total annual incentive ($I_{a,tot}$) evaluated on the basis of the Italian mechanism so-called "Conto Termico".

4. Results and description

In terms of primary energy saving (PES) the AS shows a certain reduction of the energy demands for all the configurations considered, Figure 3. However, great benefits do not derive from the aperture area growth when flat plate collectors and the new desiccant material is considered, in fact the best PES takes place for 27 m^2 of aperture area (Figure 3-a). With evacuated tube collectors greater savings appear with the widest solar field but the second better option is that of 21 m^2 (Figure 3-b). On the contrary, with silica gel DW an improvement in the PES always corresponds to the increase of the solar field aperture area (Figure 3-c-d). Evacuated tube collectors with both desiccant materials determine better energy performance, only the configuration with 20° tilt angle, flat plate collectors and MILGO DW shows worse PES than all the solutions with silica gel desiccant wheel.
The reason of the irregular trends, obtained with the new material, can be explained by considering the solar fraction indices:

- Total solar fraction

$$SF_{Total} = \frac{E_{th}^{TS-DWreg} + E_{th}^{TS-preheat} + E_{th}^{TS-postheat}}{E_{th}^{DWreg} + E_{th}^{preheat} + E_{th}^{postheat}} \tag{12}$$

- Cooling mode solar fraction

$$SF_{Cooling} = \frac{E_{th}^{TS-DWreg}}{E_{th}^{DWreg}} \tag{13}$$

- Heating mode solar fraction

$$SF_{Heating} = \frac{E_{th}^{TS-preheat} + E_{th}^{TS-postheat}}{E_{th}^{preheat} + E_{th}^{postheat}} \tag{14}$$

Where $E_{th}^{TS\text{-}DWreg}$ is the thermal energy contribution of the solar subsystem to the regeneration of the DW; $E_{th}^{TS\text{-}preheat}$ and $E_{th}^{TS\text{-}postheat}$ are the thermal energy shares of the solar subsystem respectively to the pre-heating and post-heating phases; E_{th}^{DWreg} is the regeneration energy for the DW; $E_{th}^{preheat}$ and $E_{th}^{postheat}$ are the thermal energy for the pre and post heating processes.
For all the configuration a decreasing trends with the tilt angle are recorded for the SF in cooling mode, whereas a the SFs grow with solar collectors inclination in heating mode (Figure 4). Furthermore, in cooling operation, SFs are always greater than 85% except for the configuration with 20 m^2 of flat plate collectors (Figure 4-a): this implies that not significant benefits derive from the bigger collecting surfaces, contrariwise, the solar thermal energy dissipated increases.

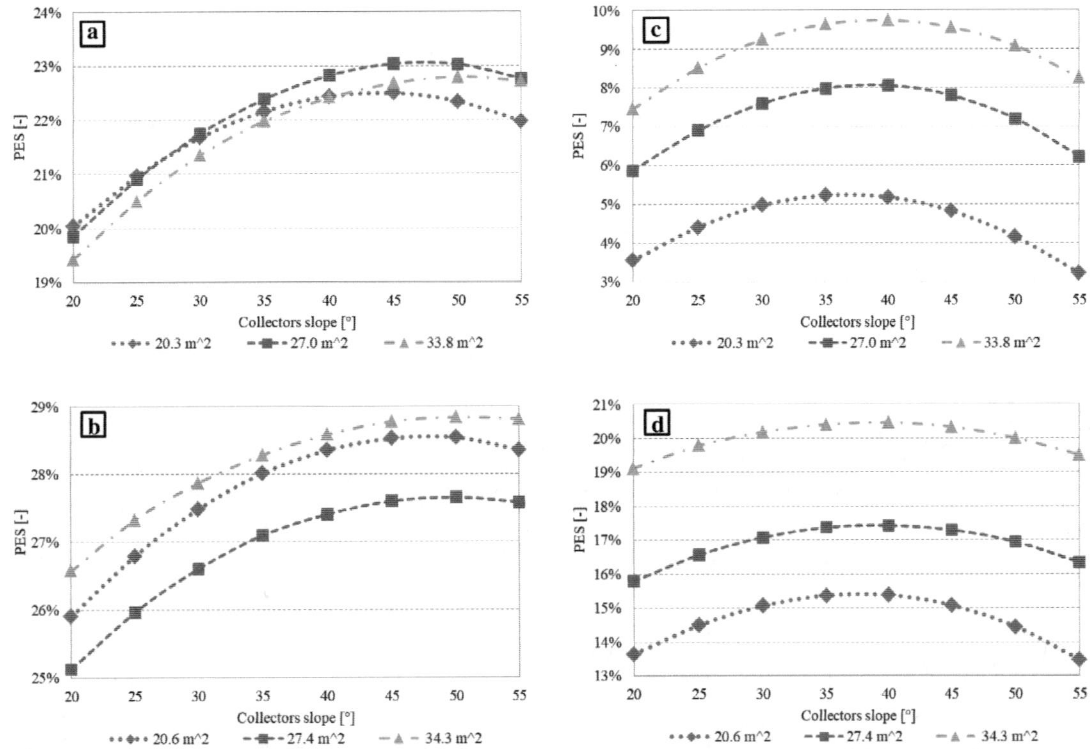

Figure 3. Primary energy savings as a function of the tilt angle and the solar field aperture area for: (a) MILGO DW and flat plate collectors, (b) MILGO DW and evacuated tube collectors, (c) silica gel DW and flat plate collectors and (d) silica gel DW and evacuated tube collectors.

Within the range of variation of the tilt angle (20°-55°) greater improvements are observed in the winter solar fractions; instead, the total SFs have a quite flat trends for each solar field surface and collector type, the optimal tilt angle appear at about 46°. In the plants with MILGO DW the maximum mean increase (about 7 percentage points) takes place between 20 and 27 m^2 (only 5 percentage point between 27 and 34 m^2).
Taking into consideration the SFs reached in the plants with silica gel DW (Figure 5) the same trends can be observed (SFs in cooling decrease with tilt angle whereas SFs in heating increase) but wider variations take place in the solar fraction evaluated in cooling mode rather than in heating especially with the solar fields of 20 and 27 m^2 and with flat plate collectors (Figure 5-a-b-d-e). These results highlight that there is a higher possibility to exploit solar radiation both in heating and in cooling operation when the desiccant wheel filled with silica gel is considered; as expected the new material allows to reduce the regeneration energy in summer and therefore small collector fields are necessary. Bigger surfaces are justified only by the need to better meet the thermal energy demand during the heating operation. Finally, looking at the total solar fractions with MILGO hygroscopic material one can see that, unlike the PES, they increase with the solar field aperture area, this because SF does not consider the electric energy used to run the fan of a dry cooler to dissipate solar thermal energy surplus (it replaces the HW-HX when hot water production is not considered).
As concern the environmental analysis, Figure 6 shows the calculated ΔCO_2. In terms of equivalent CO_2 emissions it is emphasized the condition for which better results take place with smaller collecting surfaces when the air handling unit is equipped with MILGO DW (Figure 6-a-b). Also for this index it is preferable to enlarge the solar field that drives the innovative AHU in its initial configuration (Figure 6-c-d).

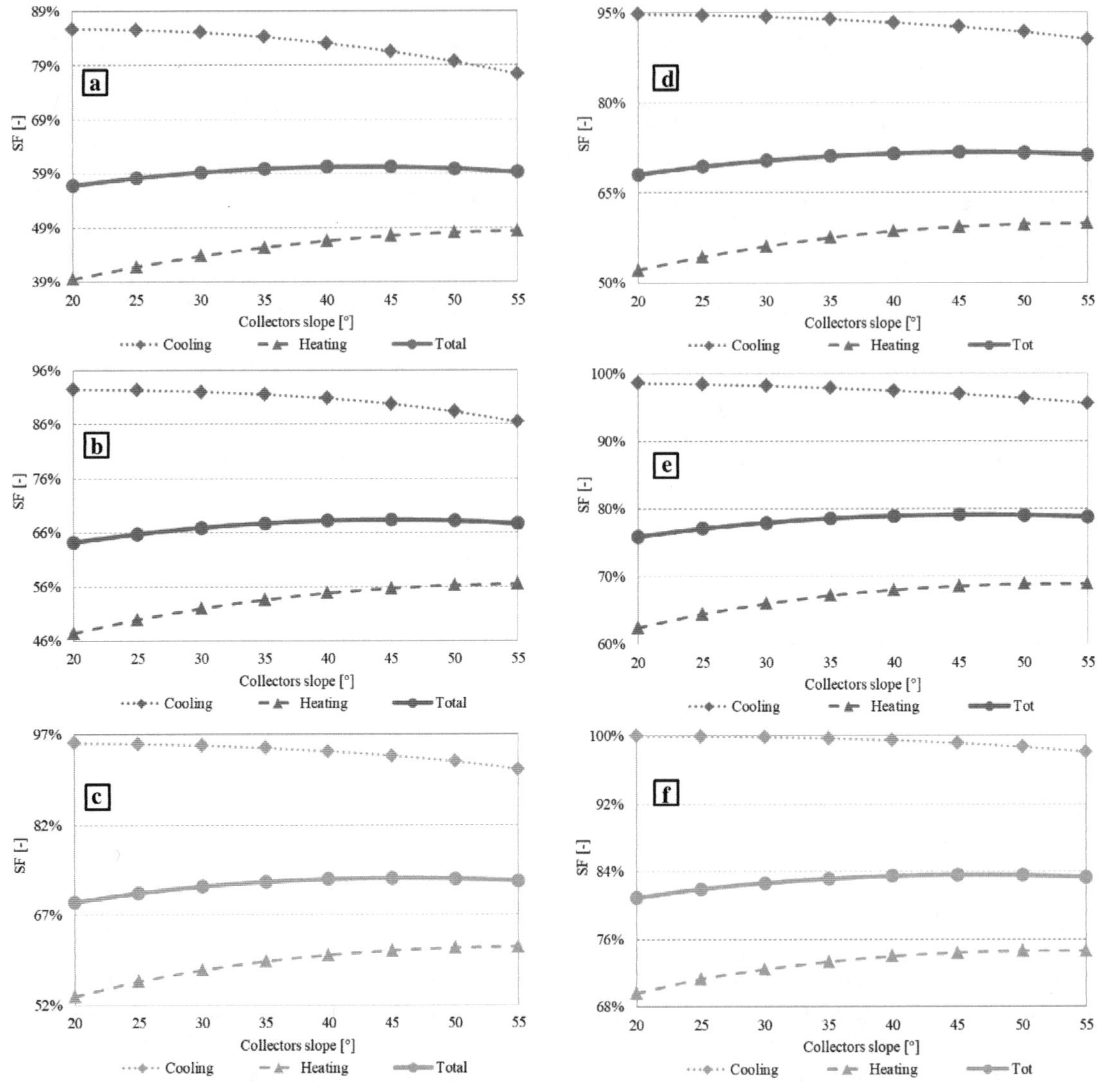

Figure 4. Solar fraction for the plant equipped with MILGO DW in cooling operation, heating operation and total considering: (a) 20 m^2 of flat plat solar collectors, (b) 27 m^2 of flat plat solar collectors, (c) 34 m^2 of flat plat solar collectors, (d) 20 m^2 of evacuated tube solar collectors, (e) 27 m^2 of evacuated tube solar collectors and (f) 34 m^2 of evacuated tube collectors.

Despite the new desiccant material achieves better energy (and environmental) results with respect to the silica gel the investment costs are too high and the simple pay back period is much longer that the useful life of the plant. Interesting SPB periods take place only if the solar thermal energy over-productions are considered for other thermal uses at low temperature. The results for the partial (50%) and total surplus exploitation are reported in Table 4 and Table 5.
The use of solar thermal energy extra-production has sensible effects also in improving the PES ad ΔCO_2. In Figure 7, Figure 8, Figure 9 and Figure 10 the diagrams of PES and the ΔCO_2 indices when the solar thermal energy surplus is recovered for half or totally, are reported. It is evident that a strong improvement in the exploitation of the renewable energy source happens: the best results are achieved with evacuated tube solar collectors and the largest solar field surface.

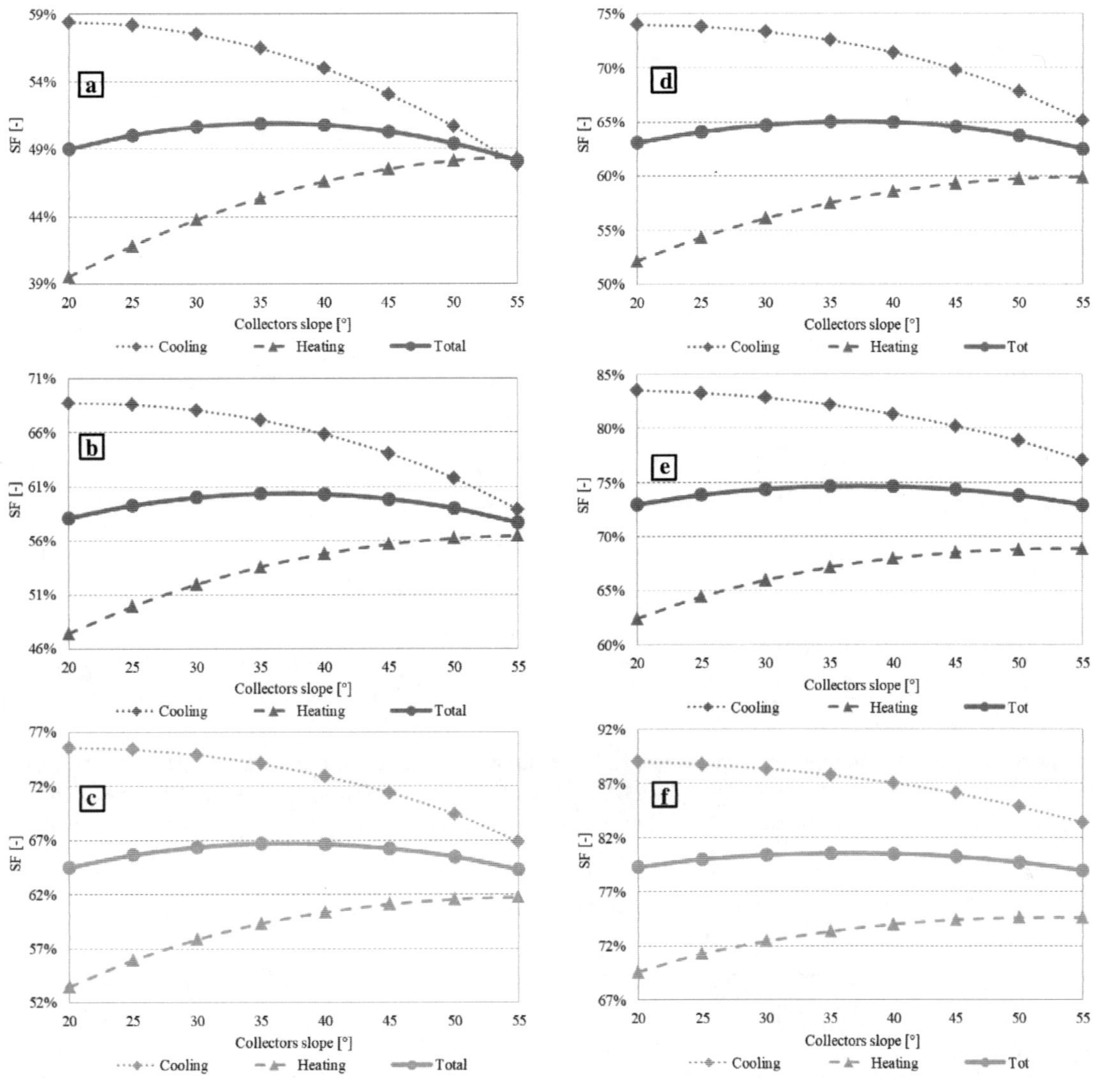

Figure 5. Solar fraction for the plant equipped with silica gel DW in cooling operation, heating operation and total considering: (a) 20 m^2 of flat plat solar collectors, (b) 27 m^2 of flat plat solar collectors, (c) 34 m^2 of flat plat solar collectors, (d) 20 m^2 of evacuated tube solar collectors, (e) 27 m^2 of evacuated tube solar collectors and (f) 34 m^2 of evacuated tube collectors.

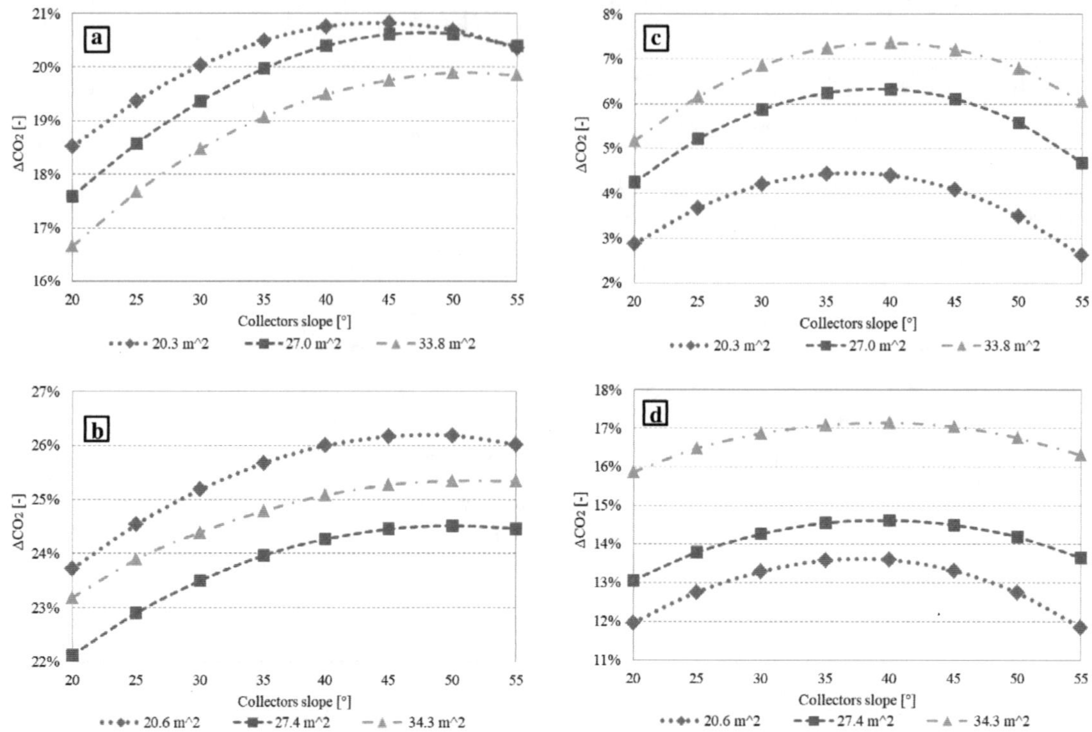

Figure 6. Equivalent CO_2 emissions avoided as a function of the tilt angle the solar field aperture area for (a) MILGO DW and flat plate collectors, (b) MILGO DW and evacuated tube collectors, (c) silica gel DW and flat plate collectors and (d) silica gel DW and evacuated tube collectors.

Table 4. SPB period for the AS in different configurations when 50% of the solar thermal energy surplus is used for additional low temperature purposes.

	Flat plate		**Evacuated tube**	
Aperture area (m^2)	MILGO (y)	Silica gel (y)	MILGO (y)	Silica gel (y)
20	15	17	17	16
27	12	11	14	12
34	9	8	12	10

Table 5. SPB period for the AS in different configurations when the solar thermal energy surplus is totally used for additional low temperature purposes.

	Flat plate		**Evacuated tube**	
Aperture area (m^2)	MILGO (y)	Silica gel (y)	MILGO (y)	Silica gel (y)
20	10	12	10	11
27	7	8	8	9
34	5	6	6	7

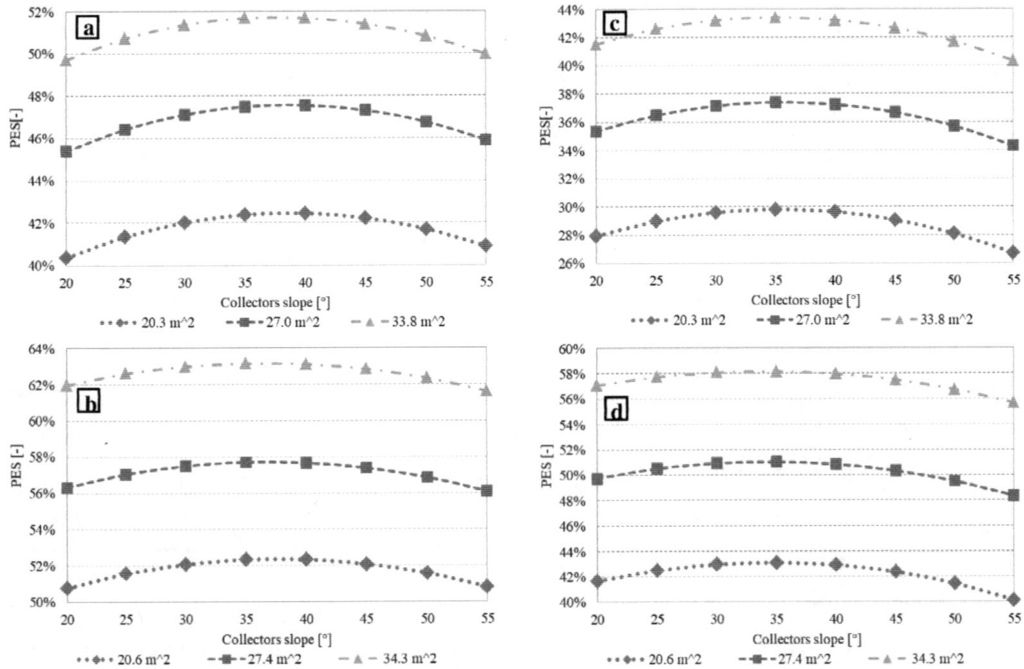

Figure 7. Primary energy savings as a function of the tilt angle and the solar field aperture area when half of the solar thermal energy surplus is used, in the following configuration: (a) MILGO DW and flat plate collectors, (b) MILGO DW and evacuated tube collectors, (c) silica gel DW and flat plate collectors and (d) silica gel DW and evacuated tube collectors.

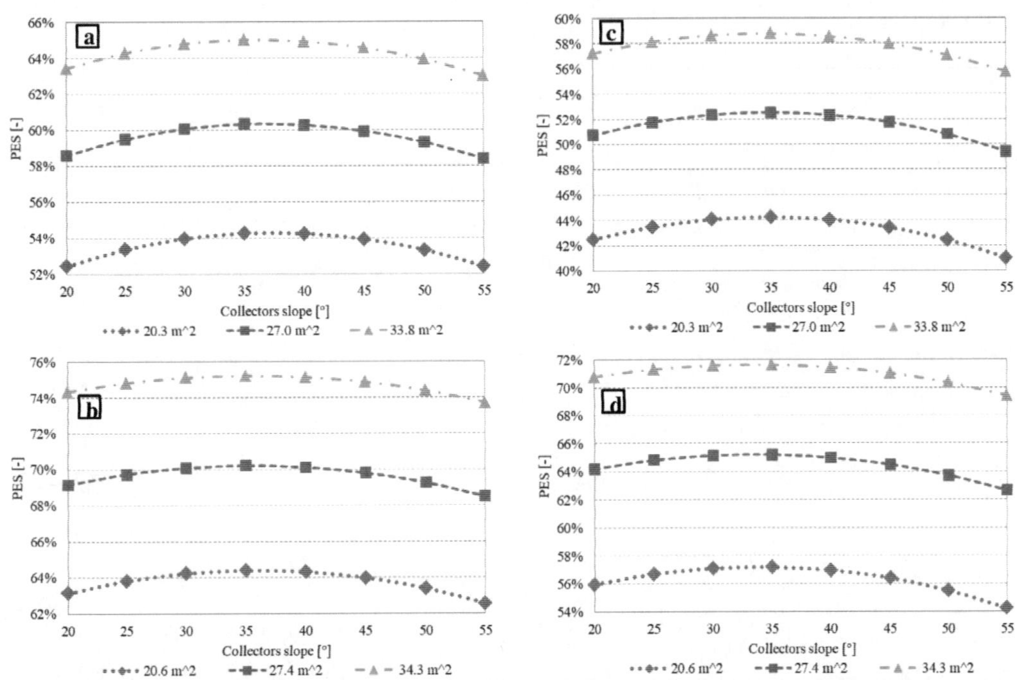

Figure 8. Primary energy savings as a function of the tilt angle and the solar field aperture area when the solar thermal energy surplus is totally used, in the following configuration: (a) MILGO DW and flat plate collectors, (b) MILGO DW and evacuated tube collectors, (c) silica gel DW and flat plate collectors and (d) silica gel DW and evacuated tube collectors.

IOP Conf. Series: Journal of Physics: Conf. Series **1224** (2019) 012025 doi:10.1088/1742-6596/1224/1/012025

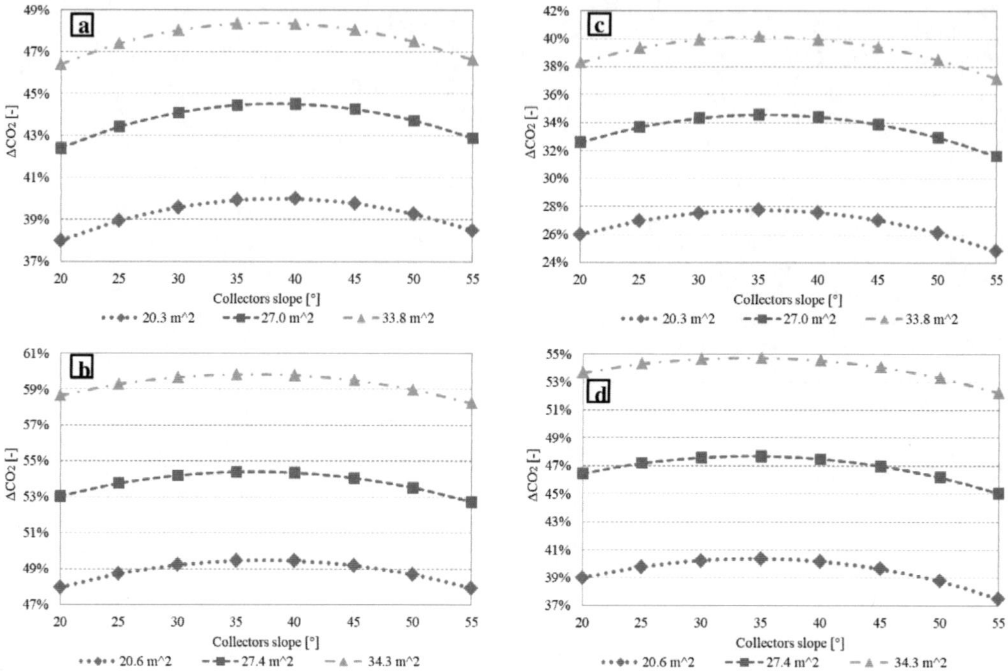

Figure 9. Equivalent CO_2 emissions avoided as a function of the tilt angle and the solar field aperture area when half of the solar thermal energy surplus is used, in the following configuration: (a) MILGO DW and flat plate collectors, (b) MILGO DW and evacuated tube collectors, (c) silica gel DW and flat plate collectors and (d) silica gel DW and evacuated tube collectors.

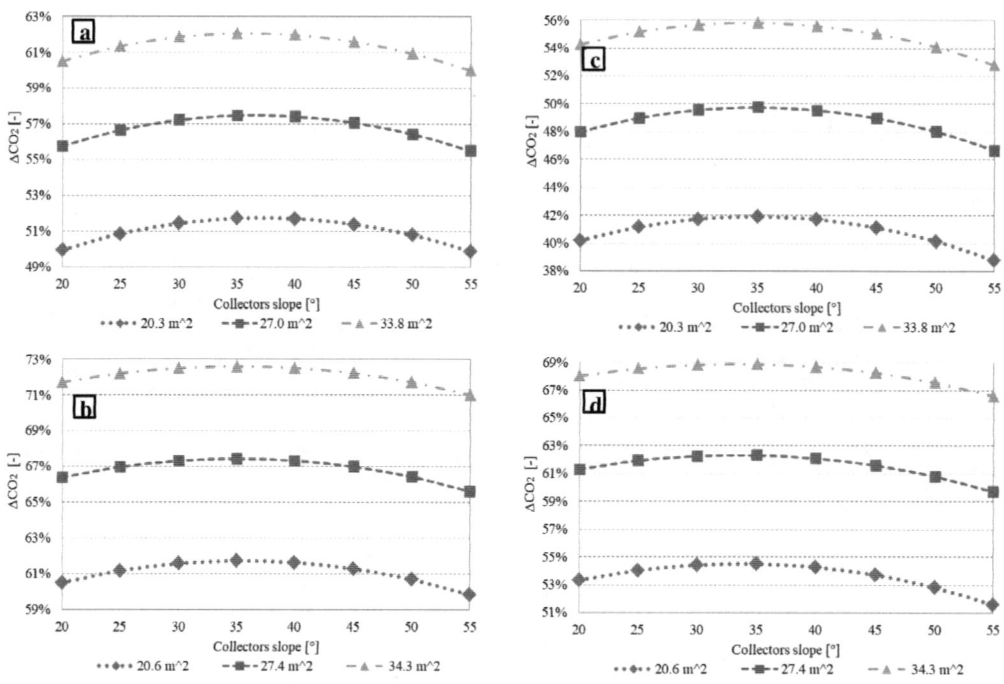

Figure 10. Equivalent CO_2 emissions avoided as a function of the tilt angle and the solar field aperture area when the solar thermal energy surplus is totally used, in the following configuration: (a) MILGO DW and flat plate collectors, (b) MILGO DW and evacuated tube collectors, (c) silica gel DW and flat plate collectors and (d) silica gel DW and evacuated tube collectors.

5. Conclusion

In this paper an innovative solar-driven hybrid AHU equipped with a desiccant wheel is compared with a conventional HVAC system that serves a university classroom located in Benevento (Southern Italy). Two alternative materials for the DW are considered and both show positive results in terms of primary energy saving and equivalent CO_2 emissions reduction. A significant improvement in the performance takes place with the new MILGO material instead of the silica gel, actually used. Furthermore, a parametric analysis was performed considering two solar collector types, three solar field area and finally tilt angles variable in the range 20-55°. The best result is achieved for the MILGO–based AHU: the percent primary energy saving reaches the value of about 23% and 29% with 27 m^2 of flat plate solar collectors and 34 m^2 of evacuated tube solar collectors, respectively. As concern the environmental analysis similar considerations can be derived, the maximum percent reduction in the CO_2 equivalent emission are about 21% and 26% with flat plate and evacuated tube collectors but in both cases with 20 m^2 of aperture area. Simple pay back period less than the useful life of the plant can be achieved only if other low temperature energy uses of the solar thermal energy over-productions are considered; in this case the best pay back periods are 5 and 6 years with 34 m^2 of flat plate collectors for MILGO and silica gel DW, respectively.

6. Nomenclature

c	Unitary Cost (€/Sm^3), (€/kWh), (€/m^2)
CO_2	Equivalent CO_2 emission (kg/y)
E	Energy (MWh/y)
EC	Extra Cost (€)
F_1, F_2	Isopotential lines
F_k	Cash Flow (€/y)
I_a	Annual Incentive (€/y)
mc	Specific Maintenance Cost (%)
MC	Maintenance Cost (€/y)
N	Number of year (y)
OC	Operating Costs (€/y)
PES	Primary Energy Saving (%)
S	Gross solar collectors area (m^2)
SF	Solar Fraction (-)
SPB	Simple Pay Back period (y)
t	Temperature (°C)
V	Volume (Sm^3)

Greek symbols

α	Specific emission factor of electricity drawn from the grid ($kgCO_2/kWh_{el}$)
β	Specific emission factor of primary energy related to natural gas combustion ($kgCO_2/kWh_{Ep}$)
ΔCO_2	Equivalent CO_2 avoided emission (%)
η	Efficiency (-)
ω	Humidity ratio (g/kg)

Superscripts

AS	Alternative System
CS	Conventional System
postheat	Post-heating phase
PP	Power Plant
Preheat	Pre-heating phase
TS	Thermal Storage
DWreg	Desiccant Wheel regeneration

Subscripts

aux	Auxiliaries
B	Boiler
CH	Chiller
Cooling	Cooling mode

EG	Electric Grid
el	Electric
F_1, F_2	Isopotential lines
Heating	Heating mode
NG	Natural Gas
p	Primary
SC	Solar thermal Collector
th	Thermal
tot, Total	Total

Acronyms

AS	Alternative System
AHU	Air Handling Unit
B	Boiler
CC	Cooling Coil
CF	Cross-Flow heat exchanger
CH	Chiller
COP	Coefficient Of Performance
CS	Conventional System
DW	Desiccant Wheel
EC	Evaporative Cooler
GHG	Greenhouse Gas
HC, HC2	Heating Coils
HVAC	Heating, Ventilation and Air Conditioning
HW-HX	Hot Water Heat exchanger
LHV	Lower Heating Value
MILGO	Hygroscopic material, consisting graphite oxide dispersed in the MIL101 metal organic framework network structure
SCOP	System Coefficient Of Performance
SC	Solar thermal Collector
SF	Solar Fraction
SPB	Simple Pay Back period
TS	Thermal Storage

7. References

[1] IEA, International Energy Agency. Energy Efficiency, 2017. Available online: http://www.iea.org/publications/freepublications/publication/Energy_Efficiency_2017.pdf [last access: 5 June 2018].

[2] Eurostat. Energy balance, January 2017. Available online: http://ec.europa.eu/eurostat/web/energy/data/energy-balances. [last access: 05 June 2018].

[3] Chaudhary G Q, Ali M, Sheikh N A, Gilani S I H and Khushnood S, Integration of solar assisted solid desiccant cooling system with efficient evaporative cooling technique for separate load handling *Appl Therm Eng* **140** pp 696–706.

[4] Abbassi Y, Baniasadi E and Ahmadikia H, Comparative performance analysis of different solar desiccantdehumidification systems 2017 *Energ Buildings* **150** pp 37–51.

[5] Merabti L, Merzouk M, Merzouk N K and Taane W, Performance study of solar driven solid desiccant cooling system under Algerian coastal climate 2017 *Int J Hydrogen Energ* **42** pp 28997–29005.

[6] Jani D B, Mishra M and Sahoo P K, Performance analysis of a solid desiccant assisted hybrid space cooling system using TRNSYS 2018 *J Building Engineering* **19** pp 26–35.

[7] Enteria N, Yoshino H, Mochida A, Satake A, Yoshie R, Takaki R, Yonekura H, Mitamura T and Tanaka Y, Performance of solar-desiccant cooling system with silica–gel (SiO2) and titanium dioxide (TiO2) desiccant wheel applied in East Asian climates 2012 Sol Energy **86** pp 1261–1279.

[8] Angrisani G, Roselli C, Sasso M and Tariello F, Dynamic performance assessment of a solar-assisted desiccant-based air handling unit in two Italian cities 2016 *Energ Convers Manage* **113** pp 331–345.

[9] Angrisani G, Roselli C, Sasso M and Tariello F, Dynamic performance assessment of a micro-trigeneration system with a desiccant-based air handling unit in Southern Italy climatic conditions 2014 *Energ Convers Manage* **80** pp 188–201.

[10] Bareschino P, Diglio G, Pepe F, Angrisani G, Roselli C and Sasso M, Numerical study of a MIL101 metal organic framework based desiccant cooling system for air conditioning application 2017 *Appl Therm Eng* **124** pp 641–651.

[11] Solar Energy Laboratory, TRNSYS 17, a TRaNsient System Simulation program 2010 University of Wisconsin, Madison.

[12] T.E.S.S. Component Libraries v.17.01 for TRNSYS v.17.0 and the TRNSYS Simulation Studio, Parameter/Input/Output Reference Manual 2004 Thermal Energy System Specialists,LLC.

[13] Calise C, Dentice d'Accadia M, Roselli C, Sasso M and Tariello F, Desiccant-based AHU interacting with a CPVT collector: Simulation of energy and environmental performance 2014 *Sol Energy* **103** pp 574–594.

[14] Maclaine-Cross I L, Banks P J,. Coupled heat and mass transfer in regenerators– predictions using an analogy with heat transfer 1972 *Int J Heat Mass Tran* **15** **(6)** pp 1225–1242.

[15] Howe R R, Model and Performance Characteristics of a Conditioning System Which Utilizes a Rotary Desiccant Dehumidifier 1983 MS thesis, University of Wisconsin, Madison, USA.

[16] Jurinak J J, Open cycle solid desiccant cooling: component models and system Simulations 1982 PhD thesis, University of Wisconsin, Madison, USA.

[17] Banks P J, Prediction of Heat and Mass Regenerator performance using nonlinear analogy method: part 1– basis 1985 *ASME J Heat Transf* **107** pp 222–229.

[18] Angrisani G, Roselli C and Sasso M, Experimental validation of constant efficiency models for the subsystems of an unconventional desiccant-based Air Handling Unit and investigation of its performance 2012 *Appl Therm* Eng **33–34** pp 100–108.

[19] Sabiha MA, Saidur R, Mekhilef S, Mahian O, Progress and latest developments of evacuated tube solar collectors 2015 *Renew Sust Energ Rev* **51** pp 1038–1054.

IOP Conf. Series: Journal of Physics: Conf. Series **1224** (2019) 012026 doi:10.1088/1742-6596/1224/1/012026

Comparison of isothermal and isoflux *g-functions* for borehole-heat-exchanger fields

C Naldi and E Zanchini

Department of Industrial Engineering, University of Bologna, Viale Risorgimento 2, 40136 Bologna, Italy

enzo.zanchini@unibo.it

Abstract. A new numerical approach to determine the *g-function* of a borehole field with the boundary condition of uniform temperature and time-constant mean heat flux at the surface of the boreholes is presented. The method is employed to compare the *g-functions* obtained by this boundary condition with those obtained by the usual condition of uniform and constant heat flux, for a single borehole and for a field of six boreholes placed in two lines. Boreholes with length 150 m, diameter 15 cm, buried depth 1.5 m and mutual distance 7.5 m, for the field, are considered. The results show that the difference between the two kinds of *g-functions* is less than 1.5 % for a single borehole, while it reaches 8.7 % for the borehole field, at high values of time. Finally, we show that the superposition of the effects of the single boreholes yields correctly the *g-function* of the field in the case of uniform heat flux, but overestimates the *g-function* in the case of uniform temperature and constant mean heat flux.

1. Introduction

Ground-source heat pumps (GSHPs) are becoming a rather diffuse technology for building heating and cooling, especially in countries with a cold climate, where air source heat pumps are not very efficient. Self et al. [1] compared GSHPs and conventional heating systems in terms of costs, CO_2 emissions and other parameters. They concluded that GSHPs are not only the most efficient and the most eco-friendly system, but also the most convenient system for building heating in Canada. The most widely used GSHPs are ground-coupled heat pumps (GCHPs), that commonly employ fields of vertical closed-loop ground heat exchangers, often called borehole heat exchangers (BHEs).

The design of BHE fields is usually performed by employing dimensionless thermal response factors, called *g-functions*, which are often embedded in programs for the simulation of BHE fields, such as EED [2], GHLEPRO [3], and EnergyPlus [4].

The *g-function* of a BHE field is the time-dependent mean dimensionless temperature of the surface of the BHEs produced by a time-constant mean heat flux per unit length applied at that surface. For a single BHE, *g-functions* are usually evaluated not only on the BHE surface, where the heat flux is applied, but also on cylindrical surfaces placed at different radial distances from the BHE axis. In fact, the *g-function* for a BHE field can be often calculated by the spatial superposition of effects in space.

G-functions are usually evaluated by employing the boundary condition of uniform and constant heat flux per unit length at the surface between BHEs and ground. The simplest scheme adopted is the Finite Line-Source (FLS) scheme, where each BHE is considered as a line-source with finite length. The analytical solution of the temperature field under this scheme was determined by Claesson and Eskilson [5] and by Zeng et al. [6]. Since the solution has the form of an integral, the temperature averaged along the BHE length has the form of a double integral. Simpler expressions of this temperature, in the form of a single integral, were developed by Lamarche and Beauchamp [7] and by Bandos et al. [8]. An extended solution, valid also for BHEs whose top is buried under the ground surface, was determined by Claesson and Javed [9]. The FLS scheme at uniform and constant heat flux allows determining the *g-function* for a BHE field by the superposition of the effects of single BHEs.

Another simple scheme is the Finite Cylindrical-Source (FCS) scheme, where each BHE is considered as a cylinder with a finite length that delivers a uniform and constant power per unit length to the surrounding solid medium. This scheme is suitable for numerical evaluations of the *g-functions*, and was employed, for instance, by Zanchini and Lazzari [10, 11] and by Monzó el al. [12]. It allows employing the superposition of the effects of the single BHEs.

A different FCS scheme for the numerical calculation of the *g-functions* was employed by Eskilson [13]. He imposed a uniform temperature and a time-constant mean heat flux per unit length at the surface between the BHE field and the ground, and evaluated the *g-functions* for many BHE fields with this boundary condition. The results were obtained by finite-difference numerical simulations and were reported in graphical form. Lamarche and Beauchamp [7] and Fossa et al. [14] compared *g-functions* obtained by the FLS model and the superposition of the effects of single BHEs with the *g-functions* obtained by Eskilson [13], for some BHE fields. They found that the FLS scheme slightly overestimates the *g-functions* with respect to the boundary condition employed by Eskilson.

Eskilson's boundary condition is not easily applied, since the time-dependent heat flux distribution that yields a uniform surface temperature and a time-constant mean heat flux is unknown. Moreover, this boundary condition does not allow applying the superposition of the effects of single BHEs. Nevertheless, Eskilson's boundary condition has been employed in recent scientific papers, because it is closer to the conditions that occur in practical applications.

Cimmino and Bernier [15] developed a semi-analytical method, based on the FLS solution, to determine *g-functions* with Eskilson's boundary condition. In this method, BHEs are divided into segments releasing different time-dependent heat fluxes per unit length, such that the surface temperature of the BHE field is uniform and the heat flux averaged on the surface of the BHE field is constant. An improvement of the method of Ref. [15], that takes into account the effects of the BHE thermal resistance, was developed by Cimmino [16]. A technique to reduce the computation time in the application of the method of Ref. [15], by reducing the number of numerical computations of the FLS solutions, was presented in Ref. [17]. Lamarche [18] proposed a modification of the method of Ref. [15], which allows reducing the number of segments for each borehole.

Priarone and Fossa [19] presented a numerical method to determine *g-functions* with Eskilson's boundary condition for a single BHE. To obtain a uniform surface temperature, they imposed a uniform and constant heat flux at the inner surface of a superconductive layer adjacent to the BHE surface. Monzó et al. [20] proposed a more general numerical method to evaluate *g-functions* with Eskilson's boundary condition, valid for any BHE field. In this method, the BHEs are supposed to be filled with a superconductive material that also interconnects them. The authors compared their *g-function* for a 3 × 2 BHE field with that obtained in Ref. [15], and found a good agreement.

In this paper, we present an alternative numerical method to obtain the *g-function* of a BHE field with Eskilson's boundary condition, that we will call isothermal *g-function*. The method is based on a recursive procedure that starts by imposing, as a first-trial isothermal boundary condition, the isoflux *g-function*, namely the uniform time-dependent surface temperature determined by applying the usual boundary condition of uniform and constant wall heat flux. The time averaged mean heat flux per unit length obtained is then used to determine a correction factor that yields the first-trial isothermal *g-function*. The latter can be used as isothermal boundary condition in another iteration. By employing this method, we analyze the difference between isothermal and isoflux *g-functions* for a single BHE and for a 3 × 2 BHE field, with the top of BHEs placed at a typical buried depth employed in Italy, namely 1.5 m. Moreover, we show that the superposition of the effects of the single BHEs cannot be applied to determine the isothermal *g-function* of a BHE field.

2. Numerical method

The *g-functions* were evaluated by accurate numerical simulations performed through the software COMSOL Multiphysics. Each borehole was schematized as a finite cylindrical source, buried at some distance from the ground surface. The ground was considered as a semi-infinite solid medium, with constant thermophysical properties. A single borehole and a 3 × 2 BHE field were considered, with the following geometric characteristics: BHE diameter D = 15 cm, BHE length L = 150 m, buried depth B_d = 1.5 m, spacing between adjoining BHEs in the 3 × 2 field S = 7.5 m.

The simulations of the single borehole were performed by employing the 2D axis-symmetric domain illustrated in figure 1. The selected domain is a cylinder coaxial with the borehole, with a radius equal to 2500 BHE diameters and a height equal to 2510 BHE diameters. The axis of the vertical coordinate, z, is directed downwards, with z = 0 at the BHE top.

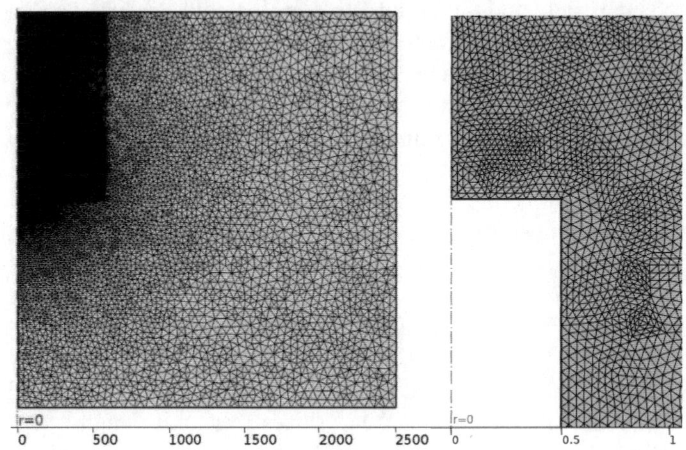

Figure 1. Computational domain and mesh for a single borehole (left) and zoom on the borehole top (right).

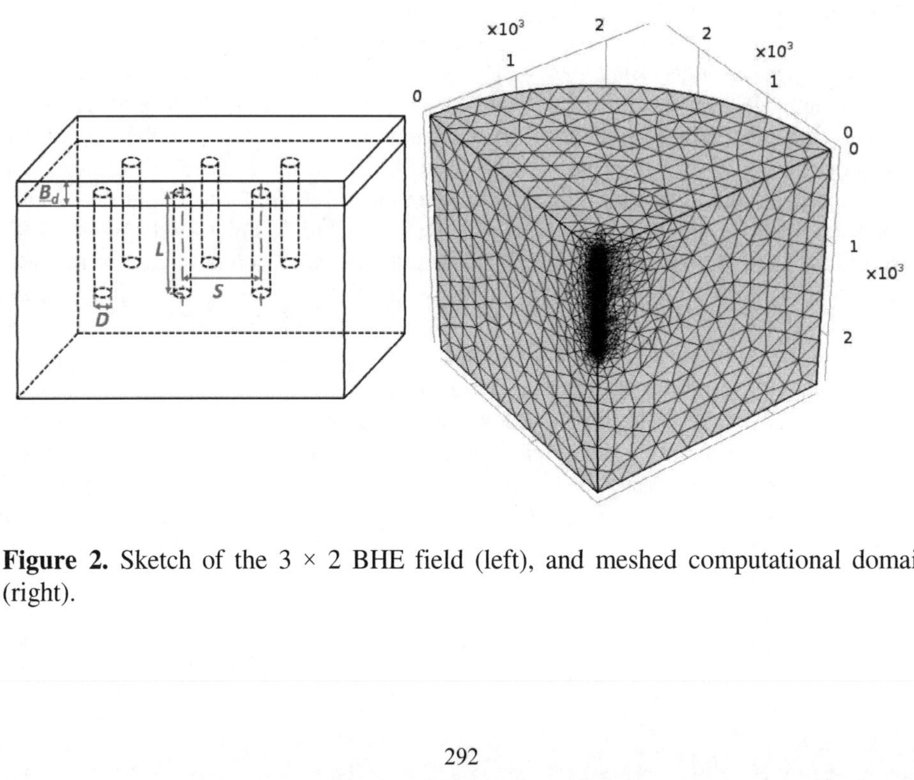

Figure 2. Sketch of the 3 × 2 BHE field (left), and meshed computational domain (right).

Figure 2 illustrates a sketch of the 3 × 2 BHE field and the 3D computational domain employed for the simulations. Thanks to the double symmetry of the 3 × 2 field, only a quarter of the field was simulated, by adopting as computational domain a quarter of cylinder that contains one borehole and a half. The cylinder has radius 2500 BHE diameters and height 2510 BHE diameters; the z-axis is directed downwards, with $z = 0$ at the BHE top.

The differential equation of transient heat conduction is

$$\frac{\partial T}{\partial t} = \alpha\, \nabla^2 T\,, \tag{1}$$

where T is temperature, t is time and α is the ground thermal diffusivity.

Both in the computational domain of the single borehole and in that of the 3 × 2 BHE field, the outer vertical surface and the bottom horizontal surface of the domain were set as adiabatic. The upper horizontal surface (ground surface) was considered as isothermal, with a constant temperature equal to the undisturbed ground temperature, T_g. The BHEs top and bottom were set as adiabatic. At the vertical surface between BHEs and ground, either a uniform and time-constant heat flux, or a uniform and time-dependent temperature were imposed, to evaluate the isoflux or the isothermal *g-function*, respectively. At the initial time instant, a uniform temperature equal to T_g was set in the whole domain.

The isoflux *g-function*, g_q, was evaluated by imposing a uniform and time-constant heat flux per unit area, q, at the BHE-ground interface:

$$q = -k\left(\nabla T \cdot \mathbf{n}\right)\Big|_{BHE-ground} = \frac{q_l}{\pi D}\,, \tag{2}$$

where k is the ground thermal conductivity, $\mathbf{n}$ is the outward unit normal and q_l is the linear heat flux applied at the BHE wall.

Results were reported in dimensionless form by employing the following dimensionless quantities

$$x^* = \frac{x}{D},\quad y^* = \frac{y}{D},\quad z^* = \frac{z}{D},\quad r^* = \frac{r}{D},\quad \nabla^* = D\nabla,\quad t^* = \frac{\alpha\, t}{D^2},\quad T^* = k\frac{T - T_g}{q_l}\,, \tag{3}$$

where (x, y, z) are the coordinates of the domain for the 3 × 2 BHE field, (r, z) are the coordinates of the domain for the single borehole, and ∇ is the nabla vector differential operator.
In dimensionless form, the boundary condition (2) becomes

$$q^* = -\left(\nabla^* T^* \cdot \mathbf{n}\right)\Big|_{BHE-ground} = 1/\pi\,, \tag{4}$$

and the corresponding dimensionless heat flux per unit length is

$$q_l^* = \pi q^* = 1\,. \tag{5}$$

The dimensionless size of the computational domain for the single BHE was $0 \le r^* \le 2500$, $-10 \le z^* \le 2500$, and that for the 3 × 2 field was $0 \le x^* \le 2500$, $0 \le y^* \le 2500$, $-10 \le z^* \le 2500$. The dimensionless geometric parameters of the boreholes were $L^* = L/D = 1000$, $B_d^* = B_d/D = 10$, $S^* = S/D = 50$ (for the 3 × 2 field).

An unstructured mesh with 440 109 triangular elements was employed for the simulation of the single BHE (see figure 1), and an unstructured mesh with 1 350 358 tetrahedral elements was adopted for the simulation of the 3 × 2 BHE field (see figure 2). In both cases, finer elements were employed in proximity of the boreholes and coarser elements towards the external boundaries of the domain.

The simulation-time interval was $10^{-6} \le t^* \le 10^6$, namely $-6 \le \log_{10}(t^*) \le 6$, divided in time steps of $\log_{10}(t^*) = 0.05$. With a borehole diameter of 15 cm and a ground diffusivity of 10^{-6} m²/s, the simulation-time interval is larger than 713 years.

Both for the single BHE and for the field, the simulation with uniform and time-constant heat flux per unit area at the BHEs surface yielded the isoflux *g-function*, g_q, which was then used as

IOP Conf. Series: Journal of Physics: Conf. Series **1224** (2019) 012026 doi:10.1088/1742-6596/1224/1/012026

time-dependent isothermal boundary condition at the surface between the BHE field and the ground in the second simulation. The inverse of the time averaged mean dimensionless heat flux per unit length at the surface between the BHE field and the ground was used as a correction factor of g_q, to determine the first-trial isothermal *g-function*, g_{T1}. Then, g_{T1} was employed as isothermal boundary condition at the surface of the boreholes in the third simulation, to determine a new mean dimensionless heat flux per unit length at the surface between BHEs and ground, which yielded the correction factor for g_{T1}, and thus the second-trial isothermal *g-function*, g_{T2}, and so on. The accuracy of the isothermal *g-function* obtained was finally checked by performing an additional simulation with this *g-function* as boundary condition, and comparing the mean time-dependent dimensionless heat flux per unit length at the BHEs surface to the desired value, 1.

In order to ensure that results are independent of the domain size, the simulation for the isoflux *g-function* of the single BHE and that for the isoflux *g-function* of the 3 × 2 BHE field were repeated, by replacing the adiabatic boundary condition at the outer vertical surface and at the bottom horizontal surface of the domain with the isothermal boundary condition $T^* = 0$. The maximum relative error was 0.03 % for the single BHE, and 0.06 % for the 3 × 2 field, both occurring at the last time instant.

In order to ensure that results are independent of the selected mesh, the simulation for the isoflux *g-function* with the adiabatic boundary condition was repeated by employing a different mesh. A mesh with 315 877 elements was used for the single BHE, and a mesh with 2 980 542 elements was used for the 3 × 2 BHE field. In each case, the relative discrepancy between the results obtained by different meshes decreases with time. For the single borehole, it is equal to 0.94 % at $\log_{10}(t^*) = 1$ and to 0.32 % at the last time instant; for the 3 × 2 BHE field, it is equal to 0.17 % at $\log_{10}(t^*) = 1$ and to 0.11 % at the last time instant.

3. Results

The evaluation of isothermal *g-functions* requires as boundary condition at the surface between BHE field and ground both a uniform temperature and a time-constant surface averaged dimensionless linear heat flux, $\bar{q}_l^*$, equal to 1. By employing the recursive procedure described in Section 2, for each iteration a uniform time-dependent temperature at the surface of the BHE field was imposed as a boundary condition, and the corresponding time evolution of $\bar{q}_l^*$ was determined.

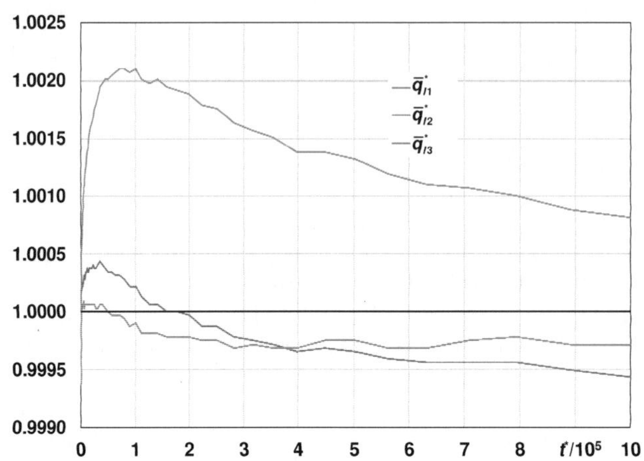

Figure 3. Plots of $\bar{q}_l^*$ versus dimensionless time, for the *g-functions* g_{T1}, g_{T2}, g_{T3}, single BHE.

For the single borehole, figure 3 illustrates the time evolution of $\bar{q}_l^*$ yielded by the isothermal *g-functions* obtained through the recursive procedure. In particular, $\bar{q}_{l1}^*$ is the surface averaged dimensionless linear heat flux produced by the first-trial isothermal *g-function*, g_{T1}, $\overline{q}_{l2}^*$ is that produced by the second-trial isothermal *g-function*, g_{T2}, and $\overline{q}_{l3}^*$ is that produced by the last isothermal *g-function*, g_{T3}. The figure shows the convergence of $\overline{q}_l^*$ to the constant value 1.

The values of the isothermal *g-functions* for the single BHE obtained by each iteration are reported in table 1, for selected time instants, together with the percent difference between the values of the second and third *g-function.* The maximum value of this percent difference, in the range $0 \le \log_{10}(t^*) \le 6$, is 0.37 %. The maximum absolute difference between the values of the last two *g-functions* is 0.0008 and occurs at $\log_{10}(t^*) = 0.25$. Even the difference between the first and the third *g-function* is rather small: the maximum relative difference, in the range $0 \le \log_{10}(t^*) \le 6$, is 0.56 % and occurs at $\log_{10}(t^*) = 0$. Consequently, even the first *g-function*, obtained by two simulations, is sufficiently accurate for technical purposes.

Table 1. Values of g_{T1}, g_{T2}, g_{T3} at selected time instants, and percent difference between g_{T2} and g_{T3}, single BHE.

$\log_{10}(t^*)$	g_{T1}	g_{T2}	g_{T3}	$100 \times (g_{T2} - g_{T3})/g_{T3}$
0	0.1933	0.1951	0.1944	0.37
1	0.3560	0.3564	0.3559	0.15
2	0.5335	0.5335	0.5333	0.03
3	0.7107	0.7105	0.7104	0.02
4	0.8747	0.8739	0.8737	0.03
5	0.9965	0.9947	0.9943	0.03
6	1.0348	1.0334	1.0337	-0.03

In the following, the isothermal *g-function* obtained by the last iteration will be called simply isothermal *g-function*.

A comparison between the isoflux and the isothermal *g-function* for the single BHE is illustrated in figure 4. For $\log_{10}(t^*) < 3.5$, the isothermal *g-function*, g_T, is graphically indistinguishable from the isoflux *g-function*, g_q. At the highest values of the dimensionless time, g_q is slightly higher than g_T, and the maximum relative difference, which occurs at the last time instant, is 1.4 %.

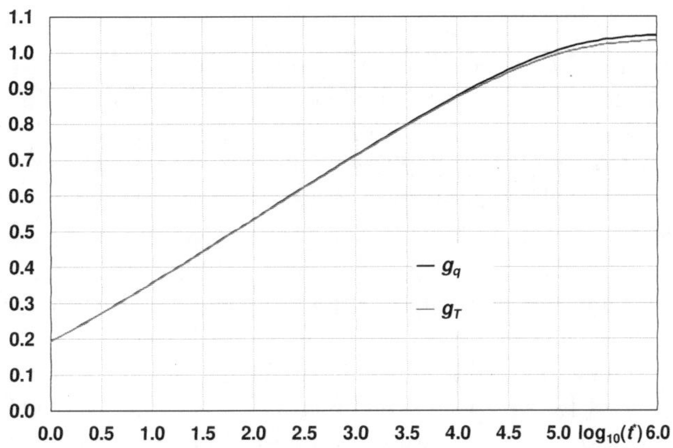

Figure 4. Isoflux *g-function*, g_q, and isothermal *g-function*, g_T, single BHE.

IOP Conf. Series: Journal of Physics: Conf. Series **1224** (2019) 012026 doi:10.1088/1742-6596/1224/1/012026

In figure 5, a comparison is made between the isoflux and the isothermal *g-functions* evaluated on cylindrical surfaces placed at different radial distances from the single-BHE axis. The isoflux and isothermal *g-functions* evaluated at 50, 100 and 400 diameters from the single-BHE axis are reported in the figure for the time interval $2 \leq \log_{10}(t^*) \leq 6$. For $\log_{10}(t^*) < 2$, all these *g-functions* are practically equal to 0. At low values of the dimensionless time, the isothermal *g-functions* are graphically indistinguishable from the isoflux *g-functions* evaluated at the same radial distance. At high values of the dimensionless time, the isothermal *g-functions* are slightly lower than the isoflux *g-functions* evaluated at the same radial distance. The absolute discrepancy between the isoflux and the isothermal *g-function* decreases with the distance from the BHE axis. In fact, the absolute difference between the two *g-functions* at the last time instant is 0.01 at a radial distance of 50 diameters (see black and red solid lines in figure 5), whereas it is equal to 0.003 at 400 diameters (see black and red dashed lines in the figure). The relative difference between the isoflux and the isothermal *g-function* is nearly independent of the distance from the BHE axis. At the last time instant it ranges from 3.15 % (at 50 diameters) to 3.82 % (at 400 diameters).

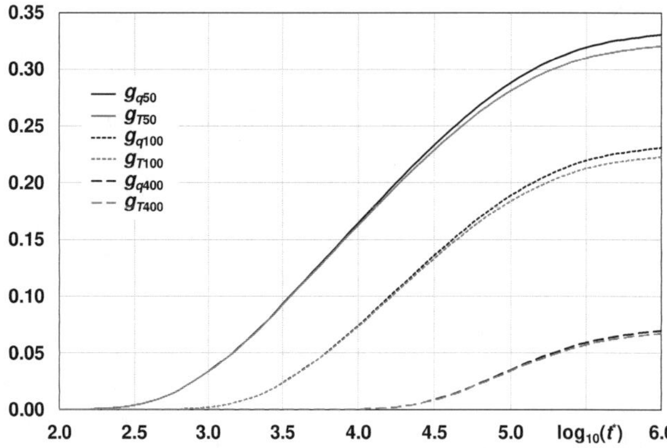

Figure 5. Isoflux *g-functions*, g_q, and isothermal *g-functions*, g_T, at 50, 100 and 400 diameters from the single-BHE axis.

The isoflux and isothermal *g-functions* were evaluated also for the 3 × 2 BHE field, by employing the recursive procedure described in Section 2. Five iterations were performed, and the convergence of the mean dimensionless linear heat flux on the BHE-field surface was checked.

Plots of the isoflux *g-function*, g_q, and of the isothermal *g-functions* g_{T1}, g_{T2}, g_{T3}, g_{T4}, g_{T5} are illustrated in figure 6, for the time interval $0 \leq \log_{10}(t^*) \leq 6$.

For $\log_{10}(t^*) < 4$, the isothermal *g-functions* are graphically indistinguishable from each other, whereas slight differences are visible at the highest values of the dimensionless time, as highlighted by the zoom on the graph of figure 6, which refers to the time interval $5.7 \leq \log_{10}(t^*) \leq 6$. The maximum relative difference between the values of the last two isothermal *g-functions* is 0.057 % and occurs at $\log_{10}(t^*) = 5.60$. Moreover, the maximum absolute difference between the values of the last two isothermal *g-functions* is 0.0013 and occurs at $\log_{10}(t^*) = 5.65$. The maximum relative difference between the first and the last isothermal *g-function*, occurring at $\log_{10}(t^*) = 5$, is 1.36 %, which confirms that even the first isothermal *g-function* is sufficiently accurate for technical purposes.

From figure 6, one can note that the isoflux *g-function*, g_q, is graphically indistinguishable from the isothermal *g-functions* for $\log_{10}(t^*) < 3.5$. At the highest values of the dimensionless time, on the contrary, g_q is significantly higher than g_{T5}, with a maximum relative difference of 8.67 %, occurring at the last time instant. As a consequence, adopting an isoflux boundary condition at the BHE-field surface, instead of an isothermal boundary condition, yields a significant overestimation of the *g-function* of the field at high times. In the following, the isothermal *g-function* obtained by the last iteration will be called simply isothermal *g-function* and will be denoted by g_T.

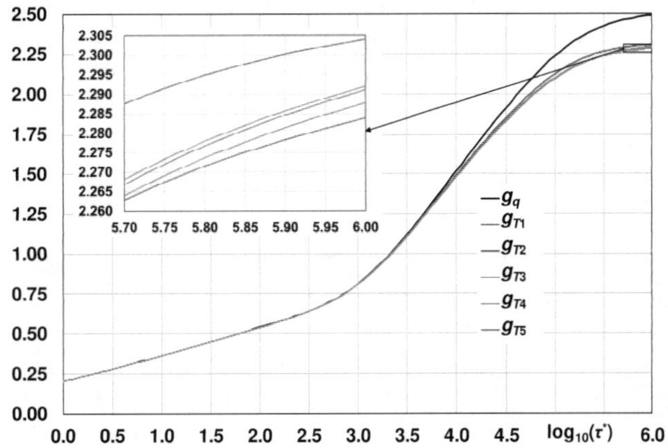

Figure 6. Isoflux *g-function*, g_q, and isothermal *g-functions* g_{T1}, g_{T2}, g_{T3}, g_{T4}, g_{T5}, 3 × 2 BHE field.

The difference between g_q and g_T is because employing an isoflux boundary condition at the BHE-field interface yields a non-uniform temperature distribution at the BHE surface. Figure 7 illustrates the vertical distribution of the dimensionless surface temperature, averaged on the circumference of the central BHEs, obtained at selected time instants by employing the isoflux boundary condition. For $\log_{10}(t^*) = 0$, the temperature distribution is almost uniform. For $\log_{10}(t^*) = 3$, the temperature distribution is still uniform, except close to the BHE top ($z^* = 0$) and bottom ($z^* = 1000$), where the temperature decreases. For $\log_{10}(t^*) = 6$ (i.e. at the last time instant), the temperature distribution has an inverted-U profile, with values at the borehole top lower than those at the bottom.

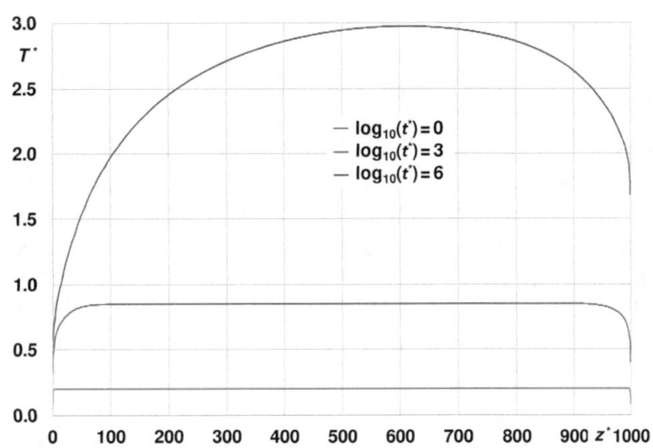

Figure 7. Dimensionless surface temperature of the central BHEs, versus z^*, isoflux boundary condition, 3 × 2 BHE field.

To check the applicability of the superposition of the effects of single BHEs, this method was applied to evaluate both the isoflux *g-function*, g_{qse}, and the isothermal *g-function*, g_{Tse}, of the 3 × 2 BHE field.

Figure 8 illustrates a comparison between g_{qse} and g_q, and between g_{Tse} and g_T. While g_{qse} is practically coincident with g_q (blue and green lines in figure 8 are overlying), g_{Tse} is sharply greater than g_T for high values of time (compare red and purple lines in the figure). The maximum relative difference between g_{Tse} and g_T, which occurs at the last time instant, is 6.0 %. The results confirm the applicability of the superposition of the effects of single BHEs to evaluate the isoflux *g-function* of a BHE field. On the opposite, employing the superposition of the effects of single BHEs to evaluate the isothermal *g-function* of the field yields a significant overestimation of the *g-function* at high values of time.

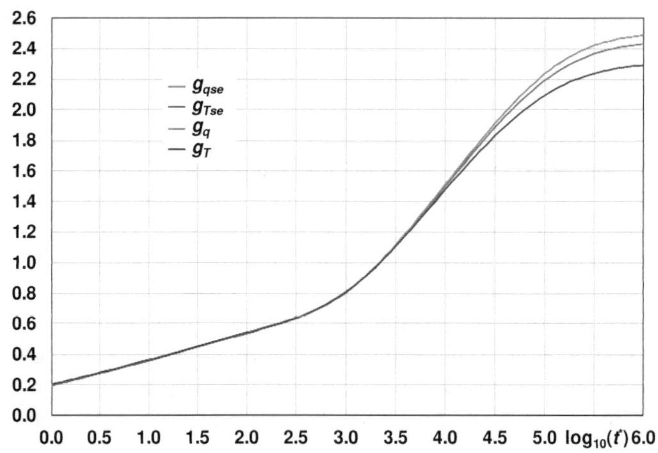

Figure 8. Isoflux and isothermal *g-functions* by superposition of the effects (g_{qse}, g_{Tse}), and isoflux and isothermal *g-functions* by 3D simulations (g_q, g_T), 3 × 2 BHE field.

4. Conclusions

We have presented a new numerical approach to determine the isothermal *g-function* of a borehole-heat-exchanger (BHE) field, i.e. the *g-function* yielded by the boundary condition of uniform temperature and time-constant mean heat flux at the surface between BHEs and ground. By this method, we have analyzed the difference between the isothermal *g-function* and the isoflux one, namely that yielded by the boundary condition of uniform and constant heat flux at the interface BHEs-ground. Moreover, we have checked the applicability of the superposition of the effects of single BHEs to determine the *g-function* of the field. The analysis has been performed for a single BHE and for a 3 × 2 BHE field, for BHEs with length 150 m, diameter 15 cm, top placed at 1.5 m under the ground surface, and spacing from each other 7.5 m in the case of the field. The following results have been found.

The difference between the isoflux and the isothermal *g-function* is positive and becomes appreciable only for values of the dimensionless time defined here greater than $10^{3.5}$, i.e. about 2.3 years for BHEs with diameter 15 cm placed in a ground with thermal diffusivity 10^{-6} m^2/s. For a dimensionless time equal to 10^6, i.e. about 713 year in the conditions stated above, the difference is about 1.4 % for a single BHE, and about 8.7 % for the 3 × 2 BHE field considered here.

The superposition of the effects of the single BHEs yields correctly the isoflux *g-function* of the field; on the contrary, it cannot be applied to determine the isothermal *g-function* of a BHE field.

References

[1] Self SJ, Reddy BV and Rosen MA 2013 Geothermal heat pump systems: Status review and comparison with other heating options *Applied Energy* **101** 341–8

[2] Hellström G and Sanner B 1994 Software for dimensioning of deep boreholes for heat extraction *Proc. Calorstock 1994* (Espoo/Helsinki, Finland) pp 195–202

[3] Spitler JD 2000 GLHEPRO-A design tool for commercial building ground loop heat exchanger *Proc. 4th Int. Conf. on Heat Pumps in Cold Climates* (Aylmer, QC, Canada) pp 1–16

[4] Fisher DE, Rees SJ, Padhmanabhanc SK and Murugappan A 2006 Implementation and validation of ground-source heat pump system models in an integrated building and system simulation environment *HVAC&R Research* **12**, 693–710

[5] Claesson J and Eskilson P 1987 Conductive heat extraction by a deep borehole, analytical studies (Technical report, Lund University, Sweden)

[6] Zeng HY, Diao NR and Fang ZH 2002 A finite line-source model for boreholes in geothermal heat exchangers *Heat Transfer–Asian Research* **31** 558–67

[7] Lamarche L and Beauchamp B 2007 A new contribution to the finite line-source model for geothermal boreholes *Energy and Buildings* **39** 188–98

[8] Bandos TV, Montero A, Fernandez E, Santander JLG, Isidro JM, Perez J, Fernandez de Cordoba PJ and Urchueguía JF 2009 Finite line-source model for borehole heat exchangers: effect of vertical temperature variations *Geothermics* **38** 263–70

[9] Claesson J and Javed S 2011 An analytical method to calculate borehole fluid temperatures for time-scales from minutes to decades *ASHRAE Transactions* **117** (2) 279–88

[10] Zanchini E and Lazzari S 2013 Temperature distribution in a field of long borehole heat exchangers (BHEs) subjected to a monthly averaged heat flux *Energy* **59** 570–80

[11] Zanchini E and Lazzari S 2014 New g-functions for the hourly simulation of double U-tube borehole heat exchanger fields *Energy* **70** 444–55

[12] Monzó P, Acuña J, Fossa M and Palm B 2013 Numerical generation of the temperature response factors for a borehole heat exchangers field *Proc. European Geothermal Congress 2013* (Pisa, Italy)

[13] Eskilson P 1987 Thermal analysis of heat extraction boreholes (Ph.D. Thesis, University of Lund, Lund, Sweden)

[14] Fossa M, Cauret O and Bernier M 2009 Comparing the thermal performance of ground heat exchangers of various lengths *Proc. Effstock 2009* (Stockholm, Sweden)

[15] Cimmino M and Bernier M 2014 A semi-analytical method to generate g-functions for geothermal bore fields *International Journal of Heat and Mass Transfer* **70** 641–50

[16] Cimmino M 2015 The effects of borehole thermal resistances and fluid flow rate on the g-functions of geothermal bore fields *International Journal of Heat and Mass Transfer* **91**, 1119–27

[17] Cimmino M 2018 Fast calculation of the g-functions of geothermal borehole fields using similarities in the evaluation of the finite line source solution *Journal of Building Performance Simulation* 655–68

[18] Lamarche L 2017 G-function generation using a piecewise-linear profile applied to ground heat exchangers *International Journal of Heat and Mass Transfer* **115** 354–60

[19] Priarone A and Fossa M 2016 Temperature response factors at different boundary conditions for modelling the single borehole heat exchanger *Applied Thermal Engineering* **103** 934–44

[20] Monzó, Mogensen P, Acuña J, Ruiz-Calvo F and Montagud C 2015 A novel numerical approach for imposing a temperature boundary condition at the borehole wall in borehole fields *Geothermics* **56** 35–44

IOP Conf. Series: Journal of Physics: Conf. Series **1224** (2019) 012027 doi:10.1088/1742-6596/1224/1/012027

Ventilated pitched roof with forced ventilation and flow homogenizer device: testing and performance assessment

C Ferrari, A Muscio

EELab – Dip. di Ingegneria "Enzo Ferrari" – Univ. di Modena e Reggio E., Italy

Corresponding author: alberto.muscio@unimore.it

Abstract. Ventilated tile roofs are common objects in the construction sector of Italy. A preferred type is characterized by a ventilated air space between the upper tile layer and the roof structure below. The air enters through openings at the gutters level, flows up below sheathing in the air space along the pitch and is finally discharged through openings along the ridge. This setup, which leaves the room below the roof sealed and habitable, allows removing the heat brought by the absorbed solar radiation thanks to the flow that is established by natural ventilation in the air space. However, its actual performance is often weak and also unpredictable due to continuously changing buoyancy forces. Nonetheless, a permanent and adequate flow can be ensured even through a relatively thin air space and for whichever irradiance and wind conditions by forced ventilation: a properly designed fan can provide the desired flow rate by extracting the air after this is collected along the ridge by a manifold. A thorough design and manufacturing of the manifold is needed, however, to avoid a highly inhomogeneous flow, which would follow the easiest path and leave most of the pitch practically unventilated. As an alternative, a throttling shutter parallel to the ridge has been proposed to progressively choke the flow entering the manifold through parallel climbing ducts as the fan is approached, possibly allowing onsite adjustment after installation. In this work the solution, developed by means of a small scale test bed, is illustrated in details and some methods to assess the performance are presented.

1. Introduction

Ventilated tile roofs are common objects in the construction sector of Italy. A common type seems that characterized by a ventilated air space between the upper tile layer above, which rests on a waterproofing membrane and a wooden slab, and the lower remaining layers of the roof structure. The air enters through openings at the gutters level, flows up along the pitch in the airspace below sheathing and is finally discharged through openings along the ridge (Figure 1). This setup, which leaves the room below the roof sealed and habitable, allows removing heat brought by the absorbed solar radiation thanks to the natural airflow that is established in the ventilated air space. Moreover, moisture that may condense in the roof can be evaporated and removed by the flowing air. An airspace as thick as 7-10 cm is needed, however, to allow an adequate airflow, but this is seldom the case. Generally speaking, buoyancy forces that induce the airflow depend on warming up of the flowing air as resulting from the complex and continuously changing thermal balance of solar absorption and other heat transfer processes, and such forces can also be countered by wind counter-pressure, so that the actual performance of such type of ventilated roof is often unpredictable.

An extensive analysis of parameters affecting the performance of ventilated roofs can be found in [1-5] and, with regard also to the roof dynamic behavior, in [6-9], considering both above and below sheathing ventilation. Studies have been aimed to improve natural flow in pitched roofs, for example by using novel tile geometries [10], proper design of the insulation layer [11], radiant barrier inside the gap [12], thermal breaker to prevents heat conduction through the air space [13], combination with cool roofs [9,14]. See also [15,16] for reviews of double skin building elements.

Forced ventilation can ensure an adequate airflow for whichever irradiance and wind conditions and even through a relatively thin air space: a properly designed fan can provide the desired flow rate by extracting the air after this collected along the ridge by a manifold (Figures 2-3). Compared to a conventional ventilated roof with natural circulation (Figure 1), the main advantage of the investigated approach based on forced circulation consists of the accurate control of the airflow, which can be given a flow rate always adequate and independent of the surrounding environmental conditions. Moreover, the airflow can be easily modulated by controlling the fan speed and possibly suppressed when it is not desired, for example at night or during the winter season. Electrical energy could also be supplied by a small photovoltaic panel installed on the roof, which will ensure the maximum power under the maximum solar irradiance (*i.e.* when ventilation is mostly needed) and for which an inverter may not be needed if a DC fan is used. A solar chimney like in [12] or a solar turbine could also be used.

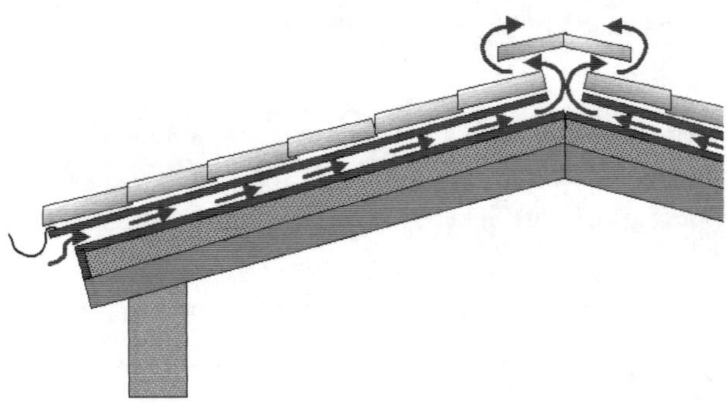

Figure 1. Pitched roof with natural ventilation: cross section and ridge outlet.

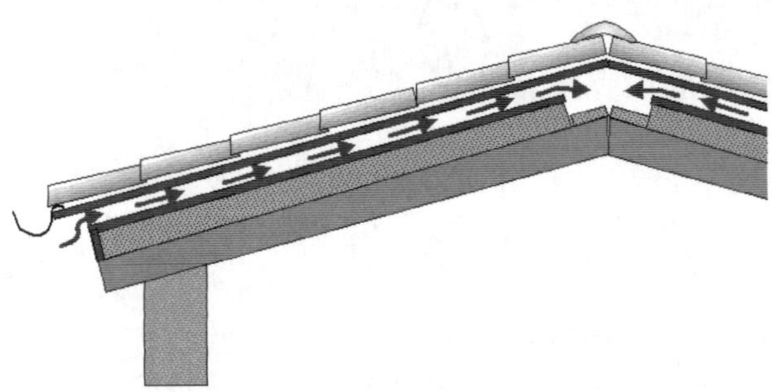

Figure 2. Pitched roof with forced ventilation:
cross section with climbing ventilation ducts and ridge manifold.

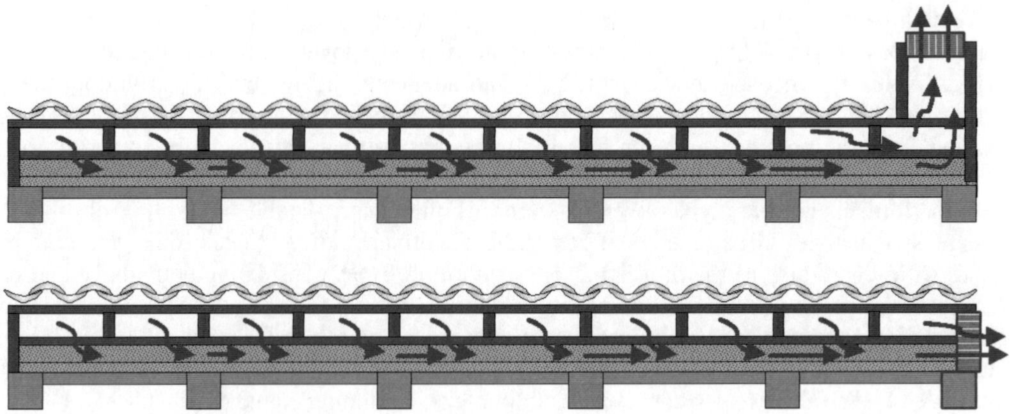

Figure 3. Ridge manifold for a pitched roof with forced ventilation: cross section along the duct outlets, layouts with chimney (top) or side (bottom) extraction fan.

A thorough design and manufacturing is needed for the ventilation system and, above all, for the manifold that collects the air along the ridge and passes it to the fan to avoid a strong non-homogeneity of the airflow, which would follow the easiest path and leave most of the pitch practically unventilated. Therefore, a throttling shutter parallel to the ridge has been proposed [17-19] to progressively choke the flow through the outlets of the parallel climbing ducts as the fan is approached (Figure 4). This approach to forced ventilation of pitched roofs, which can allow onsite adjustment after installation and, in principle, can be applied also with flat horizontal roofs [12], was already experimented through a small scale test bed, with positive results [17-19]. In this work, the concept is illustrated in details and the experimental procedure to assess the airflow distribution and adjust the throttling shutters is tested by the small scale test bed.

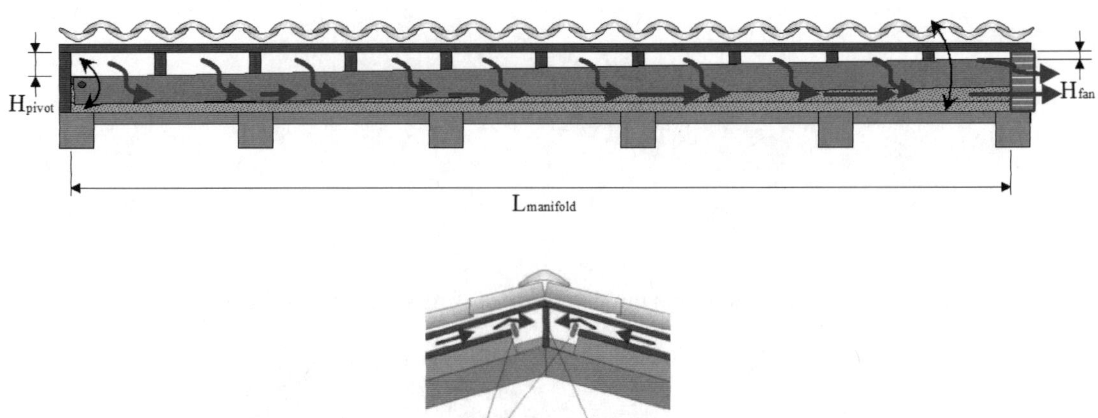

Figure 4. Ridge manifold for pitched roof with forced ventilation: cross section along the duct outlets and pivoted throttling shutters (in red).

2. Pitched roof with forced ventilation

The proposed layout of pitched roof with forced ventilation has been sketched in Figures 2-4, where a single pitch is considered for sake of simplicity. The air is sucked through openings at the gutters level, which can be made according to current practice for ventilated roofs with natural circulation; thereafter it flows along the roof pitch through a series of side-by-side ascending ducts. These ducts are built below sheathing, between the upper wooden slab that supports the tile mantle and a lower wooden slab that protects the thermal insulation layer below, the latter represented in Figures 2-4 by the pale blue layer. The side walls of the climbing ducts can be made by parallel joists that connect the upper and lower slabs; this is needed to make the flow in each duct independent of that in the others. The air flowing in the ducts subtracts and carries away heat from the tile mantle and the wooden slab above, thus limiting overheating when the roof is subjected to direct insolation. Rather than being directly discharged into the atmosphere, as it occurs in naturally ventilated roofs (Figure 1), the air is collected at the outlets of the ducts by a manifold that follows the ridge and eventually discharges the warm air into the external environment thanks to an extraction fan (Figures 2-3). The fan can be installed in a chimney close to the ridge or in an outlet hole at one end of the ridge, in the side wall, and it may serve parallel manifolds for opposite pitches of a two-pitched roof. In this case, a separation septum may be appropriate between the parallel manifolds (Figure 4) to avoid wind-induced cross air circulation. The fan can also be installed in a chimney at the midpoint of the ridge and serve symmetrical manifolds, to limit the manifold length. In order to allow a continuous tile mantle from the gutters to the ridge, the manifold can be incorporated in the roof thickness at the expense of the insulation layer, for which a lower thickness of a high performance yet expensive materials such as aerogel (represented in grey in Figures 2-4) could be afforded in view of the low quantity of material to be used.

A main issue of a forced ventilation system as outlined above is to obtain a homogeneous distribution of air: the airflow would generally follow the easiest path, which is that through the ducts closest to the fan, and it would thus leave the farthest ducts practically unventilated. Of course, the situation would be even worse without joists that divide the ventilated air space in parallel ducts. Such issue can be avoided by a proper design and manufacturing of the manifold that collects the air along the ridge and brings it to the fan, for example using a continuously varying cross-section of the manifold itself rather than a homogeneous one (as common for manifolds of parallel tube heat exchangers). This may be unpractical, however, since it would require a targeted design for each roof, made by a skilled designer with proper design tools. Moreover, a very accurate manufacturing would be needed for the manifold and the whole ventilation system, something that is not easy to ensure in the construction sector. An alternative and more viable solution can be obtained through a simple and effective regulation mechanism, which can be adjusted after the roof has been built on the basis of simple thermal and/or fluid flow measurements. The mechanism consists of a throttling shutter (showed with red color in Figure 4) that follows the ridge line and progressively chokes the flow through the outlets of the ducts as the fan is approached. It is pivoted at one end of the manifold and adjusted up or down at the other end, with a chopping motion, in order to reduce the outlet height from a maximum value H_{pivot} close to the pivot down to a minimum value H_{fan} at the end of the manifold length $L_{manifold}$.

The verification of the airflow distribution can be carried out by infrared thermography, measuring the temperature distribution of the covering surface (tiles, shingles, etc.) when this is subjected to solar radiation [17-19]: areas affected by inadequate ventilation will tend to overheat, while areas affected by an excessive flow rate will show a temperature closer to that of the cooling air. The throttling shutter can thus be adjusted, and the outlet cross section of the ducts increasingly reduced close to the fan, until an adequate homogeneousness of the surface temperature is achieved. However, time could be needed to let the temperature to reach a new steady state distribution after each adjustment, moreover changing solar irradiance may disturb the process.

An alternative and much less time consuming approach to verify the flow distribution consists of measuring the air velocity along the gutters, at the center of the duct inlets, by means of an anemometer. A hot wire or hot sphere anemometer could be used to have a small probe size and not interfere significantly with the airflow, which would also reach a new steady state distribution very quickly after each adjustment.

3. Adjustment of flow distribution

The flow distribution was tested on a small scale test bench representing a single pitch of a two-pitched roof with about 1:10 scale. The length of the manifold $L_{manifold}$ is 958 mm, the length of the pitch and the climbing ducts is 520 mm, the height of the duct cross section is 11 mm, the width is 40 mm gross and 35 mm net of spacer joists. A DC axial fan with 60 mm x 60 mm size is installed in a vertical chimney and extracts the air at one end of the manifold. A flat rod pivoted at the opposite end is used as a throttling device, with the pivot positioned in order to leave the local height H_{pivot} of the duct cross-section at the maximum value of 11 mm. The rod movement allows to change the height H_{fan} of the duct cross-section below the fan from the maximum value of 11 mm, equivalent to absence of the throttling device, down to almost 0. The air velocity is measured at each duct inlet by means of a Testo hot bulb thermal anemometer integrated in a flow, temperature and humidity probe, connected to Testo 400 portable data acquisition device with accuracy ±(0.03 m/s + 5% of the measured value).

Some laboratory experiments aimed at validating further the effectiveness of the throttling device and verifying the usefulness of the approach to assess and optimize the airflow distribution were carried out by means of the above described test bench. More specifically, the air velocity distribution was thus determined for different settings of the throttling device, as shown in Figure 5.

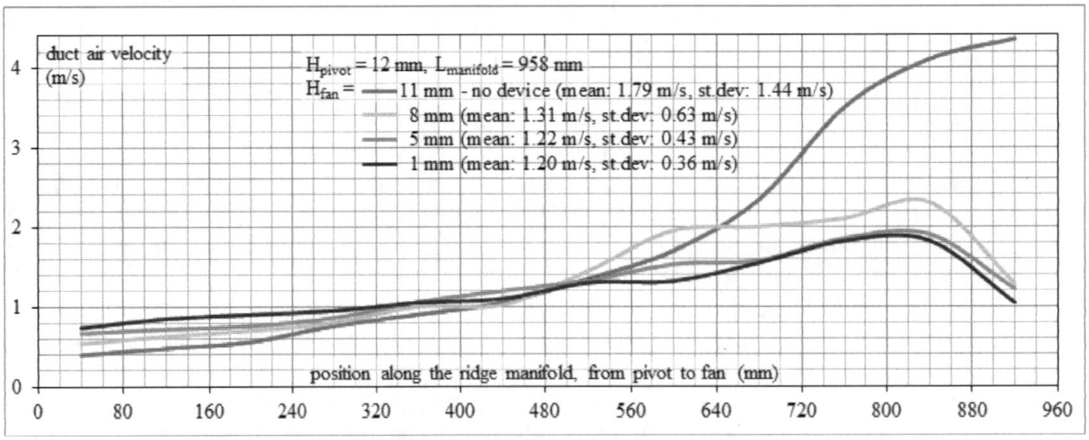

Figure 5. Air velocity distribution for different settings of the throttling device, measured at the duct inlets from pivot (left side) to fan (right side).

As expected, without the throttling device the distribution is completely non-homogeneous, with most of the air flowing in the ducts close to the fan, and a very low air velocity occurring on the opposite side. Lowering the cross-section height below the fan, H_{fan}, progressively improves the homogeneousness of the airflow, as shown by the relatively flat distribution of air velocity and the lower value of its standard deviation shown in Figure 4, with a relatively low penalization in terms of average velocity and, consequently, of total flow rate. The air velocity in the ducts farthest from the fan is almost doubled. At one point, decreasing further H_{fan} does not seem to yield significant improvements.

4. Assessment of performance of the ventilated roof

The efficiency of roof ventilation can be assessed in different ways. A parameter that will be analysed in future work is the efficiency of removal of absorbed solar heat, that is the ratio η_A (%) of the heat rate removed by the airflow and the heat rate brought by the absorbed solar radiation:

$$\eta_A = \frac{\dot{Q}_{removed}}{\dot{Q}_{absorbed}} \tag{1}$$

The heat rate removed by the airflow, $\dot{Q}_{removed}$ (W), can be calculated by the relationship:

$$\dot{Q}_{removed} = \rho_{out} \cdot c_{p,out} \cdot w_{out} \cdot A_{out} \cdot (T_{out} - T_e) \tag{2}$$

where

ρ_{out} density of the extracted air (kg/m^3)
$c_{p,out}$ specific heat at constant pressure of the extracted air (J/(kg·°C))
w_{out} average velocity of the extracted air (m/s)
A_{out} outlet cross section area (m^2)
T_{out} temperature of the extracted air (°C)
$T_e = T_{in}$ temperature of the external ambient air/entering air (°C)

The heat rate brought by the absorbed solar radiation, $\dot{Q}_{absorbed}$ (W), can be calculated as follows:

$$\dot{Q}_{absorbed} = A_e \cdot \alpha \cdot I_{sol} \tag{3}$$

where

A_e external/insolated surface area of the roof pitch (m^2)
α solar absorptance ($0 < \alpha < 1$)
I_{sol} solar irradiance (W/m^2)

Some outdoor experiments were carried out by the small scale test bed to verify the approach. A constant volume flow rate of 3.6 m^3/h was measured by a fan anemometer with probe installed in the chimney duct, downstream of the extraction fan. The temperatures of the ambient air and the extracted air were measured by type T thermocouples with 1 mm shielded probe, connected to a PicoLog TC-08 USB data acquisition system with cold junction compensation. The solar irradiance was measured by a Delta Ohm pyranometer model LP 471-PYRA-02-10 operating in the 305÷2800 nm range, with probe placed coplanar with the insolated surface. A solar absorptance $\alpha = 0.69 \pm 0.05$ was measured by a solar reflectometer D&S SR1 v6 according to the ASTM C1549 test method [20] with beam normal solar spectrum as reported in ASTM E891 [21]. Sample data for this analysis approach are shown in Figure 6.

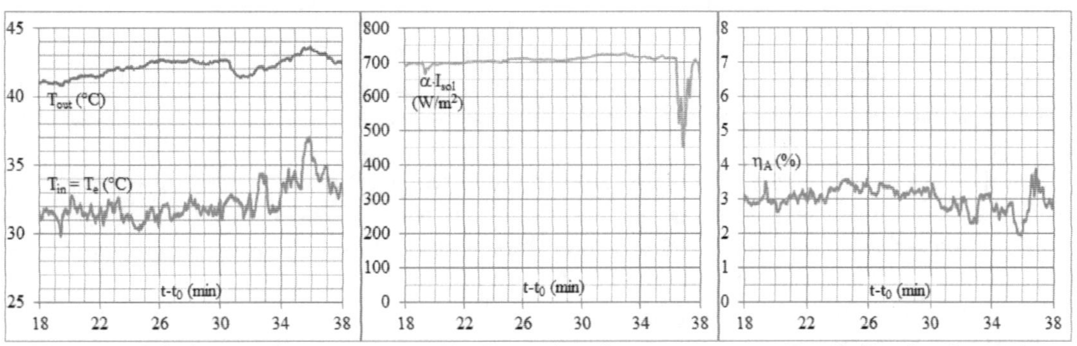

Figure 6. Inlet/external temperatures, $T_{in}=T_e$, and outlet temperatures, T_{out} (left); absorbed solar irradiance, $\alpha \cdot I_{sol}$ (middle); efficiency of removal of the absorbed solar heat, η_A (right).

An alternative and more used approach, analogous to that considered in [1], is to evaluate the reduction of transmitted heat rate η_B (%) allowed by roof ventilation with respect to absence of ventilation:

$$\eta_B = \frac{\dot{Q}_{removed}}{\dot{Q}_{transmitted}} \tag{4}$$

For a lightweight roof structure with negligible inertia and thermal transmittance U (W/(m^2°C)), the transmitted heat rate without ventilation can be estimated as

$$\dot{Q}_{transmitted} = U \cdot A_e \cdot \left(T_e + R_{se} \cdot \alpha \cdot I_{sol} - T_i\right) \tag{5}$$

where

R_{se} external surface resistance (m^2°C/W)
T_i indoor temperature (*e.g.* 27°C in summer)

R_{se} can be evaluated from the wind velocity v_{wind} (m/s) and the mean external temperature T_{me} (°C), *i.e.* the mean of external ambient temperature and external surface temperature. With typical summer values of T_{me} between 35°C and 45°C, the following semi-empirical formula can be obtained [22]:

$$R_{se} \approx 1/\left(10 + 4 \cdot v_{wind}\right) \tag{6}$$

In the considered test case, in which $T_e = T_i$, $v_{wind} \approx 1.5$ m/s, and $U \approx 1/(2 \cdot R_{se})$, the percent reduction of transmitted heat rate η_B is shown in Fig. 7.

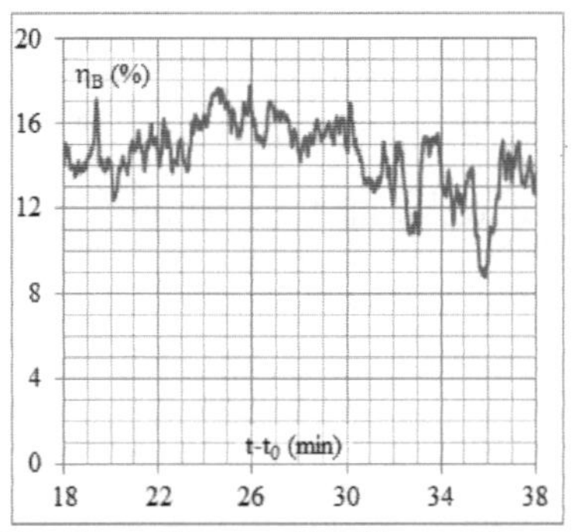

Figure 7. Percent reduction of transmitted heat allowed by roof ventilation with respect to absence of ventilation.

Generally speaking, rather than instantaneous values, averages over a given time period could be considered to take care of fluctuations of ambient conditions coupled with the thermal inertia of the roof layers, whatever low this is.

It must be underlined that the results obtained through the small scale bed are only qualitatively representative of a full scale roof system since some fluid dynamic phenomena don't linearly depend on size, yet they clearly show the effectiveness of the proposed throttling device. On the other hand, the experimental approach can be easily scaled up to a full size roof system.

5. Conclusive remarks

A novel approach to roof ventilation that relies upon forced flow rather than buoyancy force was investigated. In fact, exploiting a fan makes the airflow independent of the randomly varying environmental conditions and may limit the efforts to design and build the roof system. The considered layout is partially based on components of conventional naturally ventilated roofs (gutter inlets, ventilation cavity) and it may even permit to simplify the remaining components, consisting of a simple extraction manifold with fan instead of extraction outlets along the ridge, and also allow a continuous tile cover from gutters to ridge. Complexity of roof design and building is limited by an original regulation device, which improves flow homogeneousness in the ventilated air space and can be adjusted on site by simple thermographic or fluid velocity measurements. The effectiveness of the device was demonstrated by small scale experiments.

The fan extracting the airflow from the roof air space requires electrical energy, nevertheless a passive scheme can be obtained by adopting a small PV system installed on the roof, or a solar chimney, or a solar turbine. Moreover, the air ventilation can be easily reduced or even stopped when it is not desired, for example during the nighttime to allow radiative cooling of the tile mantle, or in winter to limit heat loss.

Approaches to assess the performance of the ventilation system are enabled by its conceptual simplicity and can be based on simple temperature and fluid velocity measurements. Such approaches were eventually tested on the small scale test bed, resulting easy to implement. In order to scale up to a full scale roof system, the test bed will be exploited to validate a CFD model, which will then be adapted to full scale simulation.

Acknowledgements
The authors wish to thank student Giorgia Paltrinieri for her support work.

References

[1] Ciampi M, Leccese F, Tuoni G 2005 *Sol. Energy* **79** 183
[2] D'Orazio M, Di Perna C, Principi P, Stazi A 2008 *Energ. Buildings* **40** 911
[3] D'Orazio M, Di Perna C, Di Giuseppe E 2010 *Energ. Buildings* **42** 1619
[4] Susanti L, Homma H, Matsumoto H 2011 *Energ. Buildings* **42** 1337
[5] Tong S, Li H 2014 *Build. Environ.* **81** 296
[6] Li D, Zheng Y, Liu C, Qi H, Liu X 2016 *Sust. Cities Soc.* **22** 8
[7] Villi G, Pasut W, De Carli M 2009 *Build. Simul.* **2** 215
[8] Zingre KT, Yang E-H, Wan MP 2017 *Energy* **133** 900
[9] Zingre KT, Wan MP, Wong KW, Toh WBT, Lee IYL 2015 *Energy* **82** 813
[10] Bottarelli M, Bortoloni M, Zannoni G, Allen R, Cherry N 2017 *Energy* **137** 391
[11] Gagliano A, Patania F, Nocera F, Ferlito A, Galesi A 2012 *Energ. Buildings* **49** 611
[12] Dimoudi A, Androutsopoulos A, Lycoudis S 2006 *Energ. Buildings* **38** 610
[13] Lee S, Park SH, Yeo MS, Kim KW 2009 *Build. Environ.* **44** 1431
[14] Di Giuseppe E, Sabbatini S, Cozzolino N, Stipa P, D'Orazio M 2018 *J. Bldg. Phys*, in press
[15] Ascione F 2017 *Solar Energy* **154** 34
[16] Zhang T, Tan Y, Yang H, Zhang X 2016 *Applied Energy* **165** 707
[17] Tarozzi L, Caffagni E, Muscio A 2007 *Proc. 9th AITA* 373
[18] Caffagni E, Libbra A, Muscio A, Tarozzi L 2008 *Proc. Terza giornata di studio UIT su 'Tecniche ottiche e termografiche in termofluidodinamica"*
[19] Caffagni E, Libbra A, Muscio A, Tarozzi L 2018 *Adv. Civil Engineering Technology* **1**
[20] ASTM 2016 *ASTM C1549-16*
[21] ASTM 1995 *E901-82(1995)e1*
[22] CEN 2018 *EN ISO 6946:18*

IOP Conf. Series: Journal of Physics: Conf. Series **1224** (2019) 012028 doi:10.1088/1742-6596/1224/1/012028

Experimental analysis of steam condensation patterns within a pressure suppression system operating at sub-atmospheric pressure conditions

R Lo Frano, D Aquaro, D Mazed, D Del Serra and G Giambartolomei
DICI- University of Pisa, Pisa- Italy

E-mail: rosa.lofrano@ing.unipi.it

Abstract
For some specific applications, the pressure suppression systems should be maintained at sub-atmospheric pressure, such as in fusion plants systems. This operation conditions differ considerably from those experienced in the suppression pools of BWR operating at atmospheric pressure.
The steam condensation in water at sub-atmospheric conditions is not sufficiently known, if not at all; therefore its effectiveness has to be investigated experimentally. In fact the steam condensation at sub-atmospheric pressure requires a subcooling capacity 2.5 times higher than that at atmospheric pressure.
In consideration of that, a low scale experimental facility was designed and built at the University of Pisa. More than 350 steam condensation tests were performed in broad influencing parameters range (e. g. pressure within the condensation tank, water temperature and steam mass flux) to study of effectiveness of the steam condensation. In doing that, different sparger configurations (with single or multiple holes) were considered to check possible interference of condensing steam jets and bubble coalescence.
Results indicate a high efficiency of condensation for all the examined conditions. The 3D map of the condensation regimes demonstrate they are dependent only on the downstream exit pressure, the water temperature and the steam mass flux per hole.

1. Introduction

The ITER Vacuum Vessel (VV) operates in normal condition at 0.8±0.3MPa and in any case the maximum internal pressure shall be limited, e.g. in case of loss of coolant accident from the in-vessel components (coolant coming from the Tokamak Cooling Water System-TCWS) or LOVA (loss of vacuum accident) event, to 0.15 MPa absolute. In particular, to limit the VV internal pressure, in case of loss of coolant from the in-vessel components, VV has been directly connected to an additional sub-system: the Vacuum Vessel Pressure Suppression System (VVPSS).

Figure 1 shows the ITER Tokamak cooling system; VVPSS is the yellow colored system on the left side.

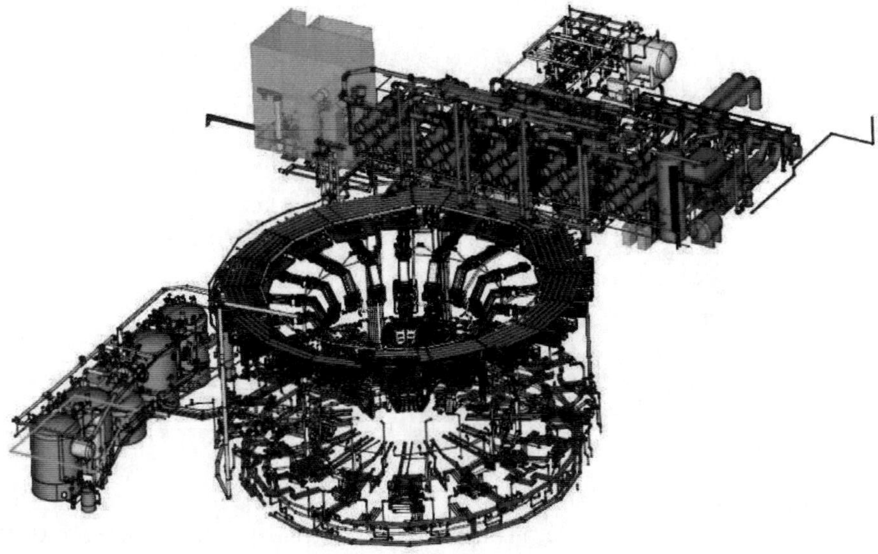

Figure 1- ITER Tokamak Cooling Water System

During an in-vessel coolant leak (Ingress of Coolant Event (ICE)) the VVPSS acts in concert with the VV drainage system: the former by condensing efficiently the hot steam in dedicated water suppression tanks allows to limit the maximum pressure to 0.2 MPa; the latter facilitates drainage of water from the VV. As a consequence of that, it appears that the steam direct contact condensation into the VVPSS suppression tanks water is the thermal-hydraulic phenomenon that guarantees the pressure safety limit are not exceeded.

In this study, the steam condensation in cold water at sub-atmospheric pressure is investigated: steam condensation regimes are determined along with the identification of main parameters influencing possibly the design of VVPSS. It is important to highlight that neither experimental nor analytical investigations of steam condensation at sub-atmospheric conditions, like those foreseen in ITER, have been yet reported in the available literature [1]÷[10].

In what follows a brief description of the VVPSS will be given. In section 3, we describe the experimental facility built at "B. Guerrini" Laboratory of the University of Pisa to study the steam condensation at sub-atmospheric condition along with the experimental work done. Critical discussion of the obtained experimental data is provided in section 4.

2. Description of VVPSS

The VVPSS is made of 4 Vapour Suppression Tanks (VSTs) of 100 m^3 each, with an inner diameter of 6.2 m and an overall height of about 4.7 m, mounted as two stacked assemblies (see Figure 2).

Each tank is designed to limit the final water temperature, after a loss of in-vacuum coolant event, to 95°C, to guarantee the structural integrity, withstanding all loads and load combinations that are within the design basis, the safe operation of the suppression system, and the confinement in the event of H_2 deflagration.

Figure 2 shows the arrangement of the VVPSS with the three Large LOCA Tanks (LLTs), and the Small LOCA Tank (SLT) that, despite the similar design, are filled with a different amount of water:

- 40 m^3 for the SLT;
- 60 m^3 for each of the LLT tanks 1, 2 and 3;

Moreover, the 3 LLTs are supposed to manage bigger LOCA events (cat. III and IV), while the SLT is supposed to manage smaller LOCA events (cat. II). VSTs are connected to the VV through two different relief lines, as shown in Figure 3, describing the VVPSS process flow diagram configuration.

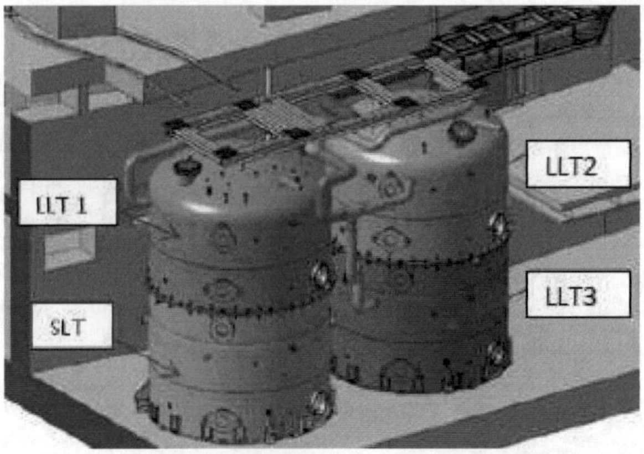

Figure 2- Overview of VVPSS with the LLTs and SLT tanks

By analyzing the VVPSS process flow diagram configuration, it appears that the SLT is connected by means of a DN300 relief line to one pipeline. A fully redundant set of double vacuum isolation bleed valves DN250 is provided to allow the trapped volume evacuation and the leak checking of the valves, as in the current VVPSS.

As to the LLTs, they are instead connected in parallel to a DN500 relief line, along which are installed two sets of rupture disks (full redundancy). When accident occurs, rupture disks opening will ensure the excess of steam is released to the DTR and condensed inside the VSTs.

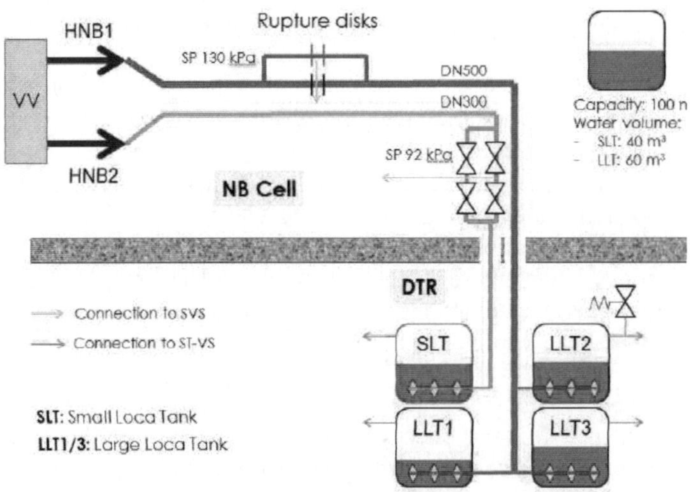

Figure 3- Schematic process diagram of the VVPSS

2.1 Experimental test facility

The experimental facility of the DICI-University of Pisa, even if a scaled model if compared to the ITER VVPSS, has been designed in order to provide a good representation of the operational conditions of the VVPSS in terms of initial pressure in the tank free space (close to vacuum conditions), initial water temperature, water head level inside the tank, sparger holes dimension and steam mass flow rate per hole (minimum and maximum steam mass flow rate per hole of the VVPSS has been covered during experimental tests).

The facility consists of eight main sub-systems, which are:

- The Superheated Steam Supply System (SSSS);

IOP Conf. Series: Journal of Physics: Conf. Series **1224** (2019) 012028 doi:10.1088/1742-6596/1224/1/012028

- The Flow Rate Control System (FRCS);
- The Condensation Tank System (CTS);
- The Auxiliary Tank (AT);
- The Vacuum System (VS);
- The Cooler System;
- The Degassed Water Supply System;
- The Data Acquisition and Control System;
- The Visualization and video recording system (VVRS).

The SSSS is constituted by an electrical steam generator of 130 kW provided in series by a heater (6 kW), able to deliver a maximum steam mass flow rate of 45 g/s at a pressure of 0.16 MPa and 150° C superheated steam temperature. During the operation, the steam generator automatically adjusts the power required to produce the requested steam mass flow.

The CTS is the main sub-system composed by the condensation tank (CT) and the sparger connected to it. This system, shown in Figure 3, has been designed in order to allow the experimental tests on DCC phenomena and monitor all the most affecting parameters.

The CT is a stainless steel cylindrical vacuum tight vessel (with thermal insulation) of 4.55 m^3 with 1.40 m ID, 3.2 m height. Moreover it is provided with a pressure safety valve set at 0.149 MPa (abs) and a manhole located at the bottom (same level of the sparger hole exit) to allow its inside inspection and maintenance.

The water inside the tank can be heated-up directly with the injection of the superheated steam from the electrical generator and cooled down through the Chiller (operated at atmospheric pressure) or indirectly flowing cold water in the coil piping located all around the internal wall of the Vessel.

In order to monitor the most affecting parameters, the CT has been deeply instrumented

Figure 4 - CTS with all the measurement sensor before insulation (left) and after walls insulation

IOP Conf. Series: Journal of Physics: Conf. Series **1224** (2019) 012028 doi:10.1088/1742-6596/1224/1/012028

The CT volume may be thought as subdivided in eight control layers (L2÷L9), whose data are provided in Table 1.

To measure temperature and pressure throughout the water volume the CT was instrumented with 28 temperature sensors (TE) (± 0.1 °C) and 8 pressure transducers (PE) (± 0.1 kPa). The water head level is measured (± 1 mm) by means of a differential pressure sensor level.

Table 1: Layer dimensions and sensors' position

Layer ID	Height (m)	Free Volume (m^3)	Located Sensors		Height Position (m)	W.H. H1.0	W.H. H1.3	W.H. H1.6	W.H. H2.0
			Temperature	pressure					
L2	0,425	0,5353	TE102	PE102	- 0.35	in Water (13 T) (4 P)	in Water (17 T) (5 P)	in Water (21 T) (6 P)	in Water (25 T) (7 P)
L3	0,400	0,6863	TE103,TE113,TE123,TE133	PE103	0.00				
L4	0,400	0,6120	TE104,TE114,TE124,TE134	PE104	+ 0.40				
L5	0,400	0,6120	TE105,TE115,TE125,TE135	PE105	+ 0.80				
L6	0,350	0,5352	TE106,TE116,TE126,TE136	PE106	+ 1.20	in VS (13 T) (4 P)			
L7	0,350	0,5352	TE107,TE117,TE127,TE137	PE107	+ 1.50		in VS (9 T) (3 P)		
L8	0,350	0,5380	TE108,TE118,TE128,TE138	PE108	+ 1.90			in VS (5 T) (2 P)	
L9	0,425	0,4790	TE109	PE109	+ 2.15				in VS (1 T) (1 P)
Total	3,100	4,5330							

The steam flows along an inner 2" pipe (electrically heated and thermally insulated) having a bottom cap drilled with holes of 10 mm diameter. The insulated outside surface is equipped with an electric heater system such to avoid steam condensation within the tube.

Figure 5 shows the bottom ends of sparger system.

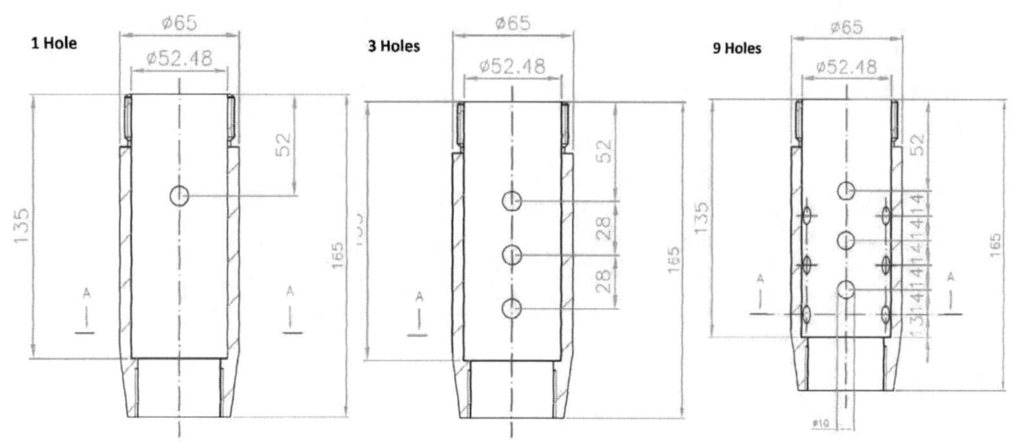

Figure 5: Technical drawing of the three configuration of the removable bottom part of the sparger: from left to right the patterns with 1H, 3H, and 9H are shown respectively.

Inside the condensation tank, a visualization and video recording system has been made with 4 high speed video cameras (GOPRO Hero4 model). Three of them were positioned at the same level of the sparger exit: the front camera is at about 1.00 m in front of the sparger exit; the left and right cameras are at 0.72 m left and 0.72 m right from the sparger exit.

The upper camera is fixed on the top of the condensation tank (and aligned along the axis) at about 2.40 m above the sparger exit. The data acquisition and control system program is coded upon LabVIEW© platform.

2.2 Description of test matrix

The test matrix was properly designed and organized in order to investigate extensively a wide range of technical parameters affecting the condensation of steam in water at sub-atmospheric pressure. In developing this matrix the following variables were investigated:

- Three sparger's configurations: 1, 3 and 9 holes (indicated with notation 1H, 3H and 9H respectively) of 10 mm diameter each. The latter sparger pattern was designed to check the jets interaction effects.

- Two thermodynamic states of the steam: superheated and saturated.

- Three steam mass flow rates for each sparger configuration:
    - 1H: 1.5, 2.5 and 5 g/s (55 tests);
    - 3H: 4.5, 6.5 and 15 g/s (25 tests);
    - 9H: 13.5, 22.5 g/s and 45 g/s (25 tests).

- Five initial water temperature (T_w) values: 10, 20, 30, 40 and 50°C.

- Water head levels available in the CT.

In particular, the attention was focused on the possible influence of the sparger configuration, basically associated to the number of holes available for each sparger, on the condensation efficiency.

Experimental tests have been performed for each sparger configuration as follows:

- High temperature tests as first in order to handle water with low air solubility rates;
- Low temperature tests as second ones.

Finally it is to remark that the condensation tests sequence has been such to move progressively from the sparger configuration, with one hole to that with the nine holes.

3. Results discussion

Results presented and discussed therein following are mainly oriented to demonstrate the efficiency of the steam condensation at sub-atmospheric condition and/at condensation regime (definition and construction of steam condensation map) in order to demonstrate the capability of VVPSS to mitigate the accident scenario, e.g. ICE IV, and fulfil design criteria.

Figure 6 shows the time-dependent behaviour of the axial increment of water temperature in case of 1H, 3H and 9H sparger configurations, and for initial water temperature of 30°C.

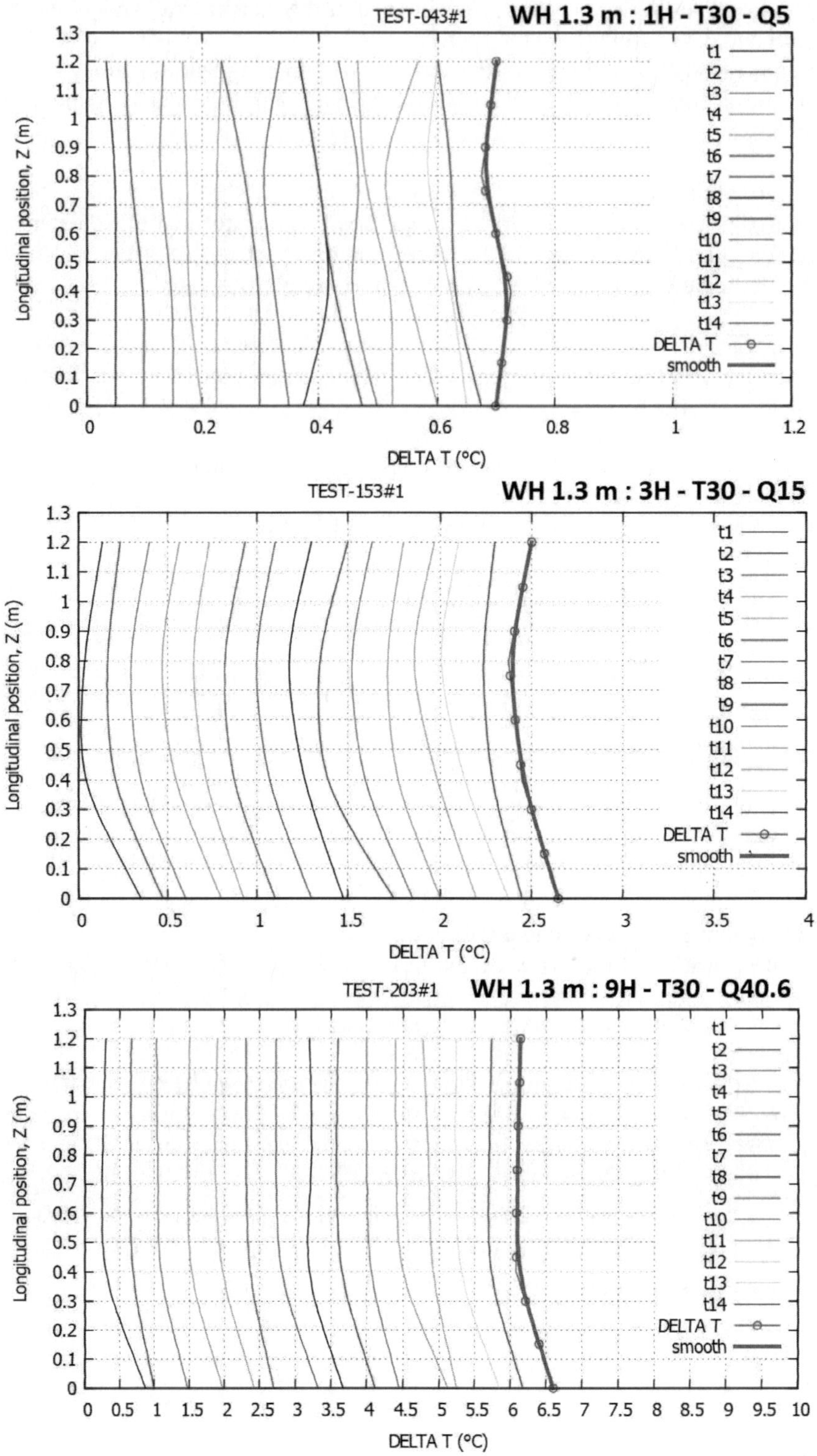

Figure 6 - Axial temperature profile (1 min time step): each colored curve represent temperature trend acquired thorough the 15 min. test duration.

Because of direct contact condensation (DCC- heat energy transferred from steam to water at vapour-water interface), it is possible to determine the steam condensation rate and the condensation efficiency, at the prevailing conditions, on the basis of the accurate measurement of water average temperature increase (ΔT_W) before and after the event of steam discharge.

The efficiency of the steam condensation was in all test conditions greater than 99.7 %. This value has been determined as follows:

$$PSF(\%) = 100 \cdot \left(1 - \frac{\Delta P_{net}}{P_{st-eq}}\right) \tag{1}$$

where $P_{st\text{-}eq}$ is the pressure-equivalent of the discharged steam (calculated based on the perfect gas law) while ΔP_{net} is the net pressure change in the vacuum space.

For any test, the increase of the water temperature (longitudinal and radial profiles), of the pressure in the vacuum space above the water head and the jet shape allow to characterise the condensation regimes and the heat transfer processes is taking place into the water mass.

It was observed that for water temperature ranges from 10 to 50°C and steam mass flux form 12 to 64 kg/(m^2 s), four different steam condensation regimes have been identified (see Figure 7). In this condensation map, it is possible to recognize the following regimes:

-Bubbling Condensation Oscillation (BCO);
-Transitional Chugging (TC);
-Stable Condensation (SC);
-Condensation Oscillation (CO).

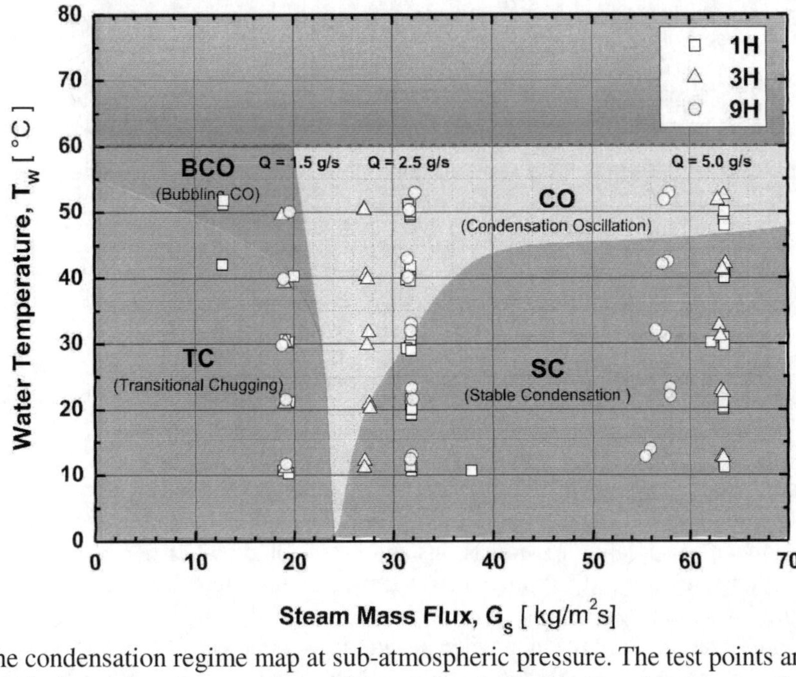

Figure 7- The condensation regime map at sub-atmospheric pressure. The test points are shown with different symbols for each sparger patterns. The pressure in free space volume was maintained slightly above saturation pressure of the water at the prevailing temperature to avoid boiling at water head interface.

In identifying the condensation regimes (CR), an important contribution is given by the image analysis of the thermal mixing (steam jet recorded by the video cameras), induced by the steam discharge in the water tank. Figure 8 shows some photographs of the typical shape of the condensing steam plume observed through the video cameras installed within the CT.

Chugging (C)	Transitional Chugging (TC)	Condensation Oscillation (CO)
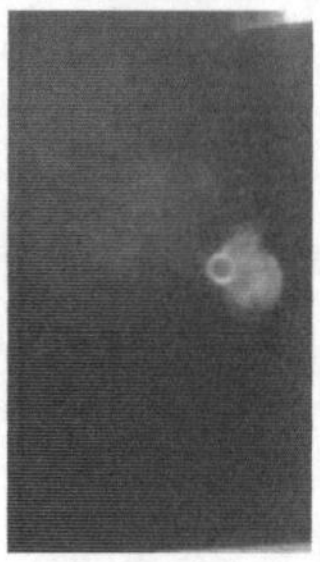	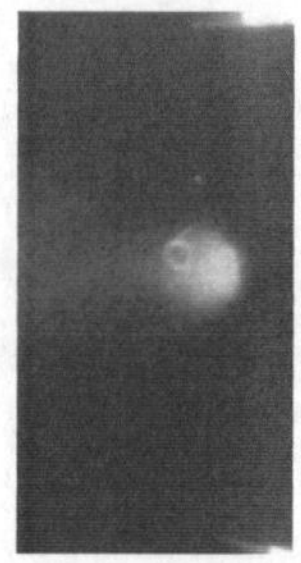	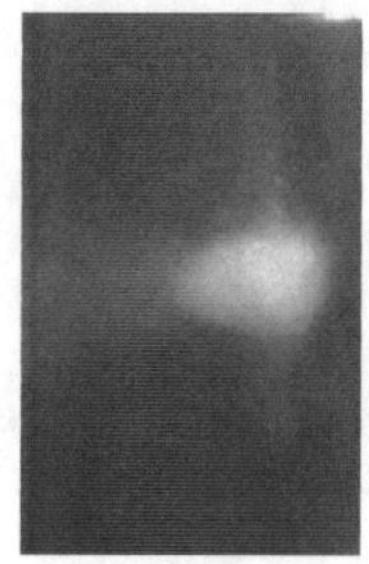
Stable Condensation (SC)	Bubbling Condensation Oscillation (BCO)	Interfacial Oscillation Condensation (IOC)
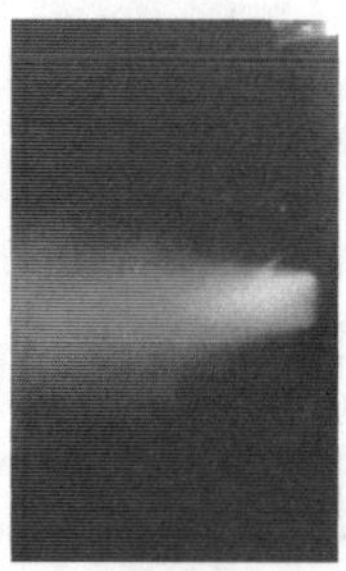	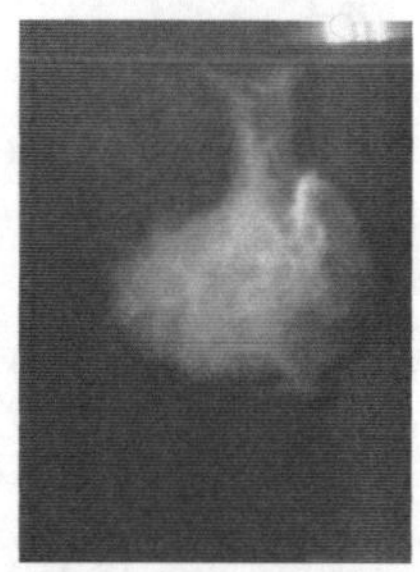	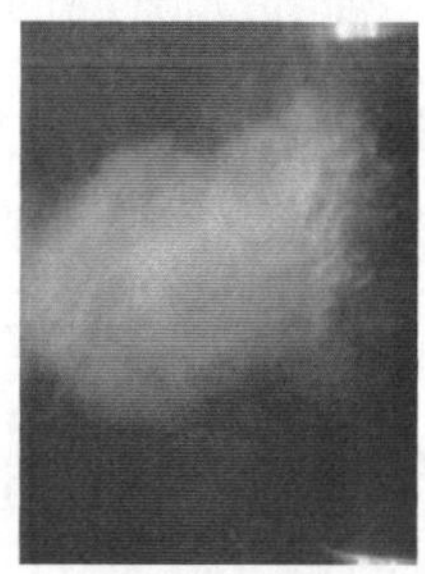

Figure 8 - Photos of the characteristic shape of the steam jets at exit hole for each identified condensation regime [13].

At sub-atmospheric pressure the CR resulted mainly governed by the water temperature (T_w) and steam mass flux to downstream pressure ratio (G_s/P_W): the *condensation map*, as shown in Figure 9, comprises six main condensation regimes (together with regions' boundaries), experimentally identified throughout the investigated parameters ranges. They are:

- Chugging (C): when the vapour-water interface locates alternatively inside and outside the sparger
- Transitional Chugging (TC): when the vapour-water interface locates outside the sparger in tank water, and the vapour core is characterized by short varying dimensions.
- Bubbling Condensation Oscillation (BCO): when the steam condenses in form of bubbles with varying dimeters, oscillating in dimensions and extension directions.
- Condensation Oscillation (CO): characterized by a condensing steam jet variable (unstable) in the longitudinal extension.
- Stable Condensation (SC): reached when a stationary condensing steam jet plume is established and is characterized by its stability. The jet plume does not change neither in length nor in radial extension.
- Interfacial Oscillation Condensation (IOC): characterized by intermittent destruction (disappearance) of the separation interface between the steam core and water. It is characterized by a random changing plume shape.

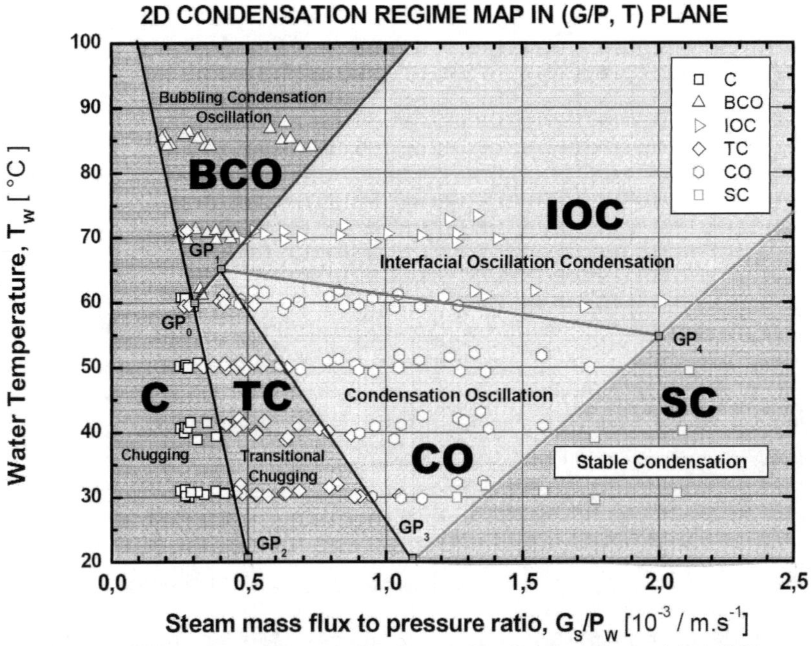

Figure 9 - Experimental CR map: each condensation region is identified with a different colour a part form the proper nomenclature. GP_i points identify the intersections of straights associated to each CR domain. Stable condensation appears at high G_s/P_w ratio for T_w from 30°C to 50°C.

The analysis of the steam condensation regime maps as a function of the downstream pressure shows that when P_W increases, the observed regions of the six condensation regime extend from the low G_S towards that with high G_S values [11]. As shown in Figure 9, the stable condensation appears at high G_s/P_w ratio for T_w from 30°C to 50°C, the reverse for unstable regimes.

Finally it has to remark that the identified condensation regimes are classified by analogue with those provided by Kuo-Shing Liang in [12] and by Chul-Hwa Song et al. [1]: consistency (and a good agreement was observed) among the compared experimental results was made by comparing the steam condensation data obtained at atmospheric pressure condition.

4. Conclusion

Results of the experimental campaign highlights:

- The water temperature increase rate for the same mass flow rate depends on the initial water temperature and the total mass of water in the tank.
- The total water temperature increase ΔT_W depends on the water mass in the condensation tank, the total mass of steam discharged, and the initial water temperature. In particular, it is independent from the water head level, the steam mass flow rate and the number of sparger holes (no evidence of hole steam jet coalescence).
- TW and GS/PW are the two governing parameters of the steam condensation which determine completely the condensation regime in any given VVPSS operation state, as far as we deal with pure steam
- No steam jet interference has been observed. The steam bubble coalescence for the sparger configuration examined (1, 3 and 9 holes) is thus very negligible.
- The condensation of steam through Direct Contact Condensation within the volume of cold water of the suppression tank is enhanced.

Finally the ITER VVPSS seems to be very effective in limiting the maximum pressure in the VV to 0.2 MPa during in-vessel coolant leak events and in maintaining the VV long-term pressure below atmospheric pressure during cat. IV event.

DISCLAIMER

The views and opinions expressed herein do not necessarily reflect those of the ITER Organization.

References

[1] J. Huang et al., Review of vapor condensation heat and mass transfer in the presence of non-condensable gas, Applied Thermal Engineering 89 (2015), pp. 469-484
[2] S.J. Hong et al., Condensation dynamics of submerged steam jet in subcooled water, International Journal of Multiphase Flow 39 (2012), pp. 66–77
[3] ITER/IO Internal Report. Preliminary Design Review, L8EUP5, 15 November 2013.
[4] R. Lo Frano et al., Fluid dynamics analysis of loss of vacuum accident of ITER cryostat, Fusion Engineering and Design, 109–111 (2016), pp. 1302–1307.
[5] C-H. Song et al., Steam Jet Condensation in a Pool from Fundamental Understanding to Engineering Scale Analysis, J. Heat Transfer, Transactions of the ASME-031004-2, 134, march 2012.
[6] J.C. Weimer et al., Penetration of vapour jets submerged in subcooled liquids, AIChE J., 19 (1973), pp. 552-58
[7] S. Cho et al., Effect of multiple holes on the performance of sparger during direct contact condensation of steam, Exper. Therm. and Flui. Sci., 28 (2004), pp. 629-38.
[8] R. Redlinger et al., 3D-analysis of an ITER accident scenario, Fusion Eng. and Design, 75–79 (2005), pp. 1233-1236.
[9] K. Takase et al., Numerical study on direct-contact condensation of vapor in cold water, Fusion Engineering and Design 63-64 (2002), pp. 421-428.
[10] D. Mazed et al., Experimental study of steam pressure suppression by condensation in a water tank at sub-atmospheric pressure, Proceedings of ICONE24, June 26-30, 2016, Charlotte, NC, USA
[11] D. Mazed et al., Experimental investigation of steam condensation in water tank at sub-atmospheric pressure, Nuclear Engineering and Design, Vol. 335 (2018), pp- 241-254.
[12] K.S. Liang, Experimental and analytical study of direct contact condensation of steam in water, Ph.D. Thesis, Massachusetts Institute of Technology, 1991.
[13] R. Lo Frano et al., Experimental investigation of functional performance of a vacuum vessel pressure suppression system of ITER, Fusion Engineering and Design, 122, (2017), pp. 42-46.
[14] R. Lo Frano et al., Investigation of vibrations caused by the steam condensation at sub-atmospheric condition in VVPSS, Fusion Engineering and Design, Volume 136, Part B (2018), pp. 1433-1437

IOP Conf. Series: Journal of Physics: Conf. Series **1224** (2019) 012029 doi:10.1088/1742-6596/1224/1/012029

LCT, PIV and IR Imaging Detection in Selected Technical and Biomedical Applications

Jan Stasiek, Marcin Jewartowski

Gdansk University of Technology, Faculty of Mechanical Engineering, Department of Energy and Industrial Apparatus, ul. G.Narutowicza 11/12, 80-233 Gdansk, Poland

jstasiek@pg.edu.pl

Abstract. Thermochromic liquid crystals (TLC), Particle Image Velocimetry (PIV) and Infrared Imaging Themography (IR) have been successfully used in non-intrusive technical, industrial and biomedical studies and applications. These four tools (based on the desktop computers) have come together during the past two decades to produce a powerful advanced experimental technique as a judgment of quality of information that cannot be obtained from any other imaging procedure. The brief summary of the history of this technique is reviewed, principal methods and tools are described and some examples are presented. With this objective, a new experimental technique have been developed and applied to the study of heat and mass transfer and for biomedical diagnosis. Automated evaluation allows determining the heat and flow visualisation and locate the area of suspicious tissue of human body.

1. Introduction

In order to demonstrate the feasibility of thermochromic liquid crystals techniques, true-colour image processing, particle image velocimetry and infrared thermography in practical heat transfer, flow visualization and biomedical contexts several experiments has been performed. The first set was carried out to investigate temperature and then heat transfer coefficient distributions (in Nusselt number form) on film-heated models in channels with ribs/turbulators. Two experimental techniques were employed in the research for comparison: firstly, the steady-state and secondly transient technique. The flow pattern produced by transverse vortex generators (ribs) was visualized using planar beam of double-impulse laser tailored by a cylindrical lens and oil particles. Sequential images of the particles in a cross sectional plane taken with CCD video camera from the downstream side of the flow were stored on a personal computer to obtain distributions of velocity vectors by means of the PIV method.

Also in experimental investigation of the film cooling effectiveness for model turbine blades using TLC technique is presented. The film cooling and visualization of the few different holes configuration in steady-state and transient performance were analyzed experimentally.

In biomedical situations, e.g. skin diseases, breast cancer, blood circulation and other medical application, TLC, IR and image processing are successfully used as an additional non-invasive diagnostic method especially useful for screening large groups of potential patients. Biomedical data presented in the paper is based on results taken from literature and from medical researchers collaborating with authors. Most of described studies and obtained data (liquid crystals experiments, pressure gradient, flow visualization and PIV) provide useful information for better understanding of thermal performances of many industrial installations and medical examination.

2. Liquid Crystals

The original use of the term 'liquid crystal' is credited to Lehman, a German physicist, however they were first observed more than 130 years ago by an Austrian botanist, Friedrich Reinitzer [1]. In many publications of the time, Lehman use terms like Fliesende Krystalle (flowing crystals) and Flussige Krystalle (fluid crystals) to describe the unusual phase that existed between the two melting points of not only cholesteryl benzoate, but also a large number of other compounds he studied.

Liquid crystals are temperature indicators that modify incident white light and display colour whose wavelength is proportional to temperature. They can be painted on a surface or suspended in the fluid and used to make visible the distribution of temperature. Normally clear, or slightly milky in appearance, liquid crystals change in appearance over a narrow range of temperature called the 'colour-play interval' (the temperature interval between first red and last blue reflection), centered around the nominal 'event temperature'. The displayed colour is red at the low temperature margin in the colour-paly interval and blue at the high end. Within the colour-play interval, the colours change smoothly from red to blue as a function of temperature [2]. The effect is reversible which, in practical therms, means that a single application of liquid crystals to a model can be used repeatedly (usually until coating is oblated or otherwise damaged). Both the colour play interval and the event temperature range of a liquid crystal can be selected by adjusting its composition, and materials are available with event temperatures from -30°C to 120°C and with colour play bands from 0.5°C to 20°C [2-3] although not all combinations exist of event temperature and colour play band widths. Widths of 1°C or less will be called narrow band materials, while those whose band width exceeds 5°C will be called wide band. The types of material to be specified for a given task depend on the type of image interpretation technique to be used [4].

Liquid crystals are classified into three categories according to their molecular structures: smectic, nematic and cholesteric. However, from a structural viewpoint, it can be argued that there are only two basic types of liquid crystals, chiral-nematic and smectic, with cholesteric being regarded as a special kind of nematic. Some of their structures are shown schematically in figure 1 [5].

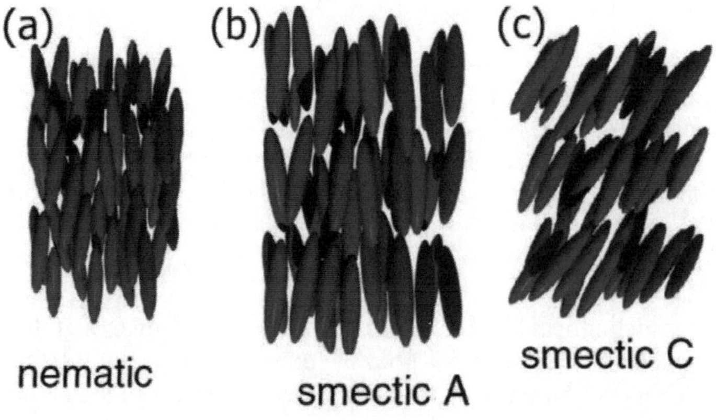

Figure 1. A different view of the mesophase structure. The orientation of the molecular director is represented by the arrows: a) nematic, b) smectic A, c) smectic C [5]

For flow analysis the the microcoencapsulated or usealed of thermochromic slurries of liquid crystals can be used to make visible the temperature and velocity fields in liquids. By dispersing the liquid crystal material into the liquid they become not only classical tracers used for flow visualisation but simultaneously small thermometers monitoring local fluid temperature [6]. The liquid crystals can be also used as temperature sensors in the case of investigations in strong magnetic field [7-8]. Magnetic field was the source of additional force, which was able to enhance heat transfer of paramagnetic fluids connected directly with temperature changes indicated by the liquid crystals. To get the qualitative information about temperature field an image analysis is necessary. Some publications described the suitability of color space transformations for such optical temperature measurements [9].

Before the execution of a thermal mapping or flow visualization experiment, we should recognize the characteristics of the overall combination of TLC, the light source, the optical camera system (ordinary, CCD black and white or RGB camera), and make a rational plan for the total measurement system. The relationship between the temperature of the crystal and the measured Hue (in the HSI – hue, saturation, intensity) of the reflected lights defines the calibration curve for the liquid crystal. The result is a curve relating the Hue of the reflected light to the surface temperature. A known temperature distribution exists on a 'calibration plate' (brass plate) to which is attached the liquid crystal layer. In order to maintain a linear temperature distribution with desired temperature gradients, one end of a brass plate was cooled by stabilized water and the other end controlled electrically to give a constant temperature [10-11]. The brass plate with the liquid crystal layer is calibrated in place in the wind tunnel with the same lighting level and viewing angle used during the data acquisition phase of the experiment [11]. The distribution of the colour component pattern on the liquid crystal layer was measured by RGB colour camera and series of images at a different temperatures defines the calibration.

Two main methods of surface temperature measurement are performed involving steady state and transient techniques which requires measurement of the elapsed time to increase the surface temperature of TLC-coated test specimen from a known initial temperature predetermined value [12-13].

Liquid crystals can be also used to make visible temperature and velocity fields in liquid by the simple expedient of directly mixing the liquid crystal material into the liquid in the small quantities. As opposed to all of the preceding researchers, who used encapsulated liquid crystals the team of Kowalewski and Stasiek [6] dissolved encapsulated chiral-nematic material in ether and sprayed the mixture into the air above a free-surface of glycerol. The ether evaporated in mid-air, leaving small drops of liquid crystal material which fell into the glycerol forming an 'almost mono-dispersed' suspensions of particles approximately 50 microns in diameters or less. The concentration was kept below 0.02 % by weight.

3. Particle Image Velocimetry

Particle Image Velocimetry technique is another well-established experimental method in fluid mechanics, that allows quantitative measurement of two-dimensional flow structure. More recently, this technique has been applied to study of heat and fluid flow in complex geometries with highly non-uniform boundary conditions [14-19].

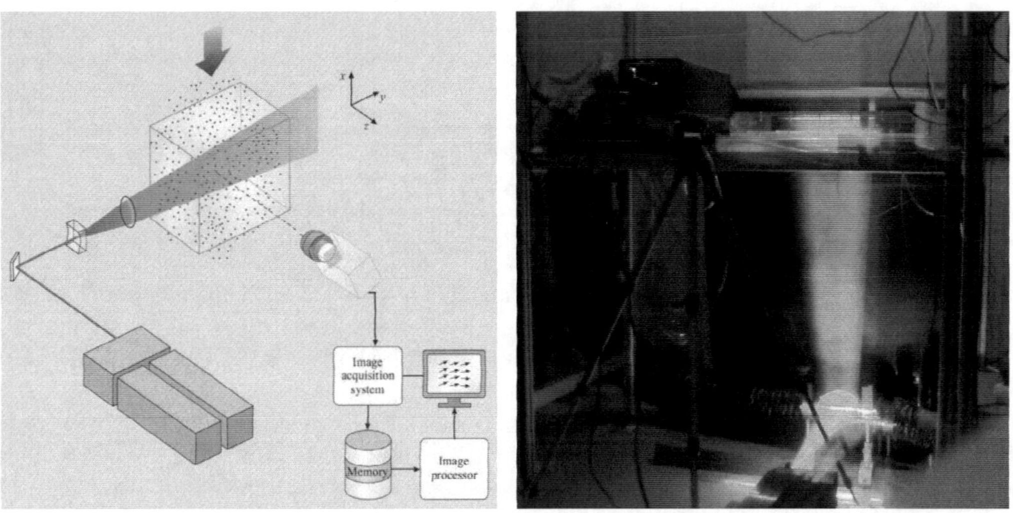

Figure 2. Schematic of a typical PIV measurements (left) and "laser light sheet" (right)

It enables measurements of the instantaneous in-plane velocity vector field within a planar section of the flow field and allows to calculate spatial gradients, dissipation of turbulent energy, spatial

IOP Conf. Series: Journal of Physics: Conf. Series **1224** (2019) 012029 doi:10.1088/1742-6596/1224/1/012029

correlations, and the like. In PIV technique selected cross-section of the investigated seeded flow is illuminated by laser light formed in thin "light sheet" (figure 2). Images of the flow are recorded by CCD camera and correlated to calculate instantaneous velocity fields. At presents the PIV measurements were performed for pure air seeded with small droplets (few micrometers in diameter) of synthetic oil DEHS (Di-Ethyl-Heksyl-Sebacat).The oil drops volumetric concentration was very low hence they did not affect the flow structure.

4. IR-Photography

Infrared photography is a non-contact measurement method, which enables visualisation and registration of temperature distribution on the surface. Every object with temperature above 0 K emits radiation as an electromagnetic wave. Radiation in the range of 0.8 μm up to 1000 μm is non visible by human eye and is called infrared radiation (IR). Its intensity is a function of temperature and emissivity of the object surface. IR-photography measures from the distance intensity of infrared radiation emitted by the body and converts it into digital signal. That signal is enhanced and transformed, which leads to obtaining a temperature field of measured object as a thermogram (map of temperature field). It can be later analysed using computer software. This technique has wide applications in industrial [20-21], scientific [22-24] and medical thermography [25].

5. Experimental facility for heat and flow visualisation

The experimental study was carried out using an open low-speed wind tunnel consisting of entrance section with fan and heaters, large settling chambers with diffusing screen and honeycomb, and then working sections (figure 3).

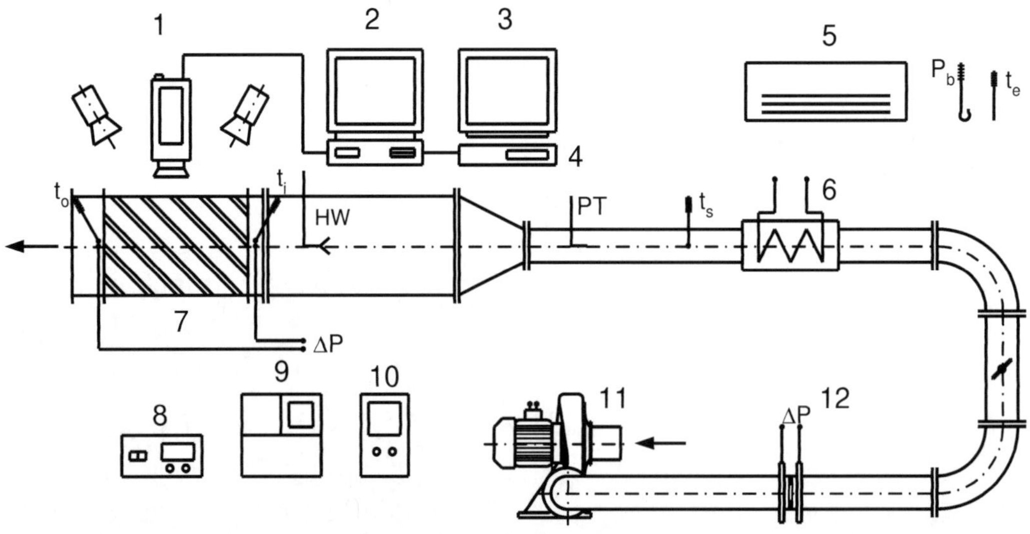

Figure. 3. Open low speed wind tunnel, 1 – RGB camera (TK-1070), 2 – PC, 3 – monitor (RGB-VMR 200), 4 – S-VHS recorder, 5 – air conditioning system, 6 – heater, 7 – LC mapping section, 8 – digital micro manometer FCO 12, 9 – DISA hot wire system, 10 – variac, 11 – fan, 12 – orifice

Air is drawn through the tunnel using a fan able to attain the Reynolds numbers in the range from 500 to 40 000. Local and mean velocity are measured using conventional Pitot tubes and DISA hot-wire velocity probe. The liquid crystals used here, manufactured in sheet form by Hallcrest [26], had an event temperature range from 30.0 to 35.0°C. In this particular experiment uncertainty for temperature measurement was estimated at about ± 0,05°C by considering only the section of the surface used in the experiment, span-wise non-uniformities in Hue value are minimized.

In many cases, as mentioned above, remarkable enhancement of local and spatially averaged surface heat transfer rates are possible with rib turbulators, in spite of the lower local Nusselt number at certain

IOP Conf. Series: Journal of Physics: Conf. Series **1224** (2019) 012029 doi:10.1088/1742-6596/1224/1/012029

locations along the ribbed surfaces. The test surface that is analyzed contains a collection of rib turbulators that are perpendicular and angled with respect to the flow stream (figure 4) [27].

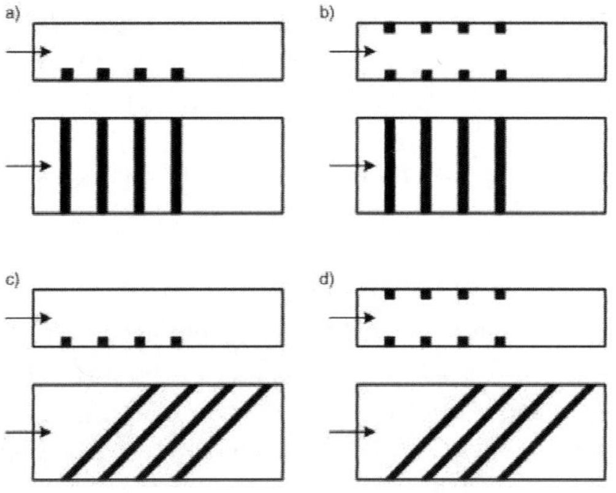

Figure 4. Schematic view of four types of transverse vortex generators (rectangular ribs)

To determine the surface heat flux (used to calculate heat transfer coefficients and Nusselt numbers), the convective power levels provided by the thermofoil heaters are divided by flat test surface area. Spatially resolved temperature distributions along the bottom rib turbulator test surface are determined using liquid crystals thermography (LCT) and true-color image processing system commercially available from Hallcrest [26]. Experimental facility is also equipped with Infrared Camera FLIR T440bx. T-Series camera offer superior sensitivity, high resolutions, excellent infrared image quality and reliable temperature measurement accuracy at an affordable price, which is ideal for scientific applications. FLIR T440bx has been specifically designed to have 8 times continuous digital zoom with a temperature range from -20°C to 1200°C, MSX Thermal Image Enhancement and many other features.

6. Examples of experimental results

In the following examples the TLC's plastic sheet and capsulated TLC's tracers have been applied to measure both temperature and velocity flow fields. For human body screening or biomedical situations few images using TLC and IR-Photography techniques are also presented.

6.1. Full field surface temperature measurements

Liquid crystals can be used to determine the distribution of the surface temperature, and if the surface heat flux can be found, this allows evaluation of the heat transfer coefficient or the Nusselt number.

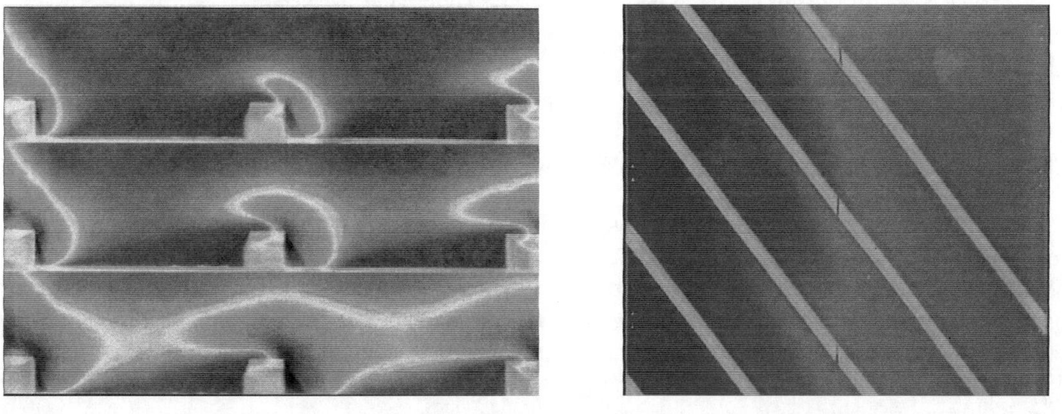

Figure 5. True-colour image from TLC for an endwall surface with columns (left) and visualization of model heat exchanger with vortex generators: ribbed surface - true-colur image (right)

Left image in figure 5 shows photographs of the colour distribution of the liquid crystal layer around a square section column.

Improvements in flow and thermal characteristics of heat exchanger do not require any demonstrations and would substantially reduce fuel and production costs. The measuring technique comprising the use of LC flexible sheets and true-colour processing may also be used for a great variety of applications and should be of considerable use in improving the design of all types of compact heat exchanger. Experimental procedures cover full field flow patterns in classic heat exchanger elements (flat plate with fine-tubes in-line, staggered and with vortex generators) describing local heat transfer coefficient and Nusselt number on the surfaces. An examples of such results are presented on right image in figure 5.

Nusselt number distribution for channel with vortex generators (rib turbulators) for one geometry with ribbed bottom and top walls (geometry b in figure 4) for different Reynolds numbers (9,000; 16,000; 26,000; 35,500) in the boundary layer is shown in figure 6. The analysis of the boundary layer was carried out to assess the impact of the channel wall on heat transfer. More information can be found in [27].

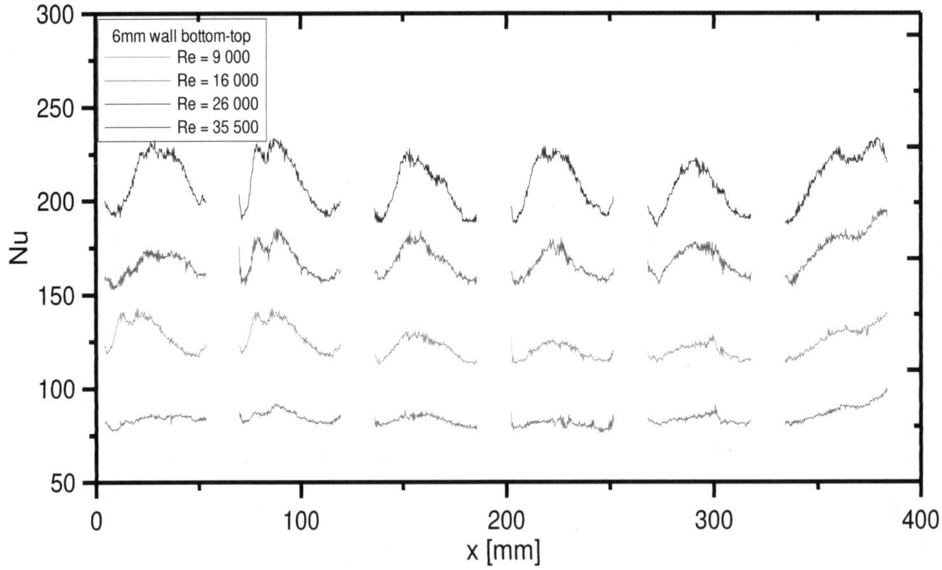

Figure 6. Local streamwise Nusselt number distribution between ribs in a boundry layer for ribbed bottom and top walls and ribs perpendicular to the flow

6.2. Natural convection in a closed cavity

Vertical temperature gradients are mainly responsible for the atmospheric or oceanographic fluid motion. Development of the nocturnal stable stratification over flat areas, inversion of temperature gradient and break-up of the convective boundary layer developed at the ground levels can be simulated using small-scale laboratory experiments.

One of the typical features characterizing instabilities generated by vertical temperature gradient are plumes or ejections appearing, when the thermal boundary layer breaks up. Figure 7 shows temperature and velocity visualization in glycerol-filled cavity under free convection in horizontal and vertical position. The photographs were made perpendicular to the line of the illumination. The specimen was illuminated using Xenon flash lamp collimated by a cylindrical lens to forma sheet which could be adjusted by 1 to 2 mm in width. The flashing of lamp was controlled by computer to flash with specified intervals between flashes from 3 seconds to 60 seconds.

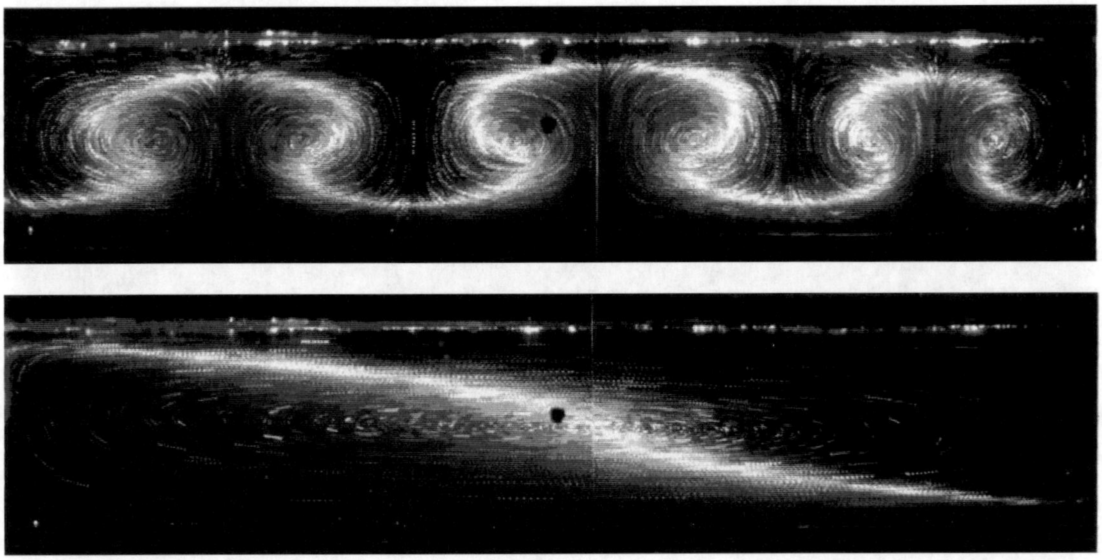

Figure 7. Visualisation of Benard cells in the horizontal (top) and vertical (bottom) cavity using TLC

6.3. Visualization of film cooling of model turbine blades

Internal cooling technique for gas turbine blades has been studied for many years. The designers need detailed data for heat transfer and temperature distributions along with the flow and streamwise pressure gradient to understand the flow physics and to improve the current internal cooling designs. Examples of modified gas turbine blades are shown in figure 8, left side [28]. Within this objective, TLC technique also has been applied to study of aerodynamic and aerothermal blade turbine design with effusive cooling concept. View of test section of low-speed wind tunnel is presented in figure 8, right side. Examples of steady-state and transient LCT measurements and visualization are shown in figure 9.

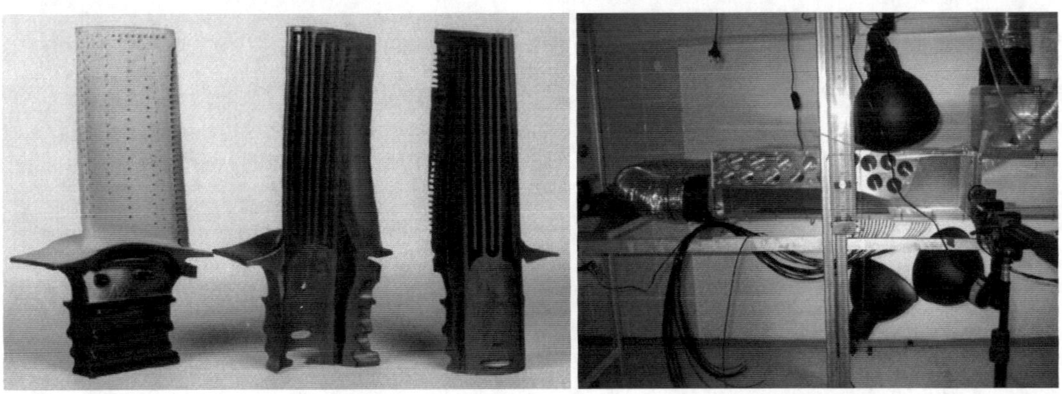

Figure 8. Gas turbine blades with cooling holes (left) and view of test section for model turbine blade cooling (right)

IOP Conf. Series: Journal of Physics: Conf. Series **1224** (2019) 012029 doi:10.1088/1742-6596/1224/1/012029

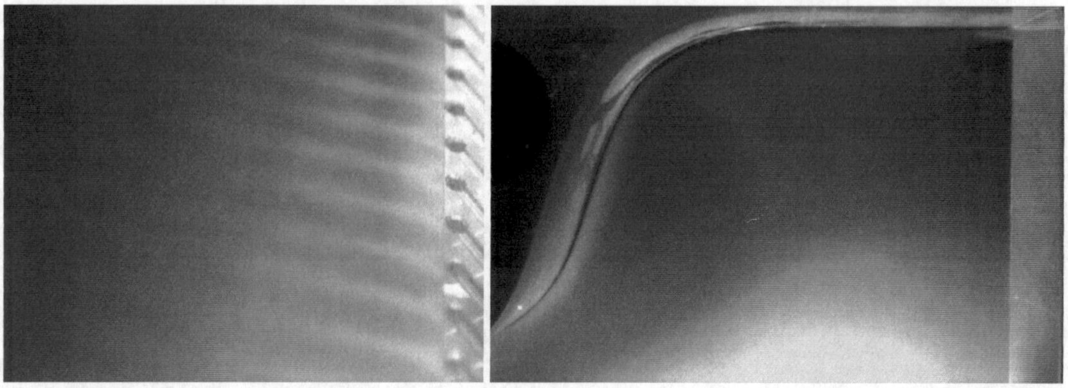

Figure 9. Steady-state LTC visualization of air flow on a blade surface with air jets (left) and end-wall temperature distribution by transient TLC method (right)

6.4. PIV measurements

Velocity fields in case of ribbed channel were obtained from the PIV measurements for different ribs geometries (figure 4) and different Reynolds numbers (Re = 9000, 16,000 and 26,000). The area scanned by the PIV method was in all cases located in the mid-vertical-plane between side walls (one position of the laser light sheet).

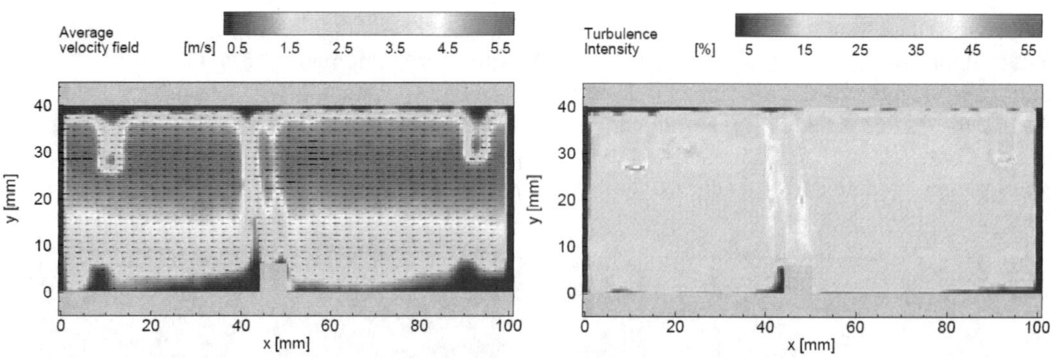

Figure 10. PIV measurements for ribbed channel: averaged velocity field (left) and turbulence intensity (right) for geometry with ribbed bottom wall

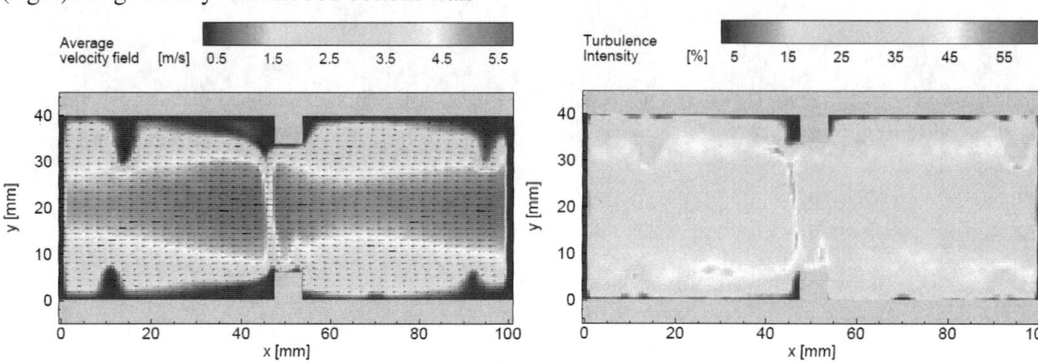

Figure 11. PIV measurements for ribbed channel: averaged velocity field (left) and turbulence intensity (right) for geometry with ribbed bottom and top walls

Figure 10 shows an average velocity field (averaged over 100 instantaneous velocity fields) for a geometry with ribbed bottom wall only (ribs height is 6 mm) and flow perpendicular to the ribs ((geometry a in figure 4)), while figure 11 show geometry with ribbed top and bottom walls (geometry b in figure 4). Reynolds number is 16,000. Maximum velocity for this cases is about 5.5 m/s and is located in the top part of the channel. Figures 10-11 show also a turbulence intensity computed from 100 instantaneous velocity fields.

6.5. Medical applications

Both, liquid crystal thermography and infrared photography, as non-invasive diagnostic techniques, can be used for skin surface temperature measurements. Obtained images can give information about pathological phenomena in the patient organism.

Liquid crystals in the mode of thermographic examination, can be also used in the case of diagnostics and monitoring of different diseases e.g.: diagnosis of cancers, tumours and inflammatory states (liver, lung, etc.), evolution of skin tests (allergy, tuberculin), traumatology and forensic medicine. Those methods are especially useful for children because they are non-invasive, fast, reliable and visually attractive. Because the temperature of human skin depends upon the metabolic level of neighbouring tissues, heat transfer can occur via muscular and adipose tissues, blood circulation, muscular tensions, body position or effect of drugs. This kind of medical diagnosis should be confirmed by other independent methods in the case of diseases suspicion [13].

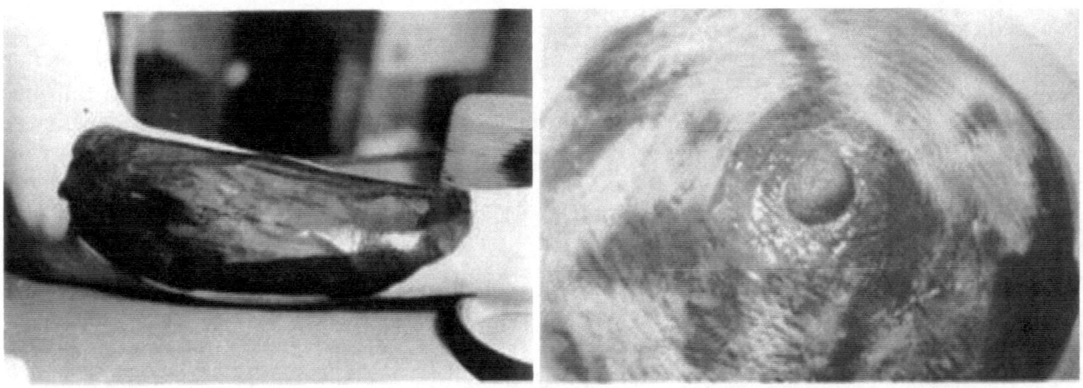

Figure 12. LC thermogram of forearm (left) and breast cancer (dark region) (right)

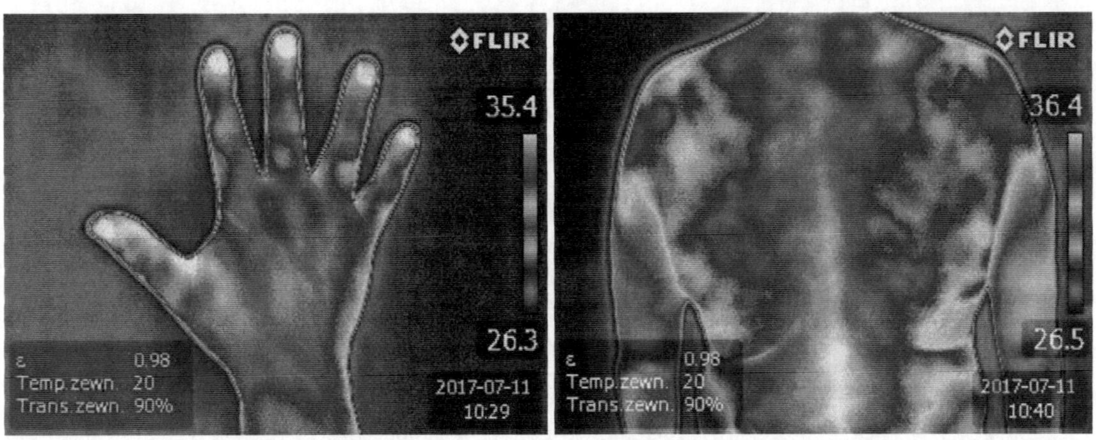

Figure 13. IR thermogram of human hand and human back

Requirements for thermographic examination are well-known [29]. The minimum area of a testing room has to be 9 square metres. All devices emitting heat should be removed, if possible. Air movements have to be minimised. The preferred temperature of the testing room should be 25 to 27°C, while humidity should be 50 to 70 %. Both these parameters should be carefully controlled during measurement. The effect of sunlight should be minimised. Light sources have to secure scattered light of low intensity to avoid light reflections. Patients should abstain from physical activity during 20 minutes before measurement. The skin area studied should be exposed for at least 15 minutes before a test to ensure thermal equilibrium with the environment [13].

The patient's position should be similar to their physiological (normal) one to avoid temperature changes due to enforced changes of blood circulation. Body areas studied should not contact other body areas. Medicines, drugs (alcohol, nicotine and so on) and treatments affecting blood circulation should not be taken, if possible, for 24 hours before measurement. Also patients should not have cold and hot drinks and not eat large meals before measurement. Cosmetics can also affect thermal properties of skin. All subsequent measurements have to be done at the same hour of the day. The thermological report should contain all conditions of measurement, including the states of the patient and environment. The patient should be allowed to see the thermographic images because their colours have a positive psychological effect especially in the case of children. Various cases have been studied and typical results of LCT are shown in figure 12 [29], while figure 13 shows examples of infrared images of a human body taken by FLIR T440bx camera.

7. Conclusions

Advanced experimental techniques, in this case true-colur image processing of liquid crystal patterns, particle image velocimetry and infrared imaging, allow new approaches to old problems and open up new areas of research and applications. Image processing data make available quantitative, full field information about the distribution of temperature, heat transfer and adjunctive medical pathogenesis and diagnostics. It is evidence from our research that liquid crystal and infrared imaging technique have many applications, not only in testing, but also in industry, medicine and the home.

The presented overview provides an introduction to thermochromic liquid crystals, digital infrared imaging and particle image velocimetry, how this measurements techniques and methods work and how they are used.

References

[1] Reinitzer R, 1899 *Monatschr, Chem, Wien* 9 pp 421-441
[2] Moffat R J, 1991 *Proc. 9th Intl. Heat Transfer Conference* Jerusalem Vol. 1 pp 308-310
[3] Jones T V, Wang Z, Ireland P T, 1992 Proc. Of the First I.Mech.E. *Seminar on optical methods and Data Processing in Heat and Fluid Flow* City University London pp 51-65
[4] Stasiek J, 1997 *Heat and Mass Transfer* 33 pp 27-39
[5] Dierking I, 2014 Chiral Liquid Crystals: Structures, Phases, Effects, *Symmetry 6*(2), 444-472
[6] Stasiek J A and Kowalewski T A, 2002 *Optoelectronics Review* 10 No.1 pp 1-10
[7] Fornalik E, Filar P, Tagawa T, Ozoe H and Szmyd J S, 2006 *Intl Journal of Heat and Mass Transfer* 49 pp. 2642-2651
[8] Fornalik E, 2007 *Journal of Theoretical and Applied Mechanics* 45 pp 557-568
[9] Ziemba A and Fornalik E, 2016 *Journal of Physics Conference Series* 745 032108
[10] Akino N, Kunugi T, Ichimiya K, Mitsuchiro K and Ueda M, 1989 *ASME J.Heat Transfer* 111 pp 558-565
[11] Stasiek J, Collins M W, 1996 *Atlas of Visualization* Vol 2 CRC Press Inc USA
[12] Stasiek J, Jewartowski M and Kowalewski T A, 2014 *Journal of Crystalization Process and Technology* 4 pp 46-59
[13] Stasiek J, Stasiek A, Jewartowski M and Collins M W, 2006 *Optics & Laser Technology* 38 pp 243-256
[14] Ambrosini D, Paoletti D, Galli G, 2015 *World Congress on Mechanical, Chemical and Material Engineering* Barcelona, Spain 325
[15] De Leo C, Paoletti D, Ambrosini D, 2018 *The European Physical Journal Applied Physics* 82(3) 30501

IOP Conf. Series: Journal of Physics: Conf. Series **1224** (2019) 012029 doi:10.1088/1742-6596/1224/1/012029

[16] Lindken R, Gui L, Merzkirch W, 1999 *Chemical Engineering and Technology* 22(3) pp 202-206
[17] Adrian R J, 2015 Experiments in Fluids 39(2) pp 159–169
[18] Merzkirch W, 2018 *Particle Image Velocimetry. In: Optical Measurements* pp 341-357 Springer, Berlin, Heidelberg
[19] Stabile L, Arpino F, Buonanno G, Russi A, Frattolillo A, 2015 *Building and Environment* 93 pp 186-198
[20] Bagavathiappan S, Lahiri B B, Saravanan T, Philip J and Jayakumar T, 2013 *Infrared Physics & Technology* 60 pp 35-55
[21] Nardi I, Lucchi E, de Rubeis T, Ambrosini D 2018 *Building and Environment* 146 pp 190-205
[22] Mikielewicz D, Wajs J, Glinski M and Zrooga A-B R S, 2013 *Experimental Thermal and Fluid Science* 44 pp 556-564
[23] Wajs J, Mikielewicz D, 2017 *Archives of Thermodynamics* 38(1) pp 123-139
[24] Raiola M, Greco C S, Contino M, Discetti S, Ianiro A, 2017 *International Journal of Heat and Mass Transfer* 108, 199-209
[25] Polidori G, Renard Y, Lorimier S, Pron H, Derrau S and Taiar R, 2017 *Int. Journal of Surgery Case Reports* 34 pp 56-59
[26] HALLCREST, 2014 *Handbook of Thermochromic Liquid Crystal Technology* LCR Hallcrest
[27] Mikielewicz D, Stasiek A, Jewartowski M and Stasiek J, 2012 Applied Thermal Engineering 49 pp 61-72
[28] Giampaolo A, 2006 *Gas turbine handbook: principles and practices* The Fairmont Press, Inc.
[29] Zmija J, Klosowicz S and Borys S, 1989 *Cholesteric liquid crystals in a detection of radiation* (Warsaw: Poland/WNT in Polish)

IOP Conf. Series: Journal of Physics: Conf. Series **1224** (2019) 012030 doi:10.1088/1742-6596/1224/1/012030

Numerical simulation of a turbulent Lead Bismuth Eutectic flow inside a 19 pin nuclear reactor bundle with a four logarithmic parameter turbulence model

Andrea Chierici*, **Leonardo Chirco***, **Roberto Da Vià***† and **Sandro Manservisi***

*University of Bologna, DIN, Lab. of Montecuccolino, Via dei Colli 16, Bologna 40136, Italy

E-mail: †roberto.davia2@unibo.it

Abstract. Computational Fluid Dynamic allows scientists and engineers to investigate fluid flow in complex geometries and evaluate heat transfer between a solid body and a fluid. In the present paper we study the heat transfer of a Lead Bismuth Eutectic (LBE) turbulent flow in a bare 19 pin nuclear reactor bundle. When dealing with low Prandtl number fluids, like LBE for which Pr = 0.025, proper turbulence models are needed to improve the prediction of heat transfer. We use here a four logarithmic parameter turbulence model in order to calculate Reynolds stresses and turbulent heat flux. In particular, an equation for temperature fluctuations and one for their dissipation are solved. These variables are used to model thermal characteristic time scales. The results are reported for different values of the Peclet number and a fixed value of the pitch to diameter ratio. The obtained values of the Nusselt number are compared with experimental correlations, that can be found in literature, and with the ones obtained using a Simple Eddy Diffusivity model, where the eddy thermal diffusivity is calculated as proportional to eddy viscosity through a modeled turbulent Prandtl number.

1. Introduction

Liquid metals are gaining more and more interest as coolant fluids in engineering applications [1]. In particular they are being considered in the design process of IVth generation nuclear rectors with the aim of building nuclear power plants with increased economic sustainability and safety levels. Liquid metals are classified as low Prandtl number fluids. They are in fact characterized by low viscosity and high thermal diffusivity values, leading to molecular Prandtl number values $Pr \simeq$ 0.01-0.025. As a consequence, more sophisticated tools need to be developed in order to obtain accurate evaluation of heat transfer in the condition of turbulent flows. Standard modeling, i.e. constant turbulent Prandtl number $Pr_t \simeq 0.85$, that is normally used for simulating turbulent flows of fluids having Pr close to one, leads to overestimate turbulent heat flux [2, 3, 4]. Various alternatives have been developed, involving different algebraic models for turbulent Prandtl number, see [5, 6, 7] and references therein. Also isotropic four parameters turbulence models in which the turbulent thermal conductivity α_t is modeled by different characteristic thermal time scales and anisotropic models for turbulent heat flux have been proposed [8, 9, 10, 11, 12]. In the present work we use a four parameter logarithmic turbulence model to study a 19 pin nuclear reactor bundle whose cross section is sketched in Fig. 1 a). Many works involving numerical simulations of turbulent heat transfer in nuclear reactor bundles can be found in literature for triangular [13, 7] and square [14, 7] lattices.

Nomenclature

C_p	Fluid specific heat capacity
D_h	Hydraulic diameter
f_h	Heated to wet perimeter ratio
k	Turbulent kinetic energy
K	Logarithmic value of k, $\mathrm{K} = \ln(k)$
k_θ	Mean squared temperature fluctuations
K_θ	Logarithmic value of k_θ, $\mathrm{K}_\theta = \ln(k_\theta)$
Nu	Nusselt number, $Nu = qh/(T - T_{in})\lambda$
P	Mean pressure
p	Pitch, distance between adiacent rods
Pe	Peclet number, $Pe = Re \cdot Pr$
Pr	Prandtl number, $Pr = \nu/\alpha$
Pr_t	Turbulent Prandtl number, $Pr_t = \nu_t/\alpha_t$
q	Imposed heat flux
Re	Reynolds number, $Re = U_b h/\nu$
T	Mean temperature
$\mathbf{u}$	Mean velocity

Greek symbols

α	Fluid thermal diffusivity
α_t	Eddy thermal diffusivity
χ	Pitch to diameter ratio
λ	Fluid thermal conductivity
μ	Fluid dynamic viscosity
ν_t	Eddy kinematic viscosity
Ω	Logarithmic value of k specific dissipation rate
Ω_θ	Logarithmic value of k_θ specific dissipation rate
ϕ	Hot-spot factor
ρ	Fluid density
Θ	Non dimensional temperature

Miscellaneous

$(\cdot)_{bun}$	Reference to whole bundle cross section
$(\cdot)'$	Fluctuation of a quantity
$(\cdot)_{sub}$	Reference to triangular sub-channels

Property	Symbol	Value	Unit
Dynamic viscosity	μ	0.00184	Pa s
Density	ρ	10340	Kg/m^3
Thermal conductivity	λ	10.76896	W/(m K)
Specific heat capacity	C_p	146	J/(Kg K)

Table 1. Lead Bismuth Eutectic physical parameters used in the present study.

Hexagonal bundle, in lattices of 19 or more rods, are studied in bare, wire-wrapped and with spacer grid configurations [15, 16, 17, 18]. We study a fully developed turbulent flow occurring in a bare rod lattice configuration, in order to provide results in a simplified geometry that could be used as a starting point for future analysis. The simulated coolant fluid is Lead Bismuth Eutectic with a molecular Prandtl number $Pr = 0.025$. In the following Section we introduce the mathematical model used to study the fluid turbulent behavior. In Section 3 we report some of the obtained results regarding the simulated section and the triangular sub-channels.

2. Numerical Modeling

In the present work we deal with the simulation of a LBE turbulent flow. The physical properties are reported in Table 1 with a resulting Prandtl number $Pr = 0.025$. In order to study the flow and heat transfer behavior of this configuration, we solve the system of Reynolds Averaged Navier-Stokes equations as

$$\nabla \cdot \mathbf{u} = 0\,, \tag{1}$$

$$\rho\frac{\partial \mathbf{u}}{\partial t} + \rho(\mathbf{u}\cdot\nabla)\mathbf{u} = -\nabla P + \nabla\cdot\left[\mu\left(\nabla\mathbf{u} + \nabla\mathbf{u}^T\right) - \rho\overline{\mathbf{u}'\mathbf{u}'}\right] + \rho\mathbf{g}\,, \tag{2}$$

$$\rho C_p\left(\frac{\partial T}{\partial t} + (\mathbf{u}\cdot\nabla)T\right) = \nabla\cdot\left[\lambda\nabla T - \rho C_p\overline{\mathbf{u}'T'}\right] + Q\,, \tag{3}$$

where $\mathbf{U}$, P and T stand for mean velocity, pressure and temperature. The system of equations (1-3) is closed with a four parameter logarithmic turbulence system of equations in order to model the Reynolds stresses $\rho\overline{\mathbf{u}'\mathbf{u}'}$ and the turbulent heat flux $\rho C_p\overline{\mathbf{u}'T'}$.

$$\frac{\partial K}{\partial t}+\mathbf{u}\cdot\boldsymbol{\nabla}K=\boldsymbol{\nabla}\cdot\left[\left(\nu+\frac{\nu_t}{\sigma_k}\right)\boldsymbol{\nabla}K\right]+\left(\nu+\frac{\nu_t}{\sigma_\varepsilon}\right)\boldsymbol{\nabla}K\cdot\boldsymbol{\nabla}K+\frac{P_k}{e^K}-C_\mu\, e^{\Omega}\,, \tag{4}$$

$$\frac{\partial \Omega}{\partial t}+\mathbf{u}\cdot\boldsymbol{\nabla}\Omega=\boldsymbol{\nabla}\cdot\left[\left(\nu+\frac{\nu_t}{\sigma_\varepsilon}\right)\boldsymbol{\nabla}\Omega\right]+2\left(\nu+\frac{\nu_t}{\sigma_\varepsilon}\right)\boldsymbol{\nabla}K\cdot\boldsymbol{\nabla}\Omega+$$
$$+\left(\nu+\frac{\nu_t}{\sigma_\varepsilon}\right)\boldsymbol{\nabla}\Omega\cdot\boldsymbol{\nabla}\Omega+\frac{c_{\varepsilon 1}-1}{e^K}P_k-C_\mu\left(c_{\varepsilon 2}f_{exp}-1\right)e^{\Omega} \tag{5}$$

$$\frac{\partial K_\theta}{\partial t}+\mathbf{u}\cdot\boldsymbol{\nabla}K_\theta=\boldsymbol{\nabla}\cdot\left[\left(\alpha+\frac{\alpha_t}{\sigma_\theta}\right)\boldsymbol{\nabla}K_\theta\right]+\left(\alpha+\frac{\alpha_t}{\sigma_\theta}\right)\boldsymbol{\nabla}K_\theta\cdot\boldsymbol{\nabla}K_\theta+\frac{P_\theta}{e^{K_\theta}}-C_\mu\, e^{\Omega_\theta}\,, \tag{6}$$

$$\frac{\partial \Omega_\theta}{\partial t}+\mathbf{u}\cdot\boldsymbol{\nabla}\Omega_\theta=\boldsymbol{\nabla}\cdot\left[\left(\alpha+\frac{\alpha_t}{\sigma_{\varepsilon\theta}}\right)\boldsymbol{\nabla}\Omega_\theta\right]+2\left(\alpha+\frac{\alpha_t}{\sigma_{\varepsilon\theta}}\right)\boldsymbol{\nabla}K_\theta\cdot\boldsymbol{\nabla}\Omega_\theta+$$
$$+\left(\alpha+\frac{\alpha_t}{\sigma_{\varepsilon\theta}}\right)\boldsymbol{\nabla}\Omega_\theta\cdot\boldsymbol{\nabla}\Omega_\theta+\frac{c_{p1}-1}{e^{K_\theta}}P_\theta+\frac{c_{p2}}{e^K}P_k-\left(c_{d1}-1\right)e^{\Omega_\theta}-c_{d2}C_\mu e^{\Omega}\,. \tag{7}$$

The system of equations (4-5) consists of the transport equations for the logarithm of turbulent kinetic energy K and the logarithmic value of turbulent kinetic energy specific dissipation rate Ω, while (6-7) are solved for thermal turbulence, in particular for logarithmic temperature fluctuation variance K_θ and the logarithmic specific dissipation rate Ω_θ. With these variables characteristic time scales for thermal turbulence are modeled in order to calculate eddy thermal diffusivity α_t. A logarithmic formulation of turbulence variables allows us to obtain an increased numerical stability of the turbulence model. Further details on the model parameters can be found in [11, 19, 20, 21]. In the present case we focus our attention in the dynamic and thermal fully developed regime. If volumetric source term is set to 0 and a constant heat flux q is imposed on the wall then temperature grows linearly along axial coordinate. Temperature gradient can be calculated by integration of equation (3) on the whole domain, obtaining

$$\frac{\partial T}{\partial z}=\frac{4qf_h}{w_{mid}\rho C_p D_h}\,, \tag{8}$$

where z is the axial direction, w_{mid} is the average axial velocity on the cross section of the domain, D_h is the hydraulic diameter and f_h is the ratio between the heated and the wet perimeter of the domain cross section. In this case the temperature field T of (3) is usually decomposed as

$$T(x,y,z)=T_{w0}-\tilde{T}(x,y)+\frac{\partial T}{\partial z}(z-z_{in})\,, \tag{9}$$

where T_{w0} is an arbitrary temperature value on the inlet surface. The new temperature field $\tilde{T}(x,y)$ describes the temperature distribution on the transverse section of the domain and does not change with the axial coordinate. The energy balance equation, written with $\tilde{T}$ as state variable, becomes

$$\frac{\partial \tilde{T}}{\partial t}+\mathbf{u}\cdot\boldsymbol{\nabla}\tilde{T}=\boldsymbol{\nabla}\cdot\left[\left(\alpha+\alpha_t\right)\boldsymbol{\nabla}\tilde{T}\right]+\mathbf{u}\cdot\hat{\mathbf{a}}\frac{4qf_h}{w_{mid}\rho C_p D_h}\,, \tag{10}$$

where $\hat{\mathbf{a}}$ is a unit vector along the axial coordinate. We also neglect secondary flows and exploit geometry symmetry planes by simulating only a sector of the whole 19 pin nuclear reactor bundle, as reported in Fig. 1 a) and b). With the fully developed regime periodic boundary conditions can be imposed on inlet and outlet sections for all the solved variables. The final system of equations (1-2) - (4-7) - (10) is implemented and solved with finite element code FEMuS [22].

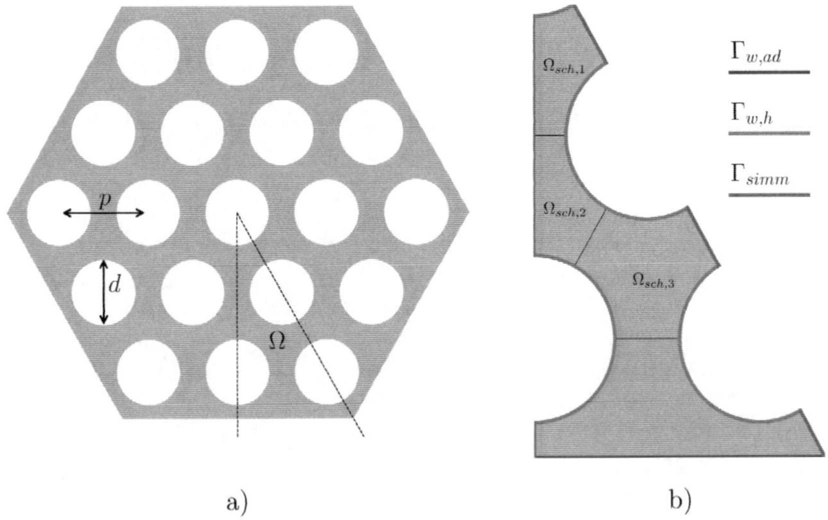

Figure 1. Schematic representation of the whole bundle cross section a) and simulated domain, boundary definition and triangular sub-channels b).

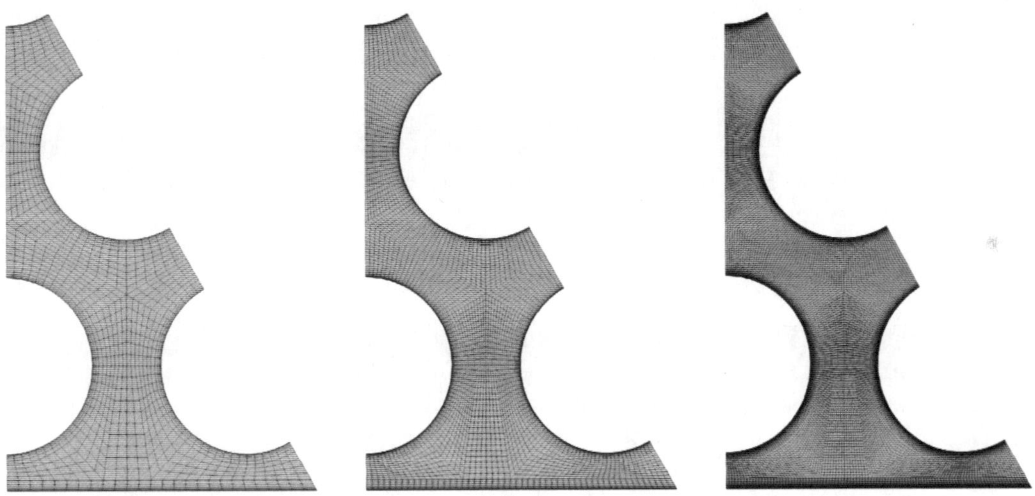

Figure 2. Meshes with three different levels of refinements used for the present work.

χ	$d_{h,bun}$ [mm]	$d_{h,sch}$ [m]	f_h
1.4	7.7	9.5	0.733

Table 2. Geometrical parameters of the simulated 19 pin nuclear reactor bundle.

3. Results

In this Section we present the results obtained from the simulation of the nuclear reactor bundle for multiple values of the Reynolds number. In particular we will refer to bulk Reynolds number, $Re_b = u_b d_{h,bun} \rho/\mu$, where u_b is the bulk velocity calculated on the whole bundle cross section and $d_{h,bun}$ is the hydraulic diameter of the bundle, and to sub-channel Reynolds

Boundary	$\mathbf{u}$	K & Ω	$\tilde{T}$	K_θ & Ω_θ
$\Gamma_{w,h}$	no-slip	near-wall b	uniform heat flux q	near-wall b
$\Gamma_{w,ad}$	no-slip	near-wall b	null gradient	null gradient
$\Gamma_{w,simm}$	null gradient	null gradient	null gradient	null gradient

Table 3. Boundary conditions imposed on boundaries represented in Fig. 1 b).

Correlation	Expression	Validity range
Gräber	$Nu = 0.25 + 6.2\chi + (0.032\chi - 0.007)Pe^{0.8-0.024\chi}$	$Pe \in [150, 4000], \chi \in [1.2, 2.0]$
Ushakov	$Nu = 7.55\chi - 20\chi^{-13} + \frac{3.67}{90\chi^2}Pe^{0.56+0.19\chi}$	$Pe < 4000, \chi \in [1.3, 2.0]$
Mikityuk	$Nu = 0.047(1 - e^{-3.8(\chi-1)})(Pe^{0.77} + 250)$	$Pe \in [30, 5000], \chi \in [1.1, 1.95]$

Table 4. Experimental correlations for Nusselt number calculation in triangular lattices.

number, $Re_{sch} = u_b d_{h,sch}\rho/\mu$, where $d_{h,sch}$ is the hydraulic diameter of the triangular lattice sub-channel. Rod lattices are usually classified using the pitch-to-diameter ratio χ, where pitch p is the distance between two adjacent rods center lines and diameter d is the rod diameter, as sketched in Fig. 1 a). Numerical values of the bundle geometrical properties are reported in Table 2. These values are relative to the experimental setup studied in [15, 18]. A sketch of the simulated domain is reported in Fig. 1 b), together with the definition of the three different boundary regions and of the triangular sub-channels $\Omega_{sch,1}$, $\Omega_{sch,2}$ and $\Omega_{sch,3}$.

Several meshes with quadratic elements and increasing levels of mesh refinements have been used, as described in Fig. 2. In particular these meshes are used to check the numerical solutions by their L_2 norm values. For given different meshes $\mathcal{T}_{h1}$ and $\mathcal{T}_{h2}$, with corresponding solutions u_{h1} and u_{h2}, a convergence criteria is defined as

$$||u_{h_{n-1}} - u_{h_n}||_{L^2} \leq 10^{-4}, \tag{11}$$

assuring full convergence for all the solutions within two or three digits.

The boundary conditions used for the present simulations are reported in Table 3, where "near-wall b" means that the imposed boundary conditions are in accordance with the near wall behavior that can be obtained from a Taylor series expansion in the near wall region [11]. We investigate turbulent heat transfer at four different values of Re_b, namely CASE A : $Re_b = 1.4 \times 10^4$, CASE B : $Re_b = 2.2 \times 10^4$, CASE C : $Re_b = 3.1 \times 10^4$ and CASE D : $Re_b = 5.4 \times 10^4$. For each case we study the behavior of the three triangular sub-channels by computing local Nusselt number Nu values and hot-spot factor ϕ. In order to compare the results with reference data we follow the methods used in [15, 18]. Nusselt number is thus calculated as

$$Nu = \frac{q d_{h,sch}}{\lambda(\langle T_w \rangle - T_b)}, \tag{12}$$

where λ is the fluid thermal conductivity, $\langle T_w \rangle$ is the sub-channel mean wall temperature and T_b the bulk temperature. Hot-spot factor is calculated as

$$\phi = \frac{T_{w,max} - T_b}{(\langle T_w \rangle - T_b)}, \tag{13}$$

where $T_{w,max}$ is the maximum wall temperature. This factor is used to determine how non-homogeneous the temperature field is on wall surface, as $\phi = 1$ would mean that the maximum temperature value is equal to the mean wall temperature. Nusselt number values are compared with experimental correlations derived for triangular lattices. These correlations are usually defined as a function of Peclet number $Pe = Re\,Pr$ and of pitch-to-diameter ratio χ. In Table 4 we report the experimental correlations of Gräber, Ushakov and Mikityuk [23, 24, 25]. Nusselt

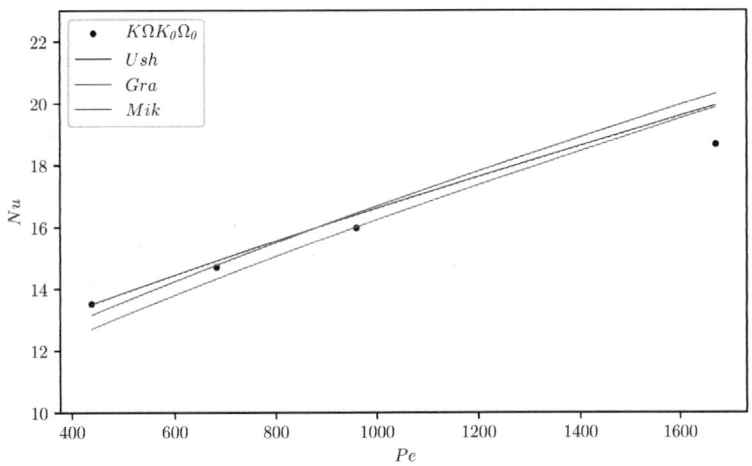

Figure 3. Comparison of computed Nusselt values with experimental correlations of Gräber, Ushakov and Mikityuk.

Pe	$\Omega_{sch,1}$	$\Omega_{sch,2}$	$\Omega_{sch,3}$
438	1.0855	1.1835	1.1905
682	1.0939	1.2158	1.2221
960	1.0965	1.2218	1.2266
1670	1.0938	1.2027	1.2048

Table 5. Hot-spot factor ϕ values for the three triangular sub-channels as a function of the Peclet number for the different simulated cases.

number values of triangular sub-channel $\Omega_{sch,1}$ are reported in Fig. 3 and compared with the experimental correlations of Table 4. The Peclet number values used here are calculated using the sub-channel Reynolds number Re_{sch}. For $Pe < 1000$ we observe that the computed Nusselt number values is between the values of Gräber and Ushakov experimental correlations, while for CASE D Nu value is smaller than correlation values. If compared with experimental values reported in [15] we observe a general under-estimation of Nu for all simulated cases. The Nu values of sub-channels $\Omega_{sch,2}$ and $\Omega_{sch,3}$ are not reported in Fig. 3 as they are very close to those calculated for $\Omega_{sch,1}$. Hot-spot factor values are reported in Table 5 for all simulated cases and for the three triangular sub-channels. Triangular sub-channel $\Omega_{sch,1}$ has the most uniform wall temperature distribution with a ϕ value in the range of $[1.085, 1.096]$, while $\Omega_{sch,2}$ and $\Omega_{sch,3}$ have $\phi \in [1.18, 1.22]$ for all simulated cases. Hot-spot factor values of sub-channel $\Omega_{sch,1}$ are in accordance with the values calculated in [7] for the case of triangular rod bundle with pitch-to-diameter ratio of 1.4. A non-dimensional temperature field Θ, calculated as

$$\Theta = \frac{(T - T_b)\lambda}{d_{h,bun} q}, \tag{14}$$

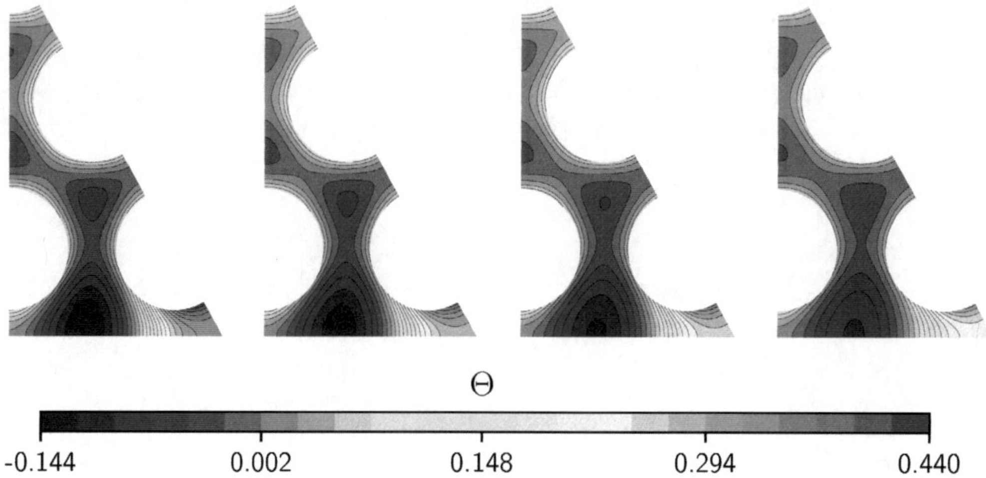

Figure 4. Non-dimensional temperature distribution Θ. From left to right CASE A, CASE B, CASE C and CASE D.

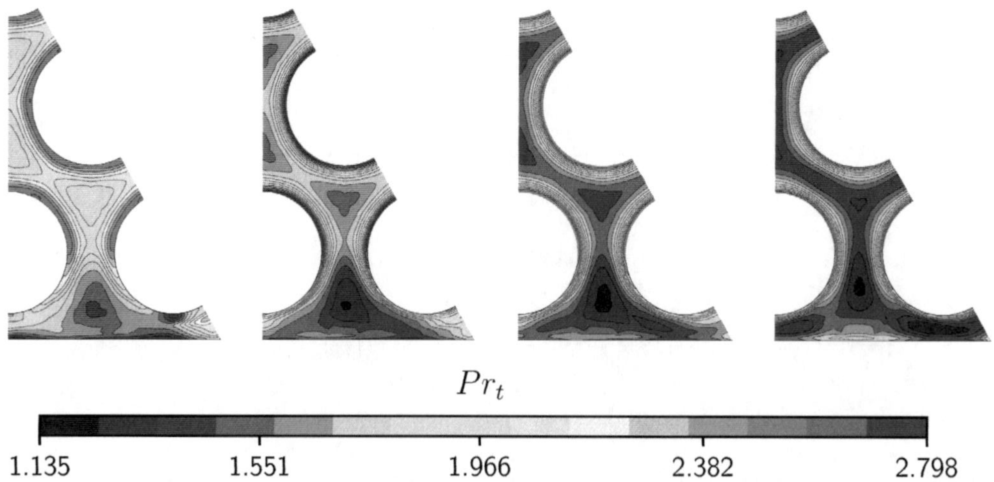

Figure 5. Turbulent Prandtl number distribution on the simulated domain for increasing Reynolds number values. From left to right CASE A, CASE B, CASE C and CASE D.

with T_b bulk temperature of the cross section, is reported in Fig. 4 in order to compare temperature behavior between the simulated cases. The results agree with the computed values of ϕ. Moreover we see that highest temperature values are located in the corner region. In Fig. 5 we show the computed turbulent Prandtl distribution on the bundle cross section for increasing values of the Reynolds number. We see that high Pr_t values are obtained in the region close to heated walls. Maximum values in those regions are in the range of $[2, 3]$ and tend to decrease as the Reynolds number increases. It is commonly suggested to set $Pr_t = 1.5$ [7] when using an algebraic model. This assumption can be a good compromise as in center of bundle cross section the computed Pr_t is approximately equal to 1.5, even if still it would be under-estimated in the near wall region.

4. Conclusions

In the present work we have studied turbulent heat transfer occurring in a 19 pin hexagonal lattice of a nuclear reactor bundle where Lead Bismuth Eutectic is used as a coolant fluid. A simplified model has been here considered, under the assumption of fully developed dynamical and thermal turbulent flow. In this way symmetry planes can be detected and only a sector can be simulated. A transformation on temperature field has been operated in order to obtain an energy balance equation where periodic boundary conditions can be applied on inlet and outlet sections. The study of turbulent heat transfer has been then transformed in a two-dimensional numerical simulation. A four parameter logarithmic turbulence model has been here used to close the system of Reynolds Averaged Navier-Stokes equations. Simulations for four different values of the Reynolds number have been performed. Three different triangular sub-channels have been considered on the simulated domain in order to compare our results with literature data. The Nusselt number values, calculated for the three different sub-channels, have been compared with experimental correlation formulated for triangular nuclear reactor bundles as a function of Peclet number [23, 24, 25]. We observe a good agreement between our results and reference data for three of the simulated cases, in particular for $Pe < 1000$, while for high Peclet numbers the Nusselt number value is under-estimated. Hot-spot factor values for the central triangular sub-channel are in agreement with the ones calculated for triangular sub-channels [7], while higher values are obtained for the other two sub-channels, denoting a non uniform temperature field on wall surfaces. Finally a comparison of turbulent Prandtl number distribution between the four simulated cases is shown. The present study is a starting point for more accurate numerical simulations of the 19 pin nuclear reactor bundle. In particular thermal field development will be considered in order to evaluate the influence of buoyancy forces on the flow field. Moreover more complicated geometries, with wire-wrapped configurations or with the presence of grid-spacers, will be analyzed to see the impact of secondary flows on heat transfer.

References

[1] Heinzel A *et al.* 2017 *Energy Technol* **5** 1026–1036
[2] Cheng X and Tak N i 2006 *Nucl. Eng. Des.* **236** 385–393
[3] Cheng X and Tak N 2006 *Nucl. Eng. Des.* **236** 1874–1885
[4] Grötzbach G 2013 *Nucl. Eng. Des.* **264** 41–55
[5] M Kays W 1994 *J. Heat Transfer* **116** 284–295
[6] Schumm T, Niemann M, Magagnato F, Marocco L, Frohnapfel B and Fr J 2015 1–12
[7] Ge Z, Liu J, Zhao P, Nie X and Ye M 2017 *Nucl. Eng. Des.* **314** 198 – 206
[8] Nagano Y and Shimada M 1996 *Phys. Fluids* **8** 3379–3402
[9] Abe K, Kondoh T and Nagano Y 1995 *Int. J. Heat Mass Tran.* **38** 1467–1481
[10] Manservisi S and Menghini F 2014 *Int. J. Heat Mass Tran.* **69** 312–326
[11] Da Vià R, Manservisi S and Menghini F 2016 *Int. J. Heat Mass Tran.* **101** 1030–1041
[12] Hattori H, Morita A and Nagano Y 2006 *Int J. Heat Fluid Fl.* **27** 671–683
[13] Manservisi S and Menghini F 2016 *Nucl. Eng. Des.* **273** 251–270
[14] Manservisi S and Menghini F 2015 *Nucl. Eng. Des.* **295** 251–260
[15] Pacio J *et al.* 2014 *Nucl. Eng. Des.* **273** 33 – 46
[16] Martelli D, Marinari R, Barone G, di Piazza I and Tarantino M 2017 *Ann. Nucl. Energy* **103** 294 – 305
[17] Jeong J H, Yoo J, Lee K L and Ha K S 2015 *Nucl. Eng. Des.* **47** 523 – 533
[18] Pacio J *et al.* 2015 *Nucl. Eng. Des.* **290** 27–39
[19] Ilinca F, Hetu J F and Pelletier D 1998 *Computers and Fluids* **27** 291–310
[20] Manservisi S and Menghini F 2014 *Int. J. Heat Mass Tran.* **69** 312–26
[21] Manservisi S and Menghini F 2014 *Nucl. Eng. Des.* **273** 251–70
[22] Code FEMuS URL `https://github.com/FemusPlatform/femus/`
[23] Graeber H, Rieger M and Dwyer O 1973 *Atomkernenergie* **19** 23–40
[24] Ushakov P, Zhukov A and Matyukhin N 1978 *High Temp. (USSR)* **15** 1027–1033
[25] Mikityuk K 2009 *Nucl. Eng. Des.* **239** 680 – 687

IOP Conf. Series: Journal of Physics: Conf. Series **1224** (2019) 012031 doi:10.1088/1742-6596/1224/1/012031

Parametric thermal analysis for the optimization of Double Walled Tubes layout in the Water Cooled Lithium Lead inboard blanket of DEMO fusion reactor

Pietro ARENA[1], Alessandro DEL NEVO[2], Pietro Alessandro DI MAIO[1], Ruggero FORTE[1], Giulio NEVONE[1]

[1] Università di Palermo, Dipartimento di Energia, Ingegneria dell'Informazione e Modelli Matematici, Viale delle Scienze, Edificio 6, I-90128 Palermo, ITALY.
[2] ENEA C.R. Brasimone, I-40032 Camugnano (BO), ITALY.

Corresponding author's e-mail: pietroalessandro.dimaio@unipa.it

Abstract. Within the roadmap that will lead to the nuclear fusion exploitation for electric energy generation, the construction of a DEMOnstration (DEMO) reactor is, probably, the most important milestone to be reached since it will demonstrate the technological feasibility and economic competitiveness of an industrial-scale nuclear fusion reactor. In order to reach this goal, several European universities and research centres have joined their efforts in the EUROfusion action, funded by HORIZON 2020 UE programme. Within the framework of EUROfusion research activities, ENEA and University of Palermo are involved in the design of the Water-Cooled Lithium Lead Breeding Blanket (WCLL BB), that is one of the two BB concepts under consideration to be adopted in the DEMO reactor. It is mainly characterized by a liquid lithium-lead eutectic alloy acting as breeder (lithium) and neutron multiplier (lead), as well as by subcooled pressurized water as coolant. Two separate circuits, both characterized by a pressure of 15.5 MPa and inlet/outlet temperatures of 295 °C/328 °C, are deputed to cool down the First Wall (FW) and the Breeder Zone (BZ). The former consists in a system of radial-toroidal-radial C-shaped squared channels where countercurrent water flow occurs while the latter relies in the use of bundles of poloidal-radial Double Walled Tubes (DWTs) housed within the breeder.

A parametric thermal study has been carried out in order to assess the best DWTs' layout assuring that the structural material maximum temperature does not overcome the allowable limit of 550 °C and that the overall coolant thermal rise fulfils the design target value of 33 °C. The study has been performed following a theoretical-numerical approach based on the Finite Element Method (FEM) and adopting the quoted Abaqus FEM code. Main assumptions and models together with results obtained are herewith reported and critically discussed.

1. Introduction

In the next future, nuclear fusion power reactors might represent a very interesting alternative to the conventional power plants using fossil fuels for electric energy generation. Within the roadmap leading to the realization of net electric power from nuclear fusion reactions on an industrial scale [1], the DEMOnstration (DEMO) fusion reactor will have the role of proving the technical feasibility of such a power plant assessing, at the same time, its economic competitiveness on the worldwide energy market. Aiming at the achievement of this goal, the major European research centres and universities have joined their effort under the EUROfusion project, funded by HORIZON 2020 UE framework programme.

Within the framework of the Work Package Breeding Blanket (WPBB) of the EUROfusion action promoting the Breeding Blanket (BB) research activities [2,3], ENEA and University of Palermo are deeply involved in the design of the Water-Cooled Lithium Lead (WCLL) BB concept, that is currently one of the two options under consideration for the DEMO reactor. In particular, it foresees the adoption of the liquid lithium-lead eutectic alloy Pb-15.7Li as breeder and neutron multiplier and the use of pressurized subcooled water as coolant, flowing at the average pressure of 15.5 MPa with an inlet temperature of 295°C and an outlet one of 328°C.

According to the DEMO Baseline 2015, the WCLL BB is foreseen to be toroidally segmented into 18 sector of 20°, each one housing 3 BB outboard segments and 2 BB inboard segments [4] directly facing the plasma and mechanically coupled by an upper poloidal plug. As to each segment, both Multi Module System (MMS) and Single Module System (SMS) have been investigated by the WCLL BB design team as potential options for its configuration, the former one being based on the adoption of poloidal stacks of modules welded to a Back Supporting Structure (BSS) while the latter one being characterized by a poloidal-oriented "banana-shaped" structural box, housing the breeder under quasi-stagnant conditions. In particular, numerical simulations performed during 2015-2017 [5,6,7] have pointed out the low efficiency of the MMS configuration, especially in terms of breeder drainage, helium extraction, hot spot temperatures and stress intensity mainly under electro-magnetic loads. The research efforts have, hence, been focused onto the inboard and outboard SMS segments, that are currently under investigation to finalize the design of their structural box and cooling system.

In particular, as a consequence of both its position and shielding function, the WCLL BB is heavily exposed to both surface and volumetric heat loads due to the high heat flux arising from plasma and to the intense heat power deposited inside its volume as a consequence of the interactions with neutrons and photons, respectively. The WCLL BB cooling system plays, hence, a pivotal role in the reactor operation and its design results particularly demanding since it has to effectively extract the overall heat power deposited into the module while keeping the structure temperature below its prescribed limit, to avoid the occurrence of thermal and structural crisis, without incurring in an unduly pumping power. Furthermore, it has to maintain the breeder temperature beyond the solidification point to avoid local formation of breeder solid grains that might endanger its normal circulation, to be considered as propaedeutic to tritium extraction. Therefore, a significant effort has been made since 2014 by the WCLL BB design team in order to assess an effective design for the cooling system of this blanket concept, considering several cooling system solutions and assessing their thermal-hydraulic performances for both the inboard and outboard SMS segments.

Within this framework finds its place the research activity reported in the present paper, that has been focussed on the optimization of the thermal-hydraulic performances of the cooling circuit conceived to remove nuclear-deposited heat power from the breeder of a typical WCLL BB SMS inboard segment. The optimization has been performed following a theoretical-numerical approach based on the Finite Element Method (FEM) and performing parametric FEM analyses by means of the Abaqus v6.14 commercial code. Main assumptions and models together with the results obtained are herewith reported and critically discussed.

2. WCLL breeding blanket inboard segment

As reported in [7,8], the WCLL BB segment design is currently oriented towards the SMS concept, which seems to show better performances than the MMS one, especially in terms of mechanical performances. According to this concept, the WCLL BB inboard segment is mainly composed by a EUROFER steel [9] actively-cooled banana-shaped structure named Segment Box (SB) that delimits the so-called Breeder Zone (BZ) where neutron multiplication and tritium breeding take place (Fig. 1).

In particular, the SB is articulated in a First Wall (FW), two Side Walls (SWs) and a Back Plate (BP) and it is reinforced against over-pressurization accidents and electro-magnetic loads by a system of poloidal-radial (vertical) and toroidal-radial (horizontal) Stiffening Plates. Conversely, the BZ houses the breeder flowing, under quasi-stagnant conditions, according to a U-shaped path along radial-poloidal-radial directions.

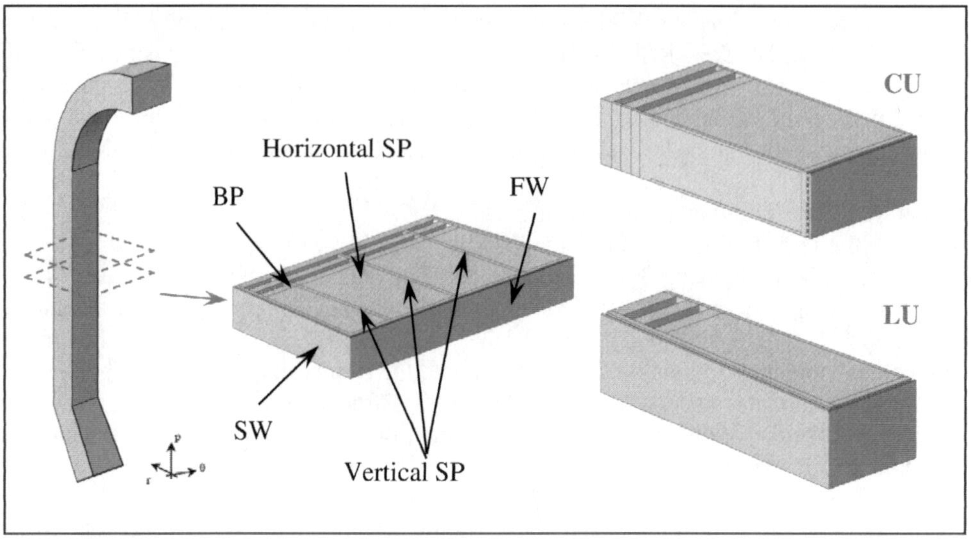

Fig.1. WCLL BB inboard segment.

The whole segment may be considered as composed of a poloidal sequence of elementary toroidal-radial cells, each one composed of two Central Unit (CU) and two Lateral Unit (LU) (Fig. 1).

The WCLL BB inboard segment cooling system is mainly articulated in two independent cooling circuits devoted to separately cool the SB and the BZ. The former relies in a set of radial-toroidal-radial C-shaped squared cooling channels (7x7 mm), poloidally arranged with a pitch of 13.5 mm, where water coolant countercurrent flow occurs with a thermal rise of 33°C from the inlet temperature of 295°C to the outlet one of 328°C. The latter is based on the use of bundles of poloidal-radial Double Walled Tubes (DWTs) housed within the breeder, where water coolant flows experiencing the same thermal rise of the SB cooling circuit. In particular, DWTs [10], made in EUROFER and characterized by an internal diameter of 8 mm and a thickness of 5.5 mm, have been widely adopted for the design of the WCLL BB concepts with the aim of reducing the likelihood of possible liquid metal-water interaction.

3. Double Walled Tubes layout optimization

The main assignment of the BZ cooling circuit is the removal of the heat power deposited within the breeder as a consequence of the interactions of its nuclei with neutrons and photons. It has to be achieved under steady state conditions while keeping the maximum temperature of SB below the limit of 550°C, prescribed by safety codes for the EUROFER, and the breeder minimum temperature beyond the limit of Pb-15.7Li eutectic alloy solidification point (≈300°C).

The design of BZ cooling circuit results particularly demanding and a research campaign has been, hence, launched in close cooperation between ENEA and University of Palermo with the specific aim to assess an optimized DWTs layout, suitable to effectively reach the target goal and meet the safety requirements while minimizing the number of tubes to be adopted.

To this purpose, the WCLL BB inboard segment design equipped with a "beer box" SP grid configuration has been considered, that relies on the use of both horizontal and vertical SPs extending for the whole radial length of the module, from FW to BP, and delimiting several toroidal-radial cells articulated in central and lateral units (CUs and LUs), each provided with its own bundle of DWTs. Accordingly, the assessment of the optimized DTWs layout has been carried out for either a CU and a LU by running a specific set of parametric thermal FEM analyses. Finally, these optimized DWTs layouts have been implemented into an equatorial elementary toroidal-radial cell of the WCLL BB inboard segment and its overall thermal-hydraulic behaviour has been numerically assessed by running a specific thermal FEM analysis.

IOP Conf. Series: Journal of Physics: Conf. Series **1224** (2019) 012031 doi:10.1088/1742-6596/1224/1/012031

3.1. Central Unit

As far as the CU is concerned, three different DWTs layouts have been investigated (Fig. 2) composed of 3, 4 and 5 DWTs, respectively (Fig. 2). In particular, all of the configurations considered are characterized by poloidal-radial DTWs equally spaced along the toroidal direction, so to pursue a more uniform cooling of the SPs. The DWTs have been placed at a distance of 1 cm from both the horizontal and vertical SPs and 1 cm far from the FW internal surface.

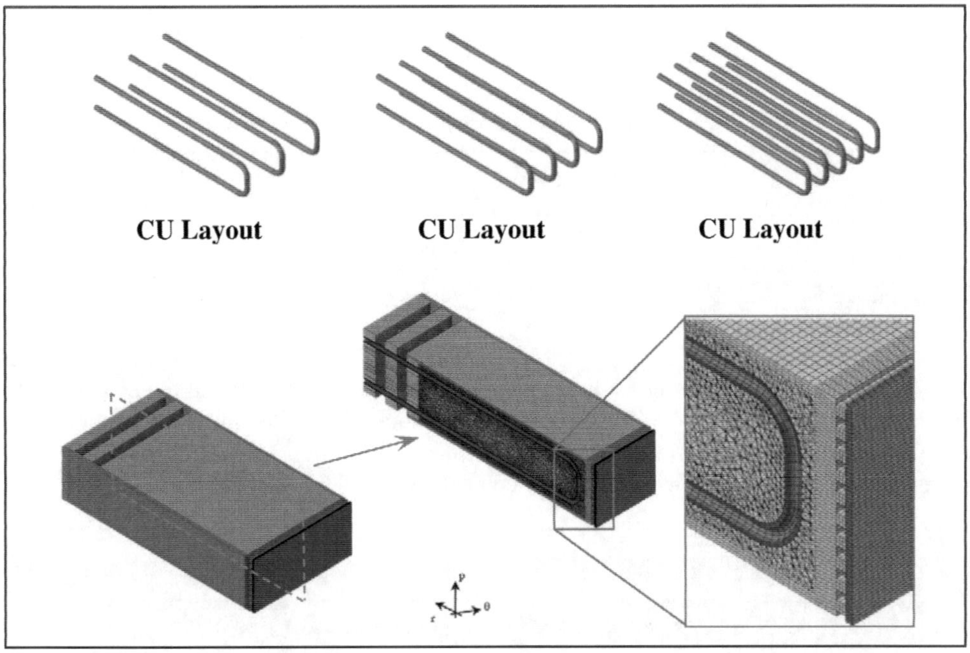

Fig.2. CU layouts and FEM models.

A detailed FEM model including structure and breeder has been set up for each layout considered, typically composed of ~$0.45 \cdot 10^6$ nodes connected in ~$1.1 \cdot 10^6$ linear elements (Fig. 2), according to the mesh settings deduced from sensitivity analyses carried out as to the outboard blanket units [11]. As to the thermal loads, a radial-dependent volumetric density of nuclear-deposited heat power, drawn from [10], has been implemented into the model while a heat flux of 0.5 MW/m^2 has been imposed on the model plasma facing surface. Moreover, convective boundary conditions have been implemented at both the FW cooling channels and the DWTs interfaces with coolant. To this purpose, heat transfer coefficients (h) have been computed by means of the Dittus&Bölter correlation [12], taking into account a preliminary estimate of coolant mass flow rate distribution per channel or tube based on that assessed for the WCLL BB outboard segment and reported in [11]. The bulk temperature has been uniformly imposed to 311.5 °C, as the average value between the coolant inlet (295 °C) and outlet (328 °C) reference temperatures. A summary of the boundary conditions applied to the models is reported in Table 1, where the mass flow rate G is relevant to a single tube or channel.

Table 1. CU cooling circuit parameters.

	G [kg/s]	h [W/(m^2 °C)]
CU Layout 0	0.0643	17441
CU Layout 1	0.0483	13855
CU Layout 2	0.0386	11590
FW	0.0776	21239

Since the breeder is expected to flow at few mm/s within the BZ, due the simultaneous action of natural convection and magneto-hydrodynamic forces, it has been considered as stagnant and a pure diffusive heat transfer has been, hence, assumed within the BZ domain, neglecting convective transport phenomena as widely accepted and already done in previous analyses [10,13,14,15]. A thermal contact model has been implemented at all of the breeder-wetted surfaces of the model, characterized by a conservative thermal conductance of 100 kW/(m^2 °C).

A steady state thermal analysis has been performed for each configuration under investigation and its thermal response has been numerically assessed. The maximum temperatures predicted in the SB as well as in the BZ are reported in Table 2.

Table 2. Maximum temperatures predicted for CU layouts.

	$T_{MAX, SB}$ [°C]	$T_{MAX, BZ}$ [°C]
CU Layout 0	711.0	760.2
CU Layout 1	577.2	654.6
CU Layout 2	533.4	598.3

As may be deduced from the analysis of Table 2, the results obtained show that the layout equipped with 5 DWTs (CU Layout 2) is the only one, among those investigated, suitable to cool the SB structure while complying with the EUROFER maximum temperature requirement. In fact, according to the thermal field distribution reported in Figure 3, the CU Layout 2 allows to obtain a maximum temperature of 533.4 °C, reached in correspondence of the vertical SPs.

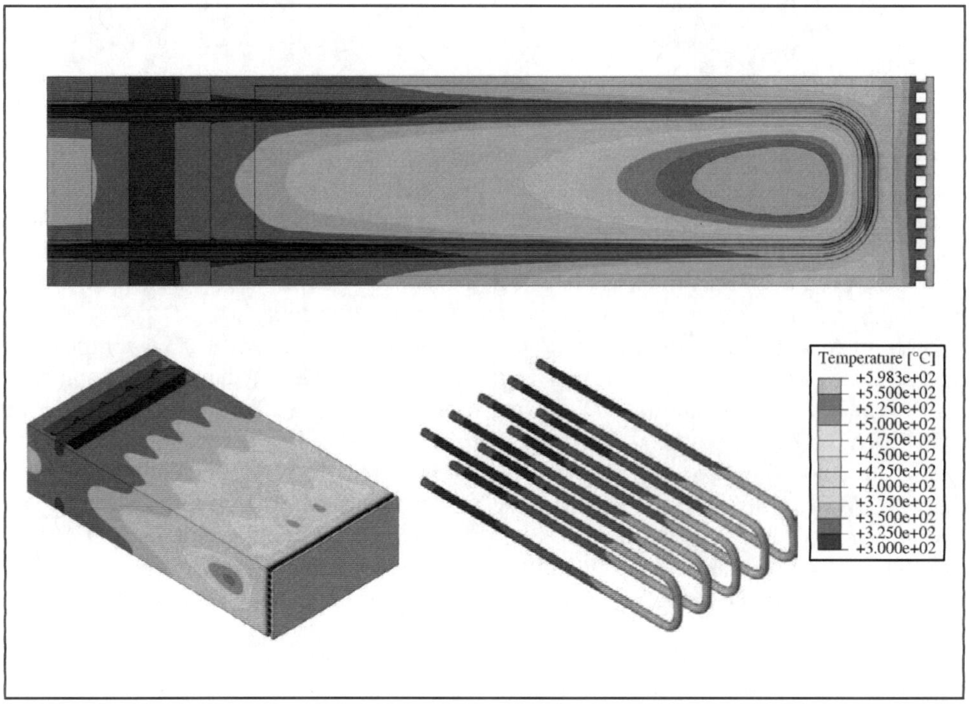

Fig.3. CU Layout 2 thermal field.

Furthermore, it has to be underlined that a minimum breeder temperature of 321.0 °C has been calculated for all the configurations analysed. It overcomes the limit value of 300 °C and, hence, allows the breeder solidification process to be reasonably excluded under normal operation.

3.2. Lateral Unit

The same procedure followed for the CU has been adopted for the thermal analysis of the LU. In particular, only two DWTs layouts (LU Layout 0 and LU Layout 1) have been taken into account in this case (Fig. 4), due to the reduced toroidal space available to accommodate them.

Both the layouts maintain the same characteristic distances from the other components already mentioned in the CU case, but they are characterized by a different orientation of the most external tube, which is rotated of 3.33° around the poloidal axis in order to follow the SW direction, remaining at the same toroidal distance of 1 cm from it.

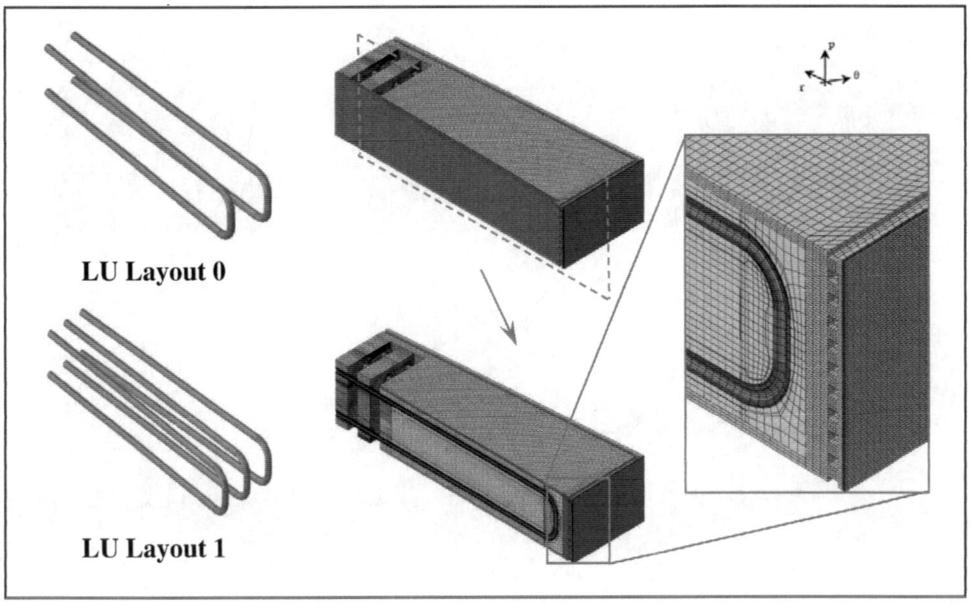

Fig.4. LU layouts and FEM model.

The same volumetric and surface heat loads adopted for the CU analysis have been imposed to the models, but different convective boundary conditions, according to the data reported in [11] and reported in Table 3, have been implemented.

Table 3. LU cooling circuit parameters.

	G [kg/s]	h [W/(m^2 °C)]
LU Layout 0	0.0597	16418
LU Layout 1	0.0398	11869
FW-SWs	0.0776	21239

A steady state thermal analysis has been performed for each configuration under investigation and its thermal response has been numerically assessed. The maximum temperatures predicted in the SB as well as in the BZ are reported in Table 4.

Table 4. Maximum temperatures predicted for LU layouts.

	$T_{MAX, SB}$ [°C]	$T_{MAX, BZ}$ [°C]
LU Layout 0	551.8	609.9
LU Layout 1	530.3	577.9

They clearly show that a minimum of 3 DWTs (LU Layout 1) is needed in order to allow the maximum SB temperature to line up at 530.3 °C, well below the required limit of 550 °C. Moreover, the thermal field of the LU Layout 1 is reported in Figure 5, showing that the maximum SB temperature of 530.3 °C is reached in correspondence of the vertical SP.

Finally, it has to be stressed that a minimum breeder temperature of 314.9 °C has been calculated for all the configurations analysed. It overcomes the limit value of 300 °C and, hence, allows the breeder solidification process to be reasonably excluded under normal operation.

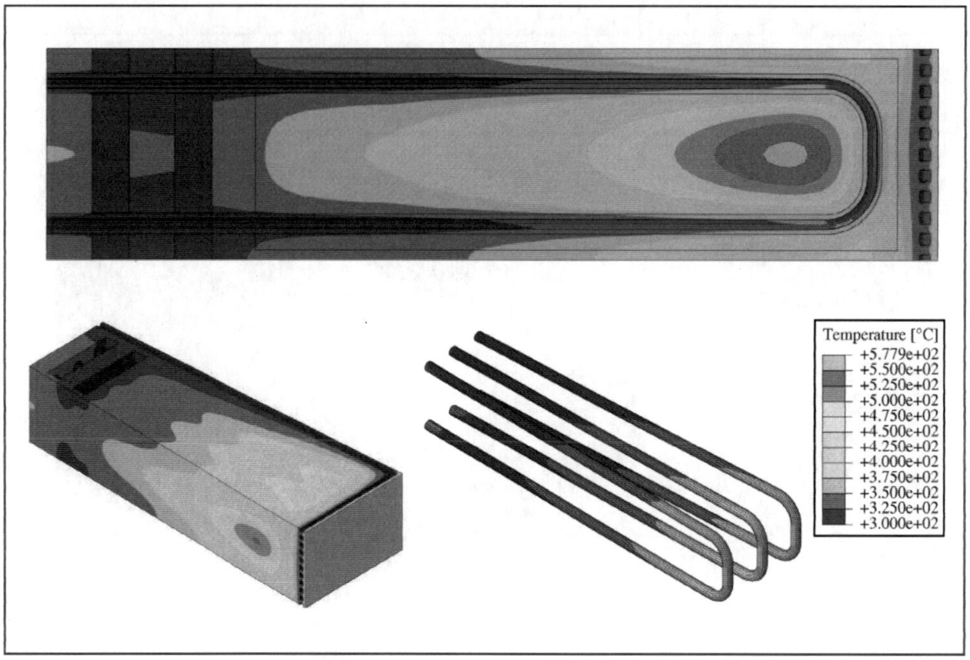

Fig.5. LU Layout 1 thermal field.

3.3. WCLL-BB inboard segment equatorial elementary cell

According to the planned activity, the optimized DWTs layouts assessed by means of the parametric analysis campaign for both CUs and LUs have been implemented into an equatorial elementary cell of the WCLL BB inboard segment to assess its overall thermal-hydraulic behaviour by running a specific thermal FEM analysis.

To this purpose a FEM model of the "optimized" WCLL-BB inboard segment equatorial elementary cell has been set up, endowed with $\sim 2.4 \cdot 10^6$ nodes connected in $\sim 2.5 \cdot 10^6$ linear elements. In order to obtain a more realistic temperature distribution, the elementary cell water domains housed within both the DWTs and the SB cooling channels have been additionally modelled. To this purpose, the same procedure already adopted in [14] has been followed in order to simulate the convective heat transfer between the structure and the coolant. In particular, a simplified "frozen" flow field approach has been adopted modelling the water domains by means of hexahedral forced convection/diffusion elements, which hold the heat transfer convective-diffusive transport equation. The coolant velocity field has been assumed to be known and, typically, the mass flow rate for unit area has been imposed to each material point, so that the velocity could be automatically code-calculated by means of the mass density. The heat flux entering the coolant domain at the steel walls interface has been constrained to be functionally dependent on the thermal gap between steel and coolant local temperatures (T_i^{wall}, $T_j^{coolant}$) by means of the following equation:

$$q''_{ij} = h\left(T_i^{wall} - T_j^{coolant}\right) \quad (1)$$

where q''_{ij} is the heat flux flowing through the coupled nodes i and j, belonging to the steel walls and the water coolant, respectively, and h is a fixed heat transfer coefficient.

The BZ cooling circuit inlet and outlet have been supposed to be placed at the bottom and the top of the elementary cell, respectively. An inlet temperature of 295 °C has been assumed for the coolant and an iterative analysis procedure has been adopted to determine the updated mass flow rate distribution per channel/tube suitable to reach the reference coolant outlet temperature of 328 °C. In particular, observing that the DWTs optimized configurations share almost the same mass flow rate per tube (Tables 1 and 3) it has been possible to determine a unique mass flow rate value to be adopted for all the DWTs, ensuring the coolant outlet mixing temperature to be equal to 328 °C with a coolant maximum outlet temperature of 332.1 °C, well below the saturation temperature of 344.8 °C (@15.5 MPa).

The mass flow rate values implemented into the model have been reported in Table 5. The heat transfer coefficients have been again calculated at the average temperature of 311.5 °C by means of the Dittus&Bölter correlation.

Table 5. WCLL BB inboard segment equatorial elementary cell cooling circuits parameters.

	G [kg/s]	h [W/(m^2 °C)]
DWT	0.0295	9347
FW-SWs	0.0679	19089

A steady state thermal analysis has been carried out and the thermal field distribution within the model has been obtained (Fig.s 6-8). As it may be observed, the results show a good agreement with those obtained by the parametric analyses of the Central and Lateral units. In particular, a maximum temperature of 531.4 °C has been predicted for the SB in correspondence of the central vertical SP, well below the limit value prescribed for the EUROFER (550°C). This value matches quite well the corresponding value of 533.4°C predicted by the parametric analyses for the CUs. Moreover, as to the BZ, minimum and maximum temperatures 304.6°C and 611.4 °C have been predicted, respectively, allowing to exclude any risk for the considered configuration to incur in breeder solidification issues.

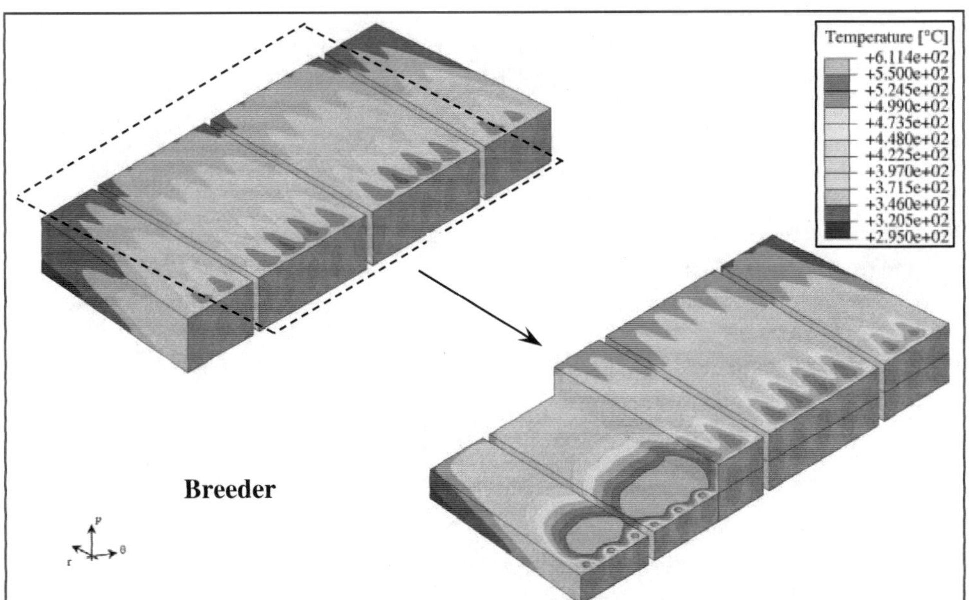

Fig.6. WCLL BB inboard segment equatorial elementary cell. BZ thermal field.

IOP Conf. Series: Journal of Physics: Conf. Series **1224** (2019) 012031 doi:10.1088/1742-6596/1224/1/012031

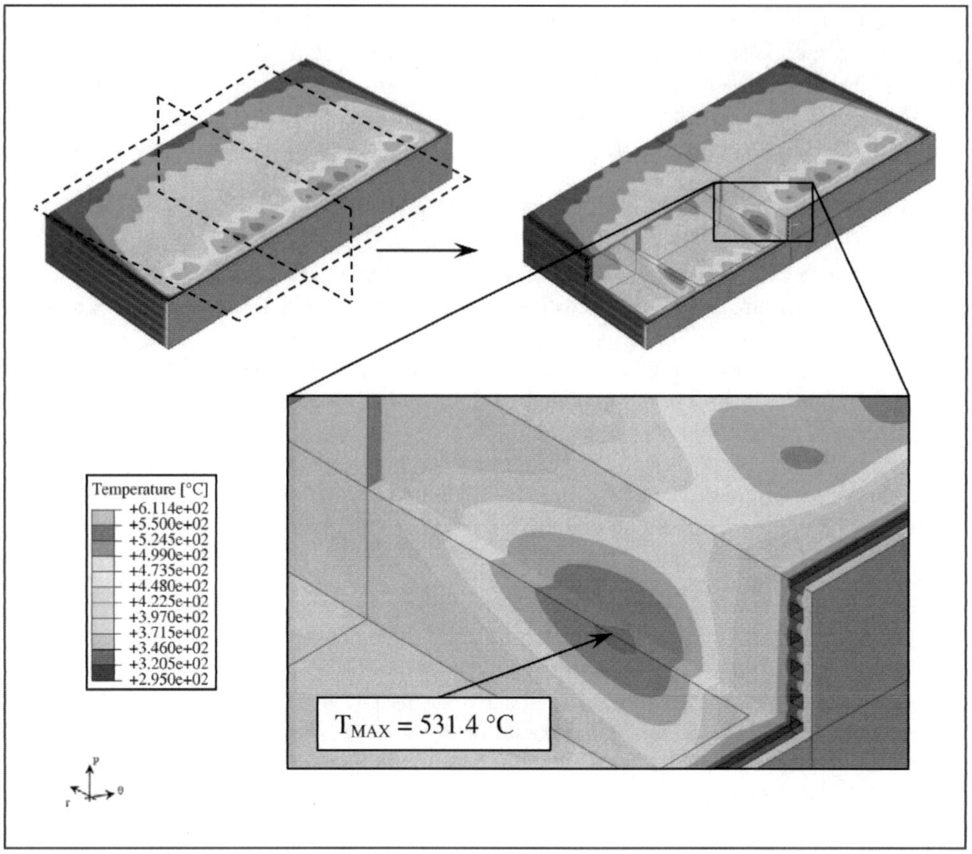

Fig.7. WCLL BB inboard segment equatorial elementary cell. SB thermal field.

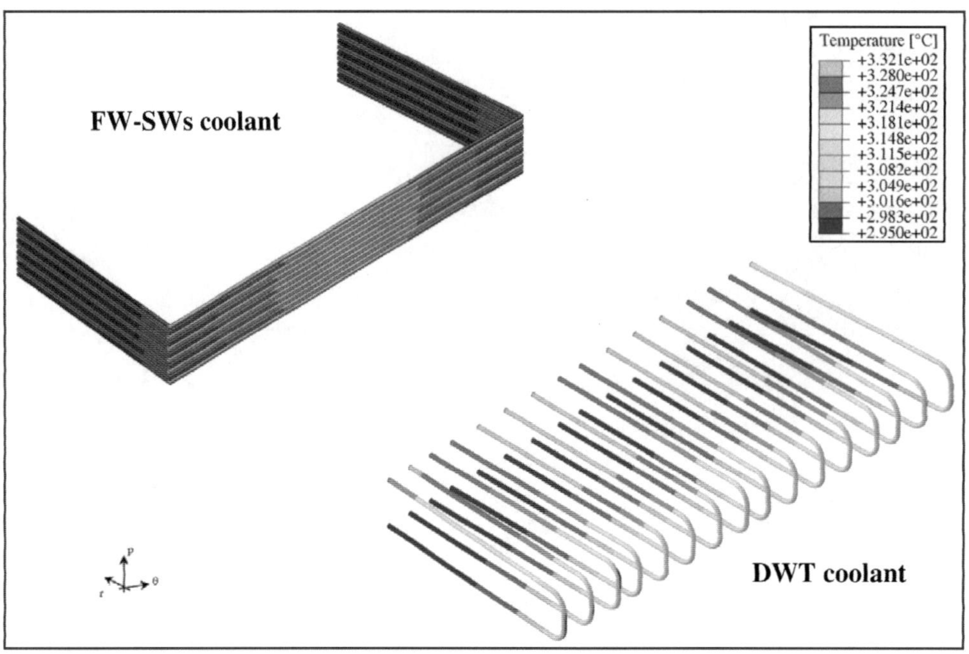

Fig.8. WCLL BB inboard segment equatorial elementary cell. Coolant thermal field.

4. Conclusions

Within the framework of the DEMO WCLL BB research activity, a parametric thermal study has been carried out by ENEA and University of Palermo, with the aim of determining an optimized configuration of the BZ cooling circuit of the WCLL BB SMS inboard segment, suitable to effectively remove nuclear-deposited heat power and meet the safety requirements in terms of maximum temperature of the structural material (< 550°C) and minimum temperature of the breeder (> 300 °C), while minimizing the number of tubes to be adopted.

The thermal-hydraulic performances of several DWTs layouts of both the CUs and LUs of the WCLL BB inboard segment equatorial elementary cell have been numerically investigated by means of parametric thermal analyses finding out that the optimized lay-outs have to include 5 and 3 DWTs for each CUs and LUs, respectively.

These layouts have been finally implemented into an equatorial elementary cell of the WCLL BB inboard segment to assess its overall thermal-hydraulic behaviour. The elementary cell steady state thermal response has been numerically predicted, showing a maximum structure temperature of 531.4 °C, computed in correspondence of the central vertical SP, and a minimum breeder temperature of 304.6 °C. The former is well below the EUROFER limit value of 550 °C, allowing to exclude any thermal crisis for the structural material, while the latter results slightly beyond the breeder limit value of 300 °C, allowing to exclude any risk to incur in breeder solidification issues under normal operation.

Disclaimer

This work has been carried out within the framework of the EUROfusion Consortium and has received funding from the Euratom Research and Training Programme 2014–2018 under grant agreement no. 633053. The views and opinions expressed herein do not necessarily reflect those of the European Commission.

References

[1] Romanelli F. et al. "Fusion Electricity-A roadmap to the realisation of fusion energy", 2012.
[2] Federici G. et al. "Overview of the design approach and prioritization of R&D activities towards an EU DEMO", *Fusion Eng. Des.*, Vol. **109-111** (2016), pp. 1464-1474.
[3] Boccaccini L.V. et al. "Objectives and status of EUROfusion DEMO blanket studies", *Fusion Eng. Des.*, Vol. **109-111** (2016), pp. 1199-1206.
[4] Tassone A. et al. "Recent Progress in the WCLL Breeding Blanket Design for the DEMO Fusion Reactor", *IEEE Trans. on Plasma Science*, Vol. **46** Issue 5 (2018), pp. 1446-1457.
[5] Del Nevo A. et al. "WCLL breeding blanket design and integration for DEMO 2015: status and perspectives", *Fusion Eng. Des.*, Vol. **124** (2017), pp. 682-686.
[6] Martelli E. et al. "Thermo-hydraulic analysis of EU DEMO WCLL breeding blanket", *Fusion Eng. Des.*, Vol. **130** (2018), pp. 48-55.
[7] Bongiovì G. et al. "Multi-Module vs. Single-Module concept: comparison of thermomechanical performances for the DEMO Water-Cooled Lithium Lead breeding blanket", *Fusion Eng. Des.*, https://doi.org/10.1016/j.fusengdes.2018.05.037.
[8] Di Maio P. A. et al. "Numerical assessment of the thermomechanical behaviour of the DEMO Water-Cooled Lithium Lead inboard blanket equatorial module", *Fusion Eng. Des.*, https://doi.org/10.1016/j.fusengdes.2018.04.098.
[9] Tavassoli A., Fusion Demo Interim Structural Design Criteria (DISDC)/Appendix A Material Design Limit Data/A3. S18E Eurofer Steel, EFDA TASKTW4-TTMS-005-D01, CEA Report DMN/DIR/NT/2004-02/A, December 2004.
[10] Chiovaro P. et al. "Optimization of the breeder zone cooling tubes of the DEMO Water-Cooled Lithium Lead breeding blanket", *Fusion Eng. Des.*, Vol. **109-111** (2016), pp. 227-231.
[11] WCLL Design Report 2017, IDM Ref EFDA_D_2MYHNE v1.0.
[12] Incropera F.P., Dewitt D.P. "*Fundamentals of Heat and Mass Transfer*", 2nd edition, John Wiley

& Sons, 1985.
[13] Di Maio P. A. et al. "On the optimization of the first wall of the DEMO water-cooled lithium lead outboard breeding blanket equatorial module", *Fusion Eng. Des.*, Vol. **109-111** (2016), pp. 335-341.
[14] Di Maio P. A. et al. "On the thermo-mechanical behaviour of DEMO water-cooled lithium lead equatorial outboard blanket module", *Fusion Eng. Des.*, Vol. **124** (2017), pp. 725-729.
[15] Arena P. et al. "Thermal optimization of the Helium-Cooled Lithium Lead breeding zone layout design regarding TBR enhancement", *Fusion Eng. Des.*, Vol. **124** (2017), pp. 827-831.

IOP Conf. Series: Journal of Physics: Conf. Series **1224** (2019) 012032 doi:10.1088/1742-6596/1224/1/012032

Application of RELAP5/Mod3.3 – Fluent coupling codes to CIRCE-HERO

N Forgione[1], M Angelucci[1], C Ulissi[1], D Martelli[1], G Barone[1], R Ciolini[1] and M Tarantino[2]

[1] Dipartimento di Ingegneria Civile e Industriale, Università di Pisa, Largo Lucio Lazzarino n. 2, 56122 Pisa, Italy.
[2] ENEA, Cento Ricerche Brasimone, Località Brasimone, 40032 Camugnano (BO), Italy

nicola.forgione@unipi.it

Abstract. This paper presents the work ongoing at the DICI (Dipartimento di Ingegneria Civile e Industriale) of the University of Pisa on the application of coupled methodology between Fluent CFD code and RELAP5/Mod3.3 system code. In particular, this methodology was applied to the LBE-water heat exchanger HERO, with the aim to analyse the performances of this component. The test section object of this study is installed inside the vessel S100 of the CIRCE facility, built at ENEA Brasimone Research Centre.
In the proposed methodology the CFD code is adopted to simulate the LBE side of the HERO heat exchanger, whereas the secondary side (two-phase flow, water-vapour) is simulated by the STH code. In this procedure, the variables exchanged between the boundaries of the two codes are: the bulk temperature and heat transfer coefficient of the ascending water (in two-phase flow) obtained from RELAP5 and transferred to Fluent code; the wall temperature at the water side surface of the pipes is calculated by Fluent and passed to RELAP5 code.
The coupling procedure was verified by comparing the obtained results with the analogous ones achieved with the RELAP5 stand-alone calculation, proving that the developed coupling methodology is reliable. Further, the coupled simulation allows to obtain more accurate information on the LBE side.

1. Introduction

System Thermal-Hydraulic codes (STH) are widely adopted to support the design of components and thermal-hydraulic systems of nuclear power plants (NPP). In particular, these codes are fundamental for nuclear safety analysis as they provide accurate predictions of transients and postulated accidental scenario in NPPs. RELAP5 [1], CATHARE [2], ATHLET [3], TRACE [4], GOTHIC [5], etc. are some of STH codes available to this aim. These codes are also named Best-Estimate (BE) codes because they analyse an accidental scenario as realistically as possible.

STH codes are generally based on one-dimensional form of mass, momentum and energy balance equations, for two-phase flow, solved in Eulerian coordinates, including models based on empirical correlations (e.g. heat transfer, frictional pressure losses, etc.). Currently, their validation has reached a high level of accuracy that allows predicting NPP behaviours within reasonable computational time. However, the STH codes are not suitable to be applied to the study of more complex 3D phenomena such as, for example, the mixing and thermal stratification phenomena in large pool systems. Indeed, the use of computational fluid dynamics (CFD) has widen its extension in Nuclear Reactor Safety (NRS) field in the last years. These codes reached a satisfactory degree of maturity for single phase flow, particularly in the study of complex 3D phenomena; they combine a refined discretization of the domain with the implementation of models to simulate turbulence, heat transfer, multiphase flows and chemical reactions. However, these advantages involve higher computational efforts compared with those required by the STH codes when applied to an equivalent domain.

In this situation, coupling techniques between STH and CFD codes for thermal-hydraulic analysis purposes appear a good solution to accurately predict the behaviour of NPP components. This methodology allows to obtain different levels of detail in multiscale phenomena. The STH code should be used to simulate system characterized by 1D components (e.g., pipe) and to model multiphase flow, while the CFD code should be used to analyse a smaller part of the domain where 3D effects are significant and/or detailed flow information is demanded.

In this frame, the University of Pisa (UniPi) developed an in-house coupling methodology between the Fluent CFD code and RELAP5/Mod3.3 STH code using a ''non-overlapping scheme". In particular, the coupling tool was applied to the NACIE experimental loop [6, 7, 8] and the CIRCE experimental pool [8, 9]. These activities were realized in collaboration with ENEA Brasimone RC (Italy), where the facilities have been designed and built.

In the present work, the coupling methodology was applied to the HERO test section, a LBE-water heat exchanger, installed in the CIRCE facility.

2. Coupled codes calculation for thermal-hydraulic analysis of the HERO test section

The ongoing activity is finalized to analyse the heat transfer performance of HERO SGBT (Heavy liquid mEtal – pRessurized water cOoled Steam Generator Bayonet Tube), which is composed by seven double-walls bayonet tubes with a stainless steel powder filling gap. The unit consists of a hexagonal wrap (6795 mm in length) whose inner and outer transversal heights are, respectively, 126 and 132 mm. Six lateral slots are realized in the wrap at the top of the active length (6000 mm) to feed the liquid metal side of the SGBT unit [10].

The performed activity is accomplished using RELAP5 system code coupled with CFD ANSYS Fluent code. In particular, the RELAP5/Mod3.3 is a version modified at the University of Pisa to employ liquid metals, like lead, lead-bismuth and lead-lithium, as working fluids [11]. This coupling approach allows to supply updated and more realistic heat transfer boundary conditions to the CFD model and to obtain more accurate information on heat transfer for this component with a detailed distribution of the LBE temperature inside the hexagonal wrapper.

The adopted methodology for the computational domain is "non-overlapping". Indeed, the primary side of the heat exchanger and the heat structure are simulated with the CFD code, while the secondary side containing water-vapour two-phase mixture is modelled with the RELAP5 system code.

2.1. CFD preliminary analysis

The CFD preliminary analysis was carried out by ANSYS Fluent solver based on the finite volume method. A 3D model of the LBE side and of the HX solid structures were developed and the mesh was created using the native tool of ANSYS Workbench. Actually, the computational domain is one sixth of the entire geometry in the transversal section, exploiting the symmetry of the section. This simplification reduces the number of elements and as a consequence, the required computational time is decreased. Figure 1(a) shows the overall geometry used in the CFD code calculation, while the details of initial and final regions are displayed respectively Figure 1(b) and Figure 1(c). Figure 2 illustrates the cross section with the description of the different fluid and solid regions reproduced in the model: the LBE part is represented in blue, while the solid structures of AISI-304 steel are depicted in red (the one in contact with LBE side) and green (the one in contact with water side). The AISI-316 powder between the two

IOP Conf. Series: Journal of Physics: Conf. Series **1224** (2019) 012032 doi:10.1088/1742-6596/1224/1/012032

steel tubes is coloured in orange. The LBE lateral inlet section is represented in yellow in Figure 1(a) reproducing one of the six entrance slots of the HERO test section.

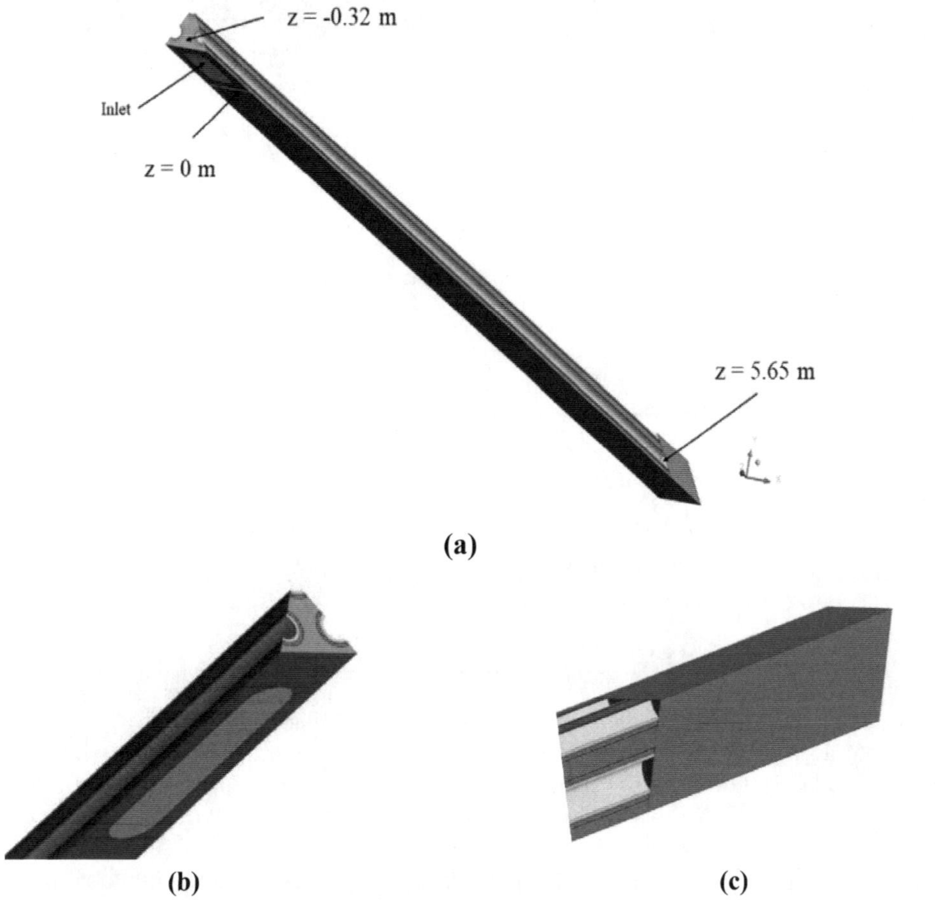

Figure 1. HERO periodic geometrical domain: overall geometry (a), LBE inlet (b) and outlet (c) region.

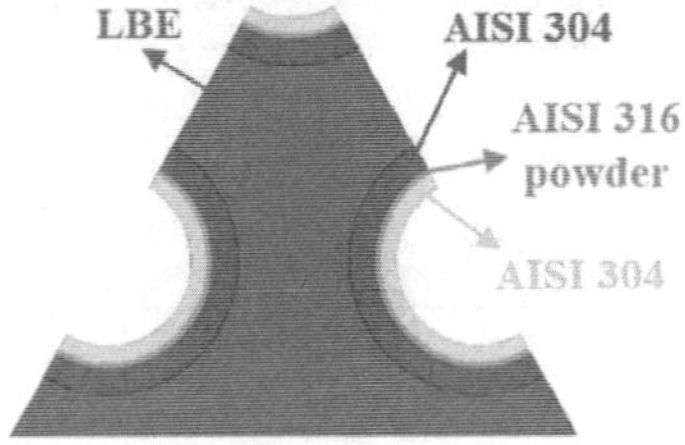

Figure 2. HERO periodic cross section.

The total length was considered including also an LBE zone below the end of the seven water tubes, which was mainly modelled to avoid the "reversed flow" occurrence at the "pressure outlet" boundary condition. This zone is noticeable in the final part of the geometry in Figure 1(c). The spatial discretization, displayed in Figure 3(a) and in Figure 3(b), is characterized by primarily structured hexahedral elements, for a total of 731454 nodes and 688213 elements; additional mesh statistic data are reported in Table 1.

IOP Conf. Series: Journal of Physics: Conf. Series **1224** (2019) 012032 doi:10.1088/1742-6596/1224/1/012032

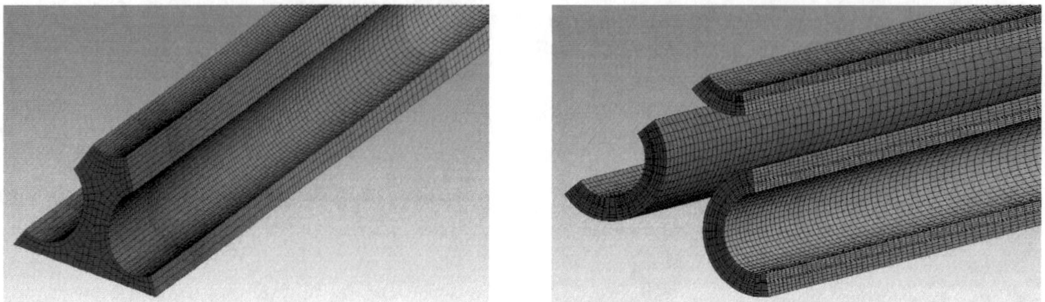

(a) (b)

Figure 3. HERO spatial discretization: LBE side (a) and tubes (b).

Table 1. Mesh static data.

Nodes	731454
Elements	688213
Max Aspect Ratio	38.94
Max Skewness	0.79
Min Orthogonal Quality	0.999
Avg Orthogonal Quality	0.983

The LBE properties, such as the density, molecular viscosity, thermal conductivity and specific heat, were taken into account as polynomial functions of the temperature in agreement with OECD/NEA Handbook of 2007 [12]. For AISI-304 steel and AISI-316 powder, the Fluent default values of density and specific heat were adopted, whereas the thermal conductivity of the AISI-304 steel was changed using a polynomial function of the temperature. A constant value of 3.41 W/(m K) was used for the AISI-316 powder, according to Ref. [13]. The thermal properties of solid structures materials are summarized in Table 2 and Table 3.

Table 2. AISI-304 steel thermal properties.

Density, [kg/m^3]	8030
Specific heat, [J/(kg K)]	502.48
Thermal conductivity, [W/(m K)]	9.437+0.0154 T

Table 3. AISI-316 powder thermal properties.

Density, [kg/m^3]	8030
Specific heat, [J/(kg K)]	502.48
Thermal conductivity, [W/(m K)]	3.41

To model heat transfer, the energy equation was activated and the "convection" condition was fixed on the pipe walls. The axial profile of the water side heat transfer coefficient and the two-phase fluid bulk temperature were imposed as thermal wall boundary conditions. The numerical values to apply as boundary conditions were obtained from a RELAP5 stand-alone simulation. The thermal wall boundary conditions were applied at the inner walls of the ascending water pipes, named Wall 1, Wall 2 and Wall 3, as shown in Figure 4. Different wall boundary conditions between the central pipe (Wall 3) and the lateral pipes (Wall 1 and Wall 2) were adopted, with the aim to obtain more realistic results.

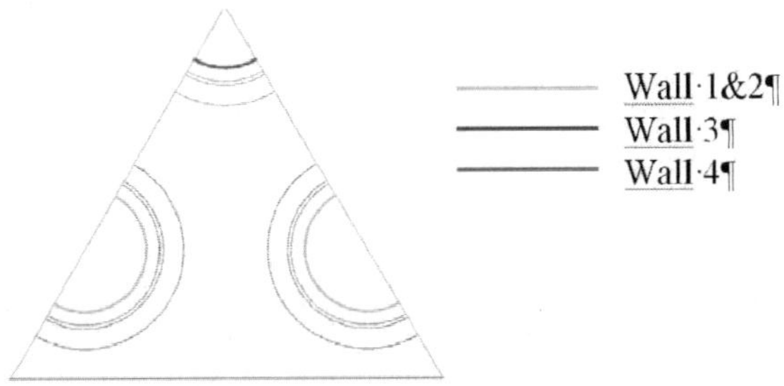

Figure 4. HERO tubes spatial discretization.

In Figure 4, Wall 4 is also shown, which represents the hexagonal wrapper, where no slip and adiabatic conditions were assumed. In addition, the LBE mass flow rate and temperature were imposed at the inlet section, while pressure outlet boundary condition was imposed at the outlet section.

The mesh was developed in agreement with the request of the standard wall function model resulting in $y^+ > 30$ at the wall boundaries. The influence of the adopted turbulence model was also investigated. In particular, three different models were attempted:

- RNG k-ε Model;
- Realizable k-ε Model;
- SST k-ω Model.

The parameters set as boundary conditions at the inlet of the domain summarized in Table 4:

Table 4. Reference boundary conditions.

Inlet LBE mass flow rate, [kg/s]	6.225
Inlet LBE temperature, [°C]	480

The obtained results of LBE temperature, velocity and turbulent kinetic energy (TKE) for the cross section placed at z = 4.1 m are reported in Figure 5, Figure 6 and Figure 7 respectively.

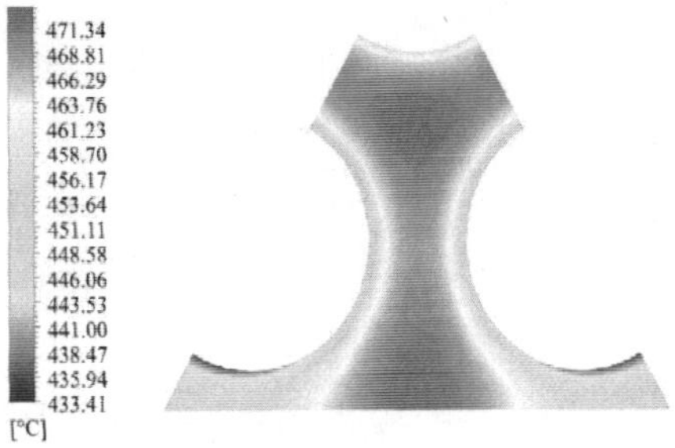

(a)

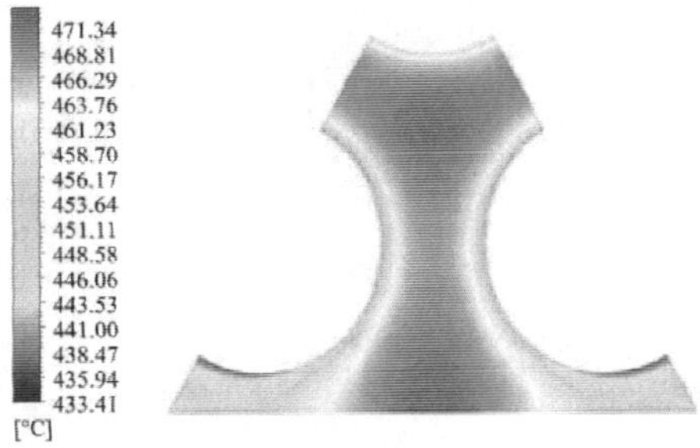

(b)

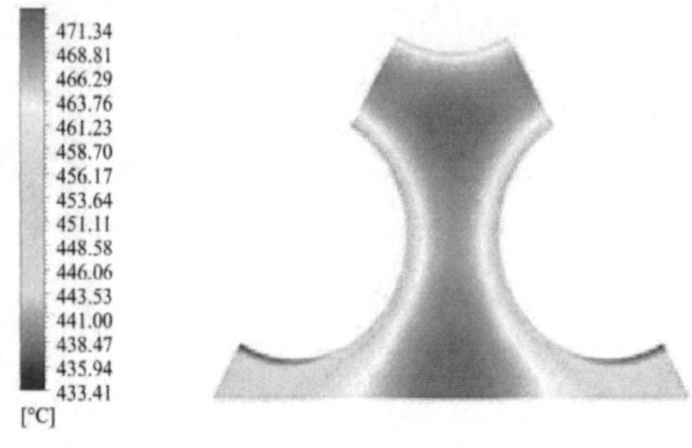

(c)

Figure 5. LBE temperature comparison between the three turbulence models: RNG k-ε Model (a), Realizable k-ε Model (b) and SST k-ω Model (c).

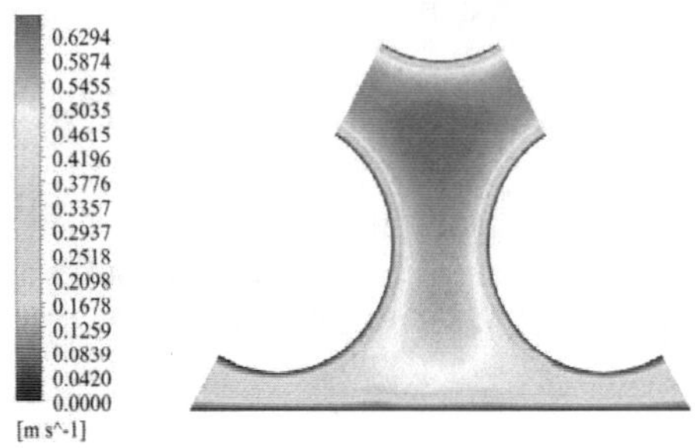

(a)

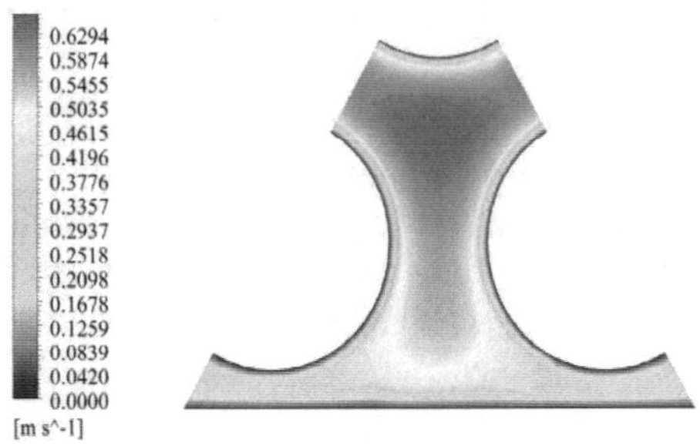

(b)

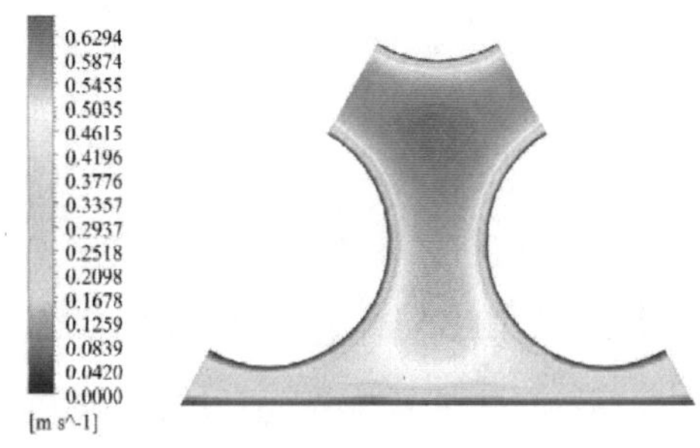

(c)

Figure 6. LBE velocity comparison between the three turbulence models: RNG k-ε Model (a), Realizable k-ε Model (b) and SST k-ω Model (c).

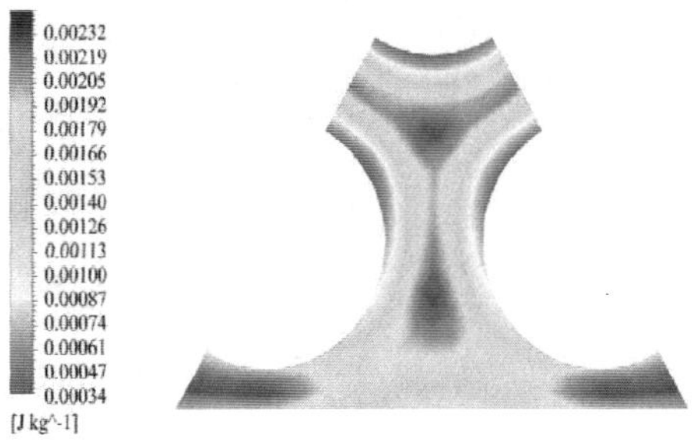

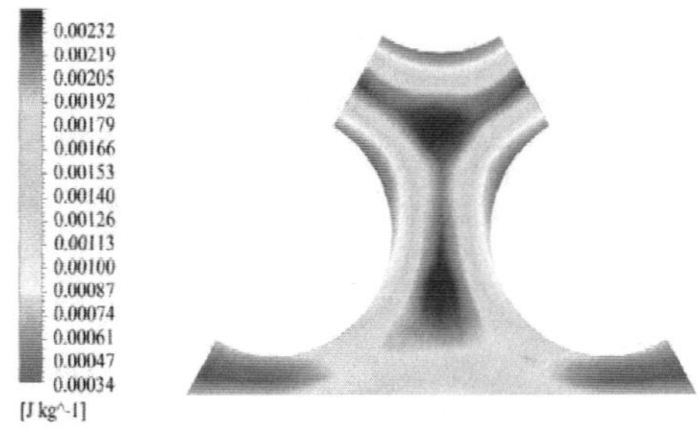

IOP Conf. Series: Journal of Physics: Conf. Series **1224** (2019) 012032 doi:10.1088/1742-6596/1224/1/012032

(a) **(b)**

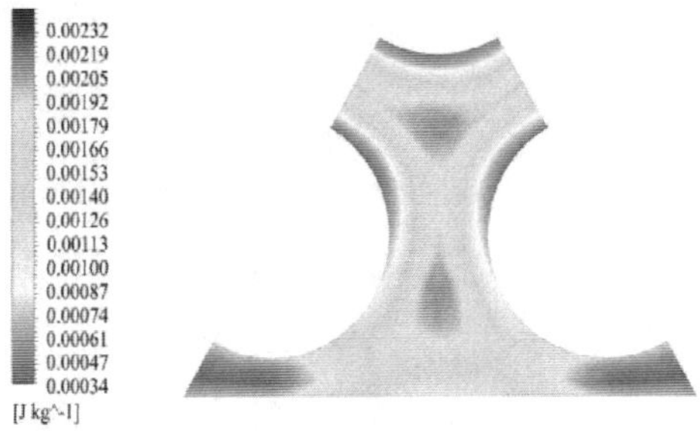

(c)

Figure 7 - LBE turbulent kinetic energy comparison between the three turbulence models: RNG k-ε Model (a), Realizable k-ε Model (b) and SST k-ω Model (c).

From Figure 5, Figure 6 and Figure 7, it can be observed that LBE temperature, velocity and turbulent kinetic energy (TKE) distributions are not highly influenced by the adopted turbulence model. In the temperature and velocity profiles, the RNG and Realizable k-ε models show a similar trend while the SST k- ω model results are slightly lower. Instead, the TKE profiles display a more noticeable difference among the models, especially in the SST k-ω model, which generally estimated smaller turbulent parameters. Nevertheless, the choice of the model was based on the comparison of the residual behaviour of the different turbulence models. The model that shows the lowest residuals is the Realizable k-ε model.

Since the turbulent Prandtl number is one of the major parameters affecting sensitively the turbulent heat transfer behaviour in Liquid Metal, additional CFD calculations with Realizable k-ε model were carried out using different turbulent Prandtl number values: 0.85, 1, 1.5 and 2. Changing the turbulent Prandtl number, the LBE temperature profile is slightly different; in particular, higher turbulent Prandtl number gave higher LBE temperature profile over the chosen line.

In absence of experimental data to compare the numerical results and to decide which turbulent Prandtl number is the most appropriate, the choice was based according to the following correlation [14]:

$$Pr_t = \begin{cases} 1.5 & Pe \leq 2000 \\ 2.5 - 0.0005Pe & 2000 \leq Pe \leq 3000 \\ 1.0 & Pe \geq 3000 \end{cases}$$

In the simulated test cases, the Péclet number is generally smaller than 2000, then a Pr_t=1.5 was chosen for the CFD-STH coupled code simulations reported in the next section.

2.2. Adopted coupling procedure

The scheme used for the coupling model is represented in Figure 8, where the thermodynamic variables exchanged at the interfaces between the two domains are highlighted. Basically, the exchange of the variables takes place at the walls of the ascending water pipes along the entire tube length. As already explained, exchanged data are differentiated for the central and the lateral pipes. This differentiation is shown in Figure 8, where the RELAP5 nodalization of the HERO secondary side is reported: the components from 420 to 452 are representative of the central pipe, whereas components from 520 to 552 represent the six lateral tubes. This separation of the central pipe from the lateral ones allows to obtain distinct boundary conditions on the pipes wall from the detailed temperature distribution in the primary side of the HX (LBE side), calculated by the CFD code. In the RELAP5 model, the water is injected through the time-dependent junction 415 inside the inlet plenum 418 and then redistributes inside the two pipes (420 and 520).

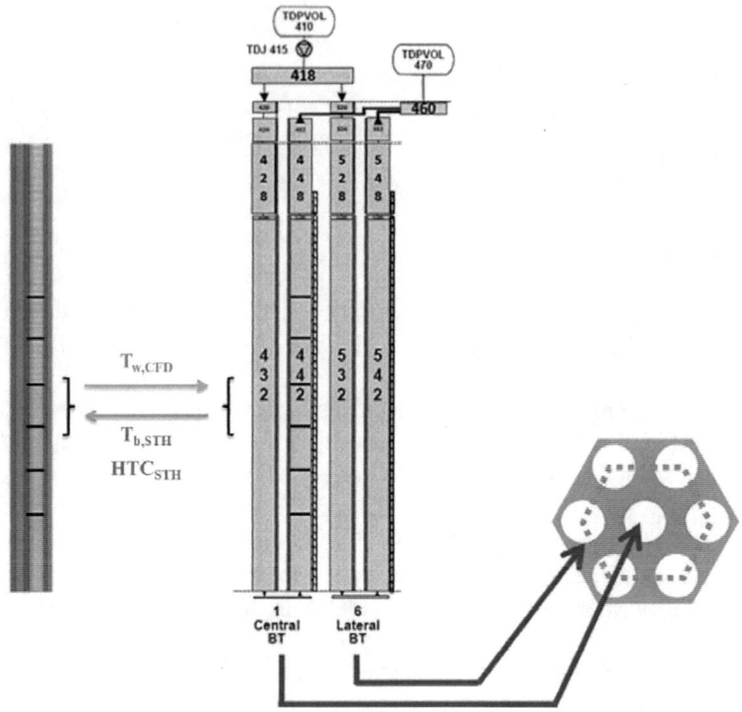

Figure 8. HERO tubes spatial discretization.

60 artificial surfaces have been created on each boundary wall of the CFD domain, in order to exchange data with the system code. The height of these surfaces is equivalent to the respective volume length of pipes 442 (542) and part of pipes 448 (548) of RELAP5. Actually, the divisions are operated on the Wall 2 (the division of Wall 1 is not necessary for symmetry) and Wall 3 because these boundaries are located where the data exchange is carried out between CFD and STH codes. The schematic representation of the artificial subdivision for the data exchange is showed in Figure 8. At each iteration, wall temperature results from Fluent are averaged over the tubes surface (axially and azimuthally) in each of the 60 above-mentioned zones; then these data are transferred to RELAP5. In turn, RELAP5 gives back 60 data points, which are used to build an axial profile (with linear trend between two consecutive points) of both the two-phase water mixture bulk temperature and the Heat Transfer Coefficient (HTC), which are given to Fluent as boundary conditions.

The numerical method adopted for the coupled simulation is the semi-implicit scheme. At each step, for both outer and inner iterations, the variables exchanged between the boundaries of the two codes are:

- the water bulk temperatures;
- the HTC at the pipes wall of the ascending water;
- the wall temperature of the ascending water (in two-phase flow).

The percentage error between the values of each variable at two consecutive iteration is chosen as convergence criteria. More detailed information on the coupling numerical algorithm can be found in Ref. [8].

IOP Conf. Series: Journal of Physics: Conf. Series **1224** (2019) 012032 doi:10.1088/1742-6596/1224/1/012032

2.3. Numerical results of HERO coupled calculations

The model for HERO coupled calculations described in section 2.2 was applied to four test cases with the aim to investigate the different performance of the component by changing any of the relevant parameters.

The hydraulic Boundary Conditions (BC) of the HX primary side (LBE) were set in the CFD model and the BCs of the secondary side (water/vapour mixture) were set in the RELAP5 model. The relevant parameters of the four cases are reported in Table 5.

Table 5. Boundary conditions for HERO coupled calculations.

Variable	Case 1	Case 2	Case 3	Case 4
LBE mass flow, [kg/s]	37	9	37	9
LBE inlet temperature, [°C]	480	480	480	480
LBE outlet gauge pressure, [bar]	0	0	0	0
Water mass flow rate, [kg/s]	0.33	0.33	0.08	0.08
Water inlet temperature, [°C]	335	335	335	315
Secondary outlet pressure, [bar]	180	180	180	180

Two different values of the LBE mass flow rate were chosen: the higher one is representative of the normal operating condition of CIRCE, the lower is representative of the natural circulation flow condition typical of CIRCE. Two different levels of water mass flow rate were also investigated, where the higher one is related to the reference design condition.

The characterizing results of the four test cases are summarized in Table 6.

Table 6. Resulting parameters from HERO coupled calculations.

Variable	Case 1	Case 2	Case 3	Case 4
LBE outlet temperature, [°C]	397.8	345.9	455.2	369.9
Water/steam outlet temperature, [°C]	381.3	357.0	478.3	475.0
Outlet static quality, [-]	1	0.38	1	1
Total removed power, [kW]	440	175	132	142

Following the design operating condition, the total power removed by HERO is about 440 kW and this value generally decreased moving from that parameters. Lower LBE outlet temperatures were obtained whenever the LBE mass flow decreased, and almost in all cases the vapour reached superheated conditions.

In the following discussion, Case 1 was chosen for extensive analysis of the results, through the comparison with the analogous RELAP5 standalone calculation. Nevertheless, the RELAP5 standalone model employees a single equivalent water tube to model the 7-tubes of the HERO bundle, which has a total flow area equal to the sum of the individual 7 tube area.

Figure 9 shows the axial trend of temperatures along the HERO test section obtained by the coupling calculation for the central pipe 442 (Figure 9(a)) and lateral pipes 542 (Figure 9(b)). The two images can be compared with Figure 10, which displays the analogous trends obtained with the RELAP5 standalone calculation. The reported profiles depict the trends of LBE temperature, wall temperature on LBE side, the wall temperature on water side and the water temperature both in the descending and ascending tubes.

At first sight, the trends of the temperatures of the central and lateral pipes are similar and there are not evident differences with the RELAP5 stand-alone calculation. Notwithstanding, thanks to the use of coupled calculation it was possible to individuate a slightly different behaviour between the central and the lateral tubes. Indeed, the two-phase flow mixture in the central tube reached the saturated steam condition earlier with respect the lateral tube and the vapour outlet temperature was higher (408.8°C in central pipe versus 378.6°C in lateral pipes). In particular, the CFD portion of the model allows to compute more detailed temperature distribution in the transversal area of the HERO bundle and is able to give to RELAP5 two distinctive boundary conditions respectively to the central and lateral pipes of the bundle.

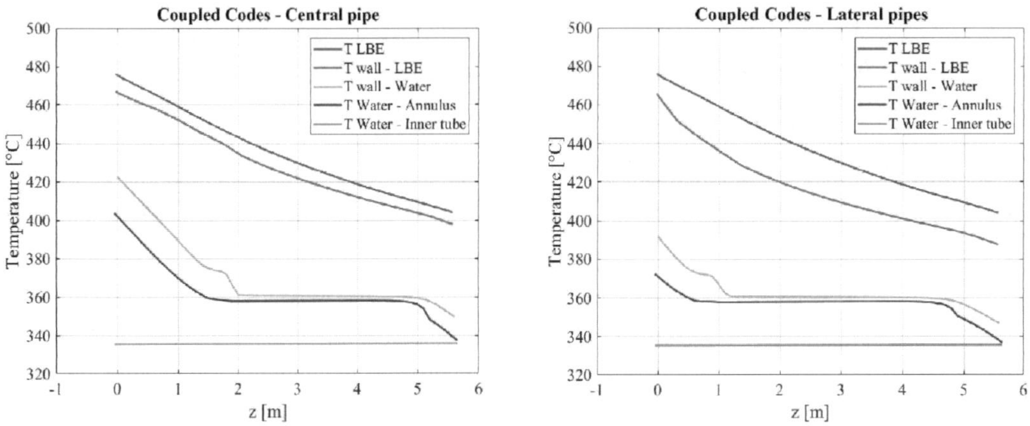

(a) (b)

Figure 9. Axial trend of temperatures in the central (a) and lateral (b) pipes from coupled calculation for Case 1.

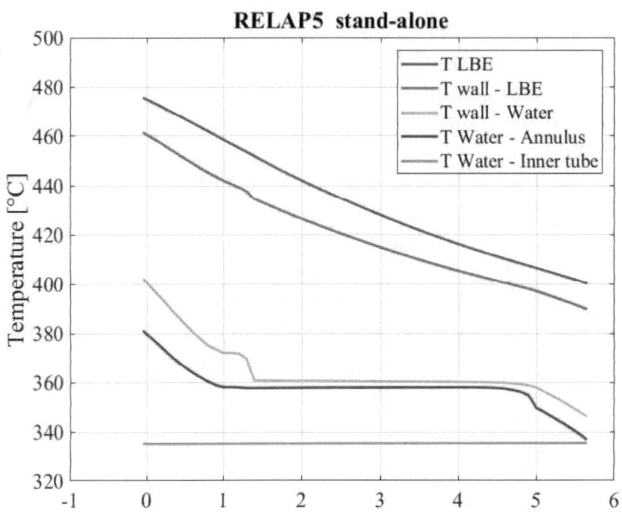

Figure 10. Axial trend of temperatures from RELAP5 standalone calculation for Case 1.

The trends of the HTC of the two-phase flow mixture in the ascending tubes are shown Figure 11, where the results achieved with the coupled calculation (both in central and lateral pipes) are compared with those obtained from the RELAP5 standalone simulation. The trend of the three curves is very similar; however, the changes in slope occurred in slight different axial positions, according to the temperature trend of the mixture. In particular, it occurs closer to the entrance region in the central pipe than the lateral ones, confirming that the coupled calculation could lead to more accurate information on heat transfer.

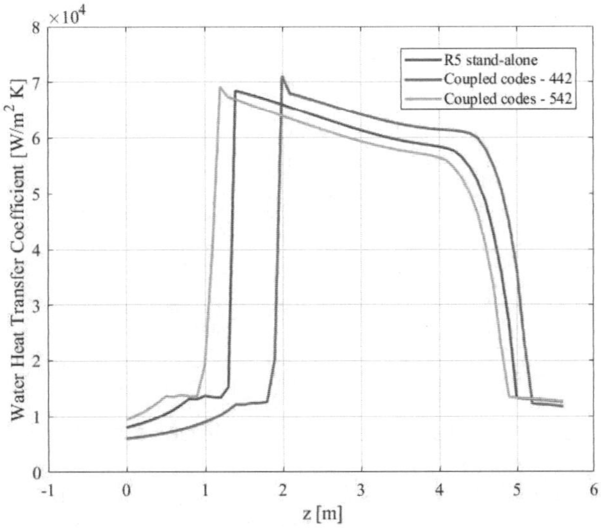

Figure 11. Comparison of the axial trend of HTC – water side for Case 1.

Concerning the value of the HTC in the LBE side, it was possible to obtain the axial trend of HTC directly from the Fluent domain of the coupled model, using the following equation:

$$HTC_{LBE}(z) = \frac{q''_w\,(z)}{\bar{T}_w\,(z) - T_b\,(z)} \tag{1}$$

where q''_w is the wall heat flux, $\bar{T}_w$ is the wall temperature at LBE side and T_b is the LBE bulk temperature. These values were extracted from artificial axial planes specifically created in the CFD domain. In each plane, q''_w and $\bar{T}_w$ were averaged over the total perimeter of both central and lateral walls. In Figure 12, the obtained axial trend is compared with the HTC in the LBE side from the RELAP5 standalone simulation, which was computed using one of the available correlations in the code (Ushakov in this case). The qualitative difference between the two trends is linked to the different way of computing the HTC. In particular, in the CFD calculation it is possible to compute a more detailed variation along the axial coordinate. The initial sharp trend of the curve is related to the entrance region, where the flow and the temperature profile are not developed and the difference between the average wall temperature $\bar{T}_w$ and the bulk temperature T_b is small.

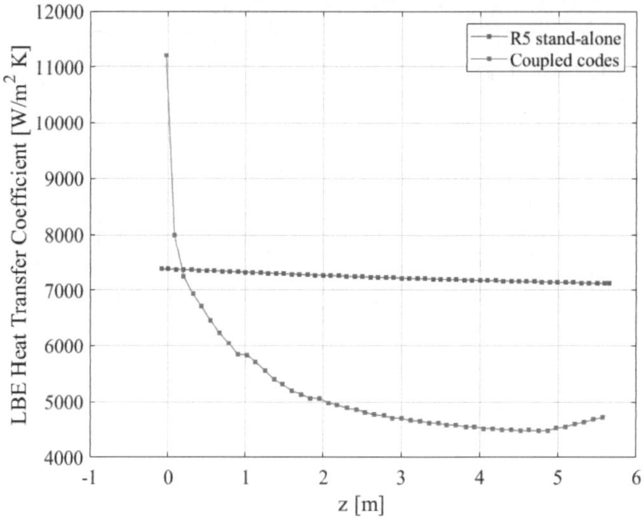

Figure 12. Comparison of the axial trend of HTC – LBE side for Case 1.

As explained in section 2.2, the exchanged heat transfer variables in HERO coupled calculations are the wall temperature (from Fluent to RELAP5) and water/vapour HTC together with the two-phase mixture bulk temperature (from RELAP5 to Fluent). Consequently, the wall heat flux at the boundary walls is computed independently by the two codes. The two heat flux trends, extracted from the Fluent and the RELAP5 domains of the coupled simulation, were compared a posteriori and are illustrated in Figure 13. The two trends are practically coincident over most part of the active length, with exception of few points where the curves have local minimum or maximum. The discrepancy was attributed to limits of the discretization and not to the physical inconsistency. Figure 13(a) and Figure 13(b) show respectively the curves for the central and the lateral tubes, which are put also in comparison with the trend from the RELAP5 standalone calculation.

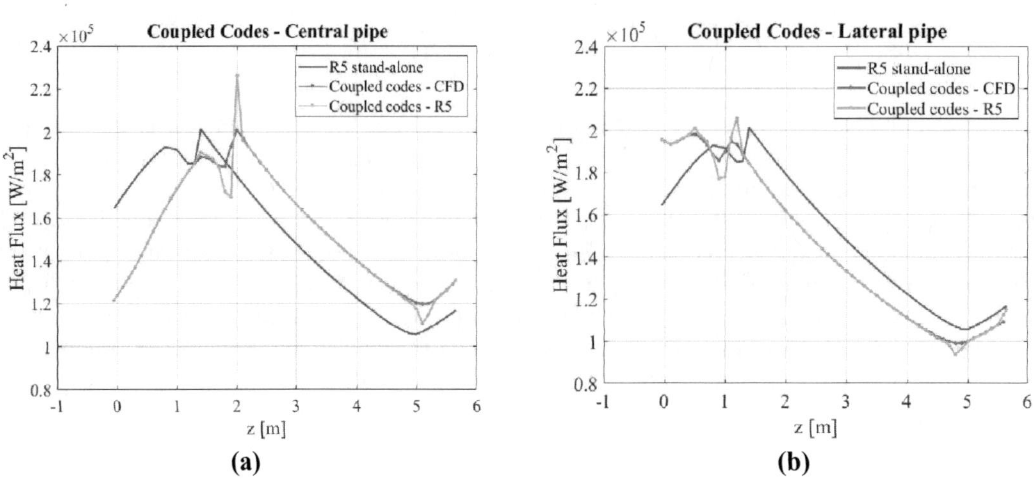

Figure 13. Comparison of the axial trend of the wall heat flux in the central (a) and lateral (b) pipes for Case 1.

The other advantage of using the coupled calculation is the opportunity to analyse more in detail the flow in the portion of the domain modelled by the CFD code. As a showing example, Figure 14 reports the LBE contour plots for temperature (left) and velocity (right) in three different section, respectively at about z=1.6, 3.6 and 4.1 m. The same scale was used for the three images associated to the same variable. In the contours of temperature, the solid domains (AISI 304 structures and AISI 316 powder) are plotted too. The LBE temperature is colder close to the pipes, where the heat transfer occurs, but also in the outer region between the lateral pipes and the hexagonal wrapper since the LBE velocity is smaller. In the latter region the heat is manly transferred by conduction. Using the same scale range, it is possible to see the different temperature levels during the LBE cooling. The velocity field in the first monitored section (z=1.6 m) appears slightly different with respect to the contour in z=3.6 m and z=5.6 m. Indeed, the first section is still affected by the entrance region and the flow is not completely developed. For the sections z=5.6 m, the flow is finally developed, and the small differences are related to temperature effects.

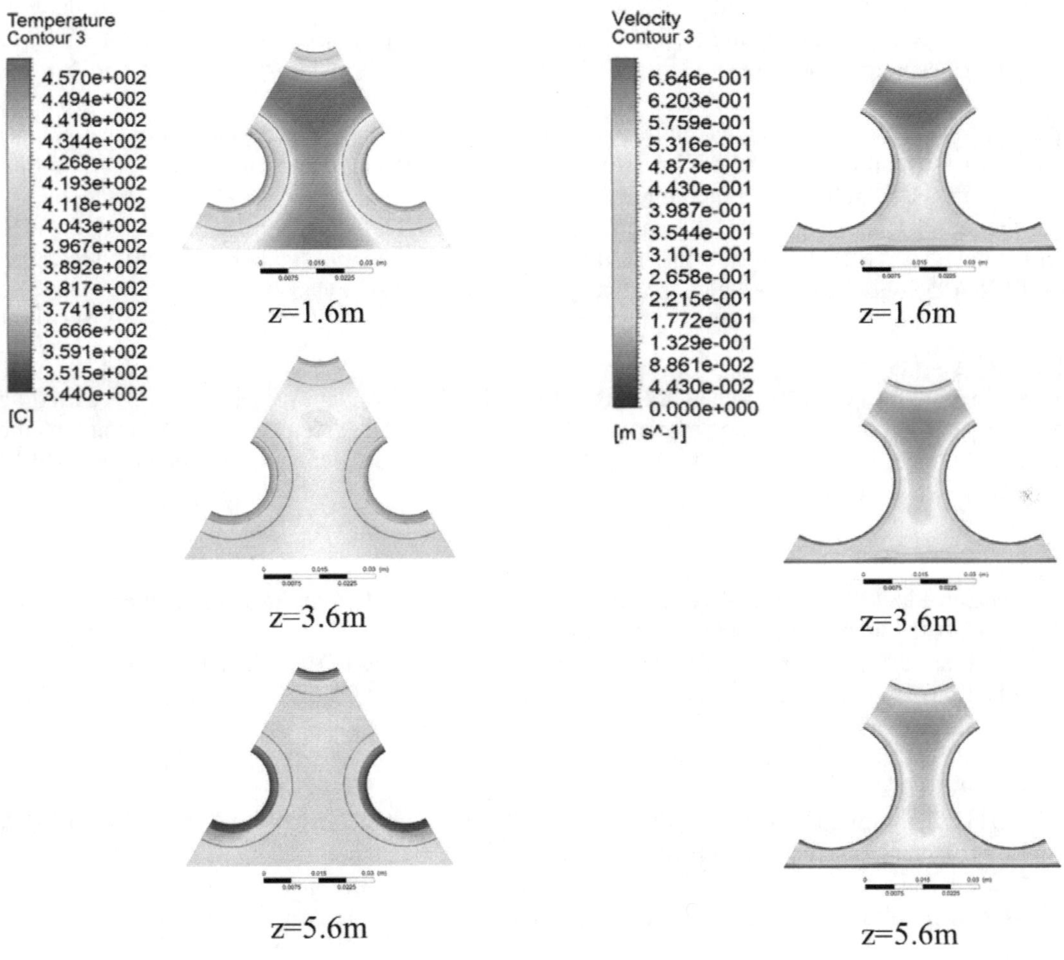

Figure 14. Temperature and velocity magnitude in the LBE side (CFD domain) of HERO at different axial position for Case 1.

3. Conclusions

The present work describes the numerical analysis performed in support of the HERO heat exchanger experimental campaign foreseen at the ENEA Brasimone R.C..

To this end, a coupling methodology between CFD code and system code already developed at the

University of Pisa is improved accounting for its application to thermal boundary conditions.

In the first part of the work, CFD sensitivity analysis were performed in order to select the proper turbulent model and to investigate the effects of the chosen turbulent Prandtl Number. A symmetrical 3D CFD model representative of one sixth of the entire geometry of the LBE side and of the HX solid structures was build.

Boundary conditions, obtained from a previous RELAP5 standalone calculation, were imposed for the CFD standalone calculations (one-way coupling). In particular, the thermal wall boundary conditions were applied at the inner walls of the ascending water.

In the second part of the work, the coupled methodology was applied to the entire HERO section: the primary LBE side and the pipe structures were modelled with the CFD Fluent code whereas the water side was modelled with the STH code RELAP5. The coupling procedure was verified by comparing the obtained results with the analogous ones achieved with the RELAP5 standalone calculation. Similar temperature, heat flux and HTC trends were obtained with the two calculations, proving that the developed coupling methodology is reliable. Further, the coupled simulation allows to obtain more accurate information on the LBE flow and on the heat transfer performance for this component.

As next step, this methodology could be adopted to analyse the performances of the HERO component in several operating conditions (by varying the water side operating pressure, LBE and/or water inlet mass flow rates and temperatures).

At the end of the experimental campaign on the CIRCE-HERO facility, collected detailed experimental data will be use for the verification and preliminary validation of the developed Fluent/RELAP5 coupling mythology in predicting the involved phenomena.

Acknowledgments

This work was performed in the framework of H2020 MYRTE project. This project has received funding from Euratom research and training program 2014-2018, under grant agreement No 662186, and by the Programmatic Agreement (AdP) between the Italian Ministry of the Economic Development (MiSE) and ENEA.

References

[1] Nuclear Safety Analysis Division 2001, RELAP5/Mod3.3 code manual Volume I: Code Structure, System Models, and Solution Methods., vol. 1.

[2] Lavialle G 2006, CATHARE 2 V2.5_1: User's Manual, SSTH/LDAS/EM/2005-035.

[3] ATHLET Mod 3.0 Cycle A. User's Manual, (2012). GRS-P-1/ Vol. 1 Rev. 6 – GRS, Garching bei München, Germany.

[4] TRACE V5. 0, 2007.Theory Manual - Field Equations, Solution Methods, and Physical Models, USNRC, Washington DC.

[5] GOTHIC Thermal Hydraulic Analysis Package User Manual, Version 8.0 (QA), NAI 8907–02, Rev. 20, Numerical Applications Inc., 2012.

[6] Martelli D, Forgione N, Barone G and Di Piazza I 2017, *Coupled simulation of the NACIE facility using RELAP5 and ANSYS FLUENT codes*, Annals of Nuclear Energy 101, 408-418.

[7] Martelli D, Forgione N, Barone G, Del Nevo A, Di Piazza I and Tarantino M, *Coupled simulations of natural and forced circulation tests in NACIE facility using RELAP5 and Ansys Fluent codes*, Proceedings of the 2014 22nd International Conference on Nuclear Engineering ICONE22, July 7-11, 2014, Prague, Czech Republic.

[8] Angelucci M, Martelli D, Barone G, Forgione N and Di Piazza I, *STH-CFD codes coupled calculations applied to HLM loop and pool systems*, Science and Technology of Nuclear Installations, Volume 2017, Article ID 1936894.

[9] Narcisi V, Giannetti F, Tarantino M, Martelli D and Caruso G 2017, *Pool temperature stratification analysis in CIRCE-ICE facility with RELAP5-3D© model and comparison with*

experimental tests, Journal of Physics: Conf. Ser. 923 012006,.
[10] Rozzia D, Del Nevo A and Tarantino M, *HERO-CIRCE Configuration*, Report ENEA CI-I-R-166.
[11] Barone G, Forgione N, Martelli D and Ambrosini W 2013, *System codes and CFD codes applied to loop- and pool-type experimental facilities*, RL 1530/2013, CIRTEN Technical Report, University of Pisa.
[12] OECD/NEA 2007, *Handbook on Lead-bismuth Eutectic Alloy and Lead Properties*, Materials Compatibility, Thermal-hydraulics and Technologies.
[13] ENEA 2016. Private communication.
[14] Cheng X and Tak N I 2006, *CFD analysis of thermal–hydraulic behavior of heavy liquid metals in sub-channels*, Nucl. Eng. Des. 236, pp. 1874-1885.

IOP Conf. Series: Journal of Physics: Conf. Series **1224** (2019) 012033 doi:10.1088/1742-6596/1224/1/012033

On the thermal dynamic behaviour of the helium-cooled DEMO fusion reactor

Ivo MOSCATO*, Luciana BARUCCA, Sergio CIATTAGLIA***, Pietro Alessandro DI MAIO*,**

* Dipartimento di Energia, Ingegneria dell'Informazione e Modelli Matematici, Università di Palermo, Viale delle Scienze, I-90128 Palermo, ITALY
** Ansaldo Nucleare, Corso Perrone 25, 16100 Genova, Italy
***EUROfusion Consortium, Boltzmannstr. 2, Garching, Germany

Corresponding author e-mail: ivo.moscato@unipa.it

Abstract. The EU-DEMO conceptual design is being conducted among research institutions and universities from 26 countries of European Union, Switzerland and Ukraine. Its mission is to realise electricity from nuclear fusion reaction by 2050. As DEMO has been conceived to deliver net electricity to the grid, the choice of the Breeding Blanket (BB) coolant plays a pivotal role in the reactor design having a strong influence on plant operation, safety and maintenance. In particular, due to the pulsed nature of the heat source, the Primary Heat Transfer System (PHTS) becomes a very important actor of the Balance of Plant (BoP) together with the Power Conversion System (PCS). Moreover, aiming to mitigate the potential negative impact of plasma pulsing on BoP equipment, for the DEMO plant is also being investigated a "heat transfer chain" option which envisages an Intermediate Heat Transfer System (IHTS) equipped with an Energy Storage System (ESS) between PHTS and PCS. Within this framework, a preliminary study has been carried out to analyse the thermal dynamic behaviour of the IHTS system for the Helium-Cooled Pebble Bed (HCPB) BB concept during pulse/dwell transition which should be still considered as the normal operating mode of a fusion power plant. Starting from preliminary thermal-hydraulic calculations made in order to size the main BoP components, the global performances of DEMO BoP have been quantitatively assessed focusing the attention on the attitude of the whole IHTS to smooth the sudden power variations which come from the plasma. The paper describes criteria and rationale followed to develop a numerical model which manages to simulate simple transient scenarios of DEMO BoP. Results of numerical simulations are presented and critically discussed in order to point out the main issues that DEMO BoP has to overcome to achieve a viable electricity power output.

1. Introduction

The EU-DEMO conceptual design is being conducted among research institutions and universities from 26 countries of European Union, Switzerland and Ukraine. Its mission is to realize electricity from nuclear fusion reaction by 2050. The recent European roadmap, has established that several hundred MW of electricity must be produced by DEMO plant which has to ensure an adequate availability and reliability of operation over a reasonable time span [1],[2]. Due to this reason the EU-DEMO plant design has to be strongly oriented to the Balance of Plant (BoP).

The BoP in fact is made up of all those systems which scope is to convert the fusion power into electricity, the electrical power supply, the cryogenic plant and several auxiliaries providing specific fluids and gases to the various apparatus according to their need [3].
The systems that are the core of the DEMO BoP, representing the fundamental energy "chain" devoted to the extraction of thermal power and to its conversion into electricity, are the Primary Heat Transfer Systems (PHTSs) of Breeding Blanket (BB), Divertor (Div) and Vacuum Vessel (VV) and the Power Conversion System (PCS). The former are in charge of extracting the thermal power generated by the tokamak and delivering it to the latter in order to be converted first in mechanical and then into electrical power by means of the turbine and the alternator, respectively.
The DEMO pulsed operation may lead to unacceptable fatigue stress on essential PCS components, therefore it is being developed a "heat transfer chain" configuration which foresees a third coolant system aimed at decoupling the BB PHTS to the PCS and mitigating the potential negative impact of plasma pulsing on the turbine and the other main equipment of the PCS: the Intermediate Heat Transfer System (IHTS). [3][4][5]. This intermediate cooling circuit is equipped with an Energy Storage System (ESS) that, during the pulse operation, collects a portion of the thermal energy removed from the blanket in order to give it to the PCS during the dwell time; in this way the turbine is allowed to work in steady-state condition both during the pulse and the dwell time of the tokamak.
It has been chosen to connect the IHTS only to the blanket heat removal system because it has to extract about 85% of the power generated by the tokamak, hence it might be intended as the main hub of the heat transfer chain.
Another element that makes the DEMO BoP design more complex respect to other power plants is the presence of separated PHTSs. Such choice is necessary because the three main systems that extract the thermal power from the machine - BB, Div and VV - have different functions and requirements making almost impossible the use of a single coolant operating at the same thermal-hydraulic (T/H) conditions. It is agreed that water will be used for the Div and VV cooling, while the selection of the BB coolants is still open; four blanket options are, in fact, under developments [6], nevertheless the DEMO BoP activities are currently focused on the two concepts whose feasibility seems achievable in shorter term: the Helium Cooled Pebble Bed (HCPB) and the Water Cooled Lithium Lead (WCLL) BB [5][7].
Within this framework, a preliminary study has been carried out to analyse the thermal dynamic behaviour of the intermediate cooling system for the HCPB BB concept during pulse/dwell transition which should be still considered as the normal operating mode of DEMO power plant. Starting from preliminary thermal-hydraulic calculations carried out to size the main BoP components, the global performances of DEMO BoP have been quantitatively assessed focusing the attention on the attitude of the whole IHTS to smooth the sudden variations of the plasma generated power. The paper describes criteria and rationale followed to develop a numerical model which manages to simulate simple transient scenarios of DEMO IHTS.

2. The HCPB BB BoP

The DEMO HCPB BoP is depicted in figure 1, where the main systems for the heat transport and power conversion are outlined, namely: the PHTSs, the IHTS with the ESS and the PCS which integrates the low temperature sources of Div and VV as feedwater pre-heaters.

2.1. Primary Heat Transfer Systems

DEMO presents four independent PHTSs. The bigger PHTS is devoted to remove the thermal power from the BB, two PHTSs are necessary to extract heat from the Div while the last PHTS is used to cool the VV. Table 1 summarises the main parameters of DEMO PHTSs.

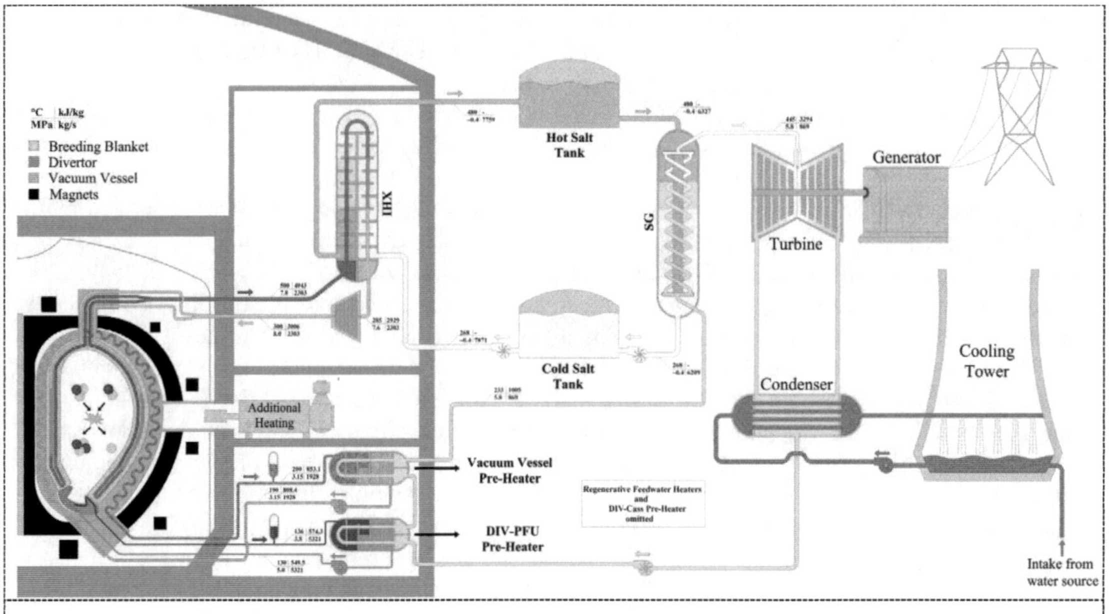

Figure 1. Conceptual scheme of helium-cooled BB BoP with ESS.

Table 1. DEMO PHTSs main data.

	BB	DIV-PFCs	DIV-Cass	VV
Coolant	Helium	Water	Water	Water
Thermal Power (Pulse/Dwell)[MW]	2389.1/21.4	136.0/1.4	115.2/1.1	86/1.0
Inlet temperature (Pulse) [°C]	300	130	180	190
Outlet temperature (Pulse) [°C]	500	136	210	200
Operating Pressure [MPa]	8	5	3.5	3.15

2.1.1. Breeding Blanket. The DEMO blanket is subdivided in 18 sectors, each one of 20°. Each sector includes two main zones, the Inboard Blanket (IB) and the Outboard Blanket (OB). These two areas are, in turn, subdivided in two sub-components, namely the First-Wall (FW) and the Breeding Zone (BZ): the FW has to withstand the plasma heat radiation and the additional heat loads due to impacting particles while the BZ contains lithium that is used as breeder material to produce tritium by means of nuclear reactions. FW and BZ are cooled in series by helium. The BB PHTS foresees a highly degree of segmentation of its cooling loops which are 9. These circuits are completely independent, from mechanical and functional point of view, in order to limit some common mode failures. In particular, there are 3 loops which are designed to cool the IB portion of the tokamak while the other 6 loops are in charge of removing the power from the OB zone. One IB cooling circuit provides helium to 6 blanket sectors while an OB loop cools the segments of 3 blanket sectors. Each IB/OB loop has 3 main parallel hot/cold legs, one heat exchanger and a couple of circulators which allow helium to circulate along the whole circuit [3][4].

2.1.2. Divertor. The Divertor is primarily a high heat flux component. As for the blanket it is subdivided in 18 sectors of 20°. Div is separated in Plasma-Facing Components (PFCs), which represent the Div targets where are collected most of the particles and the energy exhausted by the plasma, and the Cassettes that act as support for the PFCs. Both sections of the Divertor are cooled by water but at different T/H conditions , thus two separated cooling circuits are needed. Each circuit is constituted of two loops which are directly coupled to the PCS via the heat exchanger that works as feed-water preheater [3].

2.1.3. Vacuum Vessel. The Vacuum Vessel is a torus-shaped double-walled pressure vessel that provides the primary vacuum (at very low pressure and at very high purity in order to optimize the D-T fusion reaction) and shields the magnet system from neutrons. The VV supports all the in-vessel components including BB and Div. Also the VV is made up of 18 sectors and its cooling is provided by two independent circuits that give their contribution in the feed-water heating [3].

2.2. Intermediate Heat Tranfer System

The pulsed nature of the currently considered DEMO operation imposes unique design problems on the energy conversion system. In DEMO, energy is generated in the reactor for 120 minutes (pulse time) then the reactor is shut down for 30 minutes (dwell time) to recharge the central solenoid and to allow the vacuum inside the plasma chamber to be restored. However, several studies are on-going aimed at improving central solenoid recharge time and vacuum pumps performances in order to achieve a shorter dwell time which should last about 10 minutes [8]. During the dwell time the BB thermal power sharply drops to almost 1% of the nominal power due to the residual decay heat.

Since the PCS requires stable thermal operating conditions throughout the plasma operational phases, the IHTS equipped with an ESS, where a molten salt is used as heat transfer and storage medium, is being included in the BoP with the aim to mitigate the impact of plasma pulsing on main components, with particular care of steam turbine, steam generators, and the electrical grid.

The ESS of the DEMO is an industrial 2-tanks design commonly used in concentrated solar power plants. The adoption of an ESS, where part of the thermal energy is accumulated as sensible heat, allows to operate the turbine at almost continuous load equal to about the 81% of the blanket power. However, since the energies of Div and VV are not stored, to furtherly limit thermal transients inside the PCS caused by the DEMO pulsed operations, the feed-water heat-up strategy foresees variable steam extractions from turbine and by-pass lines for Div and VV heat exchangers thus allowing continuous plant operations and flexibility in the range of turbine power scale of ±20%. [9].

After a proper literature review of the molten salts technology [10][11], it has been decided to use the HITEC salt [12] as heat transfer fluid because its good thermal stability within the range of DEMO operative temperatures and its relatively low freezing point give more margins to withstand normal and off-normal transients. Table 2 summarises the operating condition of molten salt.

Table 2. IHTS molten salt operating condition.

Coolant	HITEC
Hot tank temperature [°C]	480
Cold tank temperature [°C]	268
Pressure [MPa]	0.1

2.3. Power Conversion System

The preliminary conceptual design of PCS for DEMO envisages a classical superheated steam Rankine cycle operating at about 58 bar with superheated steam at 445°C. It is equipped with steam generator, reheater - in between high and low pressure turbine stages heated by using a stream of hot molten salt -, deaerator, condenser and feed-water pre-heaters, [5].

The pre-heaters are of condensing type or single phase fluid in both tube and shell side. In particular, the latter are represented by the heat exchangers of Div and VV PHTSs which are integrated in PCS as additional (low temperature) heat sources to improve system efficiency [3][9].

3. Preliminary sizing and thermal behaviour of IHTS components

A preliminary assessment of sizes and performances for the main equipment of the IHTS has been carried out in order to identify potential technical feasibility issues. The attention has been then focused on the IHTS thermal dynamic behaviour during pulse/dwell transition and on its attitude to smooth pulsed operations.

3.1. Intermediate Heat eXchanger and Steam Generator sizing

The thermal-hydraulic design of the IHXs has been performed starting from the basic methods commonly adopted to this purpose. The study has been focused on the "tubes and shell" heat exchanger technology for both the IHXs and the steam generators. In particular, the former foreseen a typical straight tube bundle while for the steam generator a helical coiled tube bundle layout has been selected. These configurations have been chosen in order to adopt components easily available on the market since they have been widely used in nuclear and conventional industrial applications for several decades.

3.1.1. Working fluids. The main properties of Helium and HITEC have been extracted respectively from [13] and [12], while as regard as water, its properties have been estimated according to IAPWS-97 functions [14]. Table 3 reports Helium and HITEC properties functions used in design calculations.

Table 3. Helium and HITEC properties.

Coolant	Helium (T in [K] and p in [bar])	HITEC (T in [°C])
Specific heat [J/(kg*K)]	5187.6	1560.0
Density [kg/m^3]	$48.14\frac{p}{T}\left[1+0.4446\frac{p}{T^{1.2}}\right]^{-1}$	$2080-0.7324T$
Conductivity [W/(m*K)]	$2.682\cdot10^{-3}\left(1+1.123\cdot10^{-3}p\right)T^{\left(0.71\left(1-2\cdot10^{-4}p\right)\right)}$	$e^{\left\{-3.8661-2.3903\left[\ln(T)-4.8892\right]\right\}}$
Viscosity [Pa*s]	$3.674\cdot10^{-7}T^{0.7}$	$0.4465+0.1788\cdot10^{-3}T-1.1486T^2$

3.1.2. Intermediate Heat eXchanger. A two-pass tubes, two-pass shell configuration has been selected (TEMA NFN type) [15] to design this component which has been sized to withstand the thermal load that occurs during the pulse time. It has been also decided to let helium circulate through the tube bundle while HITEC salt flows on the shell side. Such choice has been made mainly due to the helium higher pressure and to mitigate the effects of loss-of-coolant accident. The tube side T/H conditions have been selected to be compliant with the blanket requirements, in fact it has been taken into account the helium temperature at the IHX outlet that reaches about 285°C before helium is compressed by the circulators reporting its temperature at the BB target value of 300 °C. Moreover, the huge pumping power needed to circulate the coolant –177 MW for the whole BB PHTS - has been also considered in the IHXs design thermal inputs [4]. Therefore, the IHX T/H performances have been numerically investigated in detail by means of in-house developed code based on the log-mean temperature difference method [16]. In particular, the tube side heat transfer model of this tool relies on the use of the widely accepted correlation by Gnielinski [17] refined by correction factor which considers the influence of surface roughness on heat transfer as suggested by Shah and Bhatti in [18]; on the shell side, instead, to take into account the effects of various leakage and bypass streams on heat transfer coefficient and pressure drop, the Bell-Delaware procedure has been followed as discussed by Taborek in [16]. The main data of heat exchanger are shown in Table 4.

3.1.3. Steam Generator. A helically coiled tube layout has been chosen to transfer the power from HITEC to water in order to produce superheated steam at about 450 °C. Due to its low operating pressure (close to atmospheric pressure), the molten salt flows downward outside the tubes, whereas the water/steam is circulated inside the pipe bundle upward generating a counter-current, cross-flow pattern. The T/H models and procedures applied to design the steam generator are mainly based on the Fenech's and Smith's methodologies [19][20] however more recent works [21][22] are also taken into account to characterise the different heat transfer regions of the water during its phase change, namely the liquid single-phase region, the subcooled boiling region, the saturated boiling region, the post-dryout region and the superheated steam single-phase region. As mentioned in sections 2.2., the power input of the steam generators is about 81% of the heat transferred during the pulse at the IHXs and it has been evaluated considering compressors being run at nominal speed and constant power during both pulse and dwell periods and according the following formula (1):

$$\frac{\int_{pulse}(Q_{BB}+W_{Compr})dt+\int_{dwell}(Q_{BB}+W_{Compr})dt+\int_{ramp-up/down}(Q_{BB}+W_{Compr})dt}{\left(t_{pulse}+t_{dwell}+t_{ramp-up}+t_{ramp-down}\right)} \tag{1}$$

where Q_{BB} and W_{Compr} are thermal and pumping power of the BB, respectively.
The resulted thermal power of 2092.3 MW is distributed among six steam generators in order to limit the size of these equipment. The main parameters of the component are reported in Table 5.

Table 4. IHXs main data

Parameter	IB IHX	OB IHX
Thermal Power [MW]	245.2	304.9
T_{in}/T_{out} helium [°C]	500.0/284.5	500.0/285.6
T_{in}/T_{out} HITEC salt [°C]	268.0/480.0	268.0/480.0
Tubes active length (per pass) [m]	17.0	16.5
Tube number (per pass) [-]	6779	8427
Tube external diameter [mm]	19.05	19.05
HITEC pressure drop [kPa]	163.4	156.1
HITEC volume [m^3]	107.4	127.9

Table 5. Steam generator main data

Parameter	Value
Thermal Power [MW]	348.7
T_{in}/T_{out} HITEC [°C]	480/268
T_{in}/T_{out} water [°C]	233/445
Tubes active length [m]	88.7
Tube number (per pass) [-]	885
Tube external diameter [mm]	22.225
HITEC pressure drop [kPa]	19.3
HITEC volume [m^3]	35.7

3.2. Storage tanks and IHTS thermal behaviour

Considering the requirement of a continuous PCS load, the characteristic time DEMO pulsation (e.g. pulse and dwell time) and the BB power, it follows that the minimum amount of energy that must be stored is about 950 MWh. Taking into account the properties of the HITEC salt and its temperature cycle, it derives the need of at least 10309 tons of fluid and thus a volume of the tanks (hot + cold) of around 12000 m^3. However, during the dwell period the energy source is no longer available at the maximum temperature due to the sharp drop of the fusion power which implies a fast decrease of the helium temperature owing to the high volumetric flow displaced by the compressors. As consequence of the helium temperature reduction, the HITEC temperature at the IHXs outlet decreases too and, in cascade, the hot tank might be affected by this cooling effect making difficult the operation of the steam generators as well as other PCS components. Such cooling process could be hypothetically avoided decreasing the mass flow rate of helium at about 1% of the nominal value operating on the compressor speed but such a huge difference from nominal condition in these heavy turbomachinery seems not to be easily achievable [23][24][25], moreover continuous and quick big changes of compressor angular velocity might decrease its qualified lifetime. For these reasons it has been preliminary chosen to keep the compressors at nominal speed assuming, as first approximation, no changes in power consumption, regardless from their inlet conditions. Feasibility studies on the design of variable speed compressors, working within speed range suitable for the main operative conditions of the BB PHTS, are currently under development with the involvement of industry [8].

3.2.1. The IHTS model and assumptions. In order to analyse the thermal behaviour of the IHTS and its attitude to flatten out the oscillating thermal input coming from the BB, a theoretical model has been set-up adopting a lumped parameter approach. This model is based on simplified time-dependent forms of the mass (2) and energy (3) equations which are herewith reported:

$$\frac{dm}{dt} = \sum_i G_i^{in} - \sum_i G_i^{out} \tag{2}$$

$$\frac{d(mh)}{dt} = \sum_i (Gh)_i^{in} - \sum_i (Gh)_i^{out} \tag{3}$$

Where m is the mass, G is the mass flow rate and h is the enthalpy.
Integrating simultaneously in MATLAB environment [26] these equations for both cold and hot reservoirs and taking into account the input and output powers which have to cross the whole simulated intermediate loop - depicted in figure 2 - it has been possible to perform a parametric assessment of the evolution of mass inventory and fluid bulk temperature inside the two storage tanks.

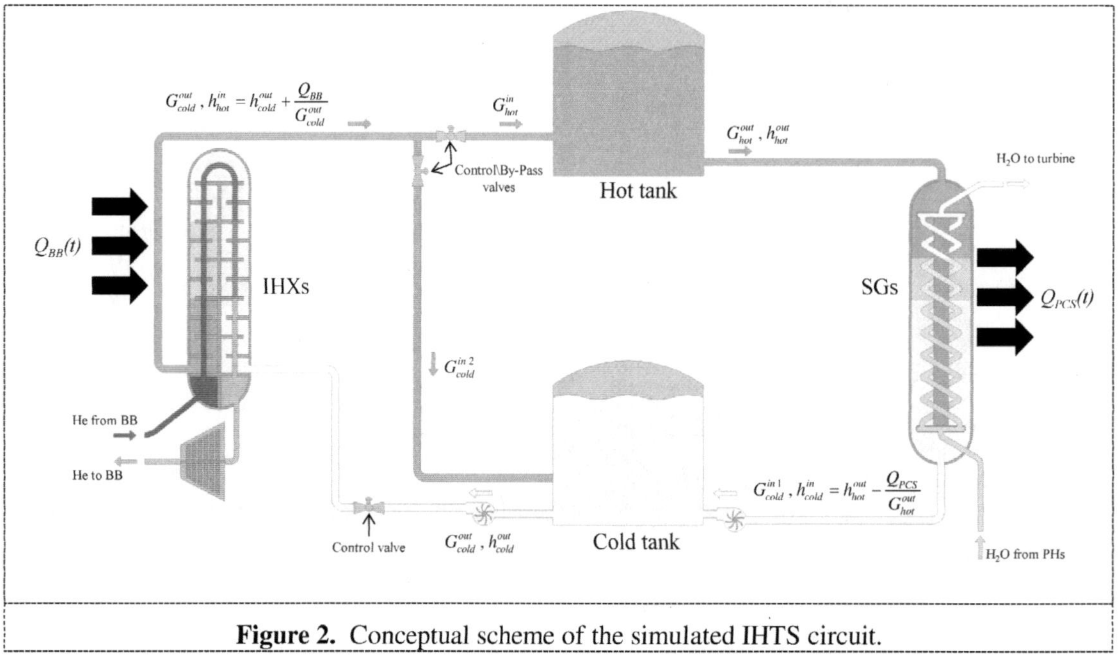

Figure 2. Conceptual scheme of the simulated IHTS circuit.

As can be gathered from the figure 2, it has been assumed that, thanks to huge IHXs heat transfer area, the power generated in the BB is instantaneously transferred to the intermediate circuit even during the pulse/dwell transition, neglecting any potential delays due to the thermal inertia of the PHTS; this assumption relies on the fact that helium thermal inertia is several order of magnitude lower than the total one of the intermediate coolant hence the time to reach its asymptotic conditions is supposed to be quite shorter compared to that of the HITEC. On the PCS side, it has been postulated that this system must always work in (quasi) steady-state condition, therefore the molten salt mass flow rate flowing through the steam generators is kept at nominal value in order to deliver a constant power to the water/steam; such assumption can be considered valid until the hot molten salt temperature does not differ too much from its nominal value allowing the steam generators to operate properly, otherwise it is likely to fall into error. The pulse/dwell transition has been analysed for three different plasma ramp-up/down scenarios: the first is a borderline case where a sudden variation of the power from dwell to pulse is simulated by means of a step function, while the other two, representative of more realistic situations [27][28][29], consider a linear increase of the BB power from dwell to pulse values, lasting 100s and 200s, respectively; the ramp-down is treated in specular way. In all three cases, during the dwell time, the molten salt charging flow (from cold tank to hot tank) is supposed to be ideally regulated in order to the average helium temperature at 400°C, by using control and by-pass valves as well as decreasing the pump speed. These preliminary studies have been performed starting with the minimum amount of HITEC stored into the tanks. Then, a sensitivity analysis has been performed with the aim of evaluating the suitable inventory of molten salt which should be stored to avoid huge variations in the coolant temperatures. Table 6 shows the main input parameter adopted in the developed transients.

Table 6. Transients input data.

Cases	$Q_{BB\ Pulse}/Q_{BB\ Dwell}$ [MW]	Q_{PCS} [MW]	t_{Pulse}/t_{Dwell} [s]	$t_{Ramp\text{-}up}/t_{Ramp\text{-}down}$ [s]	Initial mass stored [t]
Case 1	2566.1/198.4	2092.6	7200/1800	0/0	10309
Case 2	2566.1/198.4	2077.1	7200/1800	100/100	10676
Case 3	2566.1/198.4	2062.4	7200/1800	200/200	11032

3.2.2. Results. The time "0" of each simulation have been properly chosen in order to reach the beginning of the dwell period after 2 hours of the transient. Figure 3 shows the mass flow rate distributions in the main heat transfer equipment of the IHTS circuit, namely the IHXs and the steam generators, and in the by-pass duct which allows the HITEC coming from the IHXs to be recirculated toward the cold tank during dwell and ramp-up/down times. Such operation is necessary to achieve a complete balance of the masses between the two storage tanks. The molten salt mass flow rate in IHXs and steam generators have been calculated to be compliant with the input and output powers of the system. As can be observed in figures 4, 5 and 6, the calculated profiles of temperature in the main IHTS components show very common trends for all three scenarios. However, when dwell time is running out, disagreements can be recognised in the temperatures of hot tank and, as effect, at the steam generators outlet. This can be ascribed to the different mass inventory presents in the hot tank throughout the final phases of the dwell. In particular, it could be noted that the sharper is the pulse/dwell transition, the lower is the quantity of salt remained in hot reservoir at the end of dwell, finding as extreme situation the one related to the Case 1 where the tank gets empty just before the beginning of the pulse and the HITEC temperature drop to the same value reached at the steam generators outlet. With regard to the Case 1, it is evident that the assumption of providing constant power to the PCS becomes very shaky since it is hard to think that steam generators can keep the same heat transfer performances when the temperature of its power source decreases of more than 60°C. Cases 2 and 3 are less affected to temperature reduction because the mass of HITEC which is in the hot tank at the end of dwell – 225 and 443 tons, respectively – provides a thermal inertia sufficient to avoid a dramatic temperature drop maintaining it within 20° C. Nevertheless, operating continuously the steam generators inside this temperature window might intensify the ageing of the components caused by a pronounced thermal fatigue [30].

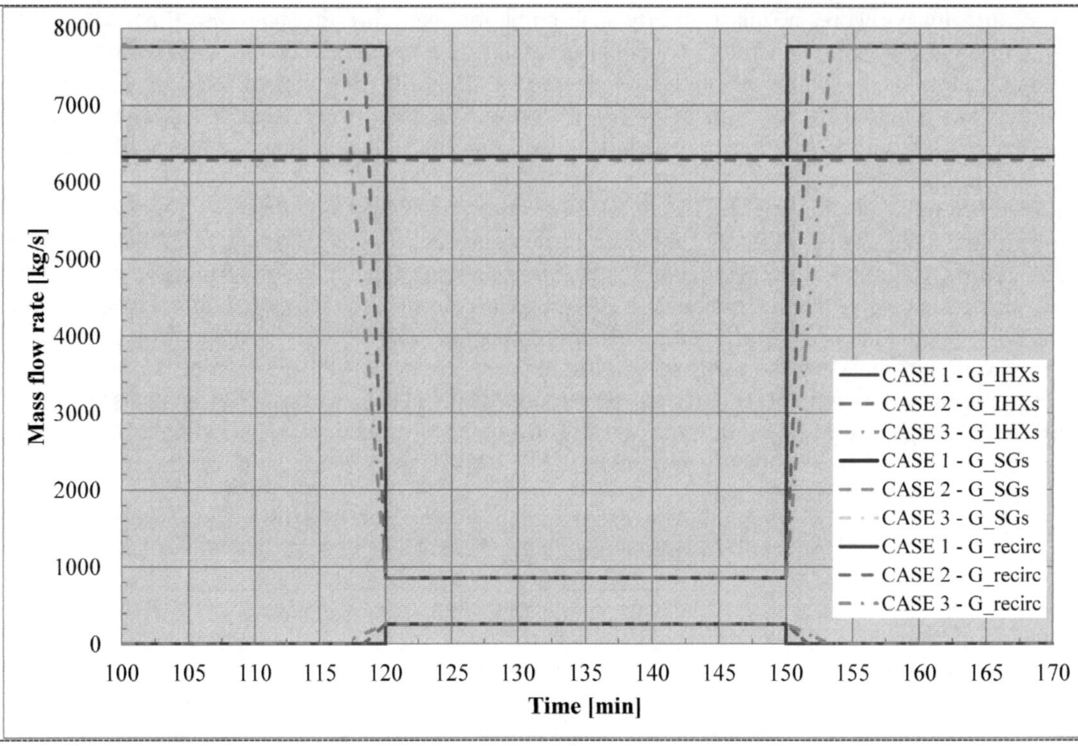

Figure 3. Mass flow rates distribution in IHXs, steam generators and by-pass duct.

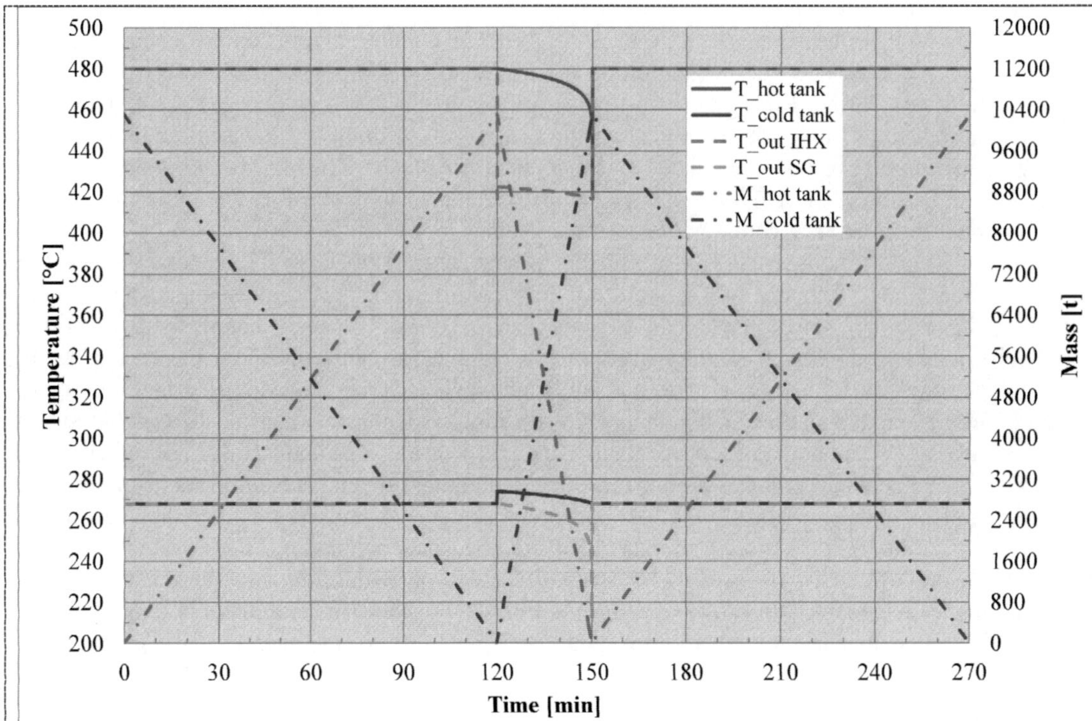

Figure 4. CASE 1 - Main IHTS temperature profiles and mass inventory of hot and cold tanks.

IOP Conf. Series: Journal of Physics: Conf. Series **1224** (2019) 012033 doi:10.1088/1742-6596/1224/1/012033

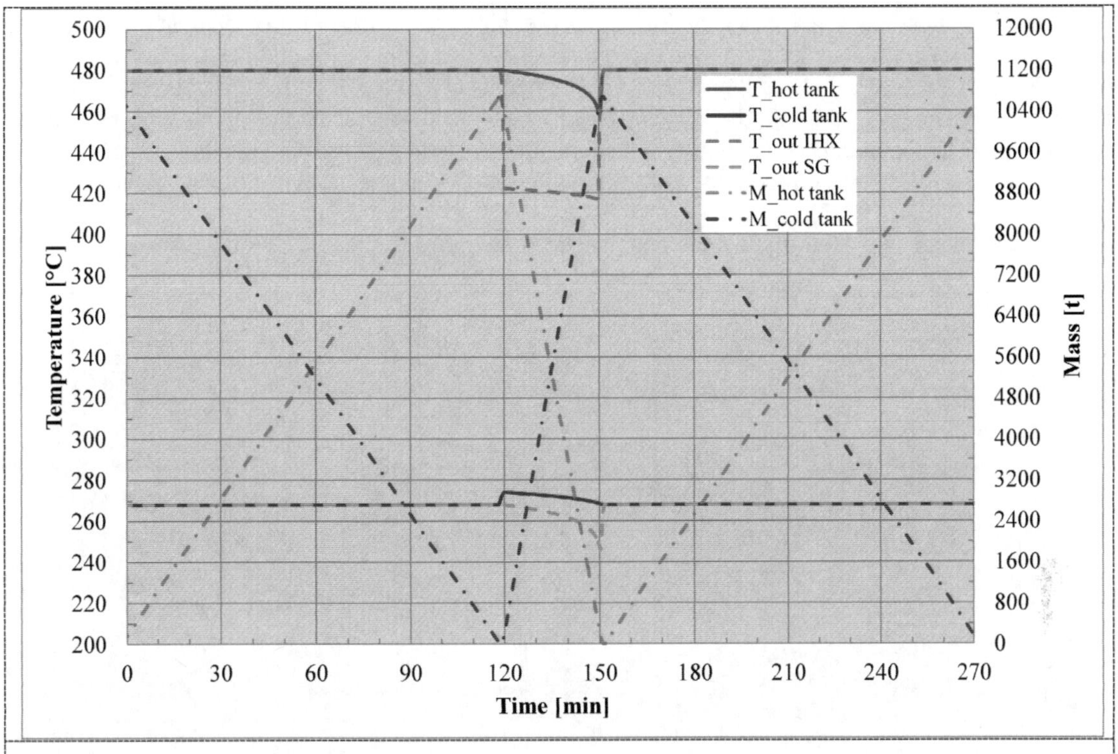

Figure 5. CASE 2 - Main IHTS temperature profiles and mass inventory of hot and cold tanks.

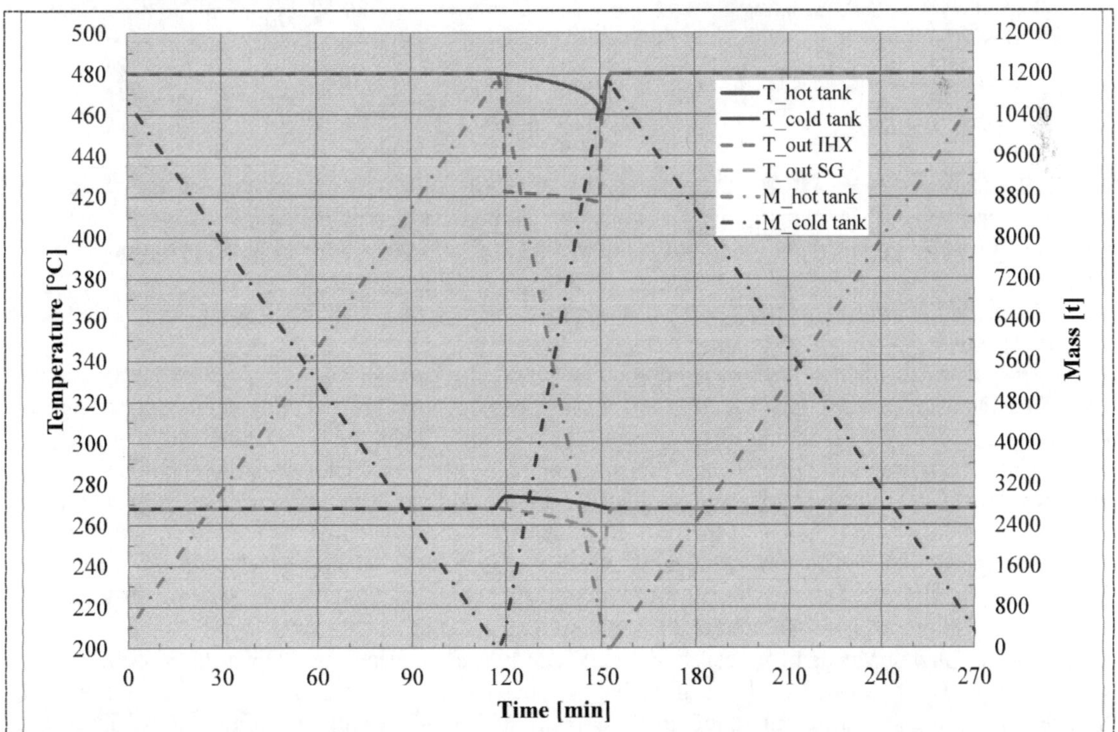

Figure 6. CASE 3 - Main IHTS temperature profiles and mass inventory of hot and cold tanks.

For the abovementioned reasons, starting from the previous cases, further analyses have been performed increasing the quantity of stored molten salt with the aim of finding the amount of HITEC which should be foreseen to keep the temperature variations in a range of 5°C. To obtain this objective it has been finally estimated that the molten salt mass should be augmented of about 75% respect to the minimum hypothetical value. Figure 7 reports the evolution of the coolant bulk temperatures during the pulse/dwell transition inside both hot and cold tank and at the IHXs outlet. Each group of three curves is referred to the Case 1#, Case 2# and Case 3# which differ from the earlier scenarios owing to the initial salt inventory that, as said, has been increased of 75%.

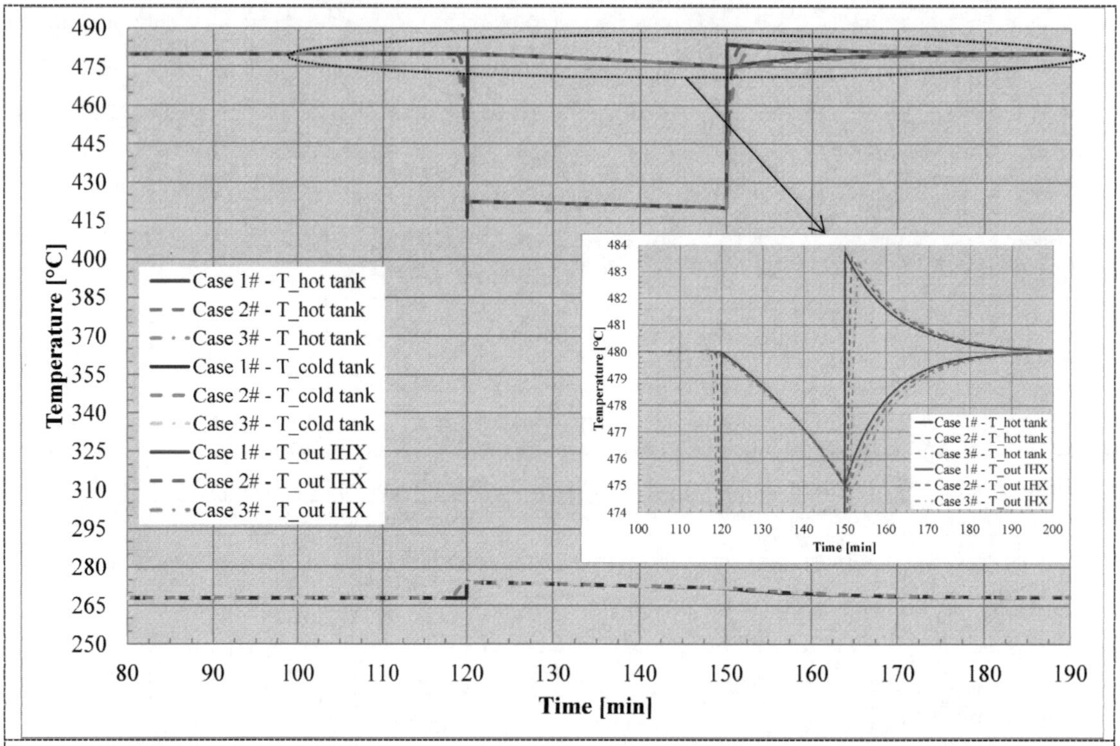

Figure 7. HITEC bulk temperatures in the hot and cold tanks and at the IHXs outlet.

4. Conclusion

A preliminary assessment of the main component sizes of the IHTS and its connected ESS for the HCPB BB concept of the DEMO fusion power plant has been carried out. The conceptual design of the main components belonging these systems have been developed focusing the attention on equipment configurations widely used in nuclear and conventional industrial applications. In particular, the choice of the "tube and shell" architecture for the design of the heat exchangers has been mainly driven by the degree of maturity that such type of layout offers as well as by its relatively simple design which usually shows great flexibility to meet almost any service requirement. The results have highlighted that the overall dimensions of these components are huge however their size and design T/H are still comparable to those which can be commonly found in steam generators of many nuclear plants. Mechanical design and verification are on-going to confirm the viability of the solution. Further refinements of the heat exchangers design are on-going dealt to optimize their heat transfer area by using different tube bundle layouts and/or increasing the temperature difference between coolants.

Thermal dynamic analyses on the IHTS have been also performed to study the global behaviour of the system during pulse/dwell transition and its attitude to smooth the sudden power variations which come from the plasma. A theoretical model based on a lumped parameter approach has been set-up to assess the evolution of both mass inventory and fluid bulk temperature in the hot and cold tanks during the main operation phases of DEMO. According to the hypotheses and the simplification made, results have shown that storing the minimum amount of molten salt could be not sufficient to reach a stable PCS operation since, in all cases analysed, the coolant is experienced by moderate to high reductions of temperature at the steam generators inlet during the final stages of the dwell time. Therefore further parametric studies have been carried out to estimate the mass needed to be stored into the two ESS tanks in order to keep the temperature variations of the HITEC in a range of 5°C. It has been found that, for the given input data, to pursue such goal about 18000 tons of molten salt would be necessary Additional analyses will be performed in the future to study the effects caused by: 1) a reduction of the dwell time from 30 to 10 minutes, 2) different BB PHTS temperature control strategies, 3) different BB PHTS compressor control strategies.

Acknowledgments
This work has been carried out within the framework of the EUROfusion Consortium and has received funding from the Euratom research and training programme 2014-2018 under grant agreement No 633053. The views and opinions expressed herein do not necessarily reflect those of the European Commission.

List of Acronyms

BB	*Breeding Blanket*	**IHX**	*Intermediate Heat Exchanger*
BZ	*Breeding Zone*	**OB**	*Outboard Blanket*
BoP	*Balance of Plant*	**PCS**	*Power Conversion System*
Div	*Divertor*	**PFC**	*Plasma Facing Component*
ESS	*Energy Storage System*	**PHTS**	*Primary Heat Transfer System*
FW	*First Wall*	**T/H**	*Thermal-Hydraulic*
HCPB	*Helium-Cooled Pebble Bed*	**TEMA**	*Tubular Exchanger Manufacturers Association*
IB	*Inboard Blanket*	**VV**	*Vacuum Vessel*
IHTS	*Intermediate Heat Transfer System*	**WCLL**	*Water-Cooled Lithium Led*

References

[1] F. Romanelli et al., Fusion Electricity: A roadmap to the realization of fusion energy, EFDA, 2012
[2] G. Federici et al., Overview of EU DEMO design and R&D activities, Fusion Eng. Des. 89 (2014) 882–889.
[3] S. Ciattaglia et al., The European DEMO fusion reactor: Design status and challenges from balance of plant point of view, Proceedings of 17th IEEE International Conference on Environment and Electrical Engineering and 1st IEEE Industrial and Commercial Power Systems Europe, 2017, 10.1109/EEEIC.2017.7977853
[4] I. Moscato et al., Preliminary design of EU DEMO helium-cooled breeding blanket primary heat transfer system, Fusion Eng. Des. (2018) https://doi.org/10.1016/j.fusengdes.2018.05.058, in press.
[5] L. Barucca et al., Status of EU DEMO Heat Transport and Power Conversion Systems, ISFNT-13 Conference (2018).
[6] L.V Boccaccini et al., Objectives and status of EUROfusion DEMO blanket studies, Fus. Eng. Des. 109–111 (2016) 1199–1206.
[7] G. Federici et al., Overview of the design approach and prioritization of R&D activitiestowards an EU DEMO, Fusion Eng. Des. 109–111 (2016) 1464–1474
[8] G. Federici et al., DEMO design activity in Europe: Progress and updates, Fusion Eng. Des. (2018) https://doi.org/10.1016/j.fusengdes.2018.04.001, in press.

[9] E. Bubelis, W. Hering, S. Perez-Martin, Conceptual designs of PHTS, ESS and PCS for DEMO BoP with helium cooled BB concept, https://doi.org/10.1016/j.fusengdes.2018.02.040, in press.
[10] A.B. Zavoico, Solar power tower design basis document, Sandia National Laboratories Report, SAND2001-2100 (2001)
[11] R. Serrano-López et al., Molten salts database for energy applications, Chemical Engineering and Processing: Process Intensification 73 (2013) 87-102
[12] L.L.C. Coastal Chemical Co., HITEC® Heat Transfer Salt.
[13] H. Petersen, The properties of Helium: Density, Specific Heats, Viscosity, and Thermal Conductivity at Pressures from 1 to 100 bar and from Room Temperature to about 1800 K, Danish Atomic Energy Commission, Risö Report No. 224, 1970.
[14] W. Wagner, H.J. Kretzschmar, International Steam Tables - Properties of Water and Steam Based on the Industrial Formulation IAPWS-IF97, Springer, Berlin, Heidelberg, 2008
[15] TEMA, Standards of the Tubular Exchanger Manufacturers Association (9th Edition), Tubular Exchanger Manufacturers Association Inc., 2007.
[16] G.F. Hewitt, Heat Exchanger Design Handbook, Hemisphere Publishing Corp., ISBN 0891161252, 1983
[17] V. Gnielinski, New equations for heat and mass-transfer in turbulent pipe and channel flow, Int. Chem. Eng. 16 (1976) 359–368.
[18] Kakac, S., R. K. Shah, and W. Aung: Handbook of Single-Phase Convective Heat Transfer, Wiley (1987), New York.
[19] H. Fenech, Heat Transfer and Fluid Flow in Nuclear Systems, Pergamon Press, eBook ISBN: 9781483150789, 1981.
[20] E. M. Smith, Advances in Thermal Design of Heat Exchangers – A Numerical Approach: Direct-sizing, step-wise rating and transients, John Wiley & Sons; 2005.
[21] J. Gou at al., An assessment of heat transfer models of water flow in helically coiled tubes based on selected experimental datasets, Ann. Nucl. Energy 110 (2017) 648–667
[22] K.W. Hwang et al., Experimental study of flow boiling heat transfer and dryout characteristics at low mass flux in helically-coiled tubes, Nucl. Eng. Des. 273 (2014) 529–541.
[23] International Atomic Energy Agency, Advances in Nuclear Power Process Heat Applications, IAEA-TECDOC-1682, IAEA, Vienna, (2012)
[24] International Atomic Energy Agency, Gas-cooled reactor coolant circulator and blower technology, Proceedings of a specialists meeting, IAEA, Vienna 1988
[25] C. F. McDonald, Helium turbomachinery operating experience from gas turbine power plants and test facilities, Applied Thermal Engineering 44 (2012) 108-142
[26] MATLAB and Mathematical Toolbox Release 2017b, The MathWorks, Inc., Natick, Massachusetts, United States
[27] P. Vincenzi et al., EU DEMO transient phases: main constraints and heating mix studies for ramp-up and ramp-down, Fusion Eng. Des. 123 (2017) 473–476
[28] C. Bustreo et al., Economic assessment of different operational reactor cycle structures in a pulsed DEMO-like power plant, Fusion Eng. Des. 124 (2017) 1219–1222
[29] J. Morris, M. Kovari, Time-dependent power requirements for pulsed fusion reactors in systems codes, Fusion Eng. Des. 124 (2017) 1203–1206
[30] International Atomic Energy Agency, Assessment and Management of Ageing of Major NPP Components Important to Safety, Steam Generators, IAEA-TECDOC-1668, IAEA, Vienna, (2011)

IOP Conf. Series: Journal of Physics: Conf. Series **1224** (2019) 012034 doi:10.1088/1742-6596/1224/1/012034

Use of multispectral and thermal imagery in precision viticulture

G Tanda[1] and V.Chiarabini[2]
[1]DIME, Università degli Studi di Genova, Genova, Italy
[2]KIM-RemoteSensing GmbH, Klagenfurt am Wörthersee, Austria

[1]Corresponding Author. E-mail: giovanni.tanda@unige.it

Abstract. The increasing demand for higher quality and yield of wine production has led to a growing interest in precision viticulture, i.e., practices of monitoring and managing spatial variations in variables related to productivity within a vineyard. This paper presents a few applications of optical measurements, in combination with monitoring systems making use of geolocation and remote/proximal sensing, to calculate vegetation indices related to plant vigour and water stress in vineyards. Measurements were performed on vineyards in Burgenland, Austria, by both aerial and proximal (terrestrial) sensing techniques. A remote-sensing, four-band multispectral sensor, placed on an Unmanned Aerial Vehicle (UAV), has been used to detect the spectral signature of the vineyard and to calculate the NDVI index, useful to selectively address the harvest on the basis of quality and quantity of grapes. Proximal, thermal infrared imaging complemented the investigation providing information about the water status of the vegetation through the CWSI index. Examples of vigour maps are provided, showing, inside a given parcel, the presence of canopies at different level of vegetation characteristics. Results provide a range of information useful to make the optimal choice in management strategies of vineyards.

1. Introduction

Precision viticulture aims at optimizing vineyard performance, in particular by maximizing grape yield and quality while minimizing costs, environmental impacts and risk. This is accomplished by measuring local variation in factors that influence grape yield and quality, and by applying appropriate viticulture management practices (e.g. timing of harvest, irrigation, etc.). Nowadays, remote and proximal sensing sensors efficiently allow to detect the vineyard status, that is, plant health, water and nutrient availability, or soil conditions.

Techniques for investigating the physiological status of vegetation include the detection of reflectance of light (solar radiation) in the visible and near infrared wavelengths and the use of infrared thermography. The spectral signature of vegetation is distinctive and characterized by a low reflectance in the visible spectral region (400–700 nm), especially in blue or red wavelengths, because of strong absorption by chlorophylls, while exhibits a relatively high reflectance in the near-infrared band (700–1300 nm), to which the human eye is insensitive, because of internal leaf scattering and no absorption [1]. The sharp contrast between reflectance in the near-infrared and visible bands has been exploited in the construction of reflectance spectral indices. In particular, the most widely used indicator of plant vigour is the normalized difference vegetation index (NDVI), related to the difference between the light reflected in the near infrared and in the red regions; healthy vegetation is characterized by large NDVI values, whereas low NDVI values correspond to stressed vegetation [2]. In the context of precision viticulture, the alternation of vegetation and soil (with the need to separate the grapevine canopy from other image features) and the spatial variability in terms of canopy characteristics within the vineyard require the monitoring of large areas with relatively high spatial accuracy; unmanned aerial vehicle (UAV) applications in remote sensing combine the requirements of gathering multispectral data at high spatial ground resolution and monitoring vineyards of small to medium size (1 to 10 hectares) in a relatively short time [3].

Infrared thermography is a diagnostic tool useful to record the surface temperature of canopy on the basis of aerial and/or terrestrial monitoring systems. The use of thermal infrared imaging of vegetation dates back to 60s (e.g., [4]). The crop water stress index (CWSI), introduced in [5] and calculated from a combination of the canopy surface temperature and environmental factors (such as net radiation, vapor pressure deficit, wind speed), can be efficiently used to map the spatial variability in water deficits. The recent development of portable thermal imagers and remote sensing technology has greatly extended the opportunities to analyse thermal properties of plant canopies and widened the available information on the growth and condition of plants.

2. Analysis

2.1 Reflectance: definitions and spectral signature of vegetation

The angular reflectance pattern of vegetation canopies is of particular interest for diagnosing plant physiological status. Several definitions of reflectance can be found in the literature (e.g., [6]) based on the feature of incident and reflected radiation. Figure 1 provides some pictorial examples of different reflectance quantities ranging from the bidirectional condition (a single direction for the incident and reflected radiances) to the bihemispherical condition (both illumination and reflection are integrated over the hemisphere). As outlined in [7], in remote sensing the most common measurement condition on field is the hemispherical-conical configuration. This definition assumes the surface incident radiance consisting of a mixture of solar direct and diffuse components creating an illumination field that is hemispherical in angular extent. On the other hand, the field-of-view of the measuring instrument implies a reflected radiance acquired within a finite solid angle.

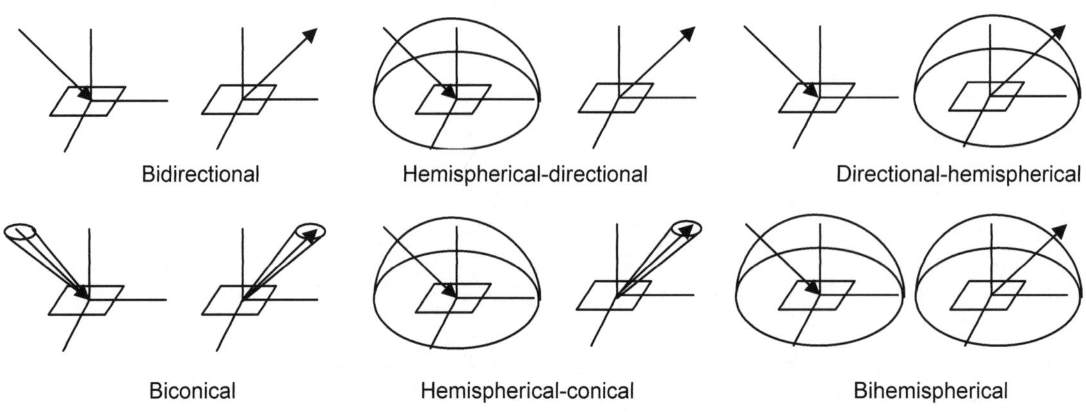

Figure 1. Reflectance definitions.

The reflectance variation with wavelength of vegetation, in the visible and near infrared ranges, can be associated with the plant physiological response to growth conditions and adaptations to the environment. As shown in Fig.2, in the visible spectrum (400–700 nm), plant reflectance is low because of absorption by photosynthetic pigments (mainly chlorophylls and carotenoids). In the near-infrared region (700–1300 nm), the structural discontinuities encountered in the leaf are responsible for high reflectance values both at plant and canopy levels. The red-edge (680–750 nm) is the wavelength of maximum slope in the increase of reflectance from red to near-infrared, while the middle infrared region (1300–3000 nm) presents a reflectance profile affected by the absorption characteristics of water and other compounds.

Of particular relevance is the difference between the high reflectance of near infrared radiation and the low reflectance (high absorption by chlorophyll) of red radiation, since it can be related to the vegetative status, as outlined in the figure.

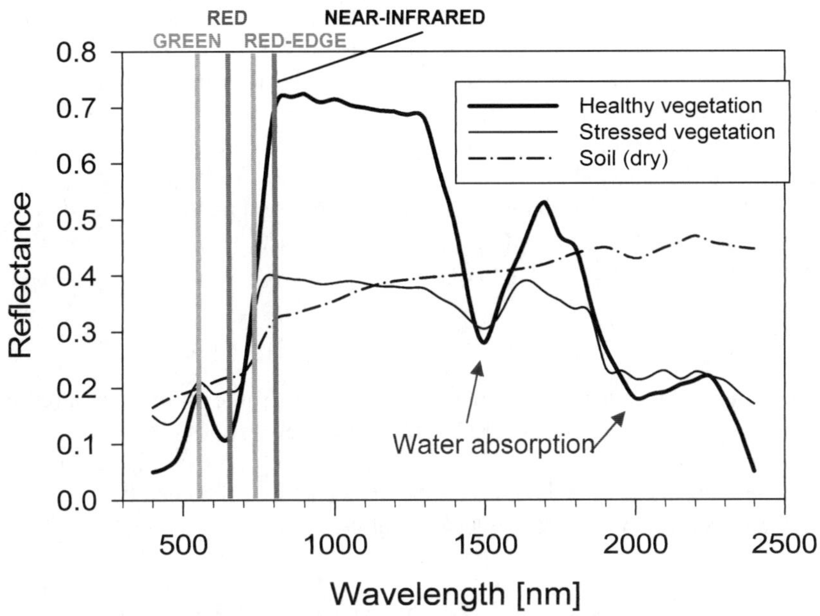

Figure 2. Typical spectral signatures (from visible to middle infrared) of vegetation and soil.

2.2 Reflectance-based vegetation indices

The accurate description of the spectral signature of target requires the use of sensors, termed hyperspectral, able to record the reflectance in a large number of wavebands, typically greater than 10. Multispectral instruments gather information in a small number of wavebands (from 2 to 10). Despite this important distinction, it is worth noting that the most widely used indicator of plant vigour, the normalized difference vegetation index (NDVI), utilises only two reflectance values, taken at the same time instant and for the same target area, according to the relation:

$$NDVI = (\rho_{NIR} - \rho_R) / (\rho_{NIR} + \rho_R) \quad (1)$$

where ρ is the spectral reflectance of the target and the subscripts *NIR* and *R* denote the near-infrared (typically 790−890 nm) and the red (typically 610−680 nm) wavelengths, respectively (see, for instance, [8], [9]). The NDVI is a number between -1 and +1 and quantifies the relative difference between the near infrared reflectance 'peak' and the red reflectance 'trough' in the spectral signature [2]. For highly vegetated targets, the vigour level is high and the NDVI value is close to unity, while for non-vegetated target the vigour level is low and the NDVI value is close to zero (negative values rarely occur in natural targets). Information regarding the spatial variations of the vigour level has many applications for improving the management of the vineyards: more vigorous grapevines produce a larger yield than stressed ones but the vine quality is lower. Therefore, NDVI maps within the vineyard can suggest a selective grape harvest, dividing parcels into zones of different grape quality and processing the grapes separately [10], with significant benefits for winegrowers since the assessment of the vineyard productivity with traditional methods (i.e. measuring basic fruit quality and yield parameters) is generally more expensive and time-consuming than remote sensing [2].

2.3 Thermal imaging and water stress indices

An important factor influencing both quality and quantity of wine grapes is water: excessive water leads to increased vegetative growth and yield, but grape quality parameters, such as sugar content, pigment formation, acidity, and wood maturation of the wine are negatively affected. Conversely, severe water stress induces stomatal closure, causing strongly reduced or no assimilative activity and shoot growth with negative effects on grape yield. Maintaining a slight-to-moderate water deficit and thus inducing a certain level of stress can be very beneficial in grapevine cultivation as it stimulates optimal quality parameters without significantly compromising yield. In order to achieve this, crop water status must be measured accurately and reliably with the aim of maintaining a predetermined level of mild stress [11]. Reflectance in the middle infrared region presents a shape mainly linked to the absorption characteristics of water and other compounds and can be indicative of the plant water status. At the same time, indices based on the surface temperature of foliage or canopy are useful to evaluate the accuracy of water stress detection based on aerial and/or terrestrial infrared thermography. The crop water stress index (CWSI) indicator, used to map the spatial variability in water deficits, is deduced from thermal imaging according to the relation:

$$\mathrm{CWSI} = (T_s - T_{wet}) \,/\, (T_{dry} - T_{wet}) \qquad (2)$$

where T is the surface temperature and subscripts s, dry and wet denote values measured by infrared thermography, upper and lower values, respectively. More precisely, T_{dry} is the surface temperature of a fully-stressed, non-transpiring plant, whereas T_{wet} is the surface temperature of a well-watered, transpiring plant. Upper and lower bounds of T_s can be theoretically calculated by the energy balance for a completely dry leaf and for a well-watered leaf, as follows

$$T_{dry} = T_{air} + (r_{HR}\, R_{ni}) \,/\, (\delta_{air}\, C_{p,air}) \qquad (3)$$

$$T_{wet} = T_{air} + (r_{HR}\, r_{aW}\, R_{ni}\gamma\,) \,/\, [(\delta_{air}\, C_{p,air})\, (r_{aW}\,\gamma + r_{HR}\,\Delta)] \;-\, (r_{HR}\,\mathrm{VPD}) \,/\, (r_{aW}\,\gamma + r_{HR}\,\Delta) \qquad (4)$$

where T_{air}, δ_{air}, and $C_{p,air}$ are the temperature (°C), density (kg m^{-3}) and specific heat (J $kg^{-1}K^{-1}$) of air, respectively; R_{ni} is the net radiation (W m^{-2}), γ is the psychrometric constant (Pa K^{-1}), VPD is the vapour pressure deficit (Pa), Δ is the slope of the curve relating saturation vapour pressure to temperature (Pa K^{-1}), while r_{HR} (s m^{-1}) and r_{aW} (s m^{-1}) are parameters related to the canopy heat transfer and water vapour resistances, respectively (see, for instance, [5], [11],[12]). Upper and lower bounds of surface temperature can either be evaluated as a function of environmental factors such as net radiation, vapor pressure deficit, and wind speed (affecting both the heat transfer and water vapour resistance) measured on field or be deduced by direct measurements over reference, dry and wet surfaces, either natural or artificial ([13]). For a well-watered crop, CWSI is typically lower than 0.2, while values of 0.8–0.9 indicate water stress conditions.

The usefulness of CWSI to assess the relative water stress is still debated. CWSI requires the input of several parameters and is strongly related to the quality of thermal images; moreover, CWSI typically exhibits diurnal changes (even without irrigation or rain) and its use in irrigation scheduling schemes should be based on daily measurements taken always at the same time of the day (for, instance, at noon, as indicated in [14]). Practical aspects of thermal imaging of vegetation are discussed in [15] and include the view angle (side view, as occurs in terrestrial inspections or top view, as occurs in aerial applications) and the sun light conditions (direct radiation, especially in the central hours of the day, increases the sensitivity whereas diffuse radiation reduces the spread of temperature between shaded and sunlit leaves). Moreover, thermal images can be taken at leaf level (or groups of leaves), as occurs in the proximal view at short distance or at canopy level, as occurs in the aerial view from large distance. The sensor quality affects accuracy and resolution of measured data; a given type of camera may provide accurate data over the foliage of a single plant or for rows of a vineyard during a proximal inspection and coarse or misleading information in remote sensing application if the resolution of the sensor is not adequate (with the single pixel incorporating fraction of soil and canopy).

2.4 Monitoring technologies

Precision viticulture has received a significant boost from remote sensing technological development. Satellite, aircrafts and more recently UAV systems have been used to record, at a distance with different scales of resolution, data over a large area in a relatively short time [3]. In particular, UAVs, remotely controlled from the ground, permit to gather synchronized and georeferenced visible, multispectral and thermal infrared images, with reduced planning time and an adequate ground spatial resolution (typically 8-10 cm/pixel for reflectance and 15-20 cm/pixel for thermal measurements at a distance of 100m). Proximal sensing is the most widely used monitoring technique for thermal imaging to inspect grapevines at leaf level; at the same time proximal monitoring is not suitable to infer the spatial variability of the vegetative status unless a very large number of measurements are taken.

3. Applications

Multispectral and thermal measurements were performed on vineyards (total monitored area about 50 hectares) in Burgenland, Austria, during two consecutive days of August 2017, a couple of weeks before the harvest, making use of both aerial and proximal (terrestrial) sensing techniques. Experiments have been performed from 10 a.m. to 4 p.m., in the presence of a clear sky. Air temperature and relative humidity were continuously recorded on ground by a portable weather station, while further meteorological data (hourly values of barometric pressure, solar irradiation and wind velocity) were obtained from a nearby automatic meteorological station. No events of rain or irrigation occurred on the last four days before and during the present investigation. The two-day period of investigation was sufficient to infer significant information concerning the plant vigour due to the long term response of multispectral measurements; conversely, water stress data based on thermal images can show daily or even hourly variations and thus are more useful in the short term.

3.1 Multispectral measurements

A remote-sensing, four-band multispectral sensor, placed on an Unmanned Aerial Vehicle (UAV), has been used to detect the spectral signature of the vineyard, at a distance of 70 m from ground. UAV (shown in Fig.3) was equipped with a multispectral sensor (Parrot Sequoia) collecting data of reflected and incident light in four spectral bands (green 550 nm, red 660 nm, red-edge 735nm, and near-infrared 790nm) at a ground spatial resolution of 9 cm/pixel. Multispectral inspection was integrated with a Zenmuse X5S digital camera to capture visible imagery. An in-built geographic positioning system (GPS) and a photogrammetric software were used to create digital orthophotographs and spatial maps of reflectance data within the target area.

Figure 3. UAV equipped with multispectral sensor, GPS device and digital camera.

Spectral reflectance ρ_λ of the target surface area has been reconstructed by the ratio of reflected and incident radiances ($R_{r,\lambda}$ and $R_{i,\lambda}$, respectively) measured at the same time instant and corrected to take into account the different solid angles subtended by the two sensors, i.e., $\rho_\lambda = k \cdot R_{r,\lambda} / R_{i,\lambda}$. The correction factor k has been determined by using a calibration sample (a 25cm x 25cm Teflon plate placed on field) of known (and high) reflectance. An example of spectral reflectance over isolated spots (two canopy surface regions at different vigour and the soil) is provided in Fig.4 (left-hand side). The right-hand side of the figure, where the selected spots are indicated, is a visible, aerial image clearly showing the typical heterogeneity of a vineyard, with sequences of narrow vine rows and poorly vegetated background soil between them. For this reason, aerial inspection requires the best spatial resolution to differentiate the canopy from the soil (whose spectral signature is completely different) and to correctly identify the response of each canopy and the great variability in the vegetative status that typically characterizes the vineyards.

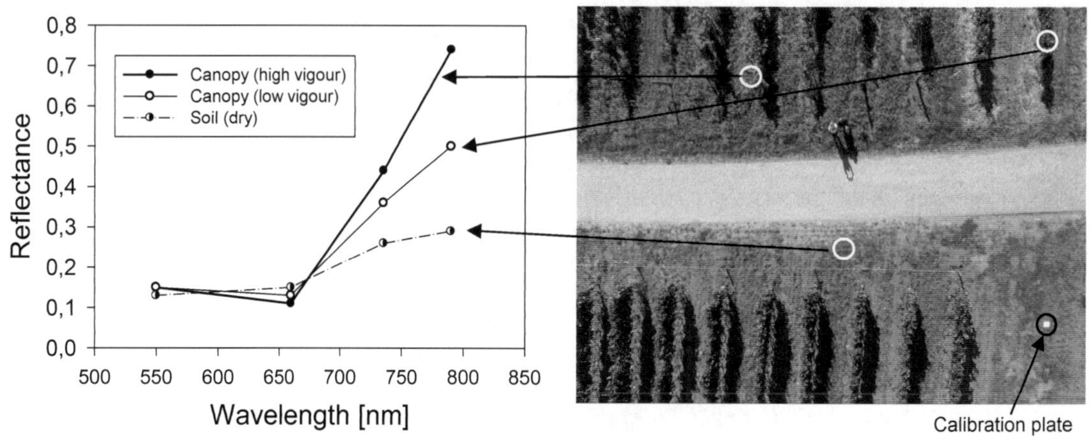

Figure 4. Reflectance measurements over isolated spots encompassing vegetation at different vigour level and soil.

A conventional representation of multispectral data condensed in a single image is the CIR (Colour Infrared Imaging) imagery, where false colours are used to highlight, for each pixel, the dominant reflection wavelength. According to the conventional code colours, the red colour indicates a strong near-infrared reflection and the red variation (from bright to light) indicates the vegetative status (from vigourously growing or healthy to mature or stressed). CIR imaging conventionally uses the green colour for the reflected red wavelength and the blue colour for the reflected green wavelength. As a consequence, regions encompassing soil, sand, dead vegetation and building or manmade materials will appear as white, blue, green or black. Figure 5 shows the standard and the CIR images of a parcel (3.5 hectares) within the vineyard: the corresponding orthophotograph (the inspected field as it would appear to human eyes) is shown of the left, while the false colour image on the right (CIR image) displays the different levels of vegetation health based on the variations in red.

A detail of Fig.5 is extracted and shown in Fig.6 (left-hand side), where the canopies of adjacent grape rows are clearly distinguishable from each other. The processing of spectral reflectance data allows the reconstruction of the vigour maps, according to the definition of NDVI provided by Eq.(1). The right-hand side of Fig.6 reports the NDVI map (according to a colour scale) of the target surface shown in the left-hand side of the same figure. In order to convey a useful indication of grapevine rows, the values for the inter-row space (which would otherwise confound grapevine measurement) have been eliminated. To improve the readability of NDVI maps, the final processing step consists in applying algorithms to produce maps of the area-averaged NDVI value (with different levels of resolution) based only on individual values recorded for the grapevine rows. Figure 7 shows the area-averaged vigour maps for a portion of the vineyard parcel of Fig.5 displayed according to different levels of resolution; grapes included in the red area (at relatively low level of vigour) appear to be mature and likely to produce a wine with an overall higher quality than that potentially achieved from zones at greater vigour (yellow and green areas).

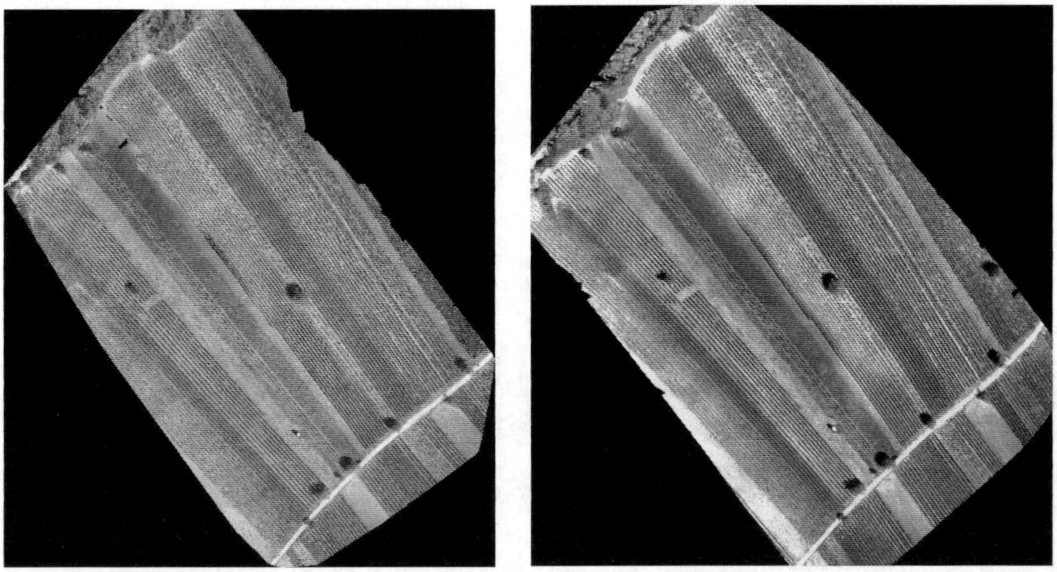

Figure 5. Orthophotograph (left-hand side) and false colour, CIR image (right-hand side) of a vineyard parcel.

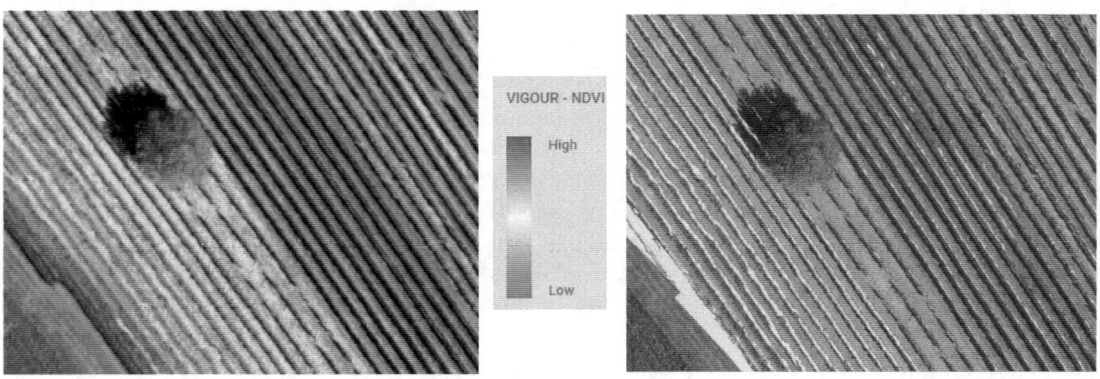

Figure 6. High-resolution CIR image of a reduced target area (left-hand side) and the corresponding NDVI map limited to the grape canopies (right-hand side).

3.2 Thermal measurements

To assess canopy temperature of the grapevine and derive the water stress status, lateral infrared thermal images (proximal or terrestrial view) were obtained using a handheld infrared camera (Model FLIR T335, Flir Systems Inc., 320 x 240 pixels, thermal sensitivity/NETD < 0.05°C), with the emissivity of vegetation fixed at 0.98.

The first example of thermal image is reported in Fig.8 for a grapevine row monitored at leaf level (at about 2m-distance). The figure shows, on the left-hand side, a mosaic image giving the standard visible view with the thermal map overlapped. The thermal map is displayed as CWSI map on the right-hand side. The picture was taken at 9:30 a.m., with T_{air}=23°C and calculated T_{dry} and T_{wet} equal to 24.7 and 19.8°C, respectively. CWSI values over the foliage are mainly in the 0.2–0.4 range,

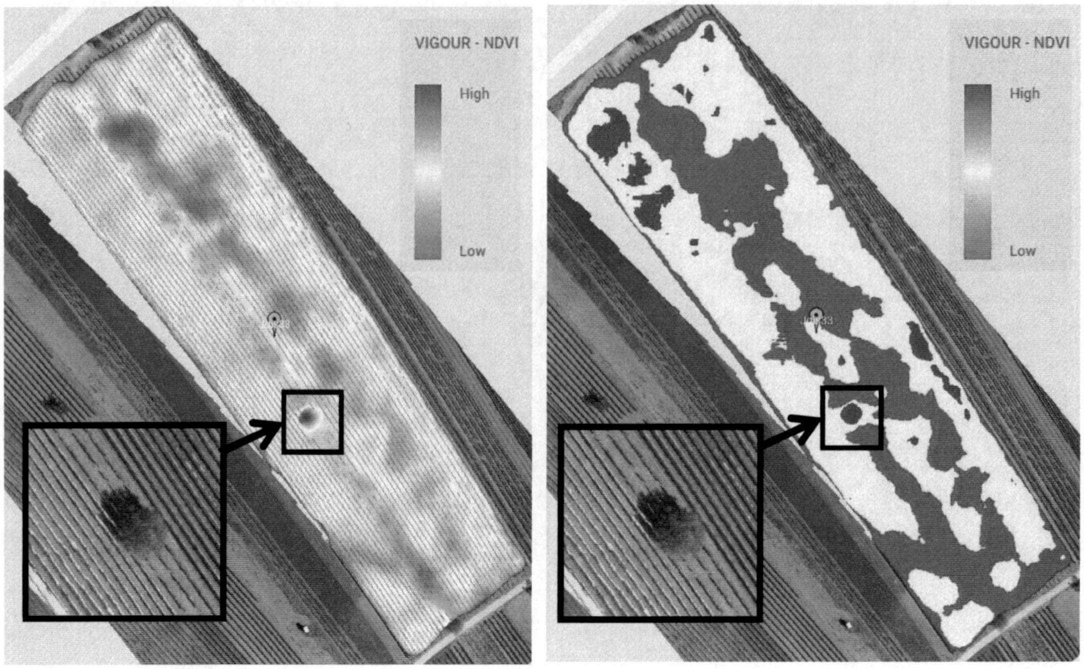

Figure 7. Maps of area-averaged NDVI values for the same vineyard parcel, processed according to different resolutions.

indicating a low to moderate degree of water stress. A different scenario is shown in Fig.9, which refers to an adjacent parcel of vineyard encompassing a larger target area. The thermal image, processed to obtain the CWSI map and overlapped to the visible image, was taken at 2:30 p.m., with T_{air}=30.5°C and calculated T_{dry} and T_{wet} equal to 35.3 and 24.9°C, respectively. The foliage of grapevine rows appears to be water stressed as CWSI is systematically higher than 0.7. This result was not surprising since the month of August 2017 was exceptionally dry and hot (with harvest anticipated by several weeks) and the vineyard was not equipped with irrigation system. Data presented in Figs. 8 and 9 show typical variations of CWSI over time during the day, with peak values during the warmer hours of the day and lower values in the first hours of the morning, when reduced sunlight allows a larger stomatal aperture and transpiration rate of foliage [14].

A final example of thermal imagery is provided in Fig.10, where a visible/thermal mosaic image is shown on the left, with the corresponding CWSI map reported on the right. This image, taken at 11:30 a.m., with T_{air}=25.5°C, T_{dry} =27.7 °C, and T_{wet} = 21.3°C, includes the lateral view of two adjacent grapevine rows and a cluster of rows displaced on a hill at about 100 m of distance (thus simulating an aerial thermal view). Again, CWSI values are relatively high but heterogeneously distributed over the two facing grapevine rows. This is due to the fact that upper and lower values of surface temperature were calculated for sunlit conditions; thus they correctly apply to the right-hand grapevine row (predominately in the sun) and not to the left-hand row (predominately in the shade). CWSI values obtained for the grapevine rows on the background are affected by the limited resolution of infrared camera unable to discriminate canopy values from inter-row values.

Figures 8−10 are not directly comparable since they refer to different hours of the day, to different scale levels (leaves, rows, rows and far canopy, respectively) and different vineyard parcels; even though they are deemed to be poorly correlated to the vigour maps previously shown, they provide some interesting and complementary indications about the water status of vegetation.

IOP Conf. Series: Journal of Physics: Conf. Series **1224** (2019) 012034 doi:10.1088/1742-6596/1224/1/012034

Figure 8. Mosaic image (visible + thermal) of a grapevine row at leaf level (left-hand side) and the corresponding CWSI map (right-hand side).

Figure 9. CWSI calculated for grape rows at medium-high water stress level.

Figure 10. Mosaic image (visible + thermal) of a vineyard parcel (left-hand side) and calculated CWSI (right-hand side).

4. Conclusions

Optical measurements and derived indices used in precision viticulture have been described and commented, together with monitoring (remote and proximal sensing) technologies. Among the wide range of techniques and data processing under investigation worldwide, attention was paid to multispectral (in the visible and near-infrared bands) and thermal imagery. Results obtained for a vineyard in Burgenland, Austria, two weeks before harvest, have been presented and discussed. Multispectral images taken from an unmanned aerial vehicle and the derived maps of vine vigour (NDVI) were capable to discriminate grape yield/quality within single vineyard blocks; for this purpose, high-resolution optical data have to be processed in order to accurately differentiate between individual grapevine rows and adjacent or inter-row soil and vegetation. At the same time, thermal imagery provided complementary information for diagnosis and quantification of plant water stress.

References

[1] Knipling EB 1970 Physical and physiological basis for the reflectance of visible and near-infrared radiation from vegetation *Remote Sensing of Environment* **1** 155−159

[2] Hall A, Lamb DW, Holzapfel B and Louis J 2002 Optical remote sensing applications in viticulture – a review *Australian Journal of Grape and Wine Research* **8** 36−47

[3] Matese A and Di Gennaro SF 2015 Technology in precision viticulture: a state of the art review *International Journal of Wine Research* **7** 69−81

[4] Fuchs M and Tanner CB 1966 Infrared thermometry of vegetation *Agronomy Journal* **58** 597−601

[5] Jackson RD, Idso SB, Reginato RJ and Pinter Jr PJ 1981 Canopy temperature as a crop water stress indicator *Water Resources Research* **17** 1133−1138

[6] Siegel R and Howell JR 1972 *Thermal Radiation Heat Transfer* (McGraw-Hill)

[7] Schaepman-Strub G, Schaepman ME, Painter TH, Dangel S and Martonchik JV 2006 Reflectance quantities in optical remote sensing – definitions and case studies *Remote Sensing of Environment* **103** 27−42

[8] Rodríguez-Pérez JR, Riaño D, Carlisle E, Ustin S and Smart DR 2007 Evaluation of hyperspectral reflectance indexes to detect grapevine water status in vineyards *American Journal for Enology and Viticulture* **58** 302−317

[9] Jones HG 2014 *Plants and Microclimate – A Quantitative Approach to Environmental Plant Physiology, 3rd Edition* (Cambridge University Press)

[10] Arnó J, Martínez-Casasnovas JA, Ribes-Dasi M and Rosell JR 2009 Review. Precision Viticulture. Research topics, challenges and opportunities in site-specific vineyard management *Spanish Journal of Agricultural Research* **7** 779−790

[11] Möller M, Alchanatis V, Cohen Y, Meron M, Tsipris J, Naor A, Ostrovsky V, Sprintsin M and Cohen S 2007 Use of thermal and visible imagery for estimating crop water status of irrigated grapevine *Journal of Experimental Botany* **58** 827−838

[12] Jones HG 1999 Use of infrared thermometry for estimation of stomatal conductance as a possible aid to irrigation scheduling *Agricultural and Forest Meteorology* **95** 139−149

[13] Testi L, Goldhamer DA, Iniesta F and Salinas M 2008 Crop water stress index is a sensitive water stress indicator in pistacchio trees *Irrigation Science* **26** 395–405

[14] Bellvert J, Zarco-Tejada PJ, Girona J and Fereres E 2014 Mapping crop water stress index in a ‘Pinot-noir’ vineyard: comparing ground measurements with thermal remote sensing imagery from an unmanned aerial vehicle *Precision Agriculture* **15** 361−376

[15] Jones HG and Leinonen I 2003 Thermal imaging for the study of plant water relations *Journal of Agricultural Meteorology* **3** 205−217

The ultrasounds as a mean for the enhancement of heat exchanger performances: an analysis of the available data

Alessandro Franco and Carlo Bartoli

Department of Energy, Systems, Territory and Constructions Engineering (DESTEC), University of Pisa, Largo Lucio Lazzarino, 56126 PISA (ITALY).

Corresponding author e-mail: alessandro.franco@unipi.it

Abstract. The aim of this paper is to investigate the potential use of ultrasounds for improving the performances of heat exchangers. In a heat exchanger, the effect of ultrasonic vibrations is important both for heat transfer intensification and for the possibility of obtaining fouling reduction: the cumulative effect can determine an interesting advantage in term of overall heat transfer coefficient increase. After a general analysis of the topic, the paper analyse the results recently obtained in the two main fields of analysis, heat transfer coefficient enhancement and fouling reduction. The perspective of the use of acoustic fields and ultrasound for optimization of heat exchangers operation is some particular fields of application are discussed.

1. Introduction

The positive effect of ultrasonic vibrations on heat transfer enhancement was reported by several authors; the topic has been recently reviewed in [1]. Free convection [2], boiling [3,4] and also melting [5] were deeply studied and the consequences of ultrasound upon these phenomena are known.

At first, the use of ultrasonic vibration concerns only convection heat transfer; but results obtained did not seem promising enough. So, the use of ultrasound was then extended to phase change phenomena. Boiling heat transfer is another important research field, where ultrasonic vibrations promote bubbles formation and therefore delay the onset of the critical heat flux. The influence of ultrasound on heat transfer has been experimentally investigated with reference to particular specific conditions and geometries: the collected results are reported in some papers, like [6–8]. One of the authors of the paper was involved during past years in the field of intensification of heat transfer in boiling and subcooled boiling using refrigerant fluids by means of ultrasounds too, [9-10]. The positive effect of ultrasound can be remarked mainly for fluids operating in subcooled conditions, while ultrasounds seem to lose their influence in saturated boiling conditions.

A possible practical application of the enhancement of heat transfer using ultrasounds is observed in the cooling of electronic components. Even if the physical problem is investigated, less attention seems to be connected with practical application and in particular to heat exchangers. The effect of ultrasonic vibrations on natural convection or phase-change heat transfer has been considered in many fundamental studies but till some years ago, only a few publications analyzed the influence of ultrasound on heat exchanger performances [11, 12].

Recently, investigations began to head toward fouling reduction for heat exchangers. An undeniable advantage of ultrasound is its cumulative effects leading to several improvements on a single system. Heat exchangers are perfectly suited examples to illustrate this doubly favourable factor, with both heat transfer enhancement and cleaning possibilities [13]. An important side-effect of ultrasound to heat exchangers is surface cleaning which essentially thanks to acoustic cavitation. This may be a promising solution to reduce the fouling process in heat exchangers. Some authors have studied specific aspects of ultrasounds connected with the operation of heat exchangers, [14, 15].

Considering this a topic of specific interest and not so much systematically investigated by various research groups, the aim of this paper is to investigate the potential use of ultrasonic waves for improving the performances of heat exchangers. In a heat exchanger, the effect of ultrasonic vibrations is important both for heat transfer intensification and for the possibility of obtaining fouling reduction: the cumulative effect determines an undeniable advantage in term of efficiency increase.

After a general analysis of the topic, the results recently obtained in the two main fields of analysis, heat transfer coefficient enhancement and fouling reduction are analyzed and the perspective of the use of ultrasound for optimization of heat exchangers operation is some particular fields of application are discussed. This paper is organized in two main parts. The first one deals with a theoretical background on heat transfer intensification by means of ultrasounds. The second part is devoted to the perspective of the use of ultrasounds with application to heat exchangers.

2. Heat transfer with ultrasounds: theoretical background

Sound, as a special form of energy, propagates through pressure fluctuations in elastic media. It can be classified as infrasound ($f < 20$ Hz), sound (audible) (20Hz $< f <$ 20 kHz), and ultrasound ($f >$ 20 kHz) according to the frequency of pressure pulsation. Over the frequency of 20 kHz, low frequency, high frequency and very high frequency ultrasounds can be observed as schematically described in figure 1. Ultrasonic applications are rigidly classified into low-and high- intensity applications. The low-intensity ultrasound is typically in the megahertz frequencies with acoustic power up to milliwatts. Figure 2 provides a qualitative summary of the various application of ultrasounds at medium power range. The power of 10 W can be considered as a reference level for distinguishing between low and high power ultrasounds. The high-intensity ultrasound is of low frequencies (from 20 to 40 kHz).

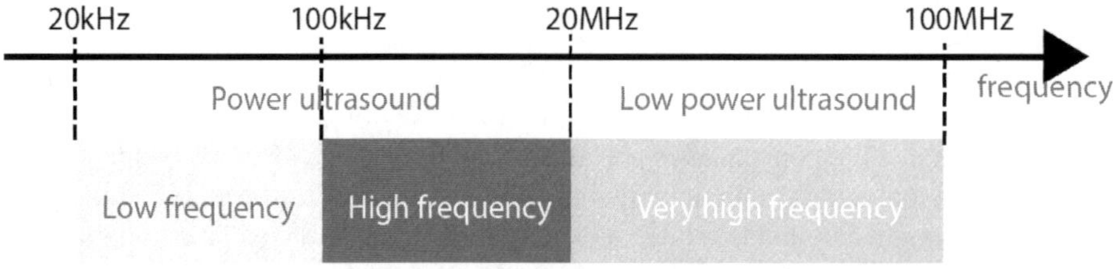

Figure 1. Classification of ultrasounds

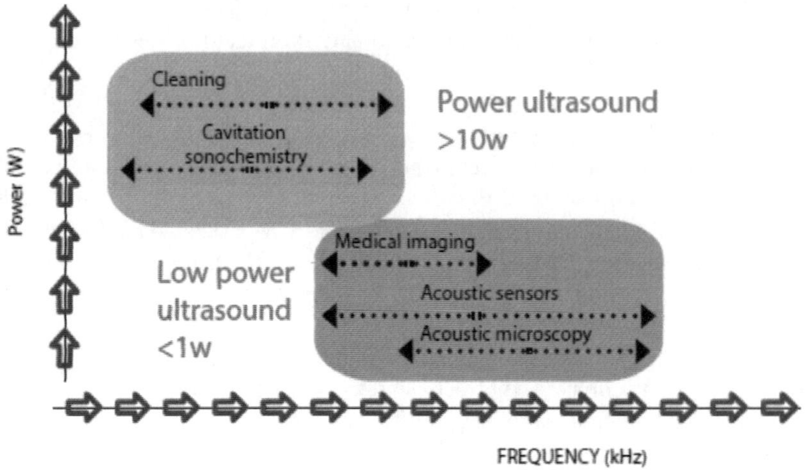

Figure 2. Utilizations of ultrasound according to frequency and power. [1]

Ultrasound are usually employed for communication, medical diagnosis, fetal imaging, under water finding (SONAR) and non-destructive testing. Some special effects are induced by power ultrasound during its transmission through media. Pressure waves caused by ultrasound transmitted in a medium cause an oscillation of the molecules around their mean position determining alternating compression and decompression of the medium. Considering the heat transfer intensification, in the specific literature on the subject, three main kinds of fluids have been taken in consideration:

1. **gas or a mixture of gasses**, for example oxygen, nitrogen, air etc.;
2. **single phase fluid**, for example liquid water in standard conditions;
3. **two phase fluid**, for example liquid and vapor (as in boiling and condensing regimes).

In the cases of liquid or two-phase fluid it is also important to take in consideration the presence of dissolved gases in the liquid (in particular air): these dissolved gases may cause the insurgence of a specific convective transfer alteration mechanism. The mechanical effects induced by ultrasound when it transmits through medium are typically cavitation in liquids and acoustic streaming in liquid or gas. But different secondary effects can be also observed. In liquid medium, when ultrasonic power attains a threshold, the rarefaction cycle may exceed the attractive forces. In such a case, cavitation bubbles could appear from existing gas nuclei.
Propagation of high-frequency acoustic wave waves, especially in a liquid results in cavitation and acoustic microstreaming which improves the heat transfer conditions substantially (figure 3).
Cavitation is the formation of small bubbles due to local pressure decrease and their subsequent growth and collapse. The implosion of these bubbles results in microscopic hydrodynamic phenomena such as micro-jet formation, shockwave propagation and thereby micro-mixing of the liquid that induce positive effects on the local heat transfer coefficient. The intensity of cavitation and its effects depend on the characteristics of the medium, such as viscosity, and process variables, like ultrasonic intensity, ultrasonic frequency or pressure. In gas media, power ultrasound can produce intense effects on the interfaces, such as pressure variations or micro-stirring, which can affect the mass transfer even if more limited effect on the intensification of heat transfer can be observed. The influence of several factors (amplitude of ultrasonic waves, external static pressure, temperature and viscosity of medium) acting, either individually or in combination can influence in an important way the amount of power transferred from the surface to the fluid. Analyzing the various studies in the literature, it might be concluded that, mainly in case of liquids, both in single phase and in two-phase conditions, the ultrasonic field often results in overall heat exchange transfer coefficient enhancement.

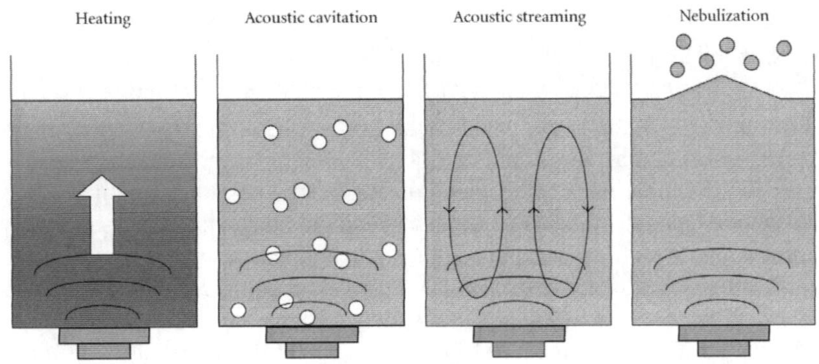

Figure 3. Four effects resulting from ultrasound propagation in a liquid

In order to quantify the effect of ultrasound on the heat transfer coefficient, an enhancement factor *EF* is usually defined. This can be considered as the ratio of the value observed with ultrasound divided by the value under silent conditions. In the case of a thermal control system, it corresponds to the ratio of the overall heat transfer coefficient in the presence of ultrasonic waves divided by the overall heat transfer coefficient without ultrasound U for the same hydrodynamic configuration:

$$EF = \frac{h_g \, with \, ultrasounds}{h_g \, without \, ultrasounds} \tag{1}$$

A theoretical prediction of the effects of ultrasound on heat transfer is not fully available and a lot of analysis in the literature are based on experimental results. In general, it can be observed that ultrasound is an efficient way to enhance heat transfer performances of liquid flow in laminar regime up to that usually observed for turbulent flow conditions. On the other hand, a theoretical analysis of the problem and the definition of the heat enhancement effect of ultrasounds appear to be really complex: the extremely fast and small-scale phenomenon have interaction with turbulence inside the fluid.
Moreover, the problem can be complicated by the fact that the acoustic field propagations could be sensitive to fluid property variations and by the complex geometry involved. Anyway, the heat transfer enhancement factor surely depends on the following elements:

- the fluid under analysis and its phase (liquid, gas, vapour);
- the operating conditions of the fluid (subcooled, saturated, subcooling degree etc.);
- the ultrasonic frequency of the generator, f;
- the ultrasonic generator power, P_{gen};
- the heat transfer surface characteristics;
- the geometry of the system;
- the material of the surface determining the possible formation of chemical substances.

3. Heat transfer enhancements by means of ultrasounds: available experimental results

Several authors have investigated the heat transfer enhancement due to ultrasonic waves using different fluids and in different operating conditions. The studies on the heat transfer enhancement by means of ultrasounds, mainly focused on measuring the heat transfer rate change with such experimental parameters as the vibration frequency, the distance between the vibration transducer and heat source. Moreover, the heat transfer enhancement depends on a great number of conditions, such as ultrasonic wave power and frequency, dimension of heated wire, chamber dimension and liquid properties, affect the heat transfer enhancement degree. In each experimental analysis, exposed in the literature since the 60s, considerable efforts to investigate the effects of individual parameter changes, like the fluid have been made. Furthermore, without generalization of those observations into reduced or dimensionless parameters, the collection of experimental data showing the influence of the factors above is severely limited unless the same conditions are realized in applications. Obviously they affect flow in a variety of ways depending on the liquid conditions. In the following sections some details about different combinations of fluid and operating conditions are discussed.

3.1. Enhancement of natural convection heat transfer

The effect of ultrasounds on heat transfer enhancement in natural convection is the first topic covered in the literature. The cases of gas and liquid can be considered. In the case of gaseous medium, the scientific literature reports three main phenomena of convective heat transfer alteration caused by an acoustic field: thermo-viscous dissipation of acoustic field, acoustic streaming and acoustic alteration of thermal boundary layer. In the case of liquid, all three phenomena described above for the gas medium are still present, with the only difference that, cause to their low compressibility, the acoustic alteration of thermal boundary layer is less important. Moreover, in this case are involved phenomena which are not present in the case of gas: vaporous and gaseous acoustic cavitation and anti-fouling effect of acoustic cavitation. General analysis on the topic are reported in [16, 18].
Even if the case of fluid in a single phase could be evaluated by means of theoretical calculation, the case is not particularly considered in experimental investigation except that in some specific papers, as [19, 20]. In the papers the authors report an average percentage increase of the heat transfer coefficient between a minimum of 18% and a maximum of 24%. This could not be considered too significant from the point of view of the heat transfer enhancement, because the calculated error on the evaluation of the heat transfer coefficient is around 12%.

3.2. Enhancement of subcooled boiling heat transfer

A paper by Li et al. [21] represents a good reference works for the topic. The experimental analyses remark that increase of the heat transfer coefficients up to 2.5-2.6 times can be observed for liquid in subcooling conditions both with water and refrigerants. Interesting heat transfer enhancements can be observed in case of evaluation of convection heat transfer of liquid flowing inside pipes, as in [22].
Yamashiro et al. studied experimentally the rapid quenching of a thin platinum wire 0.5 mm in diameter to water, in sub-cooling boiling conditions. They clarified the influence on the heat transfer rate of the sound pressure level, the cavitation phenomenon, and acoustical streaming. They varied the heat flux per unit surface, the ultrasonic generator power, the ultrasound frequency, and the working fluid level inside the ultrasonic tank, [23]. Bartoli et al. [9-10] consider the water sub-cooling degree (ΔT_{sub}) as the most important variable. Considering distilled water, the average percentage heat transfer enhancement started from 41% at ΔT_{sub} equal to 45 C, reached the maximum point, corresponding to 57% at 25 C, and then started to decrease dramatically up to 15% at 10 C. The best conditions occurred at T_{sub} equal to 25 C, f = 40 kHz and P_{gen} = 500 W: at the heat flux per unit surface equal to $1.2x10^5$ W/m^2 the heat transfer coefficient enhancement is equal to 62%. These results seem to be promising about the use of ultrasounds in thermal problems. Before of their practical application, the working fluid will be replaced with a dielectric fluid, commonly used in electronic cooling.

3.3. Enhancement of saturated boiling heat transfer

The case of boiling liquid medium is the most complicated one: with this kind of fluid all the typical phenomena are present. The tests in saturated pool boiling, exposed in some basic works like [2] and [6] shows as the ultrasounds does not really increase the heat transfer coefficient, even if the critical heat flux can be reached at higher heat flux than in the absence of ultrasonic waves. Heat transfer coefficient doesn't change significantly with or without ultrasound in saturated boiling even some effect can be observed. A detailed analysis of the problem is provided in a paper by Park and Bergles, [24] and more recently by Kim et al, [25] that compare the effects of ultrasounds in pool boiling with the same in natural convection.

4. Applications of ultrasound to heat exchangers: analysis of the available results

Heat exchangers have at least two fluids (flowing or at rest), which makes systems sometimes a more difficult task to study the effect of ultrasound as enhancement factor. The vibrational energy could be provided in the heat exchanger by two ways: directly to the fluid, that is the acoustically exited system; to the solid structure of the heat exchanger, that is the mechanically exited system.
Indeed, it must be observed that they are subjected to several constraints, and ultrasonic vibrations have influence on various parameters (e.g., heat transfer, fouling, and charge losses). But the presence of two fluids made possible a more remarkable enhancement effect as it can be observed in Table 1, showing the enhancement obtained for two particular water-water heat exchangers. The heat transfer coefficient intensification is surely the first element that need to be considered in order to evaluate the effect of ultrasound if applied to the heat exchangers. As previously reported, it was shown that concerning a single fluid, the overall heat transfer coefficient can be increased up to 2.5 times with ultrasounds.

Table 1. Enhancement factor of the heat exchangers versus cold water flow rate: influence of the hot water flow rates for a double-tube heat exchanger (a) and for a shell-and-tube heat exchanger (b). [26]

Cold water flow rate [L/min]	Enhancement factor (Double tube heat exchanger)	Enhancement factor (Shell and tube heat exchanger)
0.15		2.10 – 2.60
0.20		2.05 – 2.40
0.25	1.80 – 2.30	1.95 – 2.20
0.40	1.60 – 2.20	
0.50	1.55 – 2.00	1.60 – 1.80
0.60	1.60 – 2.05	
0.75	1.45 – 1.90	1.35 – 1.50
1.00	1.50 – 1.60	1.15 – 1.30

Considering a system like a conventional heat exchanger, in which two fluids operates, combining the increased effect on both the side an enhancement of more than two times can be easily pursued.
In the recent literature analysis of three different heat exchanger configuration can be evidenced: double-tube heat exchanger [26, 27], shell and tube heat exchanger [28] and brazed plate heat exchanger [29]. In all the cases the authors demonstrate that the increase of overall heat transfer coefficient could be remarkable, mainly if some particular conditions are maintained. It was observed that the use of ultrasonic vibrations be a powerful technique to enhance the heat transfer rate both for a double-tube heat exchanger and for a shell-and-tube heat exchanger. In order to compare heat transfer, with and without ultrasounds, energy balances are used being the tested heat exchanger in steady state condition. Without ultrasounds, the energy balance is classically given by Eq. (2), in which the pedices identifies cold (c) and hot fluid (h), taking into account an additional term q_{env} corresponding to heat losses to the environment, negligible if the heat exchanger is well insulated:

$$q_c + q_{env} = q_h \quad (2)$$

In the presence of ultrasound, the energy balance takes into account the ultrasonic power P_{US} supplied to the system to generate ultrasounds; therefore, the energy balance is written as in Eq. (3):

$$q_c + q_{env} = q_h + P_{US} \quad (3)$$

4.1. Double-tube heat exchanger
The first system of interest is the double-tube heat exchanger of the typical schematic configuration presented in figure 4. It can be operated in parallel flow configuration and/or in vertical position. This kind of configuration is object of attention of papers, like [26, 27]. According to the balance equation, typical heat transfer intensification can be illustrated by values of heat flowrates according to data of Table 2. Analyzing the data obtained by the authors, it can be observed that an increase of the heat transfer coefficient up to 100% can be obtained and this mainly for a quite low mass flow rate.

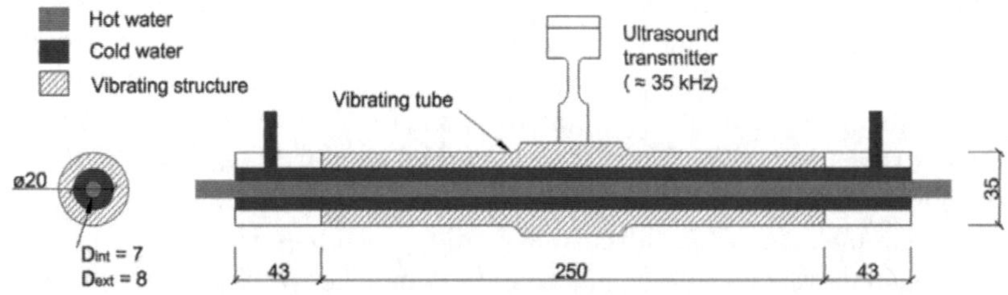

Figure 4. Sketch and dimensions of the vibrating double pipe heat exchanger (dimensions in mm)

Table 2. Examples of heat flowrates with and without ultrasound at both sides of the double-tube heat exchanger for a hot water flowrate in the inner tube of 2 L/min in a counter-flow configuration. [26]

Cold water flowrate [L/min]	q_c [W]	$q_{c,us}$ [W]	q_h [W]	$q_{h,us}$ [W]	P_{us} [W]
0.20	95	216	95	153	70
0.40	136	285	135	210	79
0.60	163	332	158	254	85
1.00	201	400	197	310	109
1.50	243	428	237	324	108
2.00	268	462	269	355	106
2.50	295	479	291	376	101
3.00	318	499	312	392	103
3.50	337	512	331	426	100
3.75	345	515	334	436	99

The comparison of the overall heat transfer coefficient without and with ultrasound indicates important increase due to ultrasounds, at least in the fluid flowing in the annular space.

4.2. Shell-and-tube heat exchanger

A special configuration of shell and tube heat exchanger using ultrasound as enhancement effect has been examined in [28]. The shell side of this exchanger vibrates at a given low ultrasonic frequency (35 kHz) resulting in what might be regarded as a vibrating shell-and-tube heat exchanger.

Experiments carried out in [28] and estimations of the overall heat transfer coefficient based upon the energy balance demonstrates that with ultrasounds, heat flow rates are increased at both sides of the heat exchanger. The authors show that the enhancement of the overall heat transfer coefficient ranges from 120 up to 250%. It is shown that the enhancement of the overall heat transfer coefficient due to ultrasound is not sensitive to the ultrasonic power supplied at the shell side. Low ultrasound amplitude results in a significant enhancement of the heat transfer. Furthermore, the enhancement factor strongly depends on the cold water flow rate at the shell side while it seems not to depend on the hot water flow rate in the tube. Figure 5 provides the results obtained in terms of overall heat transfer coefficients for both double pipe heat exchanger and shell and tube heat exchanger.

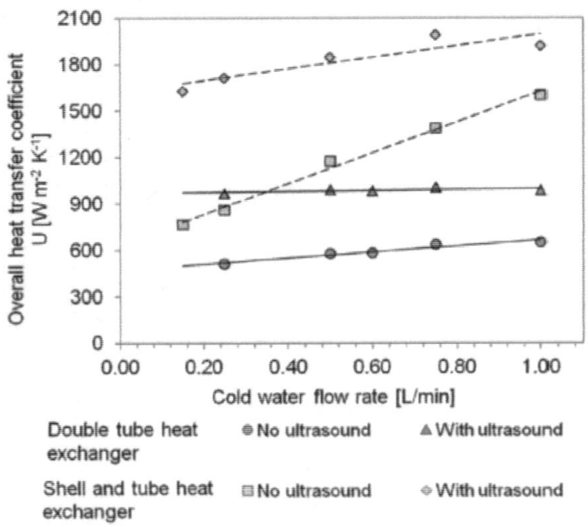

Figure 5. Overall heat transfer coefficient versus cold water flow rate: a comparison of double tube and shell and tube heat exchanger [28]

4.3 Plate heat exchanger and compact heat exchangers

A type of configuration often studied is heat transfer occurring in a channel made by two plates or beams at different temperatures with vibrations applied either to the fluid between or to one of the walls [19, 20]. This configuration has a specific application in the plate heat exchangers that has a great interest for a lot of industrial and civil applications (like a district heating systems).

The intensification of heat transfer in plate heat exchanger have been investigated in [29]. In particular, a classical double-tube heat exchanger and an industrial brazed plate heat exchanger have been considered as object of the analysis. Considering the three cases under analysis, the optimal settings for a pronounced enhancement for both configurations appear to be a slow laminar flow. Improvement of heat exchanger operation using ultrasound in compact heat exchangers is reported in [30].

5. Use of ultrasounds for anti-fouling

Fouling usually exists in industrial heat exchangers mainly when water is used as working fluid. Fouling not only decreases heat transfer performance seriously, but also reduces the life of heat transfer equipment. The anti-fouling effect of ultrasound is mainly connected with acoustic cavitation: this is interesting, being indeed an indirect convective heat transfer enhancement mechanism due to acoustic cavitation (both vaporous or gaseous). It is especially remarkable that some recent works on the

application of ultrasound in heat exchanger usually claim improvement not only for heat transfer coefficient but for pressure drop reduction or anti-fouling properties too. Moreover, the hydrodynamic actions originated from bubble implosion induce high mechanical strains in this fouling layer, making possible in some cases its disintegration and removal (and, therefore, a global increase of efficiency).
In a recent work, [31], Legay et al. reports experimental studies about fouling reduction by ultrasound inside a double-pipe heat exchanger. They tested 4 different combination of material geometry of the internal tube (5,6 mm and 6 mm of internal diameter and 10 mm or 12 mm of outer diameter of the internal tube). The results are shown in figure 6: after 20 minutes, a remarkable increase of the overall heat transfer coefficient, meaning a reduction of resistance due to fouling, can be evidenced.
In a recent paper Chen et al. [32] analysed, by means of an experimental analysis, the antifouling effect of ultrasound in a heat exchanger operating with water at different temperatures. The parameters of ultrasound were controlled in the frequency range (28–40 kHz) and power of 25 and 50 W. As exposed in Tables 3 and 4 the fouling resistance appears to be dependent on the ultrasound frequency as well as on the ultrasound power and on the initial hardness of water (expressed in mg/L). The results show that ultrasound has remarkable antifouling effects and propose similar conclusion with those exposed in [31]. Moreover in this case the authors analyse the effects of frequency: it is possible to state that antifouling rate increases with decreasing of ultrasonic frequency. Whereas a contrary trend was observed for ultrasonic power, i.e. a higher antifouling rate was reached with a power of 50W.
The use of ultrasound to prevent fouling is discussed with reference to a plate heat exchanger too [33].

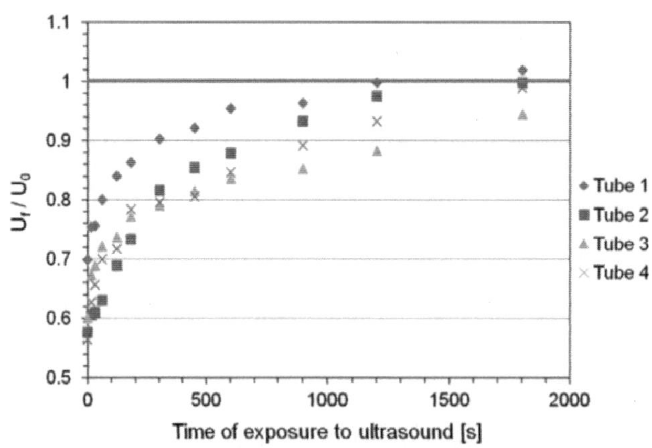

Figure 6. Ratio of overall heat transfer coefficients (fouled/clean) for four painted pipes. [31]

Table 3. Fouling resistance at different ultrasound frequencies at T=44 °C [32]

Time of exposure [hours]	No ultrasound	28 kHz	40 kHz
5	1.20 10^{-4} m^2K/W	1.00 10^{-4} m^2K/W	1.10 10^{-4} m^2K/W
10	1.80 10^{-4} m^2K/W	1.40 10^{-4} m^2K/W	1.60 10^{-4} m^2K/W
15	2.00 10^{-4} m^2K/W	1.55 10^{-4} m^2K/W	1.80 10^{-4} m^2K/W
20	2.10 10^{-4} m^2K/W	1.60 10^{-4} m^2K/W	1.95 10^{-4} m^2K/W
25	2.20 10^{-4} m^2K/W	1.70 10^{-4} m^2K/W	2.00 10^{-4} m^2K/W

Table 4. Fouling resistance variation with ultrasound power at the optimal frequency 28 kHz [32]

Time of exposure [hours]	No ultrasound	25 W	50 W
5	1.10 10^{-4} m^2K/W	0.60 10^{-4} m^2K/W	0.25 10^{-4} m^2K/W
10	1.15 10^{-4} m^2K/W	0.65 10^{-4} m^2K/W	0.30 10^{-4} m^2K/W
15	1.10 10^{-4} m^2K/W	0.65 10^{-4} m^2K/W	0.25 10^{-4} m^2K/W
20	1.05 10^{-4} m^2K/W	0.60 10^{-4} m^2K/W	0.25 10^{-4} m^2K/W
25	1.00 10^{-4} m^2K/W	0.60 10^{-4} m^2K/W	0.25 10^{-4} m^2K/W

6. Conclusions

The use of ultrasounds appears to be an effective mean to improve the heat transfer performances of fluids and can attract specific interest for some special applications like heat exchangers.

In the case of gas, current literature has identified the occurrence of three phenomena of convective heat transfer modification: the heat dissipation of the introduced vibrational energy (thermo-viscous dissipation of acoustic field); the alteration of the mean flow trajectories (acoustic streaming); the intensification of conductive heat transfer in the thermal boundary layer (acoustic modification of thermal boundary layer). With liquids, other three phenomena can be identified: the alteration and mixing of fluid flow in the thermal boundary layer, through the implosion of low pressure vapor bubbles (vaporous acoustic cavitation), a similar alteration of vaporous acoustic cavitation, but through the implosion of low pressure gas bubble (gaseous acoustic cavitation) and the removing of fouling layers by the mechanical stress inducted by acoustic cavitation (anti-fouling effect of acoustic cavitation).

In the case of two-phase fluid in boiling conditions, all the six phenomena exposed above are still present. Furthermore, a new peculiar phenomena can be identified: the destabilization of the vapor bubble on the heating surface and its facilitated detachment (acoustic release of vapor bubble).

Considering the various phenomena, it is not easy to develop a theoretical analysis of the problem; an adequate mathematical description is very difficult to be obtained due the complexity of the phenomenon and there is also a strong uncertainty about the possibility of using the theoretical results in real application. For the above mentioned reasons the authors that have considered the problem thinks that an experimental backup is highly desirable. But experimental investigation is often limited to quite specific cases and it would take a very long time to cover a meaningful extension of the results. However results obtained in specific conditions indicates that increase of the heat transfer coefficient of 200% in natural convection, 400% in subcooled conditions and of 50% in saturated pool boiling can be reached.

Considering the case of heat exchangers, a promising interesting field of application for ultrasound is represented by the control of heat transfer coefficient in consequence of a reduction of the mass flow rate. In this case the perspective of a possible regulation of the heat transfer coefficient as a consequence of mass flow rate decrease in off-design conditions and the control of fouling effects represents elements of particular interest and suggest the opportunity of further investigations, mainly in case of liquids.

7. References

[1] Legay M, Gondrexon N, Le Person S, Boldo P and Bontemps A 2011 Enhancement of heat transfer by ultrasound: review and recent advances *Int. J. Chem. Eng.* **2011** 1-17.

[2] Kim HY, Kim YG and Kang BH 2004, Enhancement of natural convection and pool boiling heat transfer via ultrasonic vibration *Int. J. Heat Mass. Tran.* **47** 2831-2840.

[3] Bonekamp S and Bier K 1997 Influence of ultrasound on pool boiling heat transfer to mixtures of the refrigerants R23 and R134A *Int. J. Refrig.* **20**, 606-615.

[4] Wong SW and Chon WY 1969 Effects of ultrasonic vibrations on heat transfer to liquids by natural convection and by boiling *AIChE J.* **15** 281-288.

[5] Oh YK, Park SH and Cho YI 2002 A study of the effect of ultrasonic vibrations on phase change heat transfer *Int J. Heat Mass Tran.* **45** 4631-4641.

[6] Iida Y and Tsutsui K 1998 Effects of ultrasonic waves on natural convection, nucleate boiling and film boiling heat transfer from a wire to a saturated liquid *Exp. Them. Fluid Sci.* **5** 108–115.

[7] Monnot A, Boldo P, Gondrexon N and Bontemps A 2007 Enhancement of cooling rate by means of high frequency ultrasound *Heat Transfer Eng* **28** 3-8.

[8] Hyun S, Lee DR, and Loh BG 2005 Investigation of convective heat transfer augmentation using acoustic streaming generated by ultrasonic vibrations *Int. J. Heat Mass Tran.* **48** 703–718.

[9] Bartoli C and Baffigi F 2012 Use of ultrasonic waves in sub-cooled boiling *App. Therm. Eng.* **47** 95-110

[10] Baffigi F and Bartoli C 2012 Influence of the ultrasounds on the heat transfer in single phase free convection and in saturated pool boiling *Exp. Therm. Fluid Sci.* **36**, 12–21.

[11] Yao Y, Zhang X, Guo Y 2010 Experimental Study on heat transfer enhancement of water-water shell-and-tube heat exchanger assisted by power ultrasonic, International Refrigeration and Air Conditioning Conference, Purdue University Purdue e-Pubs.

[12] Melkumov U and Kurbanov K 2003 Use of ultrasound for intensification of heat transfer process in heat exchangers, Proc. Int. Congr. Refrig. 1–5.

[13] Benzinger W, Schygulla U, Jäger M and Schubert K 2005 Anti-fouling investigations with ultrasound in a microstructured heat exchanger, Proc. of the 6th Int. Conf. Heat Exch. Fouling Clean, Kloster Irsee, Germany.

[14] Duan XL, Wang XY, Wang G, Chen YZ and Qiu XQ 2004 Experimental study on the influence of ultrasonic vibration on heat transfer and pressure drop in heat exchanger tubes *Petro-Chem. Equip.* **33**, 1–4.

[15] Bulliard-Sauret O, Ferrouillat S, Vignal L, Memponteil A, Gondrexon N 2017 Heat transfer enhancement using 2 MHz ultrasound, *Ultrason. Sonochem.* **39** 262-271.

[16] Fand RM, 1965 The influence of acoustic vibrations on heat transfer by natural convection from a horizontal cylinder to water *ASME J. Heat Tran* **87** 309–310.

[17] Wong SW and Chon WY, 1969 Effects of ultrasonic vibration on heat transfer to liquids by natural convection and by boiling *AlChE J* **15** 281–288.

[18] Iida Y and Tsutsui K 1992 Effects of ultrasonic waves on natural convection, nucleate boiling and film boiling heat transfer from a wire to a saturated liquid *Exp. Ther. Fluid Sci.* **5** 108–115.

[19] Nakagawa M 2004 Analyses of acoustic streaming generated by four ultrasonic vibrators in a vessel *Japan. J. App. Phys.* **43** 2847–2851.

[20] Vainshtein P, Fichman M and Gutfinger C 1995 Acoustic enhancement of heat transfer between two parallel plates *Int. J. Heat Mass Tran.* **38** 1893–1899.

[21] Li B, Han X, Wan Z, Wang X and Tang Y 2016 Influence of ultrasound on heat transfer of copper tubes with different surface characteristics in sub-cooled boiling *Appl. Therm. Eng.* **92** 93-103.

[22] Baffigi F and Bartoli C 2014 Heat transfer enhancement using ultrasonic waves in presence of Liquid: A Basic Research for Cooling Electronic *App Mech Mat* **464** 163-170.

[23] Yamashiro H, Takamatsu H, and Honda H 1998 Effects of ultrasonic vibration on transient boiling heat transfer during rapid quenching of a thin wire in water *ASME J. Heat Tran* **120** 282–286.

[24] Park KA and Bergles AE 1988 Ultrasonic enhancement of saturated and subcooled pool boiling *Int. J. Heat Mass Tran.* **31** 664–667.

[25] Kim HY, Kim YG and Kang BH 2004 Enhancement of natural convection and pool boiling heat transfer via ultrasonic vibration *Int. J. Heat Mass. Tran.* **47** 2831–2840.

[26] Legay M, Le Person S, Gondrexon N, Boldo P and Bontemps A 2012 Performances of two heat exchangers assisted by ultrasound *App. Therm. Eng.* **37** 60-66.

[27] Legay M, Simony B, Boldo P, Gondrexon N, Le Person S and Bontemps A 2012 Improvement of heat transfer by means of ultrasound: Application to a double-tube heat exchanger *Ultrason. Sonochem.* **19** 1194–1200.

[28] Gondrexon N, Rousselet Y, Legay M, Boldo P, Le Person S and Bontemps A 2010 Intensification of heat transfer process: Improvement of shell-and-tube heat exchanger performances by means of ultrasound, Chem. Eng. Proc. **49** 936–942.

[29] Gondrexon N, Cheze L, Jin Y, Legay M, Tissot Q, Hengl N, Baup S, Boldo P, Pignon F and Talansier E 2015 Intensification of heat and mass transfer by ultrasound: Application to heat exchangers and membrane separation processes, *Ultrason. Sonochem.* **25**, 40–50.

[30] Tingaud F, Ferrouillat S, Colasson S, Bulliard O, Sauret and Bontemps A 2013 Improvement of two-phase flow distribution in compact heat exchangers by using ultrasound *App. Mech. Mat.* **392** 521-525.

[31] Legay M, Allibert Y, Gondrexon N, Boldo P and Le Person S 2013 Experimental investigations of fouling reduction in an ultrasonically-assisted heat exchanger *Exp. Ther. Fluid Sci.* **46** 111-119.

[32] Chen Y, Sun S, Lai Y and Ma C 2016 Influence of ultrasound to convectional heat transfer with fouling of cooling water *App. Ther. Eng.* **100** 340–347.

[33] Hotrum NE, de Jong P, Coen Akkerman J and Fox MB 2015 Pilot scale ultrasound enabled plate heat exchanger: its design and potential to prevent biofouling *J. of Food Eng.* **153** 81–88.

Experimental study of liquid velocity profiles in large-scale bubble columns with particle tracking velocimetry

G Besagni[1], F Inzoli[1], T Ziegheneín[2] and D Lucas[2]

[1] Politecnico di Milano, Department of Energy, Via Lambruschini 4a, 20156, Milano
[2] Helmholtz-Zentrum Dresden-Rossendorf e. V. Institute of Fluid Dynamics, 01314 Dresden, Germany

E-mail: giorgio.besagni@polimi.it

Abstract. A complete knowledge of the bubble column fluid dynamics relies on understanding the global and the local fluid dynamic properties. Unfortunately, most of the previous literature focused on the "*global-scale*" fluid dynamics, whereas a limited attention was devoted to the "*local-scale*". We contribute to present-day discussion by proposing an experimental study concerning the local liquid velocity profiles within the pseudo-homogeneous flow regime. The experimental study, based on a particle-identification and particle-tracking algorithm, was conducted in a large-diameter and large-scale bubble column (height equal to 5.3 m; inner diameter equal to 0.24 m) operated in the counter-current mode. We considered gas superficial velocities in the range of 0.37–1.88 cm/s and liquid superficial velocities up to −9 cm/s. Time-averaged and transient liquid velocity fields were obtained for five superficial gas velocities and four superficial liquid velocities at two measuring heights. Subsequently, the liquid velocity observations were coupled with previously measured bubble size distributions and local void fractions, to provide a complete description of the "*local-scale*" fluid dynamics. These data would help in the validation procedure of numerical codes, to support the prediction of industrial-scale relevant conditions.

1. Introduction

In the broader framework of multiphase reactors, bubble columns are widely used as contacting devices in industrial applications, owing to their many advantage in both design and operation (i.e., low initial cost, low maintenance, high contact area between the phases, as discussed in ref. [1]). The simplest bubble column layout consists of a cylinder where the gas phase is dispersed—through a gas sparger—into a liquid phase, which is supplied in the batch mode or it may be led in either co-currently or counter-currently to the upward gas phase [2]. On the practical point of view, batch-mode bubble columns are applied in hydrogenation and fermentation processes; conversely, waste-water treatment, water ozonation, and three-phase inverse fluidized bed involve a continuous flow of the liquid phase [3]. In these cases, counter-current bubble columns are employed thanks to the higher interfacial mass transfer compared with batch-mode bubble columns [4].

The gas phase, introduced through the gas sparger, evolves in the axial direction of the bubble column either in the form of "*dispersed bubbles*" or in the form of "*coalescence-induced structures*", depending on the prevailing flow regime (please refer to the discussion proposed by Besagni et al. [2]). The complete understanding of the flow regime characteristics, which is essential to correctly design and operate bubble columns, depends on the precise knowledge of the multi-scale fluid dynamics, encompassing the "*bubble-scale*" (i.e., break-up/coalescence, swarm effects, ...) and the "*reactor-scale*" (i.e., large-scale circulation, global scale properties, ...). The fluid dynamics in bubble columns is subject to this multi-scale connection since the bubble motion drive the liquid phase, which in reverse influence the rising characteristic and break-up/coalescence of the bubbles. The most important "*reactor-scale*" parameter is the gas holdup, ε_G (defined as the volume of the gas phase divided by the total volume of the system). Conversely three parameters characterize the fluid dynamics at the "*local-scale*": (a) liquid velocity, (b) local gas fraction, and (c) bubble size distributions (*BSDs*). This concept is displayed in Figure 1, which represents a typical liquid velocity profile in batch-mode bubble columns, as outlined in the pioneering study of Ueyama and Miyauchi [5]: the liquid phase flows upward in the center of the column, it becomes zero at an intermediate radial position and, finally, it flows downward near the wall. As a consequence of such liquid velocity profile, bubbles move either towards the center of the column or towards the wall, owing to the well-known effect of the lift force [7]. Subsequently, the distribution of the gas phase determines the local void fraction profiles: for example, the local void fraction profile is center-pecked if large bubbles move towards the center of the bubble columns, as observed by Besagni and Inzoli [7]; conversely, the local void fraction profile is flat is the *BSD* is prevailing composed by small-bubbles.

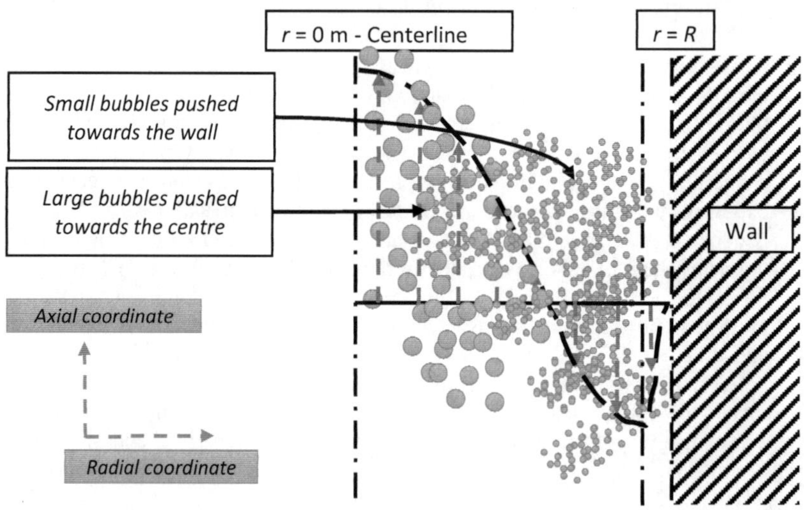

Figure 1. Radial liquid velocity profile in bubble columns (batch-mode).

In the last decades, a large amount of studies concerning bubble column fluid dynamics have been proposed; unfortunately, most of the literature investigated the batch/co-current modes and a limited attention was devoted to counter-current flows, as summarized by Trivedi et al. [6] (in particular, the reader may refer to Table 1 in ref. [6]). These studies mostly investigate the gas holdup curves [7-9], the flow regime transitions [7] and, a limited number provided *BSDs* and local flow properties measured by needle probes [7, 10]. Generally speaking, the gas holdup increases while increasing the counter-current liquid velocity, whereas the degree of this effect depends on the system design, phase properties and phase velocities [6, 7]. As stated above, three local fluid dynamic parameters (viz. liquid velocity, local gas fraction, and *BSDs*) together characterize the "*local-scale*"; however, none of the above-mentioned studies [7-10] measured the liquid velocity. Such experimental data are rare in the literature, especially if considering counter-current flows and large-scale experimental facilities; nevertheless this information is essential to completely understand the prevailing fluid dynamics in bubble columns. This concept has been stated by Trividi et al. [5], who contributed to the lack of experimental data by a lumped-parameter modeling approach. In summary, nowadays there is no experimental evidence regarding the influence of the effect of the counter-current mode on the profile proposed in Figure 1. This lack of knowledge is caused by the challenges in measuring the liquid velocity profiles, especially in dense bubbly flows in "*large-scale*" systems. In the broader group of

IOP Conf. Series: Journal of Physics: Conf. Series **1224** (2019) 012036 doi:10.1088/1742-6596/1224/1/012036

the *PIV* methods with [11] and without [12] laser illumination falls the Particle Tracking Velocimetry (*PTV*) method, which has been used in the present study. When no lasers are used for *PIV* measurements, a background light may be applied so that the shadows of particles are recorded, viz. Particle Shadowgraphy Velocimetry (*PSV*) [13]. *PSV* methods are not commonly used, but they have some advantages in bubbly flows, i.e., using a permanent light source so that no extra trigger is needed, which simplifies the experimental setup.

This study contributes to the existing discussion concerning the "*local-scale*" fluid dynamics in bubble columns, by applying a *PTV* method in a shadowgraphy setup (*PSTV*). The measurements have been conducted at the bubble column built at the Department of Energy of Politecnico di Milano. This bubble column has been developed concerning the scaling-up criteria behind the "*large-diameter*" and the "*large-scale*" concept (viz. the bubble columns has an inner diameter equal to 0.24 m, a height equal to 5.3 m, and a spider gas sparger with large openings [2, 14]). For this reason, this experimental setup can be considered as a pilot-scale experimental facility to scale-up results from "*laboratory-scale*", towards the "*industrial-scale*". We considered gas superficial velocities in the range of U_G = 0.37–1.88 cm/s and liquid superficial velocities up to $U_L = -9$ cm/s, which are within the boundaries of the pseudo-homogeneous flow regime. Time-averaged and transient liquid velocity fields were obtained for five superficial gas velocities and four superficial liquid velocities at two measuring heights. The investigated cases correspond to dense bubbly flow conditions, as the gas holdup is in the range of ε_G = 1.02 – 7.55 %. Such velocity measurements in pilot-plant scale are rarely reported in the literature, in particular for a counter-current bubble column, especially considering the "*large-scale*" concept. The liquid velocity measurements will complete a previous dataset concerning both "*global-scale*" (i.e., gas holdup) and "*local-scale*" (i.e., bubble size and shape, local volume fraction, ...) measurements [7], under the same operating conditions, aiming to provide a precise understanding of bubble column fluid dynamics and to offer a comprehensive validation database for Computational Fluid-Dynamic (*CFD*) model [15].

This paper proceeds as follows. First, the experimental setup and methods are described; subsequently, the experimental results are presented and coupled with the previously measured bubble size distributions and local void fractions. Finally, the main outcomes and outlooks are discussed.

2. Experimental setup and methods

Herein, the experimental setup, the measurements methods and the experimental matrix are presented and discussed.

2.1. Experimental setup

The experimental facility, displayed in Figure 2, is a non-pressurized vertical pipe made of Plexiglas® (ADPlast s.r.l., Milano, Italy) with a height of 5.3 m and an inner diameter of 0.24 m. A pressure reducer controls the pressure upstream of the rotameters (*(1)* and *(2)* in Figure 2), which are used to measure the gas flowrate (accuracy ±2% f.s.v., E5-2600/h, manufactured by ASA, Italy). A pump, controlled by a bypass valve, provides water recirculation and a rotameter (*(3)* in Figure 2) measures the liquid flowrate (accuracy ±1.5% f.s.v., G6-3100/39, manufactured by ASA, Italy). The gas sparger is a "*spider-gas sparger*", which is similar to the distributors used in industrial application; the gas sparger has hole-diameters in the range of d_0 = 2–4 mm. Owing to the gas sparger openings, the produced bubble size distribution (*BSD*) is polydispersed (see the measurements obtained by Besagni and Inzoli [7]). Filtered air and filtered, deionized water were used; during the experiments, the air and water temperatures were maintained constant at room temperature (T = 22 °C). The values of gas density (used to compute the superficial gas velocity), are based upon the operating conditions existing at the column midpoint, owing to the large scale of the experimental setup. The initial liquid level is set at 3.0 m above the gas sparger, thus leading to an aspect ratio (viz. the ratio of the initial liquid level to the bubble column diameter) above the critical value of 5, reported by Wilkinson et al. [14] and discussed by Besagni et al [2].

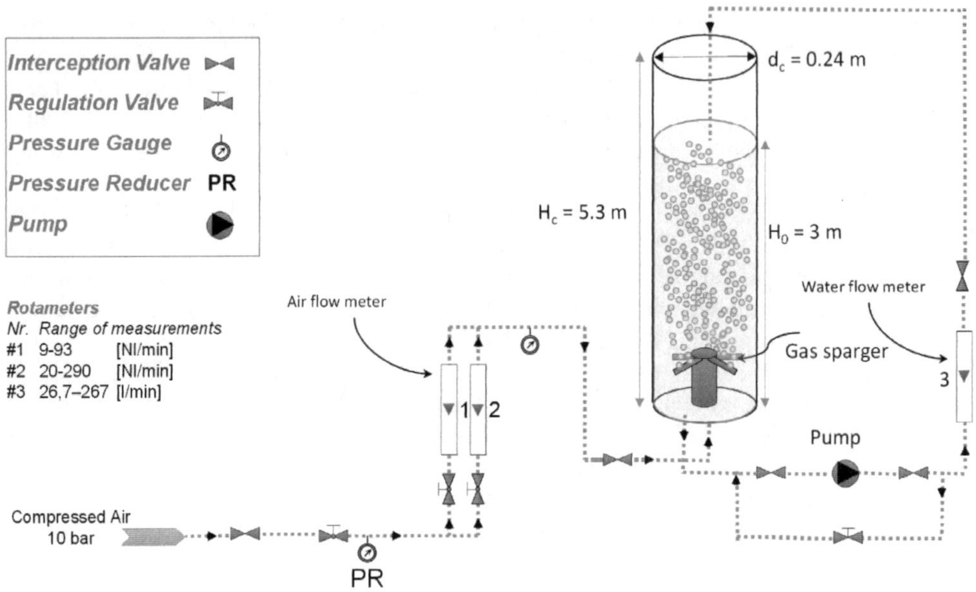

Figure 2. The experimental setup.

2.2. Experimental method

All the liquid velocity measurements were performed 0.7 m and 1.9 m above the gas sparger, to obtain void fraction profiles comparable with the local void fraction profiles reported by Besagni and Inzoli [7]. The experimental method is a shadowgraphy technique, based on a Q-VIT high-speed camera (AOS Technologies AG) with a backlight (air-cooled 400W halogen lamp); an example of the outcome of this method is displayed in Figure 3 and in Figure 4. The *PTV* method proceeds in two steps (particle identification and particle tracking); the interested reader may refer to the description proposed by Hessenkemper and Ziegenhein [13]. The sampling bias, which is found in multiphase flows owing to the coupling between the dispersed and the continuous phases, is treated with a hold processor, accordingly with the proposal of Ziegenhein and Lucas [16]. Prior to measurements the applied method and the sampling bias treatment were verified by considering integral values and comparing the *PSTV* method with *PSIV* methods in a simplified measurement setup (the flow was seeded with 2 g of 50 μm Polyamide particles - Dantec Dynamics A/S). In order to obtain a quasi-two-dimensional measuring plane, the depth of field was adjusted to 2 mm (calibrated as described in ref. [13]) by using a Samyang lens with an aperture of 2.0, a focal length of 135, and a focus distance of 0.13 m. The optical setup, was placed on a linear unit to move it stepwise along the column radius, in the range of 0.119 - 0 m (Figure 3a). With this approach, 11 measuring planes of 6x25.3 mm (Width x Height) were obtained, having a resolution of 397x1686 pixels along the radial coordinate (11.5 μm/Pixel). This method of moving the camera in radial direction is executed for the 15 flow configurations listed in the next sub-section (viz. 330 total measurements). The camera was stepwise moved until representative test prints were detected as blurred from the edge detector algorithm. It is worth noting that, despite the bubble column is operated in the homogeneous regime, a bubble plume with very long time scales exists (owing to the pseudo-homogeneous flow regime properties). Even after 1000 seconds, the averaged velocities show some changes over time. Nevertheless, the changes are rather small so that a 1000-second measuring time was a good compromise between accuracy and effort; additionally, a 1000-second measuring time has been also used for the optical probe measurements performed by Besagni and Inzoli [7]. A set of 10 images with 1000 frames per second was recorded every 2 seconds for 1000 seconds for all 11 measuring planes so that in total 55000 images in 3 hours measuring time were recorded for each operation point.

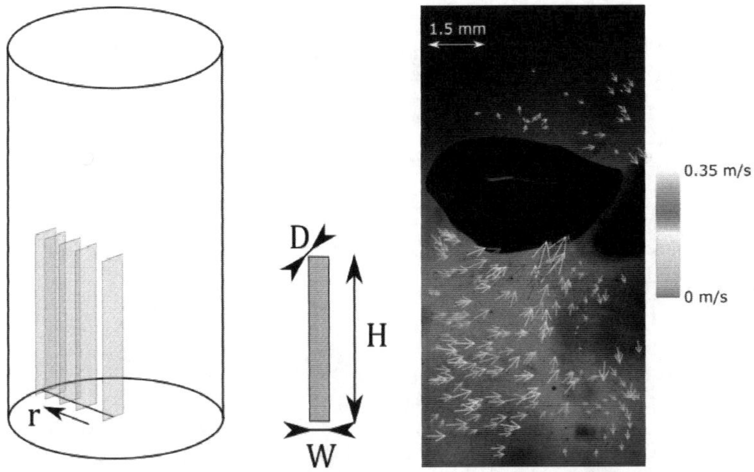

(a) Measurement locations (b) $U_G = 0.37$ cm/s, $U_L = 0$ cm/s.

Figure 3. *PVT* method: measurement method.

2.3. Experimental matrix

Taking into account the previous experimental investigations [7], this paper focuses on the description of the pseudo-homogeneous flow regime. In particular, the camera setup was installed for the particle tracking measurements at the two measuring heights of the optical probes (1.9 m and 0.7 m above the gas sparger). The flow conditions tested are listed in Table 1; in addition, Table 1 lists the flow regime transition points for the different superficial liquid velocities; it can be observed that the tested conditions falls within the pseudo-homogeneous flow regime (as also displayed in **Errore. L'origine riferimento non è stata trovata.**Figure 5) in dense bubby flows conditions (ε_G is in the range of 1.02 – 7.55 %).

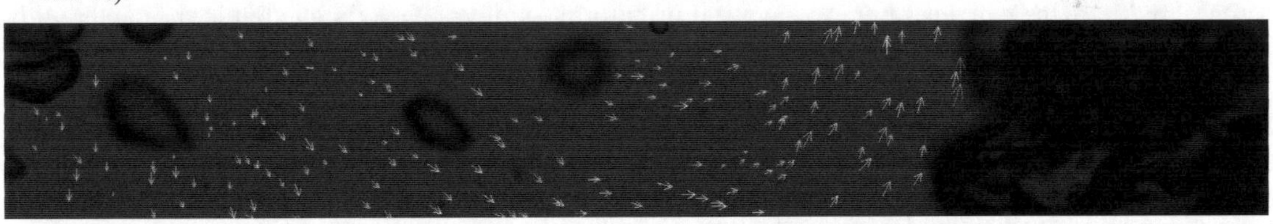

Figure 4. Local liquid velocity: example of outcome of the method.

Table 1. Operating conditions experimentally evaluated in the present study (marked as ✓) - in bracket, the corresponding gas holdup [%], measured by Besagni and Inzoli [7], is specified.

U_L [cm/s]	$U_{G,trans}$ [cm/s]	U_G [cm/s]				
		0.37	0.74	1.11	1.48	1.85
0.0	2.63	✓ (1.02 %)	✓ (1.64 %)	✓ (3.72 %)	✓ (5.30 %)	✓ (6.60 %)
-3.3	2.45	✓ (1.32 %)	✓ (2.91 %)	✓ (4.61 %)	✓ (6.63 %)	
-6.6	2.08	✓ (1.70 %)	✓ (3.62 %)	✓ (5.99 %)		
-9.2	1.73	✓ (2.44 %)	✓ (5.00 %)	✓ (7.55 %)		

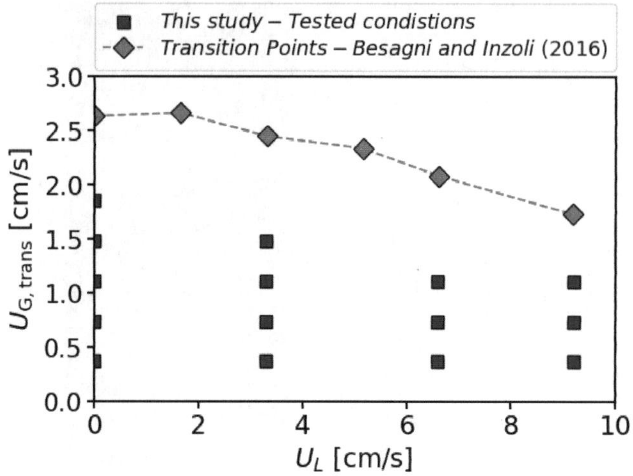

Figure 5. Operating conditions experimentally evaluated in the present study compared with the regime transition points (as evaluated by Besagni and Inzoli [7]).

3. Experimental results

Herein, the experimental results are presented and discussed in terms of both upward and angular velocities. First, the single-phase flow situations are presented; second, the two-phase flow conditions are discussed. Such presentation of the experimental results is supposed to better stress the influence of the dispersed phase on the fluid dynamic properties imposed by the liquid phase.

3.1. Single phase flow conditions

Measurements of the single phase flow situation (viz. U_G = 0 cm/s, changing U_L) were performed, to provide preliminary observations concerning the influence of the inlet and outlet boundary conditions on the flow field: the averaged liquid velocity profile for the single-phase flow is flat in the center with a steep slope towards the wall (Figure 6), owing to the boundary conditions imposed by the wall itself. Such velocity profiles are typical of turbulent single-phase flow conditions. Increasing U_L, it is observed that the profiles are just shifted to higher absolute velocities (figure 6a). Due to the asymmetrically inflow in the basin on top of the column (see Figure 7a), the water is rotating in the basin (despite a porous plate was placed at the bottom of the gas sparger, Figure 7b). This rotation can be clearly observed in the angular velocity (Figure 6b). Indeed, in the angular velocity, an asymmetry can be seen since the sign change is not in the center of the column for superficial liquid velocities above U_L = -3.3 cm/s. Also from visual observation, this circular current is preserved towards the bottom of the bubble column.

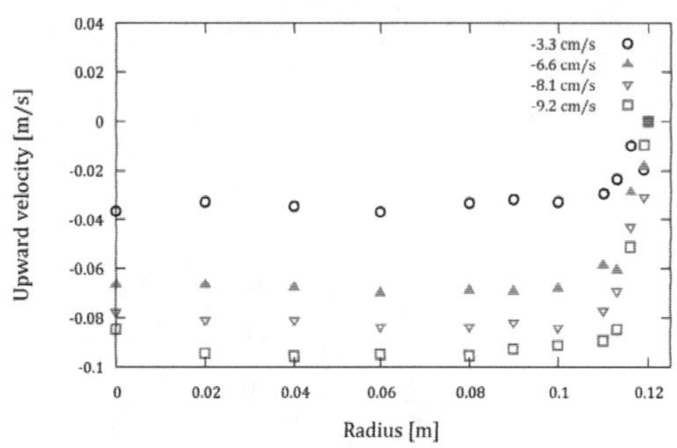

(a) Upward velocity

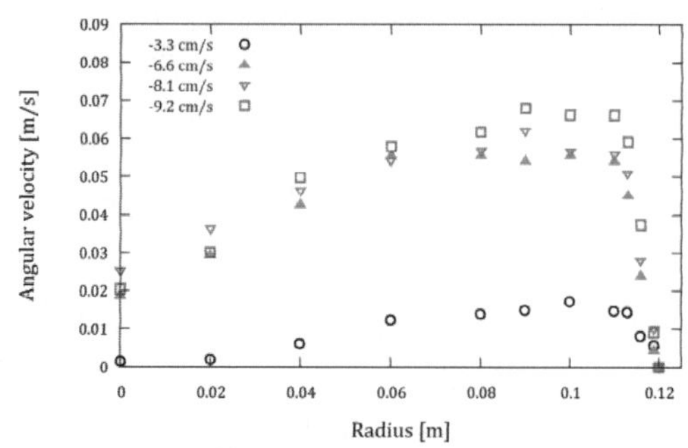

(b) Angular velocity

Figure 6. Upward and angular liquid velocity velocities profiles without gas (U_G = 0 cm/s) at different superficial liquid velocities.

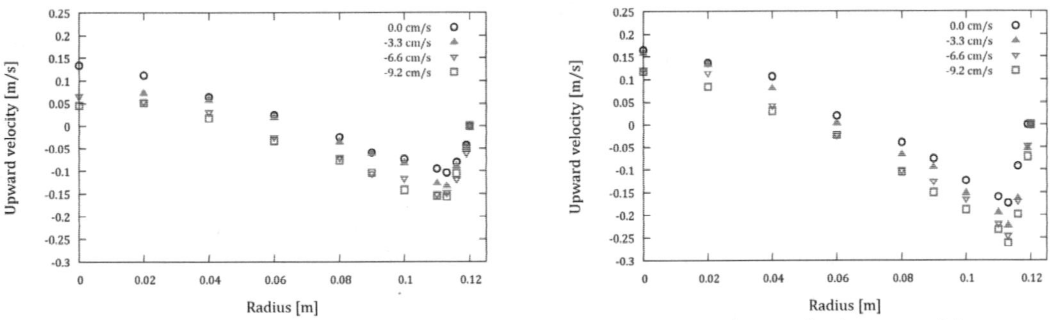

(a) Outlet conditions for water recirculation (b) Porous plate at the bottom of the gas sparger

Figure 7. Outlet conditions for the liquid phase

3.2. Two phase flow conditions

Figure 8 and Figure 9 display the time-averaged upward liquid velocities for the operating conditions listed in Table 1. In the different conditions, the liquid flows upward in the centre of the column, becomes zero at an intermediate radial position and then flows downward near the wall, which is in agreement with the liquid velocity profiles displayed in Figure 1 and discussed in ref. [5]. For all the conditions, the minimum velocity is located at approximately 7.5 mm away from the wall. Increasing U_G, at fixed U_L, the gradients in the liquid profiles get steeper, which can be nicely seen for the batch mode (Figure 8a). These findings are in qualitative agreement with the modeling approach proposed by Trividi et al. [5], despite some differences concerning the magnitude of the velocity. The upward velocity at wall increases with U_G (in agreement with ref. [5]); conversely, the influence of U_L on the upward velocity at wall is limited (which is different from the predictions of ref. [5]).

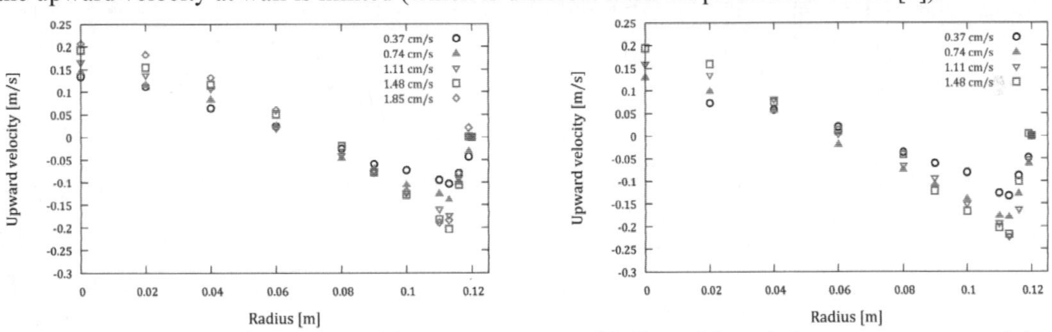

(a) U_L = 0 cm/s (batch mode) (b) U_L = -3.3 cm/s (counter-current mode)

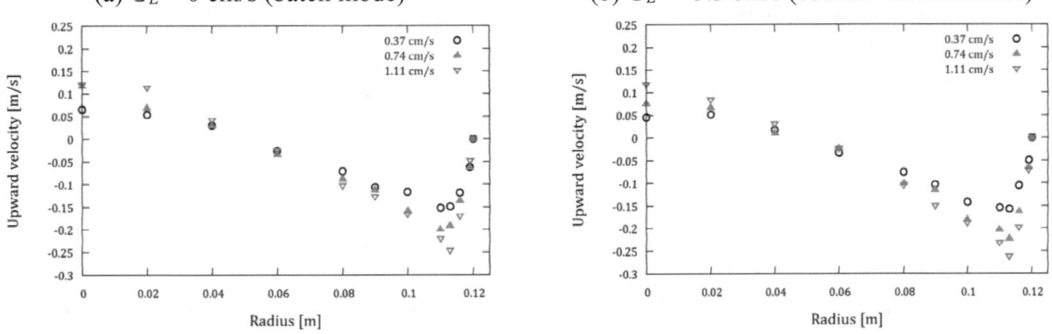

(c) U_L = -6.6 cm/s (counter-current mode) (d) U_L = -9.2 cm/s (counter-current mode)

Figure 8. Averaged upward liquid velocities for different superficial gas velocities at different counter-current flow setups 0.7 m above the gas sparger.

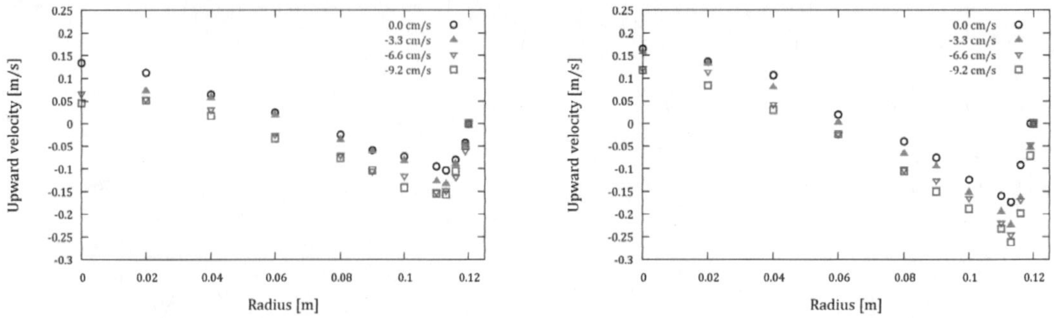

(a) $U_G = 0.37$ cm/s (b) $U_G = 1.11$ cm/s

Figure 9. Averaged upward liquid velocities for different counter-current flow setups at different superficial gas velocities 0.7 m above the gas sparger.

It is observed that the upward liquid velocity profiles are shifted downwards with no further modification, while increasing U_L at fixed U_G (Figure 8 and Figure 9). This observation is very interesting and suggests that the prevailing effect of the counter-current operation mode is to decelerate the bubbles raising the column. These experimental findings confirm the previous discussions proposed by Besagni and Inzoli [7] and provide a rational explanation regarding the prevailing effects of the liquid velocity on the bubble column fluid dynamics: the counter-current mode increases the gas holdup and the local void fraction; conversely, it decreases the local velocity. In addition, Figure 10 displays the influence of the counter-current mode on the angular velocity: the bubbly flow generates a clear circulating flow, which is changing its direction towards the center. This counter-current angular flow is slightly hindered when the liquid velocity is increased to U_L = -3.3 cm/s; conversely, owing to the asymmetrically boundary conditions, the circulating flow of the liquid inflow dominates the bubbly flow U_L = -6.6 and -9.2 cm/s

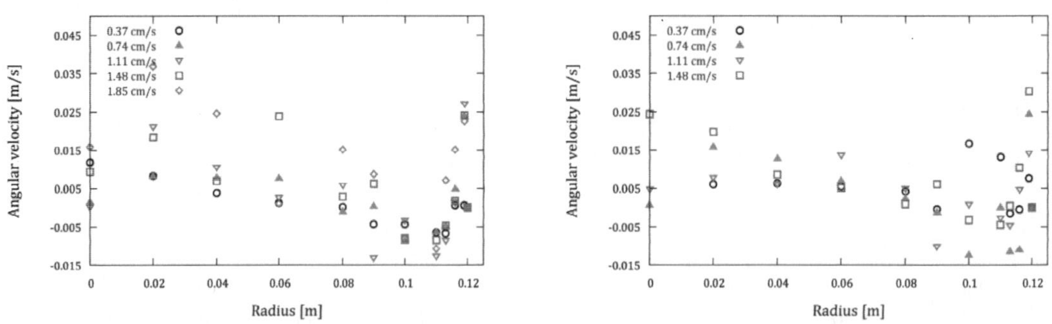

(a) $U_L = 0$ cm/s (batch mode) (b) $U_L = -3.3$ cm/s (counter-current mode)

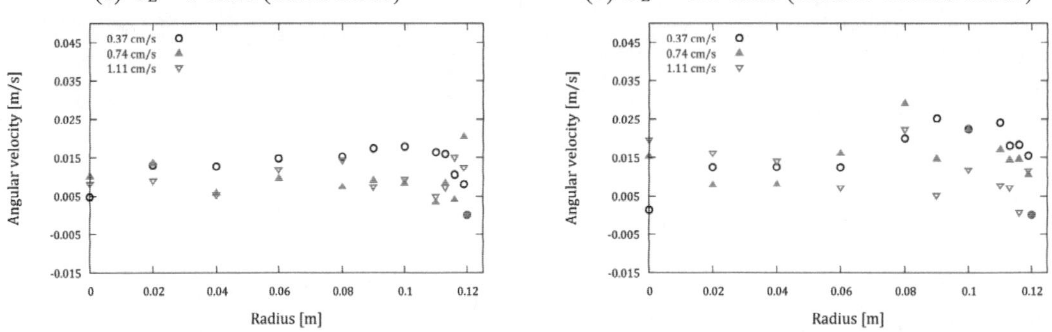

(a) $U_L = -6.6$ cm/s (counter-current mode) (b) $U_L = -9.2$ cm/s (counter-current mode)

Figure 10. Averaged angular liquid velocities for different superficial gas velocities at different counter-current flow setups 0.7 m above the gas sparger.

3.3. Description of the "local-scale" bubble column fluid dynamics

As stated above, three parameters determine the bubble column fluid dynamics at the "*local-scale*": (a) the liquid velocity, (b) the local gas fraction, and (c) the bubble size distributions (*BSDs*). To this end, Figure 11 couples the local liquid velocity profiles obtained in this study (Figure 8a), with the local void fraction profiles measured in ref. [7] and the *BSDs* reported in ref. [17] (which is a follow-up of the experimental investigation proposed in ref. [7]). The liquid flows upward in the center of the bubble column, it becomes zero at an intermediate radial position and, finally, it flows downward near the wall. The liquid velocity profiles is strictly related to the *BSDs* by the lift force. It is observed that near the wall, there is a higher number of "*small bubbles*". Indeed, it is well-known that the direction of the transversal lift force depends upon the bubble size and shape: for "*small bubbles*", the lift force acts in the direction of decreasing liquid velocity (i.e. the lift force pushes the small bubbles toward the wall); conversely, for "*large bubbles*" the lift force changes its direction (a force that can be assimilated to the lift force tends to push large and deformed bubbles towards the center of the column [18-20]). The distribution of the gas phase, imposed by the prevailing *BSDs*, determines the local void fraction profiles; owing to the poly-dispersed nature of the *BSDs* and given the local liquid velocity field, the local void fraction profile is center-pecked owing to the "*large bubbles*" moving towards the center of the bubble columns. Future studies should couple all these data by using a lift force approach and extend the present comparison to the counter-current mode.

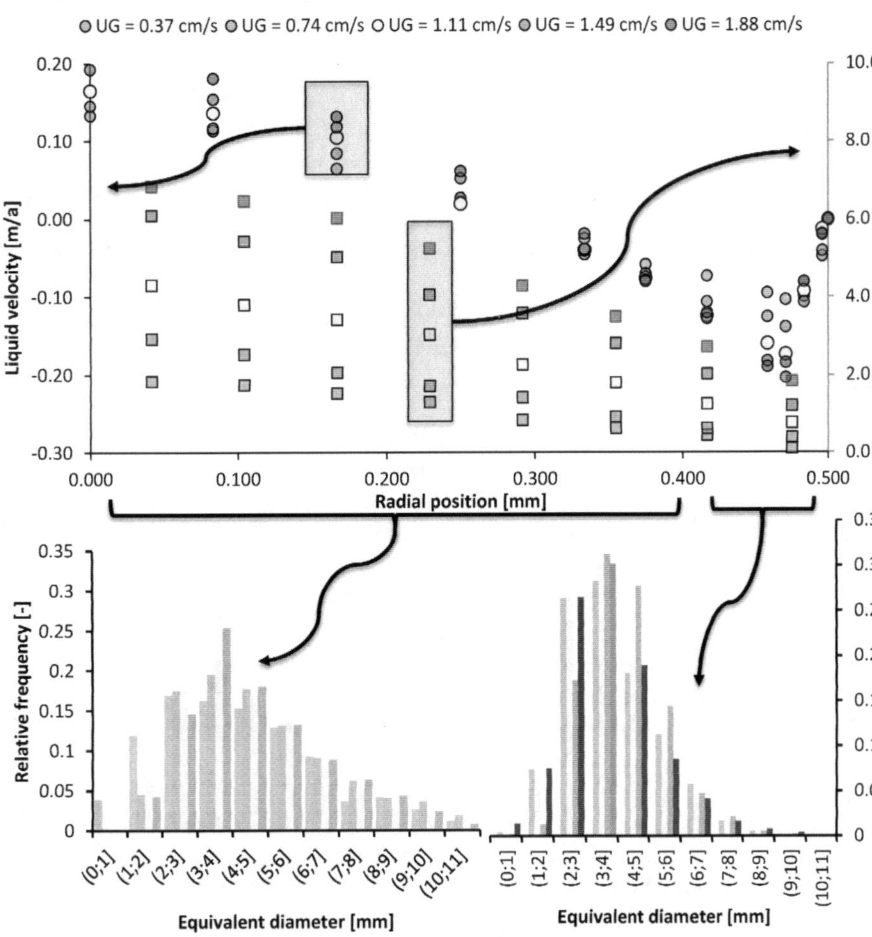

Figure 11. The "*local-scale*" bubble column fluid dynamics: liquid velocity measurements (Figure 8), local void fraction [7] and bubble side distributions [17] – batch mode.

4. Conclusions

A complete knowledge of bubble column fluid dynamics rely on understanding the global and the local fluid dynamic properties. Indeed, datasets encompassing the different scales are important to validate the capability of existing *CFD* models for upscaling. Unfortunately, most of the previous literature focused on the "*global-scale*" fluid dynamics, whereas a limited attention was devoted to the "*local-scale*". It is known that three parameters characterize the fluid dynamics at the "*local-scale*": liquid velocity, local gas fraction, and bubble size distributions (*BSDs*). The present study experimentally obtains local liquid velocity and turbulence data in a large-scale bubble column in the batch and in the counter-current flow modes. Such data complete a previous dataset concerning both global-scale (i.e., gas holdup) and local-scale (i.e., bubble size and shape, local volume fraction, ...) measurements, under the same operating conditions. It is found that, increasing the superficial liquid velocity in the counter-current mode flow only shifts the liquid velocity profiles downward. From the angular velocity, a distinct counter current circulating flow in batch mode is observed, which is dominated by the rotation of the counter-current liquid flow. The rotating counter-current liquid flow is very likely an inlet effect. The liquid velocity have been coupled with our previous measurements, to completely understand the "*local-scale*" fluid dynamics. In conclusion, the present study completes our previous studies and provides a comprehensive validation database for Computational Fluid-Dynamic (*CFD*) model validation. It has not escaped out notice that these data, coupled with the radial measurements may be used–in future studies–to provide detailed information on the local void fraction profiles and for developing a revised Drift Flux model [21] of large-scale bubble columns.

Acknowledgement
This work was partially funded by the Deutsche Forschungsgemeinschaft (DFG) under the project number 335448239.

References

[1] Rollbusch P, Bothe M, Becker M, Ludwig M, Grünewald M, Schlüter M and Franke R 2015 *Chem Eng Sci* **126** 660-678
[2] Besagni G, Inzoli F and Ziegenhein T 2018 *ChemEngineering* **2** 13
[3] Bartrand T A, Farouk B and Haas C N 2009 *Int. J. Multiphas Flow* **35**(2) 171-184
[4] Leonard C, Ferrasse J H, Boutin O, Lefevre S, and Viand A 2015 *Chem Eng Res Des* **76** 391-421
[5] Ueyama K and Terukatsu M 1979 *AIChE J* **25**(2) 258-266.
[6] Trivedi R, Renganathan T and Krishnaiah K 2018 *Chem Eng J* **338** 636-650,
[7] Besagni G and Inzoli F 2016 *Chem Eng Sci* **146** 259-290
[8] Roustan M and Wang R Y 1996 *Ozone Sci Eng* **18** 99-115
[9] Hidaka N, Mizuguchi K I and Matsumoto T 1998 *J Chem Eng Jpn* **31** 1016-1019
[10] Jin H, Yang S, He G, Wang M and Williams R A 2010 *J Chem Technol Biotechnol* **85** 1278-1283
[11] Deen N G, Westerweel J and Delnoij E 2002 *Chem. Eng. Technol* **25**
[12] Ziegenhein T, Garcon M and Lucas D 2016 *Chem Eng Sci* **153** 155-164
[13] Hessenkemper H and Ziegenhein T 2018 *Int J Multiph Flow* Accepted
[14] Wilkinson P M, Spek A P and van Dierendonck L L 1992 *AIChE J.* **38** 544-54
[15] Lucas D, Rzehak R, Krepper E, Ziegenhein T, Liao Y, Kriebitzsch S and Apanasevich P 2016 *Nucl Eng Des* **299** 2-11
[16] Ziegenhein T and Lucas D 2016 *Flow Meas Instrum* **48** 36-41
[17] Besagni G, Inzoli F, Ziegenhein T and Lucas D 2017 *Petroleum* **In press**
[18] Lucas D, Prasser H M and Manera A 2005 *Chem Eng Sci* **60** 3609-3619
[19] Tomiyama A, Tamai H, Zun I and Hosokawa S 2002 *Chem Eng Sci* **57** 1849-1858
[20] Santarelli C and Fröhlich J (2016) *Int J Multiph Flow* **81** 27-45
[21] Zuber N and Findlay J A 1965 *J. Heat Transf* **87** 453-68

IOP Conf. Series: Journal of Physics: Conf. Series **1224** (2019) 012037 doi:10.1088/1742-6596/1224/1/012037

Design and assessment of an experimental facility for the characterization of flow boiling of azeotropic refrigerants in horizontal tubes

Luigi Pietro Maria COLOMBO*, Andrea LUCCHINI*, Thanh Nhan PHAN*, Luca MOLINAROLI*, Alfonso NIRO*

* Dipartimento di Energia (Politecnico di Milano, via Lambruschini 4, 20156, Milano)

luigi.colombo@polimi.it

Abstract. An experimental facility was designed to measure pressure drop and heat transfer coefficient during flow boiling of azeotropic refrigerants in horizontal tubes. The apparatus is made of the refrigerant circuit, including the test section; the water circuit, to provide the power for evaporation; the glycol-water circuit, to fix the refrigerant pressure. The test section is a counterflow tube-in-tube heat exchanger (refrigerant inside, water outside). Plant assessment involved the measurement of pressure drop and heat transfer during R134a flow boiling in a smooth tube (outer diameter 9.56mm, inner diameter 8.92mm). Such a case study is a benchmark for the availability of a wide experimental database in the literature, which has been also summarized in many correlations. Then further tests involved a microfin tube. The experimental conditions were: evaporation temperature, 5°C; mass flux, 111÷333kg/m^{2}s; average quality, 0.15÷0.93; heat flux, 8.8kW/m^2 and 17.6kW/m^2. The uncertainty affecting the pressure drop and the heat transfer coefficient resulted lower than 1% and 5% respectively. The comparison with the literature shows satisfactory agreement with the major findings and enables using the data for the assessment of the existing models to predict both the pressure drop and the heat transfer coefficient.

1. Introduction

HVAC technologies are continually evolving to meet more and more severe requirements in terms of energy efficiency and environmental impact. The related heat transfer processes represent very often the bottleneck on improving the overall performance. In particular, if phase transitions are involved, the uncertainties related to the complex behavior of multiphase flows may significantly affect the design of a refrigeration system. A thorough characterization of boiling and condensation processes is then required to endow engineers with more and more accurate design tools.The task becomes more difficult if the smooth tube is replaced by a microfin tube. Many authors keep on working on this topic to provide experimental data [1, 2, 3, 4, 5, 6, 7] and correlations [8, 9, 10]. Accordingly, an experimental facility was designed to measure the pressure drop and the heat transfer coefficient during flow boiling of azeotropic refrigerants in horizontal tubes.Plant assessment involved the measurement of pressure drop and heat transfer during flow boiling of R134a in a smooth tube. The following step concerned the study of a microfin tube.

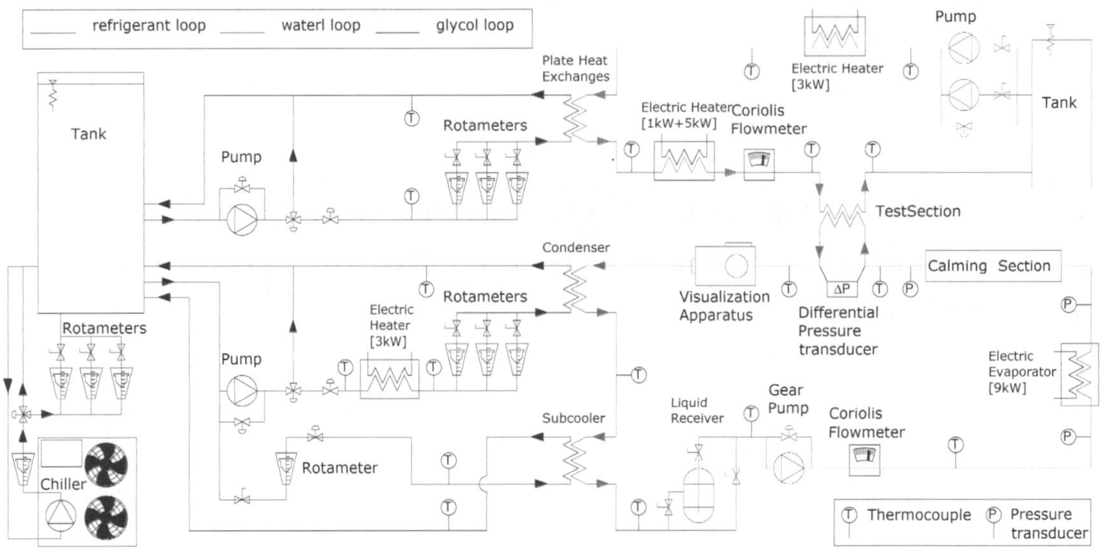

Figure 1. Experimental apparatus scheme.

2. Experimental apparatus

The experimental apparatus (Figure 1), is made of three main circuits named: the refrigerant loop (filled with R134a), the water loop (filled with demineralized water) and glycol loop (filled with a mixture of water and ethylene glycol, 30% volume concentration). They exchange thermal power each other in order to set the test condition and perform the experiments.

2.1. *The glycol loop*

The glycol loop (blue line in Figure 1) is a service circuit with two main tasks: to set the operating temperature in the test section (fixing the pressure in the condenser) and to chill both refrigerant and water. A commercial chiller (21kW cooling capacity) cools the mixture to -10°C and then it is stored in a 0.75m^3 tank. Two independent loops, one for the water and the other for therefrigerant, are connected to the tank.

The loop dedicated to the water cools down the water entering in the test section, which heats up as a consequence of the viscous dissipation.

The other loopis in charge to set the temperature in the test section and to prevent cavitation in the refrigerant pump. The former operation is accomplished setting the mass flow rate, using a manual needle valve, and the temperature (checked by a K-type thermocouple) at the condenser inlet, using a P.I.D. driven electric heater (3kW), such that the refrigerant pressure at the test section inlet (which is the sum of the pressure in the condenser and the pressure drop in between) be the saturation pressure corresponding to the test temperature. For the latter operation a bypass drains part of the cold mixture headed to the electric heater (the volume flow rate is tuned using a manual needle valve) to subcool the liquid refrigerant entering the pump.

2.2. *The water loop*

The water loop (green line in Figure 1) is designed to exchange the thermal power required for the refrigerant phase change. The heat transfer takes place in a tube-in-tube heat exchanger (water in the annulus, refrigerant in the inner duct), named test section (Figure 2), thermally insulated from the surroundings with 100 mm thick shell of rubber foam. A pump drains the water from the tank (thermally insulated with a Rockwool shell 5cm thickness, volume 0.2m^3), a bypass and a manual needle valve fix the mass flow rate, its value is measured by a Coriolis flowmeter (range [0;400]kg/h, uncertainty 0.15% of the reading). Afterwards a K-type thermocouple checks the water temperature

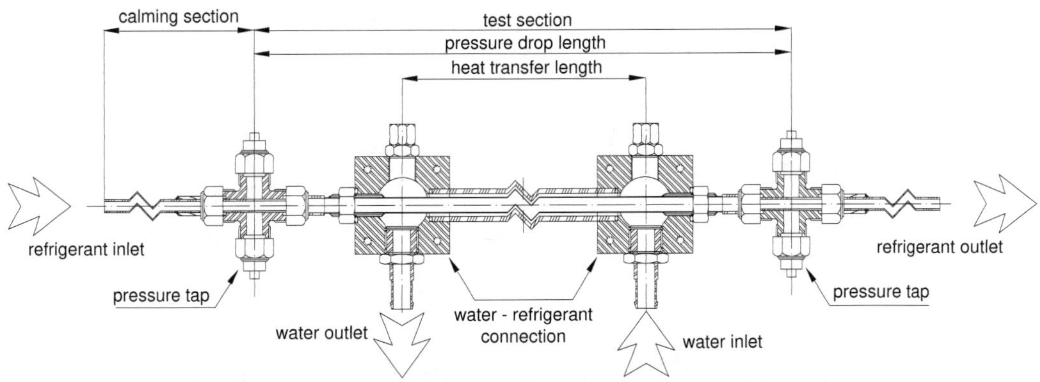

Figure 2. Test section.

and the flowenters in a plate heat exchanger (if cooling is required glycol is flushed in the other side), then an P.I.D. driven electric heater (it is made of two elements: 1kW and 5kW, it is possible to use both or only one) sets the inlet temperature in the test section (checked by a K-type) such that the power transfer causes a ±2°C temperature change of the water flow (inlet and outlet temperature are provided by two groups of 3 K-type thermocouples connected in series, uncertainty 0.1K). Finally the water returns to the tank.

2.3. *The refrigerant loop*

The main goal of the refrigerant loop (red line in Figure 1) is to provide a two phase flow of refrigerantat the operating conditions (which are defined by the mass flow rate, the inlet quality and the inlet temperature), at the inlet of the test section, where the measures take place.

A saturated liquid flow of refrigerant leaves the condenser (four plate heat exchanges and one shell end tube heat exchanger in parallel, depending on the thermal duty it is possible to use one or more of them) and enters in a plate heat exchanger (subcooler) to be chilled and to prevent cavitation in the circulation pump (gear type with inverter driven engine for the mass flow rate tuning). A Coriolis flowmeter (range [0;400]kg/h, uncertainty 0.15% of the reading) records the mass flow rate while a pressure transducer (relative, range [-1;30]bar, uncertainty ±1% of full scale) and a thermocouple (K-type, uncertainty 0.1K) check the thermodynamic state of the refrigerant as it enters in the evaporator (it is a set of 8 electric heaters, 9kW total power, driven by a software control system) which provides the power to vaporize the amount of refrigerant specified by the test section inlet quality. A straight duct (4.7 m long adiabatic duct and thermally insulated by a 5cm thick rubber foam shell), designed for the development of the two phase flow regime, is placed before the test section. Then the refrigerant enters in the test section, the visualization apparatus, for the recording and identification of the flow pattern, followsand, in the end, the refrigerant returns to the condenser.

The test section (Figure 2) is a tube in tube heat exchanger (heat transfer length L=1.11m)

Table 1. Microfin tube characteristics

Tube type			J60	smooth
inner diameter (fin root)	D_r	[mm]	8.96	8.92
outer diameter	D	[mm]	9.52	9.52
wet perimeter	S_p	[mm]	44.9	28.0
cross section area	A_c	[mm^2]	62.1	62.5
hydraulic diameter	D_h	[mm]	5.28	8.92
exchanging area ratio			1.68	1
fin number	n	[N]	60	
height	H	[mm]	0.2	
apex angle	α	[°]	40	
helix angle	β	[°]	18	

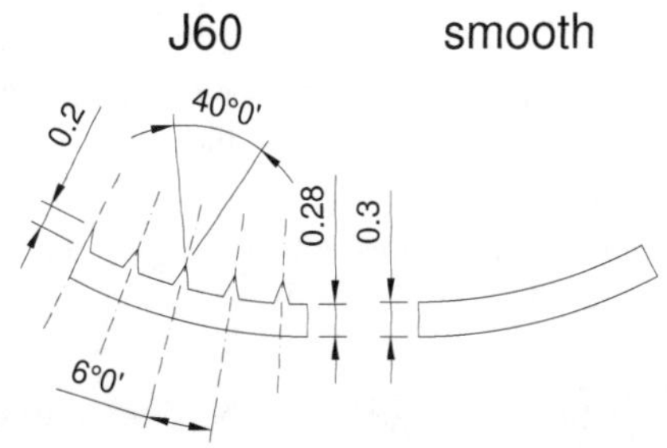

Figure 3. Microfin tube geometric characteristics

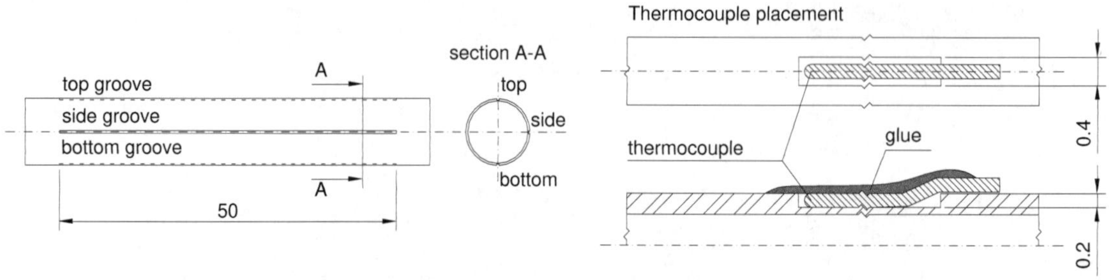

Figure 4. Thermocouple and grooves positions.

thermally insulated with 10 cm thick rubber foam shell. As different geometries are available for the inner surface of the refrigerant duct, the smooth tube is the reference to compare the microfin tube. Their geometrical characteristics are reported in Table 1 while Figure 3 shows the main differences.

To record the pressure drop a differential pressure transducer (range [-1.0;1.0]bar, uncertainty ±0.1% of the full scale), connected to two pressure taps (separated by a distance of L_p=1.3m), is used. A relative pressure transducer (range [-1;16]bar, uncertainty ±0.25% of the full scale), connected to the tap at the inlet of the test section, reads the refrigerant pressure. The refrigerant inlet and outlet temperatures are the saturation temperatures given by the pressure readings. Two groups (one at the entrance and one at the exit) of three thermocouples are glued inside grooves (length 50mm, depth 0.15mm, width 0.4mm) on the outside on the inner tube (top, side and bottom position), to measure the wall temperatures (Figure 4). The reference junction of each thermocouple (K-type, uncertainty 0.1K) is inserted in a Dewar flask filled with melting ice.

3. Data reduction

All the measuring devices are connected, via an acquisition board, to a computer which is endowed of a program to read and to store their signals. As the data acquisition ends another program performs the post processing to compute the quantities that identify the experimental operating conditions (refrigerant mass flux G, quality change in the test section Δx, average quality in the test section x_m), the pressure drop per unit length $\Delta P/L_p$ and the heat transfer coefficient h. One data point is obtained for each experimental condition, averaging 12 acquisitions; each of them is the mean value of the readings recorded during a time interval of 180s. As the sampling frequency is 1Hz, 181 samples are stored during each acquisition. Preliminary experiments were performed on single phase vapor flow to check the data acquisition and the post-processing. The results showed an agreement within ±5% with the most commonly used correlations for both heat transfer and pressure drop.

Referring to the microfin tube nominal geometrical characteristics, the net cross section area is:

$$A_c = \frac{\pi D_r^2}{4} - \frac{nH^2}{\cos(\beta)} tg\left(\frac{\alpha}{2}\right) \quad (1)$$

From the Coriolis flowmeter reading the refrigerant mass flux is computed as follows:

$$G = \frac{\dot{m}_r}{A_c} \quad (2)$$

Assuming steady state and negligible the thermal dispersion in the test section, power transfer takes place only between water and refrigerant:

$$Q = \dot{m}_a c_{pa}(T_{ai} - T_{ao}) \quad (3)$$

The quality change can be computed as:

$$\Delta x = (x_o - x_i) = \frac{\dot{Q}}{\dot{m}_r h_{lv}(p_r)} \quad (4)$$

The inlet quality can be computed performing an energy balance on the evaporator, assuming steady state and negligible thermal dispersion:

$$x_i = \frac{Q_e - \dot{m}_r c_{pr}[T_s(p_r) - T_{ri}]}{\dot{m}_r h_{lv}(p_r)} \quad (5)$$

In the end the mean quality in the test section is:

$$x_m = x_i + \frac{\Delta x}{2} \tag{6}$$

The pressure drop per unit length is immediately computed as the distance L_p between the pressure taps at the edges of the test section is known and the total (frictional and accelerative) pressure drop Δp is measured by a differential pressure transducer.

For the average heat transfer coefficient h in the test section a longer procedure is required. As it is based on the logarithmic mean temperature between wall and refrigerant at first temperatures are required. As three thermocouples read the wall temperature, both at the inlet and at the outlet of the refrigerant, their readings are averaged to get a single value for the wall temperature.

$$T_w = \frac{T_t + T_s + T_b}{3} \tag{7}$$

The inlet and outlet refrigerant temperatures are the saturation temperatures corresponding to the inlet (measured by a relative pressure transducer) and the outlet pressure (computed subtracting the pressure drop on the test section to the inlet pressure).

$$T_r = T_{sat}(p) \tag{8}$$

It follows that the logarithmic mean temperature difference is:

$$\Delta T_{ml} = \frac{(T_{wo} - T_{ro}) - (T_{wi} - T_{ri})}{ln\frac{T_{wo}-T_{ro}}{T_{wi}-T_{ri}}} \tag{9}$$

The heat transfer area of the refrigerant duct is referred to the fin rootdiameter, hence the average heat transfer coefficient in the test section is given by the following equation:

$$h = \frac{Q}{\pi D_r L \Delta T_{ml}} \tag{10}$$

Uncertainty analysis showed that the logarithmic mean temperature difference has the largest influence on the average heat transfer coefficient uncertainty. For instance, Figure 5 shows the behaviour for experiments at Q=288W (relative uncertainty 2.5%), L=1.11m (relative uncertainty 1.4%), D_r=8.96mm (the nominal fin root diameter is considered uncertainty free); for temperature readings the uncertainty is U_T=0.1K. The chart shows that, even if a single measure has a very large uncertainty (it is never lower than 3%), the average value of a single acquisition proves to be lower than 5% for all the experimental operating conditions. The same test was repeated for all the operating conditions and the same outcome followed.

4. Experimental results

The experimental conditions were identified by four quantities: the refrigerant inlet temperature in the test section, T_{ri}=5°C; the refrigerant mass flux, G ranges in the interval [110;334]kg/m^{2}s; the mean quality, x_m ranges in the interval [0.15;0.95], the heat flux referred to the fin root surface of the microfin tube, q_1=8.6kW/m^2 and q_2=2q_1=17.2kW/m^2.Heat flux values were chosen such that the quality change Δx in the test section ranges from 0.06 to 0.2.

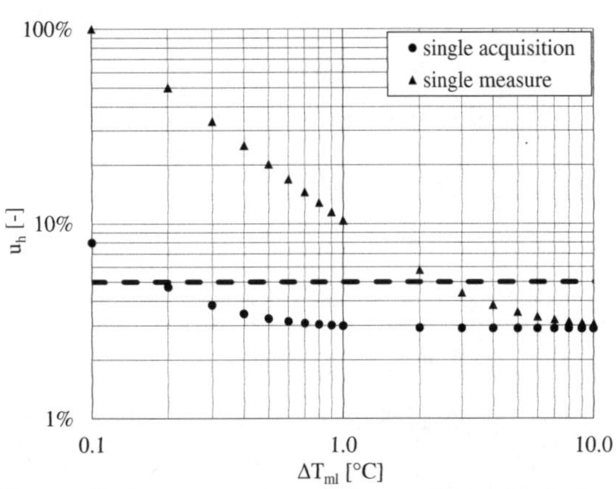

Figure 5. Average heat transfer coefficient relative uncertainty.

4.1. Smooth tube

Figure 6 reports the pressure drop per unit length for all the operating conditions tested using the smooth tube. As expected as the mass flux grows the pressure drop increases on the contrary the influence of heat flux seems to be negligible. As the mean quality increases ($0.15<x_m<0.85$) pressure drop grows then

IOP Conf. Series: Journal of Physics: Conf. Series **1224** (2019) 012037 doi:10.1088/1742-6596/1224/1/012037

($0.85<x_m<0.90$) a maximum is reached and finally ($0.85<x_m<0.95$) pressure drop falls off toward the data point for the only vapour phase pressure drop.

The heat transfer coefficient h (Figure 7) is mainly affected by the heat flux, as the latter increases the former does the same. Similar is the effect of the mass flux on the heat transfer coefficient, as the mass flux grows the heat transfer coefficient increases too.

In the end is it possible to observe, as expected, that the heat transfer increases as the mean quality increases as long as the thermal crisis takes place (x_m in the range [0.80;0.95]), then a very steep reduction of the heat transfer coefficient can be observed.

4.2. Microfin tube

For the microfin tube the effect of the heat flux, the mass flux and the mean quality on the pressure drop per unit length is the same described for the smooth tube.The only difference, as can be observed from Figure 8, is that pressure drop is larger for microfin tube, as a consequence of the fin presence.

As the heat flux increases the heat transfer coefficient h grows (Figure 9).About the connection between mass flux and heat transfer coefficienta difference, between smooth and microfin tube, can be observed.From the chartin Figure 9 two main comments arise: at low mass flux values the heat transfer coefficient raises as the mass flux grows; at high mass flux values the heat transfer coefficient is independent of mass flux. That behavior could be related to a flow regime transition from stratified flow to annular flow.In the end is it possible to observe, as expected, that the heat transfer increases as the mean quality increases, moreover the microfin tube seems to be very effective in keeping the liquid in touch with the whole perimeter of the tube because no thermal crisis was observed.

5. Comparison between experimental data and correlations

The comparison between the experimental data and the correlation prediction (reported only for the microfin tube for space constraint) is performed computing the average percentage error E and the standard deviation σ of E.

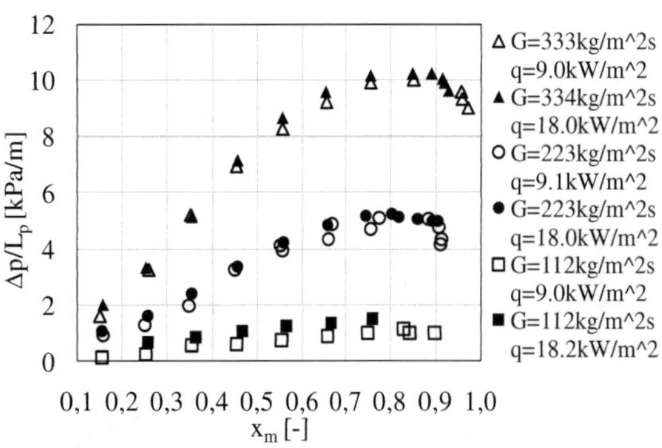

Figure 6. Pressure drop, smooth tube.

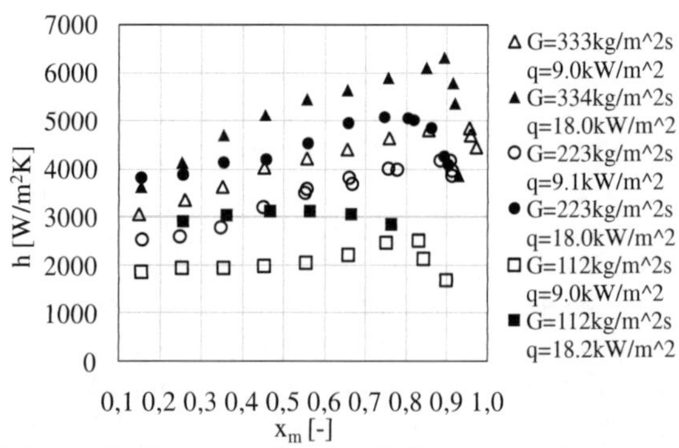

Figure 7. Heat transfer coefficient, smooth tube.

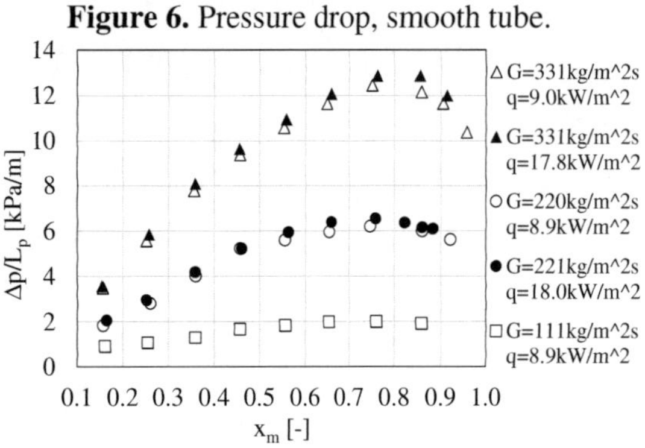

Figure 8. Pressure drop, microfin tube.

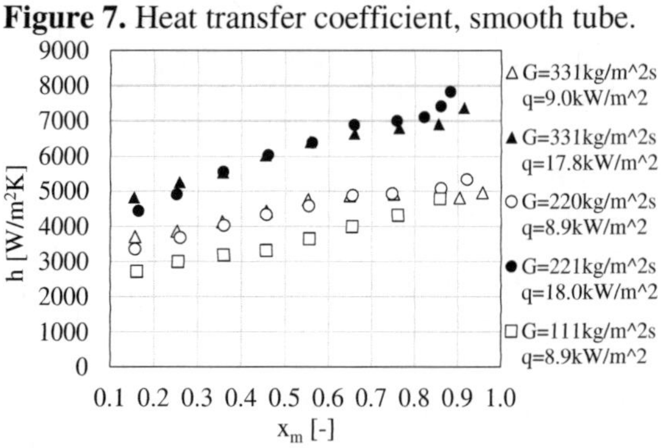

Figure 9. Heat transfer coefficient, microfin tube.

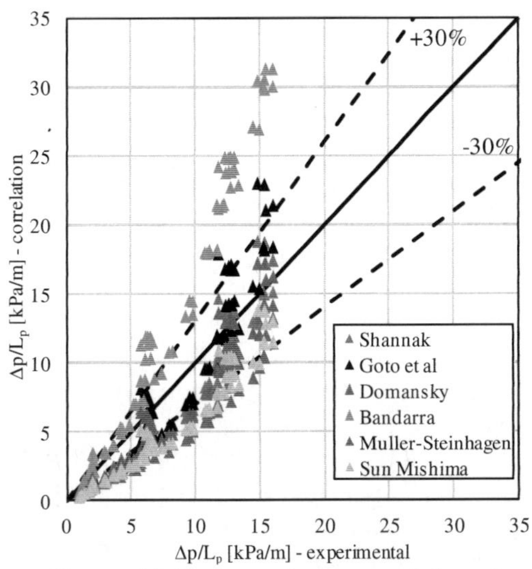

Figure 10. Pressure drop per unit length.

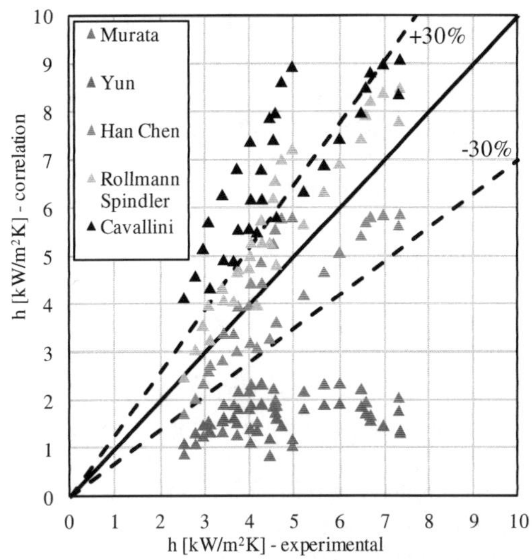

Figure 11. Heat transfer coefficient.

$$E = \frac{1}{N}\sum_{j=1}^{N}\frac{x_{j,c} - x_{j,e}}{x_{j,e}} \tag{11}$$

As reported in Table 2 and Figure 10, the majority of the correlation for pressure drop checked, but Bandarra, underestimates the experimental data. Goto et al. is the correlation with the smallest percentage error, but its standard deviation is large, and it seems unable to properly describe the trend of the pressure drop. On the contrary Sun-Mishima's and Domansky's have the smallest standard deviation but very large percentage errors, meaning that a further calibration of the parameters is required.

Table 3 and Figure 11 shows that, concerning heat transfer, the Yun's and Murata's correlations largely underestimates the data, while Cavallini's does the opposite. Han-Chen correlation shows a better agreement with the data, it has the smallest error, but the largest standard deviation.

6. Conclusions

It is known that the microfin tube exhibits a significant heat transfer enhancement, when compared to the smooth tube, butpressure drop increases as well. The experimental data shows that the latter increases slightly more (9%). Comparison between the predictions of various correlations from literature andexperimental data both on heat transfer and pressure drop was proposed. The best resultsfor the heat transfer are achieved by the Han Chen scheme of correlation, whereas, for the pressure drop, it seems that the influence ofgeometry is still not completely described in the models considered.

Acknowledgments

The financial support of MIUR through the program PRIN 2015 (Grant Number 2015M8S2PA) is greatly acknowledged. The authors thank Fabio Pulcini for the aid in taking measurement.

Table 2. Pressure drop

Correlation	E	σ
Shannak	-24%	21%
Muller-Steinhagen	-27%	20%
Bandarra	55%	34%
Sun Mishima	-39%	14%
Domansky	-41%	15%
Goto et al	-14%	31%

Table 3. Heat transfer coefficient

Correlation	E	σ
Murata	-53%	15%
Yun	-61%	9%
Han Chen	12%	32%
Rollmann Spindler	33%	22%
Cavallini	53%	25%

IOP Conf. Series: Journal of Physics: Conf. Series **1224** (2019) 012037 doi:10.1088/1742-6596/1224/1/012037

Nomenclature: Latin symbols

A_c	cross section area [m^{-2}]	p_r	refrigerant pressure [Pa]
c_{pa}	water specific heat capacity [$J \cdot kg^{-1} \cdot K^{-1}$]	Q	power exchanged in the test section [W]
c_{pr}	refrigerant specific heat capacity [$J \cdot kg^{-1} \cdot K^{-1}$]	Q_e	power provided by the electric evaporator [W]
D	outer diameter [m]	S_p	wet perimeter [m]
D_h	hydraulic diameter [m]	T_{ai}	water inlet temperature [K]
D_r	inner diameter (fin root) [m]	T_{ao}	water outlet temperature [K]
E	mean percentage error [-]	T_b	temperature in the bottom position [K]
G	refrigerant mass flux [$kg \cdot m^{-2} \cdot s^{-1}$]	T_{ri}	refrigerant inlet temperature [K]
H	fin height [m]	T_{ro}	refrigerant outlet temperature [K]
h	heat transfer coefficient [$W \cdot m^{-2} \cdot K$]	T_s	temperature in the side position [K]
h_{lv}	liquid vapour enthalpy of phase change [$J \cdot kg^{-1}$]	T_t	temperature in the top position [K]
		T_{wi}	mean wall temperature, refrigerant inlet [K]
L	heat transfer length [m]	T_{wo}	mean wall temperature, refrigerant outlet [K]
L_p	distance between the pressure taps [m]	x_i	refrigerant inlet quality [-]
$\dot{m}_a$	water mass flow rate [$kg \cdot s^{-1}$]	$x_{j,c}$	correlation value [-]
$\dot{m}_r$	refrigerant mass flow rate [$kg \cdot s^{-1}$]	$x_{j,e}$	experimental value [-]
n	fin number	x_m	average quality in the test section [-]

Nomenclature: Greek symbols

α	apex angle [°]	ΔT_{ml}	log mean temperature difference [K]
β	helix angle [°]	Δx	quality change [-]
Δp	pressure drop [Pa]	σ	standard deviation [-]

References

[1] Bandarra Filho E P, Saiz Jabardo J M, EL Barbieri P, 2004 Convective boiling pressure drop of refrigerant R-134a in horizontal smooth and microfin tubes. *International journal of refrigeration* **27(8)** 895–903

[2] Shannak B A, 2008 Frictional pressure drop of gas liquid two-phase flow in pipes. *Nuclear Engineering and Design* **238(12)** 3277–3284

[3] Han X H, Fang Y B, Wu M, Qiao X G, Chen G M, 2017 Study on flow boiling heat transfer characteristics of R161/oil mixture inside horizontal micro-fin tube. *International Journal of Heat and Mass Transfer* **104** 276–287

[4] Rollmann P, Spindler K, Muller-Steinhagen H, 2011 Heat transfer, pressure drop and flow patterns during flow boiling of R407c in a horizontal microfin tube. *Heat and mass transfer* **47(8)** 951-958

[5] Cavallini A, Del Col D, Mancin S, Rossetto L, 2009 Condensation of pure and near-azeotropic refrigerants in microfin tubes: A new computational procedure. *International Journal of Refrigeration* **32(1)** 162–174

[6] Colombo L P M, Lucchini A, Muzzio A, 2012 Flow patterns, heat transfer and pressure drop for evaporation and condensation of R134a in microfin tubes *International Journal of Refrigeration* **35(8)** 2150–2165

[7] Goto M, Inoue N, Ishiwatari N, 2001 Condensation and evaporation heat transfer of R410a inside internally grooved horizontal tubes. *International Journal of Refrigeration* **24(7)** 628–638

[8] Sun L, Mishima K, 2009 Evaluation analysis of prediction methods for two-phase flow pressure drop in mini-channels. *International Journal of Multiphase Flow* **35(1)** 47–54

[9] Muller-Steinhagen H, Heck K, 1986 A simple friction pressure drop correlation for two-phase flow in pipes. *Chemical Engineering and Processing: Process Intensification* **20(6)** 297–308

[10] Xu Y, Fang X, Su X, Zhou Z, Chen W, 2012, Evaluation of frictional pressure drop correlations for two-phase flow in pipes. *Nuclear Engineering and Design* **253** 86-97

IOP Conf. Series: Journal of Physics: Conf. Series **1224** (2019) 012038 doi:10.1088/1742-6596/1224/1/012038

Condensation heat transfer in minichannels: a review of available correlations

M Azzolin, A Berto, S Bortolin, D Del Col

Department of Industrial Engineering, University of Padova, Via Venezia 1, 35131 Padova, Italy

Corresponding author: Davide Del Col; e-mail: davide.delcol@unipd.it; tel.: +39-049-8276891, fax: +39-049-8276896

Abstract. Heat exchangers with enhanced performance are demanded in various engineering applications. Very often heat transfer devices are requested to guarantee not only high heat transfer coefficients but also small size and weight, thus limiting the charge of the operative fluid inside the heat exchanger. In order to increase the performance of condensers and to properly design new heat exchangers, it is essential to have predictive tools that are validated with experimental data. Sometimes, even well established semi-empirical correlations can be inaccurate in some microscale flow conditions or with new refrigerants. The present paper starts from the experimental database measured during condensation with different fluids (pure fluids and refrigerant blends) inside small channels (hydraulic diameter around 1 mm) at the Department of Industrial Engineering of the University of Padova. A critical review of available correlations for heat transfer during condensation in minichannels is presented. Predictions of heat transfer coefficients obtained applying selected models are compared with the experimental database that covers various refrigerants: hydrofluorocarbons (HFC, i.e. R32 and R134a), new hydrofluoroolefins (HFOs) with low global warming potential (R1234ze(E)), natural refrigerants (hydrocarbons such as propane) and zeotropic refrigerant blends of HFCs and HFOs (R32/R1234ze(E)). Refrigerant mixtures are studied because for some applications they may be a proper solution. For instance in the air-conditioning industry there are not drop-in pure fluids to replace the high global warming potential (GWP) fluids currently employed (e.g. R410A). Unfortunately, the design of condensers working with zeotropic mixtures poses the additional problem to account for the mass transfer resistance that leads to a penalization of the heat transfer coefficient. Experimental data are necessary for the assessment of predicting correlations that can be used with these new refrigerants blends.

1. Introduction

In convective heat transfer applications, flow passages sizes are shifting towards smaller dimensions in order to enhance the heat transfer coefficient and to reduce the charge of the working fluid inside the heat exchanger, limiting the risks connected to the release of potentially hazardous fluids in the atmosphere (Kandlikar et al. [1]). Minichannels devices are spreading in many applications, such as automotive condensers, air-conditioning systems, heat pipes for electronics cooling or spacecraft thermal control.

Condensation in minichannels differs from condensation in macrochannels due to the different interactions between gravity force, surface tension and gas-liquid interfacial shear stress. In

IOP Conf. Series: Journal of Physics: Conf. Series **1224** (2019) 012038 doi:10.1088/1742-6596/1224/1/012038

conventional tubes, the condensation process is dominated by vapor shear stress and gravity, while in minichannels the surface tension force can play an important role over gravitational force (Del Col et al. [2]). Depending on the operating conditions (mass velocity, fluid properties, vapor quality and channel shape), the relative role of shear stress, gravity force and surface tension can affect the transition between flow patterns, and thus the heat transfer mechanism, in a different way.

Basic attempts to develop flow regime maps for minichannels were done in adiabatic flow conditions (air-water, nitrogen-water or air-oil mixtures). Triplett et al. [3] conducted an investigation on flow regimes in adiabatic air-water flow inside circular minichannels of 1.1 and 1.45 mm diameter. As also stated by other investigators, stratified flow was not observed, while bubbly, slug, slug-annular and annular flow patterns were identified. Coleman and Garimella [4,5,6] and Garimella [7] were the first to conduct flow visualizations during condensation of pure fluid R134a inside nine different tubes of round, square and rectangular cross-sections with hydraulic diameter in the range 1-4.91 mm over the mass flux range $150 < G < 750$ kg m^{-2} s^{-1}. Four major flow regimes, including annular, intermittent, wavy and dispersed flow, were identified. The size of the intermittent region increases as the hydraulic diameter decreases, with this effect being greater at the low mass fluxes due to the fact that surface tension achieves greater significance in comparison with gravitational force at these dimensions. Furthermore, as the hydraulic diameter decreases, the effects of surface tension increasingly counteract the gravity force, promoting and extending the size of the annular film flow pattern region instead of the stratified wavy flow regime.

Correlations developed for condensation inside circular conventional tubes in some cases are even able to predict quite accurately the heat transfer coefficient inside smaller diameters. This agreement occurs mainly at high mass velocities, when the flow is mostly annular and the condensation heat transfer is dominated by the vapour shear stress on the liquid film. However, at lower mass velocities, when the interfacial shear stress is not the dominant force that affects the condensation process, models developed for greater diameter tubes tend to fail (Del Col et al. [2]). At these values of mass velocity, micro-scale characteristics of heat transfer process, and in particular the surface tension, become more important [8, 9].

In the present paper, several models for the heat transfer coefficient prediction during condensation inside minichannels are presented and compared against the experimental database obtained in the Two-phase Heat Transfer Laboratory of the University of Padova. Experimental tests were conducted with pure fluids and with a low-GWP non-azeotropic binary mixture inside a 0.96 mm inner diameter circular minichannel. The test section is a counterflow tube-in-tube heat exchanger in which the refrigerant, entering as a two-phase mixture with the desired value of the vapour quality, is cooled by the water flowing in the external annulus. A detailed description of the experimental apparatus is provided in Matkovic et al. [10]. The tested binary mixture is a blend of R32 and R1234ze(E) at 46/54% by mass composition, presenting a $GWP_{100years} = 309$ and a temperature glide of 8.2 K. Details about the database are reported in Table 1. In particular, it is worth underlining that the saturation temperature for the non-azeotropic mixture of R32/R1234ze(E) is intended as the mean value between dew and bubble temperatures.

Table 1. Condensation experimental database taken inside the 0.96 mm diameter channel.

Authors	*Fluid*	t_{sat} [°C]	$T_{sat} - T_w$ [K]	G [kg m^{-2} s^{-1}]
Matkovic et al. [10]	R134a, R32	40	5-18	100-1200
Del Col et al. [11]	R1234ze(E)	40	10-15	100-800
Cavallini et al. [12]	R245fa	40	9-16	100-500
Del Col et al. [13]	R290 (propane)	40	11-20	100-1000
Del Col et al. [14]	R32/R1234ze(E) (46/54% by mass composition)	40	10-16	150-800

2. Prediction of the heat transfer coefficient

In the present paper, several condensation models originally developed for macrochannels ($d > 3$ mm) and pure fluids have been considered for the comparison against the experimental data obtained in a 0.96 mm diameter minichannel test section. These models can be classified between shear-driven models, specifically developed for annular flow condition, and multi-regime models, for films under the simultaneous influence of shear and gravity. Furthermore, when condensation models developed for pure fluids are applied to non-azeotropic mixtures, an additional correction is needed to account for the mass transfer and diffusion phenomena that characterize the condensation process.

2.1. Condensation models for pure fluids

The shear-driven model by Akers et al. [15] was developed for annular flow condensation of pure fluids and azeotropic blends. It is based on the definition of the all-liquid flow rate, which provides the same heat transfer coefficient as an annular condensing flow. The all-liquid flow rate is expressed by an "equivalent" Reynolds number and used in a single-phase, turbulent flow equation to predict the condensation coefficient. However, the assumptions on which the equivalent Reynolds number is based result in the underprediction of many researchers' data.

Starting from the work of Akers et al. [15], Moser et al. [16] developed a condensation model for smooth tubes. A new equivalent Reynolds number approach was proposed, based on the heat-momentum analogy, which states that a relationship exists between heat transfer and wall shear stress. During single-phase flow, an increase in wall-shear stress results in a heat transfer increase. Similarly, in the two-phase flow, it is expected that, as the wall shear stress increases, the heat transfer will also increase. Therefore, if a condensing two-phase flow is replaced by an all-liquid flow with the same wall shear stress, the analogy suggests that the single-phase heat transfer coefficient would be equal to the condensation coefficient, providing that the driving temperature differences, which define the single-phase and condensing heat transfer coefficients, are the same. The two-phase multiplier for the evaluation of the pressure gradient can be predicted using several methods, such as the correlation by Zhang and Webb [17], developed for small diameter channels. The Moser et al. [16] model was validated against 1197 data points from 18 sources, regarding refrigerants R12, R22, R113, R134a, R410A and R125, inner diameters from 3.14 to 20 mm. Figure 1 shows the comparison between experimental heat transfer coefficients taken inside the 0.96 mm diameter minichannel and calculated values with the Moser et al. [16] model. On average, the Moser et al. [16] correlation shows a good agreement with the present database, with a mean absolute deviation e_{AB} equal to 13.9% and a standard deviation σ_N of 19.7%. The disagreement between experimental and calculated values is greater at low mass fluxes: indeed, at mass velocity in the range G = 100-200 kg m^{-2} s^{-1} the mean absolute deviation is equal to 15.6% and the standard deviation is of 20.5%. This should be due to the fact that the basic assumption of annular flow condensation may not be satisfied in this case. When considering the pure fluid R245fa, it is overpredicted at almost all the mass fluxes; this may be related to the different thermodynamic properties (low vapour density, high liquid density and kinematic viscosity) and to the lower operating pressures of R245fa with respect to the other tested fluids. It is interesting also to notice that the model by Moser et al. [16] is able to predict with good accuracy the experimental data of R290, except some points at low mass velocity, although it has not been developed for hydrocarbons (e_{AB} = 10.9%, σ_N =7.83%).

Cavallini et al. [18] developed a semi-empirical model to compute the heat transfer coefficient during condensation inside horizontal smooth tubes with inner diameter greater than 3 mm. This simple and easy-to-use model includes only one transition between two different regions: ΔT-dependent and ΔT-independent (where ΔT is the saturation-minus-wall temperature difference). The model includes in these two categories all possible flow regimes (annular flow, stratified-wavy flow, stratified smooth flow and slug flow). This transition criterion was obtained from an analysis of condensation heat transfer data inside macrochannels. The model was validated against a total number of 4471 data points relative to HCFCs, HFCs, HCs, carbon dioxide, ammonia and water from several independent laboratories. As can be seen from Fig. 2, the model by Cavallini et al. [18] is able to predict the experimental data inside the 0.96 mm circular minichannel with a mean absolute deviation e_{AB} equal to 9.5% and a standard deviation σ_N equal to 12.3%. Only data points with mass velocity in the range G = 100-200 kg m^{-2} s^{-1} are clearly underpredicted by more than 20% using this model; in this range of mass velocity the mean absolute deviation between calculated and experimental data is equal to 12.3% and the standard deviation is of 14.3%. This higher deviation may be related to the different flow patterns occurring in minichannel at low mass velocity as compared to macrochannels. Indeed, in the Cavallini et al. [18] model the effect of the surface tension is not taken into account in the development of the flow pattern map and this can lead to inaccurate predictions of the heat transfer coefficient at low mass fluxes.

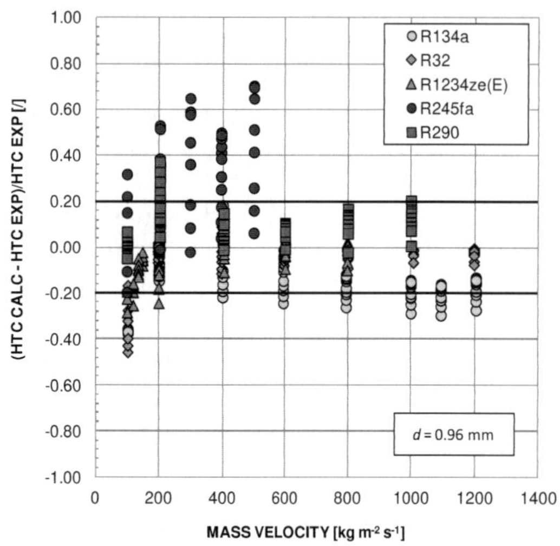

Fig. 1 Comparison between experimental and predicted values by the Moser et al. [16] model in the 0.96 mm diameter test section.

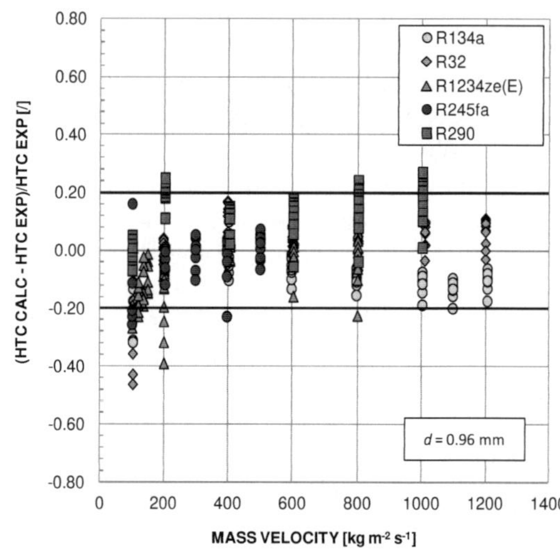

Fig. 2 Comparison between experimental and predicted values by the Cavallini et al. [18] model in the 0.96 mm diameter test section.

Shah [19] presented two comprehensive correlations for the prediction of the heat transfer coefficient, starting from the Shah [20] model developed for condensation inside macrochannels. Three different regimes were identified: Regime I corresponding to intermittent, annular and mist flow, Regime II corresponding to wavy flow and Regime III corresponding to stratified flow. The ultimate correlations proposed in Shah [19] can be applied also to minichannels, with specific adjustments to the original correlation. Indeed, at high mass flow rates, inertia force dominates the surface tension force while at low flow rates the surface tension force dominates. For this reason, an additional condition on the Weber number for all mass flowing as a vapour We_{gt} (which expresses the ratio between surface tension force and inertia force) was added to the original correlation. Furthermore, for ΔT-independent flow regime, one of the two versions of the correlation uses the equation of Cavallini et al. [18] to improve its accuracy. The Shah [19] correlations were validated by comparison with a database that includes 33 fluids, diameters from 0.10 to 49.0 mm, reduced pressures in the range 0.0008-0.946, mass velocity from 1.1 to 1400 kg m^{-2} s^{-1}, various tube shapes and orientations.

As it can be seen from Fig. 3, most of the data of the present experimental database are predicted within the ±20% error band, with a mean absolute deviation of 10.4% and a standard deviation of 13.3%. As for the previous correlations, a higher disagreement is found at G = 100-200 kg m^{-2} s^{-1} for all fluids (e_{AB} = 15.2% and σ_N = 17.7%), except for R134a. The main discrepancies between experimental and predicted values at low mass fluxes are found at low vapour qualities, mainly due to the transition criterion from Regime I (intermittent, annular and mist flow) to Regime II (wavy flow pattern) and to the increased effect of the surface tension.

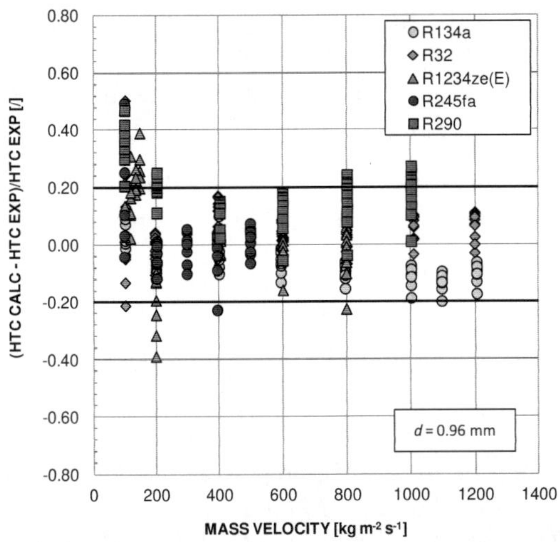

Fig. 3 Comparison between experimental and predicted values by the Shah et al. [19] model in the 0.96 mm diameter test section.

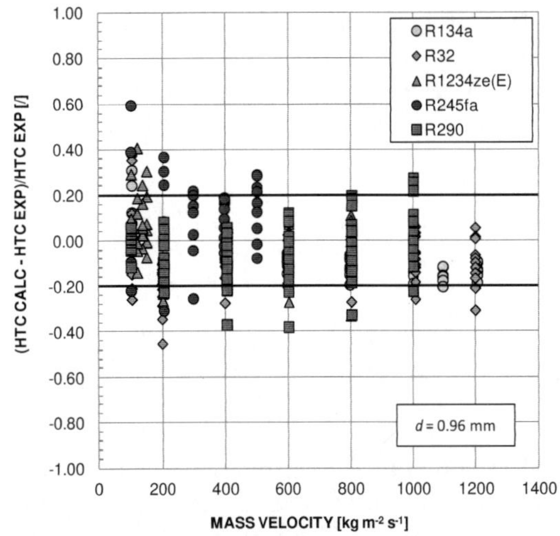

Fig. 4 Comparison between experimental and predicted values by the Thome et al. [21] model in the 0.96 mm diameter test section.

The condensation heat transfer model developed by Thome et al. [21] is based on a flow pattern map for condensation in horizontal smooth tubes, which considers fully stratified, stratified-wavy, intermittent, annular and mist flow. This model incorporates a method for the calculation of vapour void fraction spanning from low pressures up to pressures near the critical point. The proposed condensation model includes the effect of liquid-vapor interfacial roughness on the heat transfer by means of an enhancement factor. The model was validated against a very broad experimental database (fifteen fluids) including propane (R290), R134a, R32, R404A and R410A. It has been tested for mass velocities from 24 to 1022 kg m^{-2} s^{-1}, vapour qualities from 0.03 to 0.97, reduced pressures from 0.02 to 0.80 and tube diameters from 3.1 to 21.4 mm. Fig. 4 shows the comparison between experimental and calculated heat transfer coefficients with the Thome et al. [21] model. Most of the experimental data are predicted in the range ±20%, in particular the mean absolute deviation between calculated and experimental values is equal to 12.4% while the standard deviation is 14.8%. It should be stressed that, at low mass velocity (G = 100-300 kg m^{-2} s^{-1}), the predicting accuracy of the model tends to fail (e_{AB} = 15.07%, σ_N = 19.02%). In the whole range of mass velocities, the agreement between calculated and experimental data for all fluids results to be lower at very high and very low vapour qualities, in particular with hydrocarbon R290. A similar trend, but with reference to a 8.0 mm diameter macrochannel, was observed in Thome et al. [21] with hydrocarbons.

2.2. Condensation models for non-azeotropic mixtures

All the aforementioned condensation models for pure fluids and azeotropic blends are extendable to the prediction of the heat transfer coefficients of zeotropic mixtures by use of the Bell and Ghaly [22] method. This method essentially consists of assuming two thermal resistances in series for the evaluation of the heat transfer coefficient of the zeotropic mixture, as indicated in equation (1):

$$\alpha_{mix} = \left(\frac{1}{\alpha_f} + \frac{\partial q_{sg} / \partial q_t}{\alpha_g} \right)^{-1} \quad (1)$$

where the first resistance $1/\alpha_f$ is across the condensate film itself and the corresponding heat transfer coefficient is evaluated with a condensation model for pure fluid but using the thermophysical properties of the mixture. The second resistance, instead, is for the convective heat transfer in the vapor phase from the bulk temperature to the temperature at the vapour-liquid interface. In this term α_g is the convective heat transfer coefficient of the vapour flowing alone in the tube, calculated with the Dittus-Boelter equation, while $\partial q_{sg}/\partial q_t$ expresses the ratio between the sensible vapour heat flow rate and the total heat flow rate. This ratio can be calculated as a function of the vapour quality and the temperature glide of the mixture, as indicated in equation (2):

$$\frac{\partial q_{sg}}{\partial q_t} \approx x c_{pg} \left(\frac{\Delta T_{glide}}{\Delta h_{mix}} \right) \tag{2}$$

The additional thermal resistance is required to account for the different volatility of the pure components of the mixture. Indeed, the temperature glide of the blend is responsible for the creation of a mass transfer resistance owing to the fact that the less volatile component must diffuse through the vapour phase before it can condense at the liquid-vapor interface.

In the present paper, heat transfer coefficients of a non-azeotropic mixture of R32 and R1234ze(E) (46/54% by mass composition) measured during condensation inside a 0.96 mm diameter minichannel have been considered. Figure 5 reports the predicted heat transfer coefficients of the mixture calculated with the Cavallini et al. [18] model for pure fluids without the Bell and Ghaly [22] correction term, plotted against the corresponding experimental values. As can be seen, the model overestimates the experimental data with a mean absolute deviation of 11.8% and a standard deviation of 6.1%. The application of Bell and Ghaly [22] correction, as shown in Fig. 6, allows a more accurate prediction of the heat transfer coefficients with a mean absolute deviation of 6.7% and a standard deviation of 4.9%.

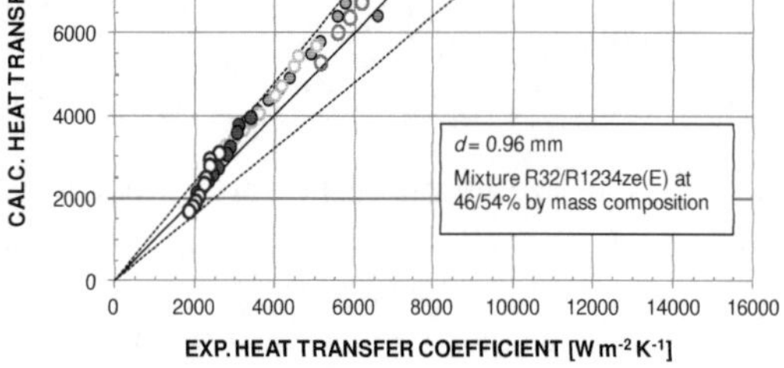

Fig. 5 Comparison between experimental and calculated heat transfer coefficients using the Cavallini et al. [18] model for pure fluids. *G* is the mass velocity in kg m^{-2} s^{-1}.

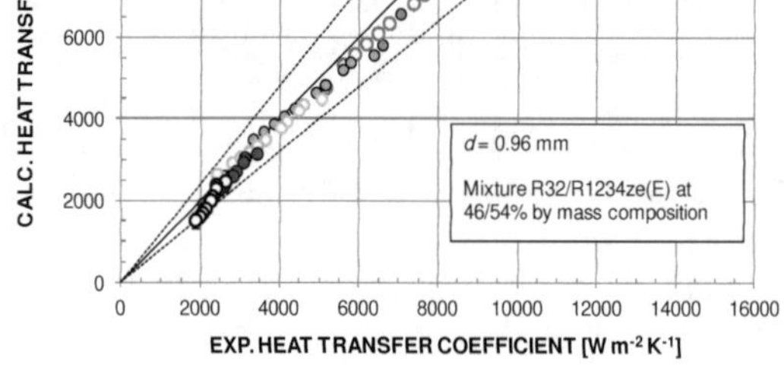

Fig. 6 Comparison between experimental and calculated heat transfer coefficients using the Cavallini et al. [18] model coupled with the Bell-Ghaly correction [22]. *G* is the mass velocity in kg m^{-2} s^{-1}.

3. Conclusions

In the present paper, condensation models for the prediction of the heat transfer coefficient inside minichannels have been compared against an experimental database taken inside a 0.96 mm diameter test section. The heat transfer models by Moser et al. [16], Cavallini et al. [18], Shah [19] and Thome et al. [21] have been considered for this comparison. Except for the model by Shah [19], the others have been developed for conventional channels, but they have proven to be applicable also to minichannels at high mass velocities. At low mass velocities, these models tend to fail because they are based on datasets obtained with conventional tubes. Above all, the surface tension effect may differ inside minichannels. At mass flux G = 100 kg m^{-2} s^{-1} the considered models display the highest disagreement with the experimental data. At these conditions they may need proper correction to predict the heat transfer mechanism. Considering the present database, the model of Cavallini et al. [18] provided the best prediction having a mean absolute error e_{AB} equal to 9.5% and a standard deviation σ_N equal to 12.3%. Furthermore, condensation models developed for pure fluids are suitable for the prediction of the heat transfer coefficient of non-azeotropic mixtures inside minichannels coupled with the Bell and Ghaly [22] correction to account for the additional mass transfer resistance.

Nomenclature

c_p	[J kg^{-1} K^{-1}]	Specific heat	q	[W]	Heat flow rate
d	[m]	Hydraulic diameter	t	[°C]	Temperature
e_{AB}	[-]	Mean absolute deviation	T	[K]	Temperature
G	[kg m^{-2}s^{-1}]	Mass velocity	x	[-]	Thermodynamic vapour quality
HTC	[W m^{-2} K^{-1}]	Heat transfer coefficient	We_{gt}	[-]	Weber number of all mass flowing as vapor

Greek symbols

α	[W m^{-2} K^{-1}]	Heat transfer coefficient	ΔT	[K]	Temperature difference
Δh	[J kg^{-1}]	Specific enthalpy difference	σ_N	[-]	Standard deviation

Subscripts

f	Film	s	Sensible	t	Total
g	Vapor	sat	Saturation	w	Wall
mix	Mixture				

References

[1] Kandlikar S G, Garimella S, Li D, Colin S and King M R 2014 *Heat transfer and fluid flow in minichannels and microchannels* (United States: Butterworth-Heinemann).

[2] Del Col D, Bortolin S and Da Riva E 2016 Prediction methods and numerical modeling of condensation heat transfer in minichannels *Encyclopedia of two-phase heat transfer and flow II - Special topics and applications,* vol 3: Special topics in condensation p 37-84.

[3] Triplett K, Ghiaasiaan S, Abdel-Khalik S and Sadowski D 1999 Gas-liquid two-phase flow in microchannels. Part I: two-phase flow patterns *Int. J. Multiphase Flow* **25**(3) 377-394.

[4] Coleman J W and Garimella S 2000a Two-phase flow regime transitions in microchannel tubes: the effect of hydraulic diameter *ASME HTD* **366** 71-83.

[5] Coleman J W and Garimella S 2000b Visualization of two-phase refrigerant flow during phase change *Proc. 34th National Heat Transfer Conference* August 20-22 Pittsburgh PA.

[6] Coleman J W and Garimella S 2003 Two-phase flow regimes in round, square and rectangular tubes during condensation of refrigerant R134a *Int. J. Refrig.* **26**(1) 117-128.

[7] Garimella S 2004 Condensation flow mechanisms in microchannels: basis for pressure drop and heat transfer models *Heat Transfer Eng.* **25**(3) 104-116.

[8] Toninelli P, Bortolin S, Azzolin M and Del Col D 2017 Effects of geometry and fluid properties during condensation in minichannels: experiments and simulations *Heat and Mass Transfer* DOI: 10.1007/s00231-017-2180-7.
[9] Bortolin S, Da Riva E and Del Col D 2014 Condensation in a Square Minichannel: Application of the VOF Method *Heat Transfer Engineering* **35**(2) 193-203.
[10] Matkovic M, Cavallini A, Del Col D and Rossetto L 2009 Experimental study on condensation heat transfer inside single circular minichannel *Int. J. Heat Mass Transfer* **52** 2311-2323.
[11] Del Col D, Bortolato M, Azzolin M and Bortolin S 2015 Condensation heat transfer and two-phase frictional pressure drop in a single minichannel with R1234ze(E) and other refrigerants *Int. J. Refrig.* **50** 87-103.
[12] Cavallini A, Bortolin S, Del Col D, Matkovic M and Rossetto L 2011 Condensation heat transfer and pressure losses of high- and low- pressure refrigerants flowing in a single circular minichannel *Heat Transfer Eng.* **32**(2) 90-98.
[13] Del Col D, Bortolato M and Bortolin S 2014 Comprehensive experimental investigation of two-phase heat transfer and pressure drop with propane in a minichannel. *Int. J. Refrig.* **47** 66-84.
[14] Del Col D, Azzolin M, Bortolin S and Zilio C 2015 Two-phase pressure drop and condensation heat transfer of R32/R1234ze(E) non-azeotropic mixtures inside a single microchannel *Sc. Tech. Built Environment* **21**(5) 595-606.
[15] Akers W W, Deans H A and Crosser O K 1969 Condensing heat transfer within horizontal tubes *Chem. Eng. Progress Symp. Series* **55**(29) 171-176.
[16] Moser K W, Webb R L and Na B 1998 A new equivalent Reynolds number model for condensation in smooth tubes *J. Heat Transfer* **120** 410-417.
[17] Zhang M and Webb R L 2001 Correlation of two-phase friction for refrigerants in small-diameter tubes *Exp. Therm. Fluid Sc.* **25**(3-4) 131-139.
[18] Cavallini A, Del Col D, Doretti L, Matkovic M, Rossetto L, Zilio C and Censi G 2006 Condensation in horizontal smooth tubes: a new heat transfer model for heat exchanger design *Heat Transfer Eng.* **27**(8) 31-38.
[19] Shah M M 2016 Comprehensive correlations for heat transfer during condensation in conventional and mini/micro channels in all orientations *Int. J. Refrig.* **67** 22-41.
[20] Shah M M 1979 A general correlation for heat transfer during film condensation in pipes *Int. J. Heat Mass Transfer* **22** 547-556.
[21] Thome J R, El Hajal J and Cavallini A 2003 Condensation in horizontal tubes, part 2: new heat transfer model based on flow regimes *Int. J. Heat Mass Transfer* **46** 3365-3387.
[22] Bell K J and Ghaly M A 1973 An approximate generalized design method for multicomponent/partial condensers *AIChE Symp. Ser.* **69** 72-79.

IOP Conf. Series: Journal of Physics: Conf. Series **1224** (2019) 012039 doi:10.1088/1742-6596/1224/1/012039

Numerical study on latent thermal energy storages with PCM partially filled with aluminium foam

Bernardo Buonomo*, Davide Ercole*, Oronzio Manca*, Sergio Nardini*

*Dip. di Ingegneria Industriale e dell'Informazione,Università degli Studi della Campania "Luigi Vanvitelli", Real Casa dell'Annunziata, Via Roma 29, Aversa (CE) 81031, IT

E-Mail: davide.ercole@unicampania.it

Abstract. A numerical investigation on the effects of the metal foam into the Latent Heat Thermal Energy Storage System, based on a phase change material, is accomplished. The geometry of the system is a rectangular box heated from below. The PCM is completely embedded in the rectangular box, it is pure paraffin wax with a low value of thermal conductivity and it melts in a range of temperatures. The metal foam is made of copper and it partially fills the rectangular box in order to understand the optimal filling ratio of the foam to have the better thermal performance. The enthalpy-porosity theory is employed to simulate the phase change of the PCM and the metal foam is modelled as a porous media that follow the Brinkman-extended Darcy model. The results are shown in term of melting time, temperature and energy at varying of time or at varying the filling height ratio.

1. Introduction

The energy thermal demand has the natural characteristic to be variable both in temporal term and in the methods of use. This variation makes difficult to match the energy demand with the energy supply, so it is important to build a thermal buffer to solve this problem and to improve the efficiency of the entire thermal energy system for various applications. A solution could the Thermal Energy Store (TES) [1]. It is a device that works as a thermal buffer in order to avoid losing of exceed energy. There are many types of thermal storage system (TES) among which the latent heat thermal energy storage systems (LHTESS) with phase change material (PCM) that represent the best choice because they work at nearly constant temperature, they are stable and they have an high energy density. The Phase Change Materials (PCMs) have the peculiarity to change their phase from solid to liquid and vice-versa during the heat transfer maintaining the operating temperature at near the melting temperature. There are many type of PCM, organic, inorganic, metal alloys etc, but the most usable PCM for thermal storage system in low temperatures is the organic paraffin. It is not corrosive, it is chemically stable, the phase change is reversible and the latent heat is very high [2]. Nevertheless the main disadvantage of the PCM is the low value of thermal conductivity, making the melting time and the solidification time very long respect to a working cycle of charge of the system. Therefore, some solution are proposed in literature, such as the usage of metal foam [3] or finned tubes [4]. Particularly, the open-cell metal foam represents the best solutions because it has a very high ratio between the area of heat exchange and its volume and the base material has high value of thermal conductivity. Some studies in literature have been accomplished, Yang et al. [5] presented a transient mathematical model to study the influence of the porosity variation on the thermal behavior of a system based on PCM, numerically. The results indicate that with the varying porosity there is a beneficial effect on the melting rate. Xiao et al [6] experimentally study the thermal behaviour between the PCM paraffin and open-cell metal foam for latent heat storage. Two type of metal foam are employed, copper and nickel with high value of porosity. They found that the copper foam is better than the nickel foam in term of heat exchange, because the effective thermal conductivity is faster for the copper foam than the nickel foam. It is worth to investigate the influence of the metal foam for not fully covered PCM because the PCM has a non-zero volume change during the melting. Some studies are accomplished to understand the filling ratio of metal foam inside the PCM, for example a two dimensional axisymmetric numerical model of PCM during the melting process in a partial thermal storage system has been analysed in Archibold et al [7]. The enthalpy-porosity method has been employed and the effects of Grashof and Stefan number have been analysed. By the results, the increment of the Grashof number enhances the heat transfer and when the Stefan number increases at constant Grashof number there is a faster melting rate. An experimental test on transient performance of a phase change material (PCM)-based heat sink filled with a copper foam has been accomplished by Zhu et al [8]. The aim has to understand the influence of the filling height ratio of the copper foam on the system. Moreover, the authors have investigated the effects of pore size and heating power. By the results they have founded that the filling height ratio might improve the thermal performance linearly. An investigation on PCM-TES with porous media at different filling ratio has been investigated in Xu et al. [9]. A new parameter has been defined to understand the melting performance and TES capacity. By the results the optimal filling height ratio is 0.7 because in this case the performance parameter is higher respect to clean case without the porous media.

In this paper a numerical investigation on the effects of the metal foam into the Latent Heat Thermal Energy Storage System, based on a phase change material, is accomplished. The geometry of the system is a rectangular box heated from below. The PCM is completely embedded in the rectangular box. The metal foam is made of copper and it partially fills the rectangular box in order to understand the optimal filling ratio of the foam to have the better thermal performance.. The results are shown in term of melting time, temperature and energy at varying of time or at varying the filling height ratio for charging phase.

2. Physical model

The model for the LHTESS consists in an rectangular box filled up with Pure PCM embedded with metal foam. The mathematical domain is 2D with a total height H equals to 20 mm and the base B equals to 75 mm. At bottom of domain, a heat flux is applied. The metal foam could completely occupies the domain, or a part of domain, therefore a filling height ratio R is defined:

$$R = \frac{h_{mf}}{H} \tag{1}$$

Where h_{mf} is the height of the metal foam inside the domain while H is the total height of the domain. In this work R is 0 - pure PCM -, 1/3, 2/3 and 1 - domain is fully occupied by the metal foam - . The other surfaces are adiabatic. The fig.1 illustrates the domain and the boundary conditions. The melting phenomenon of the PCM can be modelled with the enthalpy-porosity method [10], in which the solid-liquid interface is not explicitly traced but the method defines a mixed solid-liquid region that is described as a “pseudo” porous zone where the porosity is the liquid fraction.

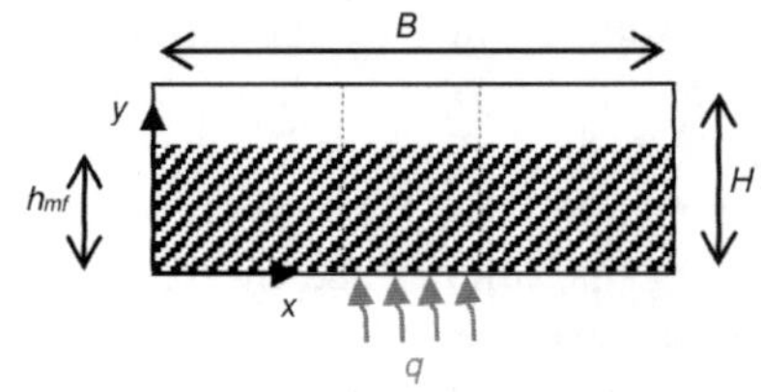

Fig.1 Mathematical model of the LHTESS.

The most important parameter of the enthalpy-porosity method is the liquid fraction, indicated with Greek letter β. The value is 0 when the zone is fully solid, 1 when it is fully liquid and between 0 to 1 in the mixed region:

$$\begin{cases} \beta = 0 & for \quad T < T_{solidus} \\ \beta = \dfrac{T - T_{solidus}}{T_{liquidus} - T_{solidus}} & for \quad T_{solidus} < T < T_{liquidus} \\ \beta = 1 & for \quad T > T_{liquidus} \end{cases} \tag{2}$$

T is the local temperature of the cell, $T_{liquidus}$ is the temperature upper which the domain is completely liquid and $T_{solidus}$ is the temperature below which it is completely solid. The similitude of the mixed region with a porous region is justify by the fact that when the liquid fraction tends to zero, the mixed region is solidifying and the velocity of the PCM tends to slow down, in fact the solid part of the mixed zone is modelled by adding a source term in the momentum equation. The metal foam is modelled with the Brinkman-extended Darcy-Forchheimer model because it behaves like a porous media; therefore another source term is used in the momentum equation to simulate its presence. The gravitational acceleration is along the y-axis and the Boussinesq approximation is considered in order to take into account the buoyancy force due to natural convection.
The governing equation are the following:

$$\nabla \cdot (\rho \vec{V}) = 0 \tag{3}$$

$$\rho \frac{D\vec{V}}{Dt} = \mu_{PCM}\left(\nabla^2 \vec{V}\right) - \nabla \vec{p} + \rho g \gamma \left(T - T_0\right)\hat{j} + \left(\frac{(1-\beta)^2}{\left(\beta^3 + 0.001\right)^3} A_{mush} + \frac{\mu}{K} + \frac{C_F}{\sqrt{K}}\rho\left|\vec{V}\right|\right)\vec{V} \tag{4}$$

$$\left((1-\varepsilon)\rho_{mf}c_{mf} + \varepsilon\rho_{pcm}c_{pcm}\right)\frac{DT}{Dt} = k_{eff}\nabla^2 T - \varepsilon\rho_{pcm}H_L\frac{\partial\beta}{\partial t} \tag{5}$$

Where V is the velocity vector of the PCM in liquid phase, t is the time, μ, ρ and T are respectively the viscosity, density and the temperature, p the relative pressure, g the gravity acceleration module, γ is the PCM thermal expansion coefficient according to Boussinesq equation where T_0 is the operating temperature, set to 310K. The first term in the square brackets is the Kozeny–Carman term that models the presence of the solid part in the mixed region during the melting, where the small number (0.001) avoids the division by zero when β is null. A_{mush} is the mushy zone constant which takes into account the damping of the velocity during the solidification [11]. Its value is set to 10^5 kg/(m^3 s) and it does not affect the rate of melting with the presence of the metal foam. The Darcy term and the Forchheimer term of the porous media are respectively the second and third term inside the square brackets in equation 4, where K is the permeability and C_F is inertial drag factor. Their values are calculated with the following relations [12]:

$$K = 0.00073(1-\varepsilon)^{-0.224}\left(\frac{d_f}{d_p}\right)^{-1.11} d_p^2 \tag{6}$$

$$C_F = 0.00212(1-\varepsilon)^{-0.132}\left(\frac{d_f}{d_p}\right)^{-1.63} \tag{7}$$

d_f is the ligament diameter of the metal foam, d_p is the pore diameter. These physical characteristics are related each other by means of the following relations [12]:

$$\begin{cases} \dfrac{d_f}{d_p} = 1.18\sqrt{\dfrac{1-\varepsilon}{3\pi}}\left(\dfrac{1}{1-e^{1(1-\varepsilon)/0.04}}\right) \\ d_p = \dfrac{0.0224}{\omega} \end{cases} \tag{8}$$

ω is the pore density of the metal foam that represents the number of pores across one inch. In this study the pore size is assigned at 15 Pore Per Inch (PPI). The heat transfer interaction between the metal foam and the PCM is governed by the Local thermal Equilibrium Assumption therefore the local temperature between PCM and metal foam is equal, so a unique Local Temperature function T in the equation (5) is defined. In equation (5) ρ_{mf} and c_{mf} are respectively the density and specific heat of the metal foam, ε is the porosity of the metal foam and its value is set to 0.80 and c_{pcm} is the specific heat of PCM. k_{eff} is the effective thermal conductivity calculated as weighted average between PCM and metal foam with the porosity:

$$k_{eff} = (1-\varepsilon)k_{mf} + \varepsilon k_{pcm} \tag{9}$$

where k_{mf} e k_{pcm} are respectively the thermal conductivities of metal foam and PCM. H_L is the latent heat of the PCM. The PCM in this study is the RT42 paraffin, the metal foam is made by copper with a density of 8978 kg/m3, thermal conductivity of 387.6 W/mK and a specific heat of 381 W/kg K. The properties of the PCM are listed in table 1. In this study the properties are assumed constant, except for the PCM density, the copper foam is isotropic and homogenous.

Table 1. Physical properties for the paraffin wax

Physical properties	Paraffin RT42 [13]
Density [kg/m^3]	840
Specific Heat [J/kg K]	2100
Thermal Conductivity [W / m K]	0.2
Dynamic Viscosity [kg/m s]	0.0269
Thermal expansion coefficient [1/K]	0.00011
Melting Heat [J / kg]	180000
Solidus Temperature [K]	321
Liquidus Temperature [K]	335

The boundary conditions are reported in the following:

- At y=0 and $0 \leq x \leq 25$: adiabatic condition.
- At y=0 and $25 < x \leq 50$: applied heat flux q=12600 W/m^2.
- At y=0 and $50 < x \leq 75$: adiabatic condition.
- For y=20 and $0 \leq x \leq 75$: adiabatic condition.
- For x=0 and $0 \leq y \leq 20$: adiabatic condition.
- For x=75 and $0 \leq y \leq 20$: adiabatic condition.

3. Numerical model

The finite volume method [14] is employed for resolving the governing equation (3), (4), (5) using the commercial code Ansys Fluent [15]. The SIMPLE algorithm is applied for coupling the pressure and the velocity; for the spatial discretization the gradient evaluation is based on green-gauss cell; Linear method is used for pressure calculation and the second order upwind scheme is utilized for momentum and energy equation. The residual convergence values are imposed to 10^{-4} for continuity equation and momentum equation while 10^{-6} is set to energy equation. A transient mode is enabled with a time step size equals to 0.1 s. The employed grid is a set of quadrilateral control volumes with a finer grid in the separation edge between the metal foam and the pure PCM at correspondence of y=h. therefore 3 different meshes are used for this study, a simple rectangular mesh for R=0 and R=1 and other two meshes for R=2/3 and R=1/3. After some tests, the mesh with R=1/3 seems more instable respect to the other height filling ratio, therefore a study to obtain a solution independent from the mesh is made with four different grids at R=1/3: 789 cells, 3840 cells, 7540 cells, 15633 cells. The mesh with 3840 cells represents the best mesh grid for the present study because the computational costs are satisfactory with the accuracy.

4. Results and discussion

In fig.2 there are the average evolution of temperature and liquid fraction for each filling height ratio.

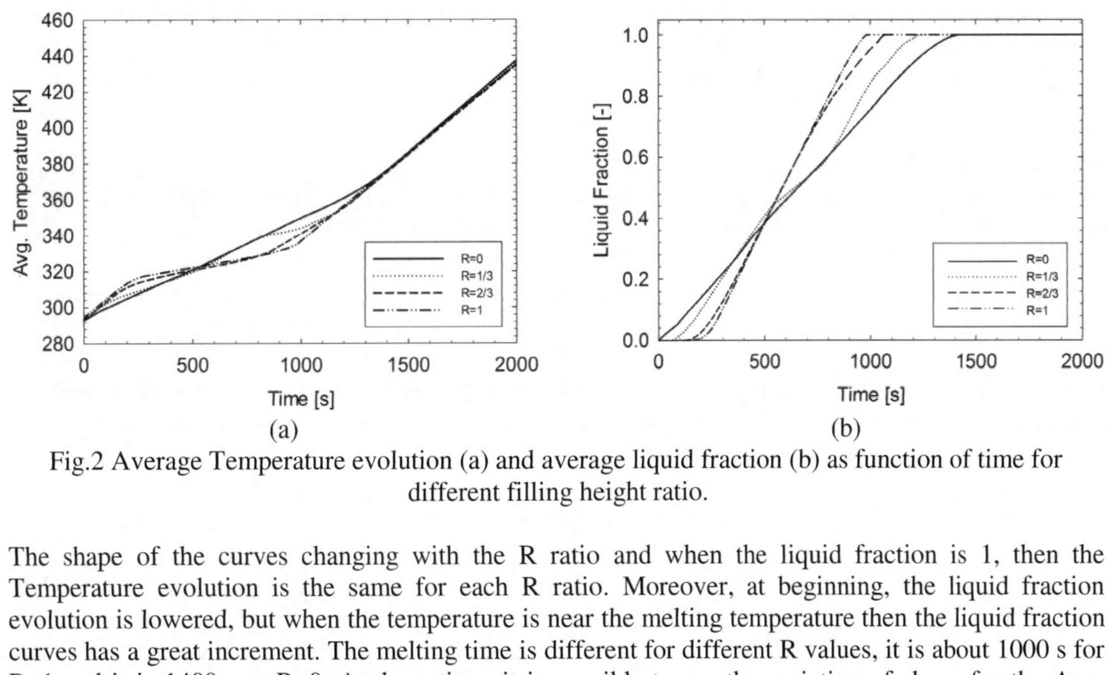

(a) (b)

Fig.2 Average Temperature evolution (a) and average liquid fraction (b) as function of time for different filling height ratio.

The shape of the curves changing with the R ratio and when the liquid fraction is 1, then the Temperature evolution is the same for each R ratio. Moreover, at beginning, the liquid fraction evolution is lowered, but when the temperature is near the melting temperature then the liquid fraction curves has a great increment. The melting time is different for different R values, it is about 1000 s for R=1 and it is 1400 s at R=0. At these time, it is possible to see the variation of shape for the Avg. Temperature curve. About the energy, in fig.3 it is possible to see that there are very few differences with the variation of height filling ratio and further studies are necessary to fully understand the stored energy inside the system.

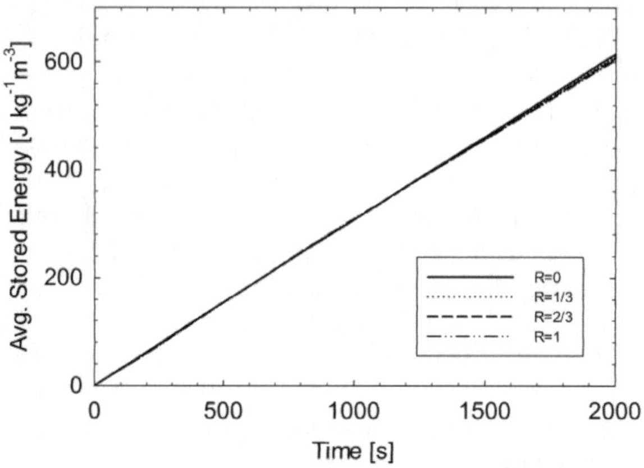

Fig.3 Average internal energy evolution per unit of mass as function of time for different filling height ratio.

Finally, in Fig. 4 the surface average bottom temperature is plotted as function of time and it can be seen that for the case without metal foam (R=0) there is a spike of temperature, because, at beginning, the PCM has a low value of thermal conductivity and therefore it behaves as a thermal insulator; but when the melting starts, the natural convection in the PCM begins and it is responsible to take the heat from the bottom lowering the bottom temperature.

5. Conclusions

A numerical investigation on the effects of the metal foam into the Latent Heat Thermal Energy Storage System, based on a phase change material, is accomplished. The geometry of the system is a rectangular box heated from below. The PCM is completely embedded in the rectangular box, it is pure paraffin wax with a low value of thermal conductivity and it melts in a range of temperatures. The metal foam is made of copper and it partially fills the rectangular box in order to understand the optimal filling ratio of the foam to have the better thermal performance. The results have showed that the melting time changing with the filling ratio but the energy is constant while for the temperature the presence of the metal foam lowers the temperature, especially in the beginning avoiding heat spikes.

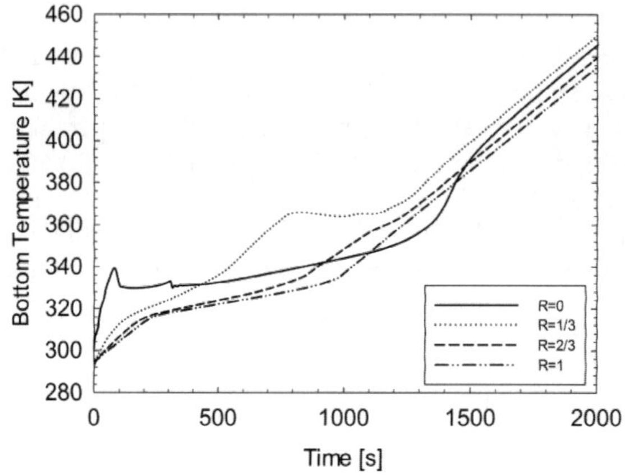

Fig.4 Surface average bottom temperature as function of time for different filling height ratio.

IOP Conf. Series: Journal of Physics: Conf. Series **1224** (2019) 012039 doi:10.1088/1742-6596/1224/1/012039

6. References

[1] Antoni G, Marc M, Ingrid M, Ana L, Pablo D, Belén Z and Luisa F C 2010 *Renew. Sust. En. Rev.* **14** 31
[2] Abhat A 1983 *Sol. En.* **30** 313
[3] Zhao, C Y ,Lu W and TianY 2010 *Sol. En.* **84** 1402
[4] Zhang Y and Faghri A 1996 *J. Enhanced Heat Transf.* **3** 119
[5] J. Yang, L. Yang, C. Xu, X. Du, Numerical Analysis on thermal behavior of solid-liquid phase change within copper foam with varying porosity, International Journal of Heat and Mass Transfer 84 (2015) 1008–1018. doi: 10.1016/j.ijheatmasstransfer.2015.01.088
[6] Xiao X, Zhang P and Li M 2014 *Int. J. Therm. Sci.* **81** 94
[7] Archibold A R , Gonzalez-Aguilar J, Rahman M M, et al. 2014 *Appl. En.* **116** 243
[8] Zhu Z Q, Huang Y K, Hua N, Zeng Y, Fan L W, 2018 *Appl. Therm. Eng.* **128** 966
[9] Xu Y, Ren Q, Zheng Z Y, He H L, 2017 *Appl. En.* **193** 84
[10] Voller and V R, Prakash C 1987 *Int. J. Heat Mass Transf.* **30** 1709
[11] Al-abidi A, Bin Mat S, Sopian K, Sulaiman M Y and Mohammed A T 2013 *Renew. Sust. En. Rev.* **20** 353
[12] Calmidi and V V, Mahajan R L 2000 *ASME J. Heat Transf.* **122**, 557
[13] Rubitherm GmBH, *www.rubitherm.de*.
[14] Pantankar S V 1980 "Numerical heat transfer and fluid flow", *Hemisphere*, Washington DC.
[15] Ansys Incorporated, Fluent 15.0.
[16] Krishnan S, Murthy J Y and Garimella, S V 2005. *ASME J. Heat Transf. 127* 995

IOP Conf. Series: Journal of Physics: Conf. Series **1224** (2019) 012040 doi:10.1088/1742-6596/1224/1/012040

Horizontal Air-Water Two-Phase Flow Measurement Using an Electrical Impedance Probe and a Venturi Flow Meter

G CHIESA[1], C BERTANI[1], N FALCONE[1], A BERSANO[1], M DE SALVE[1], B PANELLA[1]

[1] Energy Department
Politecnico di Torino
Corso Duca degli Abruzzi, 24, 10129 Torino (TO), Italy

E-mail: nicolo.falcone@polito.it

Abstract. Two-phase flow rate measurement is extremely important in many industrial sectors (nuclear, chemical, oil and gas, etc.). Nevertheless, the instrumentation required to perform these measurements is still very limited despite the huge research efforts during the last years. In the frame of the development of instrumentation for the two-phase flow rate measurement, the main aim of this study is the characterization of a spool piece consisting of a Venturi Flow Meter, and two Electric Capacitance Probes mounted in a horizontal test section, where an air-water two phase flow occurs. The probes response is analyzed versus superficial velocities of liquid and gas, flow patterns and void fraction. The test section consists of a nominal ID 80 mm horizontal transparent pipe equipped with a Venturi Flow Meter, a new capacitance probe with concave electrodes and a SIET designed electrical capacitance probe with ten long linear electrodes. In this experimental campaign, the air flow rate is in the range 0.048-0.119 kg/s (superficial velocity from 8 to 19 m/s) and the range of water flow rate is 0.0040-0.0535 kg/s (superficial velocity from 0.0006 m/s to 0.0103 m/s). These ranges of superficial velocities correspond to wavy stratified and annular-dispersed flow patterns along the test section.

1. Introduction

Two-phase flow rate measurement is still an open challenge that affects several industrial sectors such as the nuclear, the oil and gas, the chemical ones. Despite the abundancy of single-phase measurement devices relying on different physical principles, a universal device for two-phase flow rate measurement is still lacking. This is due to the fact that the flow is governed by several parameters (void fraction, phases velocities, etc.), which cannot be measured by a single instrument. For this reason, the development of a spool piece is under study. In general, a spool piece consists of a series of different instruments that are used to acquire various signals necessary to measure and to recognize two-phase phenomena. The spool piece under study couples a traditional flow rate measurement instrument such as a Venturi Flow Meter (VFM) with devices for the measurement of the void fraction. Several measurement techniques have been adopted to measure the void fraction (neutron radiography, gamma rays' tomography, optical probes, etc.) but they are very expensive, bulky and, if they involve ionizing radiations, they require particular attention and safety measures [1-5].

To measure the void fraction, another possibility is to use electrical impedance or capacitance probes (ECP) such as wire mesh sensors [6] or probes with radial electrodes [7]. Such devices have been coupled to form a "spool piece" in the studies previously carried out at Dipartimento Energia (DENERG) of Politecnico of Torino (Italy)[8].

In order to perform air-water two-phase flow measurement under different flow conditions and flow patterns, the spool piece consists of impedance probes and a VFM. An ECP designed by SIET (SIET ECP) was originally used, which allowed to get information on the phases distributions inside the probe volume, but required a very complex elaboration of 10 different signals. Moreover, the SIET ECP did not provide the average void fraction in the test section. In order to obtain directly this average void fraction, two new capacitance probes with a 24-Bit Capacitance–to-Digital Converter AD7746 [9] have been developed at DENERG (DENERG ECP) and added to the spool piece.

The spool piece is mounted on a transparent experimental apparatus to allow the visual observation of the flow patterns. The final goal is to combine the signals of the two capacitance probes and the VFM to get the void fraction, the flow pattern and the flow rate, without the need of a visual observation, such as in the case of an opaque pipe.

The two-phase flow is obtained by mixing air and water, and different flow conditions can be generated by varying the flow rate of the two fluids. In the present experimental campaign, the air flow rates are in the range 0.048-0.119 kg/s (superficial velocity in the range 8-19 m/s) and water flow rate range is 0.0040-0.0535 kg/s (superficial velocity in the range 0.0006-0.0103 m/s). The resulting flow patterns can change along the horizontal pipe and they are wavy-stratified and dispersed-annular.

This paper presents the experimental apparatus, the experimental matrix, the test procedures to characterize the signals of VFM (differential pressure), and the signals of the ECP designed by SIET and of the new ECPs with 24-Bit Capacitance–to-Digital Converter AD7746.

2. Experimental apparatus and instrumentation

The experimental apparatus (Figure 1) consists of an air supply loop, a demineralized water supply loop, a test section with the Spool Piece (SP) and two pneumatic quick-closing valves for the volume-averaged void fraction measurement, a water/air separator and the instrumentation for the measurement of single-phase fluids flow rate, pressure and temperature.

The air is fed to the test section by a blower (UNIJET500), whose rotation frequency is fixed by an inverter, while the water is supplied by a tank thanks to gravity. The mixing zone of the two fluids is located a few centimeters upstream the inlet guillotine valve. The air enters axially, whereas the water is injected radially. The water mass flow rate is measured by a rotameter, while the air flow rate is measured by a calibrated orifice flow meter.

The test section has a total length of approximately 2900 mm and consists of all the components located between the guillotine quick-closing valves; it comprises a PMMA (PolyMethylMethacrylate) horizontal pipe with a nominal inner diameter of 80 mm, which allows the visualization of the flow, the SP and an outlet pipe. The spool-piece consists of a SIET ECP with nine external long linear electrodes and one internal central electrode, two new ECPs with concave electrodes, and the VFM. The SIET ECP, whose length is 390 mm, is located approximately 700 mm downstream the inlet pneumatic valve. The pipe connecting the SIET ECP to the VFM is 995 mm long and is equipped with the two DENERG ECPs. The VFM, shown in Figure 2 a), is 340 mm long with a throat diameter of 40 mm; it is a symmetric Venturi, so both sides have the same inclination (21°) [10]. The outlet pneumatic quick-closing valve is located 495 mm downstream the VFM.

The outlet pipe has the same inner diameter of the test section and is 700 mm long. Downstream the outlet pipe a tank is installed in order to the separate the phases.

IOP Conf. Series: Journal of Physics: Conf. Series **1224** (2019) 012040 doi:10.1088/1742-6596/1224/1/012040

The test section is equipped with two pneumatic quick closing valves used to measure the average volumetric void fraction; actuating the pneumatic quick-closing valves, the liquid phase is trapped in the test section pipe and its volume can be measured by draining the trapped water. Since the pipe volume is 14.2 dm^3, the average volumetric void fraction can be determined.

The air flow rate is measured by an orifice equipped with a differential pressure transducer, having an accuracy of 0.15% of the span; the temperatures are measured by means of thermocouples (K-type, 0.5 °C accuracy) at the blower outlet, at the inlet of SIET ECP, at VFM inlet and at the test section outlet. The water flow rate is measured by an ASA rotameter with an accuracy of 0.2% of the span. The VFM is instrumented with two differential pressure transducers for the measurement of the pressure drops between the inlet and the throat (reversible pressure drop; transducer accuracy 0.2% of the span) and between the inlet and the outlet (irreversible pressure drop; transducer accuracy 0.04% of the span).

The ECP designed by SIET, shown in Figure 2 b), is 90 cm long (flange to flange) and it consists in a central electrode and 9 electrodes (39 cm long) on the external surface equally spaced of 22.5° from the upper side (0°) to the lower one (180°); their positions are shown in Figure 3. The electrodes are connected to an electronic circuit by several reed relays and two insulation transformers that prevent common mode disturbances. Each external electrode is connected, at the upper and lower extremity, to two reed relays to activate, in a predefined sequence, the excited electrode and the measuring one; the internal one is connected only in the upper extremity and it is always used as measuring electrode, when the corresponding reed relay is activated. The SIET probe can provide information about the average linear void fraction along 9 different radii and all the possible cords between the different circumferential electrodes. A complete description of the theory of the ECP has been presented in [11].

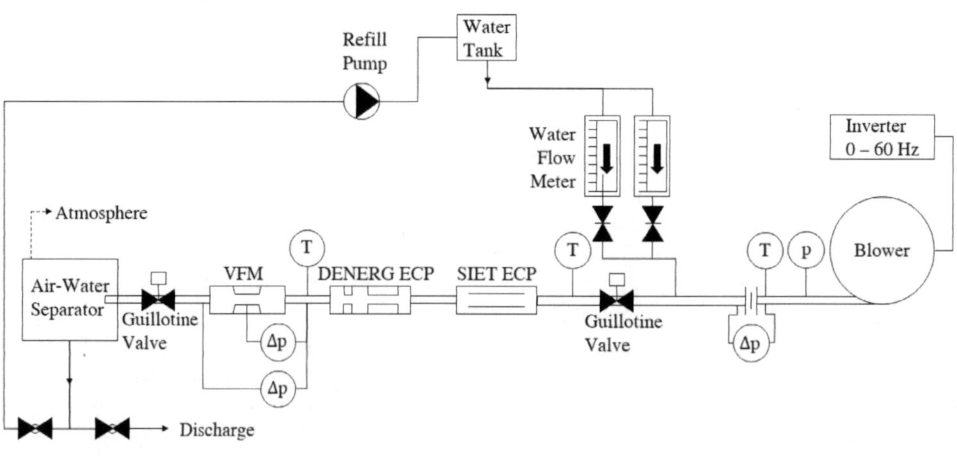

Figure 1 Schematic of the experimental apparatus.

Figure 2c) shows the new capacitance probes (DENERG ECPs), which are made by copper concave electrodes (Figure 3). The two DENERG ECPs are placed on the outer wall of the pipe and they are characterized by different surfaces S (Table 1). Two ECPs with different surfaces have been tested, since the ECP surface should be chosen taking into account the flow pattern occurring in the test section; in fact, their signal is proportional to the equivalent dielectric constant of the two-phase flow.

A 24-Bit Capacitance–to-Digital Converter AD7746 (range: ±4 pF) [9] generates the signal on one concave electrode and receives the response on the other one.

In the case of the wider electrodes the signals of the measured capacitance are out of the measurement range. For this reason it is necessary to design properly the electrodes to optimize the capacitance sensor.

The signals are acquired using three National Instruments DAQ systems (NI USB 6353, NI 9213 and NI 9205) and are managed using the LabVIEW software.

a) b) c)

Figure 2: a) The Venturi flow meter (VFM), b) SIET ECP, and c) Probes with concave electrodes designed at DENERG.

Table 1 Concave electrodes geometrical data.

Electrodes	De [mm]	D [mm]	L[mm]	g[mm]	w[mm]	S [mm^2]
A	90	80	45	20	125	5625
B	90	80	15	5	140	2100

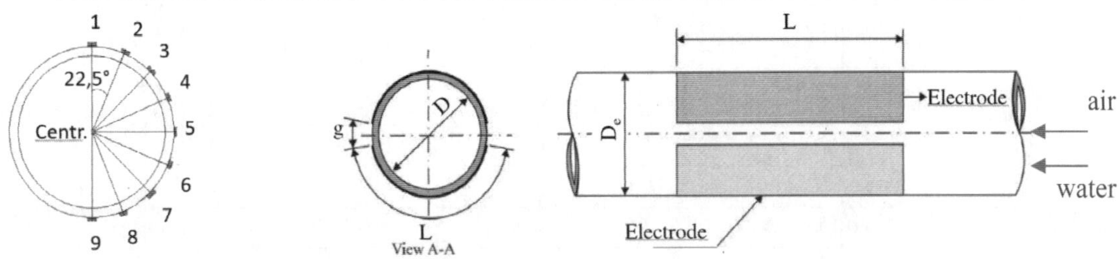

Figure 3 SIET ECP electrodes' position (left) and schematic of the concave electrodes [12] (right).

3. Test procedure and experimental matrix

All the ECPs have been calibrated in static conditions in order to characterize their output signal with liquid only, air only and known heights of water in the test section. In each run, the water and air mass flow rates are regulated respectively by changing the frequency of the inverter connected to the blower and by opening/closing a needle valve located downstream the rotameter. Since the water is provided by gravity and the water tank is located just 3 m above the test section, the possible operational range is not too wide due to the limited available gravitational head.

Pressures, temperatures and the signals from the probes are acquired when the flow in the test section has reached steady-state conditions. At the end of each run the average volumetric void fraction is measured 5-10 times, making use of the quick closing valves, to have a sufficient statistic.

As regard the ECP SIET probe, the signals are sent to the DAQ and managed by a program in the LabVIEW environment. On the other hand, the signals of the concave electrode's sensors are generated and received by an Analog Device- Evaluation board AD7746 (range: ±4 pF) and managed by an Evaluation Board Software v2.2 [13]. The measurements in the different probes and the data elaboration are synchronized by a "master" program developed in LabVIEW environment.

In the present experimental campaign, the air and water flow rates range respectively from 0.048 to 0.119 kg/s (superficial velocity from 8 to 19 m/s) and from 0.0040 to 0.0535 kg/s (superficial velocity 0.0006 - 0.0103 m/s), as shown in Table 2, which also reports superficial velocities, pressure and temperature at the test section inlet, and the average volumetric void fraction in some typical test run. . The experimental void fraction range is 0.976-0.997 and the flow quality range is 0.60-0.96. The resulting flow pattern are wavy-stratified and dispersed-annular. For the same couple of flow rates of the liquid and gas phases different separate tests have been carried out to assess the repeatability of the measurements.

Table 2 Experimental matrix.

j_g [m/s]	j_l [m/s]	x [-]	$T_{in, ECP}$ [°C]	p [barg]	α [-]
12.0	0.0016	0.901	19.6	0.091	0.992
	0.0037	0.796	20.6	0.091	0.986
	0.0054	0.724	20.9	0.091	0.983
	0.0087	0.621	20.9	0.092	0.977
	0.0106	0.572	21.0	0.092	0.973
13.5	0.0018	0.902	21.5	0.127	0.992
	0.0038	0.813	22.5	0.127	0.987
	0.0054	0.757	22.7	0.127	0.984
	0.0085	0.662	23.0	0.127	0.979
	0.0102	0.618	23.0	0.128	0.976
16.0	0.0019	0.907	24.3	0.164	0.992
	0.0038	0.828	24.7	0.165	0.988
	0.0053	0.772	24.8	0.166	0.985
	0.0086	0.677	24.8	0.167	0.980
	0.0102	0.636	25.1	0.167	0.978
19.7	0.0017	0.934	26.1	0.252	0.994
	0.0039	0.858	26.5	0.253	0.990
	0.0053	0.815	25.5	0.259	0.987
	0.0091	0.719	26.2	0.258	0.982
	0.0102	0.695	25.8	0.258	0.981

4. Results and discussion

The transparent pipes of the test section allowed to visualise the evolution of the flow pattern along the pipe. In all the experimental tests stratified-wavy and dispersed-annular flow patterns occurred along the first part of the test section, i.e. from the inlet of the test section up to the inlet of the VFM. The VFM perturbs the flow pattern, which converts to dispersed flow downstream the VFM in all the experimental tests. The experimental flow pattern along the first part of the test section have been compared with the Taitel-Duckler [14] map for horizontal flow (Figure 4) where χ is the Martinelli parameter [15][16] defined as:

$$\chi = \frac{j_l}{j_g}\left(\frac{\rho_l}{\rho_g}\right)^{0.5} \tag{1}$$

and the parameter K is defined as [14]:

$$K = \left[\frac{\rho_g j_g^2 j_f}{(\rho_f - \rho_g) g v_f}\right]^{0.5} \tag{2}$$

The agreement is good in the experimental runs with lower value of the air superficial velocity, since the flow pattern in this region is still developing; in fact the length to diameter ratio at the VFM entrance is approximately equal to 25. Downstream the VFM, the length of the pipe is 450 mm only, which corresponds to L/D equal to 5.6; therefore the flow pattern is still fully undeveloped and cannot be compared with any flow pattern map or model.

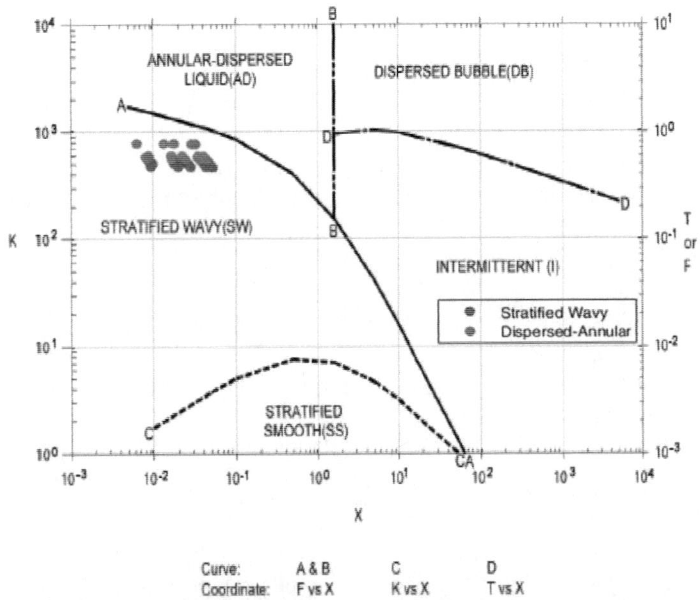

Figure 4 Taitel and Dukler map with experimental working conditions.

It is necessary to highlight that the flow pattern affects slip ratios and the distribution of interfaces between water and air and therefore also the average void fraction. The method of quick-closing valves allows to measure the average void fraction in the whole volume of the test section, i.e. the mean value between the void fraction in the first part of the test section (where annular wavy flow pattern occurs), the one downstream the VFM (where dispersed flow was seen) and the one inside the VFM.

The measured void fraction has been compared with the predictions of the Chisholm slip model [17], which is one of the more accurate for horizontal flow according to [18], and has shown a very good agreement.

As far as the responses of the VFM and ECPs are concerned, they have been analysed first of all in single-phase, following the same methodology presented in [10]. Afterwards, their two-phase response has been studied. The results are described in the following subchapters.

4.1 *Response of the VFM*

In two-phase flow the total mass flow rate through the VFM, is a function of the two-phase pressure drop Δp_{TP} measured between the inlet and the throat of the VFM and the two-phase density ρ_{TP}, and can be written as:

$$W_{tot} = K_{TP}\sqrt{2\,\Delta p_{TP}\rho_{TP}} \tag{3}$$

where K_{TP} is a proportionality coefficient that is function of geometrical data of the VFM and its single-phase discharge coefficient.

Using the approach of Lockhart and Martinelli [15] the total mass flow rate in the VFM can be estimated as:

$$W_{tot} = K_g \sqrt{2\, \Delta p_g\, \Phi_g^2\, \rho_g} \tag{4}$$

where K_g relates to the gas phase, Δp_g is the pressure drop of the gas phase estimated from the superficial velocity of the gas and from the discharge coefficient of Venturi, Φ_g^2 is the two-phase flow multiplier that considers the effect of the gas phase and ρ_g is the gas density.

The previous campaigns showed that the two-phase flow multiplier is a function of the Martinelli parameter [15][16]; on the basis of best-fit of the results of the present campaign (Figure 5) the following correlation has been developed:

$$\Phi_g^2 = \frac{\Delta p_{TP}}{\Delta p_g} = 8.3791\,\chi + 1.002 \tag{5}$$

where the pressure drop Δp_{TP} is the one measured between the inlet and the throat of the VFM and χ is the Martinelli parameter as in Equation (1).

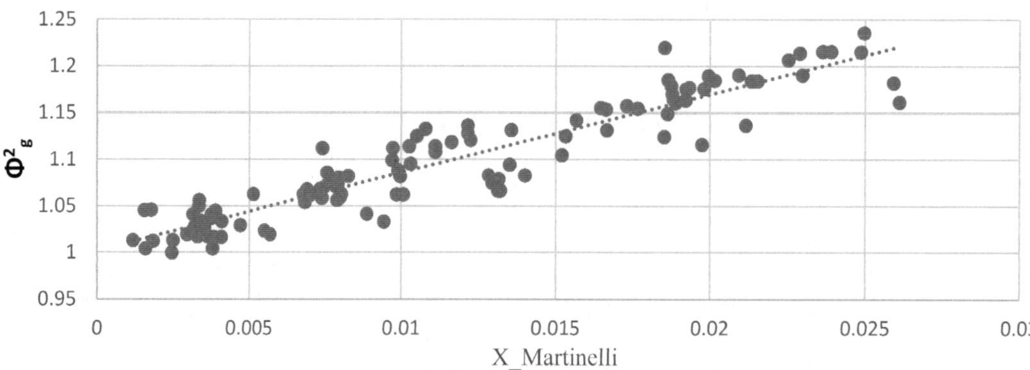

Figure 5 Experimental two-phase multiplier as a function of Lockhart-Martinelli.

Figure 6 reports the irreversible pressure drop between the VFM inlet and outlet, versus the total flow rate. Since this pressure drop is very small, a great accuracy of the transducer is necessary. The previous experimental campaigns showed that the dependency of the irreversible pressure loss on the superficial velocity of the phases and on the density of the gas phase is influenced by the flow pattern; the best-fit of the present experimental campaign allowed the development of the following equations for wavy-stratified and dispersed-annular flow patterns:

wavy-stratified flow pattern $$\Delta P_{irr} = 0.2205\,(\rho_g J_g^{1.91})\left(\frac{J_l}{J_g}\right)^{0.137} \tag{6}$$

dispersed-annular flow pattern $$\Delta P_{irr} = 0.2405\,(\rho_g J_g^{1.903})\left(\frac{J_l}{J_g}\right)^{0.147} \tag{7}$$

IOP Conf. Series: Journal of Physics: Conf. Series **1224** (2019) 012040 doi:10.1088/1742-6596/1224/1/012040

The discrepancy between the prediction of Equations (6) and (7) and the experimental data is approximately ± 10%, while 12% discrepancy was found in our previous studies which did not considered the different flow patterns.

Therefore, the knowledge of the reversible and irreversible pressure drop through the VFM could allow the evaluation of the superficial velocities of the two phases and therefore the total two-phase flow rate, but the identification of the flow pattern is fundamental in order to properly correlate the irreversible pressure drop and the two-phase multiplier to the phases superficial velocities (Figure 6).

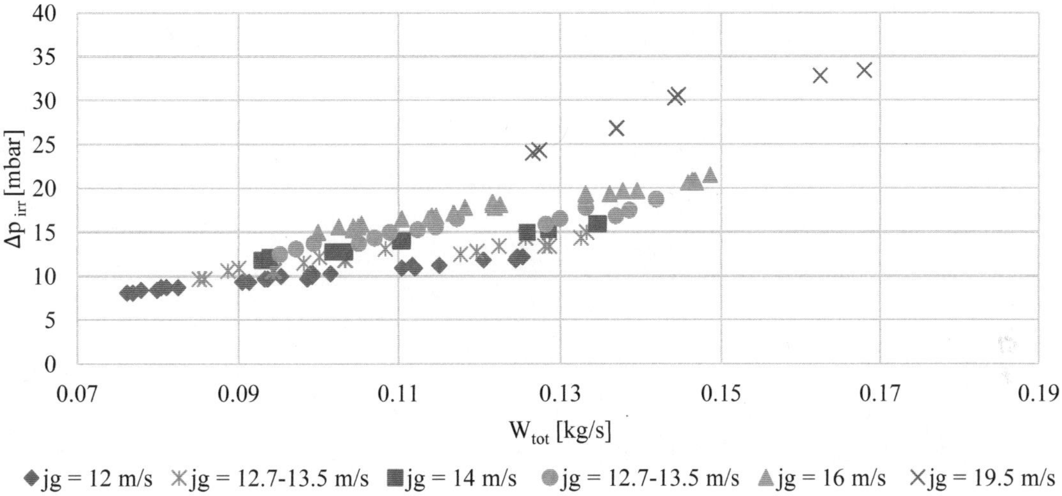

Figure 6 VFM irreversible pressure loss vs total flow rate (present experimental campaign).

4.2 *Response of DENERG ECPs*

To perform the static calibration of the new capacitance probes with concave electrodes developed at DENERG, the quick isolation valves have been closed, the test section has been gradually filled and the relationships between the equivalent electrical capacitance of the water-air mixture and the water level in the probe was found. These relationships are represented in Figure 7. In presence of a mixture of air and water, the probes behaves as capacitors in series with different dielectric constants, and they show a good sensitivity at low liquid level (level lower than the half pipe).

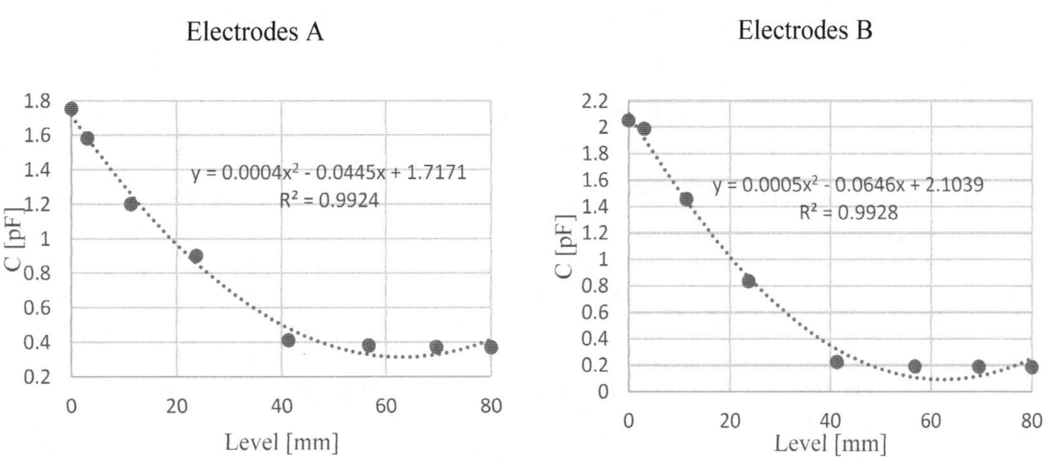

Figure 7 Calibration of the two configurations of concave electrodes' sensor.

The dynamic behavior of these two concave sensors has then been studied for different couples of values of the air and water superficial velocities. Figure 8 and Figure 9 show the measured capacitance versus the superficial velocity of the liquid phase, at different superficial velocity of the gas phase. The analysis of the signals shows that the measured capacitance increases as the void fraction increases. In particular, their sensitivity is good in case of dispersed-annular flow pattern (high capacitance values), whereas the sensitivity is poorer for the stratified-wavy flow pattern. In particular the probe A, which has a bigger surface, shows a steeper increase of the capacitance when the air superficial velocity increases; the capacitance reaches the full scale value of the measuring instrument (4 pF) for high values of the air superficial velocity (greater than 16 m/s); in this case, an ECP with lower surface (probe B) is necessary.

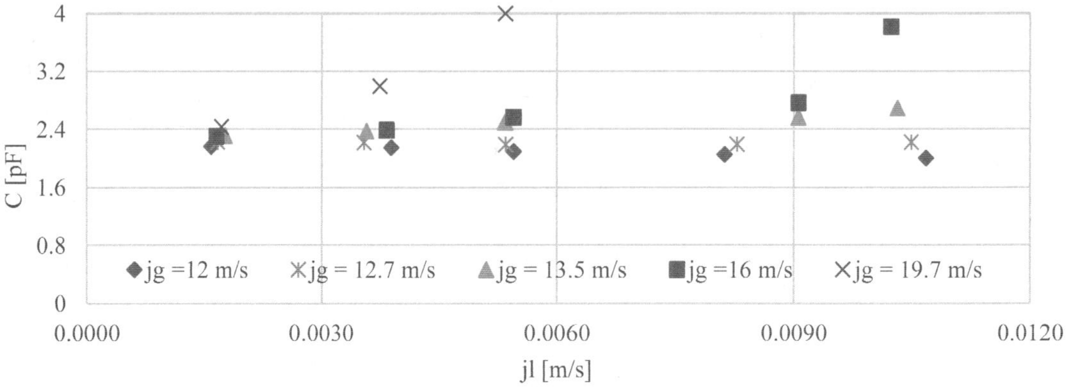

Figure 8 Capacitance values vs liquid superficial velocity, parameterized with gas superficial velocity (Electrodes A).

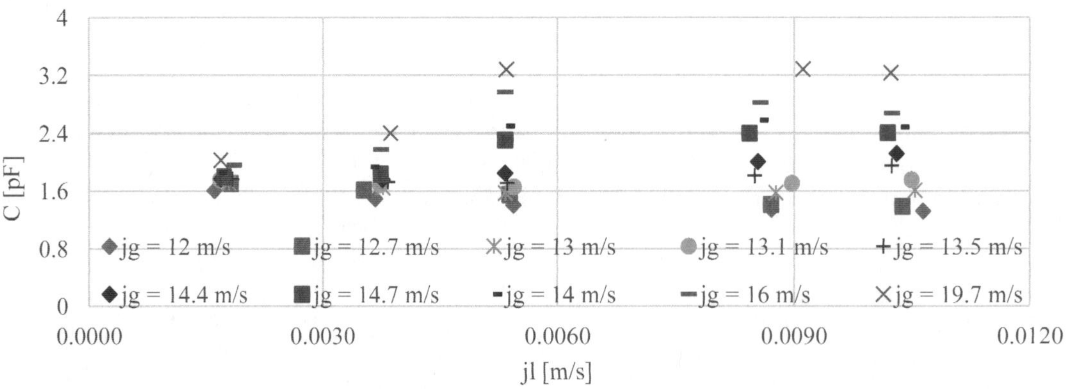

Figure 9 Capacitance values vs liquid superficial velocity, parameterized with gas superficial velocity (Electrodes B).

The relationship between the capacitance measured with the probe B and the void fraction measured by the QCV method has also been studied in the whole ranges of air and water superficial velocities. The results are reported in Figure 10, where the results have been grouped on the basis of the water superficial velocities. Four ranges of water superficial velocity have been considered; for each of them a different best-fit line has been found. Moving from the left towards the right of the void fraction axis, the flow patterns changes from wavy-stratified to dispersed-annular flow. Therefore the capacitance signal is not sufficient to determine the void fraction. The knowledge of the flow pattern is also necessary.

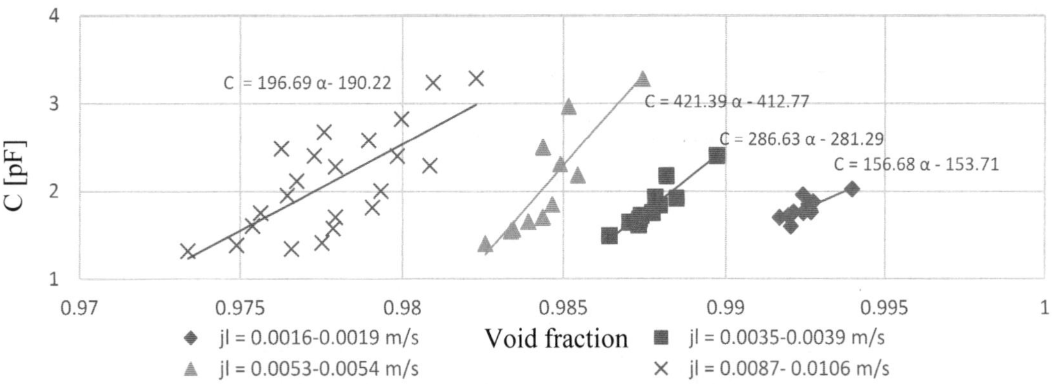

Figure 10 Capacitance values vs experimental void fraction.

4.3 *Response of the SIET ECP*

The results obtained with the SIET ECP in the present experiments are in agreement with all those of previous campaigns [10]. The curves in Figure 11 and Figure 12 show the normalized signals $V_{i,j}^*$ as a function of the position of the electrodes at different experimental values of the void fraction measured by the QCV method. $V_{i,j}^*$ is calculated as follows:

$$V_{i,j}^* = \frac{RMS_{TP-ij} - RMS_{l-ij}}{RMS_{g-ij} - RMS_{l-ij}} \tag{8}$$

where the subscript *ij* identifies the measuring electrodes combination and RMS is the root mean square of electrical signals. This value is a function of the local void fraction and of the probe sensitivity. Figure 11 reports the signal measured between electrode 1 (probe top) and all the other circumferential electrodes for two values of air superficial velocity, while figure 12 is relative to electrode 9 (probe bottom) and all the other ones.

Increasing the air flow rate, the flow pattern changes from wavy- stratified to dispersed-annular. The different distribution of the phases is identified by the ECP. In the case of higher air superficial velocities (dispersed annular flow), the measurements between the circumferential electrodes are rather insensitive to the void fraction (Figure 11b and Figure 12b), so it is necessary to consider also the measurements on the central electrode.

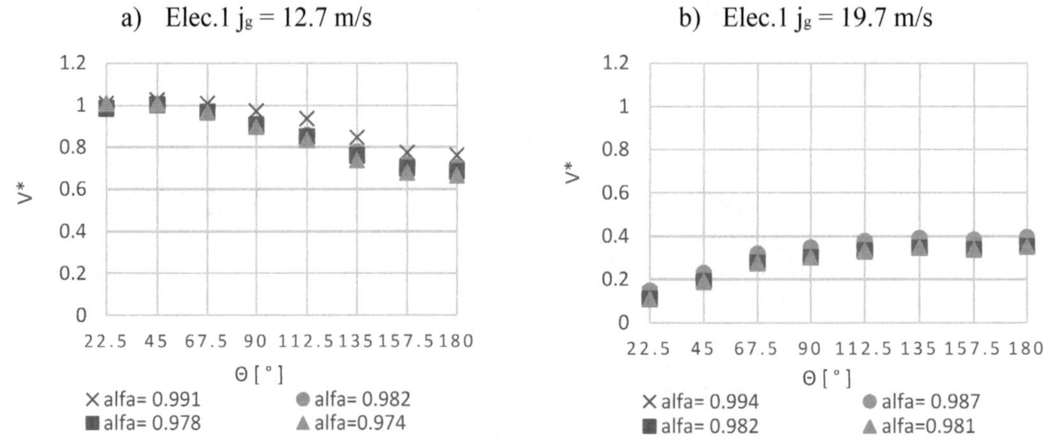

Figure 11 Dimensionless voltage drop electrode 1 (top).

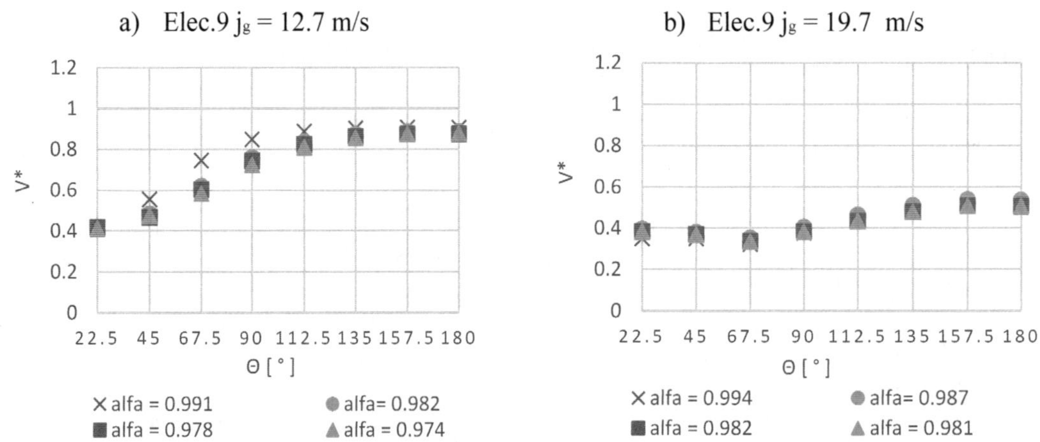

Figure 12 Dimensionless voltage drop electrode 9 (bottom).

Figure 13 shows the signals between the central electrode and the circumferential ones. The signal increases linearly with the void fraction and the trend of the curves depends on the position of the electrodes with respect to the vertical axis of the tube.

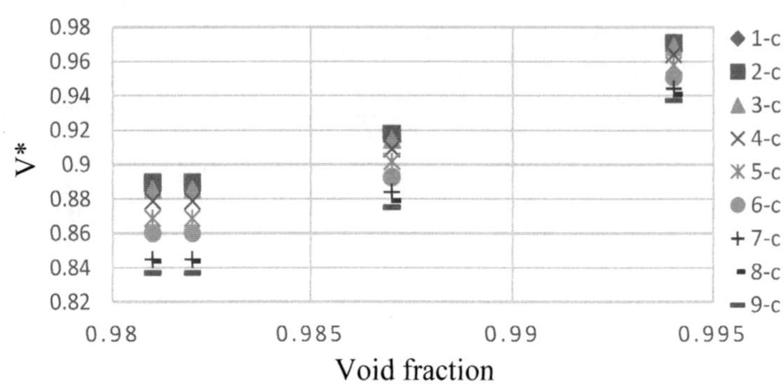

Figure 13 Dimensionless voltage drop associated with the measurements of the central electrode vs experimental void fraction.

IOP Conf. Series: Journal of Physics: Conf. Series **1224** (2019) 012040 doi:10.1088/1742-6596/1224/1/012040

5. Conclusions and perspectives

In this study a spool piece constituted by a Venturi Flow Meter, a new capacitance probe with concave electrodes designed at DENERG-Politecnico di Torino and the electrical capacitance probe designed by SIET have been characterized for wavy stratified and dispersed-annular flow pattern.

The results of the SIET ECP agree with those of the previous experimental campaigns. The DENERG ECP and the VFM demonstrated the possibility to determine respectively the average void fraction and the superficial velocity of the phases once the flow pattern is known. The elaboration of the signals from the SIET ECP can provide information on the flow pattern. In the future, taking the present experimental results into account, the design modification of the DENERG ECP will improve the performance of this probe; a second probe will also be positioned downstream the VFM and the range of the flow patterns will be extended. The final goal is the development of a two-phase flow forecasting model based on the transducers responses.

Acknowledgments: Thanks to Rocco Costantino for the technical support in the laboratory.

References

[1] Oddie G and Pearson J R A 2004 *Annu. Rev. Fluid Mech.* **36** 149-172
[2] Kumar S B, Moslemian D and Dudokovic M P 1995 *Flow Meas. Instrum.* **6** 61-73
[3] Veera U P 2001 *Chem. Eng. J.* **81** 251-260
[4] Dyakowsky T 1996 *Meas. Sci. Technol.* **7** 343-353
[5] Mishima K, Hibiki T and Nishihara H 1997 *Nucl. Eng. Des.* **175** 25-35
[6] De Salve M, Monni G and Panella B 2012 *J. Phys. Conf. Ser.* **395** 012179
[7] Monni G, De Salve M, Panella B and Randaccio C 2013 *Sci. Technol. Nucl. Ins.* **2013** 568287
[8] Monni G, De Salve M and Panella B 2014 *Prog. Nucl. Energy* **77** 167-175
[9] Analog Devices, 24-bit Capacitance-to-digital Converter with Temperature Sensor AD7746.
[10] Caramello M, De Salve M, Monni G, Panella B 2014 Modellizazione di uno spool piece per la stima dei parametri caratteristici dei deflussi bifase, Report RdS/PAR2013/095
[11] De Salve M, Monni G, Panella B, Randaccio C 2012 Electrical Capacitance Probe characterisation for vertical annular air-water flow investigation *in HEFAT2012 Conference Proceedings*
[12] Ahmed Wael H, Ismail Basel I. 2008 Innovative Techniques for two-phase flow measurements *Recent Patents on electrical engineering* **1** 1-13
[13] Analog Devices, AD7746 Evaluation Board data sheet.
[14] Taitel Y, Dukler A E 1 976 *AIChE J.* **22** 47-55
[15] Lockhart R W, Martinelli R C 1949 *Chem. Eng. Prog.* **45** 38-48
[16] Martinelli R C, Nelson D B 1948 *Trans. ASME* **70** 695
[17] Chisholm D 1973 *J. Mech. Eng. Sci.* **15** 235–236
[18] Mathure N 2010 Study of flow patterns and void fraction in horizontal two-phase flow *Master of Science, Oklahoma State University*

IOP Conf. Series: Journal of Physics: Conf. Series **1224** (2019) 012041 doi:10.1088/1742-6596/1224/1/012041

Experimental investigation on flow boiling heat transfer and pressure drop of refrigerants R32 and R290 in a stainless steel horizontal tube

B CITARELLA, G LILLO, R MASTRULLO, A W MAURO and L VISCITO

Department of Industrial Engineering, Federico II University of Naples, P.le Tecchio 80, 80125, Naples (Italy)

E-mail: alfonsowilliam.mauro@unina.it

Abstract. The purpose of this paper is to present new flow boiling heat transfer and pressure drop data in a single, horizontal smooth stainless steel tube of 6.0 mm internal diameter, in which R32 and R290 (propane) are employed as working fluids. The cross sectional average heat transfer coefficients are obtained by measuring the temperatures at the top, bottom, left and right sides of the channel. The experimental trends are analyzed for different operating conditions in terms of mass velocity (from 150 to 300 kg/(m^2 s)) and heat flux (from 10 to 40 kW/m^2). The saturation temperature is fixed to 25 °C for the heat transfer data and to 25 and 35 °C for the pressure drop experiments. The effects of the operative parameters and of the working fluids on local heat transfer coefficients and frictional pressure drop are discussed and the experimental data are finally compared with some of the available correlations taken from scientific literature.

1. Introduction

The correct design of evaporators in domestic and commercial air conditioning and refrigeration systems continuously requests flow boiling heat transfer and pressure drop data, especially in case of low-GWP refrigerants that are about to replace the conventional HFC substances [1]. Propane (R290) might be an interesting alternative thanks to its excellent thermodynamic and transport properties and miscibility with usual compressor lubricants. Nevertheless, its use is often restricted to small plants with low fluid inventory due to its high flammability. On the other hand, R32 is currently used in residential and industrial air-conditioning systems as a substitute for R410A and it is also expected to be a component of R32/HFO mixtures in the next future [2].

Although the flammability of R290 and R32 may represent a safety issue, the latent heat of these fluids is almost double of that of typical synthetic refrigerants, thus allowing the use of a considerably reduced mass flow rate required compared to HFCs fluids for the same amount of heat duty without affecting the system performance. On this regard, Mastrullo et al. [3] proposed R290 as working fluid in a light vertical freezer for commercial applications. The authors found that the refrigerant charge could be kept below the limits allowed by the current standards and also that the overall performance was increased with respect to the baseline configuration with refrigerant R410A. Later on, Fernando et al. [4] found that a 5 kW water-to-water heat pump could be run with only 200 g of propane without any penalization on the system performance when compared to a conventional design. Botticella et al. [2], in their multi-criteria optimization of a residential split system heat pump, found that refrigerant R32 is the most promising solution in terms of manufacturing costs, consumers' costs and environmental impact. Another research performed by Xu et al. [5] has pointed out that the coefficient of performance (COP) using R32 in a vapor-injected heat pump was increased up to 10% if compared to the R410A case operating under the same conditions.

Despite the wide range of possible applications for R290 and R32 as working fluids has been demonstrated, limited amount of heat transfer and pressure drop data are currently available in scientific literature. Recently, Longo et al. [6] provided flow boiling data of pure R290, propylene and R404A in a circular tube of 4.0 mm, exploring saturation temperatures from 5 to 20 °C, mass fluxes from 104 to 313 kg/(m^2 s) and heat fluxes from 15 to 31 kW/m^2. The authors found a nucleate boiling prevalent behaviour, with heat transfer coefficients not affected by a change of vapor quality especially at low mass fluxes and high heat fluxes and saturation temperatures.

Wang et al. [7] performed flow boiling experiments with R290 in a 6.0 mm horizontal tube, with low saturation temperatures (from -35 to -2 °C) and low mass fluxes (from 62 to 104 kg/(m^2 s)). The heat transfer performance was seen to be increased with mass flux at high vapor qualities and with saturation temperature at high heat fluxes, which positively affected the heat transfer coefficients in any condition.

For smaller pipes, Del Col et al. [8] studied flow boiling heat transfer and pressure drop of R290 in a 0.96 mm channel at a fixed saturation temperature of 31 °C. All the prediction methods chosen by the authors underestimated the heat transfer coefficient trends.

Flow boiling of R32 in a 4.0 mm internal diameter horizontal tube has been experimentally investigated by Longo et al. [9] for three saturation temperatures (5, 10 and 20 °C). The heat transfer coefficient was seen to be nearly independent on vapor quality and mass flux, whereas a higher sensitivity for heat flux variations was observed, especially at higher saturation temperatures.

Jige et al. [10] investigated the effect of the tube diameter on flow boiling heat transfer and pressure drop of R32 in small channels of 1.0, 2.2 and 3.5 mm internal diameters, at a fixed saturation temperature of 15 °C and by varying heat and mass flux. No differences were found on the heat transfer mechanisms with respect to conventional macro-tubes, and the heat transfer coefficients were seen to be higher with lower internal diameters under the same experimental conditions. Higher pressure drops were instead recorded for lower tube diameters and higher mass fluxes.

In this paper, new flow boiling heat transfer coefficient and pressure drop data of R290 and R32 are provided in a horizontal stainless steel tube of 6.0 mm internal diameter. Experiments were performed by investigating the effect of the operating parameters in terms of mass flux (from 150 to 300 kg/(m^2 s)) and heat flux (from 10 to 40 kW/m^2). The saturation temperature has been fixed to 25 °C for the heat transfer experiments and to 25 and 35 °C for the adiabatic frictional pressure gradient tests. Finally, the agreement with some prediction methods is also checked.

2. Experimental facility

The test facility for the flow boiling heat transfer and pressure drop experiments is schematically shown in Fig. 1, in which the refrigerant loop and the water cooling loop are displayed as black and blue lines, respectively.

The sub-cooled refrigerant passes through a magnetic gear pump, which is able to provide volumetric flow rates within a range of 1.3-2.5 dm^3/min by changing its rotating speed. The liquid mass flow rate is then measured by a Coriolis flow meter before the preheater section, which is made up of a copper tube where the heat is provided by means of four fiberglass heating tapes (each one with a nominal heat power of 900 W at 25 °C and 240 V). For the control of the local vapor quality at the test section inlet, the applied heat at the preheater section is managed thanks to a solid state relay and with the help of Arduino One controller. An adiabatic horizontal portion (40 cm) of stainless steel tube is provided in order to let the fluid be fully developed before the test section inlet. The saturated vapor is then condensed in a plate heat exchanger and finally sub-cooled in a tube-in-tube heat exchanger before the pump suction head that closes the loop. Two different throttling valves are positioned on the liquid and vapor lines, respectively, and a bypass circuit is provided at the pump outlet in order to have multiple mass flow rate control options.

Demineralized water is the cooling fluid, whose temperature is controlled by a thermostatic bath. Both the tube-in-tube and the plate heat exchangers may be independently excluded from the water flow by means of two dedicated bypass circuits.

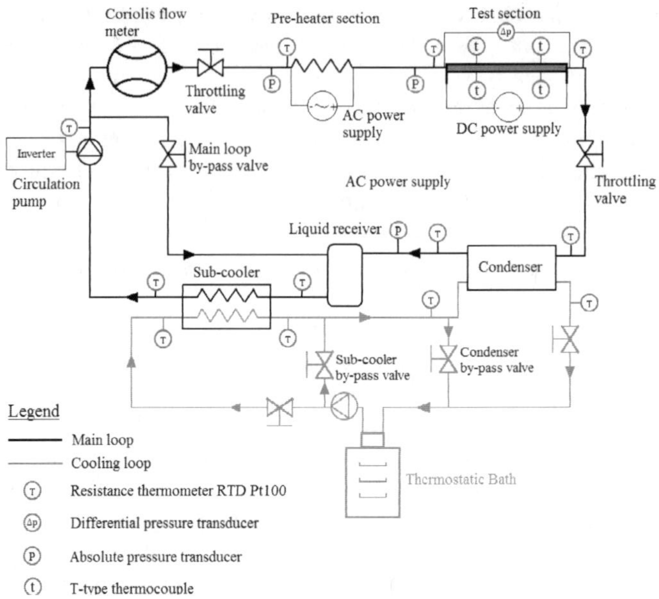

Figure 1. Schematic description of the experimental facility

2.1. Test section

All the flow boiling experiments have been conducted in a smooth, horizontal, circular stainless steel (AISI SS316) tube having an internal diameter of 6.00 ±0.05 mm and an external diameter of 8.00 ±0.05 mm. According to the tube manufacturer, the roughness of the inner surface is 10 μm. A photograph and the main characteristics of the tube test section are provided in Fig. 2. The heat flux is obtained by means of Joule effect provided by a DC power supply unit (within the range 0-8 V and 0-300 A) connected to the test section with two shorts copper wires clamped on copper electrodes welded on the tube surface (see points A and E in Fig. 2). The total measured heated length is 193.7±0.79 mm. At a distance of 237.5±0.91 mm, two pressure taps (points P and R) are provided for the measurement of the inlet absolute pressure and of the pressure drop across the test section. The local heat transfer coefficient is measured in the position C, at a distance of 146.7 ±0.64 mm from the inlet section. In this point, four T-type thermocouples are glued on the tube external surface with a high-conductivity epoxy resin, whereas a Kapton thin adhesive layer guarantees the electrical insulation from the heated channel [11]. The DC voltage for the evaluation of the imposed heat flux is measured between the position B and D in order to avoid locally concentrated tension drops in the electrodes proximity. Different layers of synthetic foam (λ = 0.040 W/(m K)) are used for the insulation of the test section and of the entire test facility, with the exception of the preheater section, which has been firstly covered with two layers of mineral wool (λ = 0.070 W/(m K)) due to the higher temperatures reached during the experimental procedure.

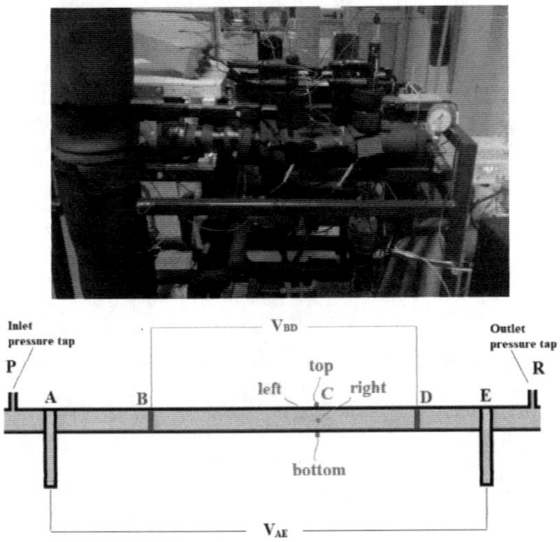

Geometrical features	
A, E	copper electrodes
B, D	voltage measurement points
C	temperature measurement point
AE	193.7 ±0.79 mm
BD	101.6 ±0.41 mm
AC	146.7 ±0.64 mm
d	6.00 ±0.05 mm (internal diameter)
D	8.00 ±0.05 mm (outer diameter)
PR	237.5 ±0.91 mm (pressure taps distance)
PC	166.9 ±0.91 mm

Figure 2. Photograph and geometrical details of the test section

2.2. Measurement instrumentation

Different RTDs are placed throughout the main and secondary loops for the measurement of the fluid temperature, carrying an overall uncertainty of ±0.180 °C. The wall temperature measurements for the evaluation of the heat transfer coefficients are instead obtained with four T-type thermocouples, calibrated *in-situ* with two resistance thermometers and a thermostatic bath. The overall estimated uncertainty for these sensors, by taking into account also the residual errors of the calibration curve, is assumed to be ±0.1 °C.

The absolute pressure at the test section inlet is measured with an absolute pressure transducer having a measuring range of 0-35 bar and an overall uncertainty of ±0.5% of the read value. The pressure drop is instead estimated with a differential pressure transducer, calibrated *in-situ* with a water liquid column and a yardstick. The instrumental uncertainty is estimated to be ±0.06 kPa.

The mass flow rate is measured by means of a Coriolis flow meter, calibrated up to 2% of the full scale, with a maximum uncertainty of ±1% of the measurement.

A digital wattmeter provides the heat applied to the preheater section, by separately measuring the voltage (100 mV-500 V) and current (1 mA-16 A). Its uncertainty is ±1.0% of the reading, as provided by the manufacturer.

The heat flux applied to the test section requires the measurement of the DC voltage (electrical voltage transducer within 0-5 V and an uncertainty of ±0.03% of the reading) and the DC current (directly measured by the DC power unit within 0-300 A and an uncertainty of ±1.0%).

2.3. Experimental procedure

The operating parameters in terms of mass flux, saturation temperature and heat flux are fixed at the beginning of each experiment, which starts from the onset of boiling and proceeds with increasing vapor quality up to the occurrence of dry-out, by means of the AC voltage applied to the fiberglass heating tapes of the preheater section. The test is stopped once the dry-out heat transfer is established (as soon as the heat transfer coefficient drops more than 25% from its previous value) and the wide wall temperature fluctuations would not guarantee high-quality experimental data.

The thermostatic bath control defines the saturation temperature of the system, whereas the inverter frequency of the electrical motor coupled with the magnetic gear pump changes the mass flow rate, which can be furtherly controlled by using the bypass circuit and/or the throttling valves on the liquid and vapor lines. Finally, the heat flux is imposed by setting the desired voltage of the DC power supply unit.

All the data are obtained in steady state conditions (with the exception of the dry-out points in which the phenomenon itself is time-dependent), with a recording frequency of 1.0 Hz. The nominal values of each sample are the arithmetic average of the measured quantities over a recording time of 90 s. The system is considered stabilized when the calculated deviation of the main parameters is sufficiently low (±2% for the mass velocity, ±0.1 °C for the saturation temperature).

3. Experimental method

3.1. Data reduction, uncertainty analysis and validation

The local heat transfer coefficient and the imposed heat flux are evaluated as follows:

$$h = \frac{q}{T_{wall} - T_{sat}} \tag{1}$$

$$q = \frac{V_{BD} \cdot I}{\pi d \overline{BD}} \tag{2}$$

where the subscripts *wall* and *sat* refer to the inner wall temperature and the local saturation temperature, respectively. The cross sectional average heat transfer coefficient shown in the results section is the arithmetical average of the four heat transfer coefficients obtained on the top, bottom, left and right sides of the test tube. V_{BD} is the voltage applied in the section *BD*, *I* is the DC current flowing in the test tube and *d* is the channel internal diameter. The fluid saturation temperature at the measurement point is obtained by considering a linear pressure drop profile from the tube inlet. The wall temperature is instead evaluated from the measured wall outer temperature T_{th} by considering 1-D heat transfer and uniform generation in the metal tube having λ_{tube} (16.26 W/(m K)) as thermal conductivity and *D* as external diameter:

$$T_{wall} = T_{th} + \frac{V_{BD} \cdot I}{4\pi\lambda_{tube} \overline{BD}} \cdot \frac{\left(\frac{D}{d}\right)^2 \cdot \left(1 - \ln\left(\left(\frac{D}{d}\right)^2\right)\right) - 1}{\left(\frac{D}{d}\right)^2 - 1} \tag{3}$$

The local vapour quality at the measurement point requires the knowledge of the local enthalpy, which is computable with an energy balance applied to the test section:

$$i = i_{in} + \frac{4 \cdot \overline{AC} \cdot q}{G \cdot d} \tag{4}$$

where *G* is the refrigerant mass flux and i_{in} is the test section inlet enthalpy, whose value is obtained from an energy balance applied to the preheater section, in which $i_{in,preh}$ is the preheater inlet enthalpy of the sub-cooled liquid, $\dot{Q}_{preh}$ is the preheater load and $\dot{m}$ is the refrigerant measured mass flow rate.

$$i_{in} = i_{in,preh} + \frac{\dot{Q}_{preh}}{\dot{m}} \tag{5}$$

All refrigerant thermodynamic properties are evaluated with the software REFPROP 9.0, whereas the whole data reduction is carried out with MATLAB software.

3.2. *Uncertainty analysis and validation*

The instrumental B-type uncertainties are composed to the standard deviation evaluated in the recording time for all the measured parameters. The uncertainty analysis of the derived results is then carried out by using the law of propagation of errors [11]-[12]. The maximum uncertainties are typically detected at the dry-out occurrence due to the higher fluctuations of the measured parameters. Lower uncertainty values are instead recorded during stable boiling. Tab. 1 shows the maximum and the average recorded uncertainty for the main operating parameters and results.

Table 1 Summary of the uncertainty analysis

Parameter	Average uncertainty	Maximum uncertainty
Saturation temperature T_{sat}	±0.05 °C	±0.08 °C
Mass flux G	±1.1%	±3.5%
Heat flux q	±0.70%	±0.76%
Vapor quality x	±0.04	±0.19
Mean heat transfer coefficient h_{mean}	±7.5%	±16%
Pressure gradient $\frac{\Delta P}{\Delta z}$	±1.8%	±11%

The preheater and test section insulation and the correct functioning of the whole experimental apparatus have been checked with liquid single-phase experiments by using refrigerant R134a. The electrical power (from 220 to 1460 W in the preheater section and from 10 to 150 W in the test section) is compared to the heat absorbed by the sub-cooled refrigerant by means of two energy balances performed in the heated sections. In all tests, the wall and ambient temperatures reached similar values than those obtained during the flow boiling experiments.

The liquid single-phase heat transfer coefficients for refrigerant R134a are also evaluated and compared to the well-known Dittus-Boelter prediction method [13]. Further detailed information on the validation procedure and results can be obtained from a previous work of the same authors with the same test facility [11].

4. Results

4.1. *Heat transfer coefficient: experiments and assessment*

The heat transfer coefficients points presented in this section are intended to be averaged over the four measurement points (top, bottom, left and right sides of the tube). The parameters and their uncertainties for all the titles and legends in the following diagrams are averaged over all the corresponding represented data.

In this experimental campaign, R32 and R290 are employed as working fluids and the saturation temperature has been fixed to 25 °C. Mass fluxes between 150 and 300 kg/(m^2 s) and heat fluxes from 10 to 40 kW/m^2 have been investigated and the effect of their variation on the local heat transfer coefficients is discussed in the following diagrams.

Fig. 3 shows the effect of the mass flux on the mean heat transfer coefficient. Fig. 3a refers to refrigerant R32 with an imposed heat flux of 10 kW/m^2 and Fig. 3b refers to propane (R290) with a heat flux of 20 kW/m^2. The heat transfer coefficients trends of R32 are different depending on the mass flux: for the lower values of 152 and 223 kg/(m^2 s), the heat transfer performance is not particularly affected by vapor quality and also by a change of the mass flux itself, even if a slight deterioration of the heat transfer coefficient with ongoing evaporation is detected for the lowest mass flux investigated. A different trend is shown for the mass flux of 301 kg/(m^2 s), in which the heat transfer coefficient increases with vapor quality (passing from 6 to 10 kW/(m^2 K)) and exposes a typical convective behavior. The dry-out phenomenon occurs at higher vapor qualities with increasing mass velocity, passing from 0.79 to 0.86 when the mass flux is changed from 152 to 301 kg/(m^2 s).

The trends of R290 are similar to those displayed for refrigerant R32 and are not affected by the mass flux and vapor quality variations for all the operating conditions investigated, showing a nearly constant value of 9-10 kW/(m^2 K) from the onset of boiling up to the dry-out occurrence, which happens at approximately $x = 0.85$ for the three operating conditions shown.

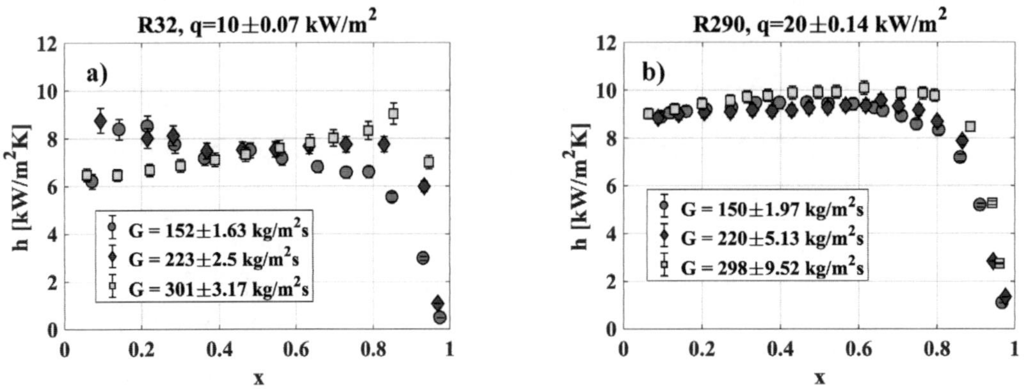

Figure 3. Effect of the mass flux on the mean heat transfer coefficient. The saturation temperature is fixed to 25 °C. (a) R32 with q = 10 kW/m^2; (b) R290 with q = 20 kW/m^2.

The effect of the imposed heat flux on the average heat transfer coefficients of R32 and R290 is shown in Fig. 4a-b, respectively. At a mass flux of 220 kg/(m^2 s) and with refrigerant R32, the heat transfer coefficient remains almost constant with ongoing evaporation when the vapor quality is included between 0.2 and 0.8 for the three curves shown. At lower vapor qualities, the heat transfer coefficient is nearly constant for a heat flux of 10 kW/m^2, whereas shows an increasing trend for q = 20 kW/m^2 and a decreasing trend for q = 40 kg/m^2. An increasing heat flux determines a significant increase of the average heat transfer coefficient (approximately +55% from 10 to 40 kW/m^2) at any vapor quality, suggesting a strong nucleate heat transfer contribution in these conditions. A higher imposed heat flux leads also to a progressive reduction of the vapor quality at the occurrence of dry-out, passing from 0.93 to 0.78.

A similar behavior is observed for refrigerant R290 at the same mass flux of 220 kg/(m^2 s), for which the dry-out occurrence is anticipated from 0.85 to 0.63 with increasing heat flux. For this fluid, the convective heat transfer is not completely suppressed due to the slight increase of the heat transfer coefficient with increasing vapor quality for all the heat fluxes investigated. The effect of the imposed heat flux is anyway more evident employing R290, with the average heat transfer coefficient that is almost doubled up when passing from 10 to 40 kW/m^2.

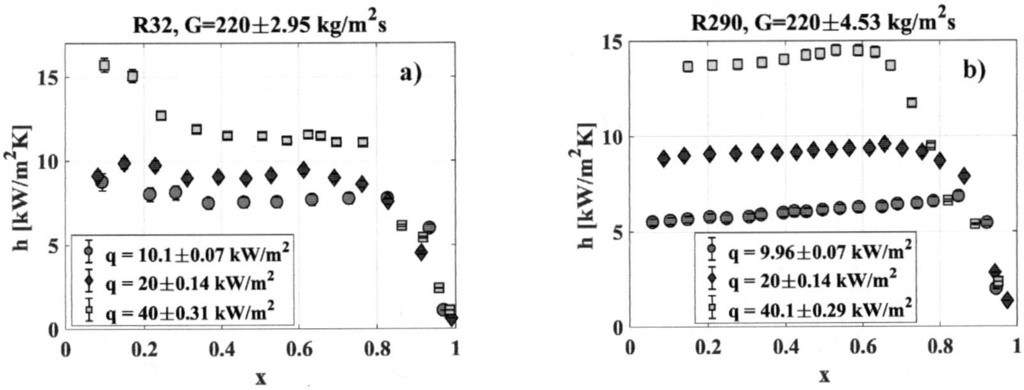

Figure 4. Effect of the heat flux on the mean heat transfer coefficient. The saturation temperature is fixed to 25 °C. (a) R32 with $G = 220$ kg/(m^2 s); (b) R290 with $G = 220$ kg/(m^2 s).

The experimental heat transfer coefficient data (including those after dry-out occurrence) are finally compared with two correlations taken from scientific literature. Particularly, the prediction methods of Wojtan et al. [14] and Del Col [15] are chosen for comparison. The statistic analysis and the graphical comparisons have been separately performed with the two refrigerants employed and are summarized in Fig. 5a-b.

The flow pattern based method of Wojtan et al. [14] was developed for macro-tubes by employing R410A as operating fluid. It works similarly for both refrigerants, providing a Mean Absolute Error (*MAE*) of 28.1% and 29.8% when applied to R32 and R290 data, respectively. The correlation of Del Col [15] is a modification of the original Gungor and Winterton [16] method, adapted for the author's data with halogenated refrigerants in a smooth horizontal tube of 8.0 mm internal diameter. From Fig. 5b, even if this method provides higher errors for higher heat transfer coefficients, it returns the best agreement with the experimental data, with a *MAE* equal to 19.6% for R32 and to 24.6% for the R290 heat transfer coefficient points.

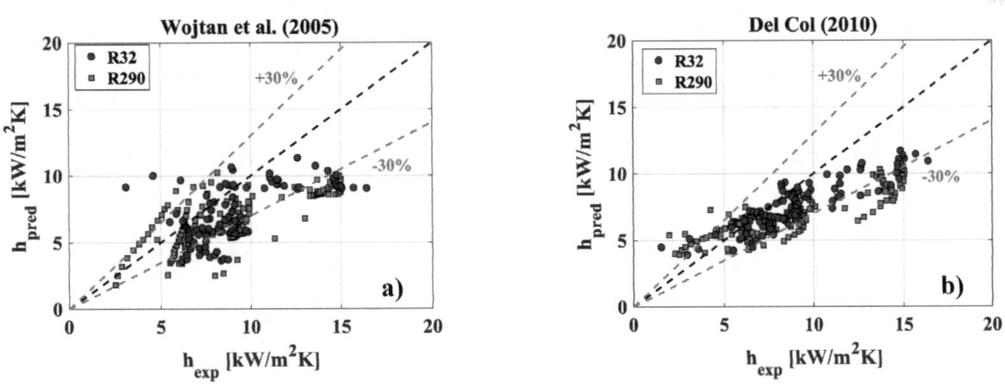

Figure 5. Experimental versus predicted heat transfer coefficients. Correlations of: (a) Wojtan et al. [14]; (b) Del Col [15]

4.2. Adiabatic pressure gradients: experiments and assessment

The effect of the saturation temperature and mass flux on the frictional pressure gradient is shown in Fig. 6a-b, respectively. The general trend is an increasing pressure drop with vapor quality, due to the flow acceleration, and then a decreasing trend after a peak value. For both fluids, an increase of the saturation temperature leads to a higher vapor phase density which in turn reduces the shear stress between the two phases, resulting in a slight reduction of the pressure drop, especially in case of high vapor qualities, where the flow velocity is more substantial. For the same reason, the frictional pressure gradients of R32 (reduced pressure of 0.29 at 25 °C) are lower than those of R290 at the same operating conditions (reduced pressure of 0.22 at 25 °C).

The pressure drop is also significantly increased with mass flux due to a higher flow mean velocity and inertia: at x = 0.8, the pressure gradient of R32 at 35 °C for G = 301 kg/m^2 s is approximately 450% higher than that obtained for G = 151 kg/(m^2 s). Similar trends with mass velocity are also found for refrigerant R290.

The experimental frictional pressure drop data are finally compared with the values obtained from the separated flow models of Zhang and Webb [17] and Friedel [18]. The graphical comparison is shown in Fig. 7a-b, with different markers for the two fluids. The correlation of Zhang and Webb [17] better predicts the R32 data, with a calculated *MAE* of 21.5%, whereas for R290 the agreement is slightly worse (*MAE* = 27.8%). On the other hand, the model of Friedel [18] provides a lower *MAE* (equal to 21.4%) for the R290 data, whereas tends to overestimate most of the R32 points (*MAE* = 41.7% and *MRE* = +38.5%).

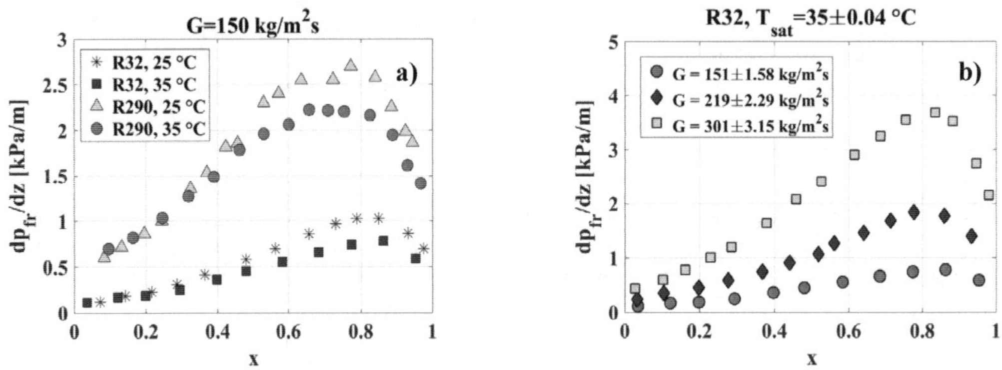

Figure 6. Experimental adiabatic frictional pressure gradients. (a) Saturation temperature effect for both fluids at G= 150 kg/(m^2 s); (b) Mass flux effect for R32 at T_{sat} = 35 °C

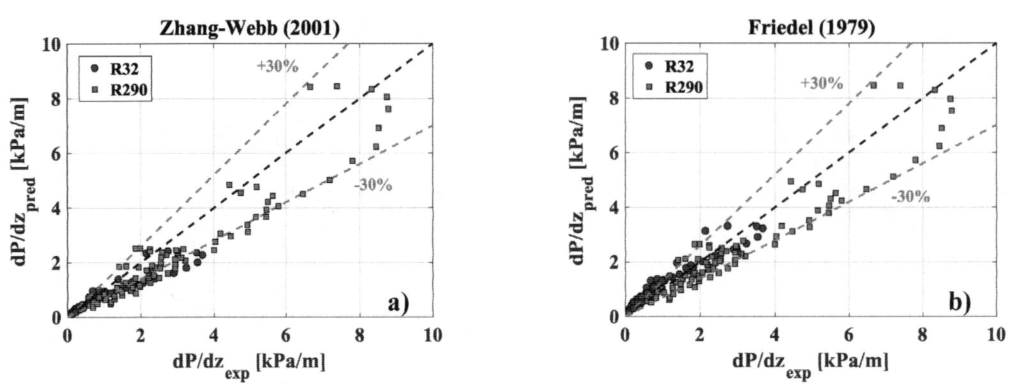

Figure 7. Experimental versus predicted frictional pressure gradients. Correlations of: (a) Zhang and Webb [17]; (b) Friedel [18]

5. Conclusions

New experimental data on local two-phase heat transfer coefficients and adiabatic pressure gradients for refrigerants R32 and R290 are obtained for a horizontal stainless steel tube of 6.0 mm internal diameter. The effects of the heat and mass flux (from 10 to 40 kW/m^2 and from 150 to 300 kg/(m^2 s), respectively) on the heat transfer coefficient and of the mass flux and saturation temperature (fixed to 25 and 35 °C) on the frictional pressure gradients are investigated. The main conclusions may be summarized as follows:

- The heat transfer coefficients of R290 present an almost constant trend with vapor quality and are not affected by a change of the mass flux, suggesting that the convective heat transfer contribution plays a minor role in these conditions. Similar trends are found for refrigerant R32, in which a typical convective behavior is observed only for the highest mass flux of 300 kg/(m^2 s). An increasing heat flux leads instead to a significant enhancement of the average heat transfer coefficients (+55% for R32 and +130% for R290 when passing from 10 to 40 kW/m^2), and also to a reduced vapor quality corresponding to the onset of dry-out.
- The two-phase heat transfer coefficient prediction method of Del Col [15] returns the best agreement when compared to the experimental data, providing a *MAE* of 19.6% and 24.6% when used with R32 and R290 points, respectively.
- When the same operating conditions are applied, the adiabatic frictional pressure gradients of R32 are lower (approximately three times) than those of R290, due to the higher reduced pressure that increases the vapor-to-liquid density ratio, reducing the shear stress between the two phases. For the same reason, the pressure drops of both fluids are lower with increasing saturation temperature. As expected, a higher mass flux provides instead a higher frictional pressure gradient, due an increased flow velocity and inertia.
- The two-phase frictional pressure drop correlation of Zhang and Webb [17] better predicts the R32 data (*MAE* = 21.5%, whereas the prediction method of Friedel [18] returns the best agreement with R290 experimental points (*MAE* = 21.4%).

Acknowledgements
Luca Viscito received a grant at Federico II University funded by "Ministero dell'Istruzione dell'Università e della Ricerca", via the project PRIN2015 "Clean Heating and Cooling Technologies For An Energy Efficient Smart Grid", which is gratefully acknowledged.

References

[1] Regulation (EU) No 517/2014 of the European Parliament and the Council of 16 April 2014 *Off. J. Union*
[2] Botticella F, de Rossi F, Mauro A W, Vanoli G P and Viscito L 2018 *Int. J. Ref.* **87** 131-153
[3] Mastrullo R, Mauro A W, Menna L and Vanoli G P 2014 *En. Conv. Manag.* **82** 54-60
[4] Fernando P, Palm B, Lunqvist P and Granryd E 2004 *Int. J. Ref.* **27** 761-773
[5] Xu X, Hwang Y and Radermacher R 2013 *Int J. Ref.* **36** 892-903
[6] Longo G A, Mancin S, Righetti G and Zilio C 2017 *Appl. Th. Eng.* **124** 707-715
[7] Wang S, Gong M Q, Chen G F, Sun Z H and Wu J F 2014 *Int J. Ref.* **41** 200-209
[8] Del Col D, Bortolato M and Bortolin S 2014 *Int J. Ref.* **47** 66-84
[9] Longo G A, Mancin S, Righetti G and Zilio C 2016 *Int J. Ref.* **61** 12-22
[10] Jige D, Sagawa K and Inoue N 2017 *Int. J. Ref.* **76** 206-218
[11] Lillo G, Mastrullo R, Mauro AW and Viscito L 2018 *Int J Heat Mass Tr.* **126** 1236-1252
[12] Moffat R J 1982 *Trans. ASME: J. Fl. Eng.* **104** 250-260
[13] Dittus F W and Boelter L M K 1930 *Univ. Calif. Publ. Engin.* **2** 443-461
[14] Wojtan L, Ursenbacher T and Thome J R 2005 *Int. J. Heat Mass Tr.* **48** 2970-2985
[15] Del Col D 2010 *Exp. Th. Fluid Sc.* **34** 234-245
[16] Gungor K and Winterton R 1986 *Int. J. Heat Mass Tr.* **29** 351-358
[17] Zhang M and Webb R L 2001 *Exp. Th. Fluid Sc.* **25** 131-139
[18] Friedel L 1979 *Proc. In: Europ. Two Phase Flow Group Meeting* **E2** 485

IOP Conf. Series: Journal of Physics: Conf. Series **1224** (2019) 012042 doi:10.1088/1742-6596/1224/1/012042

Relative importance of entrained liquid fraction and mass transfer at the interface on pressure drop of annular flows

G Lillo[1], **R Mastrullo**[1], **A W Mauro**[1] **and R Revellin**[2]

[1]Federico II, University of Naples, P. le Tecchio, 80 - 80125, Naples, Italy
[2]Université de Lyon, INSA-Lyon, CETHIL UMR5008, F-69621 Villeurbanne, France

alfonsowilliam.mauro@unina.it

Abstract. Fully developed adiabatic two-phase flows are commonly encountered in practice and their modeling is of primary importance for accurate, local calculations of pressure gradients and void fraction for the design of several industrial applications as heat exchangers and long fluid transfer piping. For large void fractions, the entrainment of liquid droplets into the vapor core can have a large effect on the pressure gradients. In this paper starting from theoretical considerations about the modeling of the pressure gradients in an annular two-phase flow, a numerical analysis as well as a discussion are proposed in order to quantify the relative importance of the entrainment on the calculations and to address further experimental studies and models on this topic.

Nomenclature

Symbols

A	cross-sectional area [m^2]
D	channel diameter [m]
e	entrainment ratio
f	friction factor
g	gravitational acceleration [$m\ s^{-2}$]
G	mass flux [$kg\ m^{-2}s^{-1}$]
J	superficial velocity [$m\ s^{-1}$]
m	mass flow rate [$kg\ s^{-1}$]
p	pressure [Pa]
R	channel radius [m]
w	velocity [$m\ s^{-1}$]
x	vapor quality

Greek symbols

β	channel inclination [°]
γ	spatial fraction
δ	liquid film thickness [m]
ε	void fraction
μ	dynamic viscosity [Pa s]
ρ	density [$kg\ m^{-3}$]
τ	shear stress [Pa]

Subscripts

i	relative to liquid/vapor interface
l	relative to liquid phase
le	relative to entrained liquid
lf	relative to liquid film
v	relative to vapor phase
W	relative to liquid/wall interface

1. Introduction

Many industrial applications such as chemical processing, petrochemical transportation, evaporators and condensers or steam generators in nuclear plants involve two-phase flows. The structure of the two-phase flow affects the performance of these applications: annular flow is one of the most important and frequent flow regimes in gas-liquid and vapor-liquid two-phase flows in tubes and channels. A thin film of liquid flows on the channel wall, surrounding a vapor core which flows in the center of the channel and may carry entrained liquid droplets in suspension. The vapor core, flowing faster than the liquid film, drags the liquid film and may atomize the tips of the waves which are formed at the liquid film surface. At the same time, entrained liquid droplets are continuously deposited back onto the liquid film after being accelerated in the gas or vapor core. Due to this continuous exchange of liquid between the liquid film and the gas or vapor core, mass and momentum exchange between the phases are strictly coupled, resulting in a complex physics of annular flows and in a particularly challenging analysis and modeling.

The optimal design and the energy efficient behavior of any two-phase flow system require accurate prediction of the pressure drop, needing the wall and the interface shear stresses as input for the modeling of mass, momentum and heat transfer in two-phase flows.

Revellin and Thome [1] developed a theoretical model for the prediction of the critical heat flux of refrigerants flowing in heated, round microchannels. In their model the basic idea was to integrate equations of mass, momentum and energy for each phase along the tube for a developing flow, in order to get a critical thickness determined by the Kelvin-Helmholtz theory, which causes the complete drying of the liquid film. The model was developed specifically for microchannels, when the relative effect of surface tension and of inertia forces guarantees both a symmetric annular flow and a marginal effect of the gravitational forces. In these conditions, the interface between vapor and liquid is smooth, the fraction of liquid entrained into the vapor core as droplets is marginal. The authors have proven that the model is able to predict experimental values with a good agreement, if the shear stresses between liquid and wall and between vapor and liquid phases are modeled with equations for smooth tubes.

Cioncolini, Thome and Lombardi [2] presented a model developed specifically for total pressure drops calculations including the actual value of the fraction of liquid entrained into the vapor core. In particular, the mass flow rate of liquid at the wall and the accelerational term reflect the effect of the liquid droplets entrained into the vapor core; consequently, the friction factor between the vapor and the liquid phases, based on the aerodynamic interaction between the phases through the Weber number of the vapor core, has been calibrated in such a way to include the effect of the entrainment.

Concerning the momentum balance for each phase, it is important to consider the momentum transfer between the phases due to the atomization of the liquid film, which goes into the vapor phase, and the corresponding deposition of liquid droplets moving from the vapor core into the liquid film.

The paper by Qu and Mudawar [3] includes this term. The model refers to a developing flow in a heated microchannel, assuming no atomization and a continuous deposition after a fully entrained boundary condition fixed at the transition between a churn and an annular flow. The friction factor for the interfacial shear stress is calculated based on the assumption of a laminar flow of the vapor core.

Even for an adiabatic developed flow with negligible compressibility effect, when the net mass transfer is zero with the equilibrium between atomization and deposition, there still is a transfer of momentum between the liquid and vapor phases due to the liquid droplets acceleration/deceleration at the interface.

In the knowledge of the authors, the case of an annular two-phase flow in a macro-channel, including the effect of the entrainment into the vapor core and the effect of the momentum transfer at the interface, has never been considered in the literature, even if it is a very common situation in different fields of application.

An issue, concerning the modeling of such a case which limits the possibility to propose and validate a model, is the difficulty to find closure equations specifically developed for this situation.

As a first step, this paper presents a discussion trying to quantify the relative importance of each effect (correction of the entrainment effect onto the mass flow at the wall and, then, of the momentum transfer) for an adiabatic, annular developed two-phase flow. Consequently, different levels of complexity into the modeling as done by work [1], [2] and [3] are introduced and generalized without any particular assumption. As a result, relative comparison of calculations with the different models are presented and discussed, in order to show their relative influence and to orient needs for further research.

2. Models description

Each of the following models assume a developed, adiabatic annular two-phase flow into a circular tube, with negligible compressibility effect (hence, no vapor quality and no pressure gradient variations).

Basic model (no entrainment) - CASE A

The conservation of momentum for the liquid and vapor phase could be expressed as follow.

$$-dp_{lf}A_{lf} - \tau_w dS_w + \tau_i dS_i + \rho_l A_{lf}\, dz\, g\cos\beta = 0 \tag{1}$$

$$-dp_v A_v - \tau_i dS_i + \rho_v A_v dz\, g\cos\beta = 0 \tag{2}$$

Where $A_{lf} = (1-\varepsilon)A_{tot}$ and $A_v = \varepsilon \cdot A_{tot}$ are the cross sections occupied by the liquid and the vapor, respectively, being ε the void fraction. The term $dS_W = 2\pi R dz$ is the elementary liquid/wall interfacial area and $dS_i = 2\pi(R-\delta)dz$ is the elementary liquid/vapor interfacial area, being R the radius of the channel and δ the liquid film thickness. β is the channel inclination angle in counterclockwise direction with respect to the gravitational acceleration vector (β=90° for horizontal flow).

The shear stress can be calculated with equation (3), where the friction factor, f_w, can be expressed according to the correlation proposed by Blasius for smooth tube, as done in [1] with no entrainment effect.

$$\tau_W = 0.5 f_W \rho_l w_{lf}^2 \tag{3}$$

$$f_W = 16\,\mathrm{Re}_{lf}^{-1} \qquad \text{for } \mathrm{Re}_{lf} = \frac{2\rho_l w_{lf}\delta}{\mu_l} < 2300$$

$$f_W = 0.079\,\mathrm{Re}_{lf}^{-0.25} \qquad \text{for } \mathrm{Re}_{lf} > 2300$$

The shear stress at the liquid/vapor interface can be defined as follows, considering the correlation proposed by Blasius for smooth tube to evaluate the friction factor f_i.

$$\tau_i = 0.5 f_i \rho_v \left(w_v - w_{lf}\right)^2 \tag{4}$$

$$f_i = 16\,\mathrm{Re}_v^{-1} \qquad \text{for } \mathrm{Re}_v = \frac{\rho_v w_v D}{\mu_v}\left(1-\frac{2\delta}{D}\right) < 2300$$

$$f_W = 0.079\,\mathrm{Re}_v^{-0.25} \qquad \text{for } \mathrm{Re}_v > 2300$$

From the momentum balance equations, the total pressure gradients for the liquid film and the vapor core can be expressed, in terms of void fraction and shear stresses as in equation (5).

$$-\frac{dp_{lf}}{dz} = \frac{2\left[\tau_W - \tau_i\sqrt{\varepsilon}\right]}{R(1-\varepsilon)} - \rho_l g\cos\beta; \qquad -\frac{dp_v}{dz} = \frac{2\tau_i}{R\sqrt{\varepsilon}} - \rho_v g\cos\beta \tag{5}$$

The average velocity of the liquid film and the vapor phases to be used for shear stress calculations are:

$$w_{lf} = \frac{\dot{m}_{lf}}{\rho_l A_{lf}} = \frac{G\left(1-x\right)}{\rho_l\left(1-\varepsilon\right)}; \qquad w_v = \frac{\dot{m}_v}{\rho_v A_v} = \frac{Gx}{\rho_v \varepsilon} \tag{6}$$

Equation (5) is valid when a symmetric annular flow exists.

Entrainment effect on the frictional term of the total pressure drop - CASE B

The entrainment ratio e is defined as the ratio between the liquid droplets mass flow rate inside the vapor core and the total liquid mass flow rate.

$$e = \frac{\dot{m}_{le}}{\dot{m}_l} \tag{7}$$

Assuming that the entrained liquid droplets flow at the same velocity of the vapor core, the spatial fraction of the total liquid occupied by the entrained liquid droplets, γ, is related to the other parameters of the flow and thermodynamic properties by the following equation:

$$\gamma = \frac{A_{le}}{A_l} = e\frac{\varepsilon}{1-\varepsilon}\frac{1-x}{x}\frac{\rho_v}{\rho_l} \tag{8}$$

The conservation of the momentum in equations (1) and (2) are formally the same when considering an annular flow with entrained liquid droplets into the vapor core. However, the entrained liquid droplets have an influence on the liquid film thickness meaning different results in the pressure gradient prediction with respect to the case without entrainment (*CASE A*, γ=0). In this approach, named *CASE B*, the equations (1) and (2) can be reduced as a function of the spatial and mass fractions for the liquid entrained and vapor as the following.

$$-\frac{dp_{lf}}{dz} = \frac{2\left[\tau_W - \tau_i\sqrt{\varepsilon+\gamma\left(1-\varepsilon\right)}\right]}{R(1-\varepsilon)\left(1-\gamma\right)} - \rho_l g\cos\beta; -\frac{dp_v}{dz} = \frac{2\tau_i}{R\sqrt{\varepsilon+\gamma\left(1-\varepsilon\right)}} - \frac{\rho_v\varepsilon+\rho_l\gamma(1-\varepsilon)}{\varepsilon+\gamma(1-\varepsilon)} g\cos\beta \tag{9}$$

The shear stresses are calculated with the same equations of the previous case (equations (3) and (4)), where the influence of the entrained liquid fraction is reflected into the velocities of phases through mass balances. In particular, the average velocity of the liquid film and the vapor phases to be used for shear stress calculations are:

$$w_{lf} = \frac{\dot{m}_{lf}}{\rho_l A_{lf}} = \frac{G\left(1-x\right)\left(1-e\right)}{\rho_l\left(1-\varepsilon\right)\left(1-\gamma\right)}; \qquad w_v = \frac{\dot{m}_v}{\rho_v A_v} = \frac{Gx}{\rho_v \varepsilon} \tag{10}$$

If the flow has no entrainment (γ=0), equation (9) is equivalent to equation (5); hence, *CASE B* includes also *CASE A*. If the flow has entrainment equal to one (γ=1), there is no more liquid at the wall and the flow is no more an annular but a mist flow. More details for the algorithm related to each case are presented in the next section.

Entrainment effect both on the frictional term of the total pressure drop and on the momentum transfer between phases - CASE C

In this case equation (5) is modified to account for the momentum transfer at the interface between vapor and liquid phases. Since for a developed flow the mass flow rate of liquid droplet atomized is in equilibrium with the one deposited (m_{atom}), equation (5) modifies in:

$$-dp_{lf}A_{lf} - \tau_w dS_w + \tau_i dS_i + \rho_l A_{lf}\,dz\,g\cos\beta + \dot{m}_{atom}\left(w_v - w_{lf}\right) = 0 \tag{11}$$

$$-dp_v\left(A_v + A_{le}\right) - \tau_i dS_i + \left(\rho_v A_v + \rho_l A_{le}\right)dz\,g\cos\beta - \dot{m}_{atom}\left(w_v - w_{lf}\right) = 0 \tag{12}$$

The total pressure gradient for the liquid film and the vapor core can be reduced as a function of the spatial and mass fractions for the liquid entrained and vapor as the following.

$$-\frac{dp_{lf}}{dz}=\frac{2\left[\tau_W-\tau_i\sqrt{\varepsilon+\gamma(1-\varepsilon)}\right]}{R(1-\varepsilon)(1-\gamma)}-\rho_l g\cos\beta-\frac{\dot{m}_{atom}\left(w_v-w_{lf}\right)}{\pi R^2(1-\varepsilon)(1-\gamma)dz} \tag{13}$$

$$-\frac{dp_v}{dz}=\frac{2\tau_i}{R\sqrt{\varepsilon+\gamma(1-\varepsilon)}}-\frac{\rho_v\varepsilon+\rho_l\gamma(1-\varepsilon)}{\varepsilon+\gamma(1-\varepsilon)}g\cos\beta+\frac{\dot{m}_{atom}\left(w_v-w_{lf}\right)}{\pi R^2\left[\varepsilon+\gamma(1-\varepsilon)\right]dz} \tag{14}$$

The shear stresses are calculated with equations (3) and (4) using the velocity in equation (10). Closure equations are needed for the atomized mass flux at the interface and for the entrained liquid fraction into the vapor core.

2.1. Closure equations

2.1.1. Atomization/deposition rate. The work of Berna et al. [4] gives a review of the extensive literature existing on the annular two-phase flow, focusing specifically on the analysis of the main phenomena which are involved. Kataoka et al. [5] obtained separate equations for the atomization and the deposition rates, also in case of non-equilibrium conditions, when the atomization and the deposition rates are unbalanced. Furthermore, a correlation for the entrainment ratio in the fully developed region is also presented.

$$\dot{m}_{atom}=\frac{\mu_L dS_i}{D}\left[6.6\,10^{-7}\left(\frac{\mu_v}{\mu_l}\right)^{0.26}(1-e)^{0.185}\mathrm{Re}_l^{0.925}We_E^{0.925}\right] \tag{15}$$

Where: $\mathrm{Re}_l=\frac{G(1-x)D}{\mu_l}$; $We_E=\frac{G^2x^2D}{\sigma\rho_v}\left(\frac{\rho_l-\rho_v}{\rho_v}\right)^{1/3}$

Ishii and Mishima [6] obtained one of the most known correlation to estimate the entrainment mass flux. The atomization rate is correlated to two dimensionless numbers, namely, a modified vapor Weber number and a liquid Reynolds number.

$$G_{atom}=G(1-x)\tanh\left(7.25\cdot10^{-7}\mathrm{Re}_l^{0.25}We_E^{1.25}\right) \tag{16}$$

2.1.2. Entrainment ratio. Kataoka et al. [5] also proposed a correlation for the calculation of the liquid droplet fraction entrained in the vapor core at the equilibrium (fully developed flow), according to Equation (17).

$$e(1-e)^{-0.25}=7.75\cdot10^{-7}\mathrm{Re}_l^{0.25}We_E^{1.25} \tag{17}$$

Cioncolini and Thome [7] presented a correlation for the entrainment ratio. The value of the liquid fraction entrained into the vapor core depends on the core flow Weber number and the core flow density. Since the core flow density depends on the entrained liquid fraction, a procedure with two steps is required, as explained in [7].

$$e=\left(1+279.6We_c^{-0.8395}\right)^{-2.209} \tag{18}$$

Paleev and Filippovich [8] proposed a correlation including the effects of droplets concentration in the vapor core. The vapor density is replaced by an average mixture density as defined in [8].

$$e=0.015+0.44\log\left[\frac{\rho^*}{\rho_l}\left(\frac{\mu_l J_V}{\sigma}\right)^2\cdot10^4\right] \tag{19}$$

3. Algorithm

Given the input parameters (fluid, mass flux, saturation temperature, vapor quality and internal diameter), the entrainment fraction can be evaluated by the selected correlation. If the entrainment fraction is less than unity, the two momentum balances have to give the same total pressure gradient (unknown) for the same void fraction (unknown), and an iterative procedure is required, as shown in the flowchart (Figure 1).

It is worth noting that if the basic model (*CASE A*) is considered, to solve equations of *CASE A* or *CASE B* with e=0 (γ=0) are equivalent approaches. Meanwhile, if entrainment is equal to unity (e=1 and γ=1), the flow pattern changes to mist flow, when the liquid film at the wall no longer exists (all the liquid phase flows inside the vapor core as droplets). In this case, the model can be solved directly, assuming γ=1: void fraction can be calculated from equation (8) and the pressure gradient can be directly evaluated from equation (9) related to the vapor phase (being the vapor in contact with the wall, while mist flow occurs).

In Figure 2 the pressure gradient evaluated by the liquid momentum balance and by the vapor momentum balance as function of the void fraction are shown, for different values of the entrainment ratio from 0 to 0.9.

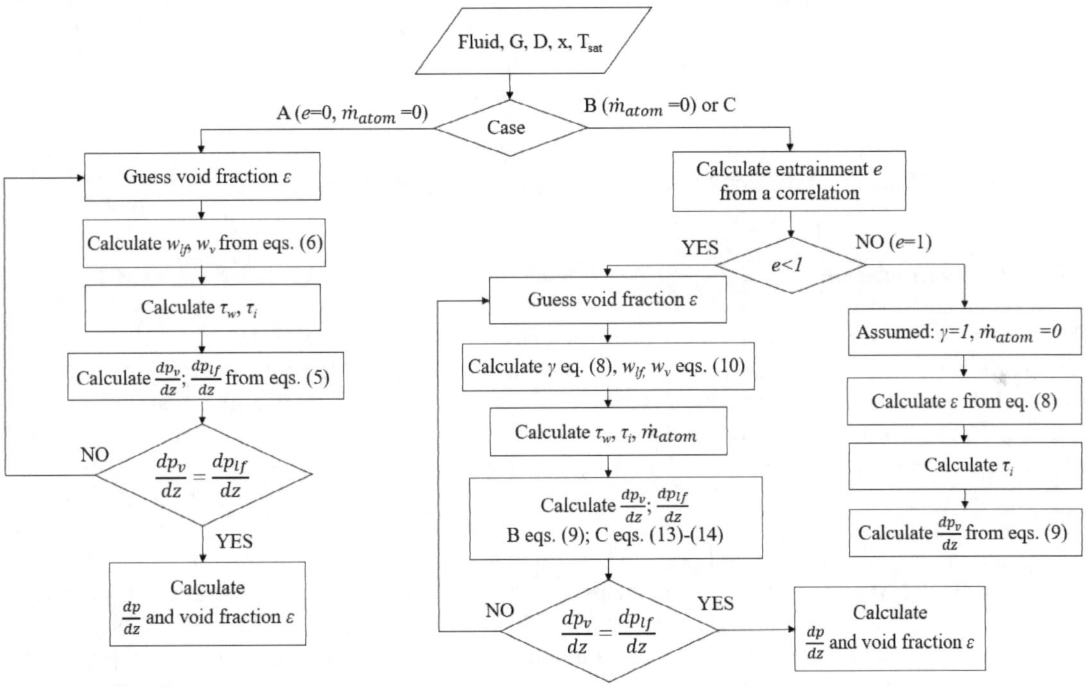

Figure 1. Iterative procedure for the calculation of void fraction and pressure gradient.

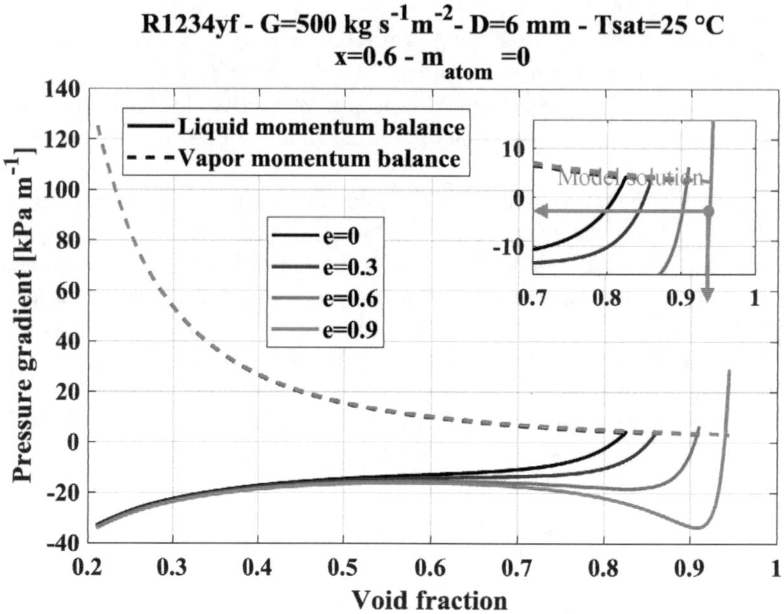

Figure 2. Liquid and vapor pressure gradient as function of void fraction at the conditions on the top of the figure, for different values of the entrainment ratio.

4. Results and discussion

4.1. Effect of the entrainment ratio on the phase velocities - CASE B vs CASE A

In this section the effects of the evaluation of the entrained liquid droplets on the pressure gradient will be studied. Figure 3 shows the entrainment ratio as function of vapor quality at given conditions. The fraction of liquid entrained as droplets increases with the vapor quality, with higher values for the correlation of Paleev and Filippovich. The model of Kataoka et al. assumes that the flow is at the thermodynamic equilibrium after x=1 (vapor single phase flow). On the contrary, the two other models (Cioncolini and Paleev) assumes a non-equilibrium flow for a vapor quality greater than unity meaning that liquid droplets at saturation coexist with the superheated vapor phase.

Figure 4 shows the entrainment ratio maps in terms of mass flux and diameter at saturation temperature of 25°C, vapor quality of 0.6 and R1234yf as working fluid. At high mass flux and internal diameter, Kataoka et al. gives entrainment ratio greater than one, which is physically impossible. As a consequence, the flow pattern is considered as mist flow: no liquid film occurs at the wall. In the investigated range of mass flux and diameter, Cioncolini and Thome calculates a maximum value of the entrainment ratio of 0.9. The correlation of Paleev and Filippovich is not sensitive to the diameter, so in the following analysis it will not be taken into consideration.

The iso-lines reported in Figure 5 show the dimensionless pressure gradient evaluated in the *CASE B* with reference to the pressure gradient evaluated in *CASE A*, using the correlation by Kataoka et al. (Figure 5a) and Cioncolini and Thome correlation (Figure 5b).

Where low entrainment is expected, the pressure gradient is near to the reference pressure gradient. On the contrary, at high mass flux and diameter, the amount of the entrained liquid by the vapor core needs to be taken into account: error up to -60% in the pressure gradient prediction could occur if the entrainment ratio is neglected, in the investigated conditions.

Depending on the boundary conditions, the relative influence of the entrainment ratio on the pressure gradient prediction could be relevant. However, different correlations give different results in terms of both entrainment ratio and pressure gradient.

IOP Conf. Series: Journal of Physics: Conf. Series **1224** (2019) 012042 doi:10.1088/1742-6596/1224/1/012042

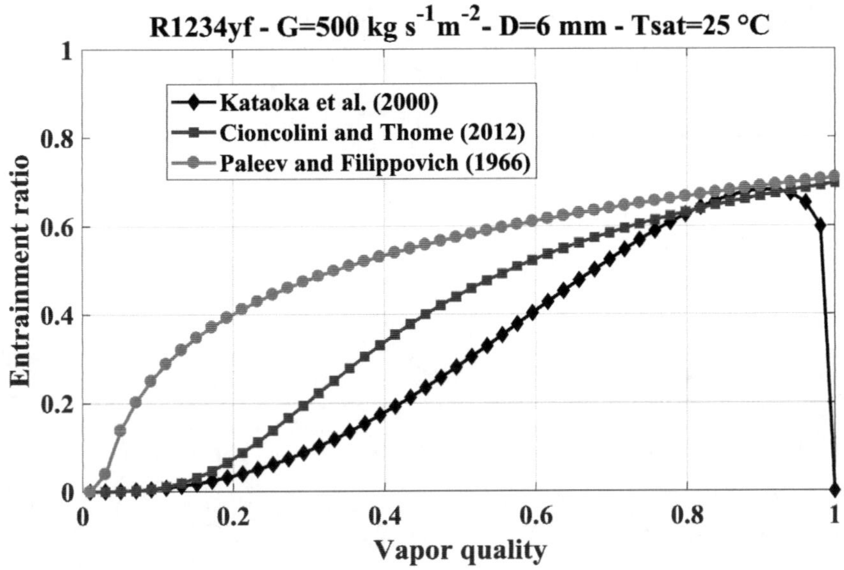

Figure 3. Entrainment ratio as function of vapor quality at the conditions on the top of the figure.

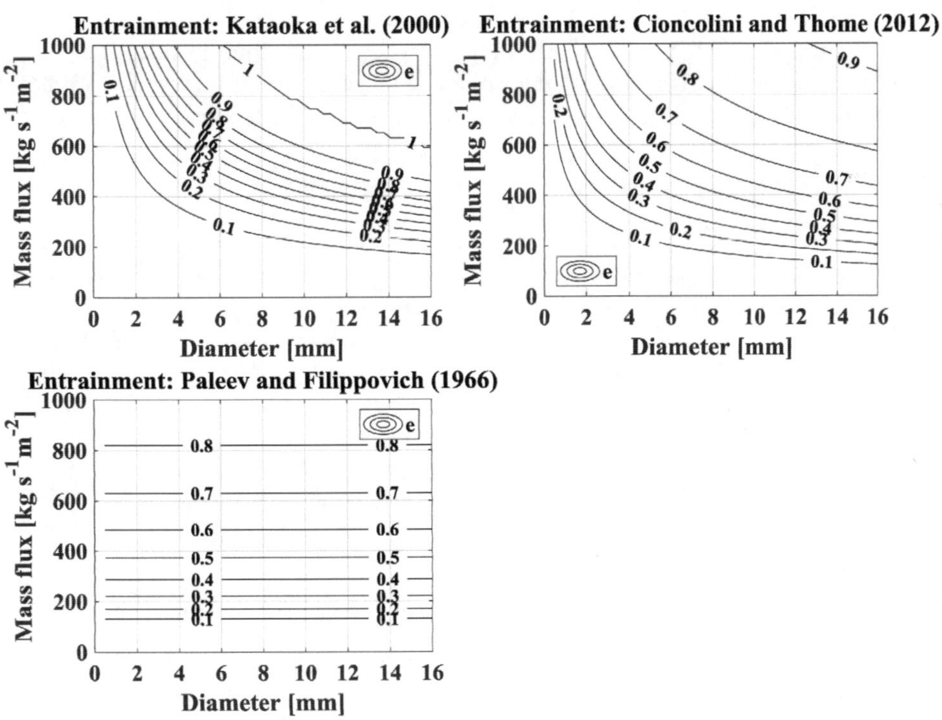

Figure 4. Entrainment ratio evaluated by Kataoka et al., Cioncolini and Thome and Paleev and Filippovich as function of mass flux and internal diameter at: T_{sat}=25°C, x=0.6.

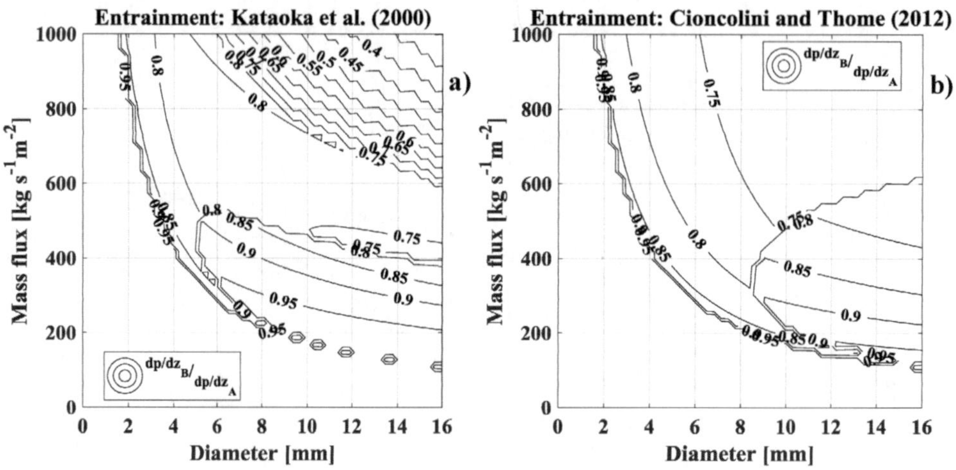

Figure 5. Dimensionless pressure gradient (CASE B) as function of mass flux and internal diameter at: Tsat =25°C, x=0.6.

4.2. Effect of the entrainment/atomization both on the phase velocities and on the momentum transfer at the interface CASE C vs CASE A

Figure 6 shows the trend of the atomization rate per unit of length as function of the vapor quality for different entrainment ratio correlations. When using Kataoka et al. correlation for the atomization rate (Figure 6a), a slight difference between the correlations of the entrainment occurs. In the investigated conditions, at vapor quality of 0.6 the atomization rate reaches the maximum value. The correlation of Ishii and Mishima (Figure 6b) gives lower values of the atomization rate with respect to Kataoka et al. Furthermore, no relevant differences between the entrainment correlations occur.

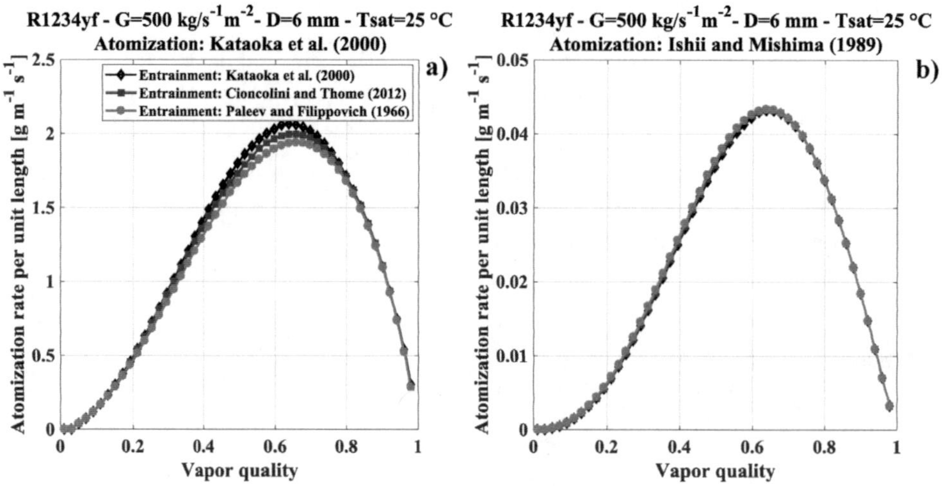

Figure 6. Atomization rate a) Kataoka et al. (2000) – b) Ishii and Mishima (1989) as function of vapor quality for different entrainment ratio correlations.

The iso-lines reported in Figure 7 shows the dimensionless pressure gradient evaluated in *CASE C* with reference to the pressure gradient evaluated in *CASE A*, using the combinations of entrainment ratio and atomization rate correlations reported on the top of the figure, at saturation temperature of 25°C, vapor quality of 0.6 and R1234yf as working fluid.

When Kataoka et al. is used to estimate the entrained liquid droplets (Figure 7a and Figure 7b), the results in terms of dimensionless pressure gradient are similar at high values of mass flux and diameter. Meanwhile, if the entrainment ratio is low, Kataoka et al. for the atomization rate over-predicts the pressure gradient respect to the *CASE A* and Ishii and Mishima under-predicts it.

Using Cioncolini and Thome (Figure 7c and Figure 7d), different results can be obtained depending on the correlation for the atomization rate. In the major of the investigated conditions, using Kataoka et al. correlation for the atomization, the model over-predicts the pressure gradient with respect to the *CASE A* (up to 50%). Instead, the prediction is under-estimated when Ishii and Mishima correlation is implemented in the investigated conditions.

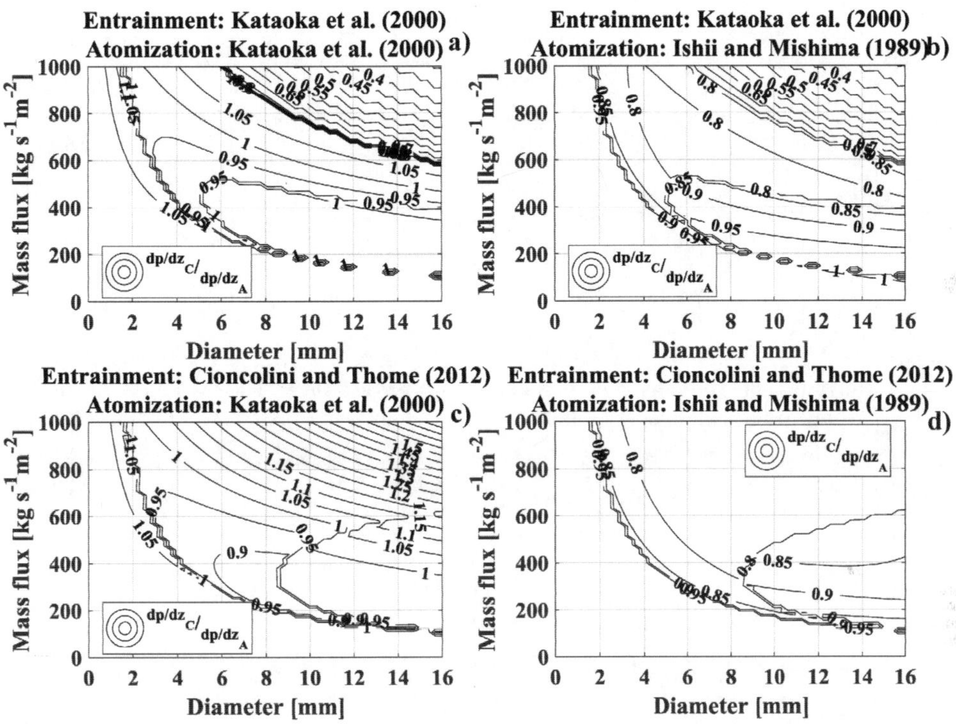

Figure 7. Dimensionless pressure gradient (CASE C) as function of mass flux and internal diameter at: T_{sat} =25°C, x=0.6.

4.3. Effect of the operating conditions

To take into account the influence of the vapor quality, an analysis varying the mass flux and vapor quality for two values of the internal diameter, 1 and 12 mm, is studied in this section. A saturation temperature of 25°C is considered. At small value of the diameter, the effect on the pressure gradient is not significant, so the results are not showed.

The iso-line at constant entrainment ratio is shown in Figure 8 as a function of the mass flux and vapor quality. at saturation temperature of 25 °C and internal diameter of 12 mm. High entrainment ratio is observed for high mass flux and vapor quality. Similarly to Figure 4, Kataoka et al. gives entrainment ratio greater than one meaning mist flow occurs in these conditions.

Figure 9 shows the dimensionless pressure gradient evaluated in the *CASE C* (including entrainment ratio and atomization rate) as function of the mass flux and internal diameter. Using Kataoka et al. for the entrainment ratio (Figure 9a and Figure 9b), at high mass flux and vapor quality the model calculates lower pressure gradient with respect to the reference state.

When Cioncolini and Thome and Kataoka et al. are used for the entrainment ratio and the atomization rate respectively, the ratio between the pressure gradient evaluated in the *CASE C* and the one evaluated in the *CASE A* assumes values up to 1.6 in the area characterized by high diameters and high vapor quality.

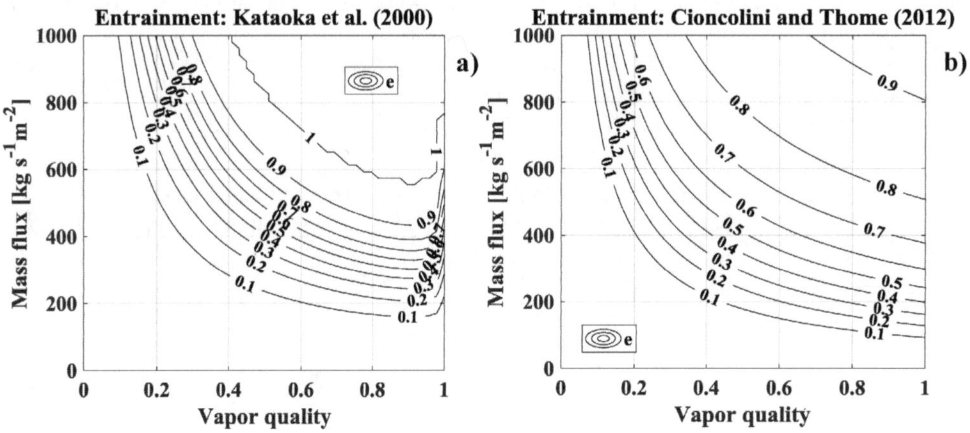

Figure 8. Entrainment ratio as function of mass flux and vapor quality at: T_{sat} =25°C, D=12.0 mm.

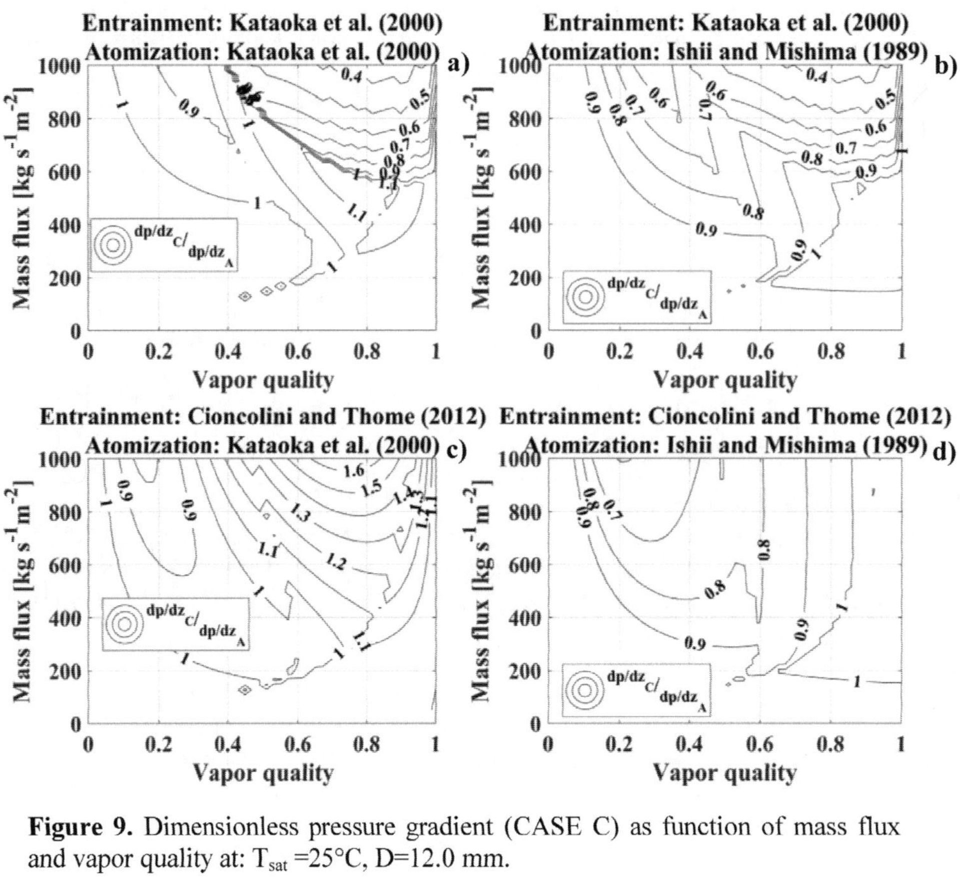

Figure 9. Dimensionless pressure gradient (CASE C) as function of mass flux and vapor quality at: T_{sat} =25°C, D=12.0 mm.

IOP Conf. Series: Journal of Physics: Conf. Series **1224** (2019) 012042 doi:10.1088/1742-6596/1224/1/012042

5. Conclusion

Starting from a basic modeling of a two-phase symmetric, fully developed annular flow with no entrainment, other two models are presented to account for the effects of the entrainment ratio on the phase velocities and on the momentum transfer at the interface between phases. The analysis was run for different correlation both for the entrainment ratio and the atomization rate, in a wide range of mass flux, channel diameter and vapor quality.

The main conclusion can be summarized as follow:

- The liquid fraction entrained into the vapor core and the momentum transfer at the interface have a large influence on pressure drop calculations and should not be neglected especially for macro-channels and large mass flux.
- The closure equations chosen for the calculations have a remarkable influence on the results: specific correlations for friction factors using equations (13) and (14) should be used to obtain more general correlations, including consistently the effects of the entrainment.

References

[1] Revellin R and Thome J R 2008 *Int. J. Heat Mass Tr.* **51** 1216-1225
[2] Cioncolini A, Thome J R and Lombardi C 2009 *Int. J. Multip. Flow* **35** 1138-1148
[3] Qu W and Mudawar I 2003 *Int. J. Heat Mass Tr.* **46** 2773–2784
[4] Berna C, Escrivá A, Muñoz-Cobo J L and Herranz L E 2015 *Prog. Nucl. En.* **79** 64-86
[5] Kataoka I, Ishii M and Nakayama A 2000 *Int. J. Heat Mass Tr.* **43** 1573-1589
[6] Ishii M and K Mishima 1989 *Int. J. Heat Mass Tr.* **32** 1835-1846
[7] Cioncolini A and Thome J R 2012 *Nucl. Eng. Des.* **243** 200-213
[8] Paleev I I and Filippovich B S 1966 *Int. J. Heat Mass Tr.* **9** 1089-1093

IOP Conf. Series: Journal of Physics: Conf. Series **1224** (2019) 012043 doi:10.1088/1742-6596/1224/1/012043

Convective and absolute instability of horizontal flow in porous media

Antonio Barletta, Michele Celli

Dipartimento di Ingegneria Industriale, Alma Mater Studiorum Università di Bologna, Viale Risorgimento 2. 40136 Bologna, Italy

E-mail: antonio.barletta@unibo.it; michele.celli3@unibo.it

Abstract. The onset of instability in flow systems has a dual nature depending on the dynamics of the growing normal modes. When the time evolution of a wave packet perturbation is tested, the growth of individual Fourier normal modes can be concealed to an observer in the laboratory reference frame. The reason is that the growing mode can be convected away by the basic flow, so that no effective growth is detected for the wave packet. The convective instability just focusses on the dynamics of each Fourier mode of perturbation, disregarding the actual amplitude growth of a wave packet, measured at a given position. When not only the Fourier modes, but all localised wave packets grow in time at a given position, then the instability becomes absolute. The two types of instability are generally distinct. This paper illustrates the transition from convective to absolute instability starting from a simple one-dimensional case. Then, this concept is employed for the stability analysis of a porous medium flow with heating from below. While the one-dimensional flow system is studied analytically, the porous medium flow stability is investigated numerically.

1. Introduction

It is well known that solutions of the governing equations of convection heat transfer may exist even if they cannot be observed in a laboratory experiment. The reason is the instability that can be developed when the dimensionless parameter or parameters describing the convection process belong to a specific domain. Such aspects are of paramount importance for the engineering community as the onset of convection cells may be undesirable when it comes to the optimisation and control of production processes. This, then, is what happens for the growth of crystals or the extrusion of plastics where the onset of convection cells should be inhibited. Likewise, the onset of thermal convection can be one of the causes of contamination for groundwater reservoirs. Conversely, many other practical situations exist where the thermal instability, and hence the emergence of cellular flow patterns, is a desirable condition meaning enhancement of heat transfer in a fluid system employed for a cooling process.

The initiation of instability is defined by the neutral stability threshold. Determining the neutral stability condition means testing the response of the flow system to quite peculiar perturbations: the normal modes. The normal modes are basically Fourier modes associated with a given wave number k and propagating either as plane waves, or as cylindrical waves, or as waves with a more complicated geometry depending on the system examined. Hence, the neutral stability threshold is a curve or a surface where the governing parameters of the flow vary as functions of k.

IOP Conf. Series: Journal of Physics: Conf. Series **1224** (2019) 012043 doi:10.1088/1742-6596/1224/1/012043

The analysis of convective instability aims to establish the neutral stability threshold. The point is that a natural perturbation of a convection process is a wave packet that forms an envelope of normal modes, i.e. monochromatic waves, with infinitely many wave numbers k. The time evolution of perturbation wave packets depends in a non-trivial way on the growth rate of the normal modes included in the envelope. Testing the response of a convection heat transfer process to wave packet perturbations is the aim of an absolute instability analysis.

The interesting aspect of the distinction between convective and absolute instability is that there can be parametric ranges where the flow is convectively unstable, but not absolutely unstable. In these ranges, normal mode perturbations can destabilise the flow while a wave packet perturbation can not. The origin of the concept of absolute instability is usually dated back to studies in the field of plasma physics, as reported by Dysthe [1].

The fundamental idea is that the onset of instability to normal modes may be concealed to an observer in the laboratory reference frame. This happens when the basic flow is so intense as to convect downstream the growing normal modes of perturbation before they turn out to be amplified at any specific position.

The aim of the study presented here is to illustrate the criterion to establish quantitatively the emergence of absolute instability in a flow system. As a simplified arena amenable for the presentation of these technical aspects, we have chosen Burgers flow [2]. The illustration of this simple case serves to build-up the method, based on the Fourier transform, to work out the analysis of the transition from convective to absolute instability of a fluid flowing in a porous medium [3]. In particular, a numerical solution is presented for the transition from convective to absolute instability in the uniform flow driven by a horizontal pressure gradient along a horizontal porous layer bounded by impermeable planes with isoflux conditions. The adopted numerical technique is based on the shooting method coupled with a Runge-Kutta solver. This procedure allows one to manage the differential formulation of the stability dispersion relation.

2. Flow stability with Burgers equation

We examine the one-dimensional Burgers equation subject to a linear force,

$$\frac{\partial W}{\partial t} + W\,\frac{\partial W}{\partial x} = \frac{\partial^2 W}{\partial x^2} + R\,(W - W_0)\,, \tag{1}$$

where $R \in \mathbb{R}$ and $W_0 \in \mathbb{R}$. The Burgers equation is just a toy model devised to study the one-dimensional flow of a fluid. In a paper by J. M. Burgers of 1939, entitled "Mathematical examples illustrating relations occurring in the theory of turbulent fluid motion", an equation similar to (1) was studied in a simplified model of developing turbulence [2].

Obviously, $W = W_0$ is a stationary solution of equation (1). In this simple model, one-dimensional flow occurs along the x-axis. To test the linear stability of this basic solution we carry out an analysis of small perturbations.

2.1. Linear stability analysis

We superpose a small perturbation to the basic solution, $W = W_0$, namely

$$W = W_0 + \varepsilon\, w, \qquad \varepsilon > 0, \tag{2}$$

where ε is a perturbation parameter such that $\varepsilon \ll 1$. On substituting equation (2) into equation (1), we obtain

$$\varepsilon\,\frac{\partial w}{\partial t} + \varepsilon\, W_0\,\frac{\partial w}{\partial x} + \varepsilon^2\, w\,\frac{\partial w}{\partial x} = \varepsilon\,\frac{\partial^2 w}{\partial x^2} + \varepsilon\, R\, w. \tag{3}$$

Then, by neglecting terms $O(\varepsilon^2)$, equation (3) reads

$$\frac{\partial w}{\partial t} + W_0 \frac{\partial w}{\partial x} = \frac{\partial^2 w}{\partial x^2} + R\, w. \tag{4}$$

Equation (4) can be solved by employing the Fourier transform. Thus, we define

$$\tilde{w}(k,t) = \frac{1}{\sqrt{2\pi}} \int_{-\infty}^{\infty} w(x,t)\, \mathrm{e}^{-\mathrm{i}\, k x}\, \mathrm{d}\, x, \qquad w(x,t) = \frac{1}{\sqrt{2\pi}} \int_{-\infty}^{\infty} \tilde{w}(k,t)\, \mathrm{e}^{\mathrm{i}\, k x}\, \mathrm{d}\, k, \tag{5}$$

were k is the wave number. We apply the Fourier transform to equation (4) so that we obtain

$$\frac{\partial \tilde{w}}{\partial t} = \lambda(k)\, \tilde{w}, \tag{6}$$

with $\lambda(k)$ defined as

$$\lambda(k) = R - k^2 - \mathrm{i}\, k\, W_0. \tag{7}$$

Equation (7) is usually called the *dispersion relation.* The solution of equation (6) is

$$\tilde{w}(k,t) = \tilde{w}(k,0)\, \mathrm{e}^{\lambda(k)\, t}. \tag{8}$$

By substituting equation (8) into equation (5), we express the perturbation as

$$w(x,t) = \frac{1}{\sqrt{2\pi}} \int_{-\infty}^{\infty} \tilde{w}(k,0)\, \mathrm{e}^{\mathrm{i}\, k x}\, \mathrm{e}^{\lambda(k)\, t}\, \mathrm{d}\, k. \tag{9}$$

The solution $w(x,t)$ given by equation (9) depends on the initial condition, $w(x,0)$, through its Fourier transform $\tilde{w}(k,0)$. We note that $w(x,t)$ represents a wave packet where

$$\omega(k) = -\Im[\lambda(k)] = k\, W_0 \tag{10}$$

is the angular frequency, and

$$b(k,t) = \frac{1}{\sqrt{2\pi}}\, \tilde{w}(k,0)\, \mathrm{e}^{\Re[\lambda(k)]\, t} \tag{11}$$

is the time-dependent amplitude of the normal mode. The single normal mode, with a given wave number k_s, is the evolution of an initial plane wave perturbation, namely

$$w(x,0) = \frac{1}{\sqrt{2\pi}}\, \mathrm{e}^{\mathrm{i}\, k_s x}. \tag{12}$$

Convective Instability

A single normal mode perturbation with a given wave number k is called convectively stable if $\Re[\lambda(k)] < 0$. It is called convectively unstable if $\Re[\lambda(k)] > 0$. The marginal condition where $\Re[\lambda(k)] = 0$ is called *neutral stability* condition.

On account of equation (7), convective instability happens when

$$R > k^2, \tag{13}$$

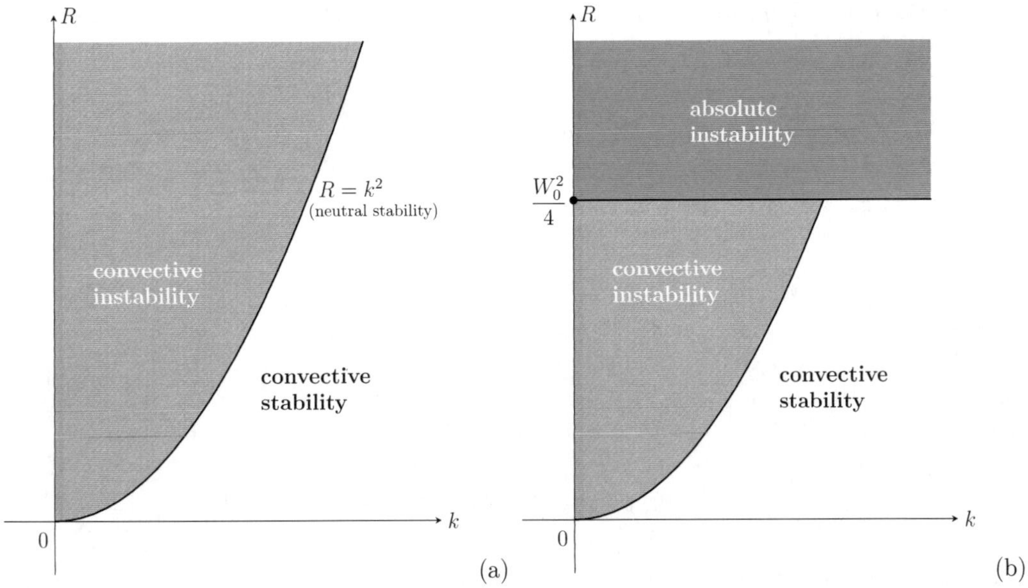

Figure 1. Drawings of the regions of convective instability (a), and absolute instability (b)

with the curve given by $R = k^2$ defining neutral stability.

A resume of the concept of convective instability and the marginal condition of neutral stability is displayed in Fig. 1(a). In this figure, only the domain of positive wave numbers is represented, as the condition of convective instability just involves k^2 and it is thus independent of the sign of k. Convective instability, to some wave numbers k, is possible only when R is larger than its *critical value*, denoted as R_c, which corresponds to the absolute minimum of R along the neutral stability curve. The corresponding value of k is the critical wave number, k_c. Thus, we have

$$k_c = 0 \,, \qquad R_c = 0 \,. \tag{14}$$

Convective instability involves quite special initial perturbations of the basic solution, having the form of plane waves with a given wave number. Such perturbations are intrinsically non-local in character. More general perturbations are given by a superposition of infinite plane waves with all possible wave numbers, as represented by the Fourier integral, equation (9). These wave packets may describe perturbations with a local support as it could be, for instance, when the initial condition $w(x, 0)$ is a Gaussian signal.

Absolute Instability

A perturbation $w(x, t)$ is defined as absolutely unstable if it is absolutely integrable over the real x-axis and if

$$\lim_{t \to +\infty} |w(x,t)| = +\infty, \tag{15}$$

for every $x \in \mathbb{R}$.

One can decide whether a perturbation $w(x, t)$ expressed through equation (9) is absolutely unstable or not by checking the large-time behaviour of the Fourier integral on the right hand side of equation (9). This task can be accomplished by employing the steepest-descent approximation.

The first step is determining the saddle points of $\lambda(k)$. In fact, equation (7) yields

$$\frac{\mathrm{d}\lambda}{\mathrm{d}k} = -2\,k - \mathrm{i}\,W_0. \tag{16}$$

Equation (16) shows that there is just one, purely imaginary, saddle point,

$$\frac{\mathrm{d}\lambda}{\mathrm{d}k} = 0 \quad \Longrightarrow \quad k = k_0 = -\frac{\mathrm{i}\,W_0}{2}\,. \tag{17}$$

One must check that the $\tilde{w}(k,t)$, expressed by equation (8), is holomorphic in the region of the complex plane $\mathbb{C}$ between the real axis and the saddle point k_0. We know that $\lambda(k)$ is holomorphic for every $k \in \mathbb{C}$. On the other hand, $\tilde{w}(k,0)$ is arbitrary, so that we must assume its holomorphy. If this hypothesis regarding the initial state $w(x,0)$ holds, we can approximately evaluate $|w(x,t)|$ for large times as [4]

$$|w(x,t)| \approx \frac{|\tilde{w}(k_0,0)|}{\sqrt{2\,t}}\,\mathrm{e}^{\Re[\lambda(k_0)]\,t}. \tag{18}$$

On account of equations (15) and (18), one can conclude that absolute instability is attained when $\Re[\lambda(k_0)] > 0$. On account of equations (7) and (17), this means

$$R > R_{\mathrm{a}} = \frac{W_0^2}{4}\,, \tag{19}$$

where R_{a} denotes the threshold for the onset of absolute instability. The condition of absolute instability is independent on the details of the initial perturbation, $w(x,0)$, inasmuch as it is absolutely integrable over the real x-axis and its Fourier transform, $\tilde{w}(k,0)$, is holomorphic.

A qualitative sketch of the concepts of convective instability and absolute instability is displayed in Fig. 1(b). This figure highlights that absolute instability is not a modal condition, meaning that its validity does not depend on the behaviour of individual normal modes, but on the asymptotic behaviour of a general class of perturbations. Interestingly enough, the condition of absolute instability turns out to be a parametric condition, given by equation (19), mostly independent of the detailed characteristics of the initial perturbations superposed to the basic stationary solution through equation (2).

3. Mathematical model of porous flow

Let us consider a horizontal porous channel having a rectangular cross-section with height H. We adopt a two-dimensional description where the coordinates are chosen so that x is the longitudinal horizontal axis and the y axis is the transverse vertical axis (see Fig. 2). We are assuming a heating from below with a uniform heat flux at $y = 0$, while the upper boundary, $y = H$, is kept isothermal with temperature T_0.

The velocity field, $\mathbf{u} = (u,v)$, and the temperature field, T, as well as the coordinates, (x,y), and time, t, can be written in a dimensionless form by adopting the transformation

$$(u,v)\,\frac{H}{\varkappa} = (u^*,v^*), \qquad (T-T_0)\,\frac{\chi}{q_0 H} = T^*, \qquad (x,y)\,\frac{1}{H} = (x^*,y^*), \qquad t\,\frac{\varkappa}{H^2} = t^*, \tag{20}$$

where $\varkappa$ is the average thermal diffusivity and χ is the average thermal conductivity of the porous medium, while q_0 is the constant wall heat flux and T_0 is a constant reference temperature. In equation (20), the asterisks denote the dimensionless quantities. Hereafter, such asterisks will be omitted in order to avoid an uselessly complicated notation. This will not induce any ambiguity as all the forthcoming equations are dimensionless.

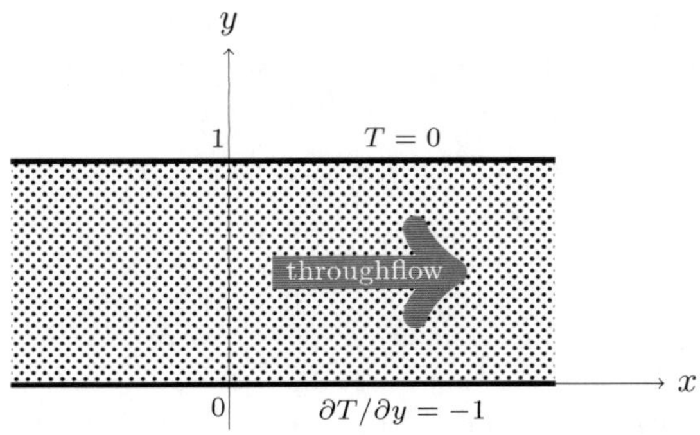

Figure 2. A sketch of the porous channel

According to the Oberbeck-Boussinesq approximation and to Darcy's law, the governing equations expressing the local balances of mass, momentum and energy can be written as

$$\frac{\partial u}{\partial x} + \frac{\partial v}{\partial y} = 0, \tag{21a}$$

$$\frac{\partial u}{\partial y} - \frac{\partial v}{\partial x} + R\,\frac{\partial T}{\partial x} = 0, \tag{21b}$$

$$\xi\,\frac{\partial T}{\partial t} + u\,\frac{\partial T}{\partial x} + v\,\frac{\partial T}{\partial y} = \frac{\partial^2 T}{\partial x^2} + \frac{\partial^2 T}{\partial y^2}, \tag{21c}$$

where the local momentum balance equation (21b) has been expressed in its vorticity formulation. In fact, the dimensionless Darcy's law comprehensive of the buoyancy force term can be written as

$$u = -\frac{\partial \Pi}{\partial x}, \qquad v = RT - \frac{\partial \Pi}{\partial y}, \tag{22}$$

where Π is a suitably defined dimensionless pressure. By evaluating $\partial u/\partial y$ and $\partial v/\partial x$ from equation (22), one obtains equation (21b).

The parameter ξ in equation (21c) is the ratio between the average volumetric heat capacity of the saturated porous medium and the volumetric heat capacity of the fluid [5]. The Darcy-Rayleigh number R is defined as

$$R = \frac{\rho g \gamma q_0 K\, H^2}{\mu \varkappa \chi}, \tag{23}$$

where γ is the thermal expansion coefficient of the fluid, g is the modulus of the gravitational acceleration, K is the permeability, μ is the dynamic viscosity of the fluid, and ρ is the fluid density at the reference temperature T_0. The boundary conditions are given by

$$y = 0 : \qquad v = 0, \quad \frac{\partial T}{\partial y} = -1,$$
$$y = 1 : \qquad v = 0, \quad T = 0. \tag{24}$$

4. Linear perturbation equations

A stationary solution of equations (21) satisfying the boundary conditions (24), which describes a parallel throughflow in the x−direction, is given by

$$(u_b, v_b) = (P, 0), \qquad T_b = 1 - y, \tag{25}$$

where subscript "b" stands for "basic solution", and the dimensionless parameter P plays the role of the Péclet number associated with the basic horizontal throughflow. We can express small-amplitude perturbations to the basic solution as

$$(u,v) = (u_{\rm b}, v_{\rm b}) + \varepsilon\,(U,V), \qquad T = T_{\rm b} + \varepsilon\,\Theta, \tag{26}$$

where (U,V) and Θ are the velocity and temperature disturbances, respectively, while ε is a perturbation amplitude such that $|\varepsilon| \ll 1$.

Taking into account that the basic solution (25) satisfies equations (21) and (24), the substitution of equation (26) into equations (21) and (24) upon neglecting terms $O(\varepsilon^2)$ yields the linearised governing equations for the perturbations. These equations can be expressed in a streamfunction formulation where

$$U = \frac{\partial \psi}{\partial y}, \qquad V = -\frac{\partial \psi}{\partial x}. \tag{27}$$

Thus, we can write

$$\frac{\partial^2 \psi}{\partial x^2} + \frac{\partial^2 \psi}{\partial y^2} + R\,\frac{\partial \Theta}{\partial x} = 0, \tag{28a}$$

$$\xi\,\frac{\partial \Theta}{\partial t} + P\,\frac{\partial \Theta}{\partial x} + \frac{\partial \psi}{\partial x} = \frac{\partial^2 \Theta}{\partial x^2} + \frac{\partial^2 \Theta}{\partial y^2}, \tag{28b}$$

$$y = 0\ : \qquad \frac{\partial \psi}{\partial x} = 0, \quad \frac{\partial \Theta}{\partial y} = 0, \tag{28c}$$

$$y = 1\ : \qquad \frac{\partial \psi}{\partial x} = 0, \quad \Theta = 0. \tag{28d}$$

Following equation (5), we apply the Fourier transform to equations (28), and express the Fourier transforms of ψ and Θ as

$$\tilde{\psi} = {\rm e}^{\lambda(k)\,t}\, f(y), \qquad \tilde{\Theta} = -\,{\rm i}\,k\,{\rm e}^{\lambda(k)\,t}\, h(y). \tag{29}$$

Therefore, the transformed equations (28) read

$$f'' - k^2\, f + R\,k^2\, h = 0, \tag{30a}$$

$$h'' - \left(k^2 + \xi\,\lambda + {\rm i}\,k\,P\right) h + f = 0, \tag{30b}$$

$$y = 0\ : \qquad f = 0, \quad h' = 0, \tag{30c}$$

$$y = 1\ : \qquad f = 0, \quad h = 0, \tag{30d}$$

where primes denote derivatives with respect to y. Equations (30) are linear and homogeneous. They define an eigenvalue problem whose solution provides us with the dispersion relation, i.e. an equation analogous to (7). However, things are significantly more complicated than in equation (7), as we are unable to find an explicit analytical expression for the complex valued function $\lambda(k)$. Then, a numerical solution of equations (30) is a sensible choice in this case. This approach can be implemented by employing the LSODE subroutine for the numerical solution of ordinary differential equations [6], and the shooting method (see, for instance, chapter 19 of Straughan [7]).

The first task to be accomplished is defining the neutral stability condition, i.e. the parametric threshold for convective instability. As illustrated for the sample case of Burgers equation, this means setting $\Re(\lambda(k)) = 0$ for $k \in \mathbb{R}$. It is readily proved that $\xi\,\Im(\lambda(k)) + k\,P = 0$ so that the principle of exchange of stabilities [8, 9] holds in a suitably defined moving reference frame.

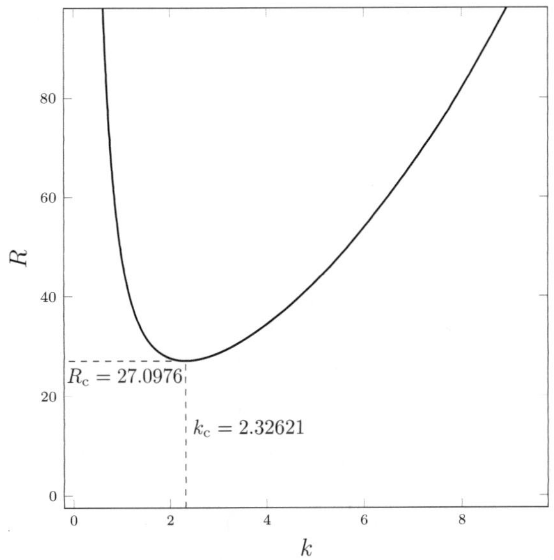

Figure 3. Neutral stability curve relative to the eigenvalue problem (31)

Thus, the evaluation of the neutral stability data involves a simplified version of equations (30), namely

$$f'' - k^2 f + R k^2 h = 0, \tag{31a}$$

$$h'' - k^2 h + f = 0, \tag{31b}$$

$$y = 0 : \qquad f = 0, \quad h' = 0, \tag{31c}$$

$$y = 1 : \qquad f = 0, \quad h = 0, \tag{31d}$$

where $k \in \mathbb{R}$. The solution of equation (31) is employed to determine the numerical function $R(k)$, i.e. the neutral stability function. In other words, R is computed as the eigenvalue of equations (31), for every prescribed value of $k \in \mathbb{R}$. The result of this computation is provided in Fig. 3. This figure shows that the point of minimum R along the neutral stability curve, i.e. the critical point for convective instability, is given by

$$k_c = 2.32621, \qquad R_c = 27.0976. \tag{32}$$

4.1. From convective to absolute instability

Absolute instability in the horizontal flow through a porous medium occurs when localised wave packets of perturbation grow unboundedly at large times t, for a given x. As illustrated in Section 2, the study of absolute instability relies on the steepest-descent approximation of the perturbation wave packet. Hence, the first step is the determination of the saddle-point in the complex plane, $k = k_0 \in \mathbb{C}$, such that $\mathrm{d}\lambda/\mathrm{d}k = 0$. The threshold of absolute instability occurs when the prescribed value of R matches the condition of zero asymptotic growth, $\Re(\lambda(k_0)) = 0$. This special threshold value of R is denoted as R_a.

The basis for the evaluation of R_a is still the eigenvalue problem (30). However, its numerical solution must be approached with the specification that $k = k_r + i\, k_i$ is a complex variable with real part k_r and imaginary part k_i. The fulfilment of the saddle-point condition can be automatically implemented by forcing the constraint $\mathrm{d}\lambda/\mathrm{d}k = 0$. One can impose this constraint

IOP Conf. Series: Journal of Physics: Conf. Series **1224** (2019) 012043 doi:10.1088/1742-6596/1224/1/012043

by doubling the order of the differential problem (30). To this end, we define

$$\hat{f} = \frac{\partial f}{\partial k}, \qquad \hat{h} = \frac{\partial h}{\partial k}. \tag{33}$$

Then, we obtain the extended eigenvalue problem

$$f'' - k^2 f + R k^2 h = 0, \tag{34a}$$
$$h'' - \left(k^2 + \xi \lambda + \mathrm{i}\, k\, P\right) h + f = 0, \tag{34b}$$
$$\hat{f}'' - k^2 \hat{f} + R k^2 \hat{h} - 2 k f + 2 R k h = 0, \tag{34c}$$
$$\hat{h}'' - \left(k^2 + \xi \lambda + \mathrm{i}\, k\, P\right) \hat{h} + \hat{f} - (2k + \mathrm{i}\, P) h = 0, \tag{34d}$$
$$y = 0 : \qquad f = 0, \quad h' = 0, \quad \hat{f} = 0, \quad \hat{h}' = 0, \tag{34e}$$
$$y = 1 : \qquad f = 0, \quad h = 0, \quad \hat{f} = 0, \quad \hat{h} = 0, \tag{34f}$$

If we denote by λ_r and λ_i the real part and the imaginary part of λ, then the solution of equations (34) must be sought by keeping $\lambda_\mathrm{r} = 0$. This means that we have four real values to get as an output of the solution process, namely

$$k_\mathrm{r}, \quad k_\mathrm{i}, \quad \xi\, \lambda_\mathrm{i}, \quad R. \tag{35}$$

In fact, the number of boundary conditions (34e) and (34f) matches exactly the order of the ordinary differential equations (34a)–(34d). However, the homogeneous nature of the differential problem (34) implies that we are allowed to introduce scale fixing extra conditions,

$$f'(0) = 1, \qquad \hat{f}'(0) = 0. \tag{36}$$

The first equation (36) introduces a normalisation, or scale gauge, for the eigenfunctions. The second equation (36) is just a consequence of the first one, obtained by employing equation (33). Since the eigenfunctions $(f, h, \hat{f}, \hat{h})$ are complex valued, equation (36) is effectively a system of four real constraints that increments the number of boundary conditions given by equations (34e) and (34f). Thus, the extra conditions are necessary and sufficient to compute the eigenvalue tuple listed in equation (35). The numerical method to achieve this task is just the same as that described for the eigenvalue problem (31). In this case, it is more demanding in terms of coding and computational time, but nothing changes on the algorithmic side. One obtains a threshold value for absolute instability, R_a, for every prescribed Péclet number, P.

4.2. Discussion of the results

In order to study the onset of absolute instability, the eigenvalue search process must be initialised in a convenient way. There is, in fact, a case where the value of R_a for a given P can be easily guessed. It is the case $P = 0$ where one expects $R_\mathrm{a} = R_\mathrm{c} = 27.0976$. Consistently, in this case, one expects also

$$k_\mathrm{r} = k_\mathrm{c} = 2.32621, \quad k_\mathrm{i} = 0, \quad \xi\, \lambda_\mathrm{i} = 0. \tag{37}$$

These expectations are grounded on the idea that a rest state ($P = 0$) is one where there is no parametric gap between convective and absolute instability, namely $R_\mathrm{a} = R_\mathrm{c}$. In fact, when there is no basic flow driving the perturbation downstream, the Fourier normal modes are non-travelling, so that they amplify or damp in place. This is a consequence of the principle of exchange of stabilities. In such situations, even a single Fourier mode which undergoes an

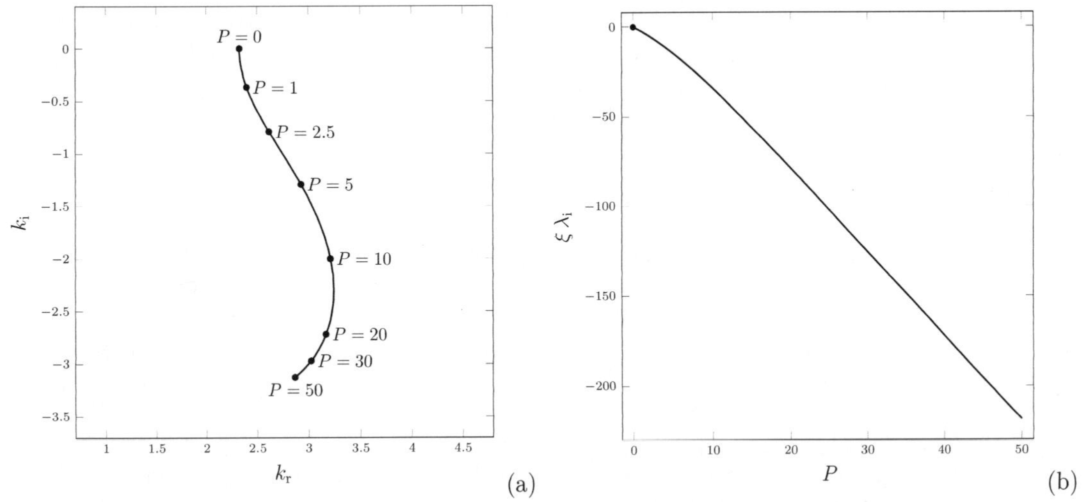

Figure 4. Migration of the saddle point in the complex $k-$plane for different values of P (a); plot of $\xi\,\lambda_i$ versus P (b)

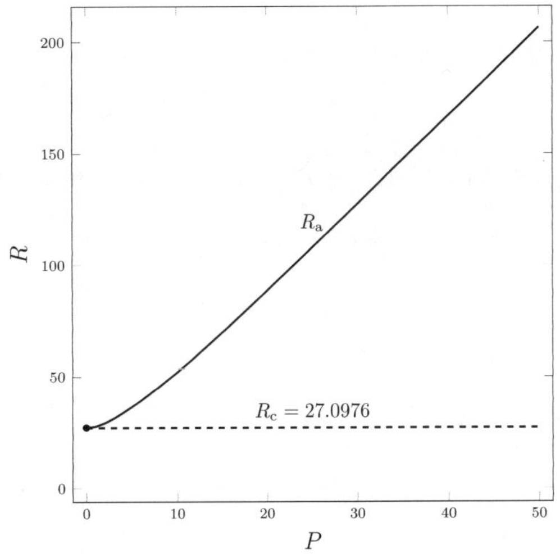

Figure 5. Plots of R_c and R_a versus P

exponential amplification is sufficient to cause an amplification of the whole perturbation wave packet, for large times.

One can track the continuous displacement of the saddle point starting from the real axis $k_0 = 2.32621$, when $P = 0$, to the complex $k-$plane when $P > 0$. Figure 4(a) displays the migration of the saddle point in the $k-$plane as P increases from 0 to 50. This figure reveals that the imaginary part of k continuously decreases from 0, while the real part undergoes a non-monotonic trend. Figure 4(b) shows that the imaginary part of the growth rate λ continuously decreases from 0 as P increases.

Figure 5 illustrates the change of the threshold values, R_c and R_a, for the onset of convective and absolute instability, respectively. The dashed horizontal line, relative to the critical value R_c, reflects the independence of P for the threshold to convective instability. On the other hand the threshold to absolute instability depends significantly on P, so that the parametric gap between convective and absolute instability gradually expands as P becomes larger and larger. We mention that qualitatively similar behaviour is observed for the case where the porous layer is bounded by isothermal boundaries [10, 11].

5. Conclusions

The transition from convective to absolute instability in a flow system has been illustrated relative to a simple example involving the one-dimensional Burgers equation. Then, this concept has been employed for the stability analysis of a porous layer with horizontal through flow. A parametric gap has been found between the threshold of convective instability and that of absolute instability. The parametric gap becomes larger and larger as the Péclet number, P, of the basic flow increases. When P tends to zero, the onset of instability has an absolute nature.

References

[1] Dysthe K B 1966 *Nuclear Fusion* **6** 215–222
[2] Nieuwstadt F T and Steketee J A 2012 *Selected papers of J. M. Burgers* (New York: Springer)
[3] Barletta A and Celli M 2017 *Fluids* **2** 1–33
[4] Ablowitz M J and Fokas A S 2003 *Complex Variables: Introduction and Applications* (Cambridge, UK: Cambridge University Press)
[5] Nield D A and Bejan A 2017 *Convection in Porous Media* 5th ed (New York: Springer)
[6] Radhakrishnan K and Hindmarsh A C 1993 Description and use of LSODE, the Livemore Solver for Ordinary Differential Equations. Tech. rep. Lawrence Livermore National Laboratory (LLNL), Livermore, CA. URL `https://e-reports-ext.llnl.gov/pdf/240148.pdf`
[7] Straughan B 2013 *The Energy Method, Stability, and Nonlinear Convection* (New York: Springer)
[8] Jeffreys H 1926 *The London, Edinburgh, and Dublin Philosophical Magazine and Journal of Science* **2** 833–844
[9] Pellew A and Southwell R V 1940 *Proceedings of the Royal Society of London A* **176** 312–343
[10] Delache A, Ouarzazi M N and Combarnous M 2007 *International Journal of Heat and Mass Transfer* **50** 1485–1499
[11] Barletta A and Alves L S de B 2017 *International Journal of Heat and Mass Transfer* **104** 438–455

IOP Conf. Series: Journal of Physics: Conf. Series **1224** (2019) 012044 doi:10.1088/1742-6596/1224/1/012044

Onset of convective instability within an inclined porous layer with a permeable boundary

Michele Celli, Antonio Barletta

Department of Industrial Engineering - Alma Mater Studiorum - University of Bologna, Viale Risorgimento 2, 40136, Bologna, Italy

E-mail: michele.celli3@unibo.it; antonio.barletta@unibo.it

Abstract. The investigation of the thresholds that identify the onset of convective instability in a fluid saturated porous layer is performed. The layer is inclined with respect to the horizontal and the boundaries are held at different uniform temperatures. The layer is also semi-permeable: one boundary is impermeable and the other one is permeable. This configuration yields a basic state characterised by a single convective cell, with zero mass flow rate, that fades for small inclination becoming completely motionless for the horizontal case. Since the central part of this cell is considered, the basic flow is parallel and the basic temperature profile is dominated by conduction. This basic state is perturbed employing small-amplitude disturbances such that the linear stability analysis can be performed. A Squire-type transformation is applied to reduce the complexity of the problem. A numerical procedure is employed to obtain the critical values of the governing parameters.

1. Introduction

The buoyancy driven instability within fluid saturated porous layers is a topic that has been driving the attention of different scientific areas for more that fifty years, from applied mathematics to geophysics or oil, civil and industrial engineering. Natural convection is, in fact, a phenomenon of a key importance when industrial processes are taken into account. The presence of convective cells may have, for instance, undesirable effects on several production processes like the extrusion of plastics or metal casting.

The classical buoyancy driven convection problem for clear fluids is the one that Rayleigh and Bénard studied at the beginning of twentieth century [1, 2]. The analogous of the Rayleigh–Bénard instability problem for fluid saturated porous layers is the so called Horton–Rogers–Lapwood (HRL) problem [3, 4]. The HRL problem involves a horizontal and impermeable layer heated from below where the local momentum balance is Darcy's law. The results obtained for the classical HRL problem together with those obtained by considering different boundary conditions are resumed in Table 6.1 of the book by Nield and Bejan [5].

The equivalent problem, compare to the HRL one, for a vertical fluid saturated porous layer was investigated by Gill [6]. He found that this configuration, a vertical fluid saturated porous layer with impermeable and isothermal boundaries, is always linearly stable. Barletta [7] found that, when permeable boundaries are considered, the vertical layer may undergo a thermal instability. The role of the porous layer inclination is thus relevant when the onset of thermal instability is studied. This role is extensively analysed by Rees and Bassom [8] for a layer

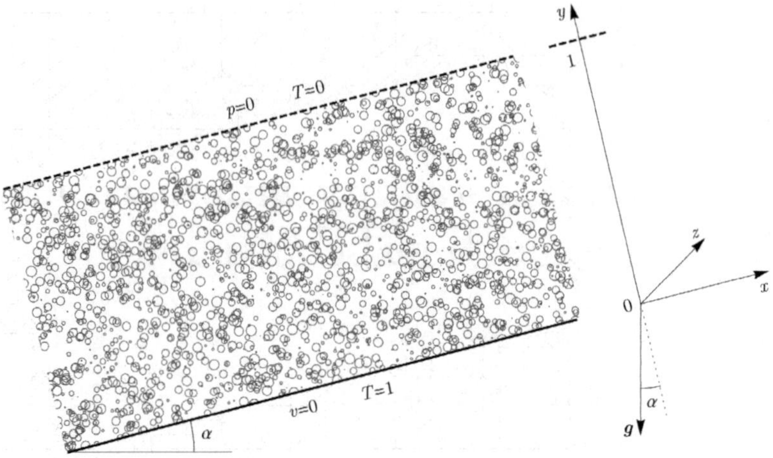

Figure 1. The inclined porous layer with its dimensionless boundary conditions.

characterised by impermeable and isothermal boundaries. The onset of thermal instability for inclined porous layers has also been studied in [9, 10, 11].

A semi-permeable porous layer tilted with respect to the horizontal is considered: one boundary is permeable, where an isobaric boundary condition is applied, the other boundary is impermeable. The two boundaries are held at two different uniform temperatures. The aim of this study is understanding the behaviour of the threshold for the onset of instability when the fluid saturated porous layer is inclined.

2. Mathematical model

The fluid saturated porous layer considered is characterised by a thickness H and infinite width in both the directions x and z. Two different temperatures are imposed at the boundaries: $T_0+\Delta T$ for $y=0$ and T_0 for $y=H$. The hydrodynamic boundary conditions are asymmetric: $y=0$ is an impermeable plane while $y=H$ is permeable, held at constant pressure p_0, as displayed in Fig. 1 in dimensionless terms. The porous layer is inclined an angle α with respect to the horizontal. Darcy's law, together with the Oberbeck-Boussinesq approximation, is employed to model the momentum transport. The fluid is incompressible and no source/sink terms are considered in the local energy balance equation. The governing equations can be written in the following dimensionless form

$$\begin{gathered}
\boldsymbol{\nabla}\cdot\mathbf{u}=0,\\
\mathbf{u}=-\boldsymbol{\nabla}p+R\,T\,(\sin\alpha\;\mathbf{e}_x+\cos\alpha\;\mathbf{e}_y),\\
\frac{\partial T}{\partial t}+\mathbf{u}\cdot\boldsymbol{\nabla}T=\nabla^2T,\\
y=0:\qquad v=0,\quad T=1,\\
y=1:\qquad p=0,\quad T=0,
\end{gathered}\tag{1}$$

where $\mathbf{x}=(x,y,z)$ is the Cartesian coordinates vector, $\mathbf{u}=(u,v,w)$ is the velocity vector, p is the total head, T is the temperature, $\mathbf{e}_x$ and $\mathbf{e}_y$ are the unit vectors for the x and y-direction, and t is the time. The following scaling is employed to obtain the dimensionless form of the governing equations

$$\frac{\mathbf{x}}{H}\to\mathbf{x},\quad\frac{\chi}{\sigma H^2}t\to t,\quad\frac{H}{\chi}\mathbf{u}\to\mathbf{u},\quad\frac{K}{\chi\,\mu}(p-p_0)\to p,\quad\frac{T-T_0}{\Delta T}\to T,\tag{2}$$

where χ is the average (between fluid phase and solid phase) thermal diffusivity, σ is the heat capacity ratio, K is the permeability of the porous layer, and μ is the dynamic viscosity of the fluid. The Darcy–Rayleigh dimensionless number R in Eq. (1) is defined as

$$R = \frac{\rho\, g\, \beta\, K\, H\, \Delta T}{\chi\, \mu}, \tag{3}$$

where ρ is the fluid density, g is the modulus of gravitational acceleration, and β is the thermal expansion coefficient of the fluid.

3. Basic state

The stationary basic solution of Eq. (1) employed here for the stability analysis is one characterised by a parallel flow with zero neat mass flow rate

$$\mathbf{u}_{\rm B} = \left(\frac{R}{2}\,(1-2y)\,\sin\alpha, 0, 0\right), \qquad \boldsymbol{\nabla} p_{\rm B} = \left(\frac{R}{2}\,\sin\alpha, R\,(1-y)\,\cos\alpha, 0\right), \qquad T_{\rm B} = 1 - y, \tag{4}$$

where the subscript "B" denotes the basic state fields. The velocity profile in Eq. (4) describes a single cell of infinite length while the temperature profile describes a pure conduction regime. This single cell arises when the layer has a non negligible inclination with respect to the horizontal, $\alpha > 0$. The horizontal case is, in fact, motionless.

The flow configuration presented in Eq. (4) can be produced by considering a porous layer of aspect ratio, width to height, sufficiently high to obtain a shallow layer. The parallel flow in Eq. (4) describes that portion of the porous layer sufficiently far from the x–direction boundaries.

4. Linear stability analysis

A pressure–temperature formulation is employed by applying the divergence to the local momentum balance equation, namely

$$\begin{gathered}
\boldsymbol{\nabla}^2 p - R\left(\frac{\partial T}{\partial x}\sin\alpha + \frac{\partial T}{\partial y}\cos\alpha\right) = 0,\\
\frac{\partial T}{\partial t} + \left[-\boldsymbol{\nabla} p + R\,T\,(\sin\alpha\;\mathbf{e}_x + \cos\alpha\;\mathbf{e}_y)\right]\cdot\boldsymbol{\nabla}\,T = \nabla^2 T,\\
y = 0: \qquad \frac{\partial p}{\partial y} = R\,\cos\alpha, \quad T = 1,\\
y = 1: \qquad p = 0, \quad T = 0,
\end{gathered} \tag{5}$$

where the y component of Darcy's law is employed to redefine the impermeability condition in $y = 0$ in terms of pressure and temperature.

The analysis of which configuration is stable/unstable with respect to the onset of thermal convection can be tackled by perturbing the basic state in Eq. (5). The pressure and the temperature fields are redefined as composed by a basic state plus a perturbation, namely

$$p = p_{\rm B} + \varepsilon\, P, \quad T = T_{\rm B} + \varepsilon\,\Theta, \tag{6}$$

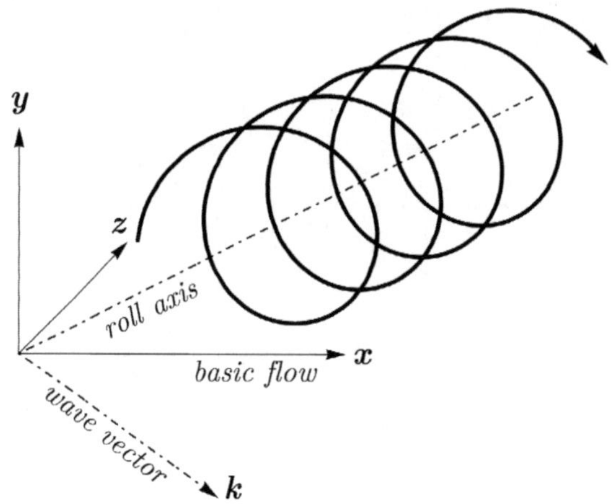

Figure 2. The inclination of the generic convective roll to the reference frame.

where the parameter ε is arbitrarily small such that, after substituing Eq. (6) to Eq. (5), one can consider only terms $O(\varepsilon)$ to obtain the linearised equations.

$$\begin{gathered}
\nabla^2 P - R\left(\frac{\partial \Theta}{\partial x}\sin\alpha + \frac{\partial \Theta}{\partial y}\cos\alpha\right) = 0, \\
\frac{\partial \Theta}{\partial t} + \frac{R}{2}(1-2\,y)\frac{\partial \Theta}{\partial x}\sin\alpha - R\,\Theta\,\cos\alpha + \frac{\partial P}{\partial y} = \nabla^2\Theta, \\
y=0: \qquad \frac{\partial P}{\partial y} = 0, \quad \Theta = 0, \\
y=1: \qquad P = 0, \quad \Theta = 0.
\end{gathered} \tag{7}$$

The flow pattern of thermal convection induced by buoyancy forces can be reasonably approximated to be roll shaped. One may thus assume the perturbations to have the form of plane waves, namely

$$\left\{P(x,y,t), \Theta(x,y,t)\right\} = \{f(y), h(y)\}\, e^{\eta\, t}\, e^{i(k_x\, x + k_z\, z - \omega\, t)}, \tag{8}$$

where η is the growth rate of the perturbation, k_x is the wavenumber in the x direction, k_z is the wavenumber in the z direction, and ω is the angular frequency. By substituting Eq. (8) into Eq. (7) we obtain two ordinary differential equations, namely

$$\begin{gathered}
f'' - k^2\, f - R\, h'\, \cos\alpha - i\, k_x\, R\, h \sin\alpha = 0, \\
h'' - \left[k^2 + \frac{i}{2}\, k_x\, R\,(1-2y)\sin\alpha - R\cos\alpha - i\,\omega\right] h - f' = 0, \\
y=0: \qquad f' = 0, \quad h = 0, \\
y=1: \qquad f = 0, \quad h = 0,
\end{gathered} \tag{9}$$

IOP Conf. Series: Journal of Physics: Conf. Series **1224** (2019) 012044 doi:10.1088/1742-6596/1224/1/012044

where the primes denote the derivatives with respect to y and $k^2 = k_x^2 + k_z^2$. Since we want to investigate the neutrally stable configuration, *i.e.* the normal modes with a null growth rate, we have set $\eta = 0$. The rolls wavevector $\mathbf{k}$ lies on the plane (x, z), as displayed in Fig. 2, and, in general, it is inclined with respect to the x direction.
One may now apply a Squire–type transformation to Eq. (9) to reduce the number of parameters, namely

$$S\,\cos\phi = R\,\cos\alpha, \qquad \tan\phi = \frac{k_x}{k}\tan\alpha. \tag{10}$$

The simplified eigenvalue problem is thus

$$\begin{gathered} f'' - k^2\,f - S\,h'\,\cos\phi - i\,k\,S\,h\,\sin\phi = 0, \\ h'' - \left[k^2 + \frac{i}{2}\,k\,S\,(1-2y)\sin\phi - S\cos\phi - i\,\omega\right]h - f' = 0, \\ y = 0: \qquad f' = 0, \quad h = 0, \\ y = 1: \qquad f = 0, \quad h = 0. \end{gathered} \tag{11}$$

The transformation in Eq. (10) yields an eigenvalue problem where the relative inclinations between basic flow direction and wavevector of the perturbations are compressed within the definition of the modified angle ϕ and the modified Darcy–Rayleigh number S.

The eigenvalue problem (11) cannot be solved analytically because of the explicit presence of the independent variable y thus it has to be solved numerically as reported in the next section.

5. Numerical method

The numerical procedure employed to solve the eigenvalue problem (11) is based on the Runge-Kutta method coupled with the shooting method. The Runge-Kutta method is employed to solve the following initial value problem

$$\begin{gathered} f'' - k^2\,f - S\,h'\,\cos\phi - i\,k\,S\,h\,\sin\phi = 0, \\ h'' - \left[k^2 + \frac{i}{2}\,k\,S\,(1-2y)\sin\phi - S\cos\phi - i\,\omega\right]h - f' = 0, \\ y = 0: \qquad f = 1, \quad f' = 0, \quad h = 0, \quad h' = \xi_1 + i\,\xi_2. \end{gathered} \tag{12}$$

In Eq. (12), two initial conditions are added to obtain a well-posed initial value problem. The eigenvalue problem (12) is homogeneous thus the initial condition $f(0) = 1$ is imposed to gauge the scale of the eigenfunctions f and h that are defined up to an arbitrary constant. The last initial condition $h' = \xi_1 + i\,\xi_2$ is fixed by introducing two unknown parameters ξ_1 and ξ_2.

The Runke-Kutta method is employed to obtain the complex valued eigenfunctions (f, h) which, in fact, depend on six real parameters $(k, \omega, S, \phi, \xi_1, \xi_2)$. The shooting method evaluates (f, h) for the four target conditions

$$y = 1: \qquad \Re\{f'\} = 0, \quad \Im\{f'\} = 0, \quad \Re\{h\} = 0, \quad \Im\{h\} = 0, \tag{13}$$

obtaining a system of four equations in the six parameters $(k, \omega, S, \phi, \xi_1, \xi_2)$. One may thus impose two of these six parameters, (k, ϕ) for instance, to obtain the remaining four unknown parameters $(S, \omega, \xi_1, \xi_2)$.

As a consequence, one can draw a neutral stability curve $S(k)$ for each given value of ϕ, as reported in Fig. 3. The minimum of each neutral stability curve defines the threshold condition

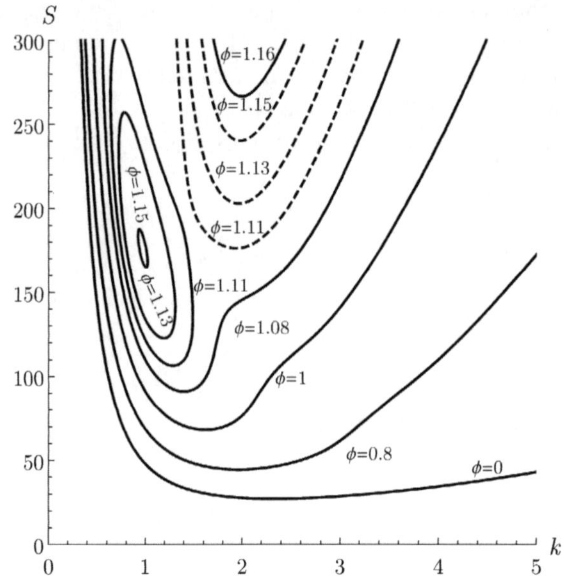

Figure 3. Neutral stability curves in the plane (k, S) for different values of the transformed angle ϕ: from $\phi = 0$, the lowest curve, to $\phi = 1.16$, the highest curve. The dashed lines are employed for those branches that are not relevant for the stability analysis since their minima are not the absolute minima for that given configuration.

for the onset of thermal instability: the critical values k_c and S_c. These minima can be obtained numerically by solving the following system of ordinary differential equations: Eqs. (12)–(13) and their derivatives with respect to the wavenumber k simplified by imposing the condition $\partial S/\partial k = 0$. This new problem is characterised by a number of equations and eigenfunctions that is doubled: four balance equations, eight initial conditions, eight target conditions, four eigenfunctions $(f, \partial f/\partial k, h, \partial h/\partial k)$. Also the number of parameters to be treated is higher compared to the original problem: nine parameters $(k, \omega, \partial\omega/\partial k, S, \phi, \xi_1, \xi_2, \partial\xi_1/\partial k, \partial\xi_2/\partial k)$. Since the target conditions are eight, one obtains the critical values of the governing parameters $(k, \omega, \partial\omega/\partial k, S)$ for each fixed angle ϕ.

6. Neutral stability curves and critical values

The neutral stability curves displayed in Fig. 3 are drawn for different values of the modified angle ϕ expressed in radians. The lowest curve is drawn for $\phi = 0$ while the highest curve, the small parabola shaped continuous line, is drawn for $\phi = 1.16$ rad. For angles higher than the threshold value $\phi = 1.10813$ rad, the curves split in two branches. The lower branch of solutions is characterised by closed curves, "islands", and by values of S that are lower compared to the other branch. These islands shrink in size for increasing values of ϕ until they are reduced to a point for $\phi = 1.15063$ rad and then disappear completely for $\phi > 1.15063$ rad. The first curves relative to higher branch of solutions are dashed because they are not relevant for the stability analysis since their minima are higher with respect to the minima of the lower branch of solutions obtained for the same value of ϕ. The higher branch of solutions becomes relevant when, for $\phi > 1.15063$ rad, the lower branch of solution does not exist anymore. This is the reason why the highest curve, drawn for $\phi = 1.16$ rad, is continuous and not dashed.

The angular frequency ω relative to both branches is different from zero, except for the horizontal case. In the horizontal case one may obtain from Eq. (10) that $\phi = 0$. One can conclude that the instability is triggered by travelling normal modes except for the horizontal case.

The minima of the neutral stability curves define the critical values of the governing parameters that identify the threshold for the onset of instability. By employing the procedure presented in Section 5, one obtains a pair (k_c, S_c) for each prescribed value of ϕ. A number of critical values are reported in Table 1 for different values of ϕ: the discontinuity for the value $\phi = 1.15063$ rad is due to fact that for this angle the lowest branch disappears. For $\phi > 1.15063$ rad the focus of the stability analysis jumps on the higher branch: for these values of ϕ this branch becomes the one with the lowest possible values of S_c. The critical values relative to the second branch keep increasing indefinitely with ϕ.

It is worth noting how the shapes of the perturbations change along with the angle ϕ. Figure 3 displays the lines at constant pressure $P(x, y, 0) = constant$, left column, and the lines at constant temperature $\Theta(x, y, 0) = constant$, right column, when the critical configurations obtained for different values of ϕ are considered. Each frame of Fig. 4 presents a whole perturbation period $2\pi/k$. The rolls shape is strongly dependent on the angle ϕ.

Table 1. Critical values of S and k for different angles ϕ expressed in radian.

ϕ	k_c	S_c
0	2.32621	27.0976
0.2	2.31021	27.7920
0.4	2.25794	30.0889
0.6	2.15446	34.8222
0.8	1.96605	44.4307
1.0	1.61979	68.1262
1.05	1.49040	80.4107
1.11	1.28671	106.347
1.15063	0.971417	175.350
1.15063	1.97285	241.505
1.16	1.98839	266.619
1.17	2.03109	302.845
1.18	2.11425	354.329
1.19	2.24545	432.855
1.2	2.47976	578.843
1.205	2.71997	735.778

7. Conclusions

A fluid saturated porous layer has been studied with respect to the onset of thermal convection. The layer is tilted to the horizontal direction and it is characterised by isothermal boundaries. Moreover, one boundary is permeable and the other one is impermeable. The linear stability analysis is performed by employing the normal modes method. A reference frame rotation plus a Squire-type transformation are applied to simplify the problem. The Squire-type transformation defines a modified Darcy-Rayleigh number S and a modified angle ϕ. The eigenvalue problem obtained is solved numerically. The most relevant results are reported here:

- The modified angle ϕ has a stabilising influence on the fluid saturated porous layer;

IOP Conf. Series: Journal of Physics: Conf. Series **1224** (2019) 012044 doi:10.1088/1742-6596/1224/1/012044

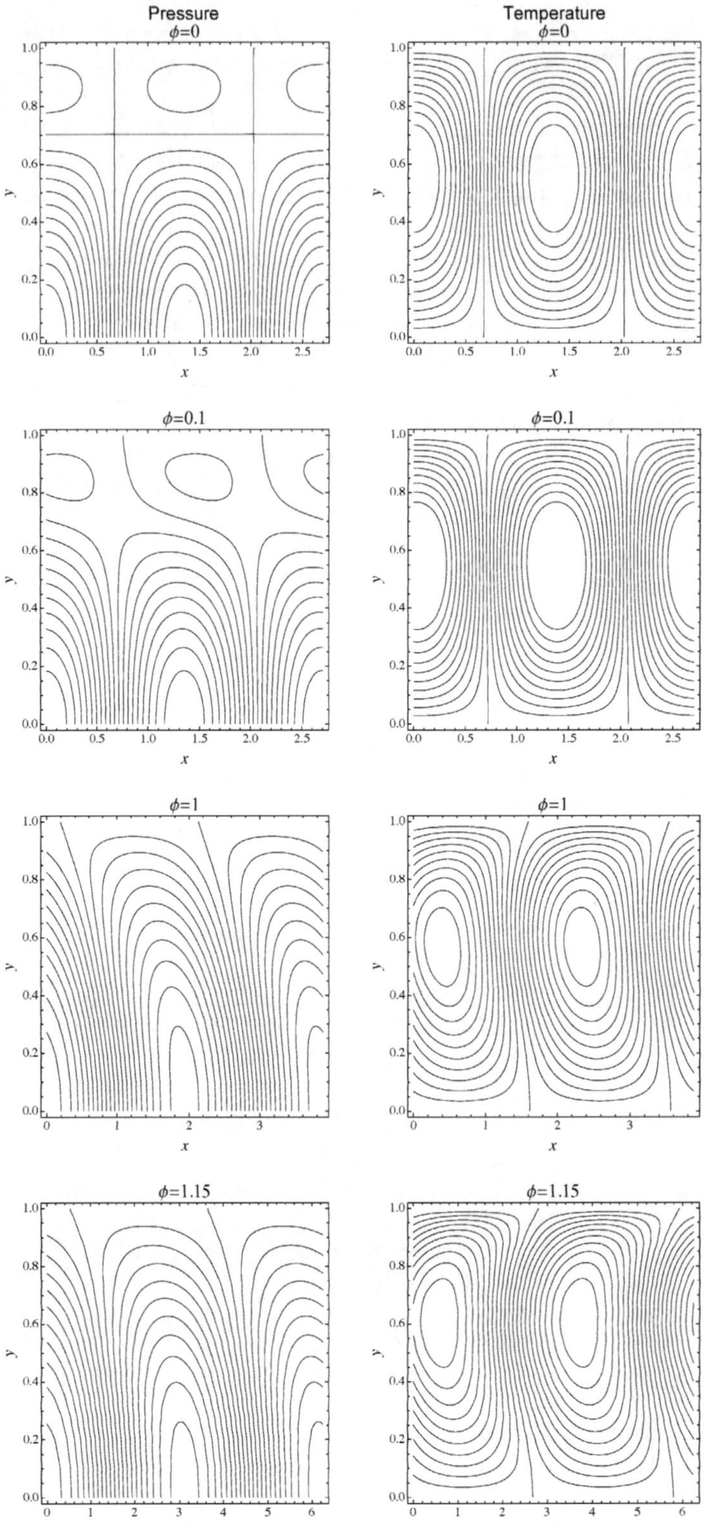

Figure 4. Perturbation pressure isolines, left column, and perturbation temperature isolines, right column, for different values of ϕ.

- The lowest values of $S_c = 27.0976$ and is obtained for $\phi = 0$. The corresponding values of $k_c = 2.32621$;
- The values of S_c increase indefinitely for $\phi \to \pi/2$;
- The instability is always activated by travelling modes, except for the case $\phi = 0$;
- The shape of the convective rolls are strongly influenced by the modified angle ϕ.

References

[1] Bénard H 1900 *Rev. Gén. Sci. Pures Appl.* **12** 1261–1309
[2] Rayleigh L 1916 *Philosophical Magazine* **32** 529–546
[3] Horton C and Rogers Jr F 1945 *Journal of Applied Physics* **16** 367–370
[4] Lapwood E 1948 *Mathematical Proceedings of the Cambridge Philosophical Society* vol 44 pp 508–521
[5] Nield D A and Bejan A 2017 *Convection in porous media* vol 5 (Springer)
[6] Gill A *et al.* 1969 *J. Fluid Mech* **35** 545–547
[7] Barletta A 2015 *Journal of Fluid Mechanics* **770** 273–288
[8] Rees D A S and Bassom A P 2000 *Acta Mechanica* **144** 103–118
[9] Barletta A and Storesletten L 2011 *International Journal of Heat and Mass Transfer* **54** 2724–2733
[10] Rees D A S and Barletta A 2011 *Transport in Porous Media* **87** 665–678
[11] Barletta A and Rees D A S 2015 *International Journal of Heat and Mass Transfer* **83** 327–336

IOP Conf. Series: Journal of Physics: Conf. Series **1224** (2019) 012045 doi:10.1088/1742-6596/1224/1/012045

The effect of PPI on thermal parameters in compact heat exchangers with aluminum foam

B Buonomo[1], A di Pasqua[1], D Ercole[1], O Manca[1*]

[1] Dipartimento di Ingegneria Industriale e dell'Informazione, Università degli Studi della Campania "Luigi Vanvitelli", Via Roma 29, Aversa (CE), 81031, Italia

*Corresponding author: oronzio.manca@unicampania.it

Abstract. A numerical analysis on a compact heat exchanger in aluminum foam is accomplished. The governing equations in two-dimensional steady state regime are written in local thermal non-equilibrium (LTNE). The physical domain under investigation is made up of a plate in aluminum foam with inside a single array of five circular tubes. The presence of the open-celled metal foam is modelled as a porous media by means of the Darcy-Brinkman-Forchheimer law. The metal foam is characterized by a porosity of 0.93 and different pores per inch (PPI), equal to 5, 10, 20 and 40 are analized. The LTNE assumption is considered to simulate the heat transfer between the fluid phase and the solid matrix of the foam. The compact heat exchanger at different air flow rates is studied with an assigned surface tube temperature. The results are shown in terms of heat and mechanical power ratio. Global parameters such as effectiveness and NTU, typical in compact heat exchangers, are showed. Furthermore, local air temperature and velocity profiles in the smaller cross section, between two consecutive tubes, as a function of Reynolds number are presented.

1. Introduction

Metal foams represent a relatively new class of materials applied in fuel cells [1], compact heat sinks [2] and solar thermal plants [3]. Metal foam heat exchangers are efficient, compact and lightweight for the open porosity, low relative density and high thermal conductivity of the cell edges and the large accessible surface area per unit volume. Aluminum foams combine low weight with good rigidity, strength, damping of vibrations and noise, shock resistance and low thermal conductivity [4]. The possible applications are dependent on many parameters such as the type of metal, the porosity (open or closed), the process by which the foam is obtained (temperature, pressure, residence time, melt flow etc.), density and overall costs.

The material has a high ratio between surface area and pressure drop, and with uniform lower density. The pressure drop is lower than in ceramic structures when considering the unit of volume [5]. Hui Wang and Liejin Guo [6] carried out an experimental study on heat transfer and pressure drop of three stainless steel foam filled tubes under convection boundary condition. The stainless steel foam filled tubes are of different pore densities (10, 30 and 70 PPI) but have the same porosity of 0.93. It was demonstrated that the inertial drag is the dominant part of the pressure drop at higher velocity. A new correlation for the pressure drop through metal foams under high velocity was showed. The effect of the boundary condition on the heat transfer performance was investigated by comparing Nusselt number obtained in the study with that obtained under constant heat flux boundary condition. It was found that the Nusselt number obtained under convective boundary condition is much lower than that obtained

under constant heat flux boundary condition. Bayomy and Saghir [7] accomplished a study on heat transfer characteristics of aluminum metal foam subjected to a water flow both experimentally and numerically. The aluminum foam was subjected to a pulsating water flow. The pressure drop was measured across the aluminum foam heat sink and the effects of the dimensionless flow frequency and the flow amplitude on the heat transfer characteristics of the pulsating flow through the aluminum foam were analyzed. An empirical correlation of the average Nusselt number as a function of the dimensionless flow frequency and flow amplitude was developed.
A comparison between the heat transfer characteristics of the steady and pulsating water flows was also conducted. Results also revealed an increase of 14% in the average Nusselt number for pulsating flow when compared to the steady flow. A numerical analysis on the heat transfer inside a open-cell aluminum metal foam was conducted by Ranut et al. [8]. For precisely recovering the microstructure of the foams, a 3D X-ray computed microtomography (μ-CT) is adopted. In this work, the results of μ-CT-based CFD simulations performed on different open-cell aluminum foams samples, for laminar flow regime, will be analized. The results demonstrate that open-cell aluminum foams are effective means for enhancing heat transfer. A convective heat transfer in metal foams was studied by Xu et al. [9] imposing the local thermal equilibrium (LTE) model and the local thermal non equilibrium (LTNE) model numerically. The results demonstrated that the heat transfer with LTE model is higher than that of LTNE model, and the LTNE influence becomes more obvious in these case: lower porosity, larger difference between the thermal conductivity of the fluid and solid phases, lower pore density and lower Reynolds values. Odabaee et al. [10] accomplished a numerical study in order to investigate the heat transfer from a metal foam wrapped solid cylinder in cross-flow. The system was compared with a finned-tube heat exchanger to discover much higher heat transfer rate with acceptable excess of pressure drop. Huisseune et al. [11] analized the performance of a metal foam heat exchanger correlated with the bare tube bundle and an existing louvered fin heat exchanger. An experimental investigation on forced convection in a compact heat exchanger made up with an aluminum foam plate was accomplished by Cicala et al. [12]. Colburn factor and Nusselt number were evaluated for different air mass flow rates and hot water inlet temperatures. A numerical investigation on a tubular heat exchanger in aluminum foam was carried out by Buonomo et al. [13] in Darcy–Forchheimer flow model and the thermal non-equilibrium energy assumption for an assigned porosity and pore density. The results were oriented to show the local behaviour of the system. The evaluation thermal and fluid-dynamic behaviours in a compact heat exchanger in aluminum foam was provided by Buonomo et al. [14] assuming the local thermal non-equilibrium hypothesis. The scope of the analysis was to identify the heat exchanger dimensions to realize a compromise between the heat transfer improvement and the increase in pressure drop.
The present numerical study is an extension of the investigation [13] because the heat exchanger is studied considering different foams characterized by various pore density. The heat transfer rate, the viscous power dissipation, their ratio and the effectiveness of the system are shown. Furthermore, the air velocity and temperature profiles inside the metal foam are presented for different PPI values.

2. Governing equations and Physical model

The 2D sketch of the tubular heat exchanger, under investigation, is showed in Figure 1. The physical model is a forced convection with a steady laminar flow in a metal foam heat exchanger. The heat exchange system is made up of an array of five tubes and they are enclosed inside the aluminum foam. The center-to-center distance between each tube, indicated as $2H$, is 0.038 m and the diameter d of the tubes is 0.012 m. The heat exchanger is characterized by the height H_{tot} of 0.19 m and the length L_{mf} of 0.038 m. The local volume averaging process is used to obtain the microscopic flow equations as indicated in [15] in order to reproduce the thermal and fluiddynamic behaviors inside the domain. The variables of the equations are written evaluating the average of the local variables on an appropriate volume, called Representative Elementary Volume (REV), using the local volume averaging technique [16]. The aluminum foam is assumed homogeneous and isotropic and the thermophysical properties of the fluid and solid phases are considered constant. Furthermore, the viscous dissipation and buoyancy force are neglected, and the thermal contact resistances between the tube surface and foam is overlooked. The Darcy-Brinkman-Forchheimer model and LTNE condition are assumed to describe the behavior of the metal foam. Using the hypothesis above indicated, the governing equations are the following:

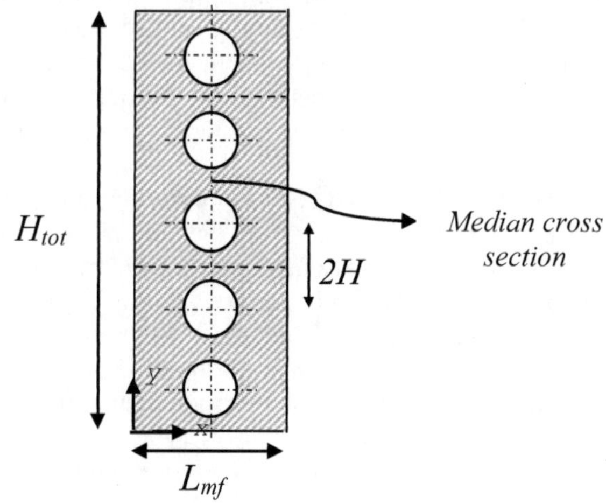

Figure 1. Heat Exchanger with five tubes (Physical Domain)

- *Continuity equation*

$$\frac{\partial u}{\partial x}+\frac{\partial v}{\partial y}=0 \tag{1}$$

- *x- momentum equation*

$$\rho_f\left(\frac{u}{\phi^2}\frac{\partial u}{\partial x}+\frac{v}{\phi^2}\frac{\partial u}{\partial y}\right)=-\frac{\partial p}{\partial x}+\frac{\mu_f}{\phi}\left(\frac{\partial^2 u}{\partial x^2}+\frac{\partial^2 u}{\partial y^2}\right)-\frac{\mu_f}{K}u-\frac{C_F}{K^{1/2}}\rho_f\sqrt{u^2+v^2}u \tag{2}$$

- *y-momentum equation*

$$\rho_f\left(\frac{u}{\phi^2}\frac{\partial v}{\partial x}+\frac{v}{\phi^2}\frac{\partial v}{\partial y}\right)=-\frac{\partial p}{\partial y}+\frac{\mu_f}{\phi}\left(\frac{\partial^2 v}{\partial x^2}+\frac{\partial^2 v}{\partial y^2}\right)-\frac{\mu_f}{K}v-\frac{C_F}{K^{1/2}}\rho_f\sqrt{u^2+v^2}v \tag{3}$$

where ϕ is the porosity, ρ_f and μ_f are fluid density and viscosity, u and v are correspondingly the velocity components in Cartesian coordinates, K and C_F are respectively the permeability and the inertial coefficient of metal foam.

-*Energy equation for the fluid phase*

$$\left(\rho c_p\right)_f\left(u\frac{\partial T_f}{\partial x}+v\frac{\partial T_f}{\partial y}\right)=\phi k_f\left(\frac{\partial^2 T_f}{\partial x^2}+\frac{\partial^2 T_f}{\partial y^2}\right)+h_{sf}\alpha_{sf}\left(T_s-T_f\right) \tag{4}$$

-*Energy equation for the solid phase*

$$\left(1-\phi\right)k_s\left(\frac{\partial^2 T_f}{\partial x^2}+\frac{\partial^2 T_f}{\partial y^2}\right)-h_{sf}\alpha_{sf}\left(T_s-T_f\right)=0 \tag{5}$$

where c_p is the specific heat, k_f and k_s are the fluid phase and solid matrix thermal conductivity, T_f and T_s are the temperature of fluid and solid phase of metal foam, respectively. In the last term in both energy equations, α_{sf} and h_{sf} are respectively the specific surface area density and the interfacial heat transfer coefficient between the fluid and solid matrix, due to the LTNE hypothesis.

The relation of Calmidi are used to evaluate K and C_F [17]:

$$K=0.00073\left(1-\phi\right)^{0.0224}\left(\frac{d_f}{d_p}\right)^{-1.11}d_p^{\ 2} \tag{6}$$

and

$$C_F = 0.00212(1-\phi)^{-0.132}\left(\frac{d_f}{d_p}\right)^{-1.63} \tag{7}$$

where d_f and d_p are respectively the ligament and pore diameter of the metal foam. These parameters are in relation with the characteristics of the aluminum foam by means of the following relations [18]:

$$\frac{d_f}{d_p} = 1.18\sqrt{\frac{1-\phi}{3\pi}}\left(\frac{1}{1-e^{-(1-\phi)/0.04}}\right) \tag{8}$$

and

$$d_p = \frac{0.0224}{\omega} \tag{9}$$

ω is the pores density of the metal foam, that is the number of pores across a linear inch. In this analysis it is variable and assumed values equal to 5, 10, 20, 40 numbers of pores per inch (PPI). The following correlations are adopted to evaluate α_{sf} and h_{sf} [19]:

$$\alpha_{sf} = \frac{3\pi d_f}{\left(0.59 d_p\right)^2}(1-e^{-(1-\phi)/0.04}) \tag{10}$$

and

$$h_{sf} = \begin{cases} \left(0.75 Re_{d_f}^{0.4} Pr_{air}^{0.37}\right)\left(\dfrac{k_f}{d_f}\right), & 1 \le Re_{df} \le 40 \\ \left(0.51 Re_{d_f}^{0.5} Pr_{air}^{0.37}\right)\left(\dfrac{k_f}{d_f}\right), & 40 \le Re_{df} \le 1000 \\ \left(0.26 Re_{d_f}^{0.6} Pr_{air}^{0.37}\right)\left(\dfrac{k_f}{d_f}\right), & 1000 \le Re_{df} \le 2\times 10^5 \end{cases} \tag{11}$$

where Re_{df} is the local Reynolds number referred to ligament diameter:

$$Re_{d_f} = \frac{\rho_f u_0 d_f}{\mu_f} \tag{12}$$

and Pr_{air} is the air Prandtl number that is evaluated as:

$$Pr_{air} = \frac{\mu_f c_p}{k_f} \tag{13}$$

The parameters of the metal foams, for different PPI, are listed in Table 1.

Table 1. Parameters of aluminum metal foams

PPI	ϕ	d_p (m)	d_f (m)	K (m^2)	C_F
5	0.93	4.48 e^{-3}	5.51 e^{-4}	2.72 e^{-7}	0.092
10	0.93	2.24 e^{-3}	2.79 e^{-4}	6.79 e^{-8}	0.092
20	0.93	1.12 e^{-3}	1.37 e^{-4}	1.69 e^{-8}	0.092
40	0.93	5.59 e^{-4}	6.89 e^{-5}	4.25 e^{-9}	0.092

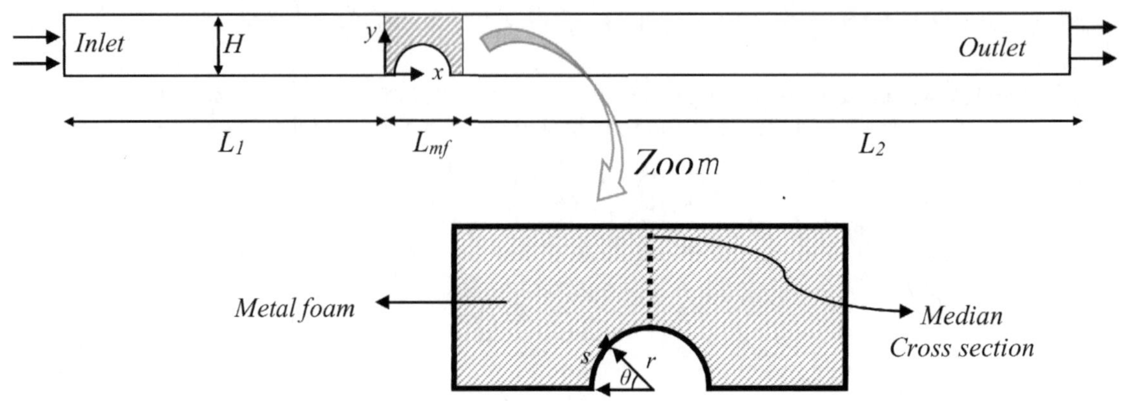

Figure 2. Numerical domain

3. Numerical model

The finite volume method is adopted to resolve the governing equations. Fluent is used to conduct the numerical simulations. Convergence criteria are imposed for the continuity and the velocity components equal to 10^{-5} while for the energy equal to 10^{-8}. As the computational domain is utilized half of a single tube, as represented in Figure 2.

The half tube is totally surrounded by the aluminum foam. The height of system is equal to H, the length L_{mf} is the same of the physical domain and the radius of the tube is equal to $d/2$. In this configuration, the heat exchanger is placed in a parallel plates channel with a rectangular cross section of 0.019 m with L_1 and L_2 equal to 0.20 m and 0.80 m, respectively.

The grid is made up of rectangular cells in the air channel and around the tube and triangular cells in the region of metal. Three different types of grids were investigated to find an independent solution from the mesh. They are constituted by 5786 cells, 23272 cells and 92702 cells, respectively. In corresponding of an inlet air velocity equal 0.12 m/s and of a metal foam characterized by 20 PPI, the evaluation of average $\overline{h}_{tot}$ on the tube surface, as showed in Table 2, points out that the grid with 5786 cells had 0.08% error than the mesh with 92702 cells. The grid adopted for the computational was the one with 5786 elements because with it a compromise between solution accuracy and convergence time was obtained.

In the analysis, the Reynolds number Re_d, evaluated as:

$$Re_d = \frac{\rho_f u_0 d}{\mu_f} \tag{14}$$

range from 100 to 1000. The inlet air velocities u_0 is calculated from the definition of Re_d. The boundary conditions of the problem are the following: the upper and bottom edges are symmetric, the tube wall is an isothermal edge at T_{tube}=323.16 K, the inlet air velocity u_0 assumes the values above indicated, the temperature at the inlet, indicated as $T_{in,air}$, is 288.16 K and to the exit is imposed the outflow condition.

4. Results

The thermal and fluid-dynamic studies are carried out for different inlet air velocities and for a fixed temperature on the external surface of tube of 323.16 K. The state regime is laminar and the LTNE model is considered to write the energy equations. In the Figure 3.a and 3.b the heat transfer rate $\dot{Q}$ and the viscous power dissipation $\dot{W}$, required to circulate the air inside the heat exchanger are reported for different PPI values:

Table 2. Grid Independence and Numerical Results

Cells Number	$\overline{h}_{tot}$ (W/m^{2}K)	% error
5786	146.88	0.08%
23272	146.79	0.02%
92702	146.76	-------

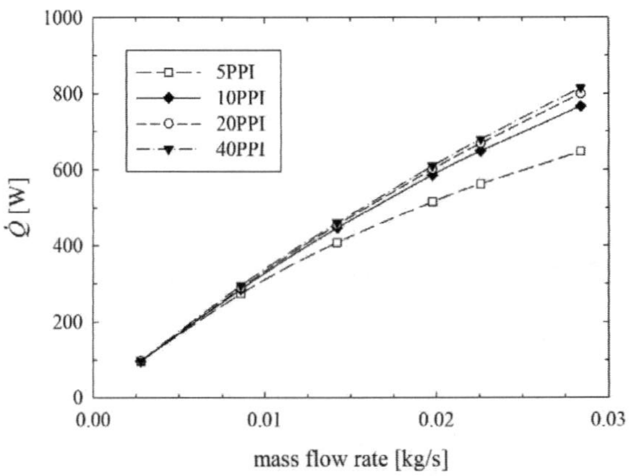

(a)

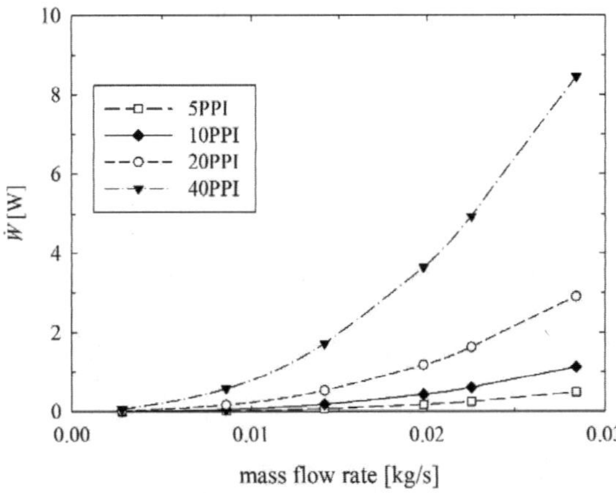

(b)

Figure 3. (a) Heat Transfer Rate $\dot{Q}$ and (b) Viscous Power Dissipation $\dot{W}$ for different PPI

It can be seen that the heat transfer rate increases with increasing pores density, but this improvement in terms of heat exchange is more visible for higher values of air mass flow rate. In addition, it can be observed as also the mechanical power increases with increasing of PPI number and of the mass flow rate. The ratio between the heat transfer rate and the mechanical power is showed, in the Figure 4, in order to know which PPI number maximizes the thermal power but at the same time does not require a high mechanical power.

The ratio between thermal and mechanical power decreases both with the increase of PPI and Re_d. This means that for high PPI values there is a high heat transfer rate but there is also a higher mechanical power required to pump the working fluid and therefore it is more convenient to use foams with a smaller number of PPI. In the Figure 5.a, the temperature profile along the median cross section into metal foam is displayed for Re_d equal to 1000. One can observe two temperature profiles, one referred to the fluid phase and another associated to the solid matrix, due to the LTNE model assumption, for all PPI values.

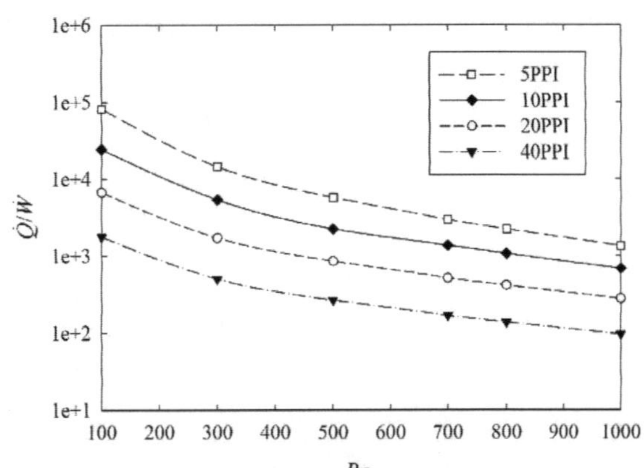

Figure 4. Ratio $\dot{Q}/\dot{W}$ for different PPI values

IOP Conf. Series: Journal of Physics: Conf. Series **1224** (2019) 012045 doi:10.1088/1742-6596/1224/1/012045

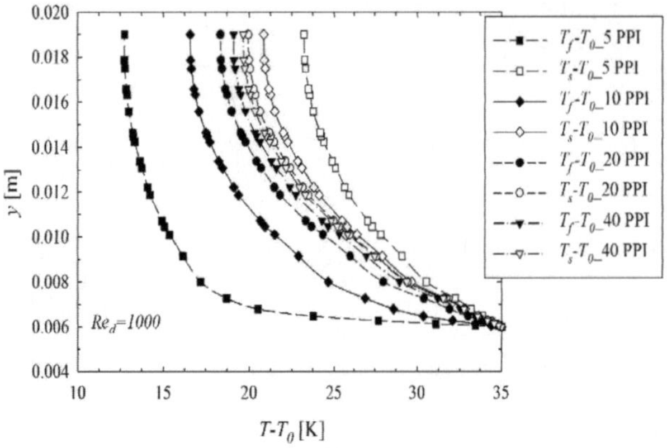

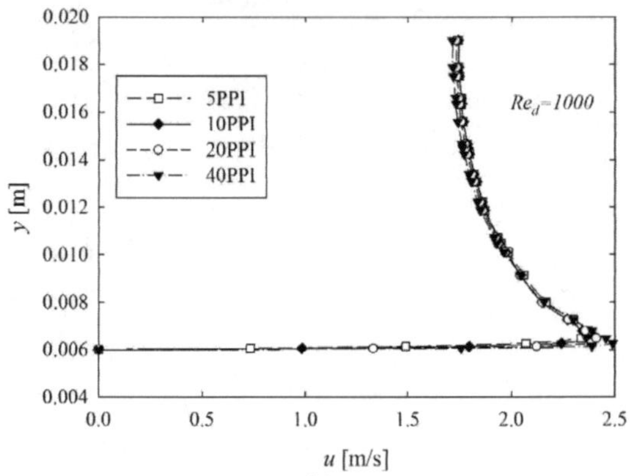

Figure 5. Temperature profile (a) and velocity profile (b).along the median section for different PPI

The difference of temperature between fluid and solid phases decreases with increasing pores density and, for this motivation, in corresponding to higher values of PPI it is possible to assume the LTE model. Furthermore, in the Figure 5.b, the velocity profiles along the median cross section in the foam are reported for Re_d=1000, showed a partial overlap of the profiles for the different PPI values.
The effectiveness ε and *NTU*, typical global parameters in a compact heat exchanger, are showed in the Figure 6:
It can be seen that the effectiveness of the system increases with increasing of PPI values and this occurs for all Reynolds numbers. These parameters are evaluated as [20]:

$$\varepsilon = \frac{\dot{Q}}{C_{\min} \cdot (T_{tube} - T_0)} \tag{15}$$

and

$$NTU = -\ln(1-\varepsilon) \tag{16}$$

with C_{min} the air-side heat capacity $\dot{m}c_p$ and T_{tube} and T_0 the free stream tube temperature and the air inlet temperature applied in the 2D simulations, respectively.

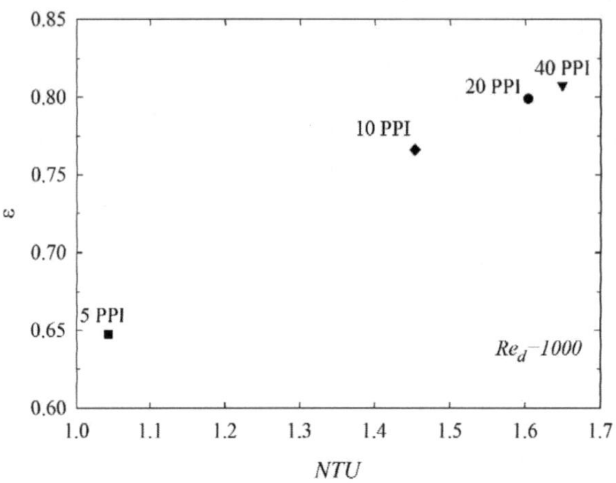

Figure 6. The effectiveness for different PPI

IOP Conf. Series: Journal of Physics: Conf. Series **1224** (2019) 012045 doi:10.1088/1742-6596/1224/1/012045

5. Conclusions

A numerical investigation on a heat exchanger with aluminum foams characterized by 5,10, 20 ,40 PPI has been accomplished in order to evaluate the thermal and fluiddynamic behaviour of the system. The results in terms of heat transfer rate show that the foam with higher pores density value improves the heat transfer than aluminum foam with a lower PPI. In addition, the results in terms of ratio between heat and mechanical power show that the use of metal foam with a lower PPI values is more efficient than one with a higher PPI value. Furthermore, the different temperature profiles in the median section into the metal foam for fluid and solid phases are showed, demonstrating that the difference temperature between two phases decreases with increasing pores density. As a final observation, the efficiency has been evaluated and it increases with increasing PPI values.

6. Nomenclature

A	Cross section	m^2
C_F	Drag factor coefficient	
c_p	Specific heat at constant pressure	J/kg K
d	Tube diameter	m
d_f	Fiber diameter	m
d_p	Pore diameter	m
f	Friction factor	
h	Heat transfer coefficient	W/m^2 K
h_{sf}	Interfacial heat transfer coefficient	W/m^2 K
H	Half pitch	m
H_{tot}	heat exchanger height	m
j	Colburn factor	
k	Thermal conductivity	W/m K
K	Porous permeability	m^2
L	Thickness of porous media	m
$\dot{m}$	Mass flow rate	kg/s
Nu	Nusselt number	
NTU	Number transfer of units	
p	Static pressure	Pa
PPI	Number of pores per inch	
Pr	Prandtl number	
$\dot{Q}$	heat transfer rate	W
r	Radius tube	m
Re	Reynolds number	
s	Curvilinear abscissa	m
T	Temperature	K
u	x-velocity	m/s
u_0	inlet air velocity	m/s
v	y-velocity	m/s
$\dot{W}$	Mechanical power	W
x	Cartesian axis direction	m
y	Cartesian axis direction	m
Greek symbols		
α_{sf}	specific surface area density	m^{-1}
Δ	Difference	
ε	Effectiveness	
μ	Viscosity	kg/m s
ρ	Density	kg/m^3
ϕ	Porosity	
ω	Number of pores per inch	m^{-1}

Subscripts

0	Inlet condition
clean	System without foam
d	Tube diameter
d_f	Fiber diameter
f	Fluid phase
mf	Metal foam
s	Solid phase of metal foam

References

[1] Yuan W, Tang Y, Yang X, Wan Z 2012 Porous metal materials for polymer electrolyte membrane fuel cells – A review *Appl. En.* **94** 309–329.
[2] Bayomy A M, Saghir Z, Yousefi T 2016 Electronic cooling using water flow in aluminium metal foam heat sink: Experimental and numerical approach *Int. J. of Therm. Sc.* **109** 182-200.
[3] Zhao W, France D M, Yu W, Kim T, Singh D 2014 Phase change material with graphite foam for application in high-temperature latent heat storage systems of concentrated solar power plants. *Ren. En.* **69** 134-146.
[4] Zhao C Y 2012 Review on thermal transport in high porosity cellular metal foams with open cells *Int. J. of Heat and Mass Transf.* **55** 3618-3632.
[5] Han X H, Wang Q, Park Y G, T'Joen C, Sommers A, Jacobi A 2012 A review of metal foam and metal matrix composites for heat exchangers and heat sinks *Heat Transf. Eng.* **33** 991-1009.
[6] Wang H, Guo 2016 Experimental investigation on pressure drop and heat transfer in metal foam filled tubes under convective boundary condition *Ch. Eng. Sci.* **22** 438-448.
[7] Bayomi A M, Saghir M Z 2016 Heat transfer characteristics of aluminum metal foam subjected to a pulsating/steady water flow: Experimental and numerical approach *Int. J. of Heat and Mass Transf* **97** 318-336.
[8] Ranut P, Nobile E, Mancini L 2014 Microtomography-based CFD Analysis of Transport in Open-Cell Aluminum Metal Foams *J. of Physics: Conference Series* **501** 012021.
[9] Xu H. J, Gong L, Zhao C. Y, Yang Y H, Xu Z. G 2015 Analytical considerations of local thermal non-equilibrium conditions for thermal transport in metal foams *Int. J. Therm. Sc* **95** 73-87.
[10] Odabaee M, Hooman K, Gurgenci H 2011 Metal Foam Heat Exchangers for Heat Transfer Augmentation from a Cylinder in Cross-Flow *Trans. Porous Media* **86** 911-923.
[11] Huisseune H D, Schampheleire S, Ameel B, De Paepe M 2015 Comparison of metal foam heat exchangers to a finned heat exchanger for low Reynolds number applications *Int. J. Heat Mass Transfer* **89** 1-9.
[12] Cicala G, Cirillo L, Diana A, Manca O, Nardini S 2016 Experimental evaluation of fluid dynamic and thermal behaviors in compact heat exchanger with aluminum foam *Energy Procedia* **101** 1103-1110.
[13] Buonomo B, di Pasqua A, Ercole D, Manca O, Nardini S 2018 Numerical investigation on aluminum foam application in a tubular heat exchanger *Heat and Mass Transfer* **54(3)** 1-9.
[14] Buonomo B, di Pasqua A, Ercole D, Manca O 2018 Numerical investigation on heat exchanger in aluminum foam *Energy Procedia* **148** 782-789.
[15] Nield D A, Bejan A 2013 *Convection in Porous Media* 4th Ed. (New York Springer).
[16] Whitaker S, 1998, *The Method of Volume Averaging* (Netherlands Springer).
[17] Calmidi V V 1998 *Transport Phenomena in High Porosity Metal Foams* Ph.D. thesis (University of Colorado, Boulder CO).
[18] Bhattacharya A, Calmidi V V, Mahajan R L 2002 Thermophysical properties of High Porosity Metal Foams *Int. J. Heat Mass Transfer* **45**: 1017-1031.
[19] Calmidi V V, Mahajan R L 2000 Forced Convection in High Porosity Metal Foams *ASME J. Heat Transfer* **122**: 557-565.
[20] Huisseune H D, Schampheleire S, Ameel B, De Paepe M 2015 Comparison of metal foam heat exchangers to a finned heat exchanger for low Reynolds number applications *Int. J. Heat Mass Transfer* **89**: 1-9.

Experimental investigation on the convective heat transfer enhancement in tubes with twisted-tape inserts

F Bozzoli[1,2], L Cattani[3], A Mocerino[1], S Rainieri[1,2] and G Pagliarini[1,3]

[1] Department of Engineering and Architecture, University of Parma, Parco Area delle Scienze 181/A I-43124 Parma, Italy
[2] SITEIA.PARMA Interdepartmental Centre, University of Parma, Parco Area delle Scienze 181/A, I-43124 Parma, Italy
[3] CIDEA Interdepartmental Centre, University of Parma, Parco Area delle Scienze 181/A, I-43124 Parma, Italy

E-mail: andrea.mocerino@unipr.it

Abstract. Passive heat transfer enhancement techniques are one of the most interesting approaches to increase the thermal efficiency of heat transfer apparatuses since they do not require external power. Between them, twisted-tape inserts are a very economical technique that can be easily installed in an already working plant in order to increase its thermal performance. The aim of the present work is to experimentally investigate the heat transfer augmentation mechanism induced by the insertion of twisted tapes. The heat transfer performance of the twisted-tape inserts is compared with that of coiled pipes, whose effect on heat transfer is based on the onset of swirl components into the flow as it happens for twisted tapes devices. The two heat transfer enhancement mechanisms have been deeply investigated by using different fluids in order to obtain a robust comparison both for the laminar and the turbulent flow regime.

1. Introduction

The heat transfer enhancement techniques constitute a very interesting research issue in the heat transfer field. In many industrial applications, such as those that involve food, chemical and pharmaceutical products, engineers have then been struggling for techniques generating enhanced heat transfer coefficients, accompanied by reduced pumping power requirements.

Today this research topic is attracting a renewed interest in the process industry due to the increase of energy and raw materials cost and this is witnessed also by the huge amount of papers available within the world scientific literature of the last decades regarding heat transfer augmentation and by the growing number of registered patents related to heat transfer enhancement technologies or devices. These techniques essentially reduce the overall thermal resistance by increasing the heat transfer coefficients with or without the increase of the heat transfer surface.

The benefits that can derive are, for instance, the reduction of the size of the heat exchanger that means costs reduction, the decrease of the temperature difference at which they work that means a reduction of the thermal stress for the product or the increase of the thermal power exchanged.

The techniques of increase of the convective heat transfer can be divided into active techniques, that require, for instance, mechanical aid or electrostatic fields, and passive techniques that do not require external power [1,2,3]. The passive techniques for the enhancement of convective heat transfer are based on changes induced on the fluid flow through a proper conformation of the surface, such as curvature of the walls or surfaces roughness or corrugation or through the insertion of devices in the main flow directly or by means of additives [1, 2].

In particular, the inserts are elements that are positioned in the flow passage with the aim of increasing the heat transfer rate. This enhancement technique results particularly attractive for the low cost, the rapid installation and the easy maintenance [4]. Within this category of passive heat transfer enhancement techniques, displaced devices, twisted tapes, wire coils are the most commonly adopted. Twisted tapes are metallic or non-metallic strips twisted with some suitable technique at desired shape and dimension and inserted in the flow with the aim of introducing swirl components into the flow and of disrupting the boundary layer at the pipe surface [5]. However, the pressure drop inside the tube will be increased by introducing the twisted-tape insert. For this reason, twisted-tape inserts have been the object of many research works that analysed the different configurations of these devices studying full-length and short-length twisted tape having a constant pitch or a variable one [6-8] in order to investigate the optimal design and achieve the best thermal performance with less friction loss.

Hong and Bergles [7] performed one of the first experimental work on twisted-tape inserts investigating both the pressure drops and the heat transfer enhancement in pipes by using water and ethylene glycol as working fluids. The insertion of twisted tapes produced an increase of the Nusselt number up to 9 times the empty tube value; in addition, the Authors proposed a correlation based on the Reynolds and Prandtl numbers.

Manglik and Bergles [8] proposed empirical correlations for the Nusselt number and friction factor, identifying a dimensionless swirl parameter. The Authors stressed the attention on this parameter, identified as a crucial factor to describe the behaviour of these devices and the interaction between viscous, inertia and centrifugal forces.

Sarma *et al.* [9] investigated the turbulent regime in pipes fitted with twisted-tape inserts; these Authors proposed a correlation for the Nusselt number as a function of the Reynolds and Prandtl numbers and the twisted tape pitch to tube diameter ratio.

Particularly interesting is the research provided by Ujhidy et *al.* [10] who investigated the laminar flow in coils and tubes containing twisted tapes and helical static elements. These Authors highlighted the similarity of the flow pattern characterised by swirl components present in coils and tubes with twisted tapes or helical static elements proposing also a modification of the Swirl number.

The present work is focused on the experimental investigation of the heat transfer augmentation mechanism induced by the insertion of twisted tapes both in the laminar and in the turbulent flow regimes. In particular, in connection with the work of [10], it has been investigated the role played by the swirl components introduced in the flow by also comparing the heat transfer performance of the twisted-tape inserts with that of coiled pipes.

2. Experimental setup

Three different stainless steel twisted tapes (figure 1) characterized by a thickness δ of 0.001 m and axial pitch p of 0.08 m, 0.15 m and 0.18 m, were considered in the present investigation.

The tape was fitted inside a straight stainless steel tube having wall thickness of 0.001 m and internal diameter d = 0.013 m, that is equal to the tape width.

The experimental setup, schematically shown in figure 2, enabled to investigate the twisted tape performances under the constant heat flux boundary condition at the fluid-wall interface.

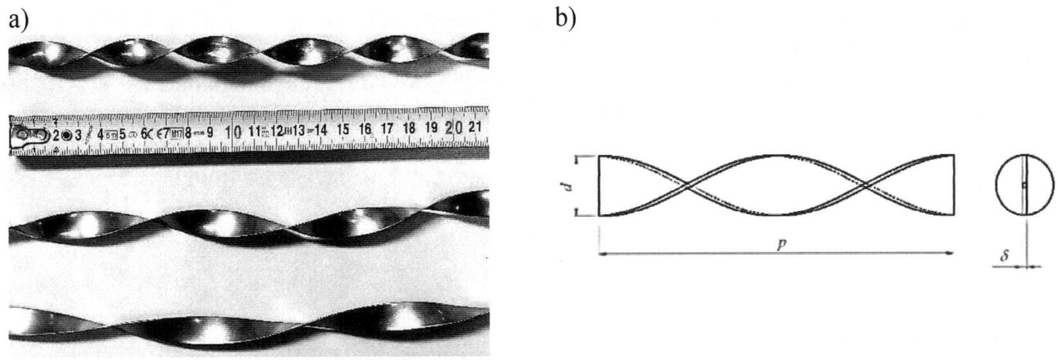

Figure 1. The twisted tapes tested (a) and a twisted tape sketch (b)

This condition was obtained by Joule dissipation in the tube wall. To this aim, the test section was fitted with steel fin electrodes which were connected to a power supply, type HP 6671A, working in the ranges 0–8 V and 0–220 A. The power supplied to the fluid was assumed uniform along the heat transfer section, while a secondary heat exchanger, fed with city water, provided a constant working fluid temperature at the inlet tube section. The outer wall of the tube was thermally insulated in order to minimize the heat exchange to the environment. The temperature of the wall and of the fluid at the inlet and outlet sections was measured by 40 type T thermocouples, previously calibrated and connected to a multichannel ice point reference, type KAYE K170-50C. The data acquisition system used to collect the measurements consists in a high precision multimeter, type HP 3458A, connected to a switch control unit, type HP 3488A, driven by a personal computer. Further details of the experimental setup are reported in [11] and, therefore, they are omitted here.

The tube wall temperature was measured by placing T-type thermocouples at different circumferential and axial locations on the external tube surface, along the heated section.

The fluid bulk temperature at the inlet and outlet sections was measured by using thermocouple sensors inside mixing chamber placed upstream and downstream the tested tube, respectively. It has to be pointed out that the mixing chamber were preceded by mixing inserts in order to reduce the fluid temperature stratification.

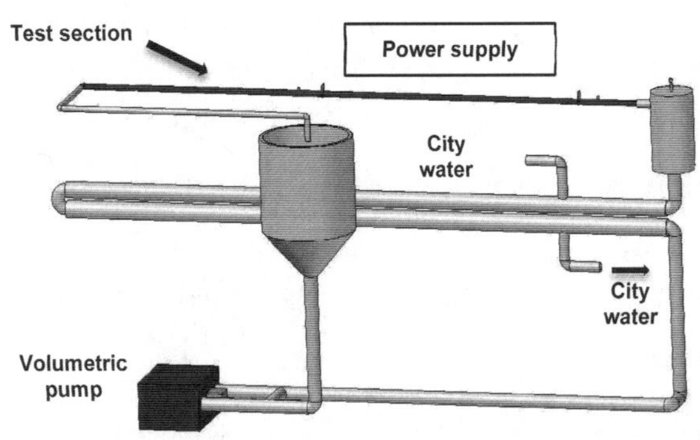

Figure 2. Sketch of the experimental setup.

The bulk temperature at any axial position was then estimated from the energy balance of the fluid, also accounting for the heat losses through the thermal insulation, whose thermal resistance was previously estimated during the calibration of the experimental apparatus. The mass flow rate was obtained by measuring the time needed to fill a volumetric flask placed at the outlet of the test rig. Both ethylene glycol and water were used as working fluid in the Reynolds number ranges 240-3500 and 1400-16100 for the two fluids, respectively. Moreover, the Prandtl number of the two working fluid, at the working mean temperature, was ranging from 40 to 55 for ethylene glycol and from 6 to 8 for water.

To assess the overall experimental procedure, the estimated and the measured fluid bulk temperature values at the outlet section were compared.

3. Results and discussion

The heat transfer performance was quantified by means of the Nusselt number, by adopting the internal tube diameter as the characteristic length in data reduction.

The local Nusselt number was computed as follows:

$$Nu_x = \frac{h_x \cdot D_{env}}{\lambda} \quad (1)$$

where h_x is the circumferentially averaged local convective heat transfer coefficient, based on the circumferentially averaged wall temperature and on the local fluid bulk temperature. In the data reduction the average bulk temperature between the inlet and outlet sections was, instead, used for evaluating all fluid properties.

The uncertainties related to the heat transfer measurements, are estimated by the classical error propagation theory [12]. According to the parameter uncertainties reported in table 1, the relative error on the measured Nusselt number is about 9%.

For the experimental configurations here considered, the fully developed conditions were always reached

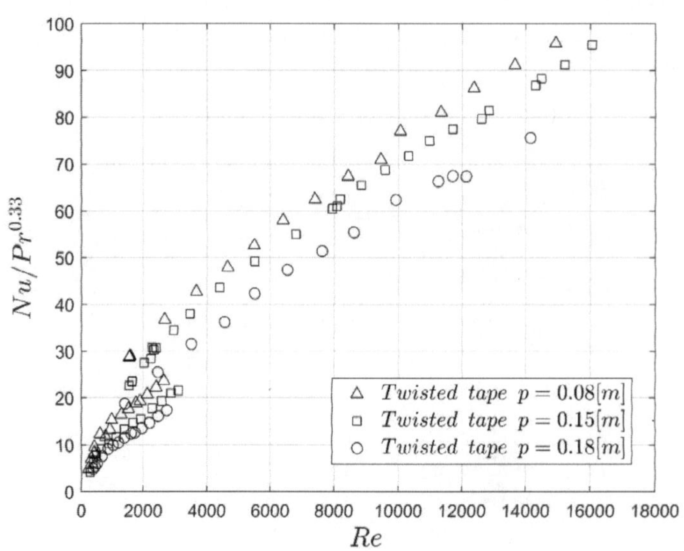

Figure 3. Asymptotic Nusselt number against the Reynolds number

Table 1. Parameter uncertainties

Parameter	$V\,[V]$	$I\,[A]$	$D[mm]$	$\Delta T\,[K]$	$\lambda\left[\frac{W}{m\,K}\right]$
Uncertainties	0.001	0.01	0.05	0.1	0.02

Therefore, the asymptotic Nusselt number was used in the discussion of the results to permit a more meaningful comparison with the coiled pipes thermal behaviour that will be discussed later. The tests were performed by varying the mass flow rate and, as a consequence, the Reynolds number.

For all the runs the asymptotic Nusselt number divided by Pr^{033} is reported in figure 3 as a function of the Reynolds number. Figure 3 points out a significant effect of the tape axial pitch. Moreover, for this particular geometry, it is also possible to identify two different flow regimes, that according to Manglik and Bergels [8], appear around $Re<2500$ and $Re>11500$ for the laminar and turbulent flow regimes respectively. Indeed, Manglik and Bergles [8] suggest that the swirl flow induced by the twisted-tape insert is characterised by the Swirl number, a dimensionless parameter which represents a balance between the centrifugal force due the helical flow, the convective inertia of the bulk axial flow and the viscous force:

$$Sw^2 \propto \frac{Centrifugal\,force \cdot Convective\,inertia\,forces}{Viscous\,forces^2} \tag{2}$$

Thus, the Swirl number can be defined as follows:

$$Sw = \sqrt{\frac{2\frac{\rho W^2}{p} \cdot \frac{\rho W^2}{d}}{\left(\frac{\mu W}{d^2}\right)^2}} = Re\sqrt{2\frac{d}{p}} \tag{3}$$

where μ and ρ are the dynamic viscosity and the density of the fluid, respectively.

The asymptotic Nusselt number divided by $Pr^{0.33}$ as a function of the above defined swirl parameter is then reported in figure 4, for both the laminar and the turbulent flow regimes.

In the same figure the correlation proposed by Hong and Bergles [7] for laminar flow regime:

$$Nu = 5.172\left(1 + 0.005484\left(\frac{Re}{p/2d}\right)^{1.25} Pr^{0.7}\right)^{0.5} \tag{4}$$

and the correlation provided by Sarma *et al.* [9] for turbulent flow regime:

$$Nu = 0.023\,Re^{0.8}Pr^{0.4}\left(1 + \frac{2\pi}{p/d}\right)^{0.87} \tag{5}$$

are reported, too.

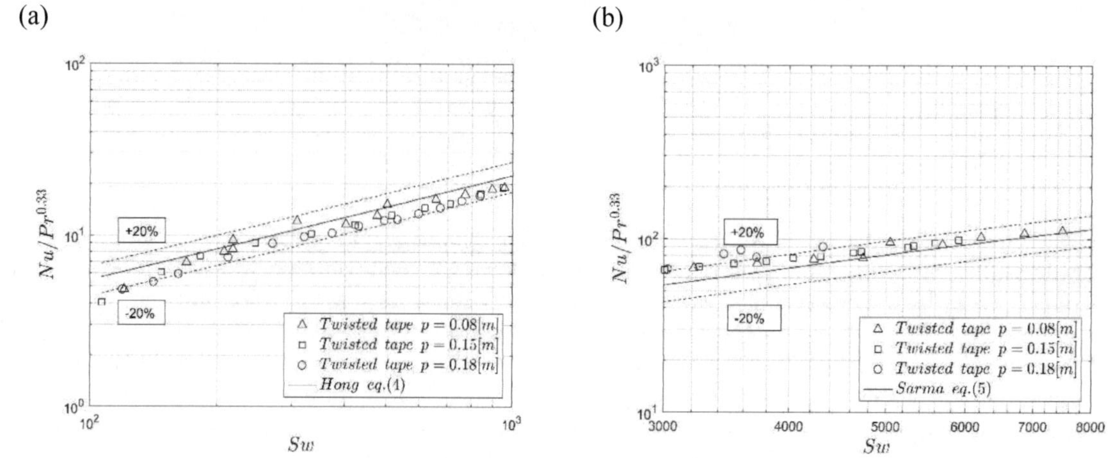

Figure 4. Asymptotic Nusselt number against the Swirl number
(a) laminar flow regime; (b) turbulent flow regime

The present experimental results confirm the prediction of the above correlations in all the considered *Sw* number range.

The convective heat transfer enhancement mechanism of the twisted-tape insert is based on the onset of a secondary swirl flow, which superimposes to the axial flow, thus reducing the thermal boundary layer thickness at the tube wall.

A similar enhancement mechanism is produced by the coiled tube [11], geometry which is schematically shown in figure 5.

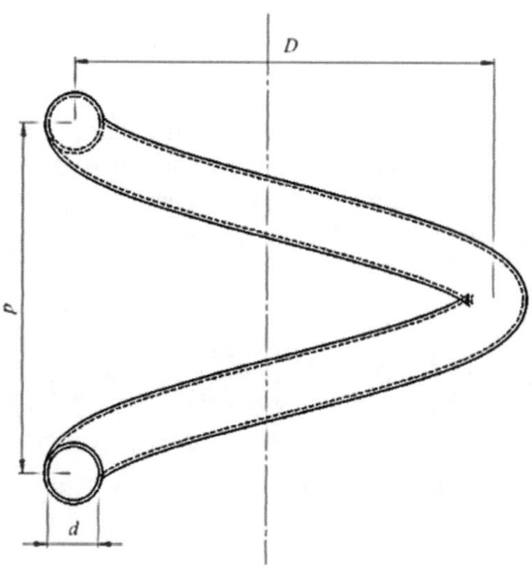

Figure 5. Sketch of the coiled pipe

For coiled tubes, Naphon and Wongwises [13] proposed the below reported correlations. For the laminar flow regime:

$$Nu = \left(2.153 + 0.31\left(Re\sqrt{\frac{d}{D}}\right)^{0.643}\right)Pr^{0.177} \tag{6}$$

which is valid for 20 < *De* < 2000, 0.7 < Pr < 175 and 0.0267 < *d/D* < 0.0884 and where *d* is the tube diameter, *D* is the coil diameter and *De* is the Dean number defined as:

$$De = Re\sqrt{\frac{d}{D}} \tag{7}$$

For the turbulent flow regime, the authors suggested the following correlation:

$$Nu = 0.328Re^{0.58}Pr^{0.4} \tag{8}$$

which is valid for 6000<*Re*<180000.

Following again the suggestion of Manglik and Bergles [8], that underlined the importance of the swirl factor, and the attempt of Ujhidy et *al*. [10] to define a dimensionless parameter that joins the swirl effects generated by different devices, it was computed an analogous Swirl number for the coiled tubes:

$$Sw = \sqrt{\frac{\frac{2\,\rho W^2}{D}\cdot\frac{\rho W^2}{d}}{\left(\frac{\mu W}{d^2}\right)^2}} = Re\sqrt{2\frac{d}{D}} \tag{9}$$

It differs slightly from the case of twisted-tape insert, since for the twisted-tape insert the centrifugal force is proportional to 1/*p*, while being proportional to 1/*D* for the coiled tube.

The correlations of Naphon and Wongwises [13] for the coiled tube are compared in figure 6 with the present experimental results obtained with the considered twisted tapes.

In the laminar flow regime correlation (6) underestimates the twisted tape performance; in the turbulent flow regime, instead, correlation (8) greatly overestimates it. This may be related to the different velocity distribution in the different flow regimes for the two considered devices.

However, the graphs point out that the Swirl number previously defined contains the information related to the swirl phenomena since both the twisted tape and the coiled tube performance has the same trend if they are graphed as a function of it. This suggest a modification of the multiplying coefficient that appear in the correlations (6) and (8) as follows:

$$Nu = 1.41\cdot\left(2.153 + 0.31\left(Re\sqrt{\frac{d}{D}}\right)^{0.643}\right)Pr^{0.177} \tag{10}$$

for the laminar flow regime, while for the turbulent flow regime:

$$Nu = 0.6\cdot 0.328Re^{0.58}Pr^{0.4} \tag{11}$$

IOP Conf. Series: Journal of Physics: Conf. Series **1224** (2019) 012046 doi:10.1088/1742-6596/1224/1/012046

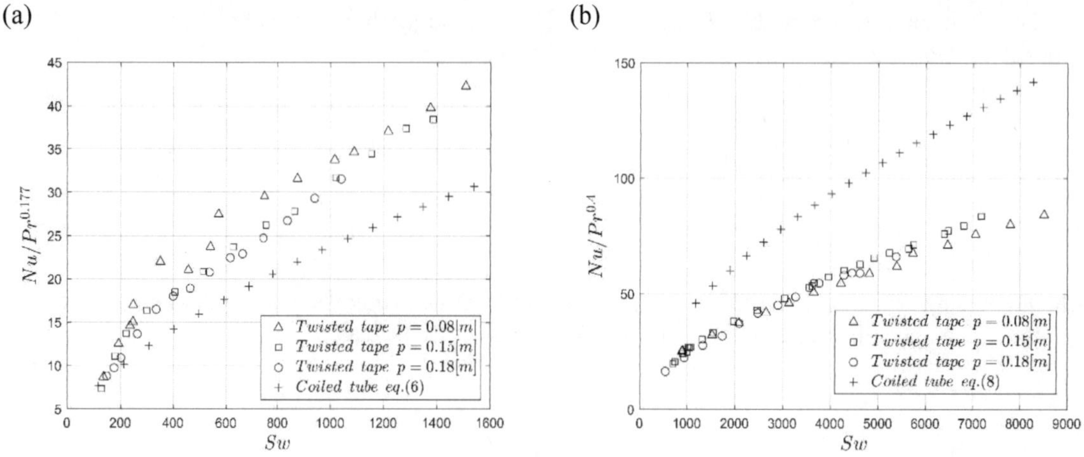

Figure 6: Nusselt number as a function of the Swirl number:
(a) laminar flow regime; (b) turbulent flow regime

The multiplicative coefficients introduced in the Naphon and Wongwises correlations were estimated by using the least square approach within the Matlab ® environment.

The modified Naphon and Wongwises correlations, expressed by equations (10) and (11), are graphed in figure 7, which confirm their ability to model the convective heat transfer mechanism of the twisted-tape insert. It has to be highlighted that the proposed correlations are valid as far as the geometrical similarity is respected.

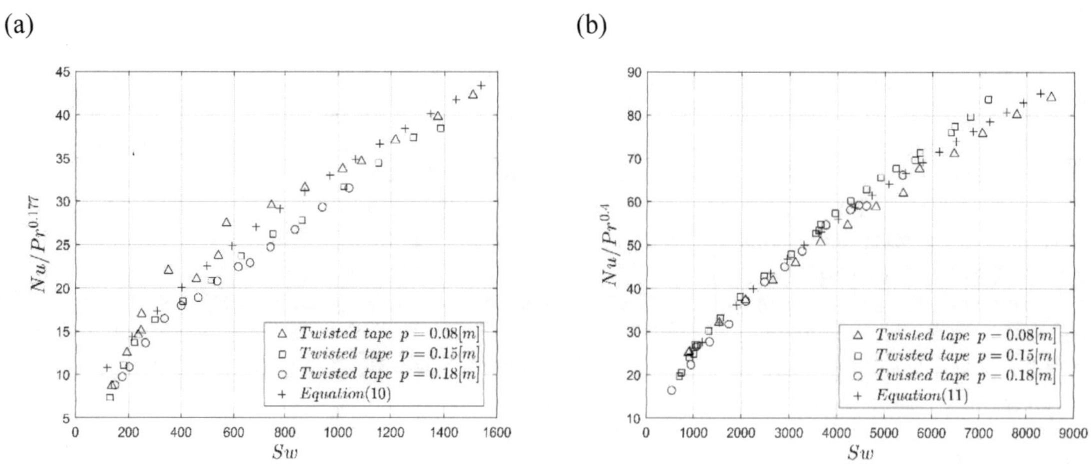

Figure 7: Nusselt number as a function of the Swirl number:
(a) laminar flow regime and (b) turbulent flow regime

IOP Conf. Series: Journal of Physics: Conf. Series **1224** (2019) 012046 doi:10.1088/1742-6596/1224/1/012046

4. Conclusion

In the present paper, the convective heat transfer performance of three different stainless steel twisted tapes, characterized by a thickness δ of 0.001 m, a width of 0.013 m and axial pitch *p* of 0.08 m, 0.15 m and 0.18 m, has been experimentally investigated. The experiments were conducted in both the laminar and the turbulent flow regimes, using ethylene glycol and water as working fluid in the Reynolds number ranges 240-3500 and 1400-16100, respectively.

The results were compared with the correlations already available in literature. Particular attention was given to the role played by the swirling flow components due to the twisted-tape inserts, by comparing the performance of this heat transfer enhancement device with that of coiled pipe, whose effect on heat transfer is based, as well, on the onset of swirl components into the flow. The results point out that a correlation obtained by modifying the correlation of Naphon and Wongwises for coiled tubes, can successfully model the convective heat transfer mechanism of the twisted-tape inserts.

Acknowledgments
This work was partially supported by the Emilia-Romagna Region (POR-FESR 2014-2020: "Applicazione delle micro/nanotecnologie per lo sviluppo di componenti innovativi per gli impianti di condizionamento " and "Progetto Nuovi paradigmi per la progettazione, costruzione ed il funzionamento di macchine e impianti per l'industria alimentare").

5. References

[1] Bergles A E 1998 *Handbook of heat transfer. Chapter 11: techniques to enhance heat transfer*, (New York McGraw-Hill)
[2] Webb R L 1994 *Principles of enhanced heat transfer* (New York John Wiley & Sons Inc)
[3] Rainieri S, Bozzoli F Cattani L and Vocale P 2014 Parameter estimation applied to the heat transfer characterisation of Scraped Surface Heat Exchangers for food applications *J Food Eng* **125** 147–156
[4] Tu W, Tang Y, Hu J, Wang Q and Lu L 2015 Heat transfer and friction characteristics of laminar flow through a circular tube with small pipe inserts *Int. J. Therm. Sci.* **96** 94-101
[5] Liu S and Sakr M 2013 A comprehensive review on passive heat transfer enhancements in pipe exchangers *Renew. Sust. Energ. Rev.* **19** 64-81
[6] Date A W and Singham J R 1972 Numerical prediction of friction and heat transfer characteristics of fully developed laminar flow in tubes containing twisted tapes *J. Heat Transfer* **94** 54
[7] Hong S W and Bergles A E 1976 Augmentation of laminar flow heat transfer in tubes by means of twisted-tape inserts *J. Heat Transfer* **98** 251-256
[8] Manglik R M and Bergles A E 1993 Heat transfer and pressure drop correlations for twisted tape inserts in isothermal tubes *J. Heat Transfer* **115** 890-896
[9] Sarma P K Subramanyam T, Kishorea PS, Rao D V and Kakac S 2002 A new method to predict convective heat transfer in a tube with twisted tape inserts for turbulent flow *Int. J. Therm. Sci.* **41** 955–960
[10] Ujhidy A, Nemeth J and Szepvolgyi J 2003 Fluid flow in tubes with helical elements *Chem. Engng and Processing* **42** 1–7
[11] Rainieri S, Bozzoli F and Pagliarini G 2012 Experimental investigation on the convective heat transfer in straight and coiled corrugated tubes for highly viscous fluids: Preliminary results *Int J Heat Mass Tran* **55** 498–504
[12] Kline S and McClintock F A 1953 Describing uncertainties in single-sample experiments *Mech Eng* **75** 3-8
[13] Naphon P and Wongwises S 2006 A review of flow and heat transfer characteristic in curved tubes *Renew. Sust. Energ. Rev.* **10** 463-490

Preliminary DNS results of friction factor at low Re inside a rectangular channel with 1:10 Aspect Ratio

F. Vignati, P. Gramazio, L. Vitali, D. Fustinoni, A. Niro

Dipartimento di Energia, Politecnico di Milano, Milan, Italy

E-mail: federica.vignati@polimi.it

Abstract. A numerical and experimental study is carried on to investigate the fluid-dynamical performanced of a smooth duct with Aspect Ratio of 10. The flow Reynolds number is between 470 and 14500, computed over the duct hydraulic diameter, ranging from laminar to turbulent flow. Direct Numerical Simulations are performed, to explore the details of all flow scales, by means of a code developed at Politecnico di Milano.

The presented numerical results include the computation of the friction factor and the identification of the transition from laminar to turbulent flow regime, and reproduce experimental and literature data fairly well. The setup of the described numerical and experimental techniques will allow to carry on a detailed investigation of the fluid-dynamical and thermal performances of ribbed duct.

1. Introduction

The investigation of the enhancement of heat transfer by means of natural convection cannot leave turbulence out of consideration [1]. Due to the complex nature of the latter, the numerical study may offer a deeper knowledge of the fluid-dynamics details [2–6]. For this reason, Direct Numerical Simulations (DNS) are performed to study the flow in a rectangular duct with smooth walls and an Aspect Ratio (AR) of $1:10$, at Reynolds number up to 20000, which has not been deeply investigated yet. The goal of this article is a first assessment on the flow features, with a particular focus on the friction factor.

This work is included in a long-term campaign carried out at the ThermALab of Energy Department of Politecnico di Milano on the enhancement of heat transfer in forced convection of air flows through rectangular ducts by means of square ribs in large variety of geometrical configurations. It represents the first numerical investigation, and it can rely on a wide experimental database obtained within the described campaign, where the Nusselt number and the friction factor were measured. The ribs are adopted to enhance heat transfer by promoting to turbulence, and this results in a larger value of Nusselt, but also of the friction factor. For each flow Reynolds number, a large

IOP Conf. Series: Journal of Physics: Conf. Series **1224** (2019) 012047 doi:10.1088/1742-6596/1224/1/012047

number of rib geometries and configurations were tested, since the program was aimed at investigating the optimal rib configuration, i.e., producing the best compromise between the heat transfer enhancement and the induced pressure drop. Comparison with experimental data obtained in a duct with the same geometry and operating conditions are also presented. For this reason, Direct Numerical Simulations should be performed, to gain a further insight in the physics of the phenomenon. This work, therefore, represents the first step in the setup of a set of the numerical tools, whose final goal is the investigation of the ribs effects on the flow Nusselt number.

2. Overview on the investigated problem

2.1. Description of the problem

Pressure drop is obtained by studying the characteristics of forced flow in a duct characterized by an Aspect Ratio AR $= L_y/2\delta$ of 10 and a total length L_x of 2400 mm, 1000 of which dedicated to the pressure measurement. The duct cross-section is $L_y = 120.0$ mm-wide in the $y-$direction and $2\delta = 12.0$ mm-high in the $z-$direction, consisting of a hydraulic diameter D_h of 21.82 mm.

The pressure drop refers only to fully developed-flow regime, either laminar or turbulent. The Reynolds number $\mathrm{Re_D}$ is computed over the hydraulic diameter D_h, the bulk velocity U_b –computed from the mass flow rate $\dot{m}$– and the kinematic viscosity ν, and it ranges from 470 to 14500. The combination between the selected Aspect Ratio and the range of $\mathrm{Re_D}$ allows to observe the transition from laminar to turbulent flow in a duct characterized by an Aspect Ratio that has been considered, for a long time, sufficiently large to produce a quasi-two-dimensional flow, but which was recently proved to suffer from three-dimensional effects [7–10]. To study the described problem, two approaches were adopted: numerical simulations, described in section 2.2, and experimental investigation, outlined in section 2.3.

2.2. Numerical simulations

The duct flow was simulated by means of a Finite Difference solver for three-dimensional, unsteady, incompressible [11] Navier-Stokes equations, which adopts a DNS approach, developed at the Department of Aerospace Science and Technology of Politecnico di Milano [12, 13] for the simulation of channel flows. The code was then adapted by the authors to simulate duct flows, too.

In the spanwise direction, zero velocity is imposed on both walls, to comply non-penetration and no-slip boundary conditions, and the derivative of the pressure in the normal direction is set to zero [14]. In the streamwise direction, periodic boundary conditions are applied to both the velocity and the pressure field, implying a fully developed simulated flow [6, 12, 15]. While periodic velocity is commonly imposed in CFD computations of developed flows, the same consideration does not usually apply to pressure, as it would result in a pressure drop equal to zero, and hence in a meaningless null friction factor. Usually two strategies are adopted: in the first one, the pressure gradient, which is assumed uniform, is imposed, and the consequent velocity is computed, under the assumption that it is streamwise periodic. In the second approach, the pressure variation between the inlet and the outlet is set to zero, i.e., periodic boundary conditions in the streamwise direction apply to pressure, too, by introducing a body force, e.g., gravity acceleration, which produces a hydrostatic pressure distribution that compensate the physical pressure drop. In this second case, either the mass flow rate or the body force is unknown. The strategy adopted in this work is derived from the latter method –which allows to impose the exact value for the mass flow rate– with the additional requirement that not only $\dot{m}$ is imposed and the pressure is assumed to be streamwise periodic, but also that the driving body force is not known. To balance the introduction of the further unknown, the solution of the Navier-Stokes equation is achieved by means of a multistep procedure, which, roughly speaking, considers only one of the two unknowns per time. This allows to assume a streamwise periodic distribution of all the variables and, at the same time, to keep the mass flow rate exactly constant, as the body force depends on this constraint. A detailed description of the procedure and its validation, for the indefinite channel case, is in [16]. In the present work, an identical technique is implemented, with the difference that the second step requires the analytical solution of a Helmholtz partial differential equation over a rectangle, i.e., the duct cross section, instead of an ordinary one.

Calculations are initialized by imposing the solution for the laminar flow, $\mathbf{u} = (u, 0, 0)$, where the normalized streamwise velocity u is known [17] and results from the analytic solution of Navier-Stokes equations. A random perturbation is added during the first time steps to the laminar solution so that, if the Reynolds number is high enough, the transitions to turbulence is triggered, whereas, for low Reynolds numbers, the flow returns to a stable, laminar condition when the perturbation ceases [12, 18–21].

Numerical simulations are performed on a fixed grid, where the node spacing and the time step are selected to resolve all the flow scales [22]. To assure good accuracy of the results and stability of the calculations, a further grid refinement is performed in correspondence of the walls and the corners. The space discretization, time step setting and numerical schemeare selected in accordance with the procedures adopted in literature, e.g., [6, 12, 13, 22]. The technique adopted to force a non-oscillating mass flow rate and to compute the friction factor, briefly recalled in this paragraph, is duly described in [16].

Figure 1 shows the profiles of the streamwise component of the velocity $\overline{u}$ –made dimensionless with respect to U_b– computed at $\mathrm{Re_D} = 1018$ in the duct with $\mathrm{AR} = 10$. The displayed, dimensionless profiles are obtained by time- and streamwise-averaging 10 instantaneous fields of $\overline{u}$, with a distance of 5 time units from each others, beginning the sample after the switching-off of the perturbation, and they overlap fairly well with the instantaneous, analytic field, as depicted in figure 1.

2.3. Experimental features

The duct, whose geometrical parameters are described in section 2.1, is made of four Plexiglas plates, with a 10 mm thickness to ensure the rigidity of the structure despite the relevant duct length. Figure 2 sketches the configuration of the test rig (not to scale).

The friction factor is estimated by measuring the pressure drop along the 1000 mm-long central section of the duct. Room air enters the test section after a 1000 mm-long section aimed at allowing the full development of the boundary layers. No discontinuity is allowed between the development and the test section, to avoid the introduction of

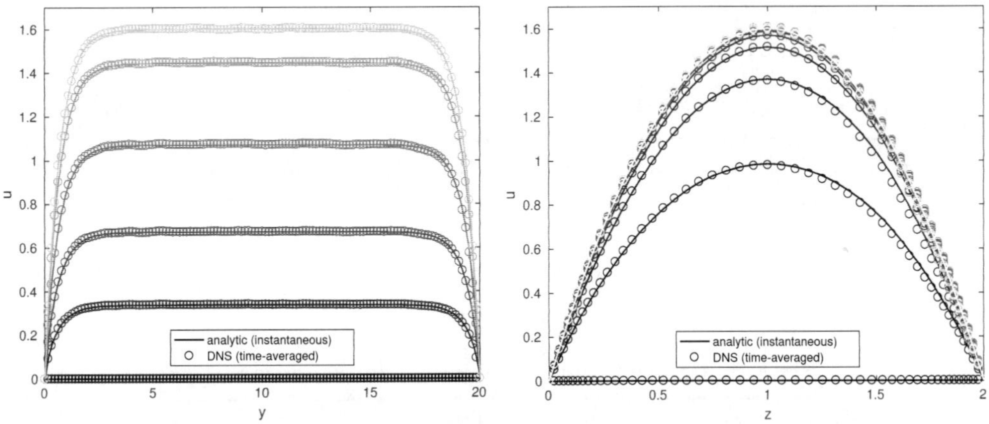

(a) (b)

Figure 1. Comparison between numerical (○) and analytic (full line) profiles of time-averaged, dimensionless $\overline{u}$ in laminar flow regime, at $\mathrm{Re_D} = 1018$. The streamwise velocity is represented (a) versus y and (b) versus z, i.e., the spanwise and height directions respectively, made dimensionless with δ. The velocity is downsampled, to better highlight the node-value of DNS results, and it is represented along (a) 10 $z-$normal and (b) 30 $y-$normal, equally spaced planes.

false perturbation. At the centerline of both the entry and the exit section of the test section, two stations for the measurement of the pressure are located, each consisting in two pressure taps, i.e. 1 mm-diameter holes, respectively on the top and on the bottom walls of the duct.

At the coupling between the entry section and the test section, and between the latter and the exit section, two stations for the measurement of the pressure are located, each consisting in a single-leg U-tube micromanometer. The micromanometer is specifically designed and produced at the ThermALab of Energy Department of Politecnico di Milano.

To achieve the reading, an electric motor is located on the top of the smaller cylinder, which moves a needle along the reservoir axis by means of a worm gear. An electric tension is imposed between the needle and the water free surface, so that the contact between them triggers the reading of the needle position.

The micrometer sensitivity is about $1\,\mu\mathrm{m_{H2O}}$, which corresponds to a pressure variation of 0.01 Pa. The overall accuracy of the test rig is of the order of $10\,\mu\mathrm{m_{H2O}}$, as computed after a statistical analysis, considering both technical features (the resolution of the electric engine and the geometrical features of the worm gear) and side effects of the measurement setup, e.g., water evaporation, mechanical vibrations and tolerances, shape of the needle, leakages and air compressibility. It is worth noticing that the order of magnitude of the accuracy is comparable with the effect of the hydrostatic pressure difference between the top and the bottom of the duct.

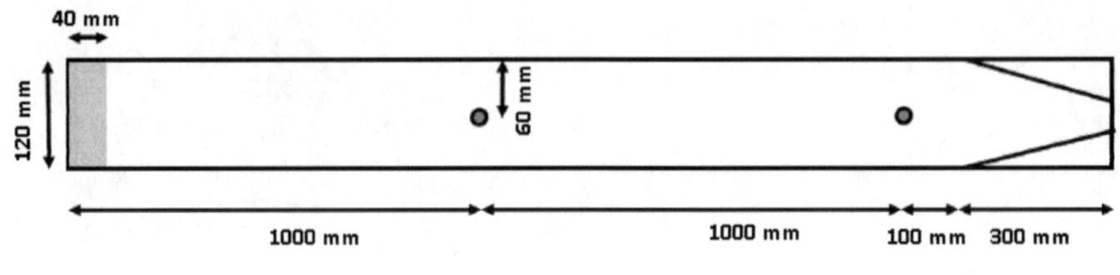

(a)

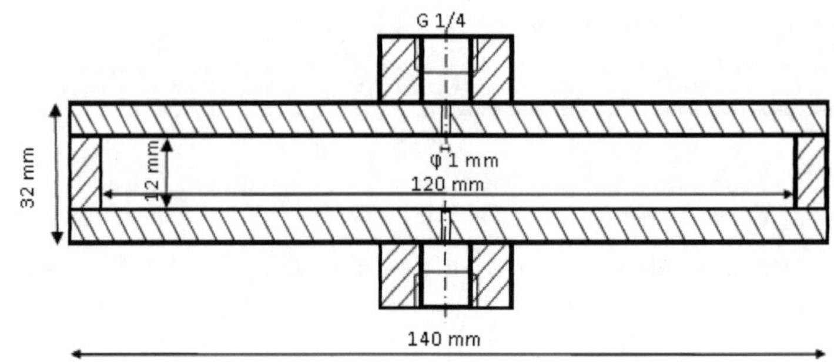

(b)

Figure 2. Sketches of the test rig for the measurement of the friction factor of a rectangular duct with AR = 10: (a) top and (b) front view (not to scale). In fig. 2(a) the flow is from left to right, and the gray dots represents the position of the pressure taps. The connection between the micromanometer taps and the duct is provided by means of threads, depicted in fig. (b).

3. Results

In the experimental framework, the estimation of Fanning friction factor f_F is performed by means of four pressure measurements: at the inlet section, the pressure is measured in correspondence of the centerplane in the spanwise direction, respectively on the top and on the bottom wall. The mean between the two results is assumed to represent the flow pressure in the considered section. The same procedure is performed at the outlet section. The friction factor is then computed by taking the slope of the straight line which interpolates between these two pressure values, and by making this result dimensionless. Unfortunately, the same procedure cannot be followed in numerical simulations, where the periodic boundary conditions impose that the pressure at every node of the outlet section is equal to the pressure evaluated at the corresponding position on the inlet surface. To estimate the friction factor, therefore, the forces across the duct are balanced, so that the force induced by the pressure drop is equivalent to the integration of non-normal stresses along lateral walls. The latter are computed as a function of the mean normal derivatives of $\overline{u}$ with respect to the walls, i.e. $\widetilde{u_z}$ and $\widetilde{u_y}$, resulting in the final expression of f_F:

$$f_F = \frac{f_D}{4} = \frac{1}{2 \cdot \mathrm{Re_D}} \left(\widetilde{u_z} + \frac{\widetilde{u_y}}{\mathrm{AR}} \right) \cdot \left(\frac{4 \cdot \mathrm{AR}}{\mathrm{AR}+1} \right)^2$$

In figure 3(a), the friction factor is displayed versus $\mathrm{Re_D}$ for the rectangular duct in laminar flow. Calculations were performed on a single core of a sixteen-core Xeon E5 2.4 GHz CPU, with 1.5 GB RAM. The time required by the $\mathrm{Re_D} = 3636$-case is of about 10 days. For the same case, the grid resolution is 150 x 500 x 150, in x, y and z, respectively. DNS results are compared with theoretical [23] and experimental results, obtained at the ThermALab of Energy Department of Politecnico di Milano, resulting in good agreement. Figure 3(b) focuses on f_F in terms of the numerator of the analytic expression of the friction factor in laminar conditions, that is $f = C/\mathrm{Re_D}$, where the numerator C depends on the duct Aspect Ratio only. The value of C is computed from DNS results, over three $\mathrm{Re_D}$, i.e., 472, 1018 and 2036, for Aspect Ratios of 1, 2, 5, 10, 20, 50 and for the indefinite channel. Since the latter is known to exhibit homogeneous features in the spanwise direction, numerical simulations of the indefinite channel are carried out with different parameters: in the spanwise direction, the no-slip condition which indicates the presence of walls is substituted by periodic conditions, which allow both slip and penetration at the boundaries. This allows to assume the channel to extend indefinitely, instead of up to AR$\cdot\delta$. According to [6], the computational domain size for the simulation of the indefinite channel is $4\pi\delta$ x δ x $2\pi\delta$, in x, y and z, respectively. DNS data agree fairly well with analytic values of C (the maximum error

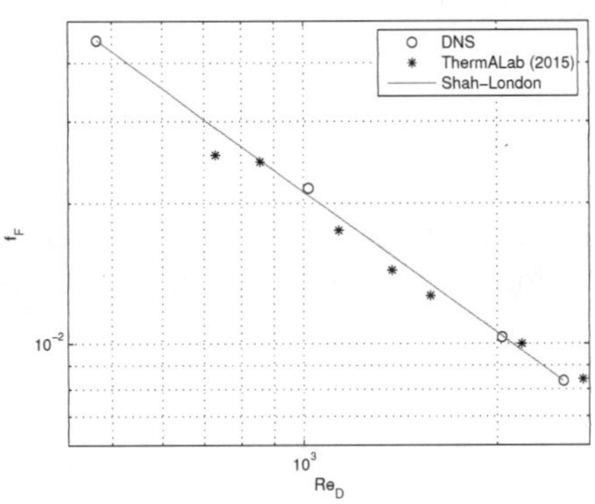

(a)

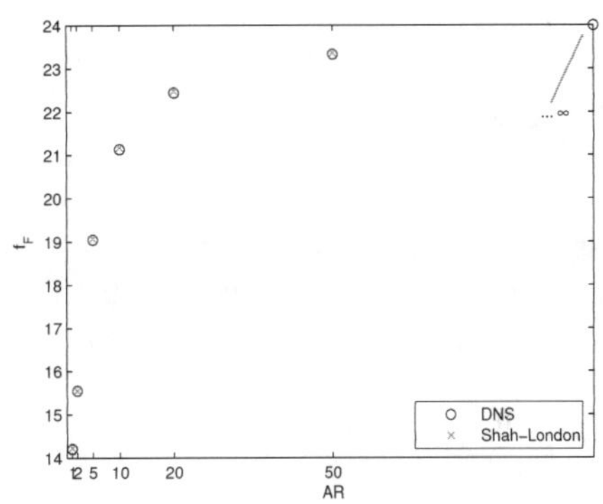

(b)

Figure 3. Comparison between numerical (o), theoretical and experimental valued of Fanning friction factor for laminar flow in rectangular ducts. (a) f_F versus $\mathrm{Re_D}$ for AR = 10; (b) C versus AR and parametrized over $\mathrm{Re_D}$. The dependence of C on $\mathrm{Re_D}$, which is theoretically null, is low enough in DNS results to be not visible in fig. (b).

between theoretical and numerical values of C is 0.2 %, for AR = 2 and $\mathrm{Re_D} = 2036$), and the variance due to $\mathrm{Re_D}$ is totally negligible.

A further method for the estimation of f_F depends on the adopted numerical technique, and consists in determining the correction to the pressure term, at each time step, required to satisfy the prescribed, constant mass flow rate. The computation of the pressure correction is detailed and duly discussed in [16].

Figure 4 compares the friction factors, as resulting from DNS simulations for diverse values of $\mathrm{Re_D}$, obtained by means of the two described methods. As expected, the numerical values are in good agreement, and indicate the values of $\mathrm{Re_D}$ between 2500 and 4500 as the range where the laminar-turbulent flow transition occurs.

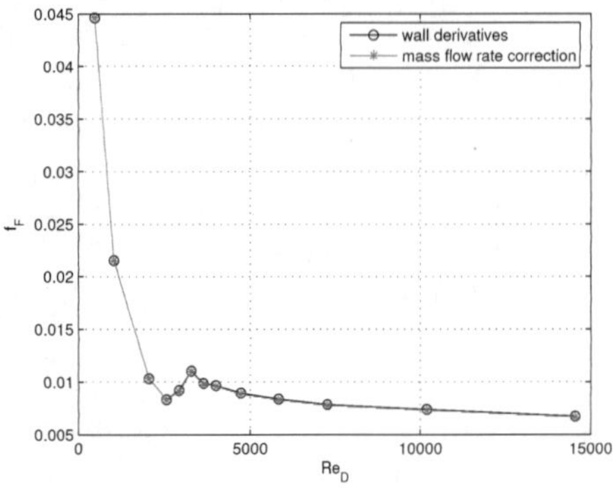

Figure 4. Fanning friction factor of a duct flow with AR = 10 at $\mathrm{Re_D}$ between 470 and 14500, obtained by means of DNS study. The two methods are compared: the wall derivatives of $\overline{u}$ ($\circ$) and the mass flow rate correction ($*$).

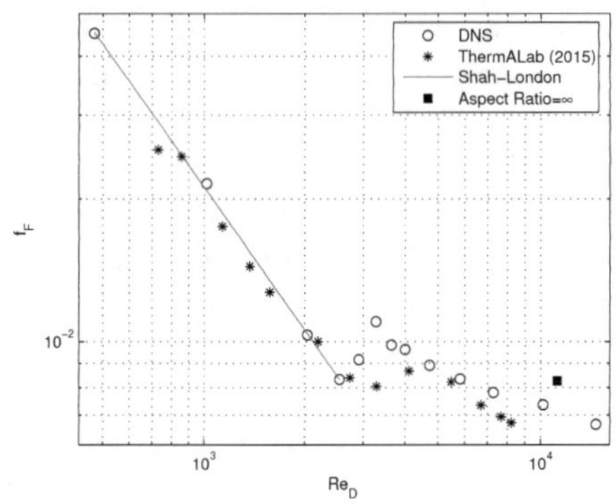

Figure 5. Comparison between DNS results and experimental data obtained at ThermaLab of Politecnico di Milano for the Fanning friction factor of a duct flow with AR = 10 at $\mathrm{Re_D}$ between 470 and 14500.

Eventually, numerical data are assessed against experimental valued for f_F obtained at ThermALab of Energy Department of Politecnico di Milano, as depicted in figure 5. The accordance between numerical and experimental data is not exact, since a deviation occurs at high Reynolds numbers. This discrepancy is probably due to the procedure adopted to estimate the experimental value of f_F, which is based on pressure measurements only along the centerline, and which therefore neglects the non-uniform distribution of the flow in the spanwise direction [7–10, 24–26]. Table 1 reports the deviations of numerical and experimental data with respect to the analytic solution (only in laminar flow) or between DNS and experimental results, computed on Fanning friction factor. The relative difference margin, even in the worst case, is below 9 %, which is in the commonly accepted range, considering also the differences between the numerical and the experimental procedures.

4. Conclusions

Direct Numerical Simulations are carried out to investigate the flow feature in a rectangular duct with Aspect Ratio of 10. The Reynolds number covers laminar flow, the transition to turbulence and "slow" turbulent flows, ranging from 470 to 14500. This work is motivated by the low number of literature on flows in correspondence of the laminar-turbulent transition in ducts with AR = 10, which is borderline to consider the internal flow to show approximately two- or fully three-dimensional features. The novel numerical code consists in a Finite-Difference solver for unsteay, incompressible, Navier-Stokes equations on structured grids, based on the code previously developed at

Table 1. Relative differences between the computed values of Fanning friction factor. The first comparison concerns the average error between analytical values, which are known in laminar conditions, and numerical (for err_{an-DNS}) or experimental (for $\mathrm{err}_{an-exper}$) data. The average relative differences between numerical and experimental data, either in laminar or in turbulent conditions, are quantified by $\mathrm{err}_{DNS-exper,L}$ and $\mathrm{err}_{DNS-exper,T}$, respectively.

err_{an-DNS}	1.1849 %
$\mathrm{err}_{an-exper}$	5.9441 %
$\mathrm{err}_{DNS-exper,L}$	8.2906 %
$\mathrm{err}_{DNS-exper,T}$	8.6916 %

Politecnico di Milano for the direct turbulence simulation of channel flows, and adapted for the simulation of duct flows.

The Fanning friction factor is computed at diverse values of $\mathrm{Re_D}$. The agreement between DNS and experimental results is good but not perfect, probably due to the different approaches adopted to compute f_F.

The provided solutions can be further improved, not only by refining the investigation tools (e.g., by adopting Compact Finite Differences or even longer computational domains, for the DNS), but in particular by unifying as much as possible the experimental and the numerical investigations. Since this work belongs to a long-time research campaign carried on at the ThermaLab of Politecnico di Milano, the final goal is to introduce in the code the simulation of more complex, non isothermal duct flows.

References

[1] Webb R L 1994 *Principles of enhanced heat transfer* (New York: Winley-Interscience)
[2] Rodi W, Mansour N N 1993 Low Reynolds number $k-\varepsilon$ modelling with the aid of direct simulation data *J. Fluid Mech.* **250** 509
[3] Mansour N N, Kim J, and Moin P 1988 Reynolds-stress and dissipation rate budgets in a turbulent channel flow *J. Fluid Mech.* **194 15-44**
[4] Blackburn H M, Mansour N N and Cantwell B J 1996 Topology of fine scale motions in turbulent channel flow *J. Fluid Mech.* **310** 269
[5] Kim J and Antonia R A 1993 Isotropy of the small scales of turbulence at low Reynolds number *J. Fluid Mech.* **251** 219
[6] Kim J, Moin P and Moser R D 1987 Turbulence statistics in fully developed channel flow at low Reynolds number *J. Fluid Mech.* **177** 133
[7] Vinuesa R, Noorani A, Lozano-Durn A, El Khoury G K, Schlatter P, Fischer P F and Nagib H M 2014 Aspect ratio effects in turbulent duct flows studied through direct numerical simulation *J. of Turbulence* **15.10** 677-706, doi:10.1080/14685248.2014.925623
[8] Vinuesa R 2013 Synergetic computational and experimental studies of wall-bounded turbulent flows and their two-dimensionality *PhD thesis, Illinois Institute of Technology*
[9] Vinuesa R, Bartrons E, Chiu D, Dressler K M, Redi J D, Suzuki Y, Nagib H M 2014 New insight into flow development and two dimensionality of turbulent channel flows *Exp. Fluids* **55** 1759 https://doi.org/10.1007/s00348-014-1759-8
[10] Vinuesa R, Schlatter P and Nagib H M 2015 Characterization of the secondary flow in turbulent

rectangular ducts with varying aspect ratio *Int. Symposium on Turbulence and Shear Flow Phenomena (TSFP-9), June 30-July 3 2015, Melbourne, Australia*

[11] Shirokoff D and Rosales R R 230 An efficient method for the incompressible Navier-Stokes equations on irregular domains with no-slip boundary conditions, high order up to the boundary *J. Comput. Phys.* **230** 8619-46

[12] Luchini P and Quadrio M 2006 A low-cost parallel implementation of direct numerical simulation of wall turbulence *J. Comput. Phys.* **211.2** 551-71

[13] Monti C M 2017 Metodo dei contorni immersi per la simulazione numerica diretta di correnti turbolente su pareti non piane *MSc thesis, Politecnico di Milano*

[14] Anders Petersson N 2001 Stability of Pressure boundary Conditions for Stokes and NavierStokes Equations *J. Comput. Phys.* **172** 40-70

[15] Quadrio M and Luchini P 2001 A $4-th$ order accurate, parallel numerical method for the direct simulation of turbulence in cartesian and cylindrical geometries *Proc. of the XV AIMETA Conf. on Theor. Appl. Mech*

[16] Quadrio M, Frohnapfel B, Hasegawa Y 2016 Does the choice of the forcing term affect flow statistics in DNS of turbulent channel flow? *European J. of Mech. B/Fluids* **55** 286-293

[17] Shah R K and London A L 1978 *Laminar Flow Forced Convection in Ducts - A Source Book for Compact Heat Exchanger Analytical Data* (New York: Academic Press, Inc)

[18] Orszag S A 1971 Accurate solution of the OrrSommerfeld stability equation *J. Fluid Mech.* **50.04** 689-703

[19] Uhlmann M and Nagata M 2006 Linear stability of flow in an internally heated rectangular duct *J. Fluid Mech.* **551** 387-404 doi:10.1017/S0022112005008487

[20] Theofilis V, Duck P W and Owen J 2004 Viscous linear stability analysis of rectangular duct and cavity flows *J. Fluid Mech.* **505** 249-286 doi:10.1017/S002211200400850X

[21] Tatsumi T and Yoshimura T 1990 Stability of the laminar flow in a rectangular duct Journal of Fluid Mechanics *J. Fluid Mech.* **212** 437-449 doi:10.1017/S002211209000204X

[22] Comini G, Croce G and Nobile E 2008 *Fondamenti di termofluidodinamica computazionale* (S.G.E)

[23] Muzychka Y S, Yovanovich M M 2009 Pressure drop in laminar developing flow in noncircular ducts: a scaling and modeling approach *J. Fluid Eng.* **131** 111105.1-9

[24] Dean R B 1978 Reynolds number dependence of skin frictin and other bulk flow variables in two-dimensional rectangular duct flow *J. Fluid Eng.* **100** 215

[25] Bradshaw P and Hellens G E 1964 The N.P.L. 59 in x 9 in boundary layer tunnel *N.P.L. Aero Report* **1119**

[26] Monty J P 2005 Developments In Smooth Wall Turbulent Duct Flows *PhD thesis, University of Melbourne*

IOP Conf. Series: Journal of Physics: Conf. Series **1224** (2019) 012048 doi:10.1088/1742-6596/1224/1/012048

Experimental results on local heat transfer coefficient inside a rectangular channel with chevron-shaped ribs

P Gramazio, L Vitali, D Fustinoni, F Vignati and A Niro

Politecnico di Milano, Department of Energy,
Campus Bovisa, Via Lambruschini 4, 20156 Milano, Italy

Corresponding author email: alfonso.niro@polimi.it

Abstract. A non-intrusive measurement technique of the heat transfer coefficient field, based on IR-thermography, is applied on a 12-mm-height rectangular channel of 1:10 aspect ratio, whose lower surface is enhanced by chevron-shaped square-cross-section ribs. The channel is operated at imposed heat flux while forced-air flows for Reynolds numbers ranging from 700 to 8000. In particular, the technique uses local temperature measurements to retrieve the radiative heat flux with a three-wavelength-band model for partially transparent enclosures, and the conductive flux towards the backside with a FEM model, which allow an edge-to-edge estimation of the local convective heat transfer. The resulting local Nusselt number maps are presented and compared to data obtained operating both at fixed temperature and inside a channel with flat surfaces.

1 Introduction

A long-term scientific study on forced convection in channels in stationary conditions has been carried out by the ThermALab laboratory at Politecnico di Milano, focusing on the enhancing effect of ribs in several configurations on the convective heat transfer. In particular, a technique for the retrieval of the local Nusselt number has been first developed [1], then critically analyzed and corrected [2], and used after more considerations on its errors and critical aspects in a transverse ribs configuration [3].

In this work the focus is on V-shaped ribs, a configuration which is also called *chevron*. On this topic, starting from the 90s, studies have been carried out in order to enhance the effect of secondary flows, see for example Han et al. [4], Krukeja et al. [5], Gao and Sunden [6]. In general, V-shaped ribs may present the apex upwards or downwards with respect to the flow direction. In the former arrangement, secondary flows move from the centre to the lateral sides of the channel, whereas in the latter the situation is reversed. It has not been yet assessed which configuration is the best in enhancing heat transfer, because controversial results have been found, probably due to the different boundary conditions taken into consideration.

This paper shows the results of the adopted technique for the estimation of local convective heat transfer for a channel with V-shaped ribs on the bottom surface, which is heated with constant heat flux, for Reynolds numbers ranging from 700 to 8000, which encompasses the transitioning regime for the reference flat channel. In particular, the adopted technique allows the measurement near the edges of the channel, a region which is mostly overlooked by other works.

2 Materials and methods

2.1 Experimental apparatus

The measurements have been carried out on an open-loop channel operated in suction, as shown in figure 1. Summarizing its main features, the channel lower surface is heated by means of two Printed Circuit Boards (PCBs) independently powered and controlled. The PCBs have a substrate made of FR-4 fiberglass-reinforced epoxy laminate material, whereas copper tracks are 0.017 mm thick, 7.8 mm wide, spaced 0.2 mm apart each other, and covered by a black lacquer which also fills up the gaps between them, making the heating surface flat, smooth and with a high emissivity. A Hendress Hauser T-Mass 65F15 flow meter is used for automatic mass-flow rate readings. A double-glazing 130 mm x 130 mm germanium window grants the optical access to a FLIR T650sc IR-camera, with a 640 x 480-pixel resolution. The main experimental features are listed in table 1.

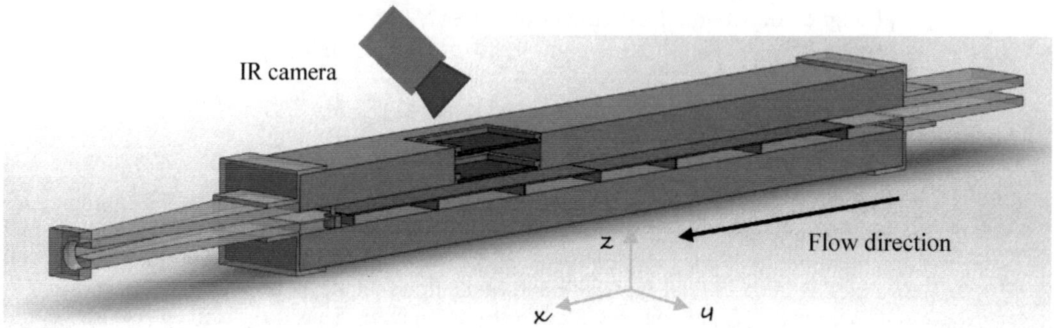

Figure 1. Section of the test channel for local heat transfer measurements, with constant heat flux

Table 1. Main experimental features

h	Channel height	12	mm
L	Channel length	840	mm
w	Channel width	120	mm
D_h	Hydraulic diameter	21.82	mm
	Height to width ratio	0.10	
s	Rib side dimension	4	mm
d	Distance rib-to-rib	80	mm
Re	Reynolds number	700 ÷ 8000	

Punctual temperature measurements are also carried out for reference by RTDs and thermocouples, as reported in table 2. The remaining temperatures, whose knowledge is needed to determine the convective heat flux distribution, are evaluated on the basis of the following assumptions, summarized in table 2 as well. The air bulk temperature $T_b(x)$ is assumed to have a linear increase along the test section, from the inlet value to the outlet one, as supposed by reference works [7]: while certainly an unrealistic assumption, as the air temperature is expected to vary in spanwise direction due to the rib configuration, since the procedure for retrieving the heat fluxes is already complex, the authors have opted for a simpler reference temperature to compute the local *h*, and therefore Nu.

The temperature of the germanium window $T_{ger}(x,y)$ is uniform and equal to the temperature T_{up} measured on its frame, as its conductivity is two orders of magnitude higher than all the other parts of the channel. Finally, the channel side wall temperature $T_{side}(x,z)$ is assumed to be constant along z-direction and equal to the mean between T_{up} and the breadboard temperature T_{down} measured at the channel side.

Table 2. Main measured and derived temperature values

Measured Temperature	**Symbol**	**Probe kind/Instrument**
Bottom wall surface	$T_w(x,y)$	IR camera
External (ambient)	T_{ext}	PT100 RTD
Inlet air bulk temperature	T_{in}	PT100 RTD
Outlet air bulk temperature	T_{out}	PT100 RTD
Channel upper side wall	T_{up}	T-type thermocouple
Derived Temperature	**Symbol**	**Method**
Air bulk temperature	$T_b(x)$	Linear interpolation between T_{in} and T_{out}
Channel side wall	$T_{side}(x,z)$	Mean between T_{up} and $T_w(x,w/2)$
Germanium window surface	$T_{ger}(x,y)$	Uniform and equal to T_{up}

2.2 Modeling

The local convective heat transfer coefficient $h_{loc,q}$ and the related Nusselt number $Nu_{loc,q}$ are evaluated, for the channel with constant heat flux, as in equations 1 and 2:

$$h_{loc,q}(x,y) = \frac{q_c''(x,y)}{T_w(x,y) - T_b(x)} \tag{1}$$

$$Nu_{loc,q}(x,y) = \frac{h_{loc}(x,y)D_h}{k_{air}} \tag{2}$$

where $q''_c(x,y)$ and $T_w(x,y)$ are the punctual values of convective heat flux and surface temperature, respectively, and $T_b(x)$ is the air bulk temperature. Since the temperatures are known, the only term which requires a further specification is $q_c(x,y)$, and it is obtained by a steady-state balance on the copper track layer

$$q_c''(x,y) = \sigma_{el} s_{Cu} - q_r''(x,y) - q_{kz}''(x,y) + q_{kxy}''(x,y) \tag{3}$$

where σ_{el} the volumetric heat dissipation due to the Joule effect, considering a uniform generation, s_{cu} the layer thickness, $q''_r(x,y)$ the heat flux radiated by the layer coating, $q''_{kz}(x,y)$ the heat flux diffusing in z direction from the rear of the copper track layer toward the substrate inside, and finally $q''_{kxy}(x,y)$ the heat flux diffusing in x-y direction through the layer itself. This flux is considered negligible in streamwise x direction, i.e. parallel to the copper tracks, due to the small values of temperature gradient in that direction; in spanwise y direction, it is calculated for each $\bar{x}$, as

$$q_{kxy}''(\bar{x},y) = k_{eq} s_{Cu} \left(\frac{\partial^2 T_w}{\partial y^2}\right) \tag{4}$$

where k_{eq} the equivalent thermal conductivity, resulting from the series of the copper tracks and the lacquer filling the gaps between them. $q''_{kz}(x,y)$ and $q''_r(x,y)$ are calculated by means of a FEM model of the section and of a custom radiative model, respectively. The solution strategy is to apply those models on N_x independent 2D y-z slices of the test section, instead of solving two 3D problems, in order to reduce computational times and complexity without losing accuracy, as discussed in [1].

The adopted radiative model is the net radiation method for diffuse and gray surfaces, which has been modified by Siegel [8] to take into account semi-transparent surfaces. In particular, the method

IOP Conf. Series: Journal of Physics: Conf. Series **1224** (2019) 012048 doi:10.1088/1742-6596/1224/1/012048

allows to compute the radiative heat fluxes for each isothermal surface in enclosures with known temperature. In our case, the system is solved for each 2D section at fixed $\bar{x}$, by discretizing the PCB surface with Npixel elements, i.e. each pixel of the IR image is considered to be an isothermal surface. The lateral (XPS wall) and top (Germanium) surfaces, even if considered at uniform temperature, are discretized too, to increase the accuracy of the view factors computed by means of Hottel's method. Finally, the model is solved for three wavelength regions, i.e., between 0 and 7 μm, 7 and 13 μm, and from 13 μm up, which are determined by the selective transmittance behaviour of the germanium window. The total transmittance of the double-glazing has been taken into account for the top surface.

The main new aspect in the present work, with respect to the previous ones, is the unavoidable presence of the ribs in the 2D slices. Within the framework of this article, the authors have followed a "ribless" approach, thus the possible fin- or insulating-effect induced by the ribs and its consequences on the temperature field has been neglected and the rib-covered regions of the thermograms are just removed from the analysis.

2.3 Analysis procedure

The analysis procedure follows the steps described below:

- Start of the experimental setup, wait for stationary conditions;
- Acquisition of a sequence of thermograms and data from temperature sensors;
- Thermogram elaboration: time-average, projective transform, cut, slicing for analysis, 2D interpolation;

Then, for each slice:

- $q''_{kz}(x,y)$: inverse conductive FEM thermal problem, given the measured temperature field inside the channel and imposing natural convection and radiation b.c. towards the external environment (see figure 2)
- $q''_{ky}(x,y)$: second derivative of the temperature field
- $q''_{r}(x,y)$: 3-band radiative model for the channel cavity
- $q''_{c}(x,y)$: Application of equation 3

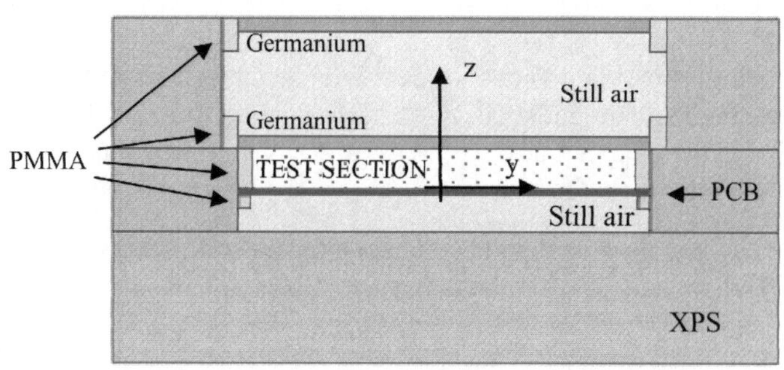

Figure 2. Geometry of the 2D FEM model of conductive heat transfer in the channel cross-section

3 Results

The local results show interesting trends. The whole spatial distribution of $Nu_{loc,q}(x,y)$ is depicted in figure 3, for three different values of Re. The selection of the three values of Re is made so that each Re would represent laminar, post-transitional and fully turbulent flow conditions in a corresponding smooth duct, respectively. It is known to Authors that the presence of the ribs affects the flow features, so that the laminar-turbulent flow transition actually occurs at much lower Reynolds numbers, but the choice of the described operational conditions allows to directly compare the performances of the ribbed channel versus the smooth duct ones. Therefore, figures 3a, 3b and 3c concern cases tested at the minimum Re, i.e., Re=700, at an intermediate value, i.e., Re=5400, and at the largest Reynolds number, i.e., Re=7550. At Re=700, the onset of relevant secondary flows is evident, due to the presence of the ribs, which affect the near-wall flow and induce a non-zero spanwise velocity. The improved-mixing effect is relevant also on the heat transfer enhancement, as $Nu_{loc,q}(x,y)$ presents a significant trend in the spanwise direction, ranging approximately from 6 to 40. Also, the amplification of the local Nusselt number in the spanwise direction follows a 3.5:1 ratio, confirming the enhancement of the heat transfer coefficient due to the flow mixing. Similarly, also for larger Re values the improved mixing due to the ribs points to an increasing of the $Nu_{loc,q}(x,y)$ value from the centerline to lateral walls, which can be estimated around 3.5-4. Two additional considerations stem from the observation of figure 3: on the first hand, the geometrical symmetry of V-shaped ribs with respect to the centerline reflects on the flow features, and hence on the local Nusselt number, which is not spanwise-uniform, but still symmetrical, too. Moreover, by comparing the three maps, figures 3b and 3c appear to be much more similar to each other than figure 3a, pointing to a possible ceiling in the maximum effect of the flow mixing on heat transfer enhancement. This effect can be further observed in figure 6, as the average rib-to-rib Nu trend decreases its slope with the increase of Re.

From the full, two dimensional distributions of $Nu_{loc,q}(x,y)$, two plots are extracted, which allow to perform a one-dimensional but more accurate analysis. Figure 4 shows the spanwise distribution of $Nu_{loc,q}(x,y)$sampled at x=40mm in the streamwise direction, so that the spanwise measurements can be carried out along the whole channel amplitude, without interruptions due to the intersection between the sampling line and the ribs. All the three cases over different Re values are represented in the same picture, to better highlight similarities and differences due to the variation in the mass flow rate (and hence on the bulk velocity and Re). Moreover, the spanwise distribution of the Nusselt number as resulting from the study of the smooth channel is added (computed in the same Re), to assess the effect of the ribs as well. The first consideration concerns the different effects on heat transfer enhancement of the ribs versus Re. The variation induced by the Reynolds number on the local Nu when the ribs are absent is way lower than the corresponding one due to the introduction of the ribs: in the smooth channel, indeed, Nu is increased by an approximate factor of 3 when passing from Re=700 to Re=7550, whereas the point-wise improvement is by a factor of approximately 5 or 6 thanks to the introduction of the ribs. A fortunate side-consideration stems from figure 4, too, concerning the quality of the curves, which confirms the ability of the model to retrieve good results near the edges.

The second one-dimensional plot that can be extracted from the full $Nu_{loc,q}(x,y)$ map is depicted in figure 5, which shows the streamwise Nusselt trend. Both pictures 5a and 5b, representing Nu(x) sampled at the centerline and at 25% spanwise, respectively, show the same Nusselt streamwise distribution, regardless of the sampling position and of Re. As expected, Nu presents a periodic trend between couples of ribs. The shape of the axial distribution depends on whether Re is considered "large" (turbulent, with reference to the ribless channel) or "small" (laminar, with the same reference case). In the first case, for both Re=5400 and Re=7550, a pseudo-skewsymmetric trend of Nu(x) can be observed. At the rib trailing edge, at the beginning of the sampling range, Nu(x) presents an intermediate value, and it grows until it reaches the peak value, which is achieved at approximately 25% of the ribs pitch. Downstream the peak, Nu(x) decreases monotonously, and it approaches the initial value at approximately 50% of the ribs pitch. In the second half of the sampling range, Nusselt duplicates the first half trend but with an opposite fashion, as it decreases until the bottom value at approximately 75% of the ribs pitch, and then it restarts growing monotonously till the leading edge of the downstream rib. In this position, Nu is close but slightly lower than at trailing edge of the upstream rib. Conversely, at low Re (Re=700), the Nusselt distribution is quasi-uniform, presenting only a low, negligible inflation in correspondence of the central region. The different behavior is probably due to the different intensity of turbulent eddies which are generated at the reattachment point downstream the ribs, i.e., the so-called "tumble vortices". In both pictures 5a and 5b, the maximum Nu oscillation with respect to the average value is around 30%. Therefore, both the absolute and the percentage streamwise variations of the local Nusselt number are significantly lower than the corresponding spanwise variations, depicted in figure 4, and for this reason the trend along the axial direction is not considered appreciable.

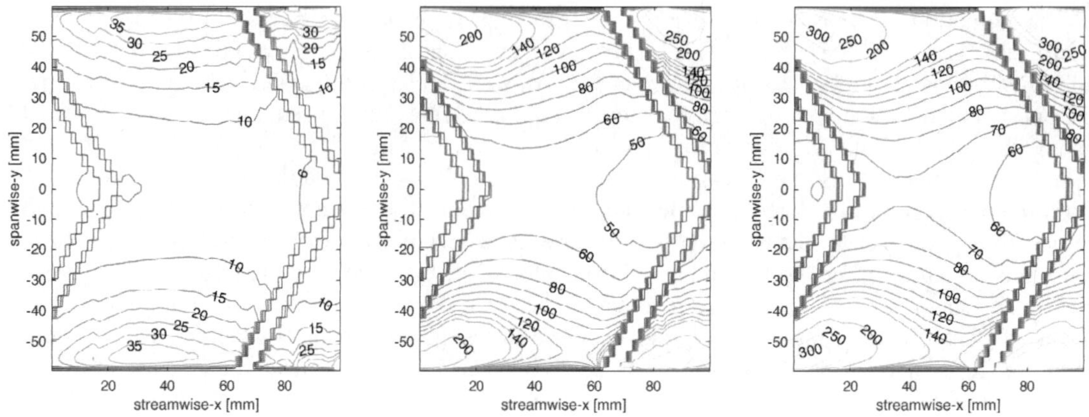

(a) (b) (c)

Figure 3. Local Nu contour maps, Re=700 (a), Re=5400 (b) and Re=7550 (c)

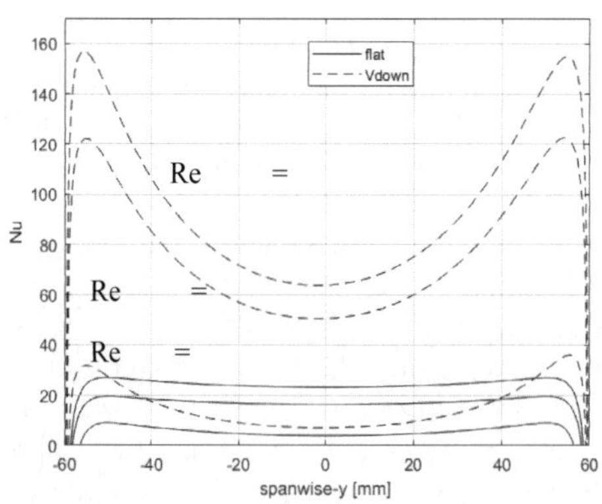

Figure 4. Spanwise Nusselt distribution, x = 40 mm, with flat channel as reference (continuous lines)

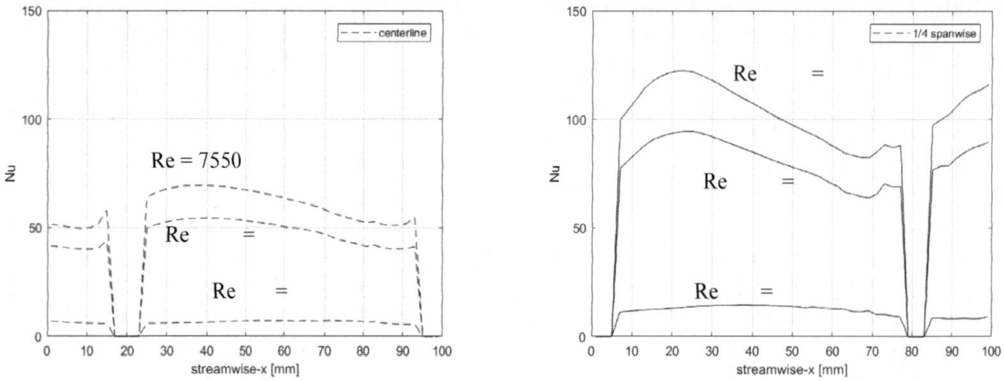

(a) (b)

Figure 5. Streamwise Nusselt distribution, centerline (a) and at y = 15 mm (b)

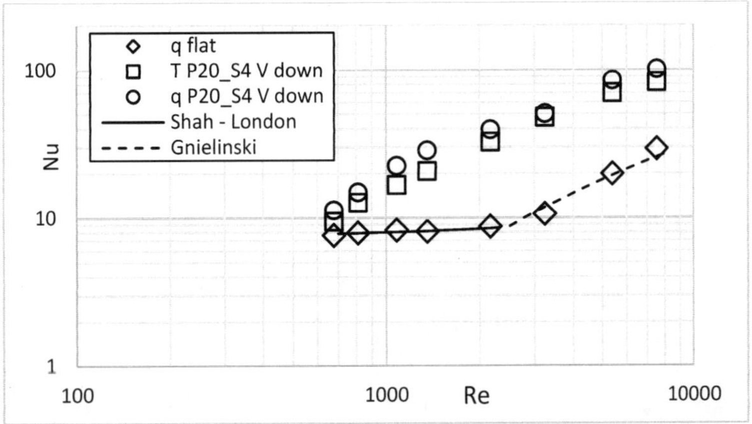

Figure 6. Average Nu on the reference flat channel (q flat, diamonds), compared with global Nu number on a channel with imposed temperature and the same V-down rib configuration (T P20_S4) and with the rib-to-rib average of Nu distribution presented in figure 3 (q P20_S4), for 8 Re. Reference correlations for laminar and turbulent flows in a flat channel are also shown as reference.

Figure 6 presents a comparison of average Nusselt numbers. The average values obtained with the method described in this paper on a flat channel, i.e. without ribs, are shown to fit well the well-known correlations by Shah and London for the laminar regime and by Gnielinski for the post-transitional regime. Moreover, the average rib-to-rib Nusselt number evaluated in this experimental campaign are compared to previous results obtained by the authors on a channel with imposed temperature and the same rib configuration, where a global Nusselt number has been measured, as described in [3]. The averaged local measurements show higher Nu values than the global ones by an average factor of 1.23. This discrepancy should be further investigated, but is probably caused by the influence of the entry region, an example of which can be appreciated on similar ducts on [9].

4 Conclusions

The present study shows the application of the developed technique for the local convective heat transfer estimation with V-shaped ribs. In particular, after a deep assessment phase described in previous works, the local trends seem superior to all the estimated sources of uncertainty, especially towards the lateral edges, were the convective heat flux ceases to be the predominant term in the energy balance. The results seem coherent with previous work in literature, and expand the limits of the analysis to the whole channel. More work is required to model the influence of the ribs, while ongoing DNS studies and flow visualization analyses are carried out to reach definitive answers on this topic.

References

[1] D Fustinoni, P Gramazio, L Vitali and A Niro 2017 *J. of Phys.: Conf. Series* **796** 012015
[2] P Gramazio, L Vitali, D Fustinoni and A Niro 2017 *J. of Phys.: Conf. Series* **923** 012052
[3] P Gramazio, L Vitali, D Fustinoni and A Niro 2018 *proc. of 16th IHTC, Beijing (China) august 10-15,* in press
[4] Han J C, Zhang Y M and Lee C P 1991 *J. Heat Trans.* **113** 590-596
[5] Kukreja R T, Lau S C and McMillin R D 1993 *Int. J. Heat Mass Tran.* **36** 2013-20
[6] Gao X and Sunden B 2001 *Exp. Therm. Fluid Sci.* **24** 25-34
[7] Han J C and Park J S 1988 *Int. J. Heat and Mass Trans.* **31** 183-195
[8] Siegel, R., "Net Radiation Method for Enclosure Systems Involving Partially Transparent Walls" NASA technical note D-7384 (1973). Technical note

[9] Peng W, Jiang P X, Wang Y P, Wei B Y 2011 *App. Therm. Eng.* **31** 2702-2708

Experimental and numerical study on a lab–scale latent heat storage prototype for cooling applications

M Caliano[1*], N Bianco[1], G Graditi[2], L Mongibello[2]

[1] Dipartimento di Ingegneria Industriale (DII) – Università di Napoli Federico II, P.le V. Tecchio, 80 – 80125 Napoli, Italy
[2] ENEA – Italian National Agency for New Technologies, Energy and Sustainable Economic Development, Portici Research Center, P.le E.Fermi, 1 - 80055 Portici (NA), Italy

*E-mail: martina.caliano@enea.it

Abstract. Latent thermal energy storage systems using phase change materials (PCMs) represent an effective way of storing thermal energy because of high-energy storage density and the isothermal nature of the storage process. In the current study, the charging and discharging characteristics of a lab-scale latent heat storage (LHTES) prototype for cooling applications are experimentally and numerically studied. Two numerical models are developed to analyse the performance characteristics of the LHTES prototype: a conductive model and a conductive-convective model. Effective heat capacity (EHC) method is implemented to consider the latent heat of the phase change material. The governing equations involved in the models are solved using the finite element based software product, COMSOL Multiphysics, and the initial and the boundary conditions are determined on the basis of the data obtained from the experimental tests. Numerically predicted temperature variations of the models during charging and discharging processes are compared with the experimental data extracted from the lab-scale LHTES prototype, and a good agreement between them is found when the conductive-convective model is used, while high deviation is observed in case of use of the conductive model. Other results are presented in terms of the performance parameters such as charging/discharging time, energy storage charge/discharge rate, and melt fraction.

1. Introduction

Latent thermal energy storage systems (LHTES) using phase change materials (PCMs) are one of the most efficient methods to store thermal energy. The use of PCM provides higher heat storage capacity and more isothermal behavior during charging and discharging processes, compared to sensible heat storage systems [1]. Therefore, many authors have evaluated experimentally and numerically the LHTES systems performance, analyzing different storage configurations and comparing different numerical models [2-11]. Charvát et al. [12] studied the effect on a solar air-based thermal system of the use of a paraffin-based PCM as thermal energy storage material. Allouhi et al. [13] numerically studied the melting and solidification processes of a PCM integrated in a solar collector. Zhao et al. [14] analyzed different operation strategies for a solar heating system, including a thermal energy storage tank based on PCM. Bejarano et al. [15] numerically studied an innovative cold energy storage system based on PCMs. Cheng and Zhai [16] numerically and experimentally studied a cold thermal energy storage system composed of a packed bed with multiple PCMs. Aljehani et al. [17] simulated a phase change composite consisting of a paraffin wax and expanded graphite for cold thermal energy storage in air conditioning applications. Niyas et al. [18] analyzed experimentally two latent thermal energy storage systems based on PCM for medium and high temperature applications. In another work [19], the same authors studied numerically the performance of a lab-scale LHTES prototype for heat storage applications. Neumann et al. [20] developed a simplified approach for simulating latent thermal energy storage systems based on PCMs consisting of a fin-and-tubes heat exchanger. They compared the results from numerical simulations to the ones of experimental tests carried out on two different storages varying their geometric configuration. Seddegh et al. [21] developed and compared two numerical models for studying a shell-and-tube LHTES system: a conduction model and a conduction-convection model. They observed that the results obtained with the conduction-convection model agreed with the experimental results.

The aim of the current work is to study the charging and discharging characteristics of a novel lab-scale latent heat storage prototype for cooling applications, both by experimental and numerical simulation. In order to analyze the performance characteristics of the LHTES prototype, two numerical models are developed: a conductive model and a conductive-convective model. Effective heat capacity (EHC) method is implemented to consider the latent heat of the phase change material. The governing equations involved in the models are solved using the finite element based software product, COMSOL Multiphysics, and the initial and the boundary conditions are determined on the basis of the data obtained from the experimental tests. Numerically predicted temperature variations of the models during charging and discharging processes are compared with the experimental data extracted from the lab-scale LHTES prototype. Results of combined conduction-convection model agree well with the experimental data, while in case of conductive model, results show a high deviation between experimental and numerical data. Moreover, results show that, in the current application, the discharging process is a natural convection dominated process, while the charging one is a conduction dominated process.

2. Description of the experiment test

The employed experimental apparatus consists of a LHTES unit, a climatic chamber, five thermocouples, a data acquisition unit, and a computer. The storage unit consists of an aluminium cylindrical container of 25.0 cm height and radius equal to 6.9 cm filled with 2.40 kg of a biological PCM, which characteristics are shown in Table 1, as given by the producer. T-type thermocouples with an accuracy of ±0.5°C are used to measure the module local temperature at five points located on the same horizontal cross section placed at 9 cm from the container bottom: on the axis, and at four points at a distance of 3.45 cm from the axis arranged to form a cross. Fig. 1 shows the LHTES prototype filled with liquid PCM, and the climatic chamber used to carry out the experimental tests. The temperature data acquisition is conducted by means the of the National Instruments NI 9213 acquisition module, using the NI cRIO 9066 controller. The experimental measurement data are recorded with sample time of 1 second.

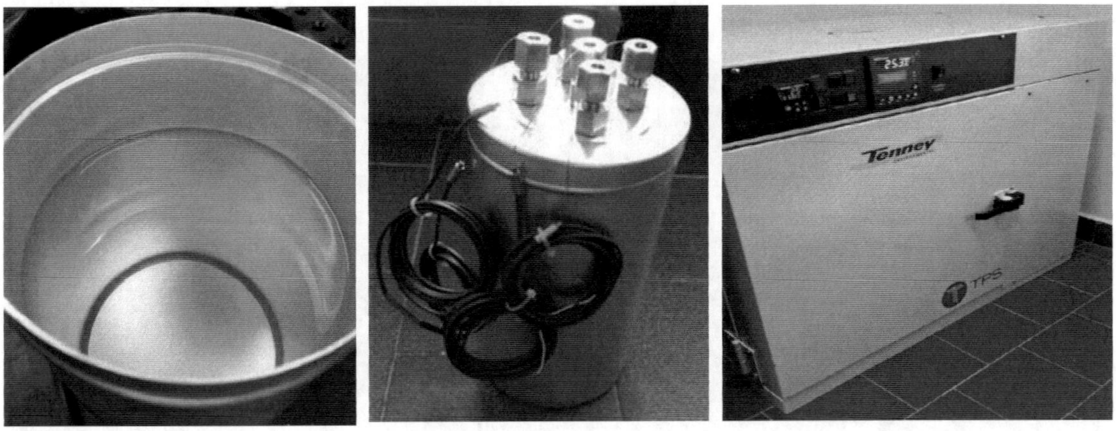

(a) LHTES unit with liquid PCM. (b) Thermocouples arrangement. (c) Climatic chamber.

Fig. 1. Pictures of the LHTES unit and the climatic chamber.

Table 1. Thermo-physical properties of the storage material.

Properties	
Melting point [T_M(°C)]	15
Latent heat [L_h (kJ kg^{-1})]	182
Thermal conductivity (W m^{-1} K^{-1})]	
Solid [k_S]	0.25
Liquid [k_L]	0.15
Density (kg/m^3)	
Solid [ρ_S]	950
Liquid [ρ_L]	860
Specific heat (J kg^{-1} K^{-1})	
Solid [$C_{p,S}$]	2250
Liquid [$C_{p,L}$]	2560

The experimental tests are carried out by varying the air temperature inside the climatic chamber. For the charging process, starting from the uniform temperature condition T_0=23.8°C, at which all the PCM is liquid, one hour temperature ramp is set up to bring the climatic chamber internal temperature to the charging temperature T_c=7°C; then the temperature inside the climatic chamber is kept at T_c for 72h. For the discharging process, one hour temperature ramp is set up to bring the climatic chamber internal temperature to the discharging temperature T_d=23°C; then the temperature inside the climatic chamber is kept at T_d until all the measured temperature inside the unit are well below to the phase change temperature.

2.1. Experimental results

Fig. 2 shows the temperature variation of the LHTES system obtained from the experiments. In detail, the variation of the local temperature at the central position and of the average of the temperatures measured with the four lateral thermocouples, for both charging and discharging process are reported. From these results, it can be seen that, during the charging process, the PCM undergoes solidification at a temperature slightly lower than 15°C. Similarly, it can be seen that, during the discharging process, the PCM starts melting at temperature lower than 15°C. From these figures, it can be also seen that the PCM temperature decrease during charging is slower than the PCM temperature increase during discharging, due to the addition of convective heat transfer model that increases the overall heat transfer rate inside the PCM during the discharging process.

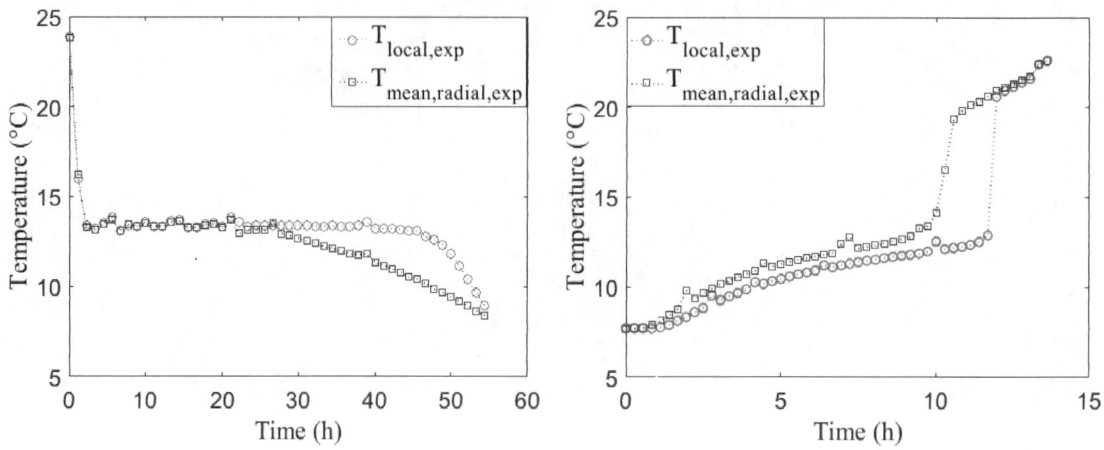

(a) Charging process. (b) Discharging process.

Fig. 2. Experimental results.

3. Numerical modelling

The geometry of the storage prototype and the symmetric boundary conditions of the experimental tests allow to implement 2D axisymmetric numerical models. Therefore, two 2D axisymmetric numerical models are developed for simulating the charging and discharging process of the considered storage unit: a conductive-mechanism-based model, and a conductive-convective-mechanism-based model.

3.1. Conductive model

In order to perform the numerical simulation of the considered storage unit, in case of conductive model, the modelling is carried out through the energy equation of eq. (1), valid for both solid and liquid phases, and the following assumptions are considered [18]: (i) PCM is homogenous and isotropic; (ii) phase change during solidification/melting occurs in a temperature range; (iii) negligible convective mechanisms.

energy equation

$$\rho_{PCM} c_{p,PCM} \frac{\partial T}{\partial t} = \nabla \cdot \left(k_{PCM} \nabla T \right) \tag{1}$$

In eq. (1), ρ_{PCM} is the density of the PCM, T is the temperature, $c_{p,PCM}$ is the heat capacity, and k_{PCM} is the thermal conductivity.

The PCM properties, such as k_{PCM}, ρ_{PCM}, and $c_{p,PCM}$ are considered constant over the entire domain, and fixed equal to the average value between the values of the liquid and solid phase.

To simulate the phase change process, the effective heat capacity method (EHC) is implemented, according to which the material effective heat capacity ($c_{p,PCM}$) depends on the latent heat of fusion (L_h) of the PCM.

$$c_{p,PCM} = c_p + L_h \frac{d\varphi(T)}{dT} \tag{2}$$

In eq. (2), $\varphi(T)$ is a non-dimensional parameter that varies during the phase transition as follows:

$$\varphi(T)=\begin{cases}0, & T<(T_M-\Delta T_M);\\ \dfrac{T-T_M+\Delta T_M}{2\Delta T_M}, & (T_M-\Delta T_M)\le T\le(T_M+\Delta T_M);\\ 1, & T>(T_M+\Delta T_M)\end{cases} \tag{3}$$

where ΔT_M defines the temperature interval in which the phase change takes place.
In detail, $\varphi(T)$ is 0 in the solid phase; $\varphi(T)$ is 1 in the liquid phase, and $0 < \varphi(T) < 1$ in the transition zone.

3.2. Conductive-convective model

In case convective mechanisms are considered, the developed thermal model is based on the following assumptions [18]: (i) PCM is homogenous and isotropic; (ii) phase change during solidification/melting occurs in a temperature range; (iii) the volume expansion during phase change is ignored; (iv) liquid PCM is Newtonian and treated with the laminar flow model. The modelling is carried out through the following equations:

energy equation

$$\rho_{PCM}c_{p,PCM}\left(\frac{\partial T}{\partial t}+\bar{v}\cdot\nabla T\right)=\nabla\cdot\left(k_{PCM}\nabla T\right) \tag{4}$$

continuity equation

$$\nabla\cdot\bar{v}=0 \tag{5}$$

momentum equation

$$\rho_{PCM}\left(\frac{\partial\bar{v}}{\partial t}+\bar{v}\nabla\cdot\bar{v}\right)=-\nabla p+\mu_{PCM}\nabla^2\bar{v}+\overline{F_b} \tag{6}$$

where p is the pressure, μ_{PCM} is the dynamic viscosity, and $\bar{v}$ is the velocity vector.

In eq. (6), $\overline{F}_b$ represents the Boussinesq approximation added to the momentum equation for including the buoyancy effects, and it is evaluated according to eq. (7).

$$\overline{F_b}=\rho_{PCM}\,\bar{g}\beta\left(T-T_M\right) \tag{7}$$

In eq. (7), $\bar{g}$ and β are the gravitational acceleration and the thermal expansion coefficient, respectively.

Also in this case, the PCM properties, such as k_{PCM}, ρ_{PCM}, and $c_{p,PCM}$ are considered constant over the entire domain, and to simulate the phase change process, the effective heat capacity method, presented in section 3.1, is implemented. As concerns the PCM dynamic viscosity μ_{PCM}, it is evaluated according to eq. (8), in order to impose a zero velocity in the solid region of the PCM.

$$\mu_{PCM}=\mu_L(1+S(T)) \tag{8}$$

$$S(T)=C\frac{\left(1-\varphi(T)\right)^2}{\varphi(T)^3+\delta} \tag{9}$$

IOP Conf. Series: Journal of Physics: Conf. Series **1224** (2019) 012049 doi:10.1088/1742-6596/1224/1/012049

In eq. (9), C and δ are arbitrary constant: δ is usually equal to 10^{-3}, and C, that defines the velocity variation into the phase transition zone, is usually between 10^3 and 10^7. δ is fixed according to ref. [18], while in order to fix the values of C, several tests are performed, for both charging and discharging process, by varying C from 10^3 to 10^{12}. Among the considered values of C, the one that allows the best overlap of the numerical simulation results with the experimental ones is 10^{10} for the charging process and $10^{3.7}$ for the discharging one.

3.3. Initial and boundary condition

Initially (t=0), the temperature is fixed equal to the temperature measured at the start-up of the experimental tests, and, obviously only in the case in which convective mechanisms are included, a no-flow condition is set, i.e. the liquid PCM velocity is given zero over the entire domain. For each t>0, the boundary conditions, for charging and discharging process, are set according to the experimental test, described in section 2. The bottom and top surfaces of the storage unit are considered adiabatic, while the boundary condition relative to the lateral surface is set according to eq (10):

$$\left\{ q_l = h_l \left(T - T_\infty \right), \qquad r = r_{\max}, 0 \le l \le l_{\max} \right. \tag{10}$$

In eq, (10), q_l is the heat flux relative to the lateral surface of the storage unit, h_l is the heat transfer coefficient relative to the lateral surface of the storage unit, T_∞ is equal to the air temperature inside the climatic chamber, and r and l are the radius and the height of the storage unit, ranging from zero to a maximum value depending on the storage unit dimension.

The convective heat transfer h_l is set at 27 W/(m^{2}K), evaluated according to the empirical correlation for vertical cylinder subjected to transverse flow under forced convection conditions of ref. [22].

3.4. Numerical treatment and mesh generation

The governing equations are solved with the finite element simulation software COMSOL Multiphysics 5.3a. The non-linearity of the problem is resolved through a segregated approach. The backward differentiation formula is adopted for the time stepping in both charging and discharging, by setting the simulation initial time step to 10^{-4} s and by using a no-fixed maximum time step. Triangle physics-controlled meshes are used, and for both developed model, a grid independence test is conducted. The simulations are performed with a Dell Precision T7610 workstation, equipped with two Intel Xeon E5- 2687 w2 processors and a RAM of 64 GB and 1866 MHz clock. Fig. 3 shows the 2D meshed computational domain of the LHTES prototype in case of conductive-convective model. Moreover, the figure shows the points at which the temperature measurements are carried out to conduct the validation, or rather the central point, placed on the container axis, and indicated with *TC* and the lateral one, indicated with *TL*.

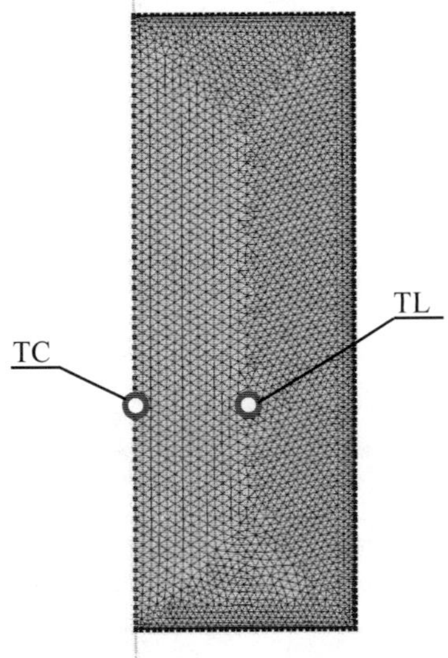

Fig. 3. Meshed computational domain of LHTES prototype and measurements points.

4. Results

In the following sub-sections, the validation, the grid independence test and various results obtained from the numerical simulations of the lab-scale LHTES prototype during charging and discharging process are presented and discussed.

4.1. Models validation

To validate the developed numerical models for the charging and discharging process, the variation of the local temperature at position *TC* in the LHTES prototype, from simulations, is compared with the corresponding temperature values obtained from the experiments, while the variation of the local temperature at position *TL* in the LHTES prototype, from simulations, is compared with the average temperature variation of the PCM obtained from the experiments.

For the charging process, the numerical simulations are carried out by setting the solidification temperature (T_S) at 13°C, while, for the discharging process, a melting temperature (T_M) of 15 °C is used. As for the parameter *C*, in order to fix the value of ΔT_M, several tests are carried out, for both charging and discharging process, by varying ΔT_M from 0.5°C to 6°C. Among the considered values of ΔT_M, the one that produces the best overlap of the numerical simulation results and the experimental ones is 2°C for the charging process and 5.5°C for the discharging one.

Figures 4 and 5 show the comparison between the experimental and numerical values for charging and discharging process, respectively. From these figures, it can be noted that the results are much better, namely closer to the experimental ones, with the conductive-convective model, especially for the discharging process, where the convective heat transfer plays a major role.

It is clear that the conductive model is not suitable for describing the heat transfer phenomena for both charging and discharging process. For this reason, only the main results obtained in the case of the use of conductive - convective model are reported in the following.

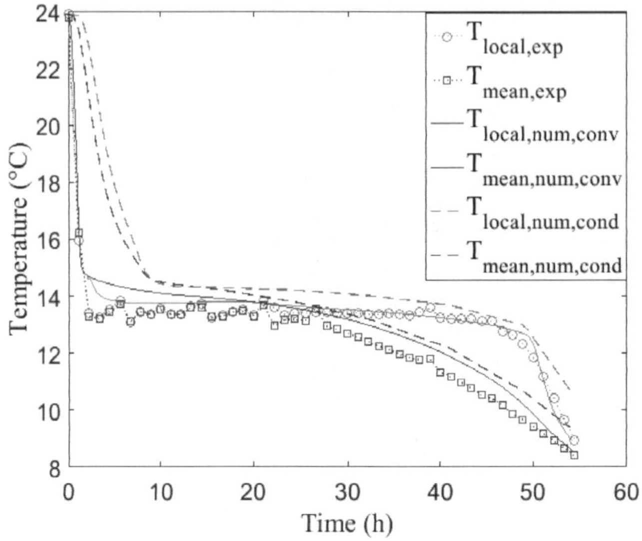

Fig. 4. Model validation: charging.

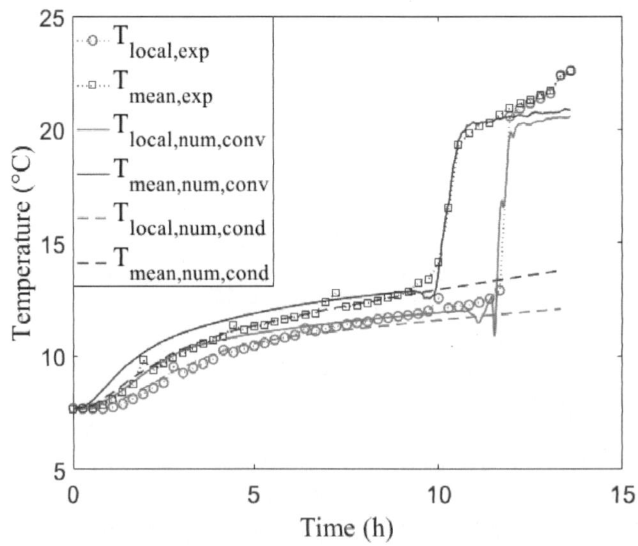

Fig. 5. Model validation: discharging.

4.2. Grid independence test

In order to test the independence of the numerical simulations results on the element size of the mesh, several simulations with different mesh element sizes are carried out for both charging and discharging process and for both the developed numerical models. In detail, meshes consisting of 948, 1780, 2788 and 5080 elements are considered for simulations. Fig. 6 shows the average temperature variation of the LHTES prototype during charging in case of conductive - convective model with the different mesh elements sizes. It is observed from Fig. 6 that the numerical simulations results are grid independent, being all the results quite overlapped. In this work, simulations are carried out considering a mesh composed of 2788 elements.

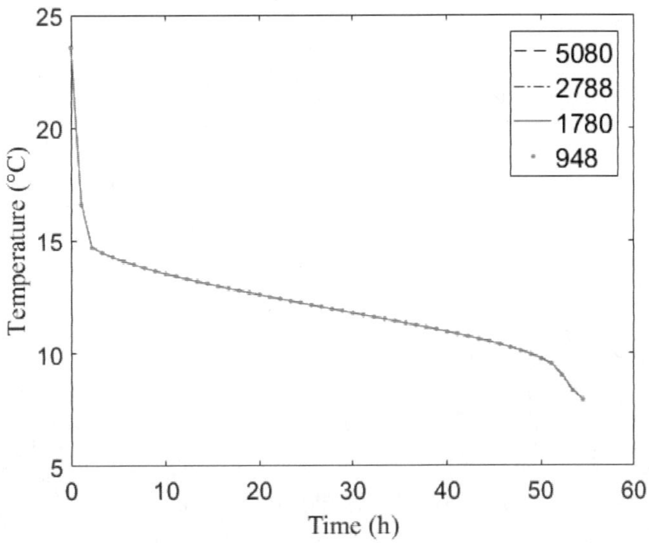

Fig. 6. Grid independence test: charging with conductive - convective model.

4.3. Melt fraction and charging/discharging time

Fig. 7 (a) and (b) shows the average melt fraction variation of the LHTES system during the charging and discharging process, respectively, evaluated as the volumetric average melt fraction of all mesh elements of the numerical model. Melt fraction is a key performance parameter which depicts the latent heat storage and discharge characteristics during the charging and discharging process. Moreover, the velocity variation of the melt fraction as function of temperature provides information about the velocity and the effectiveness of the charging and discharging processes.

It can be seen from Fig. 7 (a) and (b) that the average melt fraction of the LHTES system is 1/0 at the start of the charging/discharging processes, at t = 0 h, then it starts to decrease/increase. The solidification rate, during the charging process, is initially fast, then it decreases, whereas, due to the formation of the solid PCM on the prototype walls, that acts reducing the heat transfer rate. During the discharging, the melting rate is relatively fast for all process. The difference between the trends is mainly due to the contribution of natural convection heat transfer that occurs during the discharging process. As it can be seen from Fig. 7 (a) and (b), the charging process is much slower than the discharging one, indeed, the LHTES system is fully charged at about 50h, and discharged at about 15h.

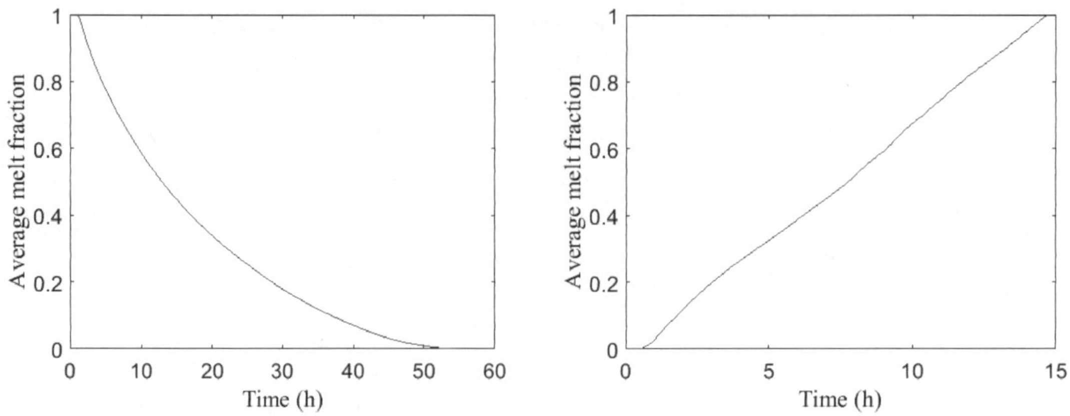

(a) Charging process. (b) Discharging process.

Fig. 7. Average melt fraction variation.

4.4. *Cooling thermal energy storage and discharge*

Fig. 8 (a) and (b) shows the total cooling energy storage rate of the LHTES system during the charging and discharging process, respectively. For convention, the incoming cooling energy is assumed positive, whereas the outcome cooling energy is assumed negative. The total cooling energy stored (E_{stored}) and discharged ($E_{discharged}$) from the PCM, and its sensible ($E_{sens,c}$ and $E_{sens,d}$) and latent ($E_{lat,c}$ and $E_{lat,d}$) contributions, are key parameters for evaluating the thermal performances of the storage unit, and are calculated as the integral of the heat flux across the prototype boundaries.

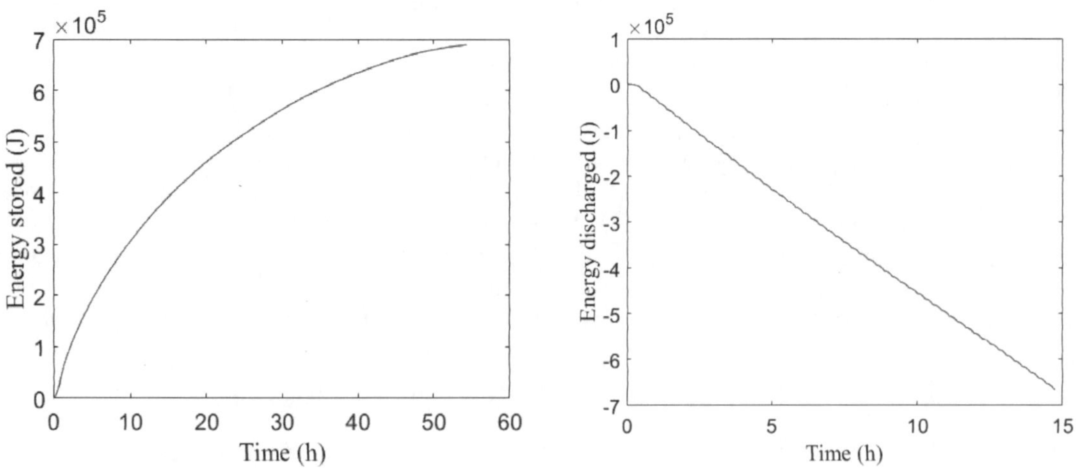

(a) Charging process. (b) Discharging process.

Fig. 8. Cooling energy storage/discharge rate.

Initially cooling energy gets stored/discharged in the form of sensible cooling energy only. Once the PCM reaches near the phase change temperature, cooling energy gets stored/discharged in the form of latent cooling energy. Similarly, after phase change of PCM, cooling energy is stored/discharged in the form of sensible cooling energy. The total cooling energy stored by the PCM is about 530 kJ, while the total cooling energy discharged by the PCM is about -488 kJ. As to the sensible and latent contributions of the total cooling energy stored and discharged by the PCM, they are shown in Table 2.

Table 2. Sensible and latent contributions of cooling energy stored and discharged.

Energy stored (kJ)	
Sensible contribution	93
Latent contribution	437
Energy discharged (kJ)	
Sensible contribution	-51
Latent contribution	-437

5. Conclusions

In this work, the charging and discharging characteristics of a lab-scale latent heat storage prototype for cooling applications are experimentally and numerically studied. Two numerical models are developed to analyse the performance characteristics of the LHTES prototype: a purely conductive model and a conductive-convective model. Performance parameters such as melt fraction, charging/discharging time, and energy storage/discharge rate are evaluated, and the following conclusions are derived:

- A better agreement with the experimental data are found with the conductive-convective model.
- Charging is a conduction dominated process, while discharging is a convection dominated process. For that, charging process is much slower than discharging one.
- The LHTES system takes about 50h to be fully charged, and about 15h to be fully discharged.
- The total amount of energy stored during charging is 530 kJ and during discharging is -488 kJ.

Future development of this work will concern the integration of the developed simulation models in a model for the simulation of cold water storage systems, in order to simulate cold storage systems using water and macro-encapsulated PCM as storage material.

6. Nomenclature

C	Constant in eq. (11)	ρ	Density (kg/m^3)
c_p	Heat capacity (J/kg/K)	*Subscript*	
h	Heat transfer coefficient (W/m^2/K)	c	Charging
k	Thermal conductivity (W/m/K)	cond	Conductive
l	Length of module (m)	conv	Convective
L_h	Latent heat of fusion (kJ/kg)	d	Discharging
r	Radius of module (m)	exp	Experimental
T	Temperature (°C)	L	Liquid
t	time (h)	l	Lateral
TL	Lateral position	lat	Latent
TC	Central position	M	Melting
ΔT	Range of fusion (°C)	num	Numerical
Greek symbols		PCM	Phase change material
δ	Constant in eq. (9)	S	Solid
μ	Dynamic viscosity (Pa·s)	s	Superior
φ	Melt fraction	sens	Sensible

References

[1] Oró E, de Gracia A, Castell A, Farid MM, Cabeza LF 2012 Review on phase change materials (PCMs) for cold thermal energy storage applications Applied Energy 99 513–533.

[2] Farid MM, Hamad FA, Abu-Arabi M 1998 Melting and solidification in multidimensional geometry and presence of more than one interface Energy Conversion and Management 39 809–818.

[3] Lacroix M 1993 Numerical simulation of a shell-and-tube latent heat thermal energy storage unit Solar Energy 50 357–367.

[4] Ng KW, Gong ZX, Mujumdar AS 1998 Heat transfer in free convection-dominated melting of a phase change material in a horizontal annulus International Communications in Heat and Mass Transfer 25 631–640.

[5] Lamberg P, Lehtiniemi R, Henell AM 2004 Numerical and experimental investigation of melting and freezing processes in phase change material storage International Journal of Thermal Science 43 277–287.

[6] Esapour M, Hosseini MJ, Ranjbar AA, Pahamli Y, Bahrampoury R 2016 Phase change in multi-tube heat exchangers Renewable Energy 85 1017–1025.

[7] Zhang Y, Du K, He JP, Yang L, Li YJ 2014 Impact factors analysis of the enthalpy method and the effective heat capacity method on the transient nonlinear heat transfer in phase change materials (PCMs) Numerical Heat Transfer Part A 65 66–83.

[8] Allouche Y, Varga S, Bouden C, Oliveira AC 2016 Validation of a CFD model for the simulation of heat transfer in a tubes-in-tank PCM storage unit Renewable Energy 89 371–379.

[9] Dutil Y, Rousse DR, Salah NB, Lassue S, Zalewski L 2011 A review on phase-change materials Mathematical modeling and simulations Renew. Sustain. Energy Rev 15 112–130.

[10] De Gracia A, Cabeza LF 2017 Numerical simulation of a PCM packed bed system: A review Renew. Sustain. Energy Rev. 69 1055–1063.

[11] Al-abidi AA, Mat SB, Sopian K, Sulaiman MY, Mohammed AT 2013 CFD applications for latent heat thermal energy storage: A review. Renew. Sustain. Energy Rev. 20 353–363.

[12] Charvát P, Klimeš L, Ostry M 2014 Numerical and experimental investigation of a PCM-based thermal storage unit for solar air systems Energy and Buildings 68 488–497.

[13] Allouhi A, Ait Msaad A, Benzakour Amine M, Saidur R, Mahdaoui M, Kousksou T, Pandey AK, Jamil A, Moujibi N, Benbassou A 2018 Optimization of melting and solidification processes of PCM: Application to integrated collector storage solar water heater, (ICSSWH) Solar Energy 171 562–570.

[14] Zhao J, Ji Y, Yuan Y, Zhang Z, Lu J 2017 Seven Operation Modes and Simulation Models of Solar Heating System with PCM Storage Tank Energies 10 21-28.

[15] Bejarano G, Suffo JJ, Vargas M, Ortega MG 2018 Novel scheme for a PCM-based cold energy storage system. Design, modelling, and simulation Applied Thermal Engeneeing 132 256–274.

[16] Cheng X, Zhai X, Wang R 2018 Thermal performance analysis and optimization of a cascaded packed bed cool thermal energy storage unit using multiple phase change materials Applied Energy 215 566–576.

[17] Aljehani A, Razack SAK, Nitsche L, Al-Hallaj S 2018 Design and optimization of a hybrid air conditioning system with thermal energy storage using phase change composite Energy Conversion and Management 169 404–418.

[18] Niyas H, Rao CRC, Muthukumar P 2017 Performance investigation of a lab-scale latent heat storage prototype - Experimental results Solar Energy 155 971-984.

[19] Niyas H, Prasad S, Muthukumar P 2017 Performance investigation of a lab–scale latent heat storage prototype – Numerical results Energy Conversion and Management 135 188–199.

[20] Neumann H, Palomba V, Frazzica A, Seiler D, Wittstadt U, Gschwander S, Restuccia G 2017 A simplified approach for modelling latent heat storages: Application and validation on two different fin-and-tubes heat exchangers Applied Thermal Engineering 125 41-52.

[21] Joybari MM, Haghighat F, Seddegh S 2017 Natural convection characterization during melting of phase change materials: Development of a simplified front tracking method Solar Energy 158 711-720.

[22] Baehr HD, Stephan K 2006 Heat and Mass Transfer, second edition Springer.

Dry deposition of particle on urban areas

M Giardina[1], P Buffa[1], A Cervone[2], C Lombardo[2]

[1]Department of Energy, Information Engineering and Mathematical Models (DEIM), University of Palermo, Viale delle Scienze, Edificio 6, 90128 Palermo, Italy

[2] Italian Agency for New Technologies, Energy and Sustainable Economic Development (ENEA), Via Martiri di Monte Sole, 4, 40129 Bologna, Italy

mariarosa.giardina@unipa.it

Abstract. Dry deposition process is recognized as an important pathway among the removal processes of radioactive pollutants in atmosphere. There is not a unique and accepted theoretical description of involved dry deposition phenomena due to the complexity of the fluid-dynamic processes that influence the deposition flux, but also because there is a lack of experimental data covering all scenarios of interest. In this paper, that is the result of a National Research Program a research activity conducted by DEIM Department of the University of Palermo and ENEA and funded by the Italian Minister of Economic Development, a new schema for parameterization of particle dry deposition velocity on urban area is proposed. The work required comparisons with some experimental data reported in literature for different particle deposition scenarios. The results show that the proposed approach can catch some aspects of phenomena involved in dry deposition processes for the examined environmental conditions with good agreements.

1. Introduction

Urbanization and industrialization are the cause of the formation of pollutants that are deposited on trees, grass, crops, water bodies, and buildings with ecological and non-ecological impacts. In the nuclear field, in the event of a severe accident, which leads to a release of radionuclides in the atmosphere, key challenges are the characterization of the specific isotopes that are released as well as studying the dispersion and deposition phenomena. This is particularly important for the definition of effective mitigation measures and actions to protect the population.

In this field, dry deposition process is recognized as an important pathway among the various removal processes of radioactive pollutants in atmosphere.

There is not a unique and accepted theoretical description of the involved dry deposition phenomena because the complexity of the fluid-dynamic processes, that influence the deposition flux, and lack of a complete experimental set of data covering all scenarios of interest. Various experimental campaigns, performed in different international laboratories, allowed evaluations of deposition velocities for different types of pollutants and deposition surfaces. Nevertheless, there is a difficulty of generalization since the velocity values differ by four orders of magnitude for gases and three orders for particles.

All the above issues limit the possibility to study the dry deposition process with a single modelling approach.

In this field research activities have been focused on the identification, among the models reported in literature, of those approaches that are capable of representing dry deposition phenomena for several categories of pollutants and deposition surfaces [1-3]. On the basis of this study, a new schema for parameterization of particle dry deposition velocity onto rough surfaces, such as an urban area, is proposed. The main aim is to develop an approach that is easy to implement within established atmospheric dispersion modeling codes as well as capable of dealing efficiently with different deposition surfaces.

The work required comparisons with some experimental data reported in literature for different particle deposition scenarios. The results show that the proposed approach can catch some aspects of the phenomena involved in dry deposition processes for the examined environmental conditions with good agreements.

2. A modified approach for dry deposition

The primary phenomena that are considered to affect the process can be described as follows:

- transport due to atmospheric turbulence in the lower part of the Planetary Boundary Layer (PBL), which is called the Surface Layer (SL). It is independent of the physical and chemical nature of the pollutant and depends only on the atmospheric turbulence level (i.e. turbulent movements of air);
- diffusion in the thin layer of air, which overlooks the air-ground interface (named Quasi-Laminar Sublayer, QLS), where the molecular diffusion for gas, Brownian motion and turbulent diffusion for particles and gravity for heavier particles become dominant;
- transfer to the ground (e.g. interception, impaction, and rebound), which exhibits a pronounced dependence on the surface type with which the pollutant interacts.

A key concept to study the dry deposition process is the deposition velocity v_d [m/s] (i.e. the deposition velocity at a given height z) that links the pollutant vertical flux to the concentration measured at quota z [m] to the ground reference level:

$$v_d = \frac{F}{C(z)} \tag{1}$$

with F pollutant flux removed per unit area and C(z) the pollutant concentration at quota z.

Considering that the reciprocal of v_d is the overall resistance to mass transfer, the influence of the above described phenomena on deposition velocity can be expressed in terms of an electric analogy, in which the resistances to the mass transfer are configured in parallel and series circuits to describe transfer factors between air and surface.

The vertical transport of particles can be modeled by assuming that the turbulent transport and particle settling can be added together as follows:

$$K_p \frac{dC}{dz} + v_s C = F \tag{2}$$

where K_p is the eddy diffusivity for the mass transfer of species with a concentration C and v_s is the settling velocity evaluated as follows:

$$v_s = \frac{d_p^2\, g\, (\rho_p - \rho_a)\, C_c}{18\, \nu_a} \tag{3}$$

being g the gravitational acceleration; ρ_p the particle density; ρ_a the air density; ν_a the air kinematic viscosity; and C_c the Cunningham slip correction factor.

By integrating the above equation, it is possible to obtain the expression of the deposition velocity as follows:

$$v_d = \frac{v_s}{1-e^{-[r(z)\, v_s]}} \quad (4)$$

where $r(z)$ is the total resistance to the transport, which can be computed as a function of d_p and height z (quota to the ground reference level), as explained in the following section.

Based on the electrical analogy, in this paper it is proposed to evaluate the resistance r for urban rough surfaces by using the schema shown in figure 1.

It is to be noted that in urban-scale dispersion models, the lowest portion of the boundary layer is often represented using surface layer similarity parameterizations. Boundary layer formulations of this type are only applicable in the inertial sublayer well above the building tops, but not in the immediate vicinity of the urban canopy elements where the flow locally depends on the particular building arrangement and thus it has a rather complex structure.

As above said, interception and inertial forces may transport particles across the various sublayers. Once a particle has traversed also the viscous sublayer, it will interact with the surface. Depending on the characteristics of the contaminant and surface, a particle may stick or bounce off. Particle deposition by any combination of these transport mechanisms largely depends on types of deposition surfaces and particle characteristics [4].

The scheme of deposition processes upon the canopy of urban surfaces proposed in this work has been modified to include the particle rebound or resuspension phenomena together with Brownian diffusion, impaction process, and turbulent transfer.

In figure 1 the aerodynamic resistance r_a (i.e. contribution to the deposition due to the atmospheric turbulence in SL) is connected in series with the resistance r_{ql} across the quasi-laminar sublayer to take into account mechanisms of diffusion by Brownian motion and impaction phenomena. The resistance r_{ql} is evaluated by considering two resistances in parallel, that is: resistance r_{bd}, which represents the Brownian diffusion, and resistance r_i, which allows to treat impaction processes.

The resistance r_i is evaluated by considering two resistances in series: resistance r_{ii} takes into account the inertial impact condition, and resistance r_{ti} considers the effects resulting from turbulent impaction.

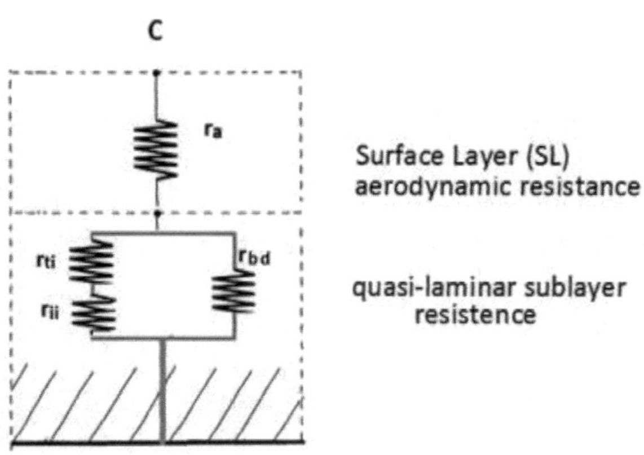

Figure 1. New schematization for parametrization of particles deposition velocity.

These last assumptions allow to take into consideration effects on particle concentration coming from both the inertial and turbulent impaction (i.e. reciprocal influence of the two impact processes on dry deposition efficiency).

Accordingly, the overall resistance r can be evaluated by using the following equations:

$$r = r_a + r_{ql} \quad (5)$$

where r_{ql} and r_i are evaluated as follows:

$$\frac{1}{r_{ql}} = \frac{1}{r_{bd}} + \frac{1}{r_i} \quad (6)$$

$$r_i = r_{ii} + r_{ti} \quad (7)$$

Following the usual methods of micrometeorology for homogeneous terrain, the pollutant concentration flux can be expressed in terms of the local flux–gradient relationship (surface-layer similarity theory) [5]. Therefore the resistance r_a can be determined by using Monin–Obukhov similarity theory as follows [6-8]:

$$r_a = \frac{1}{ku_*}\left[\ln\frac{z}{z_0} - \Psi_h\right] \quad (8)$$

being z_0 the surface roughness height above the displacement plane and k the von Karman constant (generally equal to 0.4).

Brandt et al. in [9] suggested the following relationship for calculating parameter Ψ_h in equation (8):

$$\Psi_h = -5\frac{z}{L} \qquad \text{with } \frac{z}{L} > 0 \text{ (stable atmospheric conditions)} \quad (9)$$

$$\Psi_h = e^{\left\{0{,}598+0{,}390\ln\left(-\frac{z}{L}\right)-0{,}09\left[\ln\left(-\frac{z}{L}\right)\right]^2\right\}} \quad \text{with } \frac{z}{L} < 0 \text{ (unstable atmospheric conditions)} \quad (10)$$

where L is the Monin-Obukhov length computed as follows:

$$L = \frac{u_*^3 c_p \rho \overline{T}}{kgH} \quad (11)$$

with c_p specific heat at constant pressure, $\overline{T}$ average temperature in SL, and H sensible heat.

For the resistance r_{bd} various models predict a functional dependence on Sc number such that in general:

$$r_{bd} = \frac{1}{u_*} c\, Sc^p \quad (12)$$

where c and p are constant.

The parameter p usually lies between 1/2 and 2/3 with larger values for rougher surfaces. For example, Slinn and Slinn in [10] suggested a value of 1/2 for water surfaces. Slinn in [11] suggest a value of 2/3 for vegetated surfaces. Zhang et al. in [12] used values of parameter p varying with land use categories.

In this work it is assumed in equation (12) the following relationship:

$$r_{bd} = \frac{1}{u_* Sc^{-2/3}} \quad (13)$$

The transport of particles by Brownian diffusion represented as function of $Sc^{2/3}$ in equation (12) is recommended in various works on the basis of theoretical and empirical results [13, 14].

It is proposed to evaluate the resistance for inertial impact process r_{ii} in equation (7) by using the following relationships valid for rough surfaces:

$$r_{ii} = \frac{1}{u_*\left(\frac{St^2}{St^2+1}\right)R} \quad (14)$$

Some authors suggested this formula or similar for impaction efficiency as function of smooth surfaces and surfaces with roughness elements [11]. Note that particle rebound is also included via the factor R [12].

Slinn in [11] suggested the following form for R:

$$R = e^{(-b\sqrt{St})} \quad (15)$$

where b is an empirical constant [12]. In this work it is assumed b=2 as suggested in [15].

For the calculation of resistance r_{ti}, general assumptions are reported below.

As well known, empirical relations of turbulent deposition are typically presented in terms of the dimensionless particle relaxation time τ_+:

$$\tau_+ = \tau \frac{u_*^2}{\nu} \tag{16}$$

where τ is the particle relaxation time defined, for a spherical particle, as follows:

$$\tau = \frac{d_p^2 \rho_p C_c}{18\mu} \tag{17}$$

Various models predict a functional dependence of resistance turbulent impact phenomena r_{ti} on τ_+ as follows:

$$r_{ti} = \frac{1}{u_* m \tau_+^n R} \tag{18}$$

It is to be noted that in equation (18) there is the correction factor R related to the collection efficiency for rebound evaluated by equation (15). This allows us to take into account the functional dependence of rebound phenomena on turbulent impact conditions. The constants m and n in equation (18) have been evaluated by fitting some data reported in literature for urban surfaces And they are set to 0.05 and 0.75, respectively.

3. Comparison with experimental data

The new approach for computing deposition velocity v_d is validated by comparison with experimental data reported in [16] for several meteorological conditions and deposition urban surfaces.

Figures 2 through 5 show experimental data for urban area in terms of deposition velocity as a function of frictional velocity u_*.

The datasets were taken from different experimental campaigns over a wide range of degrees of complexity: urban background, urban canopy, and industrial district or Venice lagoon surface (Italy). Summary of experimental sites in terms of measurement height (z), and roughness height z_o are reported in table 1.

In all figures the vertical bars present errors of 30%.

On the whole, a good agreement between the prediction trend obtained by using the proposed model and the examined experimental data is found.

In particular, it can be observed that there are good predictions for u_* values above about 0.3 m/s. Only limited effect of stability is observed with a slight reduction of the deposition velocities in stable conditions. This aspect is captured by the proposed approach.

Table 1. Summary of experimental sites used in aerosol sampling reported in [16].

Site	**Height z (m)**	**Roughness length z_o (m)**
Bologna	10	0.35 ± 0.02
Lecce	10	0.53 ± 0.02
Maglie	10	0.52 ± 0.02
Venice lagoon	9.6	0.11 ± 0.03 land, 0.01 ± 0.03 water

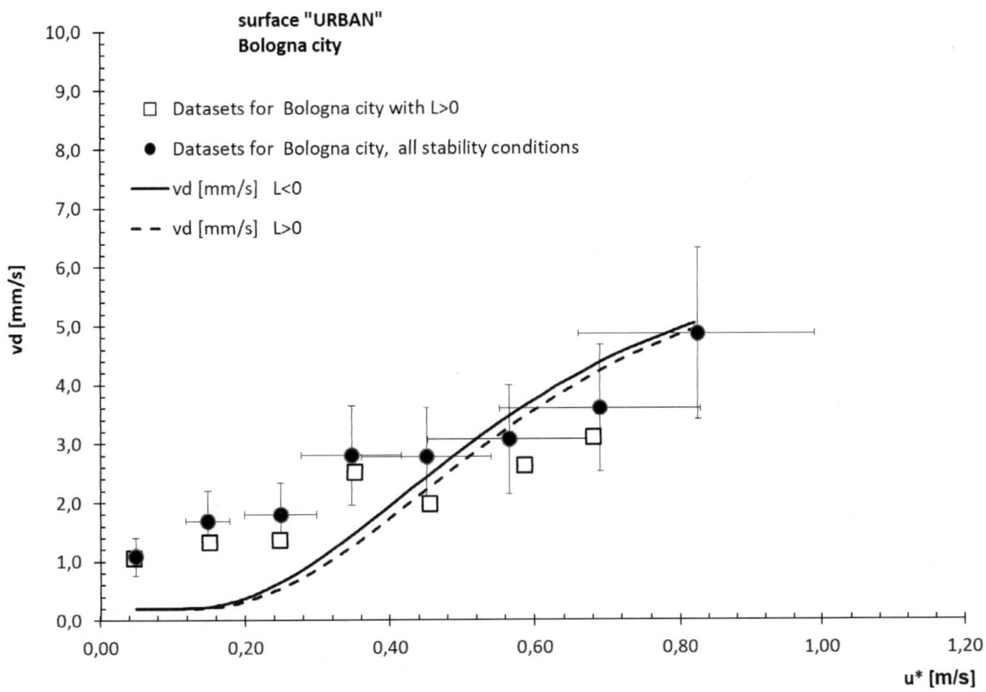

Figure 2. Deposition velocity predictions as functional dependence from friction velocity u_* and comparison with measurement datasets relevant to Bologna city, as reported in [16].

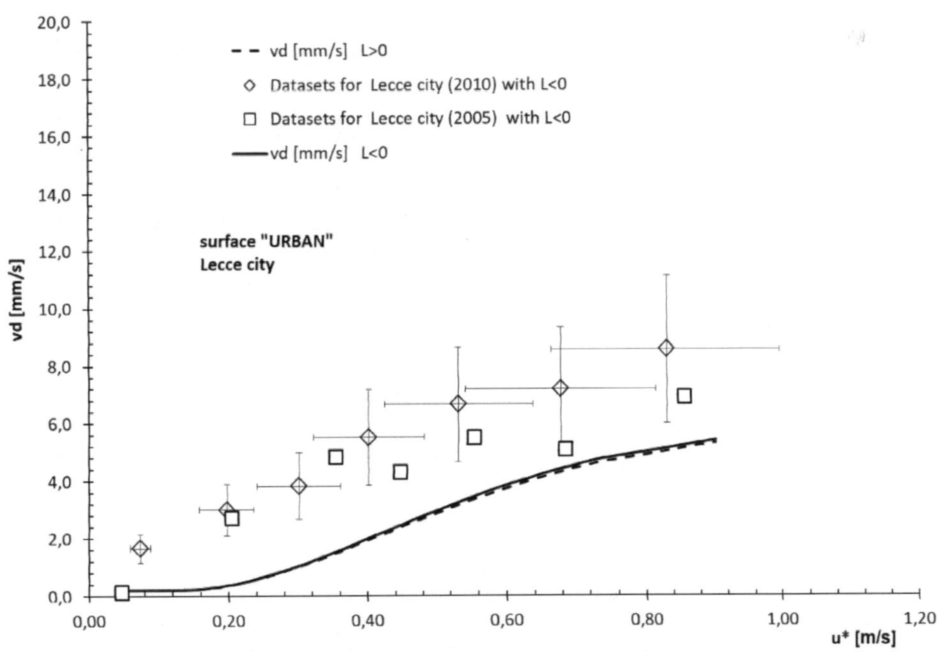

Figure 3. Deposition velocity predictions as functional dependence from friction velocity u_* and comparison with measurement datasets relevant to Lecce city, as reported in [16].

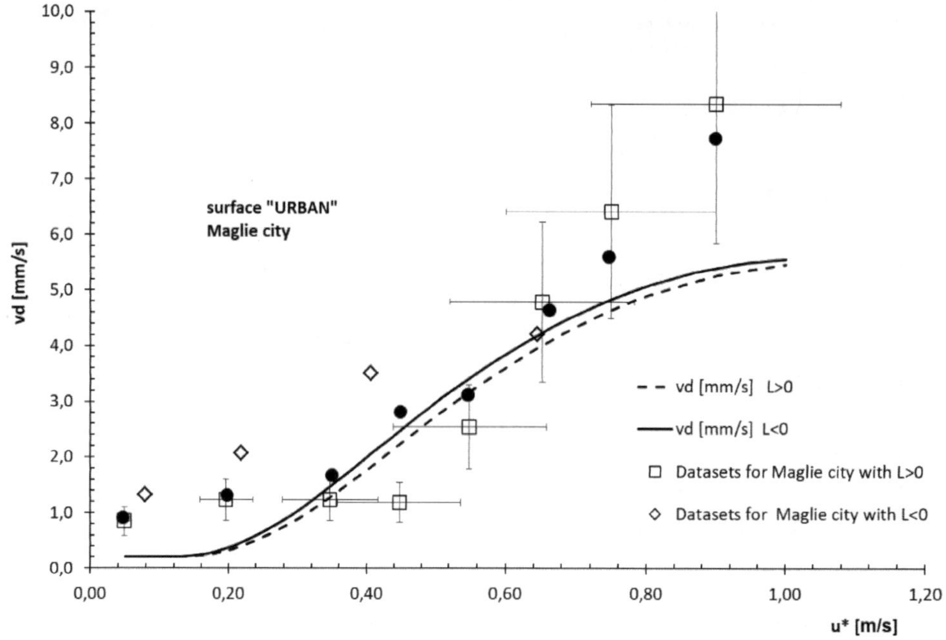

Figure 4. Deposition velocity predictions as functional dependence from friction velocity u_* and comparison with measurement datasets relevant to Maglie city, as reported in [16].

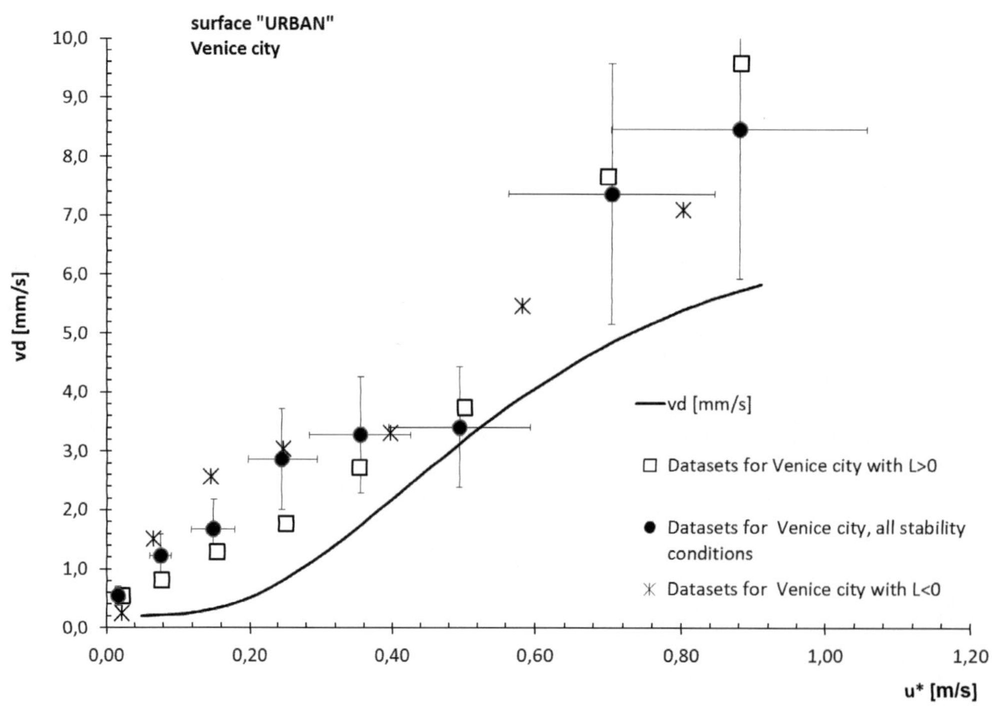

Figure 5. Deposition velocity predictions as functional dependence from friction velocity u_* and comparison with measurement datasets relevant to Venice city, as reported in [16].

4. Conclusion

The ATMES (Atmospheric Transport Model Evaluation Study) report, relevant to the study of models for the evaluation of radioactive pollutants disperse in the ambient atmosphere, highlighted that the highest uncertainties are in the parametrization of source terms and dry and wet deposition velocities.

In literature there are several models, but no one is able to treat exhaustively most of phenomenologies related to pollutants deposition because of the number and complexity of the involved processes. A review of the existing mechanistic models emphasizes the wide variety of ways captation can take place, however, a comparison of two similar scenarios provides large discrepancies between each other, which can differ by two orders of magnitude for the same particle diameter.

As highlighted in literature, the measurements of deposition velocity carried out by different international laboratories don't allow to draw general conclusions due to experimental uncertainties.

In fact, for the same typology of pollutant, experimental data show that, for gasses, the values of deposition velocity differ even by four orders of magnitude and, for particles, by up to three orders of magnitude.

The main aim of our research activity was to develop an approach easy to implement within atmospheric dispersion modeling codes as well as an approach capable of dealing efficiently with different deposition surfaces for several radioactive pollutants. Some parametrizations reported in literature are based upon a number of assumptions which may be frequently violated in practice that can strongly dependent on a number of parameter combinations such as land use classification and seasonal categories.

In this paper, a new scheme for particles deposition velocity, based on an electrical analogy, is proposed to evaluate the resistances for urban rough surfaces. It is worth to note that the correlation obtained with the new schematization is based on the hypothesis that the impact phenomena in the quasi-laminar sublayer can be affected by specific local features of the mutual influence of inertial impact processes and turbulent impact phenomena. The proposed approach is further modified to take in consideration the rebound phenomena.

The validation work, carried out by using some experimental data from literature, allowed to verify the effectiveness of the proposed approach.

Acknowledgments

The activity has been funded by Italian Minister of Economic Development

References

[1] Ambrosio G., Buffa P., Caltabellotta G., Giardina M., Palermo G., Raniolo I., Ventura F., 2015. Definizione della metodologia e degli input necessari per l'esecuzione di analisi integrate CALMET-CALPUFF ai fini della valutazione della dispersione di inquinanti radioattivi in atmosfera, rapporto di ricerca RdS PAR 2014/136, 2015

[2] Cervone I., Lombardo C., Giardina M., Buffa P., Palermo G. 2016. Modelli per calcoli di concentrazione di materiale radioattivo disperso a breve-medio raggio in aree caratterizzate da configurazioni architettoniche tipiche delle principali città italiane, rapporto di ricerca RdS PAR 2015/090, 2016

[3] Cervone I., Lombardo C., Giardina M., Buffa P. 2017. Analysis of models for the study of dry deposition processes and ECMWF forecast weather data rapporto di ricerca RdS PAR 2016/125, 2017

[4] Zufall M.J., Davidson C.I. 1998. Dry Deposition of Particles from the Atmosphere, Air Pollution in the Ural Mountains: Environmental, Health and Policy Aspects, Springer Netherlands, ISBN: 978-94-010-6192-6, pp 55-73.

[5] Businger J.A. 1973. Turbulent transfer in the atmospheric surface layer. Workshop on Micrometeorology, D. A. Haugen, Ed., Amer. Meteor. Soc., 67–100.

[6] Hanna S.R., Gifford, F.A., Yamartino, R.J. 1991. Long Range Radioactive Plume Transport

Simulation Model/Code – Phase I. USNRC Division of Contracts and Property Management, Contract Administration Branch, P-902, Washington, DC 20555. Technical report.

[7] Maryon R.H., Saltbones J., Ryall D.B., Bartnicki J., Jakobsen H.A., Berge E. 1996. An intercomparison of three long range dispersion models developed for the UK meteorological office, DNMI and EMEP. UK Met Office Turbulence and Diffusion Note 234. ISBN: 82-7144-026-08.

[8] Seinfeld J.H., Pandis S.N. 1998. Atmospheric chemistry and Physics – John Wiley&Sons, New York.

[9] Brandt, J., Christensen J. H., Frohn L.M. 2002. Modelling transport and deposition of caesium and iodine from the Chernobyl accident using the DREAM model Atmos. Chem. Phys., 2, 397–417.

[10] Slinn S.A., Slinn W.G.N. 1980. Predictions for particle deposition on natural waters. Atmos. Environ., 14, 1013-1016.

[11] Slinn W.G.N. 1982. Predictions for particle deposition to vegetative surfaces. Atmospheric Environment 16, 1785-1794.

[12] Zhang L., Gong S., Padro J., Barrie L. 2001. A size-segregated particle dry deposition scheme for an atmospheric aerosol model Atmospheric Environment, 35, 549-560.

[13] Pryor S.C. , Barthelmie R.J., Spaulding A.M, Larsen S.E., Petroff A. 2009. Sizeresolved fluxes of sub-100-nm particles over forests. Journal of Geophysical Research,114, D18212.

[14] Kumar R., Kumari K.M. 2012. Experimental and Parameterization Method for Evaluation of Dry Deposition of S Compounds to Natural Surfaces, Atmospheric and Climate Sciences, 2, 492-500.

[15] Nemitz E., Gallagher M.W., Duyzer J.H., Fowler D. 2002. Micrometeorological measurements of particle deposition velocities to moorland vegetation Q.J.R. Meteorol. Soc., 128, 2281-2300.

[16] Donateo A., Contini D. 2014. Correlation of Dry Deposition Velocity and Friction Velocity over Different Surfaces for PM2.5 and Particle Number Concentrations, Hindawi Publishing Corporation Advances in Meteorology Volume 2014, Article ID 760393.

AUTHOR INDEX